The Palms of Madagascar

THE PALMS OF MADAGASCAR

JOHN DRANSFIELD AND HENK BEENTJE

WITH ILLUSTRATIONS BY

MARGARET TEBBS AND ROSEMARY WISE

ROYAL BOTANIC GARDENS, KEW AND THE INTERNATIONAL PALM SOCIETY, 1995

Royal Botanic Gardens Kew and The International Palm Society

First published 1995

Book Design by Jeff Eden for Media Resources, Royal Botanic Gardens, Kew

Front Cover Photograph by John Dransfield

Back Cover Photograph by David Du Puy

ISBN 0 947643 82 6

Printed in Great Britain by HMSO Norwich Print Services

Contents

FOREWORD

ANYONE WHO HAS VISITED THE HUMID TROPICS CANNOT but be impressed by the number of species of palms and of their vital importance to local people around the world. As this volume shows, Madagascar is a good example of this, since it has a particularly rich palm flora. What is amazing is the poor knowledge of this flora until the field work upon which this book is based was carried out by the authors. That 75 new species have been discovered since 1986 when this study began reflects the poor state of knowledge about tropical rainforest areas. Madagascar is regarded as one of the 'hot spots' of biological diversity and it is tragic that so much of this diversity is being destroyed before it has even been named and described. This work has done a lot to rectify the information situation for at least one important component of the forests, the palms. However, it also indicates how many of these species are rare and have been reduced to extremely small populations.

This beautifully illustrated and informative volume is the result of a collaborative effort between several different and perhaps unlikely parties, the Royal Botanic Gardens, Kew, the Parc Botanique et Zoologique de Tsimbazaza, McDonald's Restaurants Ltd. (UK) and the International Palm Society. We are most grateful to our colleagues at Tsimbazaza in Antananarivo for so much help and collaboration with this project. We are also very grateful to the International Palm Society for joining Kew in publishing this work in such an attractive format; their help has allowed us to produce the large number of colour illustrations. When I see the superb quality of this book and of the research upon which it is based I am especially grateful to Paul Preston and his colleagues at McDonald's Restaurants Ltd. (UK) who agreed to sponsor what must be an unusual project for them. Their support has enabled a major contribution to the knowledge and understanding of this most important ecosystem. It is my hope that this book will stimulate further efforts to study, to conserve and to use sustainably the palms of Madagascar.

GHILLEAN T PRANCE

DIRECTOR
ROYAL BOTANIC GARDENS, KEW

ACKNOWLEDGMENTS

THIS PROJECT HAS RELIED GREATLY ON THE GOODWILL and kindness of many people. First and foremost we wish to thank our sponsor, McDonald's Restaurants Limited UK and their President, Mr. Paul Preston, who generously funded the position of Postdoctoral Research Fellow that allowed HB to work for the project. When we ran beyond our first deadline, McDonald's extended their funding by one year, allowing us to complete the project. We are immensely grateful.

In Madagascar the number of people who have helped us is very great indeed. In Antananarivo two successive Directors of the Parc Botanique et Zoologique de Tsimbazaza, Dr. Voara Randrianasolo and Dr. Albert Randrianjafy and three successive heads of the Tsimbazaza Herbarium, Armand Rakotozafy, Sylvain Razafimandimbison and Solo Hery Rapanarivo proved most supportive. Without them it would not have been possible for the project to go forward. In early stages of the project Mme Lala Rakotovao and Mme Berthe Rakotosamimanana and later Mme Adeline Zafisela helped with obtaining research clearance. Tsimbazaza staff members Armand Rakotozafy, Armand Randrianasolo, Jacky Andriantiana, Rolland Ranaivojaona, Guy Rafamantanantsoa and Jeannine Raharilala joined us in the field, not only helping to collect palms, but translating from French to Malagasy and always providing fine companionship. Dr. Bakouly Rakouth, of the University in Antananarivo gave us much encouragement. Mme Celestine Ravaoarinoromanga and M. Georges Rakotonarivo of Département des Eaux et Forêts provided permits to collect palms all over Madagascar. Alfred Razafindratsira, nurseryman, was hospitable and often provided interesting information about Malagasy palms. The manager of the Coco Beach Motel, Maroantsetra and the head waiter of the Hotel Thermale, Ranomafana provided much valued logistic support and hints about the occurrence of palms. Gerard Jean, Georges Modeste, Joubert and Noel Bemakoa, François Beantrefy, Narcisse, Jean Baptiste, Laniry, and many, many others accompanied us in the field, acting as expert guides, cooks and companions, sometimes climbing trees and often helping us to ford rivers on slippery boulders.

Mark Pigeon, Olivier Langrand, Martin Nicoll, Dominique Halleux, Nat Quansah and Dan Turk all provided very valuable information on the accessibility of forested areas and hints about where interesting palms might be found, sometimes joining one or both of us in the field.

Throughout much of the latter part of the project M. Hery R. Andriampaniry accompanied HB and, once, JD, in the field, often smoothing over international relationships.

Staff of PNUD (UNDP-United Nations Development Programme) in Antananarivo and in Mananara Avaratra helped us greatly when we visited the Biosphere Reserve at Mananara Avaratra. In particular we wish to thank Noeline Raondry (Principal National Coordinator, PNUD, Projet Biosphère), Victor Solo Rakotonirina (Directeur National Projet Biosphère Mananara-Nord), M and Mme Bruno Mananjara and Martha Klein.

Shell Exploration and Development, especially Hennie Antonissen and Henriette van der Gaag, helped HB in providing logistic support and much hospitality.

The British Embassy with three successive Ambassadors and the Netherlands Consulate in Antananarivo took great interest in the project

Missouri Botanic Garden has played a major supporting role in the project. First of all in 1986 the Director, Peter Raven, gave JD every encouragement to plan his first field work in close collaboration with the Missouri team in Madagascar. On this first trip much logistic support was provided by Dr. P.P. Lowry II who has continued to take a great interest in the project. Marion Nicoll helped in the field and dried material on the first expedition. On the second and third trips JD received a great deal of help in logistic support, field companionship and help with drying palm material from Dr. George Schatz, and on the third trip from Dr. Jim Miller. Subsequent trips have benefitted greatly from logistic support from Lucienne Wilmé and Annick Ramisamihantanirina, of the Missouri House in Antananarivo, who passed on fax messages and helped during emergencies.

In the latter part of the project two people in particular helped to make field work in Madagascar much easier and more enjoyable than it might otherwise have been. These are Rolland and Mme Victoire Raraorimanga, guardians of Kew's house in Antananarivo. Not only do they look after Kew's house, but help greatly in the pressing and drying of plants.

Our two artists, Maragaret Tebbs and Rosemary Wise have worked hard to illustrate the palms included in this book, patiently searching for often minute details that we wished to have illustrated.

Throughout the duration of the project, we have received support and encouragement from the Director of the Royal Botanic Gardens Kew, Professor Iain Prance and the Keeper of the Herbarium, Professor Gren Lucas. Kew's fund-raiser, Giles Coode-Adams, ably assisted by Ellen Wells and Kew's Finance Officer, Des Bennett, did much to make sure that the project ran smoothly. Many other Kew staff have played a role from time to time in the project. Peter Boyce has always been at hand to help in the curation of material and to look after the Palm Room when we have both been absent. Phil Cribb has taken a close interest in the progress of the project. Sasha Barrow prepared the account of *Phoenix* for us. Mark Coode helped greatly by looking after several matters during JD's absence in the field on many occasions. David Du Puy and Blaise Du Puy helped, both in the field and administratively in Madagascar and from time to time back at Kew. Both contributed some photographs for the book. Madeline Harley and Keith Ferguson have provided palynological expertise, Paula Rudall anatomical data and Margaret Johnson cytological information. Martin Cheek helped greatly on two field trips. David Cooke and Martin Staniforth have both provided enthusiastic help in the field and input into the growing of palms at Kew. In particular, David Cooke climbed many a difficult palm tree in the forest and was responsible for making the collection of the first material of several species possible; he also contributed information on growing palms at Kew for this book. Andy and the Tree Gang trained HB in tree climbing.

Milan Svanderlik, Head of Media Resources at Kew, provided a great deal of important advice. We thank Jeff Eden very much for his enthusiastic and innovative response to the challenge of converting our dry manuscript into a beautifully designed book.

Mike Lock, Soejatmi Dransfield and Juliet Beentje read the manuscript critically and we are very grateful for their many constructive comments.

JD's involvement in the palms of Madagascar began properly in 1986 with an extension of the collaboration with the L.H. Bailey Hortorium. Funding for the first crucial field trip came from the National Science Foundation grant (BSR-8407029) to Dr. Natalie Uhl. We thank NSF and Dr. Natalie Uhl for this initial stimulus. We also thank Natalie for her continued interest in the project.

In Paris we have enjoyed the close collaboration of the Laboratoire de Phanérogamie, Museum National d'Histoire Naturelle and we wish to thank the Director, Professor Philippe Morat, for his constant interest and help. We also thank Dr. Annick LeThomas, Dr. F. Badré, M. Jean Bosser, M. Jean-Noel Labat, M. Jolinon, M. Darrieulat and the late Dr. R. Hoogland.

The Keeper of the Erbario, Istituto Botanico, Firenze, provided permission to work on Beccari's palm herbarium.

The International Palm Society has been supportive throughout the project, in particular in funding David Cooke's participation in field work in 1989. We are very grateful to President Jim Cain and the members of the Executive Board for agreeing to fund the co-publication of this book.

JD wishes to thank Ken and Shirley Scrivener for the immensely valued role they played in Kingston-upon-Thames while JD was in the field.

To all these friends and colleagues we say thank you.

Finally we thank Soejatmi and Juliet, Rowan, Peter and Rosie for their great patience, their tolerance of difficult conditions in Madagascar and periods of abandonment at home, periods when HB and JD were virtually incommunicado in Madagascar and periods back in Kew when manuscript preparation took precedence over everything else. We thank them from the bottom of our hearts.

JOHN DRANSFIELD
HENK BEENTJE

ROYAL BOTANIC GARDENS, KEW
JANUARY 1995

PREFACE

"THE PALMS OF MADAGASCAR" IS THE CULMINATION of a research project formally begun at the Royal Botanic Gardens, Kew, in 1986, but the origins of the project go back further. In the early 1980s, JD began to collaborate with Dr. Natalie W. Uhl of the L.H. Bailey Hortorium in Cornell University to complete an account at the generic level of all the world's palms, a project that had been started by Professor Harold E. Moore Jr, who died in 1980 leaving the work unfinished. Between 1981 and 1986 work on this project represented the main component of JD's research at Kew. As work continued towards the completion of *Genera Palmarum,* finally published in 1987, it became obvious that most of the real mysteries remaining in the palm family related to the palms of Madagascar. Not only were Madagascar palms not well known outside Madagascar, but the material available for comparing them with other palms was often very incomplete.

On his first visit to Madagascar in 1963 Hal Moore made excellent collections of several endemic Madagascar genera, collections that provided crucial data for understanding relationships. Moore, in his engaging article in Principes (Moore 1965), described the delights of his first visit and some of the frustrations of field work and not being able to find all he wished to see. Because he had been unable to refind some of the endemic genera he returned to Madagascar in 1971 and again in 1972. However, he was still unable to refind *Beccariophoenix, Masoala* and *Louvelia.* Sadly conditions within Madagascar then became very difficult for field work by foreigners and Hal was unable to make any further visits. During the latter half of the 1970s and the early 1980s very few palm collections were made in Madagascar. Mardy Darian is one of the few palm enthusiasts to have visited the island during this period and he came back after each visit with seeds and fragments of putative new palms. One of these turned out to be the wonderful new palm, *Marojejya darianii,* a dead endocarp was the first material of *Voanioala gerardii* that we saw, and a third fragment represented the palm later described as *Lemurophoenix halleuxii.*

In the mid 1980s major changes took place within Madagascar that made it, once again, possible for scientists to visit the island and conduct field research. As the proofs of *Genera Palmarum* arrived

Forest along a river on the East Coast, **Dypsis crinita** on the left (Photo: Brian Rogers).

in the autumn of 1986, JD left Kew accompanied by David Cooke for his first nine-week period of field work in Madagascar, funded partly by The National Science Foundation as a continuation of research collaboration with Dr. Uhl in Cornell. Thanks to the generosity of Missouri Botanical Garden we were able to take advantage of the collaborative agreement they had signed with the Parc Botanique et Zoologique de Tsimbazaza in Antananarivo. During the nine weeks we visited the Masoala Peninsula and areas nearer to Antananarivo, and were able see and collect specimens representing all the then recognised genera of Madagascar, many palms being collected for the second time only. We were also able to make the first scientific collections of *Voanioala* (recounted in Dransfield 1989b) and *Lemurophoenix*. However, not surprisingly, many of the palm species we found were not fertile or were missing important stages of development. JD made a second visit in 1988 and a third visit in 1989. These three visits clarified some taxonomic problems but it became increasingly obvious that if real progress were to be made then much more field work would be required, preferably conducted continuously over at least twelve months. Furthermore, the published floristic account (Jumelle & Perrier 1945) was already seriously out of date and much in need of a revision. To this end JD prepared a project proposal for possible funding from an outside agency.

Mr. Paul Preston, the president of McDonald's Restaurants Limited (UK), approached Kew in 1990 to offer funding from McDonald's to support a research programme at Kew and the Palms of Madagascar was the project chosen. HB was the successful candidate for a postdoctoral position to work with JD, both at Kew and in Madagascar on the project for three years. By this time Kew had already signed a letter of agreement for long-term collaboration between the Parc de Tsimbazaza, the University in Antananarivo and the Royal Botanic Gardens, Kew, initially for a project conducted by Dr. David Du Puy on the Leguminosae of Madagascar. HB arrived in Madagascar in 1991 for two months to start work on the palms and to make logistic arrangements. Then between August 1991 and August 1992 he and his family lived in Madagascar, making numerous field trips throughout the island. Some of the excitement and adventure of fieldwork in Madagascar is described in three articles in Principes (Beentje 1993a, 1993b, 1994a). JD visited in April 1992 for further field work before HB returned to Britain in September 1992. A further visit was made by HB in December 1992 for two months. The concentrated fieldwork was immensely productive, in fact almost too productive, because time was insufficient to work up all the material collected. McDonald's very generously provided funding for a fourth year. In June 1994 HB and JD attended the International Palm Society's Biennial meeting in Caracas, Venezuela to invite the Board of Directors of the Society to join Kew in jointly publishing our account of the Palms of Madagascar. This was accepted and a legal agreement drawn up to cover the publishing procedure. Towards the end of the fourth year, in October 1994, we paid another visit to Madagascar to chase up some problematic palms and to try to enlarge our photographic coverage of the island's palms.

In the present book we publish the results of our work on the Palms of Madagascar, conducted since the start of the project. The major decisions on the delimitation of genera were taken after lengthy discussion. At the species level, HB and JD took on different groups and worked them up, independently at first, then discussing the results in detail. This is reflected in the attribution of authorship in the citation of new species. The descriptions were then put together, HB tackled the enormously complicated process of writing the key to *Dypsis* and then the whole manuscript was put together. We had originally hoped to publish new names before the preparation of this book. However, as time went on and we found more and more new taxa, we concluded that it would be more convenient, given the limited available time, to include the validation of new names in the monograph.

As the project draws to a close we look back on a very exciting period of research, both in the field and back in Europe. Yet we are also conscious that what we are publishing here is a mere statement of what we know about the palms of Madagascar at the beginning of 1995. Given more time to comb the forests of this extraordinary island we would undoubtedly have additions and further changes to make. We have not had time to conduct a modern phylogenetic analysis of our data and there are many other aspects of the palms of Madagascar that would make exciting and important research projects. A book is out of date the minute it is published, and that is certainly true with the present account. We hope that it will be a useful baseline of data about this rich palm flora that can be used by all interested in palms. We have enjoyed our collaboration greatly and are only sorry that it has had to come to an end.

JOHN DRANSFIELD
HENK BEENTJE

ROYAL BOTANIC GARDENS, KEW
JANUARY 1995

THE PALMS OF MADAGASCAR IN A GLOBAL CONTEXT

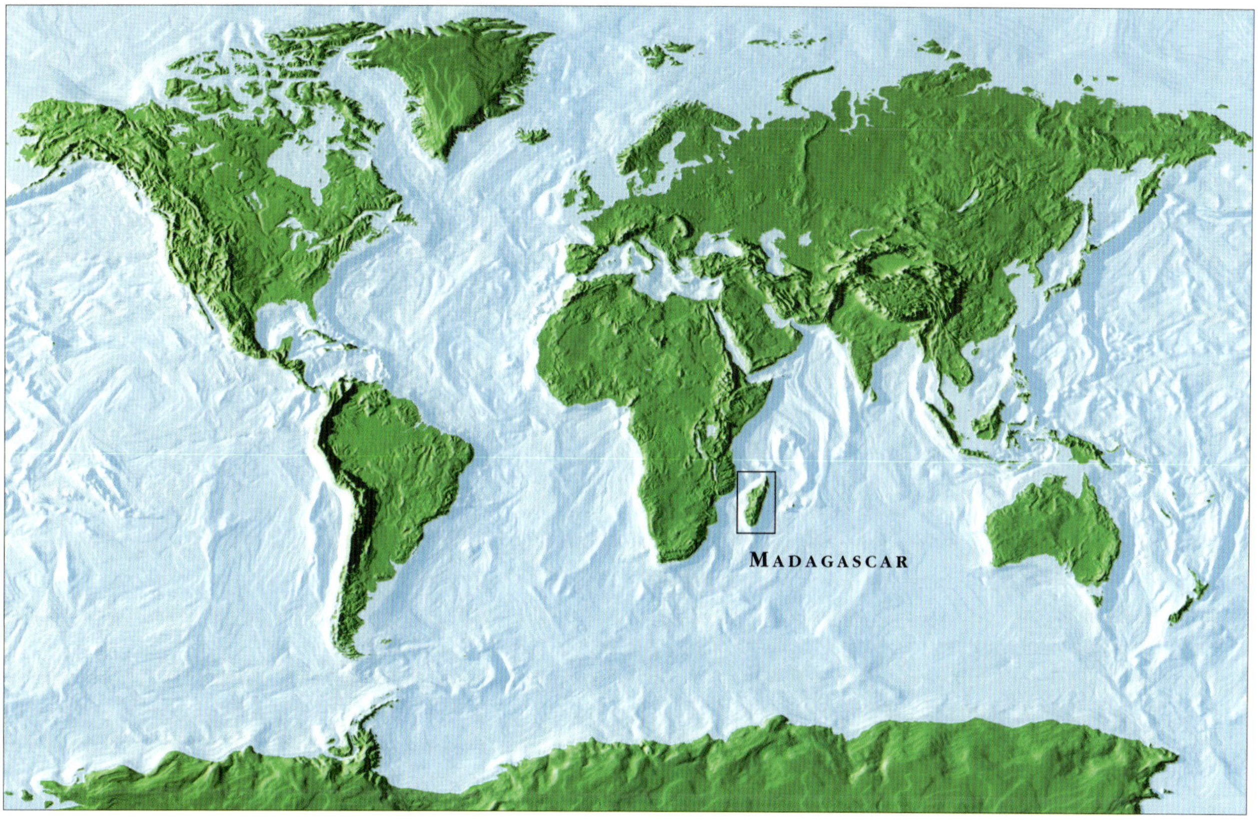

THE PALM FAMILY (PALMAE OR ARECACEAE) IS DIVISIBLE into six subfamilies (Uhl & Dransfield 1987). Recent molecular studies linked to analyses of palm phylogeny using morphological and anatomical data (Uhl et al., in press) provide some support for the demarcation of these subtribes, but also suggest some redefinition may be necessary. The subfamilies are further divided into tribes and subtribes which include what are generally regarded as being morphologically similar and closely related genera, each group representing a separate line of evolution. Representatives of four of the currently recognised subfamilies are known to occur in Madagascar; the two subfamilies that do not occur naturally on the island are *Phytelephantoideae* (endemic to northern South America) and *Nypoideae* (occurring naturally from the Bay of Bengal to the West Pacific, though with a much wider distribution in the past). One of the four subfamilies, *Calamoideae,* with 22 genera and about 650 species world-wide, is represented in Madagascar by a single species, *Raphia farinifera,* which, we believe, is probably introduced. Thus most of the palm flora belongs to three subfamilies, *Coryphoideae, Ceroxyloideae* and *Arecoideae.*

With over 170 species arranged in 16 genera, the palm flora of Madagascar is certainly diverse on a global scale. It has about three times as many palm species as does the whole continent of Africa. Of the 170 or more species, all but five are found nowhere else in the world. The five non-endemic species are *Hyphaene coriacea*, widespread in East and Southeast Africa, *Phoenix reclinata* and *Elaeis guineensis*, both widespread through subsaharan Africa, *Raphia farinifera*, an important subsistence crop palm in East Africa, and *Cocos nucifera*, the coconut, widespread throughout the coastal areas of the Indian Ocean and elsewhere in the tropics but probably native to the west Pacific. We include in our treatment the few palms that occur in the Comoro Islands, that lie between Madagascar and Africa, and one palm, *Dypsis pembana*, that is endemic to the island of Pemba off the coast of Tanzania, but belongs to a quintessential Madagascar genus. Of the sixteen genera, five are monotypic endemics (*Satranala, Bismarckia, Lemurophoenix, Beccariophoenix* and *Voanioala*), two are endemic with only two species each (*Masoala* and *Marojejya*) and two, *Ravenea* and *Dypsis* are largely confined to Madagascar but with outliers in the Comoro Islands (and (*Dypsis*) Pemba). *Orania*, with about 19 species world-wide, is largely confined to Southeast Asia but there are three species endemic to Madagascar, and *Borassus*, with about six species world-wide, occurring in Africa, India, Southeast Asia and Malesia, has two representatives in Madagascar, both endemic. Thus there is a very high level of endemism at both the generic and species level.

How do the palms of Madagascar relate to those of the rest of the world? In some cases the relationships seem obvious, while in other instances the relationships remain obscure and in need of much more study.

As currently circumscribed, Madagascar has representatives of twelve different evolutionary lines of palms, delimited at either the subtribal, or (in subfamilies which are divided only to the tribal level) tribal level. These different evolutionary lines in Madagascar may represent independent immigration events into Madagascar or may represent evolution at the subtribal or tribal level in Madagascar.

CORYPHOIDEAE

Of the three Coryphoid tribes, two are present in Madagascar. *Phoeniceae*, with the single genus *Phoenix*, is represented by the single species *P. reclinata*, mentioned above. It is tempting to suggest that its presence in Madagascar is due to relatively recent dispersal from Africa. It is unlikely to have been introduced by man, though this may be the case; long distance dispersal by birds seems more likely.

Borasseae is represented by four genera. *Borassus* (subtribe *Lataniinae*) has two species but their relationships with African species are not clear. Subtribe *Hyphaeninae* is represented by the non-endemic *Hyphaene coriacea*, and the two endemic monotypic genera *Satranala* and *Bismarckia*. *Hyphaene* has about ten species in Africa, Arabia and western India while the only other genus in the subtribe is *Medemia*, on the verge of extinction in Egypt and the Sudan. We speculate that *Satranala* and *Bismarckia* represent the remains of an early radiation of this line of evolution in Madagascar, probably predating the separation of Madagascar from Africa.

CEROXYLOIDEAE

Ceroxyloideae, as circumscribed in Genera Palmarum, is probably polyphyletic (Uhl et al., in press); however, the three constituent tribes (*Cyclospatheae, Ceroxyleae* and *Hyophorbeae*) are almost certainly each monophyletic. Only *Ceroxyleae* occurs in Madagascar, though we may mention in passing the remarkable distribution of *Hyophorbeae*, well represented in Central America, but with one genus in the Old World, *Hyophorbe*, endemic to the Mascarene Islands. *Ceroxyleae* consists of four genera: *Ceroxylon* in South America (at high altitudes), *Juania* in the Juan Fernandez Islands off the coast of Chile, *Oraniopsis* in Queensland, Australia, and *Ravenea*, with 17 species in Madagascar and the Comoros. This undoubtedly natural tribe is remarkably uniform in general morphology and is relatively unspecialised. Its wide distribution scattered through the southern hemisphere is suggestive of radiation following the break-up of the great southern continent of Gondwana.

ARECOIDEAE

Two of the six currently recognised tribes of *Arecoideae* occur in Madagascar, viz.: *Areceae* and *Cocoeae*. The former is represented by five genera belonging to four subtribes, *Oraniinae, Dypsidinae, Lemurophoenicinae* and *Masoalinae*.

Oraniinae, at first regarded as comprising two genera, *Orania* and *Halmoorea*, is now monotypic with the sinking of *Halmoorea* into synonymy. This subtribe is unusual among the *Areceae* in having a gynoecium with three locules and three ovules as opposed to the pseudomonomerous gynoecium (with a basic three-carpellate structure, but with only one fertile locule with one ovule) that is so characteristic of the tribe. *Orania* is widespread in Southeast Asia and Malesia and is particularly diverse in New Guinea. The only distinction between the three Madagascar species and the rest of the genus is in the presence of ramenta on the undersurface of the leaflets in the Madagascar species. It appears that *Oraniinae* are probably most closely related to *Manicariinae* with the single genus *Manicaria* in the New World.

In Genera Palmarum (Uhl & Dransfield 1987) *Dypsidinae* was separated from other groupings within tribe *Areceae* in subfamily *Arecoideae* on the combination of a series of characters, including a pseudomonomerous gynoecium, endocarp lacking an operculum, stigmatic remains sub-basal in the fruit, staminodes not forming a cupule, petals of pistillate flower free, not connate, staminate flowers symmetrical with free imbricate sepals, leaflet tips almost never praemorse, and stamens three or six. Thus circumscribed, *Dypsidinae* is not well defined by the presence of shared derived characters (synapomorphies), but is partially defined by absences. However, the single most useful attribute for distinguishing this group of palms is that they all occur, with a very few exceptions on nearby islands, on Madagascar. Clearly the group is rather tenuously separated from other lines of evolution within tribe Areceae. Now that all genera previously recognised within the subtribe have been included in synonymy, *Dypsis* has become an extraordinarily diverse genus, perhaps the most diverse of all palm genera, and, because of that, one of the most difficult to define. There is as great or even greater variation within this one genus as can be found within some subtribes or tribes. Most tantalising of all is the question of relationships. How is *Dypsis* related to other members of *Areceae*? Are its relationships with New World genera or with Old World genera? It might seem obvious to look at those palms which are geographically closest to Madagascar. In the Seychelles and Mascarenes, two subtribes within *Areceae* are present that are thought to be closely related to each other; *Oncospermatinae*, with two endemic genera in the Mascarenes and five endemic genera in the Seychelles (and with one more genus, *Oncosperma*, endemic to the lands bordering the eastern Indian Ocean), and *Iguanurinae* with one genus (*Dictyosperma*) endemic to the Mascarenes, and about 27 genera in Southeast Asia, Malesia and the West Pacific. However, in fruit structure members of these two subtribes are fundamentally different from *Dypsis* in the presence of an operculum in the endocarp. At the start of this project we were impressed by the many similarities between the New World genus *Leopoldinia* and *Vonitra*, the latter now included in *Dypsis*, but we are unable, as yet, to say whether these are parallelisms or representative of real relationships. There are also some superficial similarities with many other Arecoid subtribes. Until a critical phylogenetic analysis of *Dypsis* itself, and all the groupings within *Areceae*, is carried out, preferably supported by molecular studies, it seems pointless to speculate on the relationships of *Dypsidinae*.

The newly recognised *Lemurophoenicinae* with the single genus *Lemurophoenix* is superficially similar to *Dypsidinae* in inflorescence structure and external morphology of the flowers. However, the staminate flowers are multistaminate (with over 50 stamens) and the fruit morphology is unique; the pericarp is corky-warted, the endocarp is well developed, with a heart-shaped basal button reminiscent of that found in *Orania, Manicaria* and *Phytelephas*, the endosperm is shallowly and sparsely ruminate and the embryo is apical, a most unusual combination of features. The relationships of this genus are not clear but we are tempted to speculate that it represents a sister group to *Dypsidinae*.

The last subtribe is also problematic. This is the newly described *Masoalinae*, endemic to Madagascar and including the two genera *Masoala* and *Marojejya*, each with two species. In Genera Palmarum, *Masoala*, at that time known only from its type, was unplaced for lack of good data and *Marojejya* was included with the West African genus *Sclerosperma* in the *Sclerospermatinae*. *Sclerospermatinae* was defined by the following characteristics: massive short-stemmed or acaulescent palms with condensed, short interfoliar inflorescences; prophyll short; peduncular bracts one complete and several incomplete; rachis short; flowers sunken in pits in the rachillae; staminate flower with sepals free and scarcely imbricate. When we refound *Masoala* in 1986 it became obvious that it and *Marojejya* were probably closely related. The similarity in habit is striking, and the differences between the genera are based on sexuality of inflorescences and whether the inflorescences are condensed or diffusely branched. At one point we even contemplated uniting the two (see under *Masoalinae*) and it was then that we began to think that the similarities between *Sclerosperma* and *Marojejya* were superficial and represented character states linked with the parallel evolution of condensed inflorescences. In particular, we suggest that the large number of incomplete empty peduncular bracts in both *Sclerosperma* and *Marojejya* may be related to the condensation of a more diffusely branched inflorescence. It is on these grounds that we have separated the two Madagascar genera from *Sclerospermatinae* into a new subtribe. However, we still do not understand the relationships of *Masoalinae* with other members of tribe *Areceae*, and it is possible that this may only become clear when data from molecular studies become available. However, we suggest that the *Masoalinae* represents a distinct evolutionary line of palms that is unique to Madagascar.

All other Arecoid palms in Madagascar belong to tribe *Cocoeae*. Of these *Cocos nucifera*, the coconut, probably native to the western part of the Pacific, may well have been one of the first economically important plants introduced into Madagascar by seafaring people. It is now naturalised in coastal areas, particularly in the north and east of the island. Madagascar populations of the coconut show a high degree of variability. *Elaeis guineensis*, the African oil palm occurs along watercourses in the north and north-west of the island and occurs naturalised around villages through much of the East Coast lowlands. The form that commonly occurs appears to be very similar if not identical to the wild-type oil palm, found throughout the wetter parts of Africa. It is not known whether the oil palm in Madagascar is truly native or whether it was an early introduction.

The two remaining genera of cocoid palm in Madagascar are both monotypic endemics of very great interest. *Beccariophoenix* has been placed in its own subtribe, *Beccariophoenicinae*, distinctive in its

extraordinary inflorescence. *Voanioala* is included in *Butiinae*, a subtribe of ten genera, largely confined to South America, but with one monotypic genus in South Africa (*Jubaeopsis*), one genus in the western Pacific (*Cocos*) and a recently extinct palm on Easter Island (*Paschalococos*). That this subtribe, widespread through the southern hemisphere, should be present on the Gondwana fragment that is Madagascar is perhaps not altogether surprising. The relationships within this subtribe are not yet clear but are currently under study by Noblick and Sanders. Whatever they may be, it seems probable that *Voanioala* represents an ancient line of evolution that survives as a relict in Madagascar.

What makes the palm flora of Madagascar so peculiar is the very high rate of endemism, and the taxonomically isolated nature of some of its components. This seems to argue for a long and isolated evolutionary history. We may speculate that, apart from the five non-endemic species, the palms of Madagascar are representatives of evolutionary lines that were present on the island before it became separated from Africa in the late Cretaceous and from India in the early Tertiary. In marked contrast is the size of the African palm flora, yet it too has representatives of several evolutionary lines, so, despite the palms being few in number, there is nevertheless considerable diversity. There is good evidence from fossil pollen for a formerly much richer palm flora occurring in Africa, in particular in South Africa during the Tertiary, a flora that is now extinct, probably due to major climatic change. Perhaps in Madagascar we see the remnants of this once much richer southern Gondwanan palm flora, that was rafted to Madagascar at the break-up of Gondwana, and, in one line (*Dypsidinae*), has radiated spectacularly in most major habitats of the island.

Palm Habitats

Madagascar is the fourth largest island in the world, extending about 1500 km from 11° 57´ S to 25° 32´ S, covering an area of about 590,000 sq. kilometres and with an altitudinal range from sea level to 2876 m. It has a correspondingly wide range of habitats, ranging from lowland and montane rain forests to spiny desert scrub and alpine grassland.

In very general terms, the island consists of a backbone of ancient Precambian rocks, forming the high plateaux, that in the east form an escarpment and in the west slope more gently towards the Mozambique Channel. The coastal plain in the east between the escarpment and the Indian Ocean is generally rather narrow; to the west there is a much broader area of sedimentary rocks. Volcanic rock outcrops occur scattered throughout the central, north and north-western part of the island, and more rarely in the south. Limestone, sometimes occurring as spectacular karst, is confined to the western and northwestern coastal plain in areas of low rainfall, and is, sadly, of little significance for palms. The highest mountains are in the north, but throughout the high plateaux are scattered peaks, some of which reach over 1800 m above sea level. Much of the vegetation of the island has been modified, often drastically, or destroyed by man's activities over the c. 1500 years since man first colonised the island; great erosion gullies (*lavaka*) eat into the cleared lands of the plateaux and bleed red soil via the rivers, out into the Mozambique Channel.

The vegetation of Madagascar has been discussed at length in Battistini & Richard-Vindard (1972) and Koechlin et al. (1974). Humbert (1965) published a detailed vegetation map of the island with an accompanying commentary. It is not surprising that the bulk of the palm flora is confined to the perhumid rain forests of the East Coast and the northern mountains. Nevertheless there are palms specially adapted to drier areas.

Lowland Rain Forest

Lowland rain forest occurs in areas of high rainfall (more than 2400 mm a year, and sometimes greater than 3000 mm a year) at elevations from sea level up to about 800 m elevation (Koechlin et al. 1974); there is normally no clearly defined dry season, although some times

The forested island of Nosy Mangabe near Maroantsetra at the northern end of the Bay of Antongil.

The Masoala Peninsula, one of the richest areas for palms, seen from Nosy Mangabe, across the Bay of Antongil.

of the year are consistently wetter than others, and the temperature varies annually between 12 and 30 degrees Celsius. The choice of 800 m as the altitudinal limit of lowland forests may be rather arbitrary as where forest still exists at this elevation, there is usually an imperceptible transition from lowland forest to submontane forest. Nevertheless, it is usually at about this elevation that discernible changes in the palm flora take place. Lowland rain forests so defined occur in a belt running from just south of Iharana (Vohemar), south of Antsiranana in the north, southwards to Tolanaro; this is the "forêt dense ombrophile orientale" of Humbert & Cours Darne (1965). There are also areas of lowland forest away from this eastern rain forest belt around the Bay of Ampasindava, often regarded as representing a separate different floristic region, Sambirano, and in the area of Ambohitra (Montagne d'Ambre). All these areas of lowland rain forest are now reduced to a small fraction of their presumed original extent. Much of the clearance and destruction of forest has been carried out during the long history of shifting cultivation but some destruction has resulted from logging, mining and agricultural estate development. After the forest has been cleared secondary scrub (*savoka*) develops, often dominated by the traveller's palm, *Ravenala madagascariensis* (*Strelitziaceae*: not a palm at all) and introduced weeds. Very few palms survive in *savoka*, some of the few being *Dypsis mananjarensis*, the coastal form of *Ravenea sambiranensis* and, in swampy areas, the ubiquitous *Raphia farinifera*. Despite the destruction that continues, there are still significant areas of lowland forest, many of them forming the central parts of legally protected areas. Some of these blocks of rain forest remain very poorly known, their remoteness making for difficulty of access and thus partly ensuring their survival. A few significant areas of accessible lowland rain forest are described below; these descriptions emphasise the extraordinary local endemism in the palm flora of Madagascar, in particular of the undergrowth species. Although some species such as *Dypsis fibrosa*, *D. pinnatifrons* and *D. nodifera* occur throughout the lowland rain forest belt, others seem to be confined to very small areas. With forest destruction and the consequent extreme fragmentation of the remaining stands of forest, the prospects for the survival of some narrowly endemic palm species is bleak indeed. The precariousness of much of the lowland palm flora has only really become apparent during the course of this project, and emphasises the importance of detailed field surveys of threatened plants rather than mere desk exercises.

Masoala Peninsula

The most beautiful, least disturbed accessible areas of surviving lowland forest are to be found around the Bay of Antongil in the north-east of the island, in particular on the west coast of the Masoala Peninsula. In a few places around the Bay forests can be found from sea level on the coast up to the summits of the nearby mountains that range from about 400 to 1200 m above sea level. This is an area of high rainfall and high relief. Rivers are short, losing altitude rapidly, with an abundance of gorges and waterfalls and their valleys are in consequence often extremely beautiful. On slopes near the sea and further inland, *Dypsis lastelliana* and *D. fibrosa* are abundant and conspicuous, especially in partially cleared land near villages.

One of the most common and conspicuous palms in the valley bottoms is *Dypsis crinita.* It is particularly abundant on river banks, starting life as a rheophyte subjected to flood waters, but eventually reaching 10 to 15 m tall. Forested areas of Masoala are home to some of the rarest and most sought after palm species. In a narrow valley at the northern end of the peninsula grows *Lemurophoenix halleuxii*, its great trunks towering above the forest floor, the ground below carpeted with fallen fruits. With it grow other large palms such as *Dypsis bejofo, D. tsaravoasira, Orania trispatha* and *O. ravaka. Marojejya insignis* in a form with neat pinnate leaves occurs in the undergrowth along with a wealth of small dypsids such as *D. pachyramea, D. coriacea, D. procera, D. lantzeana, D. mocquerysiana, D. forficifolia* and *D. thiryana.* This area is famous also as the only substantiated locality in the world for *Marojejya darianii* (although rumours suggest it occurs elsewhere). Further to the south lies the only known locality for the forest coconut, *Voanioala gerardii,* that occurs as scattered subcanopy trees in dense very humid forest in a valley bottom, growing with the spectacular canopy palms *Dypsis tokoravina, D. tsaravotsira, D. hovomantsina, Ravenea robustior, Orania trispatha* and *O. longisquama*; in the undergrowth in the valley bottom grow two massive litter-trapping palms, *Dypsis perrieri* and *Masoala madagascariensis,* and the short sub-canopy tree palm *Ravenea dransfieldii* as well as several small species such as *Dypsis lantzeana, D. forficifolia, D. thiryana, D. mocquerysiana* and *D. pachyramea.* However, it seems that areas that are particularly rich in palms are of limited extent, and seem to occur in rather open forest that is also rich in pandans and climbing bamboo. Where the forest canopy is closed, there seem to be fewer palm species and the canopy palms, in particular, occur as only scattered individuals. What controls the distribution of these patches of the palm-rich facies of the lowland forest on Masoala is not understood, but, as in the Biosphere Reserve of Mananara Avaratra (see below) the underlying rock type may be important.

Mananara Avaratra

The Biosphere Reserve of Mananara Avaratra at the southern end of the Bay of Antongil represents, according to Nicoll & Langrand (1989), the largest remaining fragment of lowland East Coast rain forest. The reserve lies to the south and west of the small town of Mananara Avaratra and has been designated

View of the eastern boundary of the Mananara Avaratra Biosphere Reserve: fragments of more or less intact forest alternate with cleared areas, now dominated by **Ravenala madagascariensis**; clove trees, a major cash crop, in the foreground.

as a Biosphere Reserve under the United Nations Development Programme (UNDP/PNUD). Much of the forest remains rather difficult of access, but it is particularly important for palms. A total of 42 different palm species have been recorded from the reserve, making this the richest site known in Madagascar for palms. The lowland rain forest with high closed canopy is similar floristically to that of the Masoala Peninsula and many of the palms that occur on Masoala are also to be found in the reserve. In one area there is an outcrop of a dark apparently ultramafic rock that carries a most unusual palm flora (see below). Although the area is legally protected as a Biosphere Reserve, the primary forest within the reserve boundary is of limited extent and surrounding villagers of course continue to practise shifting cultivation. The management of this reserve will require an imaginative combination of sustainable development and strict conservation if the remarkable palm flora is to survive.

Île Sainte Marie

On Île Sainte Marie there are three remaining areas of forest, two consisting mainly of coastal white-sand forest typical of the East Coast (see below). The third area, typical East Coast lowland rain forest, is of historical importance to palm botany. This is the Forêt de Kalalao, a fragmented area of forest lying just to the east of the Pointe de Tafondrou, Lokintsy. We believe that this must be synonymous with Forêt de Tafondrou where Boivin collected in the early 19th Century, the type locality of several species of *Dypsis*. Although degraded and criss-crossed by paths, and with many signs of minor tree cutting activity, there is nevertheless closed and rather beautiful vegetation here, with an interesting palm flora, typical of forest developed on lateritic soils along the East Coast. Here can be seen *Dypsis lastelliana, D. corniculata, D. fasciculata, D. forficifolia, D. paludosa, D. boiviniana, D. fibrosa, Orania ravaka, Ravenea robustior* and *R. sambiranensis*. Although the forests of Île Sainte Marie may not be of much significance for the conservation of fauna (see Goodman 1993), we believe the floristic importance is great. Furthermore, the forest does have some significance for water conservation and as a tourist attraction for nearby hotels, and efforts should be made to try to protect it from further degradation.

Inside the Mananara Avaratra Biosphere Reserve; **Dypsis crinita** occurs in great abundance along river beds (Photo: Brian Rogers).

Lokobe, Nosy Be

The island of Nosy Be lies opposite to the town of Ambanja in the northwest of Madagascar. In the south of Nosy Be is a small forested peninsula near the village of Ampasindava. This is the site of the famous Lokobe Reserve, type locality for several palm species. What is remarkable about the palm flora of Lokobe, and something that is perhaps typical of the lowland forest of the Sambirano region, is the paucity of small undergrowth palmlets that are so characteristic of the forests of the East Coast. The palm flora is a rather poor one but includes a few very local species. Tree palms in the Lokobe Reserve include *Ravenea sambiranensis, Dypsis madagascariensis* (in the robust form much cultivated as *Chrysalidocarpus lucubensis*), *D. ampasindavae* and *D. nossibensis*. In the undergrowth, the only smaller palm is the very robust form of *Dypsis pinnatifrons* that was previously thought to be a distinct species, *D. sambiranensis*. The precarious state of the palms at Lokobe is described by Adany *et al.* (1994).

Betampona

Betampona is an area of lowland rain forest in the foothills of the East Coast escarpment near Toamasina. As an accessible area of forest and a protected area, Betampona has often been visited by botanists and there are a number of palm specimens from the reserve and surrounding areas. Betampona consists of primary and secondary forest; the palm flora is a rich one and includes some species not known elsewhere. Tree palms include *Orania longisquama, Ravenea dransfieldii* and *R. sambiranensis* and there is a rich assemblage of dwarf undergrowth palms such as *Dypsis schatzii, D. betamponensis, D. corniculata, D. glabrescens, D. angustifolia, D. nodifera, D. pervillei* and *D. fasciculata; Marojejya insignis* has also been recorded.

Manombo

South of Farafangana lies a small area of lowland rain forest, perhaps the last significant area of lowland forest remaining in this part of Madagascar. Although the area is protected as a Réserve Speciale, illegal cutting of timber and firewood goes on apace. Within this forest are unusual palms, some rare, others not known elsewhere, all endangered at this locality. They include tree palms such as *Dypsis mananjarensis, D. nauseosa, D. fibrosa, Orania trispatha, Ravenea robustior, R. julietiae* and *R. lakatra*, and small undergrowth palms such as *Dypsis mahia, D. angusta, D. elegans, D. digitata, D. integra, D. intermedia, D. singularis, D. simianensis* and *D. laevis*.

Lowland Forest on Unusual Substrates

Quartzite ridges

In some areas of the lowlands, outcrops of quartzite occur and on quartzite ridges, soils tend to be very poor and thin, and in such conditions an unusual palm flora is developed. On one quartzite ridge to the north of Marojejy, a curious open low canopy forest occurs with an abundance of *Masoala madagascariensis* and the climbing bamboo *Schizostachyum parviflorum.*

On quartzite ridges at about 500 m elevation near Ifanadiana, Fianarantsoa Province, we found an abundance of unusual palms including *Marojejya insignis, Masoala kona, Ravenea dransfieldii, Dypsis scandens* and a particularly beautiful densely clustering form of *D. linearis,* all growing in very low-canopy open forest of small-leafed trees, the ground carpeted with humus and moss.

Ultramafic rock

In one area of the Biosphere Reserve of Mananara Avaratra we have come across extraordinary lowland forest quite unlike neighbouring forest areas. It appears that the forest is developed on some sort of ultramafic rock (though not on a serpentine). The rock is a fine grained dark and very dense igneous one, in marked contrast to neighbouring pale granites; the rock and soil samples collected have yet to be analysed.

Where the rock is exposed, the forest has a low and rather open canopy in marked contrast to the neighbouring areas on granite with higher closed canopy, and there is an extraordinary abundance of arborescent monocotyledonous plants such as *Ravenala madagascariensis,* pandans (at least six different species), *Dracaena* sp., climbing bamboos and, of course, palms. Among the palms are many unusual species, of very restricted distribution elsewhere, and some are known only from this area. The outcrops, occurring along a deep river valley, provide a wealth of micro-habitats. In the swampy valley bottom can be found the beautiful *Dypsis beentjei* and on the wet lower slopes *Ravenea albicans, Masoala madagascariensis, Marojejya insignis, Dypsis ovobontsira, D. hovomantsina* and *D. antanambensis.*

Higher up on upper slopes and ridge tops, there tends to be a deep almost peaty humus accumulation and here grow *Satranala decussilvae, D. serpentina, D. antanambensis, Masoala madagascariensis, Marojejya insignis* and an abundance of *Orania ravaka,* all intermingled with pandans. From a palm botanist's point of view this remarkable vegetation is one of the most important areas of forest in Madagascar. Fortunately, it is within the boundary of the reserve and is also rather remote. Nevertheless, the area of forest in the Biosphere Reserve is limited and surrounded on all sides by shifting agriculture.

White sand forest

The East Coast from Tolanaro in the south up to Mahavelona north of Toamasina is remarkable for being an almost straight line running north-north-east. Throughout much of its length in this area the sandy, highly mobile beaches are pounded by Indian Ocean breakers. Behind the beach can be found a series of ancient raised beaches, marking previous shorelines. Further to the north, the coastline becomes more irregular, and there are fringing reefs that temper the effects of the Indian Ocean. At Lohatanjona Antsiraka (Pointe à Larrée) currents have deposited a sandy promontory that sticks out into the ocean opposite Île Sainte Marie. Fluctuations in currents and sea levels here have produced a remarkable striping of vegetation, highly visible from the air and on maps, marking the ancient shore lines. This coastal vegeta-

Vegetation in a swale behind a coastal white sand ridge at Ambila-Lemaitso; pandans occur in abundance along with many unusual plants.

tion developed on white sands is very distinctive but greatly threatened by development; despite the soils being generally very poor, the forest continues to be cleared for shifting agriculture, and tourist developments; mining for ilmenite (titanium ore) (near Tolanaro) and just plain pyromania add to the threat.

There are two major habitats on these coastal sands; on the higher ground on the ancient sand bars a highly characteristic forest is developed, generally with rather low canopy and with a preponderance of small-leafed trees. This is the habitat, in different areas, of *Dypsis forficifolia, D. nodifera, D. scottiana, D. saintelucei, D. fasciculata, D. lutescens, D. arenarum, D. psammophila, D. sanctae-mariae, D. dransfieldii, D. ambilaensis*, the coastal form of *Ravenea sambiranensis* and very rarely *Beccariophoenix madagascariensis*. Where this forest is disturbed, thickets of *Erica* (*Philippia*) develop and some of the palms may survive, especially *Dypsis lutescens*. Between the sand bars and old raised beaches are low-lying swales or lagoons running parallel to the coast, the largest one of which has been exploited by man as the Pangalanes Canal that runs for hundreds of kilometres behind the coast. In the swampy swales again a very characteristic vegetation is developed, dominated by sedges, pandans, ravenala and low trees. This is the major habitat of the pitcher plant, *Nepenthes madagascariensis*. There are few palms in this habitat, one particularly distinctive species being *Dypsis paludosa* that forms thickets in coastal swamp forest near Ambila-Lemaitso.

On exposed rocky shores on the East Coast north of Soanierana-Ivongo and around the Bay of Antongil, the only palm that seems to survive the effects of wind and spray in the wind-pruned vegetation is *D. lutescens*.

MONTANE FORESTS

Lowland rain forest occurs in high rainfall areas on the East Coast of Madagascar; at varying altitudes it is replaced by submontane or montane rain forest. Where rainfall is particularly high, as in the area of the bay of Antongil, submontane forest can occur at relatively low elevations and on isolated peaks montane forest can occur at surprisingly low elevations (an example of a reverse Massenerhebung effect). A good example is provided by Vatovavy, an isolated peak south of Mananjary. Here lowland forest survives on the lower slopes, but up near the summit cliffs at 400 to 500 m above sea level stunted low crown forest of small leafed trees with abundant moss in the undergrowth is found; the presence of this montane-

Montane forest on the eastern escarpment near Ranomafana (Photo: Brian Rogers).

type forest at such a low elevation may be due to a combination of factors such as poor and shallow soils, high wind and high precipitation in the summit region. The forests at the summit of Vatovavy are dominated by the elegant small tree palm *Dypsis basilonga*, that occurs in abundance. In the undergrowth of the forest here can be found *Dypsis angusta, D. eriostachys* and *D. trapezoidea*.

Analamazaotra and Mantady

The classic area of upland forest is Analamazaotra, where the old French forestry station of Perinet was founded. The area of forest near the railway station of Andasibe (Perinet) is now much degraded and shows the effects of a recent cyclone (Geralda, in early 1994) and visitor pressure. Nevertheless, as the most accessible area of submontane forest on the East Coast escarpment and type locality of many palms, this area is still of considerable interest. However, the newly gazetted national park of Mantady to the north of Analamazaotra, gives an impression of what the forest at Analamazaotra may once have looked like. There is a great wealth of palms in these escarpment forests. The most conspicuous large palm in the forest is *Ravenea robustior*, which occurs in abundance; other species of *Ravenea* also occur and include *R. madagascariensis*, the poorly known *R. latisecta, R. lakatra* (now reduced to pruned juvenile palms) and *R. louvelii*. This last species is known from a single ridge where it grows as a squat litter-trapping palm in the forest undergrowth.

Another litter-trapper, *Marojejya insignis*, is a rare palm of these escarpment forests. *Beccariophoenix madagascariensis* still occurs on ridges in the escarpment forest in the Andasibe area but it is much threatened by the depredations of seed collectors and hunters who fell them for their edible palm-heart. There are few large canopy species of *Dypsis* in this area, which seems odd when comparisons are made with forest elsewhere and it may be that such large species (such as *D. pilulifera*, have become very rare due to felling for house construction and palm-heart. *Dypsis baronii* can be abundant in the undergrowth and there are many small undergrowth palmlets such as *D. hildebrandtii, D. concinna, D. louvelii, D. jumelleana* and *D. heterophylla*. At higher elevations on the escarpment, there tend to be far fewer species. At Mandraka, for example, only two species are abundant, *Dypsis baronii* and *Ravenea madagascariensis*.

View of one of the mountain peaks in the Marojejy reserve.

Andohahela

This, the southernmost rain forest in Madagascar, lies at elevations of 200 to 1900 m above sea level. The lower regions are rather degraded but above about 500 m elevation are splendid submontane forests with a great abundance of individuals of a few species of large canopy palm, including *Ravenea robustior, R. kroctana* and *Dypsis malcomberi*; in the undergrowth grows *Marojejya insignis* and several dwarf species of *Dypsis* including *D. catatiana, D. mcdonaldiana, D. tenuissima, D. scottiana, D. pinnatifrons* and *D. linearis*. At lower altitudes and further to the west the vegetation changes dramatically over a very short distance (see below, under Transitional forest).

Marojejy

The most spectacular montane forests are to be found on the great mountain massifs of Tsaratanana, Marojejy and Manongarivo in the north of the island. The first of these areas includes the highest mountain in Madagascar, Maromokotro, 2876 m above sea level, a remote area, rarely visited and home to several palms that are known only from a few specimens (e.g., *Dypsis ankaizinensis*). Marojejy (highest peak 2133 m) and Manongarivo (highest point 1876 m) have been the subject of recent intensive collecting and sampling has included palms. Both areas are of exceptional interest to the palm botanist. Marojejy is important as the type locality of the eponymous genus *Marojejya* (though the type species *M. insignis* is now known to occur throughout the length of the East Coast rain forests, from Marojejy to Tolanaro). Palm-rich forests on Marojejy are encountered from about 700 m elevation to about 1100 m; above this, the number of species decreases. In such palm-rich submontane forests, besides *Marojejya insignis*, two other conspicuous large litter-trapping species occur, *Dypsis marojejyi* and *D. perrieri*, the latter in extraordinary abundance. Tall canopy palms include *Dypsis pilulifera, D. tsaravoasira, Ravenea robustior* and *R. sambiranensis.* In the undergrowth can be found the very robust form of *D. pinnatifrons* that was once thought to be a distinct species, *D. sambiranensis*, together with *D. cookei, D. heterophylla, D. thiryana, D. baronii, D. acuminum, D. spicata, D. lokohoensis, D. pilulifera* and *D. oreophila*. In moss forest at higher elevations, different species occur, including *Dypsis bonsai, D. andrianatonga, D. heteromorpha, D. coursii* and the enigmatic *Ravenea nana* that has not been recollected in the recent survey work by Miller. The remarkably large-fruited *Dypsis pumila* occurs above 2000 m in high altitude bogs (as illustrated in Humbert 1955), the highest recorded elevation for a palm in Madagascar.

Manongarivo

Manongarivo is a rugged mountain massif in the Sambirano region, now the focus of an integrated conservation policy in which local people are deeply involved. The absence of very small undergrowth palms in the rain forest here is surprising. Large palms include *Ravenea sambiranensis, R. robustior, Dypsis pilulifera, D. madagascariensis, D. lastelliana* (and an unusual palm very similar to this species but with grouped leaflets) and *D. ampasindavae. D. crinita, D. rivularis* and *D. pinnatifrons* are common along streams at lower elevations and in mid-altitude forest. The only other medium-sized palm recorded is *D. hiarakae.*

Tsaratanana

Since the major expeditions of Perrier in the 1920s, very few palms have been collected on the highest mountain massif of Madagascar, Tsaratanana, and as neither of us has ever been there, we have no information to add to that of Perrier. It would seem probable that the mountain forests of Tsaratanana have much in common with those of Marojejy and Manongarivo, but several species (e.g. *Dypsis ankaizinensis* and *D. tsaratananensis*) are only known from Tsaratananana and have not been found elsewhere. Malcomber refound *Dypsis montana* in 1993 after 71 years, so there is still hope for the palms of Tsaratanana.

PLATEAU FORESTS

Little remains of the forests that once covered the plateau of Madagascar. Just how extensive the forest was is not known, and there are even suggestions that the plateau may not have been covered entirely by forest before the advent of man in the middle of the first millennium. Whatever the true extent in the past may have been, the remaining forest is greatly fragmented and continues to disappear at an alarming rate. The main cause of its decrease is fire. Seasonal fires set to burn off dead grass and encourage the growth of young grass shoots for cattle grazing sweep through the plateau every year, filling the air with smoke. Such fires eat into the fragments of plateau forest, each year pushing back the forest margin. Removal of the vegetation cover then leads to soil erosion during the rainy season. The plateau landscape is scarred with huge erosion gullies (*lavaka*) and the great rivers such as the Betsiboka bleed bright red silt into the Mozambique Channel all through the year.

Dypsis decipiens in remnant plateau forest on the Itremo Massif (Photo: D. Du Puy).

The plateau forests nearest to Antananarivo lie over 100 km away at Anjozorobe and Ankazobe. A significant area of forest still exists north-west of Tsiroanomandidy at Ambohitsaratelo. These forests carry a very limited palm flora, the commonest large palms encountered being *Ravenea madagascariensis*, especially in the central and eastern part of the plateau, *R. sambiranensis*, especially in the western part of the plateau, and the beautiful *Dypsis oropedionis*. That any of these survive seems to us remarkable. *D. baronii* is the only palm to be found in the understory of the forests of the central part of the plateau. In *tapia* forests (rather low open forest composed of the Euphorbiaceous tree, *Uapaca bojeri*) *Dypsis decipiens* occurs rarely, as for example beside the main road to Mahajunga, beyond Ankazobe, and further south near Ambositra and beyond, *D. decipiens* is again encountered, and in a few places is abundant (see photo). The very rare *D. ambositrae* survives in a few gullies and open fields near Ambositra, but its future is very uncertain. There appear to be no dwarf undergrowth species in the plateau forests, which seems surprising. The fragments of mossy plateau forest south of Ambositra beside the main road near Ambatofitorahana that look so inviting are in fact devoid of palms. We have not explored the mountainous area of Andringitra, but there are early accounts by Jumelle and Perrier (1913) of extensive forests dominated by *Ravenea glauca*; whether these still survive is not known.

Transitional Forest

The southern limit of rain forest in Madagascar occurs just outside the tropics near Tolonaro in the Réserve Naturelle Integrale 11 of Andohahela. This is a large reserve that is divided into three areas; Parcelle 1 is typical of East Coast rain forests at their southern limit, Parcelle 2 is spiny forest while Parcelle 3 represents a transitional zone between the eastern rain forest and the dry spiny forests of the south. Parcelle 3 is of great interest to palm botany being the locus classicus for the triangle palm, *Dypsis decaryi*. Almost the entire wild population of this well known palm occurs within this area of transitional forest, where it has been the source of seed for export and leaves and timber for house construction. The palms are conspicuous and easily seen from the main road west of Tolanaro. *D. decaryi* has been the subject of a recent study by Joelisoa Ratsirarson, who has carried out demographic studies of the palm to act as a basis for management of the palm in the reserve.

Riverine Forest

In the drier parts of the island, forest may occur only along water courses. This riverine forest is thought by Koechlin et al. (1974) to be the primary habitat of *Bismarckia nobilis* and *Borassus madagascariensis*, but of this we are not certain. It is also the habitat of one

Burning of grassland every year is gradually destroying the last remnants of plateau forest. This area is beside the main road between Antananarivo and Mahajunga, photographed in 1992.

Riverine forest in a sandstone gorge at Isalo, habitat of **Dypsis onilahensis** and **Ravenea glauca.**

of the horticulturally most important palms, *Ravenea rivularis*. Other species of *Ravenea* that occur in riverine forest are *R. sambiranensis* (in some areas), *R. glauca*, in the sandstone gorges of Isalo and the most remarkable of them all, *R. musicalis* (see Beentje 1993b); this last species starts its life as a submerged aquatic, eventually the leaves emerging above the water and a short squat bottle trunk developing. This species is known only from a single river. *Ravenea rivularis* itself occurs on river banks and may, at least in the seedling state, behave as a rheophyte. Other palms that occur along water courses in otherwise unforested areas include *Borassus sambiranensis*, *Phoenix reclinata* and *Raphia farinifera*. *Dypsis onilahensis* occurs in riverine forest along the Analabe River in what is an otherwise quite dry area in the far north and in similar habitats in Isalo.

Spiny Forest

The only palm species recorded from the dry spiny forests is *Ravenea xerophila*, a very rare species that occurs over a small area in the south to the west of Tolanaro. Much threatened by fire, forest clearance and over-enthusiastic seed collection, this species deserves special conservation attention as one of the most unusual palms of Madagascar.

Secondary Habitats

Although some native palms will survive forest clearance and continue to grow in secondary forest, they are for the most part reproductively dead; although they flower and fruit regeneration does not occur outside the natural vegetation because seedlings are unable to survive. However, four species of palm do occur in one anthropogenic vegetation type, palm savannah, while a fifth probably introduced species, *Raphia farinifera*, is very abundant in some cleared landscapes in the east, particularly in the lowlands in high rainfall areas.

Palm savannahs

In the west of the island in the lowlands can be found palm savannahs, grasslands in which scattered or dense populations of palms occur. Four species of palms are to be found in this habitat: *Bismarckia nobilis*, *Hyphaene coriacea*, *Borassus madagascariensis* and *B. sambiranensis*, all members of Tribe *Borasseae*. Although this vegetation type is so striking and distinctive, it is generally agreed that such savannahs are anthropogenic; annual burning maintains a grassland, and the only perennial woody plants that can survive are fire-resistant species, such as these

Palm savannah with abundant **Bismarckia nobilis** (Photo: B. Du Puy and D. Du Puy).

Large areas of destroyed landscape in the eastern lowlands are covered in grassland and **Ravenala** thicket (**savoka**) (Photo: Brian Rogers).

four palm species. According to Koechlin et al. (1974), as mentioned above, the natural habitat of *Bismarckia nobilis* and *Borassus madagascariensis* is gallery forest. If this is really so, then it seems quite remarkable that these two palms have been so successful in invading the man-made grasslands. Undoubtedly they are able to resist fire. Koechlin et al. (loc. cit.) go on to suggest that as *Hyphaene coriacea* (as *H. shatan*) is not known in gallery forest at the present time, it must be one of the very few native genuinely savannah species in Madagascar. However, the same species in East Africa does occur in gallery forest and we assume that, like *Bismarckia, Hyphaene coriacea* has become much more abundant since the advent of man to Madagascar. All four species are of great ethnobotanical significance (see, for example, Audru & Andriatsarafara 1986), and the present stands may also have been influenced, in the past if not at present, by rudimentary management for their products.

NATURAL HISTORY NOTES

STUDIES OF THE NATURAL HISTORY OF THE PALMS OF Madagascar are in their infancy. The only significant detailed natural history research conducted so far is that of Joelisoa Ratsirarson, on *Dypsis decaryi* in the Andohahela Reserve near Tolanaro, and although publication of his results is imminent, we are not yet able to report his findings. He has worked on many aspects of the population biology and natural history of this rare palm, and, undoubtedly, his results will add greatly to our understanding of this species and may provide useful information for the management of the reserve.

What we present here are random notes on the natural history of palms in Madagascar, interlaced with speculation on the significance of some of the features of palm morphology. Of course, the ecology of Madagascar has been so much altered by man over the last 1500 years that it is quite possible that important interactions between palms and animals which evolved long before the arrival of man, may already be extinct. The elephant bird (*Aepyornis*) and the giant extinct lemurs (e.g., *Palaeopropithecus, Archaeoindris* and *Megaladapis*) may have played a role in the natural history of some palm species, a role that may not have been replaced by other animals. That this may be so provides endless scope for speculation that can never be tested or proved.

UNUSUAL GROWTH FORMS

Litter-trapping Plants

Several large palms in Madagascar have litter-trapping crowns. These palms are generally rather massive short-stemmed species that occur in the forest undergrowth. This growth form is exemplified by the genera *Masoala* and *Marojejya*, but there are also species of *Dypsis* such as *D. perrieri* and *D. marojejyi* which behave similarly; juveniles of *Beccariophoenix* and *Voanioala* and other palms also accumulate litter. Generally such palms have leaves that are held in a stiff "shuttlecock"; the leaves are usually marcescent, lacking petioles or with only very short petioles, and the sheaths and rachis may be deeply channelled. Leaves and other debris from the forest canopy falling into the crown are funnelled to the base of the leaves where they accumulate and rot. It is assumed that such litter accumulation may be important to the palm in capturing scarce nutrients, but no studies on this system have been conducted in Madagascar. In *Marojejya darianii* at least, the distal internodes develop adventitious roots that penetrate the accumulated litter and presumably take up nutrients. A bizarre feature of such litter-trapping palms is the presence of zigzag roots that connect the palm crown with the ground.

Roots zigzagging into the crown of **Masoala madagascariensis** in the Mananara Avaratra Biosphere Reserve.

First studied in Malaysia (see Davison 1988), these intriguing zigzag roots have also been observed in several species in Madagascar, and in *Oraniopsis* in Queensland. When the litter-trapping palm is still in its establishment phase, slender roots of neighbouring dicotyledonous trees grow into the litter in the palm crown from the adjacent leaf litter, along the margins of the sheaths, in a zigzag fashion. As they become older they gradually thicken and become woody. In the meantime, the palm continues to grow and accumulate litter. Eventually the palm stem emerges and grows vertically, leaf sheaths abscising and thus exposing the zigzag roots that appear to hang out of the palm crown. Thus it seems that the accumulation of litter by the palm may not always benefit the palm alone. In *Marojejya darianii*, the great overlapping auricles at the base of the leaf effectively produce a water-tight seal around the base of the crown; the huge entire leaves of this palm funnel rain water along with debris into the crown where it accumulates. If a leaf base of this palm is removed, a shower of water full of debris and invertebrates is released. What the role of water accumulation is in this species has not been studied. The crowns of the other species of *Marojejya*, of the two species of *Masoala* and of *Ravenea dransfieldii* are also capable of accumulating water, but not in such a spectacular way as in *Marojejya darianii.*

Climbing Palms

Until late 1994 no climbing palm was known in Madagascar, in marked contrast to Southeast Asian rain forests with their extraordinary abundance of rattan palms (belonging to subfamily *Calamoideae*), to equatorial Africa, also with rattan palms, and to the New World, where two subfamilies have climbing members (*Desmoncus* in *Arecoideae*: *Cocoeae* and one species of *Chamaedorea* in *Ceroxyloideae*: *Hyophorbeae*). Then in November 1994, as the manuscript of this book neared completion, we discovered a truly climbing palm growing on a quartzite ridge near to Ifanadiana. The palm, described here as *Dypsis scandens*, is superficially very similar to *Chamaedorea elatior* from Mexico. It has elongate internodes, flexible stems that are not self-supporting, and long leaves, with the upper leaflets reflexed with large basal pulvini, behaving as grapnels. Prior to the discovery of this new species, we had observed two other species of *Dypsis* which appear not to be entirely self-supporting, yet can scarcely be classed as climbers; these are *D. serpentina* and *D. andrianatonga*. These two species have suberect stems that branch aerially, successive branches being more slender, and sometimes flopping on surrounding vegetation. However, neither has elongate internodes or grapnel-like foliage. Two other species of *Dypsis, D. nodifera* and *D. pinnatifrons* have leaves with strongly swollen pulvini at the base of the leaflets, these being particularly conspicuous in exposed individuals at the base of the most distal leaflets which are often reflexed. However, the stems of both species are erect and self-supporting and the internodes are very short. What the adaptive significance, if any, of these pulvini is, remains uncertain.

Rheophytes and Aquatics

Rheophytes (van Steenis 1981), plants adapted to grow in the flood zone of rivers, are abundant in Eastern Madagascar, and there are rheophytic members of many flowering plant families. The familiar appearance of such rheophytes is a low bushy habit with numerous flexible stems and leaves that are narrow and willow-like or compound with very narrow leaflets, providing little resistance to flood water and thus having a greater chance of survival. Palm rheophytes have been discussed by Dransfield (1992a). Among the Madagascar palms, *Dypsis crinita* behaves very much as a rheophyte along the rocky rivers that flow out into the Bay of Antongil. Its seedling starts life in gravel at the edge of the river, and eventually the mature tree towers far above any flood level, but with the stem base still in the flood zone. Another species of *Dypsis, D. aquatilis*, known from only very incomplete material, may also be a rheophyte. The most remarkable of all palm rheophytes in Madagascar is *Ravenea musicalis* (see Beentje 1993b). This species starts its life as a submerged aquatic; eventually the leaves grow above the water level and a short squat bottle trunk develops. This species is known only from a single river. Seed germinates within the fruit wall while still attached to the infructescence. When the fruit falls into water it floats, but breaks easily and the already germinating seeds sink. The first foliar organs on the seedling are short, fusiform and hooked, and we believe that these hooked scales help to anchor the seedling to the stream bed. Subsequent leaves are pinnate, with limp leaflets that sway in the water flow. Eventually, subsequent leaves emerge from the water and have rigid leaflets, and look altogether like an ordinary *Ravenea*. *Ravenea rivularis* occurs on river banks and may at least in the seedling state behave as a rheophyte.

Spiny Palms in Madagascar

Few of the palms of Madagascar are spiny. This is in very marked contrast to the palms of Southeast Asia, Africa and South America. It is also surprising that the neighbouring island groups of the Mascarenes and Seychelles both have representatives of spiny palm groups. In Madagascar spines are confined to the following non-endemic species: *Raphia farinifera* (probably introduced), *Hyphaene coriacea* (non-endemic), *Phoenix reclinata* (non-endemic) and *Elaeis guineensis* (non-endemic). The two native endemic species of *Borassus* (genus non-endemic) have well developed irregular erose teeth on their petiole margins. *Satranala decussilvae* (endemic) has minute serrulations along the petiole margin, and in a few species of *Ravenea* (endemic) the leaf sheath margins may erode to expose hard broad fibres that are subspinescent. This is in marked contrast to the Seychelles, where five of the six endemic monotypic palm genera are densely spiny, particularly as juveniles, and the Mascarenes, where two of the five endemic genera are also strongly armed, particularly as juveniles. It appears that the strong selective pressures that have favoured the evolution of armature in palms elsewhere, have been lacking in Madagascar, and it is tempting to speculate what these might be. The very heavy armature in African palms may well have evolved as protection against large herbivores. At the present day the only extant larger mammal that might affect young palms in the undergrowth is the Madagascar bush pig (probably introduced), but we have seen none of the typical signs of bush pig damage to palms that are such a feature of Southeast Asian forests. A pygmy hippopotamus also occurred until recently in Madagascar (known in the subfossil state) (Paulian 1984), but what, if any effect, it had on palms is not known. Major herbivores are just not present in Madagascar and may never have been present. That spines have not evolved in the endemic Madagascar palm genera is almost certainly related to this absence of herbivores. In the palms of the Seychelles and Mascarenes, spinescence is particularly well developed in juvenile plants and is often (but not always) lost as the palms become adult. Here we speculate that such spinescence is in some way protection against herbivores such as the giant tortoises.

Poisonous Palms

In general secondary compounds that are poisonous (at least to man) are rare in palms, and palms are usually well protected by heavy sclerification, lignification, spines, tannins and needle crystals, rather than by alkaloids. *Orania* seems to be one of the few exceptions to this generalisation. Throughout Malesia and Madagascar, species of *Orania* are regarded as being poisonous; Burkill (1935) reports that a single fruit of *O. sylvicola* (West Malesia) will kill an elephant, though how this was discovered is not described. Fruits and palm hearts of *Orania* species are universally shunned. The palatability of the palm hearts of other Madagascar palms is well known to local people. Some species are excellent while others may smell unpleasantly (e.g. *Dypsis bovomantsina*) or be bitter and are not normally eaten. In degraded forest near to villages, it is usually the bitter-tasting palms such as *Dypsis lastelliana* that survive. The significance of the poisons in *Orania* and the bitterness of some other palms is not known, but it seems reasonable to suggest that it represents some sort of adaptation against herbivory, but not necessarily large mammalian herbivores. The interaction between wild Madagascar palms and large insect herbivores such as palm weevils that attack the stem apex would repay study. It is worth mentioning in passing that large bore holes with abundant fibrous frass formed by weevils as they bore into palm apices are of relatively frequent occurrence among larger palms (e.g. *Beccariophoenix madagascariensis* and *Ravenea madagascariensis*) in primary forest in Madagascar, but we know nothing of the beetle species involved, nor of patterns of predation.

FLOWERING AND POLLINATION

Aspects of the flowering of palms in Madagascar and their pollination present a multiplicity of topics for research. It is worth making a few general comments on flowering in Madagascar palms. There are no native hermaphroditic palm species on the island. Dioecy is confined to the coryphoid genera *Phoenix, Borassus, Hyphaene, Bismarckia* and *Satranala*, and the ceroxyloid *Ravenea*. All other palms are monoecious. All species are pleonanthic, except for *Raphia* (regarded by us as being introduced) which is hapaxanthic. In palms hapaxanthy and hermaphroditism are independently but strongly linked to particular taxonomic groupings and the absence of these two features in native Madagascar palms may be more a reflection of which evolutionary lines of palms were present or invaded Madagascar rather than the absence of particular selection pressures.

Within monoecious species in Madagascar, protandry (staminate flowers opening before the pistillate) seems to be general. So far we have only one possible record for protogyny (in *Masoala*).

Ratsirarson's study (as yet unpublished) is the only detailed study of pollination of Madagascar palms that we are aware of. Despite this we do have a few casual observations on pollination, made in passing while preparing herbarium collections or observing variation.

Flowering of palms in Madagascar may occur regularly, throughout the year (as in *Dypsis fibrosa* and *D. pinnatifrons*), it may be strongly seasonal, usually in November (e.g. *Dypsis procera* and *D. paludosa*) or it may occur at very irregular intervals. A good example of the last is provided by *Ravenea dransfieldii* which in October 1986 was to be found in full flower on the Masoala Peninsula, but in

February 1988 and October 1989 there were no signs of flowers or even fruit. In a population of *Ravenea rivularis* consisting of over 100 individuals, only staminate inflorescences were observed in February 1992 while six months later in July, only pistillate ones were seen. An extreme example is provided by *Ravenea albicans* which seems to flower at very irregular intervals; of about forty mature individuals observed in the Mananara Avaratra Biosphere Reserve, only one individual was found in April 1992 with dead pistillate inflorescences and in October 1994 one individual with dead staminate and two individuals with dead pistillate inflorescences.

Beetle Pollination

We have no direct observations on beetle pollination in Madagascar palms. However, curculionid and nitidulid beetles have been observed visiting the highly condensed inflorescences of *Ravenea louvelii* and *R. dransfieldii*. Inflorescences of both these species seem to warm up at anthesis, and the flowers emit a strong musty aroma. The flowers are partially hidden among the leaf sheaths and leaf litter accumulates among the sheaths. We have never observed anthesis in *Marojejya*. Inflorescence structure is remarkably similar to that of other, unrelated palms that are known to be cantharophilous, and we suggest as a working hypothesis that *Marojejya* is also cantharophilous. It shares with unrelated *Elaeis*, highly condensed unisexual inflorescences partially hidden among the leaf sheaths, with phases of staminate alternating with pistillate inflorescence production, with staminate and pistillate rachillae of remarkably similar form. *Elaeis guineensis* is known to be primarily pollinated by derelomid beetles, as is the dioecious calamoid genus *Salacca* in Asia in which the rachillae are also partially hidden among the leaf bases and of a form similar to that of *Elaeis* and *Marojejya*.

Bee and Fly Pollination

It is probable that bee pollination is quite widespread among the larger palms of Madagascar. We have observed honey bees and other hymenoptera visiting flowers of several large species of *Dypsis* and species of *Ravenea* such as *R. sambiranensis* and *R. madagascariensis* that have lax spreading inflorescences and sweetly-scented flowers. Bees and other insects were observed by H.E. Moore visiting the flowers of *Dypsis madagascariensis*. Bees and flies have been observed visiting the flowers of *Dypsis decaryi* (Ratsirarson, pers. com.). Many bees were observed by HB visiting flowers of *Dypsis decipiens* in the Parc Botanique et Zoologique de Tsimbazaza. Bees were observed by JD visiting the staminate flowers of *D. psammophila* on Île Sainte Marie. HB observed a Sphecoid wasp visiting the open staminate flowers of *Dypsis simianenesis* at Manombo. JD observed flies visiting staminate flowers of *Dypsis nodifera*.

Other Aspects of Pollination

The flowers of some of the smaller, three-stamened species of *Dypsis* are among the smallest of all known palm flowers. In particular, species such as *D. pachyramea* and *D. remotiflora* have minute flowers scarcely exceeding 1 mm in diameter. In such flowers, the petals hardly open, leaving a very small gap which must exclude all but the smallest insect visitors. No insect visitors have ever been observed in these flowers, yet their highly specialised structure and minute size suggest a precise and highly evolved pollination system that deserves to be studied in detail.

Flower Galls

A very common feature of the inflorescences of the smaller members of *Dypsis*, particularly those with minute three-stamened flowers, is the frequent occurrence of galled flowers. These galls appear as small horn-like structures, composed of the enlarged valvate petals of the staminate flowers. Thus the inflorescence appears to carry young fruit, or staminate flowers of two sizes and shapes: normal small rounded staminate flowers and pointed galled flowers that are about two to three times as long as the ungalled flowers. Inside the galled petals can be found a small insect grub. We know nothing more of the insect responsible for the galling, but their occurrence seems suggestive of a coevolved relationship that would be worthy of further study.

Fruit Dispersal

As with pollination, next to nothing is known about the dispersal of palm fruits in Madagascar. We may speculate that the brightly coloured fruits of many species of *Dypsis* are probably dispersed by birds. It is probable, also, that lemurs play a role in dispersal. Black lemur and *Lepilemur dorsalis* have been observed by Birkinshaw to feed on the fruits of *Dypsis madagascariensis* and black lemurs may be responsible for dispersal of this species (Birkinshaw, pers. com.). Birkinshaw also observed Madagascar Blue Pigeon feeding on the fruit of *Ravenea sambiranensis*. Ratsirarson (pers. com.) observed Black Parrot and *Lemur catta* feeding on the mesocarp of *Dypsis decaryi* and pig droppings were also found to be full of the seeds of this species at

Andohahela. Overdorff reported that fruits of *D. nodifera* are eaten by *Lemur fulvus* and *L. rubriventer* (that also feed on the leaves). Villagers in the Sahavary area reported to JD that red-ruffed lemurs feed on the fruit of *Marojejya darianii.*

Although fruit size and coloration of most species of *Dypsis* tempt one to suggest that they are most likely to be dispersed by frugivorous birds, a few species are dull-coloured, green or brown at maturity. In particular, the species that were formerly included in *Vonitra* are generally dull-coloured. Borne on the very long arching peduncles, it is tempting to suggest that these may be adapted to attract fruit-eating bats, but we have absolutely no evidence for suggesting this and guides at various sites have informed HB that fruits of these species are much liked by bush pig, the tracks of which were very evident nearby. *D. crinita,* occurring as it does in abundance along rivers in the north-east, always in flower and fruit, would make an excellent subject for a study of pollination and dispersal, with the added bonus that it appears to behave as a rheophyte (see above).

Large fruits occur in the borassoid genera (*Borassus, Hyphaene, Bismarckia* and *Satranala*). In Africa *Borassus* and *Hyphaene* are known to be dispersed by elephants, the seed sprouting in their dung. It is very difficult to imagine any extant native Madagascar animal dispersing the fruits of *Borassus,* and even *Bismarckia* which occurs by the hundreds of thousands in the palm savannahs of the west has a fruit that seems to be too large for dispersal by any extant animal. In the case of *Borassus* and *Hyphaene* fruits may well be dispersed by flood waters and this would account for their occurrence along watercourses. *Bismarckia,* on the other hand, is most abundant at the present day in anthropogenic grassland that is maintained by burning. Its extraordinary abundance could be explained by dispersal by cattle, but we have no evidence to support this conjecture. *Satranala,* as a primary forest palm of the East Coast, presents an altogether different problem. We have speculated (Dransfield & Beentje 1995) that the extraordinarily sculptured endocarp may be an adaptation to surviving passage through the crop of a giant bird (i.e. a now extinct Ostrich-like rattite such as *Aepyornis*). We use the same argument in discussing the dispersal of *Voanioala*. In the case of the latter palm, there is no effective dispersal at the present day, as far as we can see. Endocarps lie in a carpet under the fruiting trees, and all germinated seeds seem to be under the crown of the parent tree. We note that other palms with heavily sclerified endocarps are found in South America and New Guinea where living rattites occur, and, at least in New Guinea, are known to ingest whole palm fruits that have highly thickened and sculptured endocarps.

We know nothing of the dispersal of *Orania* and *Lemurophoenix* with their large fruits with endocarps that have heart-shaped basal protuberances. In the latter genus, dispersal at the present day seems, like *Voanioala,* to be very limited as the ground may be carpeted with rotting fruits and old endocarps but *Orania* is widespread throughout the forest zone.

The Cultivation of Madagascar Palms at the Royal Botanic Gardens, Kew

David Cooke

Since the start of the Madagascar palm project in 1986, approximately 155 palm collections have been introduced into the Living Collections at the Royal Botanic Gardens, Kew. Some of these palms have proved to be very difficult to grow in Kew outside their natural environment. Samples of some of these collections have also been distributed to other botanic gardens as far afield as Trivandrum in southern India, Limbe in Cameroon, the Lyon Arboretum in Hawaii, Fairchild Tropical Garden in Florida and Kebun Raya Indonesia at Bogor, Indonesia.

Most of these palm collections consisted of ripe seed, but in a very few instances where seed was immature or unavailable, seedlings were introduced. All these collections were made with permits granted by the Departement des Eaux et Forêts in Antananarivo and export permits from the same organisation. Samples of the collections were deposited in Parc de Tsimbazaza living collections. Rather few living collections were made on the first field trip to Madagascar in 1986. During the second field trip in 1988, which coincided with a fruiting season, a large number of seeds were collected. Many of those that have been collected have now been introduced successfully into the display houses at Kew.

Fruits were collected when they were colourful and ripe, though it is sometimes difficult to gauge whether the seed is completely ripe as it may take some months to mature fully. The outer fleshy fruit wall must be removed as in some species it is thought to contain growth inhibitors which reduce the rate of germination; ideally this is done in the field or as soon as possible. The seed or seed together with the enveloping endocarp is then wrapped in damp tissue paper or moss and placed in a labelled plastic bag. Back in the nursery at Kew, the seed is immediately sown in a well drained compost. High temperatures are required for best germination to take place; we have found optimum temperatures to be between 28°– 35°C with a very high humidity. The best way of achieving this is in an enclosed wooden box with a tight fitting clear glass or plastic lid, placed inside a glasshouse or in a sunny position to gain maximum heat and light. Ventilation and light watering daily is very important; too much water and humidity can cause the seed to rot. If the seed has become dry during storage it is advisable to soak it for two to three days before sowing. Germination can take from three weeks to eighteen months. We have found it best to sow seed in deep clay pots; under our growing conditions at Kew, we have found these preferable as their porous nature prevents waterlogging and provides good aeration of the seedling roots. Seed was sown and covered with a layer of compost equal to the thickness of the seed. Seedlings were potted on as soon as the second seedling leaf had opened, into individual pots just big enough for the roots to grow. If the pot is too big, the seedling may become prone to root rot. The compost used at Kew for palm seedlings consists of an equal mix of fine and coarse bark, loam, 10 mm grit, coir and 2 grams per litre of a general slow release fertiliser; some micronutrients may be added only when the young palms are about 2 years old. In Kew, situated as it is at more than 51° N, with low natural light intensities in the winter months, we have found that the best time for potting on is in the spring, every other year. It seems that the seedlings grow best if they are potbound for a while, though this may not be generally true for Madagascar palms when grown elsewhere.

Inside the Palm House at the Royal Botanic Gardens, Kew (Photo: Media Resources, Royal Botanic Gardens, Kew).

In the early days of the project, we had most difficulty in growing the small undergrowth species of *Dypsis*. The problems seemed to be related to the very limited amount of seed available in the field and their size, often as small as 6 mm to 8 mm in diameter, and their consequent susceptibility to rapid dehydration. Between 1986 and 1989 a total of 33 different collections of these small species of *Dypsis* were made, and then in the 1990s a further 12 were made. Today 20 of these are growing well in the Palm House and nursery and one species, *Dypsis mocquerysiana*, has flowered, pollination has occurred and two fruits developed. Other understory palms at Kew grow more slowly. One reason for the slow growth may be the lack of an appropriate mycorrhizal association. This will need much more investigation.

Larger types of Madagascar palm seed, ranging from 1 cm to 3 cm in diameter, have proved to be much easier to germinate and have also proved to be more adaptable when planted out in the Palm House. However, large seeds with remote germination such as are found in *Orania* and *Lemurophoenix* have proved problematic. Because they produce long "droppers" (the cotyledonary stalks), we have used deep clay pots for sowing such species. Their "droppers" can be up to 30 cm long and need special care in handling as they are very brittle; sometimes four hands were needed to transfer the seedlings into larger containers. Early growth of these species in pots may prove not to be much of a problem, but once planted out in open ground they tend to die. Perhaps it would be better to try to germinate such species where they are intended to grow on to maturity to avoid the root shock that occurs on transplanting.

Two of the most interesting palms that we have grown at Kew are *Ravenea musicalis* and *Voanioala gerardii*. In the former, the seeds are 1–1.5 cm in diameter. In the wild, the seed germinates and grows in fast flowing water, so the best way we have found to germinate them was in a shallow clay pan standing in a tray of water, with the water coming at least half way up the pan, ensuring that the compost does not dry out at any time. Once the seedlings had produced three seedling leaves they were potted individually and the pots were submerged in deep water completely covering the leaves. The water temperature was kept at a constant 23°C with the use of a thermostatic fish tank heater in the water tank and half of the water was changed every other month to keep it fresh. Within about two years a plant with a height of about 40 cm was produced. This palm grows well in still water and is on display in the Waterlily House and in the Princess of Wales Conservatory. Every January the Waterlily House closes for replanting and maintenance and the temperature is dropped from 23°C down to 10°C for four weeks. This seems to have had no ill effect on the growth of *Ravenea musicalis*.

The Forest Coconut, *Voanioala*, with its large oval endocarp up to 9 cm long posed different problems. Once the seed had germinated it produced a long "dropper" and so large pots had to be used when potting the seedlings on. This unusual palm is very slow growing and is best planted out quickly into its permanent position so as to cause minimal root disturbance. The one in the Palm House was planted out in 1989, just after the restoration, from the seed collected in 1986 and has grown to a height of one metre.

COLLECTING PALMS FROM MADAGASCAR

MADAGASCAR IS HOME TO SOME OF THE MOST BEAUTIFUL and unusual palms of the world. Several of these have proved to be easily cultivated and species such as *Dypsis lutescens* and *Bismarckia nobilis* have long been important horticultural subjects. Over the past 15 years we have also seen the introduction and rise in popularity of *Ravenea rivularis*, surely one of the most majestic palms for cultivation in the tropics and subtropics. Besides these species that are grown by the million by commercial growers, many others have been introduced in recent years. It has seemed to us that all that a palm has needed to be desirable and much sought after is a pedigree that includes an origin in Madagascar. Such has been the demand for any palm from Madagascar that commercial seed importers and amateurs alike have mounted expeditions to Madagascar to acquire any available seed. Palm taxonomists have found it difficult to keep pace with these activities and there are now many species widely grown throughout the tropics and subtropics (but particularly in Australia and southern California) that are known only by trade names. These trade names are certainly of value (as their usage proves), but rarely do they have any relationship to scientific names. Even when a trade name coincides with a validly published scientific name, there is no certainty that the cultivated palm is correctly identified by the name. "*Chrysalidocarpus mahajunga*" (or *"majunga"*), used widely for a form of *Dypsis madagascariensis* is a name that has never been validly published. *"Mahajunga"* refers to the fact that it was collected near to the town of Mahajanga on the west coast of Madagascar. The palm referred to in trade as *"Bejofo"* may well be our new species, *Dypsis bejofo*, but until the cultivated specimens flower and fruit their true identity will remain uncertain. Then there are palms referred to in the trade by local names that we have not encountered in Madagascar. Only very rarely has it been possible to guess at their identity in the absence of diagnostic features of flowers and fruit.

Many potential readers of our book will have in their collections palms from Madagascar that lack valid scientific names and we hope that our book will help to resolve some problems of identity. However, we are well aware that our keys, being based on mature palms that we have observed in the field or as herbarium specimens, are probably inadequate for determining accurately the identity of seedlings in cultivation. This seems inescapable, but is regretted. All we can advise is patience. Wait until your palms flower and fruit; then they should be identifiable if they have been described. Of course, there is always the possibility that you have a new, undescribed species in your collection. We shall always be pleased to learn of remaining palm mysteries from Madagascar and try to help to resolve them.

When faced with such a wonderfully rich palm flora, the enthusiast will want to acquire plants, especially of the more unusual and newly described species. We encourage the introduction of palms from Madagascar into cultivation but we cannot approve of, or wish to encourage, seed and seedling collecting in Madagascar that flaunts local and international laws and gives no thought to the future of the species or the well-being of local people.

Tourists who visit Madagascar may well be tempted to collect palm seed. This is not a prohibited activity but potential seed collectors should be aware that permits are required to enter all forested areas and that for protected areas, collection of plants is usually prohibited except for scientific purposes. Furthermore, permits are required for the export of all plant material, including seed, from Madagascar, and material needs to be accompanied by phytosanitary certificates and full CITES documentation where necessary. In the case of palms, only two species (*Dypsis decipiens* and *D. decaryi*) come under CITES legislation. Luggage is checked on flying out of Madagascar at the airport in Antananarivo. Recently it would appear that the Malagasy authorities are taking a more serious view of infringements of laws governing the collecting of plants and animals in Madagascar. In 1992, three foreigners were caught red-handed illegally collecting chamaeleons at night in the Ankarafantsika Réserve Naturelle Integrale, and, in the confusion of the capture, two of the foreigners were shot and mortally wounded by police officers. This case has highlighted within Madagascar the fact that foreigners have at times operated illegally and has sensitised the whole issue of collecting and exporting plants and animals.

The horticultural potential of the palms of Madagascar is certainly very great. With a wide range of latitude, altitude and climate, and varied geology, Madagascar has palms to suit gardens throughout the moister tropics and subtropics and warm temperate regions. In *Ravenea xerophila* there may even be a palm suited to the dry tropics. The horticultural potential has been realised by many growers and wild palms of Madagascar are now being exploited for the international horticultural trade. Anyone who has seen *Lemurophoenix halleuxii* and *Dypsis decipiens* growing in the wild will undoubtedly have thought of the possibility of introducing these wonderful palms to a wider public. Sadly, the only functioning botanic garden in Madagascar, the well maintained Parc Botanique et Zoologique de Tsimbazaza in Antananarivo, is at relatively high altitude (c. 1200 m above sea level), an altitude where *Ravenea madagascariensis* and *Dypsis decipiens* grow well but where the wonderful palms of the humid lowlands of north-eastern Madagascar are less likely to thrive. Thus the current potential for introducing native palms to the Malagasy themselves is limited. However, it is hoped that in the future a lowland botanic garden may be developed.

The introduction of Madagascar palms elsewhere in the world could, in our estimation, have a very significant role in conservation, but perhaps not in the way popularly conceived, of saving palms from extinction by growing them in overseas botanic gardens and private collections. Recent events such as Hurricane Andrew in Florida that in 1992 caused such damage to Fairchild Tropical Garden, home of one of the premier tropical palm collections, and periodic droughts on the West Coast of the USA that have caused problems of water availability for private palm collections, have emphasised the vulnerability of palms in collections. The need to plant populations of palm species rather than individuals, in order to maintain genetic diversity, has rarely been put into practice. Perhaps only in Townsville Palmetum in Queensland has this been carried out as a matter of policy.

Undoubtedly the best way of conserving a palm species and preventing it from becoming extinct in the wild is to maintain it in its wild habitat, which means managing the area where it grows. This, of course, often requires funds – funds for the development of reserves, for controlling boundaries and, perhaps most important of all, for education, to help create awareness of the importance of these splendid plants. At present much of the trade in palm seed from Madagascar would appear to benefit few people, most of whom live outside Madagascar. If village people realise the value of palm seed, and of maintaining wild palm populations that can provide a sustainable supply of seed for the international horticultural trade without endangering the wild populations, then there will be strong disincentives to palm felling and perhaps even to forest clearance, and Madagascar itself will gain greater benefit from trade in its native palms than it does at present.

USES OF PALMS IN MADAGASCAR

PALMS ARE A VITAL PART OF VILLAGE LEVEL ECONOMY in Madagascar. Though they do not figure on World Bank statistics or the balance of payments, for many people living in villages and small settlements they are an important part of everyday life. Palms are used as food, in construction, handicrafts, medicine, magic and so on.

Export of palm products is very restricted these days, but the commerce in piassava used to be of considerable importance. Sheath fibre from the 'Vonitra' group of *Dypsis* (especially *D. fibrosa, D. utilis, D. crinita*) was harvested all over the East Coast and exported to Europe, where it was used in making brushes. In Britain in 1915 (Rendle, 1917), Madagascar piassava was valued twice as highly as West African raphia fibre used for the same purpose, and 10–20% higher than South American piassava. In France in 1950 a kilo of clean piassava from Madagascar was worth 30–50 francs.

All following uses are taken from palm collections seen by us, or are from information we have obtained from local people.

Used in house construction: species used as timber, because of the hardness of the outer fibres, either as building poles, or as planks, with the outer part of the trunk split off in c. 3 meter long 'planks', these being flattened out and then dried. *Ravenala* planks dried in the same way are sometimes mixed with true palm planks; this is considered an adulteration, as *Ravenala* planks are of inferior quality, being much lighter.

Beccariophoenix madagascariensis, house construction (unspecified)
Bismarckia nobilis, partition wall planks
Dypsis ampasindavae (Nosy Be), house walls

Construction of a typical Betsimisaraka house at Hiaraka, Masoala Peninsula; leaves of **Dypsis fibrosa** are interwoven to produce thatching panels.

Hats bought in the market at Ifanadiana, made largely from the leaves of **Ravenea dransfieldii.**

D. madagascariensis (W Madagascar), floorboards (Nosy Be, exported to mainland)
D. malcomberi, house walls
D. mananjarensis, house planks
D. nossibensis, planks (Nosy Be, exported to mainland)
D. nauseosa, roofing beams, floorplanks
D. pinnatifrons, house beams
Orania longisquama, house wall planks
O. trispatha, house construction
Raphia farinifera, petioles used in wall construction
Ravenea julietiae, wood for construction (not specified)
R. madagascariensis, planks for house walls and floors

Coconut palms fringe the beach on the East Coast, as viewed by a Malagasy artist.

R. robustior, outer wood for floorboards, tables, house walls, said to be termite-resistant
R. sambiranensis, planks used for floorboards

Leaves used in thatching:
Bismarckia nobilis
Dypsis ceracea
D. crinita
D. decaryi
D. fibrosa (extensively used in the Masoala Peninsula)
Masoala madagascariensis
Ravenea robustior

Palm-heart eaten:
Beccariophoenix madagascariensis
Borassus madagascariensis
Dypsis ampasindavae
D. ankaizinensis
D. baronii
D. basilonga (excellent)
D. canaliculata (said by others to be poisonous)
D. decipiens
D. hovomantsina (smelly)
D. ligulata
D. madagascariensis (excellent, *Kindro*)
D. oreophila (excellent, *Kindro*)
D. malcomberi
D. mananjarensis
D. perrieri
D. pilulifera
D. prestoniana

D. tsaratananensis (excellent, *Kindro*)
D. tsaravoasira
D. utilis
Hyphaene coriacea (also sometimes used to make palm wine, tapped from the heart)
Raphia farinifera
Ravenea albicans
R. dransfieldii (though others believe this is poisonous)
R. robustior (said to be large enough to feed a whole village; but in some areas this species is not eaten)
R. sambiranensis (slightly bitter, cooked with manioc)

Palm-heart bitter or poisonous:
Dypsis canaliculata (but others say this is edible)
D. lastelliana
D. nauseosa
Orania longisquama
O. trispatha
Ravenea dransfieldii (though others believe this is edible)
Ravenea glauca

Palm fruit eaten:
Borassus madagascariensis, used in fermented fruit drink
Dypsis baronii
D. decaryi (formerly also used in making a fermented drink)
D. madagascariensis
D. utilis
Phoenix reclinata
Raphia farinifera
Ravenea sambiranensis

Palm 'pith' eaten:
Bismarckia nobilis
Borassus madagascariensis: also shoots of germinating seedlings eaten
Raphia farinifera
Ravenea robustior

Stems used to make blow-pipes, hollowed out with red-hot iron:
Dypsis confusa
D. hiarakae
D. mahia
D. nodifera
D. oreophila
D. schatzii

Leaf fibres used in basketry, hat-making etc.:
Beccariophoenix madagascariensis, for the 'Manarano' hats
Dypsis mananjarensis, "esteemed fibre", but uses not specified
Hyphaene coriacea, basketry, hat-making, rope-making
Phoenix reclinata, basketry
Raphia farinifera, basketry, cloth-making

Leaves of **Hyphaene coriacea** are woven into thatch and baskets (Photo: B. Du Puy and D. Du Puy).

Ravenea dransfieldii, hat-making
R. lakatra, hat-making
R. xerophila, hat-making, winnowing baskets

Formerly used to make salt, from the trunk pith:
Dypsis lastelliana
Ravenea robustior

Various uses:
Dypsis ceracea, used for brooms
D. crinita, piassava used as an oil filter in the Manongarivo
D. decipiens, leaves used in erosion control
D. fibrosa, inflorescence used as a broom, sold in Maroantsetra market. Formerly one of the main piassava producers
Ravenea julietiae, hollowed out trunks used as irrigation pipes
Raphia farinifera, fruits used in personal ornaments such as necklaces

Harvesting coconuts, a woodcarving from Ambositra.

Medicine:
Dypsis andrianatonga, a leaf decoction is used as a drink for convalescence, and is highly prized by the Tsimihety of Manongarivo
D. crinita, the heartwood is employed in a cure for children's cough (Manongarivo)

Magic:
Dypsis madagascariensis, used as an aphrodisiac
Masoala kona, leaf stuck in a bamboo pole, used to ward off thunderstorms
Ravenea glauca, used to make mattresses used in exorcism

Cultivation in Madagascar:
Bismarckia nobilis
Dypsis baronii, in the central plateau and especially Antananarivo
D. decipiens
D. lutescens
Ravenea madagascariensis

Common in cultivation outside Madagascar:
Bismarckia nobilis
Dypsis decaryi, Triangle Palm
D. lutescens
D. madagascariensis
Ravenea rivularis

LOCAL NAMES OF MADAGASCAR PALMS

THERE IS A GREAT VARIETY OF PALM NAMES. IN GENERAL there are a few (usually older) people in every village who have an enormous knowledge of local plants and their uses, but if you happen to hire as a guide someone who knows the forest because he goes there to hunt or trap, he may come up with imaginative names rather than correct ones. There is sometimes a risk of obtaining a 'local name' from a guide who just does not like to disappoint you; also, there is ample scope for mistakes in the name written by the plant collector, after transmission from the dialect of the guide, to interpreter, to collector. Another cause of confusion may arise from the interface between languages: we have seen a 'local name' which was Malagasy for 'tomorrow morning'.

We have been unable to gauge the plant knowledge of women; this might require a female investigator.

Some names for palms are the same throughout the island. Many species of *Ravenea* are called *Anivo* or *Anivona*, but this extends to *Orania longisquama* which may resemble *Ravenea* species. *Vonitra* is used for those *Dypsis* species with piassava such as *D. fibrosa, D. utilis, D. crinita, D. pusilla*. *Satrana* is used for fan palms such as *Hyphaene* and *Bismarckia*. *Sindro* is used for all three *Orania* species.

In other cases a single species may have various names in various dialects: *Dypsis nauseosa* is called *Mangidibe* in Antaisaka, and *Rahoma* in Antambahoaka. *Dypsis oreophila* is called *Lafaza* in Betsimisaraka and *Fitsiriky* in Tsimihety. Or the same local name might be used for different species in different dialects: *Kona* in Tsimihety is used for *Ravenea robustior*, but further south is used by Tanala and Antaimoro for *Masoala* and *Marojejya* species.

A modifier is sometimes used: *Vakaka* is used for *D. mananjarensis* and *D. malcomberi*, but *Vakakabe* for *Ravenea robustior* and *R. krociana*. *Maroala* is used for *Beccariophoenix*; *Maroalavehivavy* (female *Maroala*) denotes *Marojejya insignis*.

Other names indicate a particular use for a palm rather than a single species, e.g. *Kindro* is used for palms which produce good palm-heart, e.g. *Dypsis madagascariensis, D. tsaratananensis, Ravenea sambiranensis*. The name *sira* (meaning salt) is used for palms which, in days gone by, were used to produce salt, usually by burning the pith: *D. lastelliana, Ravenea robustior*.

Sometimes a name has a meaning describing the habit or peculiarity of the species: *Faribazo* means sugarcane tree, and describes the trunk of *Dypsis baronii*, which looks slightly like sugarcane.

Fohitanana means short hand, and probably describes the short inflorescence of *Marojejya insignis*.

Menavozona means red neck, and is used for *Dypsis lastelliana* with its furry red leaf sheath, while the opposite *Madiovozona* (clean neck) is used for *D. madagascariensis* which has a glabrous leaf sheath. *Sarimadiovozona* means 'bastard offspring of *Madiovozona*' and is used for *Dypsis rivularis*.

Besofina means big ears, and describes the large auricles on the leaf sheath of *Marojejya insignis* or *Dypsis perrieri*; an alternative name is *Menamosona*.

Mafebely means 'a bit bitter' and describes the palm-heart of *Ravenea sambiranensis*; *Hovomantsina* means 'stinking heart' and refers to the smelly palm-heart of *Dypsis hovomantsina*; *Mangidibe* means very bitter, and describes the palm-heart of *Dypsis nauseosa*.

Ravina means leaf, and is used in *Ravimbe*, big leaf (*Marojejya darianii*), *Ravimbontro* (*Dypsis fibrosa*), as well as the famous *Ravenala* (*Ravina-ala*, forest leaf).

Shihara, Sihiara, Sincare, Sinkara, Sinkiara, Tsikara, Tsinkary, Tsinkiara are all variants on the same name and are used to denote any small palm, i.e. less than 4 meters tall. Sometimes the modifier '*madiniky*' is added for really small palms, less than 50 cm high.

Laafa seems a fairly general name for large tree palms; two palms standing next to each other and patently different (one with grouped leaflets, one with regular ones) were both called *Laafa* by a guide who knew quite a few other palms by individual names. It is possible *Laafa* is used for palms with useful fibres.

Tsirika, Fitsiriky: D. concinna, D. hiarakae, D. nodifera, D. oreophila, D. pinnatifrons; these all have long thin stems and are used for making blowpipes.

Voanio is the coconut palm, and *voanio-ala* is the forest coconut, *Voanioala*, though the name is sometimes used loosely to indicate any large palm in the forest - but this might depend on the guide!

CONSERVATION, THREATS AND POSSIBLE ACTION

AT THE MOMENT, TWO PALM SPECIES FIGURE ON APPENDIX II of CITES: *Dypsis decaryi* and *D. decipiens*. It seems inadvisable to add to this number, since the identification of palm material to species would pose enormous difficulties to the authorities. The *in-situ* protection of rare species and organized, sustainable harvesting of seed by local people, to whom part of the profit would flow, would seem a better objective.

The protected area network in Madagascar encompasses a large number of National Parks and Special Reserves. Actual protection within these areas depends very much on the infrastructure present, and on the

Habitat destruction caused by shifting cultivation is the major threat to palms in Madagascar.

commitment of staff; Betampona Special Reserve is tightly run, while Lokobe Special Reserve continues to suffer the depredations of palm-fellers who sell the planks for house construction. The moist forests, in which palm species diversity is highest, are represented in the system by:

- Lokobe RNI (Réserve Naturelle Intégrale)
- Manongarivo RS (Réserve Spéciale)
- Tsaratanana RNI
- Marojejy RNI
- Anjanaharibe-Sud RS
- Nosy Mangabe RS
- Mananara Avaratra Biosphere Reserve and National Park
- Marotandrano RS
- Ambatovaky RS
- Betampona RNI
- Zahamena RNI
- Mangerivola RS
- Mantady National Park
- Analamazaotra RS
- Ranomafana National Park
- Andringitra RNI
- Ivohibe RS (montane forest)
- Manombo RS
- Kalambatitra RS (montane forest)
- Andohahela RNI

Sometimes the Reserve or Park represents the only forest left in an area (e.g. Manombo), sometimes slash-and-burn or logging is going on inside the Reserve or Park (e.g. Manombo, Lokobe), and sometimes excellent forest is not protected, and still in a good state, usually because of its inaccessibility (Rantabe hinterland, between Maroantsetra and Andapa, E of Ambositra). Some wonderful forest is not protected, and is slowly disappearing: E of Ifanadiana, Midongy area, Sainte-Luce.

A good overview of these forests and their protection can be found in Nicoll & Langrand (1989).

Many species have very restricted distribution areas; species such as *Marojejya darianii, Ravenea louvelii* and *Voanioala gerardii* are known from single populations, encompassing less than a square kilometer in the wild. With such species even minor local disturbances to the habitat can have critical results. But sadly, habitat disturbances occur on a much larger scale these days, and Madagascar is no exception. The demand for agricultural land to feed a growing population, coupled with old-fashioned food production methods due to lack of expertise and funds, has resulted in medium scale forest clearance on a national basis, and is continuing to do so. Lowland forest especially has become fragmented, and specialized habitats such as white sand forest are now restricted to only a few square kilometers.

Specific threats to palms include overharvesting for timber (such as in the Réserve Naturelle Intégrale de Lokobe), cutting for palmheart (probably a prime cause for the decline in numbers of such palms as *Beccariophoenix, Dypsis pilulifera* and *Dypsis decipiens)* and destructive collecting by palm fanatics. This last factor has affected the populations of *Marojejya darianii* and *Lemurophoenix halleuxii*, where all seed is collected and so rejuvenation is arrested; and in the population of *Beccariophoenix madagascariensis* described in an article in Principes, Journal of the International Palm Society, a tree of this species was cut down to obtain the seed, resulting in the then known population of the species to decline by 10%.

Possible Actions

Habitat conservation is the most desirable, but also the most difficult action result we can strive for. To increase food production on the current agricultural acreage is an obvious but time-consuming solution, but there is certainly considerable scope for improvement in rice-growing techniques; these techniques are already in use in countries such as Indonesia and the Philippines, and a doubling of the Malagasy rice crop seems certainly possible without increasing the rice-growing area.

Timber production and firewood harvesting are factors in forest disturbance and deterioration, and the demand for these products is increasing. Small-scale plantations of fast-growing and coppicing species, especially on a village level, would do much to channel the demand for firewood (95% of cooking is done on fuelwood and charcoal) towards reusable rather than destructive harvesting. Improvements in charcoal-making techniques are being tested in small-scale trials in Antananarivo, and such mundane matters deserve much more support from the conservation world. The introduction of more fuel-efficient charcoal stoves (such as the clay-lined ones) is a similar small-scale measure with enormous potential profits for the natural forests.

A serious problem in Madagascar is the regular burning of vast areas of grassland, both in the south and in large areas of the east coast. The purpose of burning is the stimulation of young grass shoots for cattle, but the acreages burnt seem out of all proportion to the number of cattle that benefit. The poor quality of the fodder grasses able to survive the burning is certainly a contributing factor to the ever-increasing regularity of the burning. Any alternative, either in fodder production or in management techniques, would contribute enormously to long-term solutions for many of the country's problems, since any regeneration in the vegetation is prevented by fire.

Establishing Parks or Reserves without involving local people is virtually useless, and certainly a self-defeating policy; even if local people are interested in

Typical scene in the eastern escarpment forests in October and November; burning of cut forest precedes rice planting.

View of the rolling plateau landscape north of Antananarivo; impoverished grasslands, burned every year, now cover much of the plateau while huge erosion scars (**lavaka**) eat into the hill sides.

conserving the forest, because of their use of forest products, if the reserve is seen as an alien imposition from either the capital or even from foreigners, there will be resentment which will work against long-term conservation. There are several 'paper' Reserves or Parks where local involvement is minimal, and the day-to-day need for food and money causes over-harvesting or even plunder of the resource – as in Manombo Special Reserve, for instance, a last remnant of forest in a sea of burnt grassland, where the extraction of timber and the excision of agricultural land threaten populations of several palm species, some of which are only known from this Reserve and are being described in this book for the first time.

Integrated conservation projects should be complemented by the provision of alternatives in food and firewood requirements. The establishment of forest core and buffer zones is a well-established precept, and in the WWF project in the Manongarivo Reserve the theory seems to work in practice with a large degree of success. Local people have been involved in the delimiting of the Reserve, and are able to continue their traditional activities in the forest buffer zones; to compensate them for loss of potential agricultural clearing areas there are village health projects, help with village schooling projects, and improvements in rural access. The fact that the WWF basecamp hut near the bufferzone was built with wood of *Ravenea robustior* (category: Vulnerable) at first seems shocking, until one realizes that the buffer zone and core zone help protect the majority of the populations which used to be over-harvested!

Specific palm conservation and protection projects are desperate measures – but sometimes they work. *Dypsis decaryi*, the well-known triangle palm, occurs in a small area in southern Madagascar and suffered from both burning of the habitat (dry low forest) and over-collecting of seed. Five hundred hectares, encompassing the main part of the total population (about a thousand trees) has been set aside as Parcelle 3 of the Réserve Naturelle Intégrale II d'Andohahela. The conservation goes hand in hand with education, and as the plot is adjacent to a main road tourists also visit the palms regularly.

The awareness level of the need for palm conservation, and indeed conservation in general, of the general public in Madagascar is very low – most people are more concerned with getting three meals a day and increasing their income above the level of less than half a dollar a day. Conservation education will only be digestible when it is accompanied by the prospect of increasing the standard of life: easier access to firewood, a better stove, increased rice harvests, cleaner drinking water. Most rural people in the forest areas realize the need for forests to be conserved – they obtain medicines, timber, food, and a host of 'minor' forest products from them. But there is an enormous need for investment in education and small, local improvements which will cause a cumulative effort in conservation.

SOME SPECIFIC RECOMMENDATIONS

Ravenea louvelii is restricted to an area of less than 1 square kilometre near Andasibe: specific protection measures should be relatively easy to implement, and would probably save the species from extinction.

Manombo Forest Reserve is the last remnant of lowland forest in Fianarantsoa Province, and home to several strictly endemic palm species, as well as to several critically endangered palm species. It is a small reserve (5020 hectare) and is being destroyed by commercial timber harvesting and the resulting agricultural clearance. The addition of another 5000 hectare of unprotected forest which is adjacent to the Reserve is recommended, as well as strict protection from commercial activities and agricultural conversion. This would be an excellent subject for a pilot project with local involvement with conservation, and a showcase for community buffer zones coupled with village woodlots and charcoal improvement schemes.

It is recommended that the Parc Botanique et Zoologique de Tsimbazaza receives funds to set up an ex-situ breeding program. Staff would need to be trained and a secure area set up for species that are in danger in the wild, e.g. *Marojejya darianii, Beccariophoenix madagascariensis, Ravenea louvelii, Borassus madagascariensis, B. sambiranensis, Orania trispatha, Lemurophoenix halleuxii, Voanioala gerardii, Dypsis utilis*. The Botanic Garden at Iavalohina near Toamasina (Tamatave) would be an excellent coastal counterpart in such a programme.

History of Palm Botany in Madagascar

Few botanists have actively studied the palms of Madagascar and this is probably a reflection of the paucity of material collected and the fact that, until recently, most of it was housed in a single institute, the Museum National d'Histoire Naturelle, in Paris. The first collections of palms from Madagascar to reach European botanists were made by French natural historians such as Philibert Commerson (*Dypsis commersoniana* near Fort Dauphin and *D. forficifolia* around the Bay of Antongil) in 1770–1771, Francisco de Noronha (*Dypsis nodifera*) in 1775 and Pierre Poivre (*Dypsis hirtula, D. lutescens* and *D. poivreana*, possibly on Île Sainte Marie) at some time between 1740 and 1757 and it is on these collections that the first descriptions of any Madagascar palms were based by Martius, published in Historia Naturalis Palmarum (1823–50). *Dypsis forficifolia, D. hirtula* and *D. pinnatifrons*, based on fragments, some possibly of mixed origin, were the first true Madagascar endemic palms to be described and named. The first publication of a species of *Ravenea*, the next endemic genus to be recognised, was in fact based on material cultivated in Europe from the Comoros, rather than from Madagascar.

The first person to describe more than one or two palms piecemeal was Baillon who in 1894–5 published a paper entitled "Les palmiers malgaches à petites fleurs" that included descriptions of the then known species of *Dypsis*, new collections having reached Paris from collectors such as Boivin and Aubert Du Petit Thouars. By the early 1900s, significant new collections had accumulated, important ones made by the German plant collector, Hildebrandt, and the Reverend Baron, who worked for the London Missionary Society in Madagascar, and these, together with the earlier collections in the Paris Herbarium and elsewhere, formed the basis for the first monograph of Madagascar palms (Beccari 1906). Beccari added to the information in this paper in his large folio book, Palme del Madagascar (Beccari 1912–1914). Although Beccari had not visited Madagascar, his descriptions were, as usual, meticulous and detailed and provided a firm base from which to go forward. The publication coincided with the start of Henri Perrier de la Bâthie's long career of field work in Madagascar, during which large numbers of palms were collected including material of several new genera. His collections were worked up by himself and Henri Jumelle in France and published mostly in the Annales du Musée Colonial de Marseille, in a long series of papers, spanning some 20 years. Beccari accounted for those new genera published before 1920, and despite having rather limited material at his disposal, was able to place most of them in a global context. His account of Old World Arecoid palms (edited by Pichi-Sermolli) that includes all of the genera published by Jumelle and Perrier was not published until 1956 (Beccari & Pichi-Sermolli), long after he had died.

In 1928 Humbert started his collecting in Madagascar, visiting many areas not previously visited, and accumulating a very large number of specimens, including palms. Some of these, together with the material accumulated up until then, formed the basis for the account of the palms in the Flore de Madagascar et des Comores, by Jumelle and Perrier (1945), but it is surprising that a great deal of material present in the Paris Herbarium was not included in that account. Because of the very fragmentary nature of the herbarium collections at Jumelle's disposal, many of his descriptions lack important details, and he was forced sometimes to use abstruse details of "oppositipetiolar ligules" to distinguish species, details that have proved to be of little real taxonomic value. After the appearance of the 1945 flora account, Humbert continued to collect and several of his collections represented new taxa, including the new genus *Marojejya*, described in 1955, and others, such as *Dypsis malcomberi, D. andrianatonga, D. pumila* and *D. bonsai,* not named and described until the present book. In the early 1960s Moore paid the first of three visits to Madagascar. His aim was to investigate the taxonomic limits of Madagascar palm genera, rather than to account for all species. Data from his collections and concepts were included in a series of papers, culminating in The Major Groups of Palms and their Distribution (Moore 1973), followed by Genera Palmarum (Uhl & Dransfield 1987). By the 1980s it was obvious that the account of palms in the Flore de Madagascar et des Comores was seriously out of date, and of limited value for the identification of palms from Madagascar. The history of the present project is given in the Preface.

Taxonomic and Nomenclatural problems

The purpose of taxonomy and nomenclature is to provide an ordered framework of names for organisms, to act as basic reference points. When a group of plants is well known and the differences between species well understood, a taxonomy can be proposed that reflects the natural relationships of the plants. The better the group is known, the more refined the taxonomy can be. New discoveries and new insights obtained using novel research techniques will provide evidence that can be used to refine the taxonomy further. Such refinements are inevitably reflected in the nomenclature of the group studied – name changes need to be made. There is of course an understandable resistance to changing names, but, contrary to popular belief, taxonomists usually make name changes only where it is absolutely necessary to reflect the state of taxonomy of the studied organisms.

This present study of the palms of Madagascar, started in 1986, has uncovered so many undescribed palms and so much new information about species that are already described and named that we have had to develop completely new circumscriptions to the groupings of palms in Madagascar. In other words, we have not only had to describe new palm genera and species but we have also had to sink many genera and species into synonymy where we have discovered intermediates. It is of no use to maintain historically recognised genera that can no longer be distinguished. Thus the present book includes many name changes, some of which affect palms well-known in horticulture. There may well be resistance to the use of these new or old, long-forgotten names. People will continue to refer to *Dypsis lutescens* as *Chrysalidocarpus lutescens* (as well as the quite invalid name *Areca lutescens*) and to *Dypsis decaryi* as *Neodypsis decaryi*. However, we hope to show here that neither *Neodypsis* nor *Chrysalidocarpus* can be satisfactorily defined and distinguished. That they are all species of *Dypsis* should, we think, make the situation eventually much clearer and less confused.

The names of plants are validated by types. These are usually the herbarium specimens which provided information for the first valid description of the species. Since the adoption of the Linnean binomial system for naming plants in 1753, rules have gradually been developed, refined and refined again over time, to provide a legal framework for the naming of plants. In the early part of the nineteenth century such rules had not yet been formulated; species were described, named and published but frequently with only minimal reference to the specimens on which the descriptions were based, and without indication of a unique, type specimen. In some instances, such as the coconut, no specimens were referred to in the first description, but perhaps this is not really a problem as everyone knows what a coconut looks like. There are, however, other species such as *Chrysalidocarpus lutescens,* where there is no type, and there is great potential for confusion in the correct application of the name. On the other hand, an author such as Martius in his monumental "Historia Naturalis Palmarum", that was published before the concept of types had really developed, might refer to two or three specimens that he had studied to draw up his description, without indicating which specimen was typical of the species. This provides immense scope for confusion. In interpreting the application of such a name the present day taxonomist has to find the specimens cited by the original author. In the case of Madagascar palms, these are often in the Herbarium in Paris, but they are sometimes in Geneva, or in Munich, or elsewhere. Having found them, the taxonomist can then study them in detail. If the two or three specimens indeed represent the same species then it is usually a straightforward matter to decide which of the specimens most nearly matches the original published description. This can then be selected to act as the type of the species, the lectotype, the process being referred to as lectotypification. If the two or three specimens do not represent the same species – and with early, poorly collected palm specimens, this is quite often the case – then decisions have to be taken on which specimen most nearly matches the description in the original publication. It is then chosen as lectotype, and the other elements are excluded. In the course of the present study we have had to reinterpret old names such as *Dypsis boiviniana* (that was indeed based on a mixture of two species) and even had to tussle with the typification of such an apparently well understood name as *Chrysalidocarpus lutescens.*

The interpretation of Madagascar palm species is, perhaps more than in any other palm flora, complicated by grossly inadequate type specimens. For example, several species of *Chrysalidocarpus* and *Neodypsis* were described by Jumelle based on specimens that lack fruit, and hence do not provide information on whether the endosperm is ruminate or homogeneous, the only character that separates these two historically recognised genera. How Jumelle decided whether a species belonged to one or the other genus is beyond our comprehension. The type specimen of one of the most attractive and distinctive undergrowth palmlet, *Dypsis corniculata*, consists of a single inflorescence. Beccari described his *Neophloga corniculata* in 1906 based on this fragmentary material, collected by Boivin at Tafondrou, Île Sainte Marie, in the middle of the nineteenth century. At the time Beccari described the species, there were so few palm specimens from Madagascar, that it was an easy matter to recognise that the inflorescence did not match any other palm species from Madagascar and thus represented an undescribed species. In the Flore de Madagascar, Jumelle and Perrier (1945) included *Neophloga corniculata* as an insufficiently known species. In November 1994, when we visited the Forêt de Kalalao on Île Sainte Marie, the presumed Tafondrou of Boivin, we collected an undergrowth palmlet with the characteristic inflorescences that we were able to match with the type of *Neophloga corniculata*. In fact, now that we know what the palm looks like, we have been able to establish that *Dypsis corniculata* is quite widespread in north-eastern Madagascar, and is no longer an insufficiently known species. This was one of the successes of the project, but there are still several species described in the Flore de Madagascar that we now, in turn, have to characterise as "insufficiently known species". In particular an understanding of the tall forest emergent species of *Dypsis* is bedevilled by inadequate representation in herbaria, with poorly collected type specimens and sketchy original descriptions. We may expect further changes in this group in the future as more well-prepared collections are made.

As palms are the source of important forest products used by villagers in Madagascar, it is not surprising that there are plenty of vernacular names for palms. The best local guides have been very helpful to us, not only in guiding us to new and exciting palms such as *Voanioala gerardii*, but also in pointing out the differences that they perceive between the different palms in their area. Local names, however, are only rarely consistently applied. The best local guides may apply names with uncanny precision in one area, but other villagers from the same area are rarely as consistent; poorer knowledge and careless identification are the cause of this, which would be true almost anywhere in the world. Furthermore local names in one area for one species recognised by us, may be applied to a completely different species in another area. This has provided a nagging source of uncertainty that has caused us constantly to question the limits of the species we recognise.

Having constructed a new taxonomy for the palms of Madagascar and made the necessary nomenclatural changes that reflect this, we are very much aware that what we are publishing is not the whole story. As late as October 1994, as the manuscript neared completion, we spent over one month in Madagascar, checking our descriptions and keys and taking photographs, and we discovered three more new species in an area we had previously visited. Furthermore, there are large areas which we have not yet visited. Northwest of Toamasina lies Zahamena and west of Soanierana-Ivongo, also on the East Coast, lies Ambatovaky. We have flown over these two areas of protected forest and admired from the air their wonderful pristine forests and spectacular waterfalls, and even on one occasion caught glimpses of palm crowns emerging from the forest canopy, without our being able to say to which species they belonged. Both areas are difficult of access and we have not yet visited them. Few palm collections have been made in Zahamena and, as far as we know, no collections have been made in Ambatovaky. When these and other areas are visited and scientific collections of palms made, novelties are bound to be discovered and new observations of variation made that could well lead to further name changes.

We hope that what we publish here reflects clearly the state of taxonomic knowledge of the palms of Madagascar in the mid 1990s and will be useful as a basis, a primary resource, for further work on this astonishingly rich palm flora.

OUTLINE OF CLASSIFICATION

PALMAE

SUBFAMILY	TRIBE	SUBTRIBE	GENUS	
CORYPHOIDEAE	PHOENICEAE		**I. Phoenix**	**P. reclinata** *Jacq.*
	BORASSEAE	LATANIINAE	**II. Borassus**	**1. B. madagascariensis** *Bojer ex Jum. & H. Perrier* **2. B. sambiranensis** *Jum. & H. Perrier*
		HYPHAENINAE	**III. Hyphaene**	**H. coriacea** *Gaertn.*
			IV. Bismarckia	**B. nobilis** *Hild. & H. Wendl.*
			V. Satranala	**S. decussilvae** *Beentje & J. Dransf.*
CALAMOIDEAE	CALAMEAE	RAPHIINAE	**VI. Raphia**	**R. farinifera** *(Gaertn.) Hylander*
CEROXYLOIDEAE	CEROXYLEAE		**VII. Ravenea**	**1. R. robustior** *Jum. & H. Perrier* **2. R. krociana** *Beentje* **3. R. moorei** *J. Dransf. & N.W. Uhl* **4. R. madagascariensis** *Becc.* **5. R. latisecta** *Jum.* **6. R. sambiranensis** *Jum. & H. Perrier* **7. R. julietiae** *Beentje* **8. R. rivularis** *Jum. & H. Perrier* **9. R. musicalis** *Beentje* **10. R. glauca** *Jum. & H. Perrier* **11. R. hildebrandtii** *C.D. Bouché* **12. R. nana** *Beentje* **13. R. dransfieldii** *Beentje* **14. R. xerophila** *Jum.* **15. R. lakatra** *(Jum.) Beentje* **16. R. albicans** *(Jum.) Beentje* **17. R. louvelii** *Beentje*
ARECOIDEAE	ARECEAE	ORANIINAE	**VIII. Orania**	**1. O. trispatha** *(J. Dransf. & N.W. Uhl) Beentje & J. Dransf.* **2. O. longisquama** *(Jum.) J. Dransf. & N.W. Uhl* **3. O. ravaka** *Beentje*

Subfamily	Tribe	Subtribe	Genus	
		Dypsidinae	IX. Dypsis	1. **D. bejofo** *Beentje*
				2. **D. canaliculata** *(Jum.) Beentje & J. Dransf.*
				3. **D. hovomantsina** *Beentje*
				4. **D. ceracea** *(Jum.) Beentje & J. Dransf.*
				5. **D. ampasindavae** *Beentje*
				6. **D. tsaravoasira** *Beentje*
				7. **D. nauseosa** *(Jum. & H. Perrier) Beentje & J. Dransf.*
				8. **D. oropedionis** *Beentje*
				9. **D. pilulifera** *(Becc.) Beentje & J. Dransf.*
				10. **D. mananjarensis** *(Jum. & H. Perrier) Beentje & J. Dransf.*
				11. **D. malcomberi** *Beentje*
				12. **D. prestoniana** *Beentje*
				13. **D. tokoravina** *Beentje*
				14. **D. ifanadianae** *Beentje*
				15. **D. lastelliana** *(Baill.) Beentje & J. Dransf.*
				16. **D. leptocheilos** *(Hodel) Beentje & J. Dransf.*
				17. **D. ligulata** *(Jum.) Beentje & J. Dransf.*
				18. **D. saintelucei** *Beentje*
				19. **D. ovobontsira** *Beentje*
				20. **D. tanalensis** *(Jum. & H. Perrier) Beentje & J. Dransf.*
				21. **D. ankaizinensis** *(Jum.) Beentje & J. Dransf.*
				22. **D. madagascariensis** *(Becc.) Beentje & J. Dransf.*
				23. **D. decaryi** *(Jum.) Beentje & J. Dransf.*
				24. **D. decipiens** *(Becc.) Beentje & J. Dransf.*
				25. **D. basilonga** *(Jum. & H. Perrier) Beentje & J. Dransf.*
				26. **D. ambositrae** *Beentje*
				27. **D. heteromorpha** *(Jum.) Beentje & J. Dransf.*
				28. **D. baronii** *(Becc.) Beentje & J. Dransf.*
				29. **D. andrianatonga** *Beentje*
				30. **D. serpentina** *Beentje*
				31. **D. onilahensis** *(Jum. & H. Perrier) Beentje & J. Dransf.*
				32. **D. acuminum** *(Jum.) Beentje & J. Dransf.*
				33. **D. lutescens** *(H. Wendl.) Beentje & J. Dransf.*
				34. **D. arenarum** *(Jum.) Beentje & J. Dransf.*
				35. **D. psammophila** *Beentje & J. Dransf.*
				36. **D. pembana** *(H.E. Moore) Beentje & J. Dransf.*
				37. **D. cabadae** *(H.E. Moore) Beentje & J. Dransf.*
				38. **D. humblotiana** *(Baill.) Beentje & J. Dransf.*
				39. **D. lanceolata** *(Becc.) Beentje & J. Dransf.*
				40. **D. pumila** *Beentje*
				41. **D. tsaratananensis** *(Jum.) Beentje & J. Dransf.*
				42. **D. oreophila** *Beentje*
				43. **D. ambanjae** *Beentje*
				44. **D. coursii** *Beentje*
				45. **D. rivularis** *(Jum. & H. Perrier) Beentje & J. Dransf.*

Subfamily	Tribe	Subtribe	Genus	
		Dypsidinae	IX. Dypsis	**46. D. marojejyi** *Beentje*
				47. D. commersoniana *(Baill.) Beentje & J. Dransf.*
				48. D. humbertii *(Jum.) Beentje & J. Dransf.*
				49. D. scottiana *(Becc.) Beentje & J. Dransf.*
				50. D. singularis *Beentje*
				51. D. intermedia *Beentje*
				52. D. mcdonaldiana *Beentje*
				53. D. jumelleana *Beentje & J. Dransf.*
				54. D. linearis *(Becc.) Beentje & J. Dransf.*
				55. D. bonsai *Beentje*
				56. D. caudata *Beentje*
				57. D. scandens *J. Dransf.*
				58. D. faneva *Beentje*
				59. D. fanjana *Beentje*
				60. D. boiviniana *Baill.*
				61. D. mangorensis *(Jum.) Beentje & J. Dransf.*
				62. D. sanctaemariae *J. Dransf.*
				63. D. soanieranae *Beentje*
				64. D. curtisii *Baker*
				65. D. pervillei *(Jum.) Beentje & J. Dransf.*
				66. D. elegans *Beentje*
				67. D. heterophylla *Baker*
				68. D. concinna *Baker*
				69. D. schatzii *Beentje*
				70. D. corniculata *(Becc.) Beentje & J. Dransf.*
				71. D. thiryana *(Becc.) Beentje & J. Dransf.*
				72. D. trapezoidea *J. Dransf.*
				73. D. hiarakae *Beentje*
				74. D. confusa *Beentje*
				75. D. sahanofensis *(Jum. & H. Perrier) Beentje & J. Dransf.*
				76. D. eriostachys *J. Dransf.*
				77. D. lutea *(Jum.) Beentje & J. Dransf.*
				78. D. betamponensis *(Jum.) Beentje & J. Dransf.*
				79. D. mahia *Beentje*
				80. D. pulchella *J. Dransf.*
				81. D. plurisecta *Jum.*
				82. D. andapae *Beentje*
				83. D. occidentalis *(Jum.) Beentje & J. Dransf.*
				84. D. montana *(Jum.) Beentje & J. Dransf.*
				85. D. bernierana *(Baill.) Beentje & J. Dransf.*
				86. D. poivreana *(Baill.) Beentje & J. Dransf.*
				87. D. catatiana *(Baill.) Beentje & J. Dransf.*
				88. D. coriacea *Beentje*
				89. D. minuta *Beentje*
				90. D. tenuissima *Beentje*
				91. D. lucens *(Jum.) Beentje & J. Dransf.*
				92. D. simianensis *(Jum.) Beentje & J. Dransf.*
				93. D. integra *(Jum.) Beentje & J. Dransf.*
				94. D. digitata *(Becc.) Beentje & J. Dransf.*

SUBFAMILY	TRIBE	SUBTRIBE	GENUS	
		DYPSIDINAE	IX. Dypsis	95. **D. brevicaulis** *(Guillaumet) Beentje & J. Dransf.*
				96. **D. forficifolia** *Mart.*
				97. **D. interrupta** *J. Dransf.*
				98. **D. lantzeana** *Baill.*
				99. **D. remotiflora** *J. Dransf.*
				100. **D. louvelii** *Jum. & H. Perrier*
				101. **D. zahamenae** *J. Dransf.*
				102. **D. angustifolia** *(H. Perrier) Beentje & J.Dransf.*
				103. **D. pinnatifrons** *Mart.*
				104. **D. procera** *Jum.*
				105. **D. paludosa** *J. Dransf.*
				106. **D. mirabilis** *J. Dransf.*
				107. **D. fasciculata** *Jum.*
				108. **D. lokohoensis** *J. Dransf.*
				109. **D. thouarsiana** *Baill.*
				110. **D. perrieri** *(Jum.) Beentje & J. Dransf.*
				111. **D. moorei** *Beentje*
				112. **D. dransfieldii** *Beentje*
				113. **D. nossibensis** *(Becc.) Beentje & J. Dransf.*
				114. **D. crinita** *(Jum. & H. Perrier) Beentje & J. Dransf.*
				115. **D. utilis** *(Jum.) Beentje & J. Dransf.*
				116. **D. fibrosa** *(Wright) Beentje & J. Dransf.*
				117. **D. antanambensis** *Beentje*
				118. **D. pusilla** *Beentje*
				119. **D. aquatilis** *Beentje*
				120. **D. nodifera** *Mart.*
				121. **D. thermarum** *J. Dransf.*
				122. **D. angusta** *Jum.*
				123. **D. anovensis** *J. Dransf.*
				124. **D. ambilaensis** *J. Dransf.*
				125. **D. glabrescens** *(Becc.) Becc.*
				126. **D. laevis** *J. Dransf.*
				127. **D. viridis** *Jum.*
				128. **D. hildebrandtii** *(Baill.) Becc.*
				129. **D. bosseri** *J. Dransf.*
				130. **D. furcata** *J. Dransf.*
				131. **D. lanuginosa** *J. Dransf.*
				132. **D. mocquerysiana** *(Becc.) Becc.*
				133. **D. cookei** *J. Dransf.*
				134. **D. beentjei** *J. Dransf.*
				135. **D. pachyramea** *J. Dransf.*
				136. **D. monostachya** *Jum.*
				137. **D. spicata** *J. Dransf.*
				138. **D. acaulis** *J. Dransf.*
				139. **D. canescens** *(Jum. & H. Perrier) Beentje & J. Dransf.*
				140. **D. ramentacea** *J. Dransf.*

Subfamily	Tribe	Subtribe	Genus	
		Lemurophoenicinae	**X.Lemurophoenix**	**L. halleuxii** *J. Dransf.*
		Masoalinae	**XI. Masoala**	**1. M. madagascariensis** *Jum.* **2. M. kona** *Beentje*
			XII. Marojejya	**1. M. insignis** *Humbert* **2. M. darianii** *J. Dransf. & N.W. Uhl*
	Cocoeae	Beccariophoenicinae	**XIII. Beccariophoenix**	**B. madagascariensis** *Jum. & H. Perrier*
		Butiinae	**XIV. Voanioala**	**V. gerardii** *J. Dransf.*
			XV. Cocos	**C. nucifera** *L.*
		Elaeidinae	**XVI. Elaeis**	**E. guineensis** *Jacq.*

KEY TO GENERA OF NATIVE OR NATURALISED PALMS IN MADAGASCAR

HOW TO USE THIS KEY

This key is based, as much as possible, on characters that are easily appreciated in the field. It relies heavily on vegetative features, but, nevertheless, it may be necessary to examine old inflorescences as well. Look at the palm carefully. If it's too tall to examine features of leaves and inflorescences, check the ground for fallen leaves and examine these carefully. Several exotic palms are widely cultivated in Madagascar, e.g. *Trachycarpus fortunei, Livistona chinensis, Washingtonia spp, Phoenix canariensis, P. roebelenii, Chamaedorea spp, Roystonea regia, Areca catechu, Dictyosperma album* and *Syagrus romanzoffiana*. These are not included in this key.

1. Leaves palmate 2
   Leaves pinnate or entire and pinnately ribbed 5

2. Petiole margins armed with irregular black erose teeth . . . . . . **Borassus**
   Petiole margins unarmed, or armed with regular spines, black erose teeth absent 3

3. Petiole margins with well defined upward pointing spines to 1 cm or more long . . . . . **Hyphaene**
   Petiole margins with minute teeth or unarmed 4

4. Petiole virtually unarmed; leaf segments without clear transverse veinlets; endocarp more or less smooth with very obscure ridging without, within with irregular crests and ridges that penetrate the homogeneous endosperm; palm of western savannahs, cultivated in east . . . . . . **Bismarckia**
   Petiole minutely toothed; leaf segments with clear transverse veinlets; endocarp winged and crested without, smooth within; endosperm deeply ruminate; palm of eastern rain forest . . . . . . **Satranala**

5. Petiole or base of leaf rachis armed with short or long spines 6
   Petiole or base of leaf rachis unarmed, very rarely the sheath margins with a few reflexed almost spine-like fibres 7

6. Leaflets induplicate (V-shaped in cross-section), basal leaflets modified as long spines; leaflet tips spine-like . . . . . . **Phoenix**
   Leaflets reduplicate (Λ-shaped in cross-section), fibres of sheath margin spine-like and mid ribs of basal leaflets modified as spines; leaflet tips not spine-like, more or less irregularly two-lobed . . . . . . **Elaeis**

7. Leaflet margins and midribs densely armed with short spines; massive hapaxanthic palm of village margins and disturbed habitats, apparently not in primary forest . . . . . . **Raphia**
   Leaflet margins and midribs unarmed; pleonanthic palms 8

8. Leaflet tips praemorse 9
   Leaflet tips entire or bifid, but not praemorse 10

9. Tall solitary palms of the canopy or subcanopy, with stems at least 12 cm diam. . . . . . . **Orania**
   Slender undergrowth palmlets with stems not exceeding 2 cm diam. . . . . . . **Dypsis**
   (D. thiryana, D. trapezoidea)

10. Palms with well defined crownshafts 11
Palms without well defined crownshafts 12

11. Immense canopy palm with staminate flowers with more than 50 stamens, large corky warted fruit and seeds with apical embryo . . . . **Lemurophoenix**
Palms of varying size with staminate flowers with 3 or 6 stamens, smooth fruits and seeds with basal or subbasal embryos . . . . **Dypsis**

12. Leaf with fibrous sheath that extends to the base of the lowermost leaflets, but disintegrates partly to produce a false petiole (check petiole margin for fibre remains) 13
Leaf with true petiole or no petiole 15

13. Palm of villages, village margins and naturalised in strand forest; female flowers very large, at least 3 cm diam. (the coconut) . . . . **Cocos**
Palms of natural vegetation; female flowers not exceeding 2 cm diam. 14

14. Leaflets c. 100 to 130 on each side of the rachis; peduncular bract borne at the tip of the peduncle, at least 1 cm thick, circumscissile . . . . **Beccariophoenix**
Leaflets c. 70 on each side of the rachis; peduncular bract borne near the base of the peduncle, about 5 mm thick, persistent . . . . **Voanioala**

15. Massive litter-trapping palms of the forest undergrowth, petioles absent or short, the leaf bases marcesent, forming gutters that trap debris 16
Not so 21

16. Leaflets dark green on upper surface, grey-white on under surface . . . . **Ravenea** *(R. albicans)*
Leaflets green on both surfaces 17

17. Petiole well developed, it and the sheath densely covered with shining chestnut-brown scales . . . . **Ravenea** *(R. louvelii)*
Petiole absent, or if developed with or without dull, irregular rather amorphous indumentum . . 18

18. Leaflets grouped and fanned within groups . . . . **Dypsis** *(D. marojejyi)*
Leaflets regularly arranged or if irregular, then not regularly grouped and fanned, or blade entire . 19

19. Inflorescences in bud exserted, torpedo-like, peduncular bract densely covered with red tomentum; stigmas basal in fruit . . . . **Dypsis** *(D. perrieri, D. moorei)*
Inflorescences in bud, hidden or if exserted then not torpedo-like, and peduncular bract glabrous; stigmas lateral or apical in fruit 20

20. Inflorescences highly condensed, hidden among leaf sheaths, usually unisexual . . . . . . **Marojejya**
Inflorescences diffuse, exposed, bisexual . . . . **Masoala**

21. Solitary small to very large tree palms with regularly pinnate leaves and more or less open sheaths; dioecious; inflorescences with c. 5 bracts . . . . **Ravenea**
Solitary or clustered, palmlets to tree palms with regularly or irregularly pinnate leaves, sheaths usually at least 25% tubular; monoecious; inflorescences with 2 bracts . . . . **Dypsis**

SUBFAMILY CORYPHOIDEAE
TRIBE PHOENICEAE

1. PHOENIX

BY SASHA BARROW

L., Sp. Pl. 1188 (1753); Jum. & H. Perrier, Fl. Madagascar 30: 19 (1945); N.W. Uhl & J. Dransf., Genera Palmarum 214 (1987). Type: *P. dactylifera* L.

Dwarf or creeping to large, solitary or clustered, armed, pleonanthic, dioecious palms. **STEM** often clothed with spirally arranged leaf bases. **LEAVES** induplicate, pinnate, usually marcescent; sheath forming a fibrous network; petiole very short to well developed, adaxially channelled to flattened or ridged, abaxially rounded; rachis elongate, tapering, adaxially rounded or flat to angled, abaxially rounded to flat, usually terminating in a leaflet; leaflets single-fold, acute, regularly arranged or variously grouped, the proximal few modified as spines (acanthophylls), parallel veined, midrib usually evident abaxially, often bearing scales, emergent leaves frequently with brown floccose indumentum and/or wax. **INFLORESCENCES** interfoliar, branching to 1 order, the staminate and pistillate superficially similar; peduncle flattened, short to elongate, in the pistillate frequently elongating after fertilization, bearing an often caducous, sometimes bivalved, 2-keeled, glabrous or floccose-hairy prophyll; other bracts inconspicuous; rachis flattened, usually shorter than the peduncle; rachillae unbranched, numerous, often spirally arranged along rachis, the rachillae bearing spirally arranged, low triangular bracts, each subtending a solitary flower. **STAMINATE FLOWERS** with 3 sepals connate in a low cupule; petals 3, ± valvate, acute or rounded, much exceeding the calyx; stamens usually 6 (rarely 3 or 9), filaments short, erect, the anthers linear, latrorse; pollen elliptic or circular, monosulcate, with reticulate or finely reticulate, tectate exine; pistillode absent or 3, minute. **PISTILLATE FLOWERS** globose; sepals connate in a 3-lobed cupule; petals imbricate, strongly nerved, about twice as long as the calyx or more; staminodes usually 6, scale-like or connate in a low cupule; carpels 3, distinct, follicular, ovoid, narrowed into a short, recurved, exserted stigma, ovule attached adaxially at the base, anatropous. **FRUIT** usually developing from 1 carpel, ovoid to oblong with apical stigmatic remains; epicarp smooth, mesocarp fleshy, endocarp membranous. **SEED** elongate, terete or plano-convex, and deeply grooved with intruded seed coat below the elongate raphe, hilum basal, rounded, endosperm homogeneous; embryo lateral or subbasal. **GERMINATION** remote-tubular; eophyll undivided, narrowly lanceolate.

DISTRIBUTION. About 17 species in the Old World, in the drier tropics and subtropics. One non-endemic species in Madagascar.

KEY TO PHOENIX

1. Usually occurring as clustering thickets; stem rarely more than 20 cm diam.; leaves arcuate, bright shiny green; ripe fruit not exceeding 2 × 1.5 cm. .... 1. **P. reclinata**

   Usually occurring as solitary individuals; stem usually to 40 cm in diam. or more; leaves stiff, glaucous green; ripe fruit 4–7 × 2–3 cm. .... **P. dactylifera**

1. PHOENIX RECLINATA

The wild date palm is widespread throughout the continent of Africa. It is one of the very few native palms of Madagascar that also occurs outside the island. On the basis of the very limited material available to him, Beccari considered the form occurring in Madagascar sufficiently distinct as to be recognized as a variety. However, in our view the Madagascar plants fit within the range of variation of the wild date in Africa, so that recognition of a separate variety does not seem justified.

DISTRIBUTION. NW and NE Madagascar, with reported stands in isolated sites in the SW; also in the Manakara area. *P. reclinata* is a widely distributed species growing throughout sub-tropical Africa, mostly restricted to dry, open areas, from sea-level to c.1500 m.

HABITAT. Riverine vegetation; seasonally inundated plains.

LOCAL NAMES. *Dara, Taratra, Taratsy* (Jumelle & Perrier 1913, 1945).

USES. Leaflets used in basketry; fruits more or less edible.

CONSERVATION STATUS. Not threatened.

Phoenix reclinata *Jacq.*
Fragm. Bot. 1: 27, t. 24 (1801); Jum. & H. Perrier, Fl. Madagascar 30: 19, fig. 3 (1945); J. Dransf., Fl. Trop. E. Africa, Palmae: 15 (1986). Type: t. 24, Jacq., Fragm. Bot.

SYNONYMS:
P. reclinata var. ***madagascariensis*** Becc., Bot. Jahrb. Syst. 38, Beibl. 87: 4 (1906); Jum. & H. Perrier, Ann. Inst. Bot.-Géol. Colon. Marseille sér. 3, 1 (1): 60 (1913); Jum., Cat. Pl. Madagascar, Palmae: 23 (1938). Syntypes: without locality, *Baron* 2319b (K); Nosy Be, Lokobe forest, *Hildebrandt* 3304 (K, P).

Satranala decussilvae, in lowland forest south of Mananara Avaratra.

Phoenix reclinata (from East African material). **A** habit × 1/100; **B** basal part of leaf rachis showing acanthophylls × 1; **C** mid section of leaf with grouped leaflets × 1; **D** mid section of leaf with regular leaflets × 1; **E** leaf tip × 1; **F** prophyll × 1; **G** part of staminate inflorescence × 1; **H** staminate flower, one petal removed × 5; **J** part of pistillate inflorescence × 1; **K** pistillate flower × 5; **L** fruit × 2.5; **M** seed × 2.5. **A** from a photograph, **B** – **H** from *Dransfield* JD4828, **J**, **K** from *Dransfield* JD4829, **L**, **M** from *Dransfield* JD4632. Drawn by Eleanor Catherine.

P. comorensis Becc. in Bot. Jahrb. Syst. 38, Beibl. 87: 5 (1906); Becc., Palme del Madagascar 54 (1914).
P. reclinata var. ***comorensis*** (Becc.) Jum. & H. Perrier, Fl. Madagascar 30: 20 (1945). Syntypes: Without locality, *Boivin* 2289 (K, P); Mayotte, *Boivin* 3100 (K, P), **synon. nov.**

Phoenix reclinata

Clustering dioecious palm, often thicket-forming. **TRUNK** 3 m tall, usually oblique, rarely erect, slender, to 15 cm diam., dull brown, with persistent leaf sheaths in 1–2 m below crown, irregularly marked with oblique circular leaf scars, cracked vertically; injured stem exuding clear yellowish gum. **LEAVES** 2–3 m long, arcuate; leaflets with the proximal ones modified as sharp spine-like acanthophylls, the distal regularly arranged in a single plane, median leaflets in groups of 2–5, 25–45 × 2.0–3.6 cm, lower leaflets often fanned, abaxial leaf surface with white scurfy indumentum. **INFLORESCENCES** with prophyll c. 70 cm long, greenish-yellow, covered with brown tomentum when young, two-keeled, splitting ventrally to reveal inflorescence. **STAMINATE INFLORESCENCE** erect in bud, arching when dead, peduncle not greatly elongating; rachillae to 20 cm long, to 1.5 mm diam., numerous, flexuous, arranged in irregular groups and spirals. **STAMINATE FLOWERS** creamy-white, musty scented; calyx cupule three-lobed, 1 mm high; petals 6–7 × 2–3 mm, apex thickened, acute-acuminate, margin minutely ciliate-dentate; stamens 6, sessile, 3–4 mm, linear-latrorse; pistillodes 3, minute. **PISTILLATE INFLORESCENCE** elongating beyond prophyll on fruit maturity; prophyll as in staminate; rachillae 9–26 cm long, c. 2 mm diam., thicker than those of the staminate inflorescence. **PISTILLATE FLOWERS** globose, 3–4 mm; carpels 3, usually only one reaching maturity. **FRUIT** ovoid-ellipsoid or almost obovoid, 18–20 × 9–12 mm, minutely mucronate, at the base with persistent perianth, ripening orange; mesocarp dry, becoming sweet. **SEED** 12–14 × 6 mm, deeply grooved (raphe) along ventral side; embryo centrally positioned on opposite side to raphe.

Phoenix reclinata in scrub near Vohemar (*Beentje & Andriampaniry* 4692).

NOTE. The vegetatively polymorphic nature of *P. reclinata* has led to the recognition of certain extreme forms as distinct species or varieties. A host of intermediate forms make it very difficult to delimit infraspecific taxa within the total range of variation. Despite the variation in gross vegetative morphology, all forms of *P. reclinata* are grouped together by floral and fruit characters which vary very little within the species. *Phoenix reclinata* in Madagascar shows variation consistent with that of the species in Africa. It was first described from Madagascar by Beccari (1890), who later (1906, 1914) recorded it as *P. reclinata* var. *madagascariensis* Becc. on the basis of very minor differences in persistent fruit perianth. Jumelle & Perrier (1913, 1945) considered this variety as synonymous with the species.

Beccari (1906) separated *P. comorensis* from *P. reclinata* using the same characters. Jumelle & Perrier (1945) considered this a variety of the species, but the differences are too small for the taxon to be upheld, and it is put into synonymy here.

SPECIMENS SEEN Comoro Is.: Mayotte, (y. fr.), *Boivin* 3100 (K, P); Moheli, June 1967 (fl., fr.), *L. Bernardi* 11792 (K); without any locality, Dec. 1848 (ster.), *Boivin* 2289 (K, P). Nosy Be: Lokobe forest, Dec. 1879 (fl.), *Hildebrandt* 3304 (BM, K, P)(syntype of *P. reclinata* var. *madagascariensis*). Iharana: 2 km S. of Iharana, June 1992 (infr.), *Beentje & Andriampaniry* 4691 (K, MO, TAN), (fl.) 4692 (K, MO, TAN). Belo-Tsiribihina: Belo-Tsiribihina, (fl.), *Decary* 15535 (P). Mangoky: Isahaina, near Beroroha, Oct. 1933 (infr.), *Humbert* 11298 (K, P, TAN). Mananjary: without precise locality, March/April 1909 (fl.), *Geay* 8055 (P). Morondava: Morondava, sine die, *Greve* 154 (P); Sakondry plain, July 1898 (lvs.), *Grandidier* s.n. (P). Beroha: Mangoky R basin near Beroha, *Perrier* s.n. (P).
WITHOUT PRECISE LOCALITY: North Madagascar, (fl.), *Baron* 6724 (K); Central Madagascar, (fl.), *Baron* 2319b (K); Central Madagascar, (infr.), *Baron* 4776 (K); Central Madagascar, (infr.) *Baron* 4839 (K).
CULTIVATED. Antananarivo, Tsimbazaza, Feb. 1940 (fl.), *Herb. Jard. Bot. Tana* 4267 (P).
REPORTED. SW, Menarandra River near Tranoroa, riverside trees with this name and general description reported by local people, 1992 (fide *Beentje*); Mananjary road, W of town, *Beentje* anno 1992.

PHOENIX DACTYLIFERA

Phoenix dactylifera L., the date palm, is occasionally cultivated in the larger towns.

DISTRIBUTION. According to Fl. Madagascar: Antsiranana (Diégo-Suarez), Mahajanga (Majunga), Toliara (Tuléar).
SPECIMENS SEEN. Toliara: Toliara (Tuléar), sine die (dead infl.), *Croat* 31012 (TAN).
LOCAL NAMES. *Atserijy* (Malagasy, fide Croat).

Tribe Borasseae

Key to Borasseae

1. Small, often clustering palm with (usually leaning) trunk 1–6 m high clothed in old sheaths; leaf sheath 38–40 cm long; petiole with curved spines to 1 cm high; leaf blade with conspicuous filaments in the sinuses between the segments .... **Hyphaene**

   Solitary palms with erect trunks 2–20 m high with some marcescent leaves, but an otherwise smooth clean trunk; leaf sheath > 45 cm long; petiole with erose teeth or minute spines < 3 mm long; leaf blade without filaments in the sinuses 2

2. Leaf with scattered reddish scales all over the minor veins; petiole with large erose teeth on the margins; fruit > 10 cm diam., (1–2)–3-seeded .... **Borassus**

   Leaf without small scales, though ramenta sometimes present on the midribs; petiole with minute (< 3 mm) teeth on the margins, or with smooth margins; fruit < 5 cm diam., usually 1-seeded, if more, then fruit lobed 3

3. Leaves without (or with very faint) transverse veinlets between the parallel veins; endocarp smooth, faintly ridged; palm from dry habitats in W Madagascar .... **Bismarckia**

   Leaves with clear transverse veinlets between the parallel veins; endocarp winged; palm from rain forest in E Madagascar .... **Satranala**

Subtribe Lataniinae

11. Borassus

L., Sp. Pl.: 1187 (1753); Jum. & H. Perrier, Fl. Madagascar 30: 10 (1945); N.W. Uhl & J. Dransf., Genera Palmarum 222 (1987). Type: *B. flabellifer* L.

Tall, robust, solitary, armed, pleonanthic, dioecious tree palms. **Stem** massive, covered in a lattice of leaf bases abscising cleanly in older specimens, then rough, ringed with wide leaf scars. **Leaves** induplicate, strongly costapalmate; sheath open early in development, later with a wide triangular cleft at the base of the petiole; petiole deeply channelled adaxially, rounded abaxially, surfaces smooth to minutely rough, margins of sheath and petiole armed with coarse irregular teeth; adaxial hastula conspicuous, triangular or scalloped, abaxial hastula a low ridge (?always); blade suborbicular to flabellate, divided along adaxial folds to c. half its length into regular, stiff single-fold segments, these shortly bifid, interfold filaments present or absent, surfaces smooth, ramenta or tomentum along abaxial costa and ridges of folds, midribs prominent abaxially, transverse veinlets conspicuous, short, numerous. **Inflorescences** interfoliar, shorter than the leaves, the staminate and pistillate dissimilar; staminate inflorescence branched to 2 orders; peduncle very short; prophyll 2-keeled, with long tubular base, limb short, pointed, variously split apically; peduncular bracts lacking (?); rachis also short, rachis bracts similar to the prophyll; first-order branches long, flattened, each bearing a prophyll and branched digitately into several (1–3) rachillae; rachillae large, elongate, catkin-like, bearing spirally arranged, imbricate bracts, connate laterally and distally to form large pits, each containing a reflexed cincinnus of c. 30 staminate flowers, exserted singly in succession from the pit mouth. **Staminate flowers** each subtended by a long membranous bracteole; sepals 3, asymmetrical, connate only basally or to 2/3 their length, distinct lobes keeled, elongate, membranous, stiff; corolla with a long stalk-like base and 3 short rounded lobes, ridged adaxially; stamens 6, filaments short, triangular, anthers medifixed, elongate, latrorse; pollen elliptic, monosulcate, with verrucate, tectate exine, finely reticulate between the warts; pistillode small, conical. **Pistillate inflorescence** unbranched or with a single first-order branch; peduncle short; prophyll tubular, pointed, 2-keeled, split ventrally about half its length; peduncular bracts few (2 or more?), if present as long as or longer than the peduncle; rachilla massive, bearing large cupular bracts, the first few empty, the subsequent each bearing a single pistillate flower, several empty bracts above the flowers. **Pistillate flowers** large, each bearing 2 lateral cuplike, rounded, leathery bracteoles; sepals 3 distinct, imbricate, thick, rounded; petals 3, similar to sepals; staminodes triangular, connate basally in a low cupule, sterile anthers present or not; gynoecium rounded, tricarpellate, with a central, basal septal nectary, stylar region hemispherical, stigma a low knob, carpels each with a basal orthotropous ovule and 2 lateral bodies, perhaps vestigial ovules. **Fruit** large, rounded, sometimes wider than long, bearing 1–3 seeds, stigmatic remains apical, perianth enlarged, persistent; epicarp smooth, mesocarp thick, fibrous, often fragrant, endocarp comprising 3 hard bony pyrenes. **Seed** shallowly to deeply bilobed, pointed, basally attached, endosperm homogeneous with a central hollow; embryo apical. **Germination** remote-tubular; eophyll undivided, elliptical.

Borassus sambiranensis, growing south of Ambanja (*Schatz et al.* 2415) (Photo: G. Schatz).

Distribution. About five species in the Old World; one species in Africa, one in India, SE Asia and Malesia, one in New Guinea; two endemic species in Madagascar.

Key to Borassus

1. Pistillate rachilla 25–50 cm long, with 7–20 flowers; fruit 15–17 × 15–18 cm ... 1. **B. madagascariensis**
   Pistillate rachilla 84–90 cm long, with c. 30 flowers; fruit 11–13 × 9–11 cm .... 2. **B. sambiranensis**

Note. None of the differences enumerated by other authors (Beccari; Jumelle & Perrier) seems reliable. The above were the only differences we could find, from specimens and reliable descriptions of types or unseen Perrier collections. D. Du Puy, who has seen both taxa in the field, feels that fruit and seed characters, as well as the general habit of the trees, warrant their continued recognition as species.

Beccari (1913) states *B. sambiranensis* differs from *B. madagascariensis* in the smaller inflorescences, in the larger teeth on the petiole, and in the rectangular pyrenes; and that it thereby rather resembles *B. aethiopum* var. *bagamoyensis* (= *B. aethiopum*, fide Dransfield 1986) from which it only differs in the fibrovascular entry point of the pyrene being a deep hollow.

Jumelle & Perrier (1945) state that *B. sambiranensis* differs from *B. madagascariensis* in the sheaths being golden yellow, and shot through with larger fibers, in the petiolar marginal teeth being recurved, black, and 8–15 mm long, in the staminate spikes being thinner, 2 cm thick, in the pistillate inflorescences often having 2 spikes, in the bracts being more numerous, eleven in number, all hard and acute at the apex, and in the infructescence having more fruits with the fruits being smaller and almost twice as long as wide.

Differences from *B. aethiopum* seem to be very few: the Madagascar taxa have fewer leaves in the crown (*B. aethiopum* is said to have c. 40 in the Flora of Tropical East Africa), a longer costa (c. 40 cm; *B. aethiopum* costa 10 cm); more empty bracts (*B. aethiopum*: 2–3); the fruit of *B. aethiopum* is c. 12 × 14 cm (Flora of Tropical East Africa).

1. Borassus madagascariensis

Strangely, this is one of the least known palms of the island. Despite its occurrence in the West, it has not been collected often or in a proper manner, probably because of its giant stature. Until further collections become available, the differences with both *B. sambiranensis* and mainland African *Borassus aethiopum* will remain unclear, and slightly doubtful.

Distribution. Western Madagascar.
Habitat. Along rivers, on alluvium; c. 70 m.
Local Names. *Dimaka, Marandravina, Befelatanana* (Sakalava).
Uses. The stem pith gives a slightly bitter sago-like substance which is eaten (Jumelle & Perrier 1945), as is the palm-heart; hollowed-out stem formerly used (fide Perrier) for containers. Alcohol is produced from the fruit; shoots of germinating seedlings are eaten.
Conservation status. Vulnerable. Although known from several sites, habitat destruction is continuing all over the distribution area.

Borassus madagascariensis *Bojer ex Jum. & H. Perrier* Ann. Inst. Bot.-Géol. Colon. Marseille sér. 3, 1 (1): 61, pl. 33-35, figs. 14-15 (1913); Becc., Webbia 4, 2: 349 (1913); Jum., Cat. Pl. Madagascar, Palmae: 7 (1938); Jum. & H. Perrier, Fl. Madagascar 30: 12, fig. 2 (1945). Types not mentioned, localities: "sur les bords de la rivière Marou-voai dans la baie de Bombetok, autour du village de Majungay, Madagascar; cult. à Bois Chéri; cult. au Jardin du Roi, Pamplemousses".

Synonym:
Borassus flabellifer L. var. ***madagascariensis*** Jum. & H. Perrier, Ann. Inst. Bot.-Géol. Colon. Marseille sér. 2, 5: 389, figs 2–4 (1907).

Solitary palm. **Trunk** 10–16 m high, cylindrical or with a swelling at or above the middle of the trunk, and smooth, 40–60 cm diam., to 75 cm diam. at the swelling and 50 cm diam. just below the crown (fide Perrier). **Leaves** 12–30 in the crown, spirally inserted, costapalmate; sheath with a central proximal cleft; petiole 2–3 m long, distally 3.5–7 × 3.5 cm and channelled, green with scattered brown scales on both surfaces, the margins darker with black irregular and rather thin flat spines, the largest 3–7 mm high and 11–15 mm wide; adaxial hastula 12–15 mm high, V-shaped; costa 21–35 cm long, flanked proximally by hastula to 6 cm long and c. 5 mm high; lamina 1.6–2.2 m long, 2.5–3 m wide, undulating, with

Borassus madagascariensis
(x: sight records)

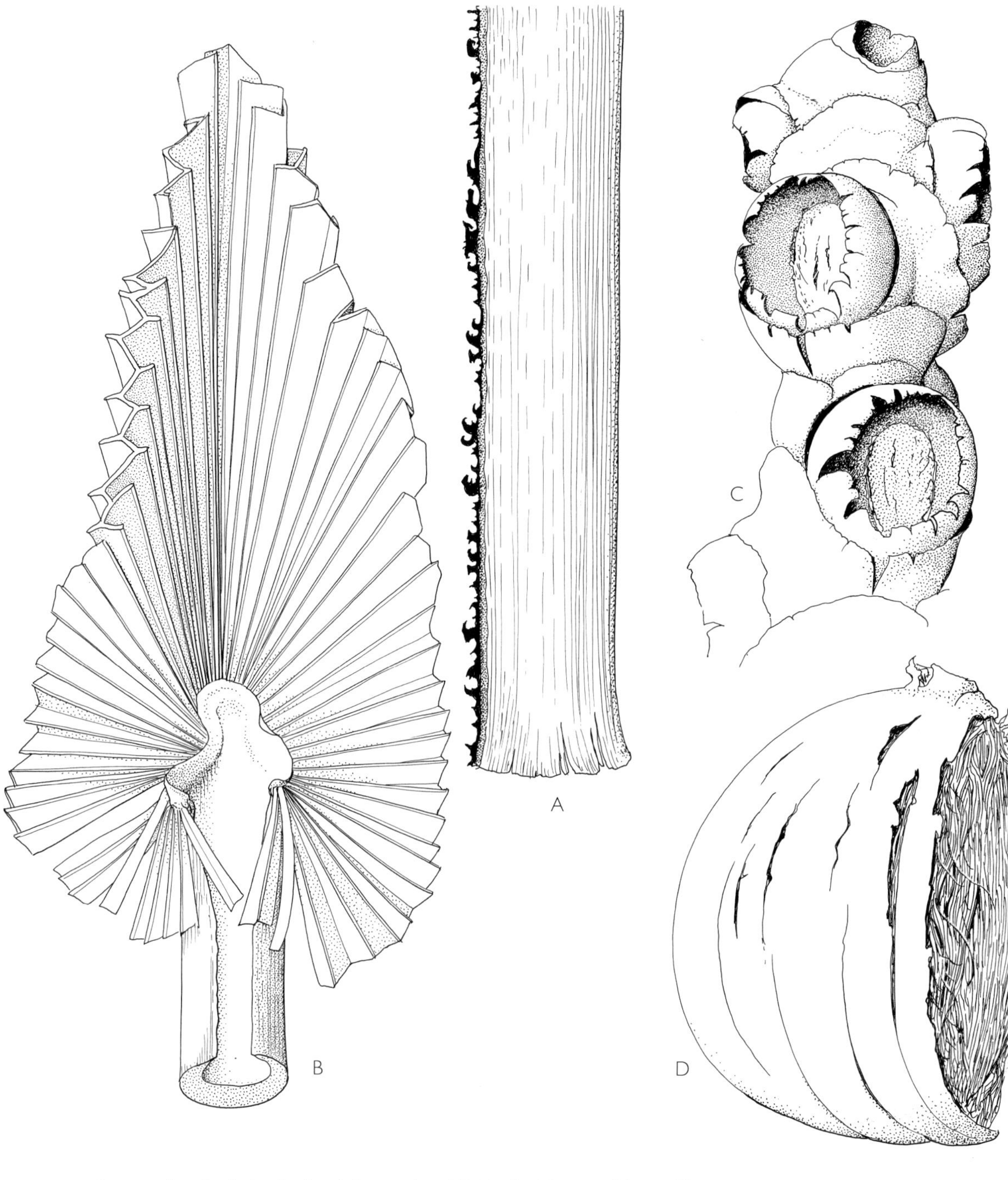

Borassus madagascariensis. **A** detail of petiole margin × 1/2; **B** adaxial view of base of lamina × 1/3; **C** detail of fruiting rachilla × 1/2; **D** part of fruit enclosing one seed × 1/2. **A** – **C** from *B. Du Puy et al.* MB791, **D** from *B. DuPuy et al.* MB256. Drawn by Rosemary Wise.

60–95 segments 1.2–2 m long, segments pale green, 4–9 cm wide, on both surfaces with reddish or faint orange scattered glands, main vein 1 per segment, with numerous fine longitudinal veinlets and faintly visible transverse veinlets, midrib proximally denticulate on the abaxial surface. **STAMINATE INFLORESCENCE** (fide Perrier, Beccari) branched to 1 or 2 orders, c. 1.2 m long, with 4–14 partial inflorescences; peduncle 20–60 cm long; prophyll 30–36 × 13–16 cm; peduncular bracts 1–4, 30–42 × 13–20 cm, the proximal keeled, all slightly beaked; rachis c. 12 cm, with 4–14 first order branches; first order branches 38–60 cm long, c. 2 cm wide, concave on one side, convex on the other, with obtuse margins, bearing up to 3 rachillae; rachis bract 36–48 × 5–15 cm; rachillae 35–40 cm long, 3–6 cm diam., subcylindrical or slightly flattened, ending in an abrupt narrowed conical part; flowers 10–12 per cavity; sepals free (Perrier) or connate for 50% (Beccari), obtuse, 5.5 × 2 mm; petals connate for 5 mm, free for 3 mm, 1 mm wide in the free part, rounded; stamens as long as the petals. **PISTILLATE INFLORESCENCE** unbranched, c. 1.2 m long; peduncle c. 37 cm long, 9 × 3.5 cm diam., slightly concave on one side; prophyll 35–48 cm long, 10–18 cm wide when flat, chestnut red-brown adaxially, pale brown with remnants of patches of indument abaxially; peduncular bracts 8–9, 28–40 × 10–15 cm; rachilla 25–50 cm long, c. 6 cm diam. (8–10 cm in fruit), with (7, fide Perrier) 16–17 (–20) flowers; bracts crowded, 2–3.5 cm high, coriaceous or almost woody, the distal margin somewhat erose; pistillate flowers solitary, with two bracteoles 2.5–3 cm high and 10–11 cm wide; petals c. 5 cm long; filaments united in a low ring with 6 thin filaments and sterile anthers, these T-shaped. **FRUIT** subglobose, 15–18 cm diam. (–25 cm, fide Perrier); exocarp smooth, mesocarp very fibrous (pulpy when fresh) and c. 1 cm thick; pyrene c. 12 × 8–12 × 5–7 cm, shaped as one-third of a globe. **SEED** (fide Beccari, Perrier) 6.5–8.5 cm long, 7–7.5 cm wide, 5–5.5 cm thick.

Note. We are not convinced of the distinctness from *B. aethiopum* Mart., though the seeds seem larger in the Madagascar taxon. Jumelle and Perrier state that the main differences are the fruit being turbinate (but then they illustrate a fruit which does not bear this out! They do admit that this is a matter of mutual compression, and solitary fruits are spheroid) and the disposition of the segments at the top of the petiole (without elaborating). They state it resembles *B. deleb*, which Dransfield (1986) considers to be a synonym of *B. aethiopum*.

Specimen seen. Antsalova: Manambolo R. gorges near Bekopaka, March 1990 (fr.), *Du Puy & Du Puy* MB791 (K, TAN). ?Miandrivazo: Sakeny R plain, July 1989 (fr.), *B. Du Puy et al.* MB256 (K, TAN).
Sight records. West coast: Ambongo-Boina, Mahajamba R., Betsiboka R, Mahavavy R; Menabe sector, Tsiribihina R, Malio R (Mangoky tributary)–all from the Flora (Jumelle & Perrier 1945).

2. Borassus sambiranensis

A majestic tree, often with a characteristic bulge in the middle of the trunk. Like *B. madagascariensis*, this species is under-collected.

Distribution. NW Madagascar.
Habitat. ?Alluvial plains; 30–50 m.
Local Names. *Dimaka* (Tsimihety).
Uses. Not recorded.
Conservation status. Endangered. HB has seen fifteen mature individuals in the Sambirano area; seedlings occur, but no young trees. Fires probably prevent establishment.

Borassus sambiranensis *Jum. & H. Perrier*
Ann. Inst. Bot.-Géol. Colon. Marseille sér. 3, 1 (1): 67 (1913); Becc., Webbia 4, 2: 346 (1913); Jum., Cat. Pl. Madagascar, Palmae: 7 (1938); Jum. & H. Perrier, Fl. Madagascar 30: 13 (1945). Type: not mentioned, "from Sambirano and Ifasy, NW", probably *Perrier* 12069, lower Sambirano valley (not seen).

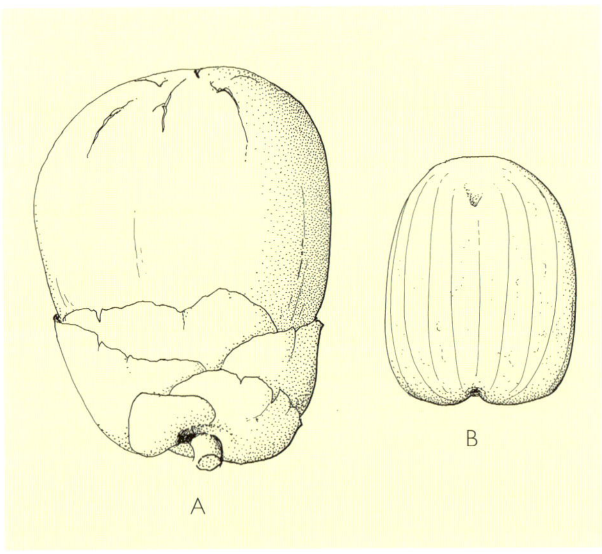

Borassus sambiranensis. **A** fruit × 1/2; **B** pyrene × 1/3. All from *Beentje & Andriampaniry* 4708. Drawn by Rosemary Wise.

Solitary palm. **Trunk** 12–18 m tall, ventricose, the wider part above the middle, c. 45 cm diam., the widest part c. 65 cm diam.; internodes 5–6 cm long, grey; nodal scars faint, 1–2 cm. **Leaves** 14–22 in the crown, spirally inserted, porrect to spreading to pendulous, forming a globose crown; sheath c. 1.1 m long, proximally 50–70 cm wide, not clearly distinct from the petiole, with triangular basal cleft 60–80 cm deep, brown, pale green, whitish ?waxy, with marginal fibers; petiole c. 1.2–1.9 m long, proximally 6–12 × 6–12 cm, distally c. 7 x 3–4 cm diam., glabrous or nearly so, orange-brown with grey margins beset with erose teeth to 15 mm; hastula adaxially continuous, 1.5–2.5 cm high, abaxially in two parts flanking the proximal costa, each 5–10 cm long and c. 3 mm high; lamina costapalmate, the costa c. 40 cm long, with 54–86 segments, the outer ones curved upwards, the inner ones curved downwards, main veins 1 per segment, ? with scattered glands/scales. **Staminate inflorescences** c. 80 cm long, erect; first order branches 4 × 1.5–2 cm diam., slightly channelled adaxially, convex abaxially, with obtuse margins, with up to 3 rachillae; rachillae c. 22 cm long, 3–3.5 cm diam.; pits c. 1 cm deep; bracteoles c. 4 mm high. **Pistillate inflorescence** unbranched, in fruit 1.2–1.4 m long, curved; peduncle (?partial) 52–60 cm long, distally 3.4 × 2.6 cm diam., glabrous; single bract seen 60 cm long, 5–6 cm diam., open in the distal half except for the 10 cm long beak, the whole bract glabrous, ?slightly waxy; rachilla 84–90 cm long, 5–6 cm diam., with 30 flower scars; bracts crowded, 2–3.5 cm high, the distal margin darker in colour and erose; pistillate flowers solitary, with two bracteoles 2.5–3 x 7–10 cm; sepals 3–4 cm high, 7–9 cm wide; petals 4.5–5 x 8–10 cm; young fruit with a staminodal ring at the base 3–4 mm high with 6 narrowly triangular teeth, terminated by a tiny linear anther rudiment. **Fruit** 11–13 × 9–11 cm, with 3 pyrenes per fruit; pyrene (fide Beccari) ± rectangular, c. 8 × 7.5 × 4.5 cm. **Seed** 7.5–8 x 6–6.5 × 4 cm, ± rectangular, truncate at base, emarginate at the apex.

Specimens seen. Ambanja: Ambanja, Oct. 1988 (fr.), *Schatz et al.* 2415 (K, MO). Analalava: Maromandia, July 1992 (dead stam.), *Beentje & Andriampaniry* 4707 (K, TAN) and (fr.), *Beentje & Andriampaniry* 4708 (K, MO, TAN).
Reported. Reported in Fl. Madagascar: Ifasy R.; Ambilobe: Nosy Mitsiou; valley of Mahavavy R.

SUBTRIBE HYPHAENINAE

III. HYPHAENE

Gaertn., De Fructibus et Seminibus Plantarum 1: 28 (1788). Type: *H. coriacea* Gaertn.

Dioecious palms. **TRUNK** often dichotomously branched, sometimes underground branching only. **LEAVES** borne spirally; leaf base sheathing, split centrally to produce a triangular cleft; petiole armed with marginal spines; hastula conspicuous adaxially, absent abaxially; blade costapalmate with single-fold segments. **INFLORESCENCES** with flattened peduncle; prophyll 2-keeled, tubular; peduncular bracts 1 or 2, tubular; rachis longer than the bracts; first-order branches compressed, semi-circular in cross-section, with rachillae borne in a fascicle at the tip; rachillae with imbricate, laterally connate bracts each surrounding a floral pit with fluffy hairs. **STAMINATE FLOWERS** in groups of 3, embedded in hairs, one flower exposed at a time; calyx tubular with 3 small lobes; corolla stalk-like at base with 3 triangular imbricate lobes; stamens 6, borne at the base of the corolla lobes, with short filaments and basifixed anthers; pistillode minute. **PISTILLATE FLOWERS** solitary, much larger than the staminate, on a densely hairy pedicel; sepals 3, free, imbricate; petals 3 free, imbricate; staminodal ring membranous, with 6 teeth, each tipped by an empty anther; ovary globose with 3 sessile stigmas, 3-locular, usually with only 1 carpel developing. **FRUIT** with persistant perianth and basal stigmatic remnants; mesocarp fibrous; endocarp hard, stony. **SEED** with homogeneous endocarp, hollow; embryo apical. **GERMINATION** remote-tubular; seedling-leaf entire, lanceolate, plicate.

DISTRIBUTION. About ten species in the Old World, mostly in Africa, also in Arabia and western India. One non-endemic species in Madagascar.

HYPHAENE CORIACEA

A common species at low altitudes in the West. This species occurs in continental Africa and some islands between Madagascar and Africa in the same habitat; it is one of the few palms which occurs both in Madagascar and elsewhere, but we believe it is native.

DISTRIBUTION. W Madagascar.
HABITAT. Littoral or inland, in grassland or wooded grassland, especially on sand; slight slope or on the flat; able to withstand fire, and sometimes locally common; 1–300 m.
LOCAL NAMES. *Satrana* (Antankarana, Sakalava); *Sata* (Sakalava, fide Hildebrandt).

Hyphaene coriacea, growing in palm savannah (Photo B. Du Puy and D. Du Puy).

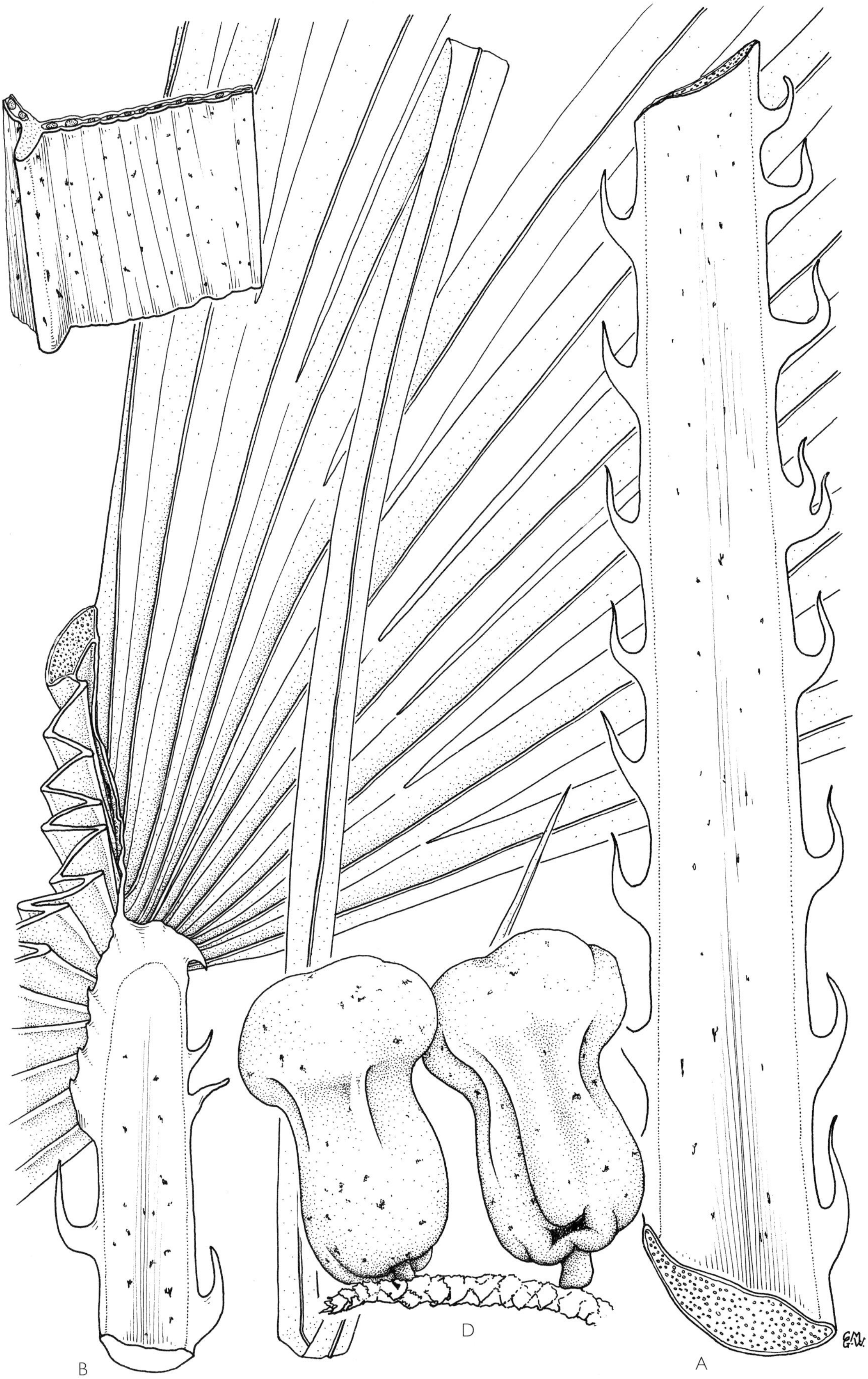

Hyphaene coriacea (from East African material). **A** petiole × 1; **B** base of blade × 3/4; **C** detail of segment × 1; **D** fruits × 3/4. **A – C** from *Dransfield* JD4817, **D** from *Dransfield* JD4848. Drawn by Christine Grey-Wilson.

USES. Leaf fibres used in basketry, hat-making, rope-making. Palm heart edible. Sometimes used to make palm wine.

CONSERVATION STATUS. Not threatened: widespread and common.

Hyphaene coriacea

Hyphaene coriacea Gaertn.
De Fruct. 1: 28, t. 10 (1788); J. Dransf., Fl. Trop. E. Africa: Palmae 29 (1986). Type: ?coll. *Berkhey* (Holotype TUB; isotype FI).

SYNONYMS:

H. hildebrandtii Becc., Bot. Jahrb. Syst. 38, Beibl. 87: 7 (1906); Jum. & Perrier, Ann. Inst. Bot.-Géol. Colon. Marseille sér. 3, 1 (1): 68 (1913). Type: near Beravi, *Hildebrandt* 3052 (Holotype at B, †; isotypes K, P).

H. baronii Becc., Bot. Jahrb. Syst. 38, Beibl. 87: 7 (1906). Type: incolis "Befelatana", *Baron* 5923 (Holotype, K; not seen at K).

H. shatan Bojer ex Jum., Cat. Pl. Madagascar, Palmae: 15 (1938); Jum. & H. Perrier, Fl. Madagascar 30: 8, fig.2 (1945), **nom. invalid.** Syntypes: *Perrier* 679 (P), 11952 (P), 11953 (not seen), 17872 (P); *Boivin* 2282 (P); *Hildebrandt* 3052 (K, P, type of *H. hildebrandtii*); *Richard* 144 (P); Mananda à Mahabo, *Grandidier* s.n. (P);

NOTE this has among the syntypes the type of *H. hildebrandtii* and is really a homotypic synonym of *H. hildebrandtii* (which is cited in the synonymy); the cited "protologue" by Bojer in Hort. Maur.: 308 (1837) has no description.

Clustering palm, often seeming solitary, in groups of 2–6. **TRUNK** 1–6 m high, 10–20 cm diam., usually unbranched but occasionally branched, clothed in old leaf-bases, with distinctive criss-cross pattern. **LEAVES** 9–20 in the crown, porrect or spreading with the rachis recurved, to 1.8 m, and to c. 9 marcescent ones; sheath open, split at the base for up to 40 cm, 38–40 cm long, waxy, brown, with fibrous margins; petiole 60–97 cm long, proximally 4–6 × c. 3 cm, mid-way flattened, c. 2 cm wide with black triangular spines to 1 cm long and curved towards the distal end, distally 1–3.5 × 1.5 cm, pale brown, waxy, with scattered scales; adaxial hastula an oblique, erose fringe to 6 mm high; costa 27–60 cm long; lamina c. 70 cm long, 112 cm wide, with 39–55 segments, the sinuses with conspicuous filaments, outer folds 31–48 × 1.2–2.6 cm, divided to the base or unsplit for 2–7 cm, more inner ones to 63–82 × 4–5 cm, unsplit for up to 20 cm, central folds 40–58 × 1.2–4 cm, unsplit for 7–9 cm, main veins 1–2, apices attenuate and bifid over 1–1.5 cm, midrib with dense to scattered dark brown and grey scales, the faint parallel minor veins with scattered reddish scales. **STAMINATE INFLORESCENCES** interfoliar, branched to 2 orders; rachillae solitary or in groups of 2–4, 9–36 cm long, 0.7–1.2 cm diam., the rachillar axes waxy and with reddish scales; bracts 2–3 × 3–3.5 mm, densely stellate-hairy. **STAMINATE FLOWERS** with sepals imbricate, 2.5–3 × 0.8–1.6 mm, narrowly obovate with a rounded apex; corolla greenish, the stalk 1–2 mm high, the lobes 2–2.8 × 1.5–1.8 mm, (slightly) spathulate and hooded, rounded; stamens with filaments 0.8–1.8 mm and thin, connate with their fleshy bases, anthers 1–1.8 × 0.6–0.8 mm, dorsifixed, versatile, yellow; pistillode not seen. **PISTILLATE INFLORESCENCES** interfoliar, 60–120 cm long, branched to 1 order with 2–5 rachillae, pendulous in fruit; peduncle 43–56 cm long with 4–5 bracts; bracts 17–19 cm long and distally scaly; rachillae with the stalk c. 20 cm long, the fertile part 14–21 cm long, 0.8–1.2 mm diam. **PISTILLATE FLOWERS** with the rachilla bract c. 2 mm high, c. 8 mm wide, inside near the base with dense hairs (?from the rachilla); pedicel 0.5–4 mm high, densely pubescent; sepals 3.5–4.5 × 2.2–3.6 mm; petals 2.5–3.7 × 2–3.2 mm, slightly obovate, obtuse with a ciliolate apex; staminodes connate at the base, 1.5–2 mm high, thin; ovary angular-globose, 3.2–3.5 × 2–3.3 mm. **FRUIT** irregularly top-shaped, 5–6 cm high, 4–6 cm diam., on a densely hairy pedicel up to 12 × 7 mm; mesocarp fibrous; endocarp hard, woody and fibrous. **SEED** c. 2.7 × 2.7 cm; endosperm homogeneous with central hollow.

SPECIMENS SEEN. Antsiranana: Bay de Rigny, Sept. 1848 (fl.), *Boivin* 2282 (P). Iharana: 30 km NW of Daraina, June 1992 (fr.), *Beentje* 4694 (K, TAN). Analalava: Maromandia, June 1923 (fl.), *Decary* 2111 (P, TAN). Marovoay: Marovoay, sine die (fl.), *Perrier* 11952 (P); idem, Dec. 1926 (fl.), *Perrier* 17872 (K, P); 16 km W of Ampijoroa, Dec. 1991 (fl.), *Beentje* 4552 (BH, K, MO, P, TAN). Ambato Boeni: Anjiajia, Aug. 1952 (fl.), *Bosser* 3506 (P, TAN). Maevatanana: Tsinjorano near Ikopa R., July 1898 (fl.), *Perrier* 679 (P). Ihosy: Isalo, Oct. 1924 (fl.), *Humbert* 2856bis (P). Manja: Beravi, June 1879 (fl., y.fr.), *Hildebrandt* 3052 (K, P; type of *H. hildebrandtii*). Betioky: Beza Mahafaly Reserve near Fitanabo, May 1987 (juv.), *Phillipson* 1767 (P, TAN).

LOCALITY UNCLEAR. NW Madagascar, Mananda to Mahabo (?), Sept. 1866 (fl.), *Grandidier* s.n. (P); N Madagascar, anno 1837 (fl.), *Richard* 144 (P). Without any locality: *Lam & Meeuse* 6135 (L, P).

SIGHT RECORDS. Ambilobe/Iharana: 50 km E of Ambilobe until Iharana (*Beentje*). Nosy Be: littoral on Nosy Be (Fl. Mad). Iharana: 15–25 km S of Iharana (*Beentje*). Antalaha: Antalaha, possibly planted (Fl. Madagascar). Boriziny: N of Boriziny (*Beentje*). Antsohihy, Analalava: common between these towns (*Beentje*). Mahajanga: Mahajanga to Ankazomborona (*Beentje*). Mananara Avaratra: Mananara, possibly planted (Fl. Madagascar). Betioky: S of Andranovory (*Beentje*); S of Betioky (*Beentje*).

Hyphaene coriacea. Close-up of crown in fruit (Photo B. Du Puy and D. Du Puy).

Subtribe Hyphaeninae

IV Bismarckia

Hildebrandt & H. Wendl., Bot. Zeit. 39: 90, 93 (1881); Benth. & Hook.f., Gen. Pl. 3: 882 (1883); Becc., Bot. Jahrb. Syst. 38, Beibl. 87: 5 (1906); Palme del Madagascar 57 (1914); N.W. Uhl & J. Dransf., Genera Palmarum 230 (1987). Type: *B. nobilis* Hildebrandt & H. Wendl.

Robust, solitary, unarmed, pleonanthic, dioecious tree palm. **Leaves** induplicate, costapalmate; sheath laterally ridged at base, with a conspicuous triangular cleft below the petiole, with rows of scales in patches; adaxial hastula often very large, distinctly lop-sided, abaxial hastula absent; blade divided to c. 1/4–1/3 of its length along adaxial folds into regular, stiff, single-fold segments, these further shortly bifid, interfold filaments conspicuous. **Inflorescence** interfoliar, solitary, the staminate and pistillate similar; peduncle ± rounded in cross-section; prophyll short, 2-keeled, included in the subtending sheath; peduncular bracts several, tubular, rather loosely sheathing, with a broad, split triangular limb, sometimes strongly keeled; rachis longer than the peduncle; rachis bracts like the peduncular, decreasing in size distally; first-order branches crescent-shaped in cross-section, longer than the subtending bract, not bearing a prophyll, branching at the tip to produce a group of 3–7 radiating, catkin-like rachillae, the group sometimes reduced to 1; staminate rachillae usually more numerous than the pistillate, slightly sinuate, bearing a tight spiral of rounded, densely hairy, striate bracts, connate laterally and partially adnate to the axis to produce pits, densely filled with hairs; pistillate rachillae usually more massive than the staminate, each bract subtending a solitary flower. **Staminate flowers** borne in a cincinnus of 3, embedded in the hairs, one flower emerging at a time, each bearing a membranous bracteole; calyx tubular, membranous, with 3 short, rather irregular lobes distally; corolla with a stalk-like base almost as long as the calyx lobes bearing at its tip 3 ovate, hooded, valvate lobes; stamens 6, borne at the base of the corolla lobes, the filaments elongate, briefly united at the base, gradually tapering, anthers medifixed, versatile, latrorse; pollen elliptic, monosulcate, with areolate, tectate exine; pistillode short, conical. **Pistillate flowers** solitary, borne on a short hairy pedicel, the pedicel greatly elongating in fruit; sepals 3, imbricate, briefly connate basally; petals 3, triangular, briefly connate at the base; staminodes united by their flattened triangular filaments to form a ring with 6 teeth, tipped with flattened, sagittate anthers; gynoecium tricarpellate, rounded, tipped with 3 low, slightly reflexed stigmas, septal nectaries present, ovule orthotropous with 2 lateral bodies, attached adaxially at the base. **Fruit** usually developing from 1 carpel, ± ellipsoid, ovoid or rounded, the stigmatic remains and abortive carpels basal, the abortive carpels often enlarging to form 2 swellings; epicarp smooth, mesocarp fibrous, ± aromatic, endocarp thick, irregularly flanged and pitted within, and with a conspicuous central intrusion at the base. **Seed** basally attached, outer surface smooth or sculptured; endosperm homogeneous but grooved to match the endocarp intrusions; embryo apical. **Germination** remote-tubular; eophyll entire, strap-shaped.

Distribution. Monotypic. Endemic to Madagascar.

Bismarckia nobilis. Sun shining through leaf.

BISMARCKIA NOBILIS

This must be the most common palm in Madagascar. It is a handsome palm, which does well in the drier tropics and subtropics and is planted in many countries. The genus is named for the German chancellor, Otto von Bismarck (1815–1898), one of the few cases where botanists have named a species after a politician. When Madagascar was colonized by the French, they included the species in *Medemia*, thus getting rid of the name *Bismarckia*, but most botanists (including the present authors) regard the genera as distinct, and therefore *Bismarckia* is the correct name for the Madagascar palm.

DISTRIBUTION. N and W Madagascar, often the only tree in the landscape; occasionally planted in E Madagascar as well.

HABITAT. Plateaux, plains, in all terrains; very common, in large numbers as the only tree on regularly burnt palm grassland on the west coast; rarely mixed with *Hyphaene*. Sometimes in flooded areas.

LOCAL NAMES. *Satra, satrabe, satrana, satranabe, satrapotsy* (Sakalava).

USES. Trunk emptied and flattened for use as planks or in partition walling; leaves used for roofing and basketry; pith serves as a slightly bitter sago. Outside Madagascar a prized ornamental for the drier (sub-) tropics.

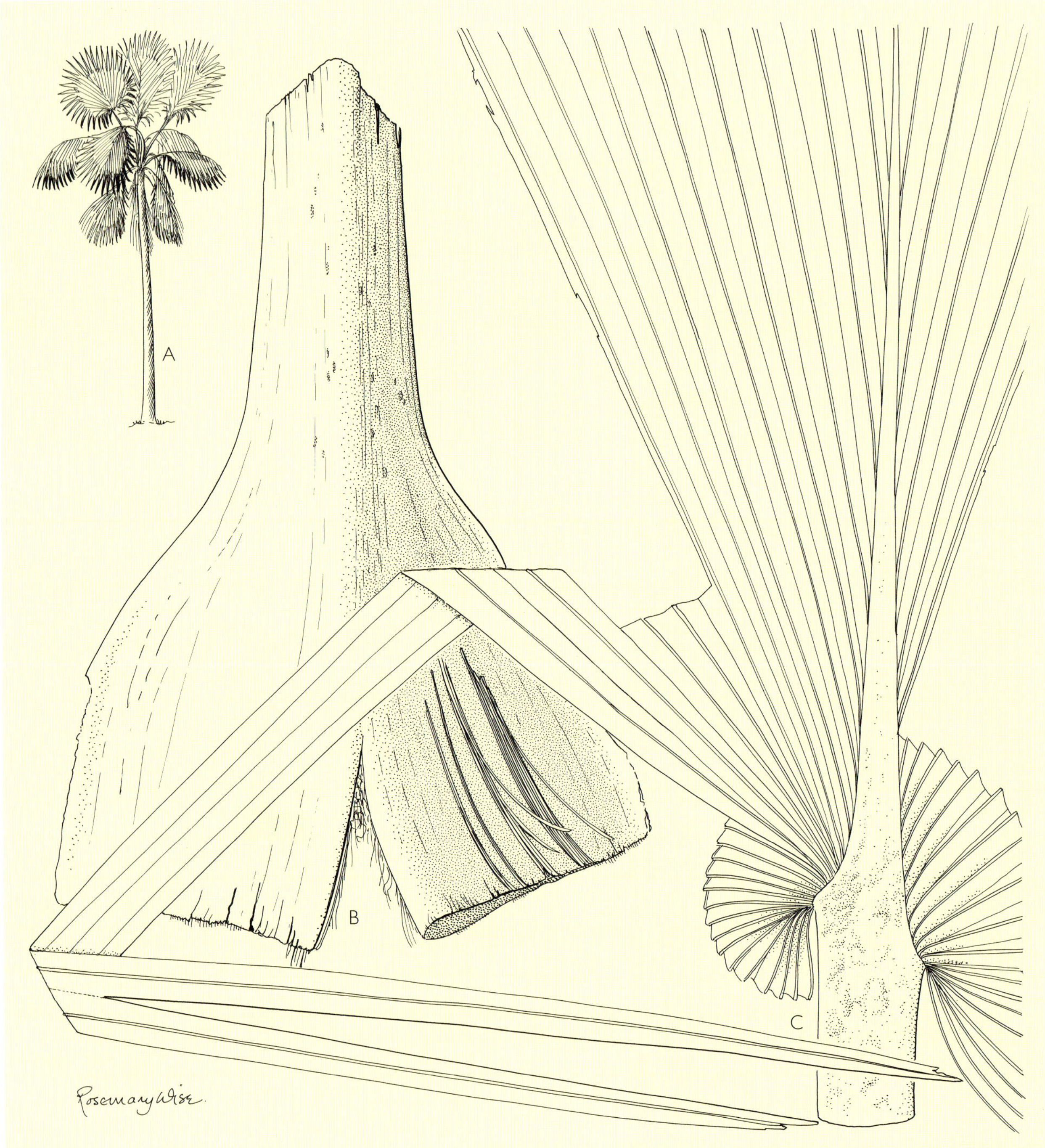

Bismarckia nobilis. **A** habit × 1/100; **B** base of petiole × 1/3; **C** abaxial view of leaf blade × 1/3. All from *Beentje* 4555. Drawn by Rosemary Wise.

CONSERVATION STATUS. Not threatened. Widespread and very common.

Bismarckia nobilis Hildebrandt & H. Wendl.
Bot. Zeit. 39: 94 (1881); Stein, Gartenfl. 35: 193, t. 1221 (1886); Becc., Bot. Jahrb. Syst. 38, Beibl. 87: 6 (1906); Becc., Palme del Madagascar 57 (1914). Type: Madagascar, Beravi, *Hildebrandt* s.n. (Holotype FI).

SYNONYM:
Medemia nobilis (Hildebrandt & H. Wendl.) Drude, Nat. Pflanzenfam. II, 3: 39 (1887); Gallerand, Compt. Rend. Acad. Paris 138: 1120 (1904); Jum. & H. Perrier, Ann. Inst. Bot.-Géol. Colon. Marseille sér. 3, 1 (1): 74 (1913); Jum., Cat. Pl. Madagascar, Palmae: 16 (1938); Jum. & H. Perrier, Fl. Madagascar 30: 6 (1945).

Bismarckia nobilis
(x: sight records)

Palm 2–20 m tall (the 30–50 m mentioned in Stein, 1886 is surely an exaggeration). **TRUNK** smooth, grey-brown, cylindrical, very straight, 20–40 cm diam., to 80 cm at very base, often marked distally with spiral twists, either clockwise or counter-clockwise; internodes c. 1.5 cm; pith abundant, white. **LEAVES** 13–30 in the crown, porrect to slightly arching, marcescent in immature individuals, neatly abscising under their own weight in mature trunked individuals; sheath c. 80 cm long, pale brown with a white-waxy covering and with scattered laciniate scales, at the interface with the petiole with 2 auricles in the shape of a 90° bend in the sheath edge; petiole 70–250 cm, proximally 12 × 3.5 cm diam., distally 3.5–5.5 × 1.3–2.5 cm diam., abaxially convex, adaxially almost flat to channelled, the surfaces greyish white, densely covered in white wax and patches of reddish, fringed caducous scales, with sharp edges or with minute, 1 mm long teeth, grey-waxy; blade c. 1.5 m diameter at greatest width, 50 cm at shortest, blade held mostly flat, grey, yellowish by the hastula, interfold filaments conspicuous, costa 43 cm, segments 50–77, inserted on one side 2–11 cm lower than on the other, the outermost undivided for c. 25–50%, 42–80 × 1.7–3 cm, the innermost undivided for c. 80%, 79–87 × 4.1–6 cm, slightly waxy abaxially with pale grey-brown ramenta on the midrib, transverse veinlets very faint or not visible. **STAMINATE INFLORESCENCES** branched to 2 orders; bracts 5–6, acute; first order branches 30–35 × 3.5 × 0.5 cm; rachillae digitate, in groups of 1–9 at the end of the branches, crimson, 15–25 cm long, 0.8–1 cm across; floral bracteoles thick, short, wide. **STAMINATE FLOWERS** with the sepals ovate, acute, c. 3.5 mm long; petals obtuse, c. 7 mm long, connate in the lower half; stamens 7 mm, with long thin filaments; ovary rudiment c. 3.5 mm high. **PISTILLATE INFLORESCENCES** interfoliar, erect, c. 120 cm, branched to 2 orders; peduncle c. 30 cm, proximally c. 3.2 × 1.5 cm; prophyll c. 28 cm; peduncular bracts inserted at c. 7 cm, c. 41 cm long (closed for c. 22 cm), inserted at c. 12 cm and c. 50 cm long (closed for c. 29 cm); rachis c. 66 cm; first order branches 36–44 cm long, proximally c. 16 × 8 mm across, distally c. 13 × 4 mm across, straw-yellow, glabrous except for the most distal part bearing some scattered scales, with 2–5 digitately arranged rachillae; rachillar bracts 27–53 cm, cylindrical, open only distally, proximally glabrous, distally with dense or scattered scales; rachillae 15–23 cm long, 1–1.5 cm across; floral

Bismarckia nobilis, growing near Isalo.

Bismarckia nobilis. Crown in full fruit (Photo: D. Du Puy).

bracteoles thick, short, 2–3 mm high, 8–16 mm wide. Pistillate flowers stalked, the pedicel hairy, elongating after anthesis; sepals c. 3 × 3.5 mm, similar to the petals, but more acute; ovary oblong, stigmas quite early in development lateral-basal. **FRUIT** dark brown, ovoid, obtuse, 40–48 × 30–35 mm, flanked at base by 2 aborted carpels, or occasionally bi- or tri-lobed, with crustaceous pericarp, spongy mesocarp and hard endocarp. **SEED** 35–38 × 22–24 mm; endosperm homogeneous but penetrated by ridges on the inside of the endocarp.

SPECIMENS SEEN. Mitsinjo: Katsepy, Dec. 1991 (fr.), *Beentje* 4555 (K, MO, P, TAN). Maintirano: Beravi, at the foot of Beturea/Rano-be between Ansahafi and Ansunaki, *Hildebrandt* s.n. (Holotype FI). Amboasary Atsimo: Anadabolava, sine die (stam.), *Humbert* 12458bis (P).
WITHOUT LOCALITY: *Lam & Meeuse* 6136 (L, P).

CULTIVATED: Ampasimanolotra, Ambila-Lemaitso, Nov. 1986, *Dransfield et al.* JD6442 (fr.; K, TAN), JD6443 (K, TAN).
SIGHT RECORDS. Jumelle & Perrier (1913) report this species in the upper basin of the Linta R. (Ampanihy Fivondranana); in large numbers in the Ambongo (Besalampy, Soalala); in dry sites in the Boina (Mitsinjo, Mahajanga); in Ampasimena (?Nosy Mitsiou); at Mananjeba R. (Ambilobe); in the Loky (= Lokia) R. valley (Antsiranana/Iharana); at the base of Mt. Andrafiamena (Iharana); and rarely in the Mahavary plain (Ambanja). HB has seen the species in Boriziny (common between Boriziny and Antsohihy), Analalava (occasional between Antsohihy and Maromandia), Mahajanga (rare in the Marovoay area), Maevatanana (occasional N of Maevatanana), Ihosy (common around the Isalo); Katsepy, in littoral forest margin; dry forest margins, Boriziny. B. Du Puy bought a hat made from leaves of this species in Maverano (Miandrivazo).

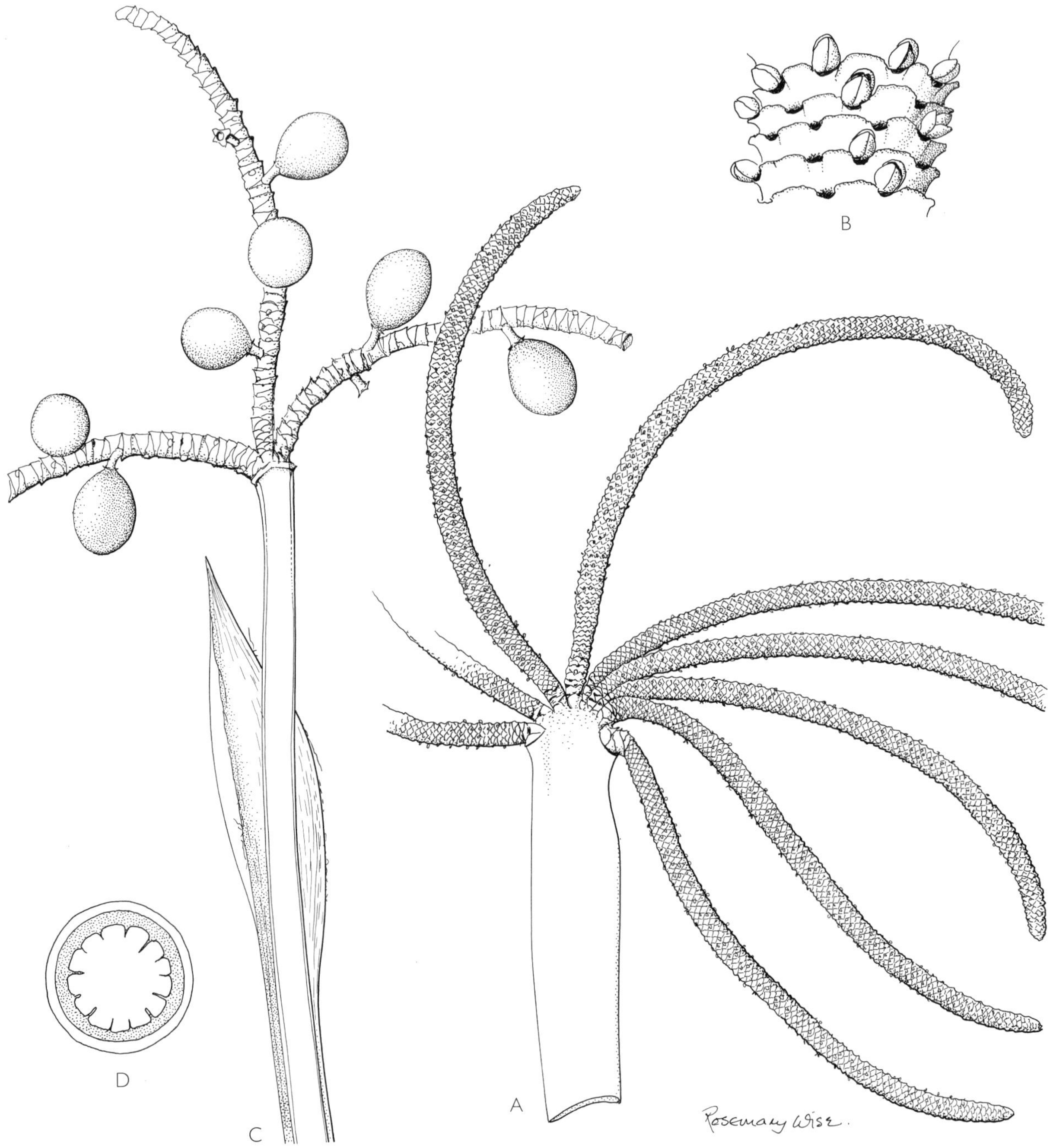

Bismarckia nobilis. **A** first order branch of staminate inflorescence × 1/3; **B** detail of staminate rachilla × 4; **C** first order branch of infructescence × 1/3; **D** fruit cross section × 4/5. **A**, **B**, from *Dransfield et al.* JD6442, **C**, **D**, from *Dransfield et al.* JD6443. Drawn by Rosemary Wise.

SUBTRIBE HYPHAENINAE

V. SATRANALA

J. Dransf. & Beentje, Kew Bull. 50 (1): 87 (1995). Type: *Satranala decussilvae* Beentje & J. Dransf.

Robust solitary pleonanthic tree palm. Stem erect, irregularly ringed with leaf scars, somewhat swollen at the base. **LEAVES** induplicately costapalmate, neatly abscising under their own weight in mature individuals; sheath lacking auricles, with a conspicuous triangular cleft below the petiole; petiole adaxially channelled near the base, distally ± flattened, abaxially rounded, the margins sharp, bearing minute irregular teeth distally, surfaces covered in patchy hairs, scales and white wax; adaxial hastula present, abaxial hastula absent; blade divided to c. 1/4 to 1/3 of its radius along adaxial folds into induplicate segments, further divided by short splits along abaxial folds, interfold filaments caducous, lamina covered with thin white wax, transverse veinlets conspicuous, close, somewhat sinuous; lamina anatomy dorsiventral. **STAMINATE INFLORESCENCE** interfoliar, solitary, shorter than the leaves, branching to 2 orders; peduncle ± rounded in transverse section; prophyll short, 2-keeled, included in the subtending leaf-sheath; peduncular bracts several; rachis longer than the peduncle; rachis bracts decreasing in size distally, tubular, rather loosely sheathing, with a broad, split, triangular limb, sometimes strongly keeled, densely covered in rusty tomentum; first-order branches crescent-shaped in cross-section, longer than the subtending bract, not bearing a prophyll, branching at the tip to produce a group of 3–9 radiating, catkin-like rachillae, rarely at the inflorescence tip the group reduced to a single branch; rachillae slightly sinuous, bearing a tight spiral of rounded, densely hairy, striate bracts, connate laterally and partially adnate to the axis to produce pits, densely filled with hairs. **STAMINATE FLOWERS** unknown; pollen (found remaining among inflorescence bracts) ellipsoid in apertural or non-apertural view; monosulcate; tectum coarsely rugulate-fossulate, but smoother around the tectum. **PISTILLATE INFLORESCENCE** similar to the staminate but with fewer rachillae in groups of not more than 3. **PISTILLATE FLOWERS** unknown. **FRUIT** developing from a single carpel, globose, stigmatic remains basal; epicarp smooth, mesocarp fleshy and fibrous, endocarp hard, woody, externally with broad anastomosing flanges, one principal flange forming a crest along the vertical axis of the endocarp, internally the endocarp smooth, lacking a basal intrusion and lacking a germination pore opposite the embryo. **SEED** globose, basally attached; endosperm ruminate, deeply and irregularly penetrated by integumental tissue, solid, embryo apical. **GERMINATION** remote-tubular, eophyll palmate with 2–3 segments.

DISTRIBUTION. Monotypic; endemic to Madagascar.

SATRANALA DECUSSILVAE

One of the most exciting discoveries made during this project. The first fan palm from the rain forests of the East Coast, looking quite like a *Bismarckia*, but with totally different fruit. The endocarp is flanged on the outside, with straight and sinuous wings. It is tempting to speculate on the adaptive significance of this extraordinary structure. We suggest the endocarp may be adapted to being swallowed by large birds (such as the now extinct *Aepyornis*) – a theory which, of course, cannot be tested. However, it is worth noting that very heavily sclerified and sculptured endocarps are found in *Ptychococcus* and *Brassiophoenix* (Uhl & Dransfield 1987) and in two species of *Licuala* (M. Forrero, pers. comm.) in New Guinea, where they appear to be adapted to dispersal by cassowaries, extant relatives of the extinct Madagascar elephant bird.

DISTRIBUTION. Mananara Biosphere Reserve, known from a single site.
HABITAT. Wet forest on shallow soils overlying ultramafic rock, in steep-sided valley rich in pandans and palms, at 250 - 285 m above sea level.
LOCAL NAME. *Satranabe* (Betsimisaraka).
USES. Not recorded.
CONSERVATION STATUS. Endangered. Known from a single site, where we have seen 30 trunked trees, 40 young ones, and many seedlings.

Satranala decussilvae. Juvenile palm in lowland forest.

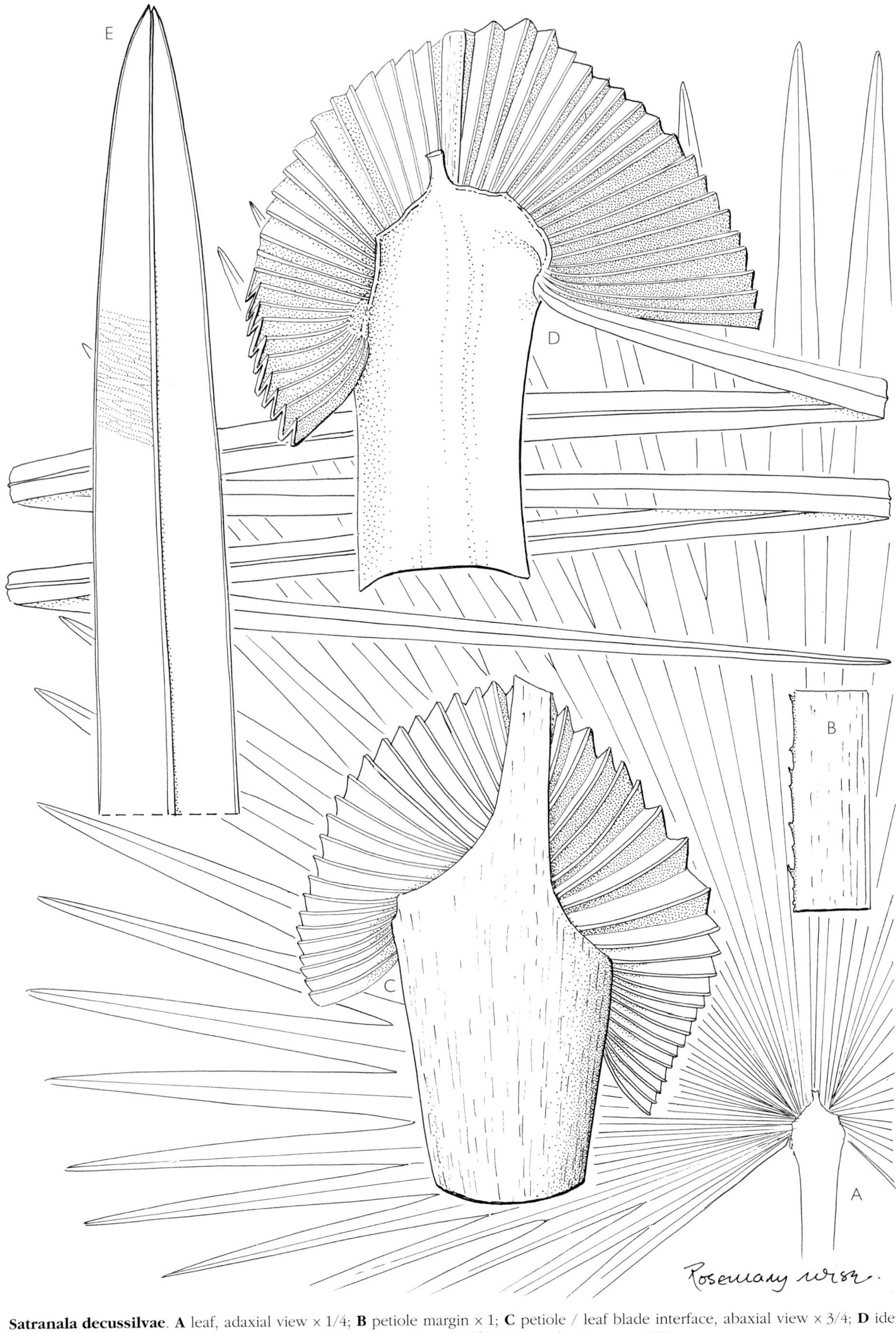

Satranala decussilvae. **A** leaf, adaxial view × 1/4; **B** petiole margin × 1; **C** petiole / leaf blade interface, abaxial view × 3/4; **D** idem, adaxial view × 3/4; **E** leaf segment apex × 1. All from *Beentje et al.* 4628. Drawn by Rosemary Wise.

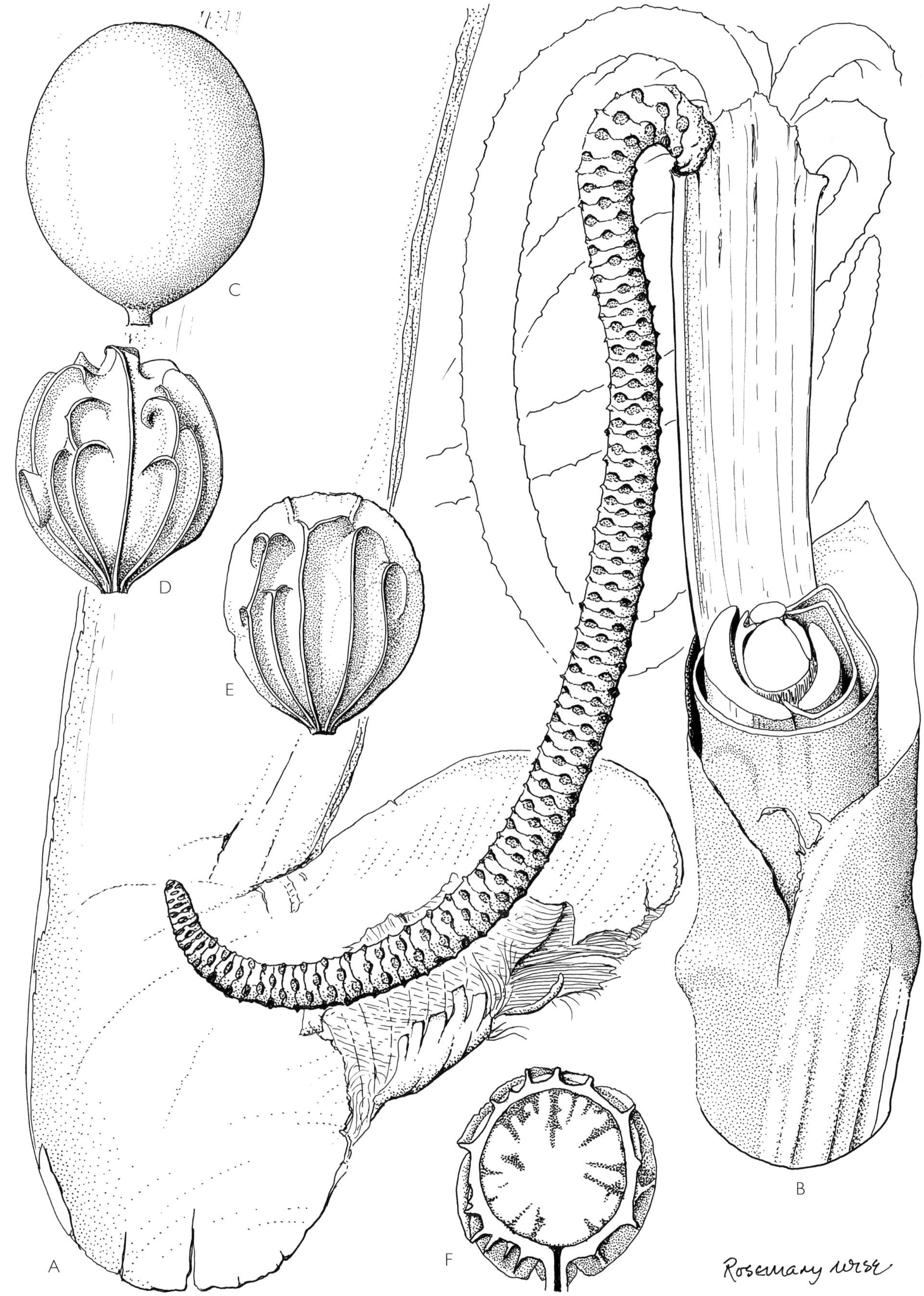

Satranala decussilvae. **A** leaf sheath base, adaxial view of lateral half × 1/2; **B** part of staminate inflorescence × 1; **C** fruit (reconstruction) × 1; **D** seed, meridional view × 1; **E** idem, but turned 90 ° through vertical axis × 1; **F** seed in cross-section × 1. All from *Beentje et al.* 4628. Drawn by Rosemary Wise.

Satranala decussilvae *Beentje & J. Dransf.*
Kew Bull. 50 (1): 89 (1995). Type: Madagascar, Mananara Avaratra, April 1992, *Beentje et al.* 4628 (Holotype K; isotypes BH, MO, P, TAN, WAG).

Satranala decussilvae

Solitary tree palm. **TRUNK** 8–15 m tall, 15–18 cm diam., hard, smooth, ± straight, obscurely ringed with scars, sometimes with aerial roots above the base of the trunk; internodes 8–10 cm, nodal scars 1.5 cm wide. **LEAVES** (9 in young plants) 20–24 in the crown, porrect, rather stiff, with up to 6 marcescent old leaves; sheath 46–60 cm long, 35 cm diam., at the very base 70 cm wide, split from 14–44 cm, abaxially chestnut-brown near the base, more distally pale brown with scattered scales, adaxially chestnut-brown, glabrous; petiole 1.4–1.5 m (to 2.7 m in young plants), proximally 7–10 × 5–6 cm, distally 5 × 1.5 cm, abaxially with thin white tomentum and wax, adaxially brown near the base, distally green with elongate scales, shallowly channelled, the margins proximally with spines to 3 mm, distally with minute spines; adaxial hastula forming a flange 3–8 mm high, with a central lobe to 15 mm long; blade costapalmate, 110–180 cm from hastula to apex, 240–260 cm wide, with 54–57 segments, costa to 33 cm, abaxially c. 2 cm wide at the base; segments almost flat, apically bifid for 1–10 cm, with three main veins and numerous close sinuous transverse veinlets, abaxially with conspicuous white wax, both surfaces with many large laciniate scales near the base, distally with few small brown scales, outer folds 88–102 × 1.3–3.2 cm, unsplit in the basal 8–9 (outermost) –20 cm, intermediate folds 104–130 × 4.4–5.5 cm, unsplit in basal 50–52 cm, central folds 114–181 × 4–6 cm, unsplit in basal 80–100 (–137 in young plants) cm. **STAMINATE INFLORESCENCE** interfoliar, arching, spreading, 230 cm long; peduncle to c. 45 cm long, 3–3.5 × 2–2.8 cm in section distally; prophyll and peduncular bracts not seen; rachis bracts rusty-brown tomentose, the proximal 42–50 cm long, c. 6 cm diam., the distal 31 cm long, c. 1.5 cm diam.; first order branches 13, the proximal 46–50 cm long, 3–3.5 × 0.6–0.8 cm diam., with 6–9 rachillae, the distal 40–45 cm long, 0.6–0.7 × 0.4–0.6 cm diam., with 1–3 rachillae; basal rachillae 29–31 cm long, 1.1–1.4 cm diam., distal rachillae to 22 cm long, 1.1 cm diam. **PISTILLATE INFLORESCENCE** similar to the staminate, rachillae 28–30 cm long, c. 1–1.2 cm diam. **FRUIT** globose to ovoid, to 5.6 × 5 cm, epicarp smooth, purple-black, shiny, mesocarp c. 1 cm thick, rather dry-fleshy, endocarp ± globose, 34–48 × 33–45 mm, the wall c. 1–2mm thick, outer surface with conspicuous, mostly vertical flanges, radiating from the base, one forming a crest along one entire vertical circumference, thus forming two faces, the flanges on the two faces c. 6 on each face, branching and anastomosing, tending to end blindly and not joined to the main crest, the flanges 3–6 mm high, the whole splitting along the main crest to allow the cotyledonary stalk to emerge; inner surface of endocarp smooth. **SEED** closely adherent to endocarp, to 30 × 32 mm, apparently considerably smaller in some individuals; endosperm solid, deeply ruminate; embryo apical, c. 6 × 2 mm. **EOPHYLL** palmate; germination remote-tubular.

NOTE. As the seed germinates the cotyledonary stalk pushes its way out of the endocarp which gapes slightly along the major crest, the endocarp thus appearing somewhat like a walnut with two valves; no other palm known to us germinates in this way.

SPECIMENS SEEN. Mananara Avaratra, Oct. 1991 (vegetative, fruit), *Beentje* 4474 (K, TAN); April 1992 (old staminate infl.), *Beentje et al.* 4628 (Holotype K; isotypes BH, MO, P, TAN, WAG); Oct. 1994 (fr.), *Beentje & Dransfield* 4810 (K, TAN), 4815 (K, TAN), 4807 (K, TAN).

Satranala decussilvae, in full fruit, south of Mananara Avaratra.

Satranala decussilvae. Close-up of fruit in section.

SUBFAMILY CALAMOIDEAE
TRIBE CALAMEAE
SUBTRIBE RAPHIINAE

VI. RAPHIA

P. Beauv., Fl. Owar. 1: 75 (1809). Type: *R. palmapinus* (Gaertn.) J. Hutchinson.

Hapaxanthic, monoecious palms. **LEAF** reduplicately pinnate; leaflets 1-fold, spiny along margins and mid-vein, with wax on the abaxial surface. **INFLORESCENCES** produced simultaneously in the axils of the most distal, reduced leaves or bracts; axis bearing a prophyll and several empty tubular bracts, followed by subdistichous or faintly 4-ranked bracts each subtending a branch; each branch with a basal prophyll and several empty tubular bracts followed by subdistichous or faintly 4-ranked bracts each subtending a rachilla; rachillae in the same plane, each with a basal prophyll and several empty bracts followed by distichous tubular bracts each subtending a flower; rachillae in the proximal part with pistillate flowers, in the distal part with staminate flowers; each flower enclosed in a 2-keeled prophyll, the pistillate flowers with a second tubular bracteole. **STAMINATE FLOWER** with a tubular calyx, scarcely 3-lobed; corolla tubular proximally, with 3 free lobes distally; stamens 6–20, briefly epipetalous and usually connate to some degree by their fleshy filaments; pistillode minute (if present). **PISTILLATE FLOWER** with a tubular 3-lobed calyx; corolla proximally tubular, distally with 3 free lobes; staminodal ring present, epipetalous, with sterile, flattened, sagittate anthers; pistil with 3 free or connate terminal stigmas; ovary covered with vertical rows of reflexed fimbriate scales; locules 3, each with a single ovule, but usually only one ovule developing. **FRUIT** tipped with the stigmatic remains, covered with the enlarged reflexed scales arranged in vertical rows, each normally with a depressed vertical line; mesocarp oily. **SEED** with a moderately thick, dry testa and ruminate endosperm; embryo lateral. **GERMINATION** adjacent-ligular; eophyll pinnate.

DISTRIBUTION. Approximately 28 species, all but two confined to the wetter parts of Africa. *R. taedigera* in W Africa and S America, and *R. farinifera* in E Africa and Madagascar.

RAPHIA FARINIFERA

All Raphia palms seen by us were closely associated with human habitation; variation within the Madagascar populations seems to be minimal, especially as regards the fruit; this gives us reason to believe the species has been introduced to the island. This same species is common and widespread in continental Africa.

Raphia farinifera. Group of individuals growing at the edge of rice fields (Photo B. Du Puy and D. Du Puy).

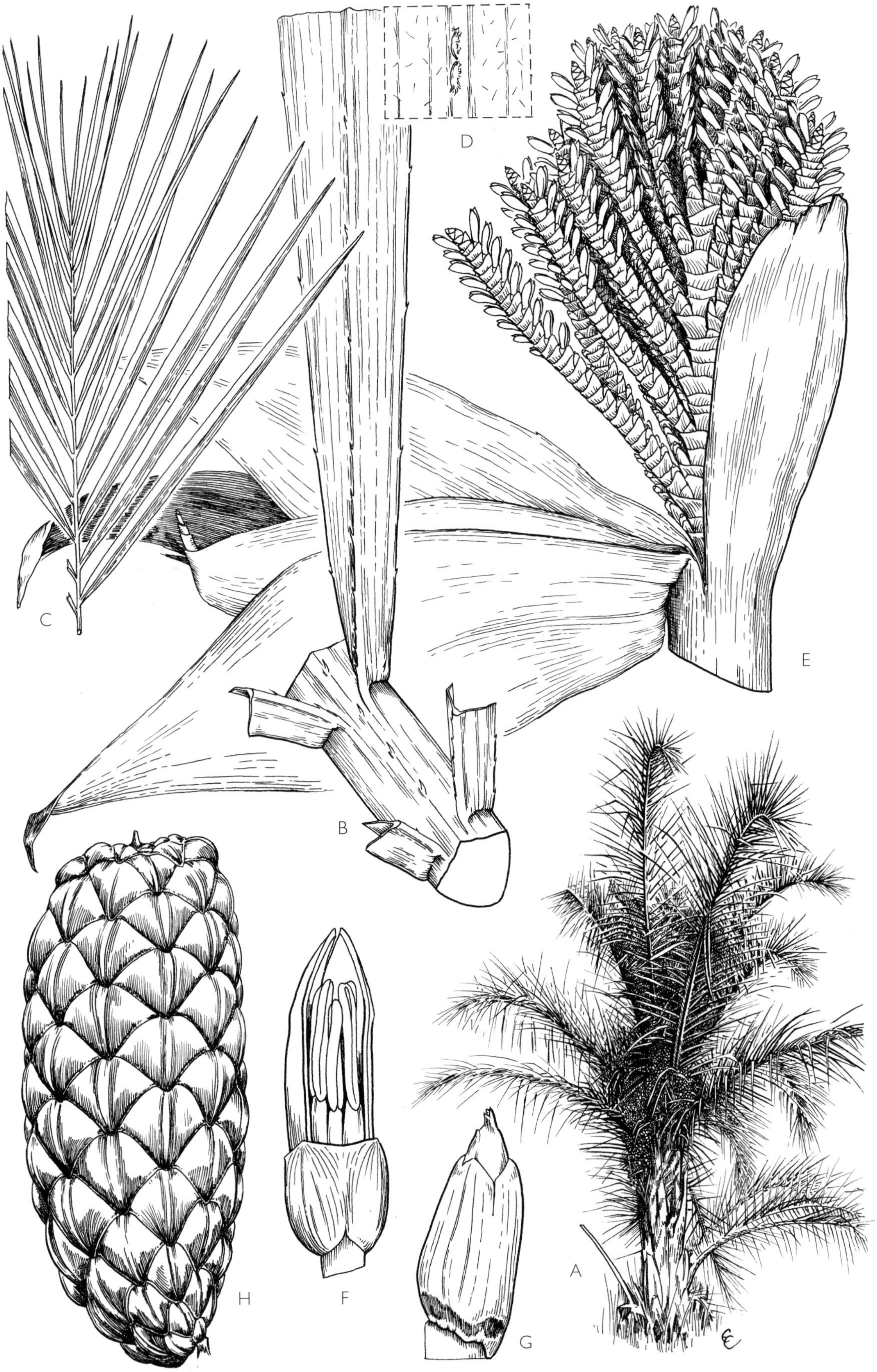

Raphia farinifera (from East African material). **A** habit × 1/150; **B** mid section of leaf × 3/4; **C** leaf tip × 1/5; **D** detail of leaflet undersurface × 2; **E** first order branch of inflorescence × 2/3; **F** staminate flower, one petal removed × 5; **G** pistillate flower × 5; **H** fruit (proportionately longer and narrower than is usual in Madagascar) × 1. **A** from a photograph, **B – D** from *Greenway* 5390, **E**, **H** from *Polhill & Paulo* 985, **F**, **G** from *McArthy & Eyles* 6678. Drawn by Eleanor Catherine.

DISTRIBUTION. Mainland Africa; in Madagascar probably introduced.
HABITAT. Moist situations (swamps, stream banks) near human habitations; 50–1000 m.
LOCAL NAMES. *Raffia* (general).
USES. Fibres from young leaves used for a variety of crafts, including hat-making, fibre-weaving for clothing and basketry; petioles used in hut construction; fruits and hearts edible.
CONSERVATION STATUS. Not threatened.

Raphia farinifera
(incl. sight records)

Raphia farinifera *(Gaertn.) Hylander* Lustgården 31: 91 (1952).

SYNONYMS:
Sagus farinifera Gaertn., De Fructibus 2: 186, t. 120 (1791). Type: t. 120 in Gaertn., De Fructibus 2 (1791).

Sagus ruffia Jacq., Fragm. Bot.: 7, t. 4/2 (1801). Type: t. 4/2, Fragm. Bot. (1801).
Raphia ruffia (Jacq.) Mart., Hist. Nat. Palm. 3: 217 (1838); Jum. & H. Perrier, Ann. Inst. Bot.-Géol. Colon. Marseille sér. 3, 1 (1): 83, t. 43 (1913); Jum., Cat. Pl. Madagascar, Palmae: 24 (1938); Jum. & H. Perrier, Fl. Madagascar 30: 14, fig. 3 (1945).

Solitary palm, though clustering in mainland Africa. **TRUNK** to 10 m, covered in persistent leaf sheaths. **LEAVES** c. 12 in the crown, porrect, slightly spreading, giving the crown a "shuttle-cock" appearance, very long, to 20 m; leaf base sheathing, with ragged ligular edge; petiole rounded in section; sheath and petiole c. 1.5 m long; rachis several meters long, reddish, distally keeled, proximally to 13 cm wide and decreasing to 1 cm, with scattered scales; leaflets up to 150 on each side of the rachis, inserted in 2 planes and thereby giving the whole leaf a feathery appearance, stiff, attenuate, the median 87–103 × 3.6–3.7 cm, the distal 16–36 × 0.4–1.7 cm, main veins 1, margins with small (1–3 mm long) yellow spines from base to apex of leaflet, midrib adaxially with similar spines to 4 mm, waxy, with many minute reddish scales/glands scattered over the abaxial surface, and sparse ramenta on the midrib. **INFLORESCENCE** pendulous from the axils of reduced leaves at the stem apex, massive, to 3 m long and 35 cm wide, branched to 2 orders; peduncle distally c. 5.5 × 4.5 cm diam., glabrous; primary prophyll c. 25 × 28 cm; peduncular bract c. 18 cm long and 8 cm diam., tubular for c. 11 cm; rachis glabrous; second order prophylls c. 9 cm long; first order branches with 13–32 rachillae packed very densely in almost one plane; rachillae 6–13 cm long, c. 8 × 5 mm diam., with dense flowers. **STAMINATE FLOWERS** with a tubular bract, 7–7.5 × 5–6 mm, broadly ovate, acute; prophyll c. 6 mm long and 3 mm diam.; calyx tubular, 4.5–5 mm high, the lobes < 0.2 mm high, slightly ciliolate; corolla with a tube 2–3 × 1.2–1.5 mm, the lobes 6–6.6 × 2.1–2.5 mm, narrowly ovate and acute, not thickened; stamens 6, inserted at the mouth of the tube, filaments slightly connate, 2–2.8 × 0.5–0.8 mm, anthers 3.2–3.6 × 1.2–1.3 mm, basifixed, locules slightly divergent and sagittate at the base; pistillode not seen. **PISTILLATE FLOWERS** with a tubular bract c. 10 × 9 mm, narrow at the base, widening in the tubular part and then narrowing to an acute apex; prophyll 7.5–8 mm, 2-keeled; bracteole 2.5–3.2 mm; calyx tubular and slightly urceolate, split, 5–6.5 mm high with a truncate apex; corolla tubular for 1–1.3 mm, the lobes narrowly triangular and acute, 2.7–3 × 1.5–1.8 mm; staminodes not seen; ovary c. 5.5 × 2.7 mm, covered in fimbriate scales. **FRUIT** ovoid, 5–6 × 4–4.5 cm with a conical base and a rounded apex with a beak to 5 mm, covered in c. 12 rows of reflexed scales, these with a median vertical groove, the largest scales c. 16 × 16 mm, chestnut-brown in colour. **SEED** ovoid, c. 3.5 × 3.2 cm; endosperm densely ruminate, the ruminations almost reaching the centre of the seed.

Raphia farinifera. Infructescence, Manambolo River (Photo B. Du Puy and D. Du Puy).

NOTE. According to Gerard Jean (pers. comm.), plants take 20–25 years from seed to flowering, and 5–6 years from flowering to ripe fruit; all the fruits mature in the same year.

Perrier stated that raffia in Madagascar varies quite a bit as regards the length of the fibres, and in the form and size of the fruit. The most striking variety, he said, is one with very large fruits in the Sambirano. No material of this has come to light.

SPECIMENS SEEN. Sambava: Sambava area, Nov./Dec. 1950 (roots), *Humbert & Capuron* s.n. (P). Maroantsetra: Maroantsetra, near the airport, Feb. 1988 (bud, fr.), *Henderson et al.* 757 (K, P). Ifanadiana: near Ifanadiana, Nov. 1994 (ster.), *Dransfield* JD7516 (K, TAN).
The Sambava specimen consists of only pneumatophores, but one has to assume that Humbert knew a Raphia when he saw one!

SIGHT RECORDS: Perrier says "very common in all the warm parts of the N of the island, in the West starting from Cap St André (now Tanjona Vilanandro), in the East starting from Mananjary; exists in the South East and in the Central Region but there only planted".

SUBFAMILY CEROXYLOIDEAE
TRIBE CEROXYLEAE

VII. RAVENEA

C.D. Bouché, Monatsschrift Vereines Beförd. Gartenbaues Königl. Preuss. Staaten 324 (1878); Hook.f., Bot. Mag. 110: t. 6776 (1884); Becc., Bot. Jahrb. Syst., Beibl. 38: 39 (1906); Jum., Ann. Inst. Bot.-Géol. Colon. Marseille sér. 4, 5 (1): 23 (1927); Jum. & H. Perrier, Fl. Madagascar 30: 166 (1945); Becc. & Pic.-Serm., Webbia 11: 178 (1956); N.W. Uhl & J. Dransf., Genera Palmarum 299 (1987). Type: *R. hildebrandtii* C.D. Bouché.

SYNONYMS:

Ranevea L.H. Bailey, Cyclopedia of American Horticulture: 1497 (1902). Type as for *Ravenea.*

Louvelia Jum. & H. Perrier, Compt. Rend. Hebd. Séances Acad. Sci. 155: 410 (1912); Ann. Inst. Bot.-Géol. Colon. Marseille sér. 3, 1 (1): 58, T.32, (1913); Jum., Ann. Inst. Bot.-Géol. Colon. Marseille sér. 4, 5: 22 (1927); Ann. Inst. Bot.-Géol. Colon. Marseille sér. 5, 1: 7 (1933); Jum. & H. Perrier, Fl. Madagascar 30: 161 (1945); Becc. & Pic.-Serm., Webbia 11: 177 (1956); N.W. Uhl & J. Dransf., Genera Palmarum 297 (1987). Type species: *L. madagascariensis* Jum. & H. Perrier [= *Ravenea louvelii* Beentje].

Subgenus ***Euravenea*** Becc., Palme del Madagascar 49 (1914).
Subgenus ***Raveneopsis*** Becc., Palme del Madagascar 51 (1914). Type: *R. madagascariensis* Becc.

Solitary, unarmed, dioecious, pleonanthic palms. **STEMS** sometimes swollen at the base. **LEAVES** spirally inserted, reduplicate, pinnate, often upward pointing, giving the crown a "shuttlecock" appearance, abscising neatly or marcescent; crownshaft absent; sheaths densely tomentose, closed at first, soon disintegrating opposite the petiole into fine or coarse fibres; petiole adaxially flattened or channelled, abaxially rounded, usually tomentose, often with sharp margins; rachis flattened adaxially, keeled abaxially in the more distal part, channelled laterally somewhat above the petiole; leaflets numerous, regularly arranged, single-fold, usually stiff, elongate, acute or acuminate, the proximal few often very slender, abaxially often with ramenta, sometimes with waxy covering, rarely with white indument abaxially, midribs more prominent adaxially, marginal veins sometimes very distinct, transverse veinlets rarely distinct, short. **INFLORESCENCES** interfoliar, axillary, 1 or several within a single prophyll in a leaf axil, erect or spreading, short and condensed to long-exserted; staminate and pistillate superficially similar but usually the staminate more slender, the staminate branched to 1 or 2 orders (to 3 orders in a single specimen of *R. robustior*) and the pistillate almost always to 1 order (to 2 orders in a single specimen of *R. robustior*); peduncle elliptic or round in cross-section, usually tomentose; prophyll short, 2-keeled, incompletely tubular, disintegrating before the peduncular bracts; peduncular bracts tubular, 4, the proximal 2 shortest, apically open and quickly tattering, the more distal ones narrow, splitting distally and longitudinally, all bracts persistent, tomentose abaxially, glabrous and usually chestnut red-brown adaxially, generally coriaceous; several non-tubular peduncular bracts sometimes present in the distal part of the peduncle, their shape broadly triangular, acute; rachis usually shorter than the peduncle, bearing numerous spirally arranged first order branches subtended by small bracts (acute or acuminate, usually adnate to the subtending branch); rachillae short (in *R. louvelii*) or more often elongate, the staminate ones shorter and more slender than the pistillate ones, usually stiff, slender, slightly zigzag, usually glabrous, the proximal portion bare, bearing numerous spirally arranged short pedicels, each subtended by a minute triangular rachilla bract, usually adnate to the stalk. **FLOWERS** congenitally open, more crowded in the distal part of the rachilla. **STAMINATE FLOWERS** symmetrical to slightly asymmetrical; sepals 3, triangular, connate in the basal part; petals 3, broadly ovate to ligulate, often fleshy, sometimes connate with the fleshy bases of the antesepalous stamens; stamens 6, filaments slender, short, basally expanded, shortly adnate to the sepals and petals, sometimes in two series and then the antepetalous stamens inserted part of the way up the petals, in *R. louvelii* forming an androecial ring by the connate fleshy filaments; anthers often somewhat twisted, basifixed, ± sagittate, latrorse; pollen ± circular, monoporate, exine tectate with supratectal spines or neatly reticulate; pistillode trifid, short, with 3 vestigial locules and ovules. **PISTILLATE FLOWERS** with sepals and petals similar to the staminate; staminodes 6 (–10 in *R. krociana*), broadly triangular with sterile anthers; gynoecium ovoid, tricarpellate; carpels 3, connate, each unilocular, uniovulate, stigmas 3, fleshy, recurved, ovules pendulous, hemianatropous. **FRUIT** globose or ellipsoid when 1-seeded (as in most species), in the 1–3-seeded species slightly 2-lobed or 3-lobed in 2–3-seeded fruits, yellow, orange or red, less often dark purple or black, with apical, lateral, or subbasal stigmatic remains; epicarp minutely pebbled, mesocarp fleshy, sometimes a sclerotic endocarp present; seedcoat brown or black, thin. **SEED** globose (in 1-seeded fruits), hemispherical (in 2-seeded fruit) or shaped as 1/3rd of a sphere (in 3-seeded fruit), hilum basal, raphe branches indistinct to coarse, sparsely anastomosing, endosperm homogeneous; embryo basal. **GERMINATION** adjacent-ligular or remote-ligular; eophyll bifid or pinnate.

Ravenea rivularis, growing beside a small river at Ilkaka.

DISTRIBUTION. Seventeen species, of which fifteen endemic to Madagascar, and two endemic to the Comoro Islands.

KEY TO RAVENEA

1. Leaflets whitish-tomentose abaxially; leaf rachis striped transversely; litter-collecting palm of moist forest .... 16. **R. albicans**

   Leaflets green and glabrous abaxially, except for midrib scales; leaf rachis not striped; palm not litter-collecting (except *R. louvelii*) 2

2. Trunk with very dense skirt of sheath remnants in the distal 1–4 m, this layer at least doubling the diameter; > 5 marcescent leaves present 3

   Trunk smooth, or with some sheath remnants parallel to the trunk; marcescent leaves absent or very few 4

CONTINUED OVERLEAF

KEY TO RAVENEA CONTINUED

3. Palm of dry bushland or low forest in southern Madagascar; leaflets on opposite sides of the rachis at a 90° angle to each other; leaf sheath 13–20 cm wide when flattened .... 14. **R. xerophila**

Palm of rain forest, only known from Moramanga area; leaflets all in one plane; leaf sheath 6.5–8 cm wide when flattened .... 17. **R. louvelii**

4. Petiole absent, the sheath margins continuing into the most proximal leaflets ...... 13. **R. dransfieldii**

Petiole present, 6 cm long 5

5. Leaflets with sinuous transverse veins connecting the longitudinal veins. Trunk with erect short woody sheath remnants along its length ....15. **R. lakatra**

Leaflets with only longitudinal veins visible (rarely visible in *R. glauca*). Trunk smooth 6

6. Leaflets 18–43 on each side of the rachis, in mid-leaf 16–46 x 1–2 cm; leaf rachis < 1 m long, < 7 mm wide in mid-leaf 7

Leaflets < 49 on each side of the rachis, in mid-leaf usually > 50 cm long; leaf rachis > 1.1 m long (except occasionally in *R. sambiranensis, R. julietiae* and *R. musicalis*, which have multiple staminate inflorescences) 8

7. Leaf rachis 33–40 cm long; median leaflets 16–35 cm long. Madagascar .... 12. **R. nana**

Leaf rachis 48–99 cm long; median leaflets 27–46 cm long. Comoro Islands ...... 11. **R. hildebrandtii**

8. Staminate trees 9

Pistillate trees 18

9. Staminate inflorescence multiple 10

Staminate inflorescence solitary 16

10. Palms growing in standing or flowing water. Trunk ventricose, 30–50 cm diam.; petiole 6–20 cm long 11

Palms of dry land. Trunk columnar, 7–29 cm diam.; petiole 10–80 cm long 12

11. Inflorescence branched to 1 order; internodes 1–2 cm long; wood soft..........................9. **R. musicalis**

Inflorescence branched to 2 orders; internodes 4–8 cm long; wood with hard layer just below the bark .... 8. **R. rivularis**

12. Petals 7.5–9 mm long; leaf sheath with floccose tomentum; palm of dry forest and canyons .... 10. **R. glauca**

Petals 1.8–6 mm long; leaf sheath with short velvety tomentum 13

13. **NOTE.** The following four species are hard to distinguish when in staminate inflorescence.

Leaf rachis > 3 m long; median leaflets 5–6.8 cm wide .... 5. **R. latisecta**

Leaf rachis < 3 m long; median leaflets < 5.2 cm wide 14

14. Leaves straight or slightly arching, often held on edge distally; leaflets < 3 cm apart in mid-leaf .... 4. **R. madagascariensis**

Leaves arching strongly, never held on edge, the leaflets held in a V, in mid-leaf > 3 cm apart 15

15. Older leaf sheath with fibrous edges; leaflet ramenta small, < 1 mm .... 6. **R. sambiranensis**

Older leaf sheath with smooth margins; leaflet ramenta large, > 2 mm .... 7. **R. julietiae**

16. Median leaflets 2–2.5 cm wide; leaf rachis 1.3–2.4 m long; inflorescence rachis < 50 cm long. Comoro Islands .... 3. **R. moorei**

Median leaflets 2.5–7.5 cm wide; leaf rachis 2.2–4 m long; inflorescence rachis > 55 cm long. Madagascar 17

17. **NOTE. 1. R. robustior** and **2. R. krociana** are indistinguishable at this stage – mainly due to the lack of good staminate material for the latter, which is only known from the Andohahela Reserve, SE Madagascar.

KEY TO RAVENEA CONTINUED

18. Pistillate inflorescence multiple .... 4. **R. madagascariensis**
    Pistillate inflorescence solitary 19

19. Peduncle 1.4–3.1 m long; seed 14–20 mm across; small graceful palm of East Coast .... 7. **R. julietiae**
    Peduncle < 1.1 m long 20

20. Palms of standing or flowing water, with ventricose trunk of 30–50 cm diam.; leaf rachis 1.2–1.8 m long 21
    Palms of dry land, either with a cylindrical trunk, or with ventricose trunk (*R. robustior, R. krociana*) and then with leaf rachis 2.2–4 m long 22

21. Inflorescence with 125–146 rachillae; fruit red, 7–9 mm .... 8. **R. rivularis**
    Inflorescence with 58–68 rachillae; fruit orange, 14–23 mm .... 9. **R. musicalis**

22. Leaves with the leaflets on opposite sides of the rachis at an angle of c. 90° with each other; petiole distally 1–3 cm wide 23
    Leaves with leaflets in one plane; petiole distally 3.7–8 cm wide 24

23. Fruit orange or red, 9–12 mm; leaf sheath short-tomentose, with marginal fibres .. 6. **R. sambiranensis**
    Fruit yellow, 20–23 mm; leaf sheath with floccose tomentum, without obvious marginal fibres .... 10. **R. glauca**

24. Leaf sheath fibres fine, < 1 mm across; leaf rachis 1.3–2.4 m long, in mid-leaf with leaflets 2–2.5 cm wide .... 3. **R. moorei**
    Leaf sheath fibres thick, > 1.5 mm across; leaf rachis 2.2–4 m long, in mid-leaf with leaflets 2.5–7.5 cm wide 25

25. Inflorescence rachis 15 cm long; fruit red, 7–10 mm .... 5. **R. latisecta**
    Inflorescence rachis 43–80 cm long; fruit orange, 8–30 mm 26

26. Petals 2.8–3.2 mm long; staminodes 10; fruit 25–30 mm ....2. **R. krociana**
    Petals 4–4.2 mm long; staminodes 6; fruit 8–18 mm ....1. **R. robustior**

1. RAVENEA ROBUSTIOR

One of the most majestic of Madagascar palms, with its slightly ventricose trunk reaching into the canopy. In the Flora (Jumelle & Perrier 1945) this species was considered to occur in only two sites; our investigations and collecting have shown that it is in fact quite widespread, occurring from sea-level up to the mountains.

DISTRIBUTION. NW, E and SE Madagascar, from Manongarivo to Marojejy and south to Andohahela.
HABITAT. Moist forest in valley bottoms, on medium or steep slopes, near water or near hill crests; in open or closed forest; often locally common; 1–1000 (-2000) m.
LOCAL NAMES. *Hovotravavy* (Tsimihety); *Manara, Tanave, Retanana* (Betsimisaraka); *Monimony, Loharanga* at Analamazaotra, *Anivona* at Andasibe (fide Dransfield, Moore); *Laafa* at Ranomafana; *Anivo, Lakabolavo* at Amby; *Bobokaomby* at Manombo; *Vakabe, Vakaky, Vakaboloka, ?Bokombio* at Andohahela.
USES. Palm-heart eaten (though the Chef de Poste at Analamazaotra informs us the palm-heart is rather bitter, and not much liked); formerly used to make salt from the ash of the trunk (fide Perrier); young leaves used to make brooms; outer wood used for floor-boards, tables and house walls.
CONSERVATION STATUS. Rare. Though widespread, the species is not common, and the continued cutting for palm-heart and construction wood might move it to the Vulnerable category in the near future. Protected populations are the ones in the Ranomafana National Park, Marojejy and Manongarivo Reserve and at Analamazaotra, though trees of this species are still being cut within the Reserves.

Ravenea robustior *Jum. & H. Perrier*
Ann. Inst. Bot.-Géol. Colon. Marseille sér. 3, 1 (1): 49, t. 27 (1913); Jum., Ann. Inst. Bot.-Géol. Colon. Marseille sér. 4, 5, (1): 40 (1927); Cat. Pl. Madagascar, Palmae: 25 (1938); Jum. & H. Perrier, Fl. Madagascar 30: 174 (1945); Beentje, Kew Bull. 49: 633 (1994). Type: not indicated, loc. Analamazaotra, "loharanga", 800 m, probably *Perrier* 11795 (Holotype P).

SYNONYM:
Ravenea robustior Jum. var. ***kouna*** Jum., Ann. Inst. Bot.-Géol. Colon. Marseille sér. 4, 5 (1): 45 (1927); Jum. & H. Perrier, Fl. Madagascar 30: 175 (1945). Type: Madagascar, Tsaratanana, *Perrier* 16069 (Holotype P).

Majestic tree palm. **TRUNK** (6–) 12–30 m high, columnar or slightly ventricose, 20–60 cm diam. at breast height, increasing to 75 cm, decreasing to 16–40 cm near crown; base bulbous, 15–50 cm high, 47–100 cm across, with surface roots to 30 cm long, 6–12 mm across, with minute adventituous side roots; outer wood hard, heartwood white and soft; leaf scars obscure, 3–7 cm; internodes 7–25 cm (near crown 4–5 cm); bark pale brown, pale reddish or grey, closely fissured to smooth. Distal part of trunk usually with remnants of sheaths. Wood extremely hard on the outside, due to many black fibres; heartwood soft, white. **LEAVES** 11–25 in crown, spiral, porrect, held in shuttlecock, straight or nearly so, often held on edge in the distal part of the leaf; leaflets stiff or curved downwards in the proximal part of the leaf; sheath grading smoothly into the petiole, ligules present or absent, c. (38–) 50–112 cm, 16–45 cm wide, bulbous or not, abaxially densely white- to grey-brown tomentose, later glabrescent, adaxially pale orange, with some stiff reflexed marginal fibres 1.5–2.5 mm across; petiole 17–134 cm, proximally 8–17 × 2–4 cm, distally 3.8–8 × 1.3–4.5 cm, channelled with sharp edges or slightly convex adaxially, convex abaxially, thickly grey-brown tomentose, glabrescent; rachis 2.2–4 m, in mid-leaf 1–4.7 cm high, 1.7–3.3 cm wide, medially sharply keeled or flat (on same tree!), abaxially with grey indument but quickly glabrescent; leaflets regular, in one plane or those at opposite sides of the rachis at a slight (up to 140°) upwards angle, dark green, (40–) 50–105 on each side of the rachis, the proximal (19–) 45–120 × 1–4 cm, median 60–126 × 2.5–7.5 cm (interval 2–5 cm), distal 13–46 × 0.3–3.5 cm, top pair often connate for up to 5 cm, ramenta large, many in young leaves, a few proximal ones in older leaves, deciduous, 1–6 main veins. **STAMINATE INFLORESCENCE** solitary, erect, interfoliar or infrafoliar among dead leafbases, branching to 2 (–3 in *Humbert* 6232) orders; peduncle 50–60 cm, proximally 2.8–3.3 × 2.5 cm, distally 2.2–2.8 × 1.3–2 cm, densely pubescent; prophyll 13–47 cm; 1st peduncular bract 18–77

Ravenea robustior. **A** cross section of part of stem × 1; **B** sheath × 1/6; **C** upper part of petiole and basal leaflets × 1/5; **D** mid section of leaf × 1/5; **E** leaf tip × 1/5; **F** staminate inflorescence × 1/12; **G** part of infructescence × 1/12; **H** rachilla in fruit × 2/3. **A – F** from *Beentje* 4576, **G**, **H** from *Beentje* 4600. Drawn by Margaret Tebbs.

× 8 cm (inserted at 3–11 cm from the base of the peduncle), 2nd (39–) 109–160 × 10 cm (inserted at 5 cm), 3rd 166–209 × 15.5 cm (inserted at 13 cm), 4th 175–209 cm (inserted at ?), all bracts abaxially with thick pale or red-brown tomentum; non-tubular peduncular bract c. 19 × 1 cm; rachis 84–131 cm long, proximally pubescent, distally glabrous, yellowish; proximal rachis bract 4–19 × 0.8–2.4 cm; rachillae many (in the lowland population with 60–140 branched and 39 unbranched first order branches), distally densely packed, porrect, straw-yellow, straight or distally sinuous, (5–) 10–47 cm, 1.5–2 mm across; pedicels quite closely set, 0.2–3 mm long; bracteoles 0.7–1 mm long; calyx with connate part 0.8–1.8 mm long and 1.4–2.2 mm across, and free sepals 0.9–1.5 mm long and 1.2–1.8 mm wide, triangular, acute; petals ovate, acute, 1.4–5 × 1.2–2.2 mm, not connate or very briefly connate by the filaments; filaments of all 6 stamens equal, 0.5–1 mm, not or only slightly attached to the petals; anthers 1.5–2.8 × 0.7–1.3 mm; pistillode 0.6–0.8 mm. **Pistillate inflorescence** interfoliar, solitary, erect, spreading or pendulous in fruit, all axes orange in fruit, branched to 1 order (to 2 orders in *Beentje* 4600); peduncle 45–100 cm, proximally 3–4.5 × 2.3–2.5 cm, distally 1.2–3.5 × 0.9–2.8 cm, pale green, proximally densely grey-white pubescent, distally glabrescent; prophyll 16–22 × 9–10 cm; peduncular bracts 21–40 × 9 cm (inserted at 5–20 cm from the base of the peduncle), 50–70 × 5–6 cm (inserted at 10–24 cm), 90–120 × 5–6 cm (inserted at c. 27 cm), 70–150 × 5–6 cm (inserted at 30–70 cm), all grey-brown tomentose abaxially; rachis 55–80 cm, with 45–100 spreading or reflexed rachillae; proximal rachis bract c. 32 × 2.2 cm; rachillae pale waxy green to orange (in fruit), 9–81 cm, 3–4 mm across, distally sinuous, proximally with bulbous bases, 1.2–2.4 × 1.2–1.8 cm, glabrous; pedicels 1–28 mm; bracteole 0.7–1.2 × 0.5–0.6 mm, connate for 0.8–6 mm; calyx connate for 0.8–2.2 mm, 2–3.5 mm across, free lobes 1.5–2.2 × 1.4–2.5 mm; petals (1.6) 4–4.2 × 1.8–2.4 mm; staminodes 0.8 mm; ovary c. 3.5 mm. **Fruit** orange, obovoid to ovoid-globose, 10–18 × 8–15 mm; usually one-seeded, then stigmatic remains subbasal; occasionally 2–3-seeded, then stigmatic remains terminal. **Seed** red-brown, hard, 9–16 × 6–13 mm; seedcoat brown, 0.2 mm thick. **Eophyll** bifid.

Ravenea robustior

Note. A population found near Sahasinaka growing in the open (*Beentje* 4677, 4717) may flower when trunks are as short as 6–8 m, but the only other differences with the rest of the populations are the cylindrical trunks, the rather long and wide leaflets (up to 7.5 cm wide, rather than up to 6 cm wide) and the rather long staminate rachillae, and the large seed (13–16 × 11–13 mm, rather than 9–13 × 6–11 mm). The more typical form of *R. robustior* was seen to grow in the same forest patch, and we feel uncertain about the taxonomic status of the form, which looks different, has a different

Ravenea robustior, growing in humid valley bottom in lowland forest, Masoala.

Ravenea robustior. Tree in the open near Sahasinaka (*Beentje & Andriampaniry* 4677).

Ravenea robustior. Swollen base of trunk, Antalavia.

local name, but in herbarium specimens is almost indistinguishable from typical *robustior*. *Beentje* 4686 from the Anjanaharibe Mts. is quite similar to the Sahasinaka population, and also grows in the open.

SPECIMENS SEEN. Ambanja: Manongarivo, Bekolosi, Feb. 1992 (old stam.), *Beentje* 4569 (BH, K, MO, P, TAN); *idem*, Feb. 1992 (green fr.), *Beentje* 4576 (BH, K, MO, P, TAN); Tsaratanana, 2000 m, April 1924 (stam., pist.), *Perrier* 16069 (type of var. *kouna*; holotype P). Andapa: Marojejy E, N of Mandena, Nov. 1986 (old pist.), *Dransfield et al.* JD6767 (K, TAN); idem (y. fr.), *Dransfield et al.* JD6768 (K, TAN); Anjanaharibe: Beamalona, June 1992 (stam.), *Beentje & Andriampaniry* 4686 (K, TAN). Maroantsetra: Hiaraka, Oct. 1986 (stam.), *Dransfield et al.* JD6379 (K, P, TAN); idem (old stam.), *Dransfield et al.* JD6381 (K, TAN); idem (old pist.), *Dransfield et al.* JD6380 (K, P, TAN). Sainte-Marie: Kalalao forest, Nov. 1994 (pist.), *Dransfield et al.* JD7518 (K, TA) and (stam.) *Dransfield et al.* JD7520 (K, TAN). Moramanga: Rahobevava to Andasibe, March 1951 (fr.), *Cours* 4384 (K, P, TAN); Analamazaotra, no date (stam., pist.), *Perrier* 11975 (Holotype P); idem, Nov. 1986, (stam.), *Dransfield et al.* JD6422 (K, P, TAN); idem, March 1988 (fr.), *Dransfield et al.* JD6491 (K, TAN); idem, Nov. 1963 (stam.), *Moore & Lefevre* 9029 (BH, TAN). Ifanadiana: Ranomafana, Ambatolahy, March 1991 (fr.), *Beentje* 4419 (BH, K, MO, P, TAN). Manakara: Amby, May 1992 (old stam.), *Beentje & Andriampaniry* 4662 (K, TAN), 4665 (K); idem, May 1992 (sd.), *Beentje & Andriampaniry* 4660 (K); idem, May 1992 (fr.), *Beentje & Andriampaniry* 4677 (BH, K, MO, P, TAN); idem, July 1992 (old stam.), *Beentje & Andriampaniry* 4717 (BH, K, MO, P, TAN). Farafangana: Manombo, Jan. 1993 (old stam.), *Beentje & Andriampaniry* 4789 (K, TAN). Midongy Atsimo: 24 km S of Midongy, May 1992 (yfr.), *Beentje & Andriampaniry* 4668 (K, TAN). Tolanaro: Andohahela, Ranohela valley, Oct. 1928 (stam.), *Humbert* 6232 (P); Manampanihy valley, W of Eminyminy, Feb. 1934 (yfr.), *Humbert* 14031 (P); Andohahela south, Dec. 1989 (old stam.), *Dransfield et al.* JD6780 (K, TAN); idem, March 1992 (fr.), *Beentje* 4600 (BH, K, MO, P, TAN) and (dead stam.) *Beentje* 4604 (K).

2. RAVENEA KROCIANA

This has very much the appearance of *R. robustior* but lacks the hard outer wood of that species; it has a very restricted distribution. The species is named for Ray Kroc, the founder of McDonald's; this organization funded a 4-year Madagascar Palm Project at Kew, during which period HB discovered *R. krociana*.

DISTRIBUTION. SE Madagascar: Andohahela.
HABITAT. Moist forest, on steep to rather flat mid slope; 420–545 m.
LOCAL NAME. *Vakakabe* (Antanosy).
USES. Not recorded.
CONSERVATION STATUS. Vulnerable. We have seen this species in only one site, where the numbers are fairly low (c. 60). The site is within the Andohahela protected area. There seem to be no specific threats.

Ravenea krociana *Beentje*
Kew Bull. 49: 636 (1994).

Typus: Madagascar, Andohahela, *Beentje* 4605 (Holotype K; isotypes BH, MO, P, TAN).

Majestic tree palm. **TRUNK** 20–30 m, ventricose, c. 26 cm diam., diameter just below crown about 18 cm (with internodes 5 cm, scars 3 cm); bark pale brown, internodes 8–9 cm, scars c. 2 cm; wood white, soft, with thin outer layer of hard wood with black fibres. **LEAVES** 14–26 in the crown, porrect, held on edge distally, with

Ravenea krociana. Mature tree showing ventricose trunk, Andohahela.

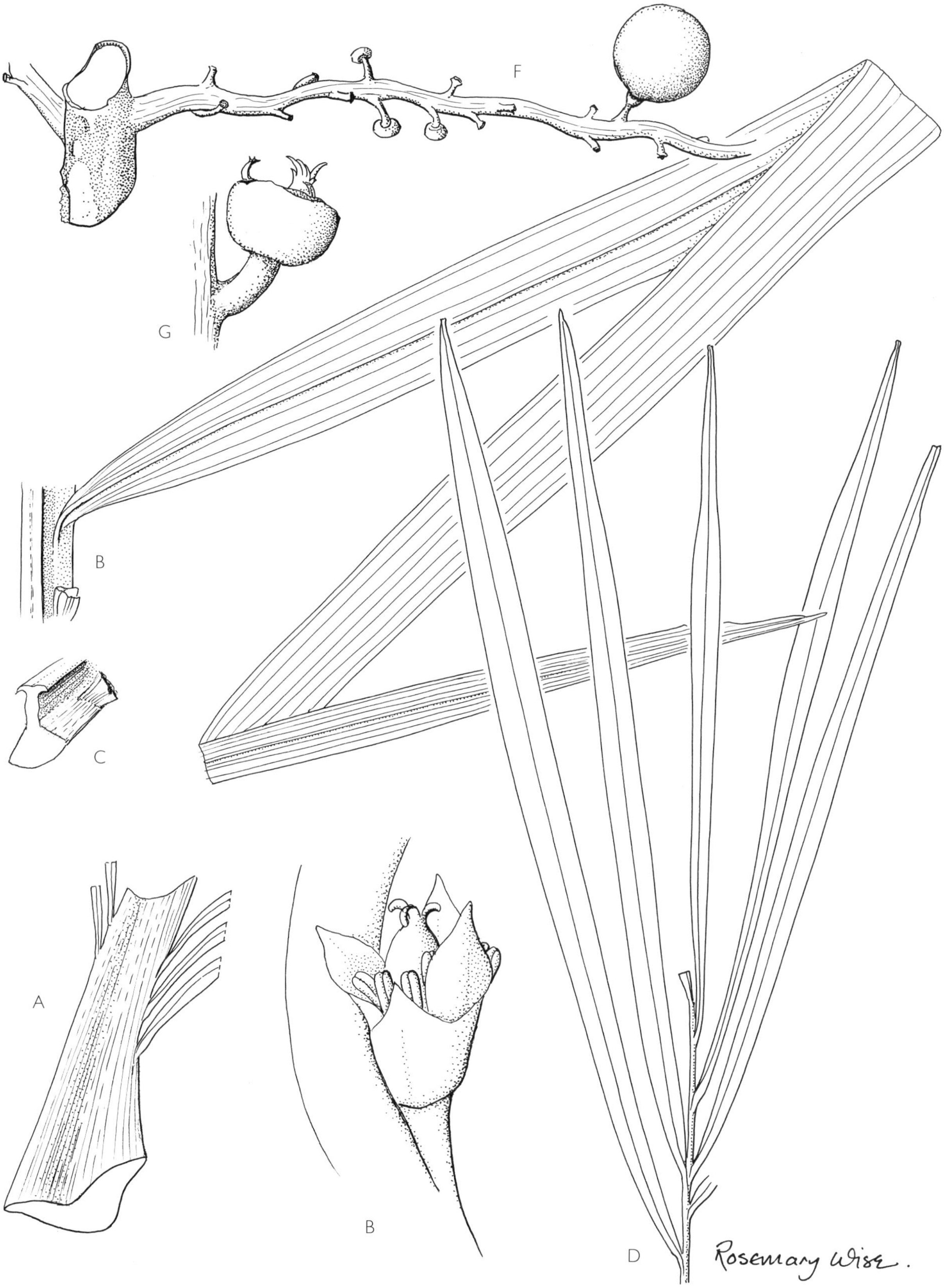

Ravenea krociana. **A** proximal portion of rachis × 1/2; **B** median portion of rachis with leaflet × 1/2; **C** cross-section of rachis in mid-leaf × 1/2; **D** leaf tip × 1/2; **E** pistillate flower bud, from an immature inflorescence × 6; **F** portion of infrutescence × 2/3; **G** fruiting calyx × 4. All from *Beentje & Andriampaniry* 4605. Drawn by Rosemary Wise.

distally pendulous leaflets; sheath c. 92 cm, pale brown tomentose, with stiff marginal fibres; petiole 30–40 cm long, c. 6.5 × 3 cm across, slightly channelled adaxially, convex abaxially, pale brown tomentose; rachis about 3.6 m, in mid-leaf 3 × 2 cm across and still flat adaxially (not yet keeled), pale brown tomentose; leaflets in one plane, distally pendulous, 101–104 on each side of the rachis, the proximal 46–62 × 0.5–2.2 cm, median 89–96 × 4.2–4.5 cm (3–3.5 cm interval), distal 17–35 × 1.5–3 cm, ramenta ? absent, main veins 3–5. **Staminate inflorescence** solitary (deduced from dead material), at least 130 cm, branched to 2 orders; peduncle > 35 cm, distally 2.2 × 1.5 cm; loose peduncular bract at least 120 cm; rachis > 55 cm; rachillae 13–27 cm, 1 mm across. **Pistillate inflorescence** solitary, pendulous in fruit, c. 150 × 85 cm, branched to 1 order; peduncle 64–77 cm, orange-green, glabrous, proximally c. 5 × 3 cm, distally 2.6 × 1.8 cm; prophyll 13–17 × 10–22 cm and thin, fibrous, white, disintegrating; peduncular bracts 20–33 × 6–8 cm (inserted at 7–13 cm from the base of the peduncle), 47–53 × 8 cm (inserted at c. 15 cm), 87–98 cm (inserted at 15–26 cm), 89–104 × 4.5 cm (inserted at 43–44 cm); rachis 43–52 cm, pale green, floccose-tomentose to glabrescent, pale green; rachillae 45–50 in number, 18–45 cm, 4 mm across but the bulbous base 15 mm across, pale green, slightly reflexed to spreading, zigzag; pedicel 2–3 mm; (the following measurements from bud material) calyx connate for 1.4 (2 in fruit) mm, 2.3 mm across, free lobes 1.1 × 1.2 mm; petals 2.8–3.3 × 1.7 mm; staminodes 10, c. 1.4 mm high; ovary 2.3 mm, 1.5 mm across. **Fruit** orange, subglobose, 27–30 × 25–28 mm, one-seeded, with the stigmatic remains subapical, calyx in fruit increasing to a sub-woody cup-like structure 4.5–6 mm across and 2–3 mm high, with faint indications of the calyx lobes, and occasionally with a remnant of a petal. **Seed** dark brown, 18–20 × 16–21 mm, with large hilar spot (?). **Eophyll** pinnate.

Note. The ten staminodes in the pistillate flower are unique within the genus.

In November 1994 we saw a population of a very large *Ravenea* species on Vatovavy, south of Mananjary. These spectacular palms appeared to have large fruits, much larger than in *R. robustior*, but we could not obtain any. It is possible that this is a second colony of *R. krociana*, although the wood seemed fairly hard.

Specimens seen. Tolanaro: Andohahela, March 1992 (fr.), *Beentje* 4605 (Holotype K; isotypes BH, MO, P, TAN); idem, 15 Dec. 1992 (dead stam.), *Beentje & Andriampaniry* 4761 (K).

Ravenea krociana. Infrutescence showing the large fruits, Andohahela.

3. Ravenea moorei

One of the two Comoro *Ravenea* species; this is the larger of the two, the other (*R. hildebrandtii*) is much smaller. Before 1986 this palm was grown under the name *R. hildebrandtii*, but that taxon is almost unknown in cultivation outside Fairchild Tropical Garden. The species is named after Professor H.E. Moore, the great expert on palms, who collected the type.

Distribution. Comoro Islands: Grande Comore.
Habitat. Secondary moist forest; 500–800 m.
Local name. *Inazi* (fide St. John, but this resembles Swahili for coconut).
Uses. Not recorded.
Conservation status. Endangered. Seed of this species was collected by DeArmand Hull in October 1993 at the north end of Grand Comore at c. 500 m, above Mitsamouli; only two female trees were seen by Mr. Hull. He considers them highly endangered due to continuing agricultural pressure on the forest in this area.

Ravenea moorei *J. Dransf. & N.W. Uhl*
Principes 30 (4): 159 (1986); Beentje, Kew Bull. 49: 637 (1994). Type: Comoro Islands, Grand Comore, *Moore & Moelevoce* 9028 (Holotype BH; isotypes K, P, TAN).

Robust tree palm. **TRUNK** to 20 m, 11–35 cm diam., with a hemispherical base up to 105 cm across and some 40 cm high, diameter near the crown c. 13 cm (from a photo), scars visible, internodes c. 20 cm long in the lower trunk, much shorter towards the apex; nodal scars not very prominent; bark grey-brown; base of crown bulbous. **LEAVES** 11–19 in crown, porrect or spreading, straight, held on edge in distal half, 2.26–3 m long; sheath up to 60 × 19.5 cm, soon splitting, margins with long fine fibers, abaxial surface covered with thick tomentum, white in unexposed areas, turning brown on exposure, becoming thinner distally; petiole 10–25 cm long, 2.4–4.5 cm wide, 1–1.5 cm thick, adaxially flattened, abaxially ± rounded, sparsely tomentose abaxially, glabrous adaxially; rachis 1.3–2.4 m long, abaxially rounded, adaxially shallowly grooved or keeled (keel to 10 mm wide, narrowing in the more distal part to a sharp narrow keel), laterally grooved, sparsely tomentose to white-waxy abaxially, in mid-leaf 1.7–2 × 1.2 cm across; leaflets 60–80 on each side of the rachis, the proximal very crowded, 15–54 × 0.8–1.7 cm, median 49–70 × 2–2.5 cm (interval 2.5–2.8 cm), distal 12–43 × 0.4–1.7 cm, main veins c. 5, the midrib the largest, prominent adaxially, abaxially bearing a few conspicuous, close, regularly arranged large pale ramenta, transverse veinlets obscure, caducous tomentum present abaxially near the base. **STAMINATE INFLORESCENCE** solitary (only known dead) c. 90 cm long, branched to 2 orders; peduncle c. 35 cm long, 1.5 cm across, rounded in cross section, bracts not known; rachillae numerous, slender, up to 15 cm 1.5 mm. **PISTILLATE INFLORESCENCE** solitary, c. 1.5 m long, spreading or pendulous, branched to 1 order; peduncle 60–90 cm, ellipsoid in cross-section, proximally 1.9–2 × 12 mm, distally c. 15 × 11 mm, green with caducous, soft brown tomentum; prophyll tubular, 2-keeled, c. 20 × 5 cm, rather thin, bearing caducous pale brown tomentum, tattering and becoming fibrous with age; peduncular bracts with the longest one c. 130 × 5 cm; rachis c. 60 cm long, glabrous and dotted; rachillae numerous, c. 100, rather stiff, straight, glabrous, green, the proximal ones 30–47 cm with flattened base 7 × 3.5 mm, the distal

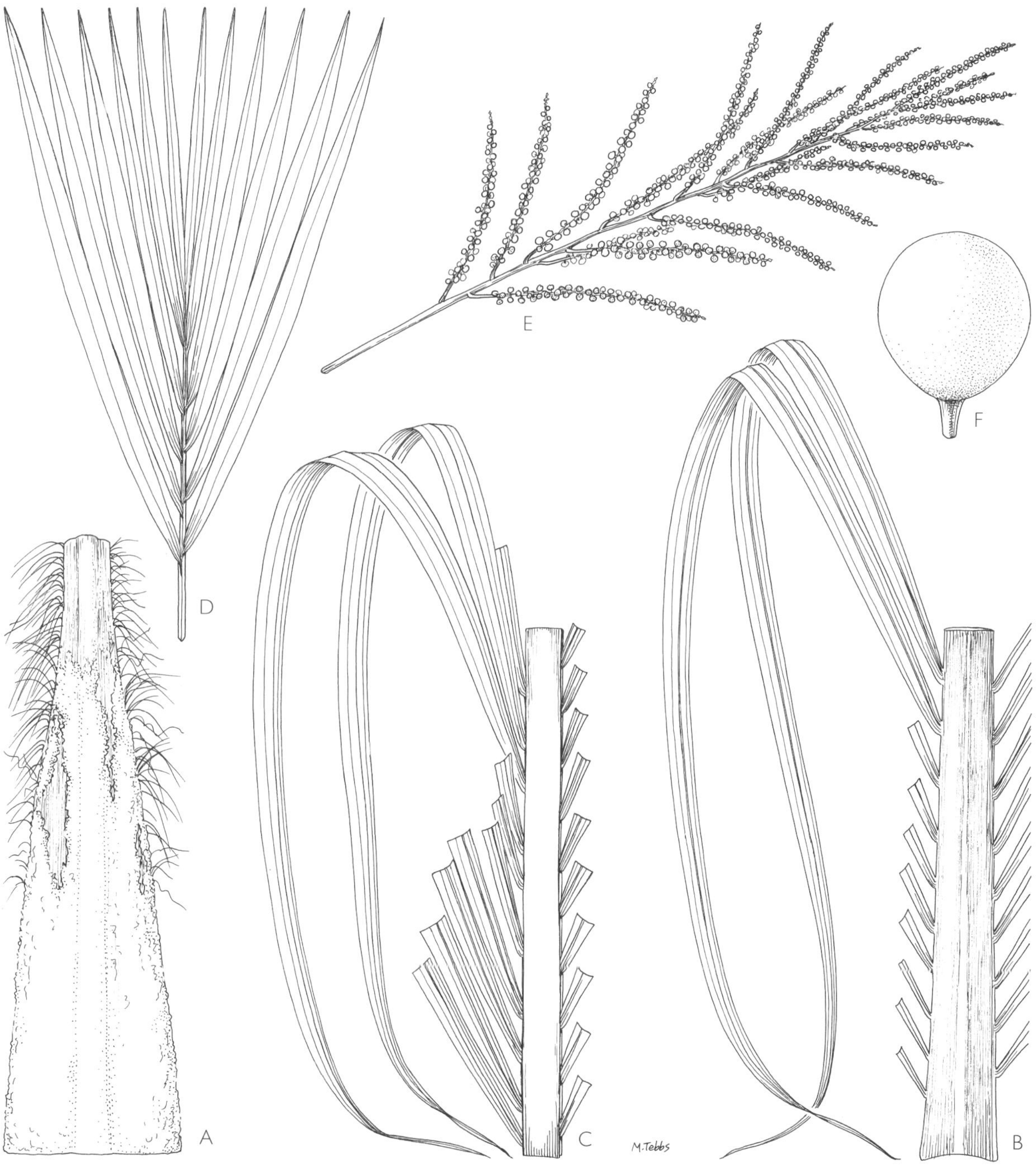

Ravenea moorei. **A** sheath × 1/6; **B** petiole and basal leaflets × 1/3; **C** mid section of leaf × 1/3; **D** leaf tip × 1/3; **E** pistillate inflorescence × 1/10; **F** fruit × 1.5. All from *St John* 26542. Drawn by Margaret Tebbs.

ones 15–20 cm; bracteole at lowermost rachilla ovate and attenuate, c. 45 × 11 mm, pedicels 0.5–5 × 1–1.5 mm; pistillate flowers not known. **FRUIT** yellow to orange, globose, c. 16 mm (when dry 11–14 mm), one-seeded, stigmatic remains lateral to sub-basal, persistent sepals triangular, c. 1.2 × 1 mm; persistent petals similar but smaller; epicarp smooth, wrinkling on drying, mesocarp fleshy, c. 1.3 mm thick, endocarp not differentiated. **SEED** spherical, 9–11 mm diameter, dark brown.

NOTE. Probably closely related to *R. robustior* of Madagascar.

SPECIMENS SEEN. Grande Comore: between Boboni & Kafeni below the chemin du Capt. Dubois, Nov. 1963 (fr., dead stam.), *Moore & Moelevoce* 9028 (Holotype BH; isotypes K, P, TAN); idem, Aug. 1961 (fr.), *St.John* 26542 (BH, K).

4. RAVENEA MADAGASCARIENSIS

A very elegant, medium-sized palm of the high plateaux; less common at lower altitudes. It is not rare in cultivation, where it is generally known as *R. madagascariensis* var. *monticola*; however, that variety does not really differ from the true *madagascariensis*.

DISTRIBUTION. Central and E Madagascar: Manerinerina and Andasibe to Kalambatitra and Midongy.
HABITAT. Moist to rather dry hill forest, on steep slopes or hill crests, occasionally on steep slopes in riverine forest; 25–1700 m.
LOCAL NAME. *Anivo*, *Anivokely*, *Anivona* (general); *Tovovoko* (Ifanadiana).
USES. Outer wood used in house walls and floors.
CONSERVATION STATUS. Rare. The species is fairly widespread, and in areas such as Ifanadiana it may be locally common. The only protected population is the one in Mantady National Park.

Ravenea madagascariensis *Becc.*
Bot. Jahrb. Syst. 38, Beibl. 87: 40 (1906); Jum. & H. Perrier, Ann. Inst. Bot.-Géol. Colon. Marseille sér. 3, 1 (1): 46, pl. 25–26 (1913); Becc., Palme del Madagascar 51, fig. 42, pl. 48 (1914); Jum., Ann. Inst. Bot.-Géol. Colon. Marseille sér. 4, 5 (1): 28, t. 111 (1927); Cat. Pl. Madagascar, Palmae: 24 (1938); Jum. & H. Perrier, Fl. Madagascar 30: 178 (1945); Beentje, Kew Bull. 49: 639 (1994). Type: Madagascar, Imerina, *Hildebrandt* 4132 (Holotype P; isotypes K, WU).

SYNONYM:
R. madagascariensis Becc. var. ***monticola*** H. Perrier, Fl. Madagascar 30:179 (1945), **synon. nov.** Type: Madagascar, Manerinerina, on the Tampoketsa between the Ikopa and the Betsiboka, *Perrier* 16834 (Holotype P).

Graceful palm. **TRUNK** 5–12 m, 16–22 cm diam., near crown 8–15 cm diam.; wood hard, with black fibrous layer just below the bark; base of trunk often thickened, the boss 10–60 cm high and 38–40 cm across, often with surface roots; trunk occasionally with remaining sheath bases in distal part; internodes 3.5–14 cm, brown or grey; nodal scars c. 2.5 cm; base of crown bulbous, 22–28 cm across. **LEAVES** 10–26 in the crown, sometimes with a few marcescent leaves; leaves porrect, straight in "shuttlecock" mode or less often slightly arching, often with the distal part held on edge; sheath 30–104 cm, the base 15–40 cm wide, dull brown with whitish brown to grey-brown tomentum, glabrescent, with fibrous edges, the fibres reflexed; petiole 20–80 cm, proximally 4.5–7 × 2–3.5 cm across, distally 3–4.5 × 1.5–2.5 cm across, adaxially channelled with sharp edges, with the same tomentum as the sheath, glabrescent; rachis 1.9–3 m, in mid-leaf 1.5–2.5 × 1.2–2 cm, keeled in the distal half, when young with whitish tomentum over green, but soon glabrescent; leaflets pale or rich green, occasionally slightly waxy, in one plane or at an angle of up to 45° with the rachis, 55–82 on each side of the rachis, the proximal 31–90 × 0.3–3.5 cm, the median 54–95 × 2.2–5.2 cm (interval 1.2–2.3 cm), the distal 11–40 × 0.3–3.5 cm, often with the terminal pair partly joined, main veins 4–6, ramenta red-

Ravenea madagascariensis. Staminate tree near Andasibe (*Dransfield et al.* JD6430).

brown, large, abundant along the midrib abaxially but quickly caducous leaving few on the basal part, occasionally with some on the smaller veins, margins often uprolled. **STAMINATE INFLORESCENCE** interfoliar, multiple in 4s–9s, the innermost opening, the outermost often remaining in bud stage until abscising; individual inflorescences branching to 2 orders; peduncle 40–119 cm, proximally 6–10 mm across, distally the same or slightly thinner, green, proximally with a dense grey-brown tomentum thinning out towards the distal end, but glabrescent; common prophyll 18–23 × 8.5 cm, tattering; peduncular bracts 26–33 cm, 53–57 cm, 120–158 cm (inserted at c. 30 cm from the base of the peduncle), 131–162 cm (inserted at c. 40 cm), densely pubescent abaxially, adaxially smooth, glabrous, red-brown; non-tubular peduncular bract 11–13.5 × 0.7–0.8 cm; rachis 37–63 cm, proximally tomentose but quickly glabrescent, with (16) 30–46 branched first order branches and 0–36 unbranched ones, all branches porrect; proximal rachis bracts 1–8 × 0.4–0.8 cm; first order branches glabrous, cream or green, with 1–3 rachillae, at the base 3–6 × 2–4 mm; rachillae 3.5–20 cm, 1 mm across, sinuous in the distal part. **STAMINATE FLOWERS** dense, yellowish, sweetly scented; pedicel 0.3–2 mm; bracteole 0.6–0.7 × 0.3 mm, acute; calyx connate for 0.8–2.4 mm, 1.4–1.8 mm across, free lobes 0.5–1.3 × 1.3–1.6 mm, triangular, acute; petals free or connate at the extreme base only, 1.8–4.2 × 1.8–2.5 mm, broadly ovate, shortly acuminate; filaments 0.3–0.5 mm, connate at the base for 0.2 mm; anthers 1.4–2.7 × 0.5–0.8 mm; pistillode 0.8–1 × 0.4–0.5 mm. **PISTILLATE INFLORESCENCES** multiple, in groups of 3–7 (only rarely solitary), erect, pendulous or spreading in fruit; individual inflorescences branching to one order; peduncle 42–100 cm, proximally 1.1–2 × 1–1.3 cm, distally 0.7–1.4 cm across, dark grey-brown tomentose, glabrescent; common prophyll 20 × 7.5 cm; peduncular bracts 11–32 cm (inserted at 0–15 cm from the base of the peduncle), 37–74 × 2.2 cm (inserted at 2–33 cm),

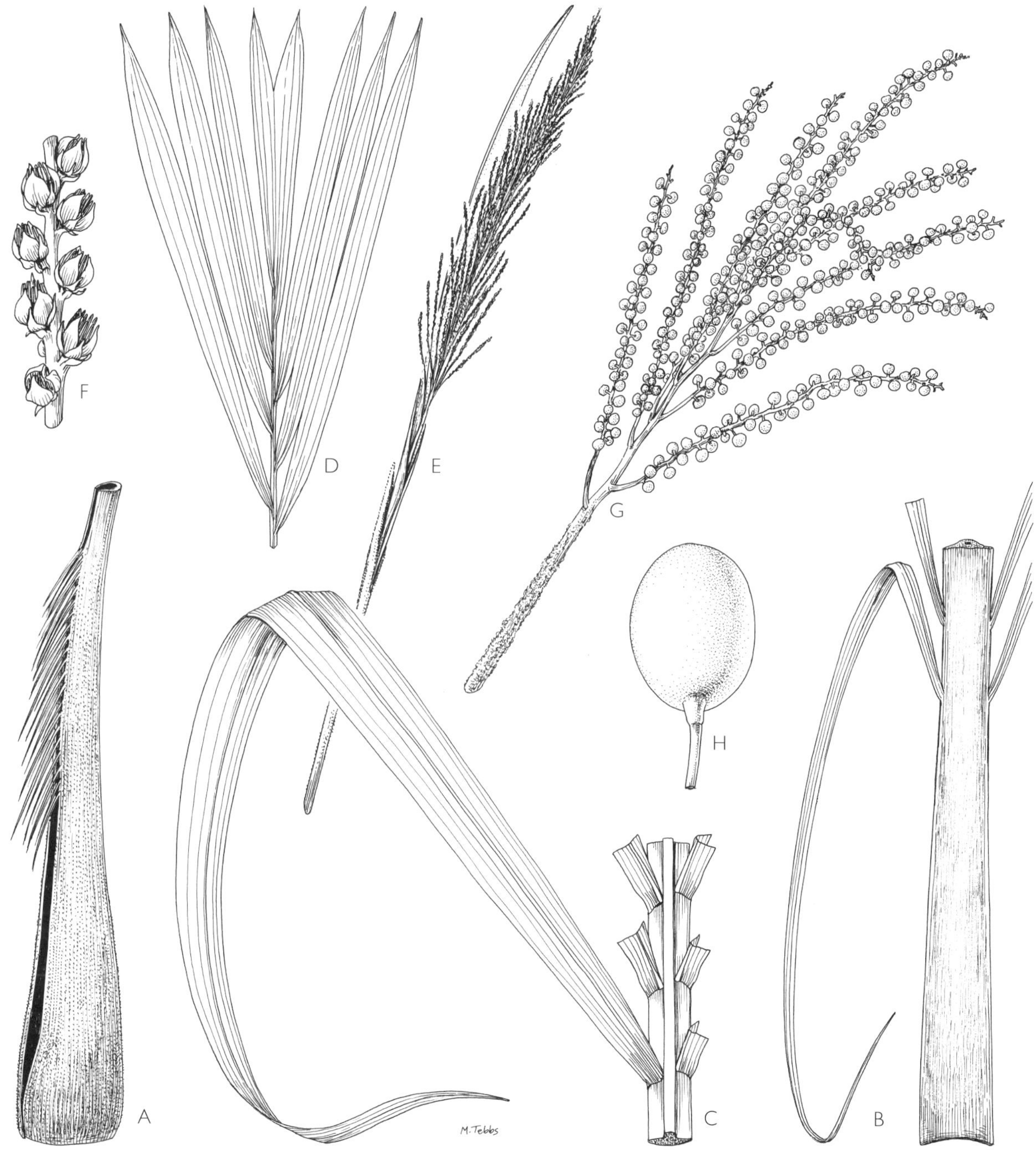

Ravenea madagascariensis. **A** sheath × 1/5; **B** petiole and basal leaflets × 1/3; **C** mid section of leaf × 1/3; **D** leaf tip × 1/3; **E** staminate inflorescence × 1/5; **F** detail of staminate rachilla × 2; **G** part of infructescence × 1/3; **H** fruit × 2.5; **A – D**, **G**, **H** from *Beentje* 4450, **E**, **F** from *Beentje & Andriampaniry* 4724. Drawn by Margaret Tebbs.

84–151 × 2.3 cm (inserted at 5–38 cm), 81–152 × 2.5 cm (inserted at 20–45 cm), densely tomentose abaxially, smooth, glabrous and red-brown adaxially; non-tubular peduncular bract 9.3–20 × 0.5 cm, proximally tomentose but quickly glabrescent; rachis 30–83 cm, glabrous, with (16) 41–78 porrect rachillae; proximal rachis bracts c. 2.6 × 1.3 cm; rachillae dense, porrect, green, glabrous, occasionally slightly waxy, 5–44 cm, 1.2–2.5 mm across, the bare base 4–9 × 3 mm. **PISTILLATE FLOWERS** not very dense, yellow; pedicel 0.5–6 mm; bracteole 0.5–3.7 1.3 mm, acute, occasionally connate with the pedicel for up to 1.5 mm; calyx connate for 1–2 mm, 1.7–2.3 mm across, the free lobes 0.4–2.3 × 1.3–2.2 mm, triangular, acute; petals fleshy, 1.7–4 × 2–2.7 mm, ovate, acute; staminodes 0.8–1.5 × 0.6–0.8 mm; ovary 2.6–4 mm, 2–2.4 mm across. **FRUIT** orange, globose or obliquely ellipsoid, 5–10 × 7.5–10 mm, one-seeded; stigmatic remains subapical to lateral. **SEED** brown, 6–7 × 5–5.5 mm.

Ravenea madagascariensis
(**x**: sight records)

NOTE. The description of *Nicoll & Abraham* 342 is suggestive of *R. robustior* (trunk 25 m, 20–40 cm across; 8–9 leaves in crown, with pendulous pinnae; fruit red) but the specimen is certainly *R. madagascariensis*. *Guillaumet* 4172 (P, TAN) from Marojejy is described as having multiple inflorescences; the specimen is pistillate, and looks quite like *R. sambiranensis*, but if it is truly the pistillate tree which has multiple inflorescences, then the specimen is true *R. madagascariensis*, and the distribution area of this species would be considerably extended.

Most publications, including the protologue, have one of the syntypes as *Chevalier* s.n., but Chevalier never collected on Madagascar, and the true specimen is a *Chapelier* collection, with Beccari's handwriting on it.

The var. *monticola* was described as different, based on the width of the leaflets (but the leaflets of the types of both varieties are of about the same width, 1.7–2.4 cm), on the proximal leaflets being longer than the median ones (a not uncommon feature in this, and other, genera, and of no taxonomic significance), and on the pistillate inflorescences being in groups of three in var. *monticola*, as opposed to solitary in var. *madagascariensis*. However, the types of var. *madagascariensis* are either staminate (*Chapelier* s.n.) or without field notes (*Hildebrandt* 4132), and there is nothing to indicate whether var. *madagascariensis* has solitary or multiple pistillate inflorescences; the confusion was probably caused by Jumelle (1927b) who states that the type is from Analamazaotra, and its pistillate inflorescences are solitary. He cites *Perrier* 11960 and 12011; we have not seen the latter, but the former has the inflorescence cut above the base. He also cites two further specimens, which are both *R. sambiranensis*, with true solitary pistillate inflorescences. Since the types of var. *madagascariensis* and var. *monticola* agree in all other respects, and the large amount of material referable to this species is rather variable, but not to such an extent that a division into varieties is warranted, the varieties are re-united.

SPECIMENS SEEN. Anjozorobe: Manerinerina, Dec. 1924 (fr.), *Perrier* 16834 (type of var. *monticola*; holotype P); Ambohitantely, April 1972 (stam., pist.), *Peyrieras* s.n. (P). Moramanga: Andasibe, Nov. 1986 (old pist.), *Dransfield* JD6425 (K, P, TAN); Mantady, Dec. 1991 (green fr.), *Beentje* 4546 (BH, K, MO, P, TAN); Anivoke, Nov.1986 (stam.), *Dransfield et al.* JD6430 (K, P, TAN); & Nov. 1986 (pist.), *Dransfield et al.* JD6431 (K, P, TAN); idem, Sept. 1991 (pist.), *Beentje* 4450 (BH, K, MO, P, TAN); Analamazaotra, no date (stam., pist.), *Perrier* 11960 (P). Ambositra: Ranomena, July 1928 (dead pist.), *Humbert & Swingle* 4868 (P); idem, July 1992 (dead pist.), *Beentje & Beentje* 4741 (K, MO, P, TAN; identification not quite certain). Ifanadiana: Tsaratanana, March 1991 (old stam.), *Beentje* 4435a (BH, K, MO, P, TAN); idem, March 1991 (fr.), *Beentje* 4435b (BH, K, MO, P, TAN); idem, Nov. 1991 (pist.), *Beentje* 4526 (BH, K, MO, P, TAN); idem, Nov. 1991 (stam.), *Beentje* 4527 (BH, K, MO, P, TAN); Ambinanindrano, Nov. 1991 (pist.), *Beentje* 4530 (BH, K, MO, P, TAN). Manakara: 9 km NW of Manakara, July 1992 (pist.), *Beentje & Andriampaniry* 4723 (BH, K, MO, P, TAN); idem (stam.), *Beentje & Andriampaniry* 4724 (BH, K, MO, P, TAN). Iakora: 2 km S of Kalambatritra Peak, Jan.1987 (pist. & fr.), *Nicoll & Abraham* 342 (K, TAN). Midongy Atsimo: 20 km S of Midongy, May 1992 (old stam.), *Beentje & Andriampaniry* 4666 (K, MO, TAN); idem, May 1992 (yfr.), *Beentje & Andriampaniry* 4667 (BH, K, MO, P, TAN).

WITHOUT PRECISE LOCALITY: Imerina, anno 1880 (pist., fr.), *Hildebrandt* 4132 (type of *R. madagascariensis*; holotype P; isotypes K, WU); Central Madagascar, (pist., fr.), *Baron* s.n. (K); no date (stam.), *Chapelier* s.n. (syntype; P).

LOCALITY NOT FOUND: Androndrombe, April 1941 (cf. stam.), *Herb. Jard. Bot. Tana* 4847 (P).

SIGHT RECORD. Mandraka (*Beentje & Dransfield*).

5. RAVENEA LATISECTA

A rather doubtful species; it is quite possible that this is really the same as *R. madagascariensis*, but female trees are needed to solve the question. The Latin name refers to the wide leaflets.

DISTRIBUTION. Only known from the Andasibe area.
HABITAT. Moist forest at 900–1000 m, steep slope near hill crest.
LOCAL NAME. Not recorded.
USES. Not recorded.
CONSERVATION STATUS. Endangered. The species only occurs in a single site, and numbers are very low (we have seen four trees).

Ravenea latisecta *Jum.*
Ann. Inst. Bot.-Géol. Colon. Marseille sér. 4, 5 (1): 35 (1927); Cat. Pl. Madagascar, Palmae: 24 (1938); Jum. & H. Perrier, Fl. Madagascar 30: 175 (1945); Beentje, Kew Bull. 49: 642 (1994). Type: Madagascar, Analamazaotra, *Perrier* 12008 (not seen, absent from P).

Medium-sized undergrowth palm. **TRUNK** 6–10 m, 15–20 cm diam., internodes 4.5–12 cm, nodal scars c. 5 cm; wood hard. **LEAVES** 6–13, porrect to spreading; sheath 60–74 x 20 cm, white-tomentose to pinkish, abaxially with many small tufts of black hairs (not in *Beentje* collections); petiole 15–50 cm, proximally 7 x 4 cm across, distally 5 x 5 cm, with a dense brown tomentum; rachis c. 3.6 m, in mid-leaf 2–2.5 cm wide, almost flat abaxially and covered there with soft brown or white tomentum, adaxially with keel; leaflets stiff to slightly drooping, in one plane, about 73 on each side of the rachis, the proximal 55–100 x 2.5–3.5 cm, median 60–105 x 5–6.8 cm, distal 12–34 x 1.5–3.7 cm. **STAMINATE INFLORESCENCES** interfoliar, according to Jumelle solitary, but multiple in 5s–7s in *Beentje* collections, pendulous when dead, 140–160 cm, branched to 2 orders; peduncle 60–90 cm, 8 x 7 mm across; prophyll not seen; peduncular bracts seen 86, 160, 176 cm long; rachis 57–84 cm, with 30–40 branched and 32–60 unbranched first order branches; rachillae 3.5–20 cm, 0.7–1 mm across, with spaced flowers; pedicels 0.3–0.5 mm. **STAMINATE FLOWERS** with the calyx connate for 0.9 mm, with free lobes 0.5–0.6 x 0.2 mm; petals free, 2.4–2.5 x 1.2 mm; stamens equal, free, with filaments 0.1–0.2 mm and anthers 1.4 x 0.4 mm, rounded at base and apex. **PISTILLATE INFLORESCENCE** (description from Jumelle) ?solitary, with peduncle 60 cm long, 3 x 1.5 cm across; bracts 20 cm (inserted at 5 cm from the base of the peduncle), 80–85 cm and 90 cm (inserted at 10–15 cm), 70 cm (inserted at c. 70 cm) respectively; rachis 15 cm or less; rachillae c. 50 in number, close, 35–45 cm long. **PISTILLATE FLOWERS** with the calyx small, with the sepals connate for about half, the free part triangular; old petals marcescent, black, ovate, acuminate, 3.6 mm; older ovary oblong, with stigmatic remains lateral. **FRUIT** (description from Jumelle) long-pedicellate, ovate or elliptic, 10 x 7 mm, yellow turning red; stigmatic remains lateral.

Ravenea latisecta

NOTE. The type is not in Paris, and could not be traced in any other herbarium. We are suspicious about the description of the pistillate inflorescence by Jumelle, and we believe he only saw part of the rachis, probably the proximal part. His statements about the inflorescence being solitary or multiple are notoriously unreliable; if the pistillate inflorescence is multiple, this taxon could be an extreme form of *R. madagascariensis*. We failed to find any pistillate trees in Mantady, and we have included the species in the key according to Jumelle's description.

SPECIMENS SEEN. Moramanga: Mantady, April 1992 (old stam.), *Beentje et al.* 4657 (BH, K, MO, P, TAN); idem, (ster.), *Beentje et al.* 4652 (K, TAN); idem, Dec. 1992 (stam.), *Beentje & Andriampaniry* 4776 (K, TAN).

Ravenea sambiranensis. Coastal form with strongly curved leaf rachises, white sand forest, Île Sainte Marie.

6. RAVENEA SAMBIRANENSIS

An elegant palm, rather like *R. madagascariensis* but with curved (rather than straight) leaves. The trees stay small on white sand on the East Coast and in populations on high mountains, but grow to canopy size in high forest on Nosy Be. The name *sambiranensis* comes from the Sambirano River in NW Madagascar, in which region the species was first found.

DISTRIBUTION. NW, W and E Madagascar, Manongarivo to Marojejy and down the East coast to Vangaindrano; outlier in Bemaraha area.
HABITAT. Littoral forest on white sand, dense moist forest or dry montane forests, steep slopes, hill crests or on almost level ground; in the West in remnant riverine forest; 1–2000 m.
POLLINATION AND DISPERSAL. Flowers visited by bees and small flying beetles (pers. obs.); fruit eaten by Madagascar Blue Pigeon, and defecated seeds germinating well (as did seed from fallen fruit) (C. Birkinshaw, pers. comm.).

LOCAL NAME. *Anivo; Anivona; Mafahely* (general); *Soindro* (Tsimihety); *Ramangaisina* (Betsimisaraka).
USES. Outer wood used for floorboards; young palm-heart cooked with manioc and eaten, but slightly bitter (hence *Mafahely*, which means a little bit bitter).
CONSERVATION STATUS. Vulnerable. Though widespread, the species occurs in fairly low numbers throughout habitats which are under threat. Littoral forests on white sand are disappearing fast. The western populations occur in high-risk areas. The submontane populations are depleted by cutting the trees for palm heart.

Ravenea sambiranensis

Ravenea sambiranensis *Jum. & H. Perrier*
Ann. Inst. Bot.-Géol. Colon. Marseille sér. 3, 1 (1): 50, t. 28, figs. 11–12 (1913); Becc., Palme del Madagascar 50, fig. 40 (1914); Jum., Ann. Inst. Bot.-Géol. Colon. Marseille sér. 4, 5: 31 (1927); Cat. Pl. Madagascar, Palmae: 25 (1938); Jum. & H. Perrier, Fl. Madagascar 30: 179 (1945); Beentje, Kew Bull. 49: 643 (1994). Type: not indicated, locality Manongarivo, 800 m (pistillate material, probably *Perrier* 12041, holotype P).

SYNONYM:
Ravenea amara Jum., Ann. Inst. Bot.-Géol. Colon. Marseille sér. 4, 5 (1): 33 (1927); Jum., Cat. Pl. Madagascar, Palmae: 24 (1938); Jum. & H. Perrier, Fl. Madagascar 30: 177 (1945). Type: Madagascar, Tsaratanana, c. 2000 m, *Perrier* 16070 (Holotype P).

Slender small palm to majestic tree palm. **TRUNK** 2–30 m, 5–29 cm diam. (near crown 5–10 cm); basal boss 25–40 cm high, 32–50 cm across, with surface roots; internodes 2–13 (–18) cm (near crown 1–4 cm), nodal scars faint or clear, 2–4 cm; rarely with sheath remnants present below the living crown; bark soft, pale grey or brown; outer wood with very hard black fibres; heartwood soft, white. Base of crown bulbous, 18–40 cm across. **LEAVES** 10–28 in the crown, porrect, arching to strongly arching; sheath 12–70 × 10–20 cm, whitish or brown-tomentose externally, in young leaves entire and fibrous, rupturing opposite the petiole when ageing and then the margins with hard fibres; petiole 13.5–76 cm, proximally 1.3–7 × 0.6–4 cm across, convex and white- or brown-grey-pubescent abaxially, slightly convex, flat or concave with sharp edges adaxially, distally 0.9–3 × 0.5–1.6 cm, glabrescent, dull green; rachis 0.6–2.5 m long, in mid-leaf 1.2–3.4 × 1–1.7 cm across, convex and white- to red- or grey-brown-tomentose to -puberulous abaxially, glabrescent, adaxially in the proximal part with a channelled keel, more distally with a flattened keel 4–5.5 mm wide, more distal becoming angular and sharp; leaflets rigid, lanceolate, oblique, straight, acute, sometimes pendulous in their distal 1/3 (rarely in one plane), dark green, arcuate, the tips somewhat bent over, 35–67 on each side of the rachis, the proximal 20–78 × 0.2–3.3 cm, median 40–100 × 1.3–4 cm (interval 3.5–5 cm), distal 6–63 × 0.3–2.7 cm, the top pair sometimes connate for up to 2 cm, abaxially on the midrib (and sometimes on the veins) with rather many small ramenta in young leaves–none, or only basal ones in older leaves, main veins 2–7. **STAMINATE INFLORESCENCE** interfoliar, multiple in 5s–9s, the central ones maturing before the outer ones, erect; individual inflorescences 30–117 cm, branched to 1 or 2 orders; common prophyll membranous, 2.5–23 × 5–12 cm, rounded, tattering, densely tomentose abaxially, glabrous adaxially; peduncle 25–74 cm, 5–9 mm across proximally, distally 3–7 mm across, green, grey-brown tomentose to glabrescent; peduncular bracts respectively 7–35 × 3.2 cm (inserted at 0–0.4 cm from the base of the peduncle), 20–60 × 2–2.2 cm (inserted at 1–2 cm), 56–125 (inserted at c. 3 cm), 68–125 × 5.3 cm (inserted at c. 7 cm), all white- to grey-brown-tomentose abaxially, adaxially cream, smooth and glabrous; non-tubular peduncular bract 1.5–3.7 × 0.4–0.6 cm, but in *Beentje 4501* a non-tubular peduncular bract inserted at 53 cm, 25 × 1.7 cm, membranous, adnate to the peduncle for 21 cm; rachis 20–54 cm, cream, densely pubescent proximally but glabrescent or with flaking patches, with 12–65 branched and 7–38 non-branched first order branches; proximal rachis bract 1.3–20 × 0.6 cm; first order branches at base 3–4 × 2 mm, with 3–6 rachillae; rachillae porrect, 3–23 cm, 1–1.5 mm across, sinuous, cream, with dense flowers above a 2 cm bare base; pedicels 0–1 mm; bracteole 0.7–1.8 × 0.5–0.8 mm, narrowly triangular, acute. **STAMINATE FLOW-**

Ravenea sambiranensis. Multiple staminate inflorescences, Ambohitsaratelo (*Dransfield et al.* JD6448)

ERS custard-yellow, scarcely to sweetly scented; calyx with connate part 0.5–1.5 × 1.3–1.8 mm, free sepals 0.9–2.5 × 0.7–1.6 mm, ovate, acute; petals free, 1.8–5 (–6.1) × 2–3 mm, broadly ovate, acute, cream to bright custard yellow; anthers equal, 1–3.7 × 0.8–1.6 mm, filaments equal, 0.3–1.5 mm; ovary rudiment 0.4–2 × 0.3–1 mm. **PISTILLATE INFLORESCENCE** interfoliar, solitary, erect, 70–175 × c. 30 cm, branched to 1 order, with green axes turning orange in fruit; peduncle 29–93 cm, proximally 1.2–5 × 0.6–3 cm across, distally 0.9–2.3 × 0.5–1.5 cm, white- to brown-pubescent, glabrescent; prophyll 5–22 × 9 cm, rounded, tattering, white-pubescent; peduncular bracts 7–34 cm (inserted at 1–4 cm from the base of the peduncle), 23–90 × 4.5 cm (inserted at 2–10 cm), 53–153 cm (inserted at 7–18 cm), 55–155 cm (inserted at 12–43 cm), glabrous and chestnut-brown adaxially, silvery-white pubescent abaxially; non-tubular peduncular bract inserted at 10 cm below rachis, 6–28 × 0.6–0.7 cm; rachis bracts 10–16 × 1.5–2 mm; rachis 14–60 cm, with 28–77 porrect rachillae; rachillae green to pale yellow, 5–50 cm, 1.5–2.5 mm across, with bulbous base 3.5–13 × 3–4 mm and 2 cm bare part, the distal part sinuous; bracteole 1.5–2.8 × 0.6–2 mm, narrowly triangular, acuminate; pedicels 0.3–6 × 1.3 mm (maximum dimensions in fruit only) long. **PISTILLATE FLOWERS** quite dense, very fragrant, sticky; calyx connate for 0.6–2 mm, 1.3–2.5 mm across, the lobes 0.6–2.6 × 1–2.4 mm, ovate to triangular and acute to acuminate; petals 2.2–5 (–10, *Du Puy* 807) 1.5–2.3 mm long, (broadly) ovate and acuminate, pale or bright yellow; staminodes c. 1.3 × 0.3 mm; ovary 2.2–3.2 mm high, 1.3–1.8 mm across. **FRUIT** orange to coral-red, ovoid to oblong, rounded at the apex, 10–12 × 9–10 mm, one-seeded, with lateral stigmatic remains; in *Birkinshaw* 136 sometimes 2- or 3-lobed (but all fruits in this collection are without seed – though they are orange in colour); fruit often with a high percentage of abortive seed. **SEED** brown, 7–8 × 5–7.5 mm. **EOPHYLL** bifid.

Ravenea sambiranensis. **A** proximal portion of rachis and tip of petiole × 1/3; **B** median portion of rachis × 1/3; **C** leaf tip × 1/3; **D** adaxial view of multiple staminate inflorescences showing incomplete prophyll × 1/5; **E** distal part of infructescence × 1/5; **F** mature fruit × 3. **A – C** from *Beentje et al.* 4572, **D** from *Beentje & Andriampaniry* 4705, **E**, **F** from *Beentje* 4582. Drawn by Margaret Tebbs.

NOTE. There were several problems to be solved around this taxon. The types of *R. sambiranensis* and *R. amara* are quite distinct, but with material HB found in the Manongarivo Mountains the distinctions blurred; only the length of the leaf rachis and the number of leaflets is less in the plants from higher altitudes, but all other characters fall within the normal range of variation. Tree size may vary from 6 m (diam. 10 cm) in montane populations, to 25 m (diam. 30 cm) in moist lowland forest.

Populations from riverine forest in the dry western Bongolava and the Bemaraha Tsingy, more than 700 km from the main population, show no differences from typical *R. sambiranensis.*

Montane populations from Marojejy differ slightly in a longer petiole, a higher number of leaflets, a longer rachis of the staminate inflorescence; but not enough to warrant distinction even at varietal level.

Eastern populations from coastal white sand forest (remnants), as well as from poor soils more inland, differ slightly in the more arching leaves and the slightly longer anthers; all differences are at one end of a range rather than absolute, and therefore we are treating this species as a variable one, rather than distinguishing geographically based subspecies with very minor morphological differences.

Beentje et al. 4625 has a hastula-like structure on the leaf rachis.

SPECIMENS SEEN: Nosy Be: Lokobe forest, near Ampasindava, Oct. 1963 (fr.), *Moore & Abdallah* 9024 (BH, TAN); idem, July 1992 (fr.), *Birkinshaw* 136 (K); idem, July 1992 (pist.), *Beentje & Andriampaniry* 4699 (K, MO, P, TAN). Ambanja: 2–4 km SW of Ambalafary, Jan. 1992 (fr.), *Beentje* 4558 (K, TAN); idem, (old stam.), *Beentje* 4560 (K, MO, P, TAN); idem, Feb. 1992 (fr.), *Beentje* 4582 (BH, K, MO, P, TAN); "Sambirano forests", without precise locality, 500 m alt., Oct. 1919 (y.fr.), *Perrier* 12041 (Holotype P); Manongarivo, Bekolosi, Feb. 1992 (old stam.), *Beentje et al.* 4570 (BH, K, MO, P, TAN); idem, Feb. 1992 (old fr.), *Beentje et al.* 4572 (BH, K, MO, P, TAN); idem, Antsatrotro, Sept. 1991 (yfr.), *Malcomber & Razafimandimbison* 883 (K, MO, P, TAN); idem, April 1992 (stam.), *Malcomber et al.* 1401 (K, MO, TAN); Tsaratanana massif, c. 2000 m, April 1924 (stam., pist.), *Perrier* 16070 (type of *R. amara*, holotype P). Analalava: 18 km NNE of Maromandia, July 1992 (stam.buds), *Beentje & Andriampaniry* 4705 (BH, K, MO, P, TAN). Andapa: Marojejy, Nov. 1989 (stam.), *Dransfield et al.* JD6757 (K, TAN); idem, Nov. 1989 (pist.), *Dransfield et al.* JD6758 (K, TAN); Mt. Beondraka, Oct. 1989 (stam.), *Miller & Randrianasolo* 4391 (K, MO, TAN). Miandrivazo: Ambohitsaratelo-Bebao, Jan. 1985 (fr.), *Dorr & Barnett* 3619 (K, P); idem, Nov. 1986 (pist.), *Dransfield et al.* JD6448 (K, P, TAN) and (stam.) *Dransfield et al.* JD6445 (K, P, TAN); idem, June 1974 (stam.), *Morat* 4591 (K, P, TAN); Manambolo R. in Bemaraha massif, 28 March 1990 (yfr.), *B. Du Puy et al.* MB 807 (K, MO, P, TAN). Maroantsetra: Maroantsetra, Oct. 1963 (stam.), *Moore* 9010 (BH, TAN); idem, Oct. 1963 (pist.), *Moore* 9019 (BH, TAN); 15 km S of Maroantsetra, Oct. 1986 (pist.), *Dransfield et al.* JD6409 (K, P, TAN); Manambia, Oct. 1986 (stam.), *Dransfield et al.* JD6408 (K, P, TAN). Mananara Avaratra: Antanambe, Oct. 1991 (stam.), *Beentje* 4456 (BH, K, MO, P, TAN); idem, April 1992 (fr.), *Beentje et al.* 4625 (BH, K, MO, P, TAN). Toamasina: Betampona, Dec. 1925 (y.fr.), *Perrier* 17473 (P); idem, Oct. 1991 (stam.), *Beentje* 4491 (K, MO, P, TAN); idem, Oct. 1991 (pist., old fr.), *Beentje* 4502 (BH, K, MO, P, TAN); Ambodiriana, Dec. 1944 (fr.), *Cours* 1945 (K, P, TAN); Andrambolahy Kely to Andranampony, April 1951 (y.fr.), *Cours* 4512 (K, P, TAN). Ampasimanolotra: Ambila, Feb. 1924 (y.fr.), *Perrier* 15991 (P). Vatomandry: Ilaka Atsipanana, Oct. 1991 (stam.), *Beentje* 4503 (BH, K, MO, P, TAN); idem, (pist.), *Beentje* 4504 (BH, K, MO, P, TAN). Ifanadiana: Ranomafana, Maharira, April 1993 (stam.), *Malcomber et al.* 2448 (K,P). Farafangana: S of Farafangana, May 1992 (old stam.), *Beentje & Andriampaniry* 4675 (BH, K, MO, P, TAN). Vangaindrano: near Ranomena, May 1992 (old stam.), *Beentje & Andriampaniry* 4674 (K).

SIGHT RECORD. Île Sainte Marie (*Dransfield*, 1994).

SIMILAR SPECIMENS. Toamasina: Betampona, *Beentje* 4501 (BH, K, MO, P, TAN); differs from typical *sambiranensis* in many sheaths remaining on trunk, full of humus and debris; sheath 28 cm; median leaflets 56–68 × 2.8–3.9 cm; staminate inflorescence multiple, non-tubular peduncular bract cylindrical, porrect from proximal first order branch, thin, green, 25 × 1.7 cm porrect part, apex fringed; 22 first order branches (branched) + 7 solitary; and Tolanaro, Ste Luce, March 1989 (part of infrutescence only), *Dumetz* 632 (K, MO, P, TAN). In Jumelle (1927) the following specimen is cited: dry forest in the Ankaizina, on sandstone, c. 800 m alt, (pist.), *Perrier* 11954. This collection is not at Paris or Antananarivo.

7. RAVENEA JULIETIAE

A graceful palm of lowland forest. The female inflorescences are unique in that they are longer than the leaves, but the male trees are quite like *R. sambiranensis.* The species is named for HB's wife Juliet, who first pointed it out in the field.

DISTRIBUTION. E Madagascar, between Mananara Avaratra and Vangaindrano.

HABITAT. Moist lowland forest; on slight to steep mid slopes; 50–285 m.

Ravenea julietiae. The type tree at Amby (*Beentje & Andriampaniry* 4783).

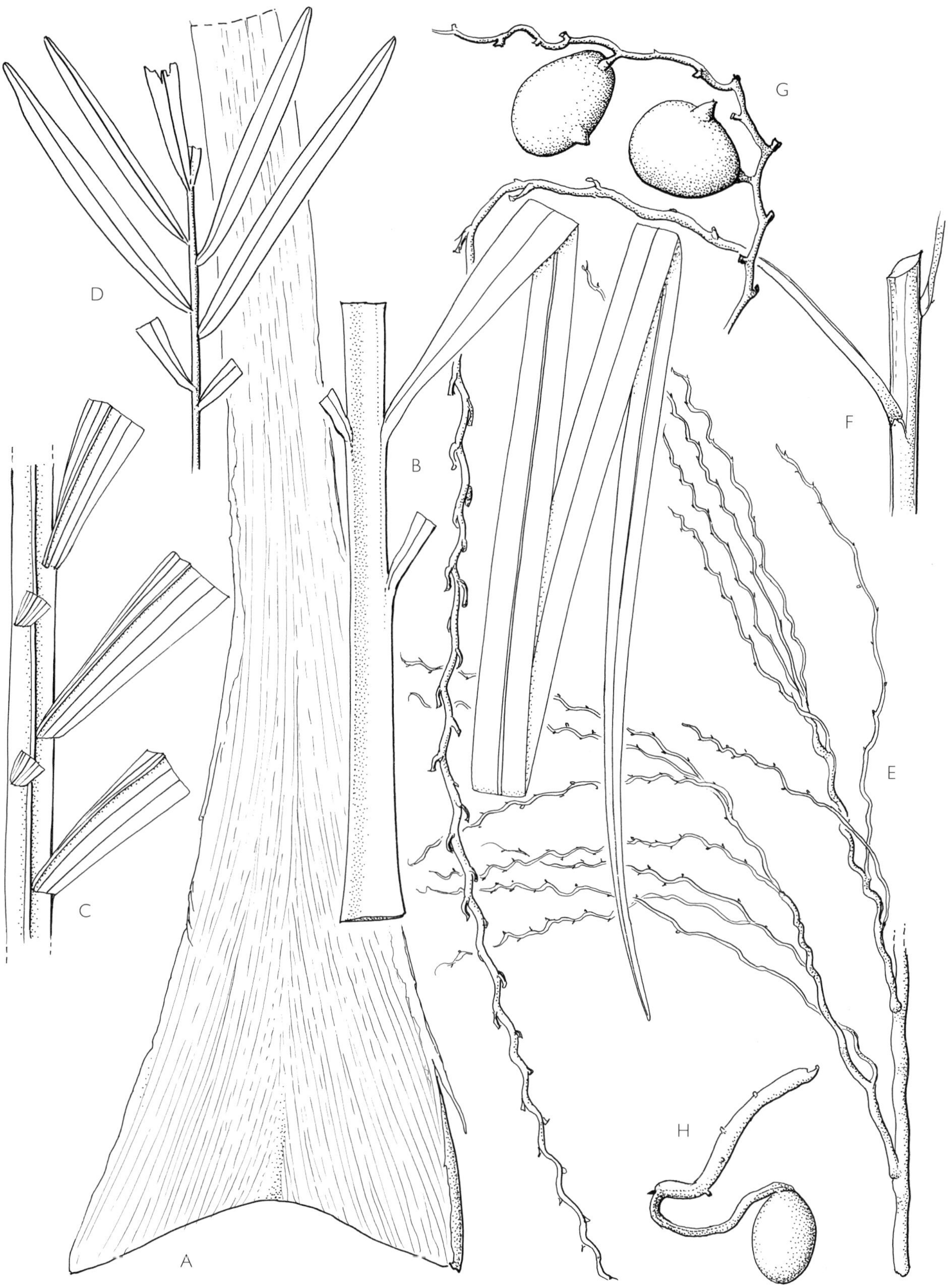

Ravenea julietiae. **A** leaf sheath × 1/2; **B** proximal leaflet × 1/2; **C** median portion of rachis × 1/2; **D** leaf tip × 1/3; **E** portion of staminate inflorescence × 2/3; **F** portion of pistillate inflorescence × 2/3; **G** portion of rachilla with immature fruit × 2/3; **H** germinating seed × 2/3. **A – D, F – G** from *Beentje & Andriampaniry* 4787, **E** *from Beentje & Andriampaniry* 4784, **H** from *Beentje* 4722. Drawn by Rosemary Wise.

Local Name. *Sindro madiniky* (Betsimisaraka, Antanambe); *Saroroira* (Betsimisaraka, Ampasimanolotra); *Vakapasy, Anive, Anivona* (Tanala/Antaisaka).
Uses. Used in construction; hollowed trunks used for irrigation pipes.
Conservation status. Endangered. Numbers are very low (less than 50 trees seen). Both southern localities (where the bulk of the population occurs) are being destroyed rapidly.

Ravenea julietiae *Beentje*
Kew Bull. 49: 646 (1994). Type: Madagascar, Amby forest, *Beentje & Andriampaniry* 4783 (Holotype K; isotypes BH, MO, P, TAN).

Graceful medium-sized palm. **Trunk** 3–10 m, 10–15 cm diam., diameter near crown c. 7 cm; basal boss to 15 cm high, 40 cm across; internodes 6–12 cm (near crown 2 cm), nodal scars c. 1 cm, bark brown-grey; wood very hard, with black fibres just below the bark; base of crown bulbous, c. 15 cm across. **Leaves** 9–23 in the crown, gracefully arching; sheath 40–80 x 12–16 cm, pale green, the base abaxially with white to pale brown tomentum; petiole 30–80 cm long, proximally 3.5–6 x 2.2–4 cm across, distally 2.5–4 x 1.4–3.5 cm, deeply channelled with sharp edges, glabrescent or with a few scattered scales; rachis 1.1–2.8 m, channelled or flat for some 40 cm, keeled for the rest, green with white scattered scales, in mid-leaf 1.7–2 cm wide; leaflets stiff or with the distal part of the leaflet pendulous, the leaflets on opposite sides of the rachis at an angle of c. 90° to each other, 34–48 on each side of the rachis, the proximal 58–110 x 1.3–3.7 cm, median 47–90 x 3.3–5.2 cm (interval 4 cm), distal 10–26 x 1.2–2.3 cm, ramenta apparently none, but in young leaves sparse large ramenta over entire length of midrib. **Staminate inflorescence** interfoliar, only known dead, multiple in 5s–7s, the individual inflorescences up to 90 cm, branched to 2 orders; peduncle 38–47 cm, distally c. 5 x 3 mm across; prophyll not seen; peduncular bracts unknown, except for one found on the ground, 84 cm long; rachis 46–55 cm, distally zigzag, with about 26 branched and 5 unbranched branches; rachillae 3.5–17 cm, distally sinuous. **Pistillate inflorescence** interfoliar, solitary, erect, spreading in fruit, 250–400 cm, branched to 1 order; peduncle 145–305 cm, proximally 2–2.2 x 1.3–1.8 cm across, distally 1.5 x 1 cm across, green with silvery-brown indument; prophyll 15–20 x 5 cm; peduncular bracts 22–36 cm (inserted at 2–6 cm from the base of the peduncle), 37–114 cm (inserted at c. 13 cm), 185–225 cm (inserted at c. 19 cm), 240–280 cm (inserted at 25–115 cm); rachis 28–70 cm long; rachillae 29–34 in number, 20 (distally) –49 (proximally) cm, 2.5–4 mm across in fruit; pedicels 1–8 mm; flowers unknown. **Fruit** colour unknown, ellipsoid, 22–27 x 17–20 mm and with lateral beak in younger stage, one-seeded. **Seed** ovoid or ellipsoid, black, 19–20 x 14–17 mm; seedcoat black, 0.2 mm thick. **Germination** remote; eophyll bifid.

Note. The staminate trees seem to have more leaves than the pistillate, with ranges from 21–23 (staminate) and 11–19 (pistillate). Male trees also seem to have longer petioles, 60–80 cm rather than 30–50 cm.

The species is distinguished from other species with solitary pistillate, but multiple staminate inflorescences, by its extraordinarily long pistillate inflorescences, its large black seeds, and the dimensions of its leaflets. It is the only *Ravenea* species apart from *R. louvelii* in which remote germination has been observed, rather than the usual adjacent-ligular germination.

Specimens seen. Mananara Avaratra: Antanambe, April 1992 (old pist.), *Beentje et al.* 4623 (K, MO, P, TAN). Ampasimanolotra: Andrambolahy Kely to Andranampony, April 1951 (fr.), *Cours* 4510 (K, P, TAN). Manakara: Amby forest, May 1992 (old pist.), *Beentje & Andriampaniry* 4661 (K, TAN); idem, July 1992 (old pist.), *H. Beentje & J. Beentje* 4719 (K, TAN); idem (fr., seedl.), *Beentje* 4722 (K). Farafangana: Manombo, Jan. 1993 (fr.), *Beentje & Andriampaniry* 4783 (Holotype K; isotypes BH, MO, P, TAN) and (old stam.), *Beentje & Andriampaniry* 4784 (K, TAN), and (yfr.), *Beentje & Andriampaniry* 4787 (K).

8. Ravenea rivularis

An aptly named species, since it only grows along riverbanks. It has recently become very popular in cultivation, and when well-watered grows at an impressive speed. The species can be admired from the main road Ihosy–Tulear, in a large stand at Ilkaka village near the Isalo National Park.

Distribution. S central Madagascar, Mangoky and Onilahy river systems.
Habitat. In shallow standing water on riverbanks, swampy valley bottoms, either in deciduous forest or in gallery forest; 350–750 m; gregarious, often forming stands.
Local Name. *Gora* (Sakalava), *Bakaly, Vakaka* (Bara), *Malio* (near Manera); *majesty palm* (in horticulture).
Uses. Seeds collected for export.
Conservation status. Vulnerable. Some 60 trees seen in two populations; they do not grow in any protected area.

Ravenea rivularis *Jum. & H. Perrier*
Ann. Inst. Bot.-Géol. Colon. Marseille sér. 3, 1 (1): 54, t. 29–31 (1913); Becc., Palme del Madagascar 51, fig. 41 (1914); Jum., Ann. Inst. Bot.-Géol. Colon. Marseille sér. 4, 5 (1): 37 (1927); Cat. Pl. Madagascar, Palmae: 25 (1938); Jum. & H. Perrier, Fl. Madagascar 30: 172 (1945); Beentje, Kew Bull. 49: 649 (1994). Type not indi-

cated, lectotype: Madagascar, upper Imaloto basin, near Isalo, *Perrier* 11958 (lectotype P).

Ravenea rivularis

Majestic tree palm. **TRUNK** 5–22 (30?) m, cylindrical or slightly inflated towards the middle, 36–50 cm diam., near crown 15–18 cm diam.; internodes 4–10 cm; wood with tough black fibre layer in the outer part; inner wood soft; bark pale brown-grey (not white, as Fl. Madagascar states!); diameter of base of crown c. 22 cm. **LEAVES** 16–25, porrect to pendulous, slightly arching, held on edge in the distal half of the leaf; sheath 30–50 x 15–16 cm, green and proximally with cottony white or grey-brown indument; margins ragged and fibrous; petiole 6–20 cm long, 4–4.5 x 1.5–1.6 cm across, adaxially flat with slightly sharp edges, white-waxy and -scaly; rachis green, white and scaly when young, 1.2–1.7 m long, 1–1.7 cm wide towards the middle, abaxially almost flat or slightly convex, densely pubescent to glabrescent, adaxially flat in the proximal half, more distally keeled with flat top and with the keel almost as wide as the rachis; leaflets regular, in one plane, stiff to pendulous, attenuate, green to slightly yellow, slightly waxy, 70–73 on each side of the rachis, the proximal 20–55 x 0.5–1.7 cm, the median 52–64 x 1.5–3.2 cm (interval c. 2 cm), the distal 16–42 x 0.4–1.9 cm, especially abaxially with the veins bearing minute whitish scales, also with a few large brown or grey ramenta on midrib abaxially. **STAMINATE INFLORESCENCE** multiple in 5s–7s (Fl. Mad. says solitary), erect, branched to 2 orders, 86–90 cm; peduncle 32–45 cm, green, proximally 0.9–1.1 cm across and densely pubescent, distally c. 0.8 cm across and glabrescent, slightly flattened; prophyll not known; peduncular bracts (based on one collection, cut just above base) 18 cm, 85 cm, 88 cm, 71 cm (the last inserted at 13 cm from the base of the peduncle), abaxially short-pubescent; non-tubular peduncular bract at 15–30 cm from the base, membranous, 35–41 x 2 cm; rachis bracts c. 6 x 0.2 cm at the base of the rachis; rachis 52–89 cm, slightly pubescent; branches 60–100, dense; rachillae 3–21 cm long, sinuous in the distal part; bracteole c. 0.7 x 0.4 mm; pedicel 0.5–1 mm long. **STAMINATE FLOWERS** with the calyx connate for 0.5–0.7 mm, 0.6–1 mm across, free lobes 1.2–1.3 x 0.7–0.8 mm, triangular, acute; petals 3.5–5.8 mm, ovate, acute; filaments 0.4 (epipetalous, adnate for 0.3–0.4 mm)–0.8 (free) mm; anthers 3.1–3.4 x 1.2–1.5 mm; pistillode 0.5 x 0.6 mm. **FLOWERS** with slight resinous smell. **PISTILLATE INFLORESCENCES** solitary, erect or spreading, branched to 1 order, 130–150 cm long; peduncle 40–60 cm, proximally 4 x 2.5 cm across, distally 1.2–2.2 x 1–1.8 cm across, green, white-scaly; prophyll 13–16 cm long, adnate for c. 2 cm; bracts narrow, 24–35 x 5–7 cm, 52–65 x 6 (inserted at c. 4 cm from the base of the peduncle), 110–130 x 5 (inserted at c. 5 cm), 124–150 x 4.5 cm (inserted at c. 6 cm), abaxially white-scaly; at c. 60 cm from the base with a non-tubular peduncular bract about 20 x 0.7 cm; proximal rachis bract 6–15 x 0.3 cm; rachis 50–72 cm, with 125–146 densely packed reflexed rachillae; rachillae 10–32 cm, 1.5 (–2.5 in fruit) mm across, with bulbous base, proximally 3.5 mm across,

Ravenea rivularis. View of crown, near Isalo (*Beentje & Beentje* 4714).

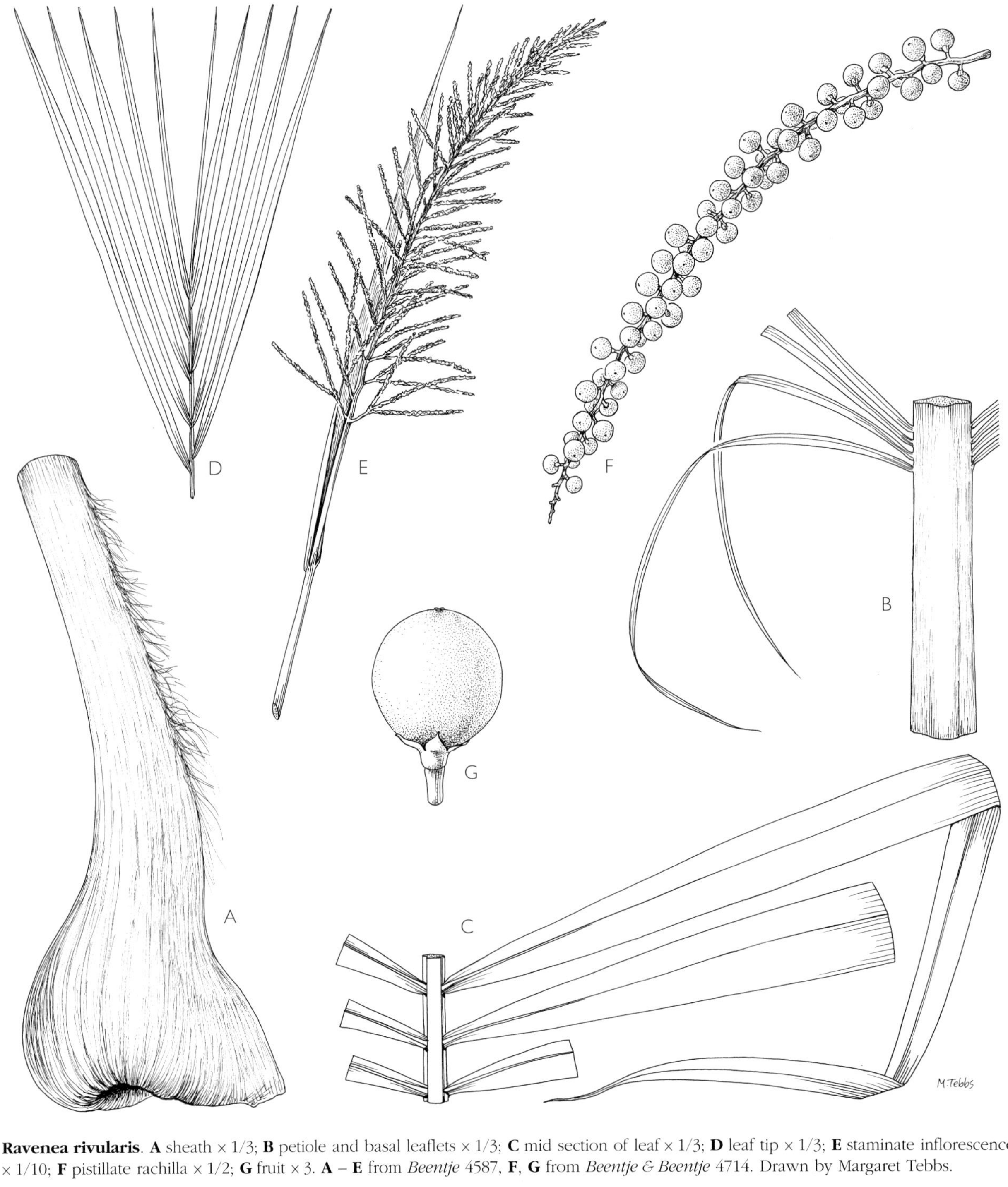

Ravenea rivularis. **A** sheath × 1/3; **B** petiole and basal leaflets × 1/3; **C** mid section of leaf × 1/3; **D** leaf tip × 1/3; **E** staminate inflorescence × 1/10; **F** pistillate rachilla × 1/2; **G** fruit × 3. **A** – **E** from *Beentje* 4587, **F**, **G** from *Beentje & Beentje* 4714. Drawn by Margaret Tebbs.

green with white bloom; pedicel 1–4 mm long. **Pistillate flower** with the calyx connate for 1–1.3 mm, free lobes 1–1.3 × 1.3 mm, triangular, acute; petals ovate/triangular, acute, 3–3.5 × 2.5 mm; staminodes 6, short; ovary ovoid-conical, slightly shorter than the petals. **Fruits** bright red, globose to slightly ellipsoid, 7.5–9 × 7–8.5 mm, 1-seeded, with bitter pulp; stigmatic remains subapical to lateral. **Seed** 5.5–6 × 5.5 mm. **Seedling** with 2 scale leaves and a bifid eophyll.

Note. The protologue mentions two populations, at Imaloto at the foot of the Isalo and at the Matitina R. near the sea (E Madagascar); the first one must have been the Perrier collection chosen as lectotype (which has staminate and pistillate flowers, as well as young fruit, just as in the protologue); the second one is only cited as *Perrier* s.n. in the Flore de Madagascar, a specimen which we have not seen; this might be an early collection of *R. musicalis*, the only eastern species looking vaguely like *R. rivularis*. HB has searched for palms at the site indicated, but failed to find a single (wild) one; older people did not know of any native palms, nor did they recognize the local name given (*Akoraka* in the Tanala language).

Jumelle (1927b) mentions populations seen on Makay sandstone, in the Mangoky basin; the Flore de Madagascar adds the Sakeny R. valley (Menabe) and

Madiovalo, just West of Ambato-Boeni, near the mouth of the Betsiboka R.; and the Bongolava (without mentioning which one; the name just means long plateau and there are several). The first three sites are indicated on the map by question marks. We consider the Madiovalo and Bongolava ones as unlikely; possibly these refer to riverside trees of *R. sambiranensis*, which has been collected in such a habitat on the (Miandrivazo) Bongolava.

SPECIMENS SEEN. Sakaraha: Manera, on the Fiheranalnal R., Oct. 1924 (fr.), *Perrier* 16580 (P). Ihosy: 40 km E of Ranohira, July 1992 (yfr.), *H. Beentje & J. Beentje* 4714 (BH, K, MO, P, TAN); idem, July 1992 (seedling), *H. Beentje & J. Beentje* 4715 (K); Ilkaka, Feb. 1992 (stam.), *Beentje* 4587 (BH, K, MO, P, TAN); upper Imaloto R., near Isalo, no date (stam., pist., yfr.), *Perrier* 11958 (Holotype P).

SIGHT RECORDS. Prof. Morat has photographs of this species from Ankazomarefo, near the Zombitse Forest.

9. RAVENEA MUSICALIS

The only true Madagascar water palm, and possibly the only palm in the world of which the seeds sprout under water; fruit and seed are adapted to dispersal by water. The fruits float, but open at the slightest bump; the liberated seed has already sprouted within the fruit, so that when it sinks to the river bottom it can quickly establish itself. The curved scale leaves might even help it to catch onto protuberances, so it would not be washed away easily. The name refers to the musical sound caused by seeds of various sizes dropping from various heights into the river.

DISTRIBUTION. S Madagascar, only known from one site.
HABITAT. At 0.5–2.5 m depth in flowing water; seen as a rheophyte on submerged rock pavement, then sterile and to 1 m only; 1–50 m.
LOCAL NAME. *Torendriky* ("submerged trunk").
USES. No uses known to local people.
CONSERVATION STATUS. Vulnerable. Occurs, as far as we know, only along a single river, where the population consists of some 450 trees. Rejuvenation at this site looks good, but as an aquatic species this is very vulnerable to habitat changes.

Ravenea musicalis *Beentje*
Principes 37, 4: 199, fig. 1–7 (1993); Kew Bull. 49: 651 (1994). Type: Madagascar, Andriambe R., *Beentje & Andriampaniry* 4611 (Holotype K; isotypes BH, MO, P, TAN).

Small to medium-sized ventricose palm growing in 0.5–2.5 m deep water. **TRUNK** 2.5–8 m, ventricose with base (at water level) to 50 cm across, (above water) 30–40 cm diam., near the crown c. 11 cm across, internodes here 0.5 cm, nodal scars 0.5 cm; bark pale brown, soft, with internodes 1–2 cm; wood soft, cream-coloured, fibrous, without hard fibres. **LEAVES** 14–16, porrect to spreading, arching, held on edge in distal half, with stiff or arching leaflets; sheath 36–41 × 13–20 cm, adaxially orange, abaxially proximally orange, distally green, with thin grey tomentum; fibres few; petiole 15–19 cm long, proximally 3.5–5 × 1.5 across, distally 2.2–2.3 × 0.5–0.6 cm across, glabrous, keeled; rachis 1.3–1.8 m, in mid-leaf 1–1.5 cm across, with little abaxial tomentum; leaflets in one

Ravenea musicalis. Submerged seedlings at the type locality, with emergent young plants at upper left.

Ravenea musicalis. **A** leaf sheath × 1/3; **B** petiole and proximal leaflets × 1/3; **C** mid-portion of leaf × 1/3; **D** leaf tip × 1/3; **E** inflorescence, schematic; **F** peduncle × 1/2; **G** portion of infrutescence × 1; **H** fruit × 1; **J** seed × 1; **K** seedling × 1. All from *Beentje & Andriampaniry* 4611. Drawn by Rosemary Wise.

plane, stiff, 59–63 on each side of the rachis, the proximal 36–47 × 0.5–1.5 cm, median 42–53 × 1.6–2.4 cm (interval 2–2.5 cm), distal 10–30 × 0.4–1.3 cm, ramenta none or few, large, basal on midrib and outer main veins, main veins 4. **STAMINATE INFLORESCENCE** multiple in 5s, the individual inflorescences to 115 cm, branched to 1 order, pendulous in later stage; peduncle 36–38 cm, proximally c. 1 cm across, distally 0.6–0.7 cm across; prophyll 29–30 cm; peduncular bracts 38 cm, 64 cm (inserted at c. 2 cm from the base of the peduncle), 84 cm (inserted at c. 5 cm), 80 cm (inserted at c. 10 cm); rachis about 54 cm, with many dense rachillae; rachillae 7–24 cm, 1–1.5 mm across; flower scars distant. **STAMINATE FLOWERS** unknown. **PISTILLATE INFLORESCENCE** solitary, spreading, 105–125 cm, branched to 1 order, the axes green; peduncle 48–52 cm, proximally 3–5 × 2–2.5 cm, distally 2–3 × 1.3–2 cm; prophyll c. 10 × 4 cm; peduncular bracts 20–24 cm (inserted at 0–2.5 cm from the base of the peduncle), 49–52 cm (inserted at 3–3.5 cm), 82–83 cm (inserted at 4–9 cm), 100–103 cm (inserted at 10–24 cm); rachis 39–55 cm, with 58–68 branches; rachillae 9–42 cm, the proximal spreading, the distal porrect, the base proximally flat, 0.6–1.5 × 1 cm, in fruit 3–3.5 mm across; pedicels c. 0.5 mm. **PISTILLATE FLOWERS** with the calyx connate for 1 mm, 1.5 mm wide, free lobes 1.7–2.6 × 1.6–2 mm, ovate, acute; petals in fruit only present as fibre remnants, c. 2.5 mm long. **FRUIT** orange, 17–23 × 14–19 mm, one-seeded; stigmatic remnants subapical to lateral. **SEED** brown, 10–14 mm across, hard, seedcoat black, 0.2 mm thick. **SEEDLING** with 3-4 scale leaves, the first small, the second, third and fourth to 9 cm long and with curving tips; eophyll pinnate.

NOTE. Differs from all other species of *Ravenea* in its habitat and habit, its extraordinary floating fruit with its spongy mesocarp, splitting at the slightest provocation and then exposing the already germinating seed, the number of its scale leaves on the seedling, and the lack of hard fibres in the outer wood.

SPECIMENS SEEN. Belavenona, R. Andriambe, March 1992 (fr.), *Beentje & Andriampaniry* 4611 (Holotype K; isotypes BH, MO, P, TAN); idem (old stam.), *Beentje & Andriampaniry* 4612 (K, MO, TAN); idem, Dec. 1992 (y.fr.), *Beentje* 4756 (BH, K, MO, P, TAN).

Ravenea musicalis. Fruiting tree at the type locality, Southeast Madagascar.

10. RAVENEA GLAUCA

A variable species, plants can be tall and slender in canyons at Isalo, but they can also flower when almost stemless in a dry river bed. The original description states that this species forms vast forests in the Andringitra mountains, between 1200 and 1800 m, to the exclusion of any other plant; it is said to be resistant to bush fires. Having seen the stands in the Isalo, HB finds monospecific forests without undergrowth hard to believe. The Latin name refers to the glaucous, waxy undersides of the leaflets.

DISTRIBUTION. Central-S Madagascar: W side of Andringitra Mts and Isalo.
HABITAT. Dry forest and in ravines and rocky or sandy gullies; altitude 670–1250 (?1800) m.
LOCAL NAME. *Anivo*; *Sihara* (Bara).
USES. Not eaten, since the heart is bitter.
CONSERVATION STATUS. Vulnerable. We are uncertain about the status of populations in the Andringitra; no plants of this species have been collected there since 1922. Elsewhere this species is only known from the Isalo, where population size is probably a few hundred.

Ravenea glauca *Jum. & H. Perrier*
Ann. Inst. Bot.-Géol. Colon. Marseille sér. 3, 1 (1): 56, fig. 13 (1913); Jum., Ann. Inst. Bot.-Géol. Colon. Marseille sér. 4, 5 (1): 24 (1927); Cat. Pl. Madagascar, Palmae: 24 (1938); Jum. & H. Perrier, Fl. Madagascar 30: 170 (1945); Beentje, Kew Bull. 49: 653 (1994). Type: none indicated, except locality, Massif d'Andringitra, W slopes, 1200–1800 m (pist. only); lectotype: Madagascar, Andringitra, W slopes, *Perrier* 13649 (P).

Slender palm. **TRUNK** 0.2–8 m tall, rarely with sheath remnants in the distal part, 7–12 cm diam.; basal boss to 10 cm high, 18 cm across; internodes 2–5 cm, brown-grey; nodal scars 0.7–2 cm light brown; wood hard. **LEAVES** 14–20 in crown, porrect and spreading, slightly arching; sheath 10–50 × 7–10 cm, white-floccose, with variously-sized remnants of more distal sheath material; petiole 10–50 cm, proximally 1.5–2.5 × 0.8–2 cm across, distally 1–1.8 × 0.8–1.4 cm, (but most of this used to have adjacent sheath material: true petiole 3–18 cm), channelled adaxially, convex abaxially, with raised sharp edges, abaxially white-tomentose; rachis 1.1–2 m, in mid-leaf 0.7–2 cm across, abaxially rounded with two blunt wings, adaxially channelled with blunt edges, densely white-tomentose when young; leaflets regular, the leaflets on opposite sides of the rachis at an angle of 90°, stiff in the proximal part, half-pendulous in their more distal part, green or glaucous with waxy covering when fresh, 49–73 on each side of the rachis, the proximal 9.5–59 × 0.1–1.1 cm, median 50–70 × 1.2–2.3 cm (interval 1–2 cm), distal 13.5–42 × 0.3–1.4 cm, ramenta sparse on the abaxial side, on the midrib only, spaced, large, white, main veins 2–4. **STAMINATE INFLORESCENCE** multiple in (1) 2s–6s, branched to 1–2 orders (in the Isalo populations the rachillae only branched at the very base of the inflorescence, and sometimes

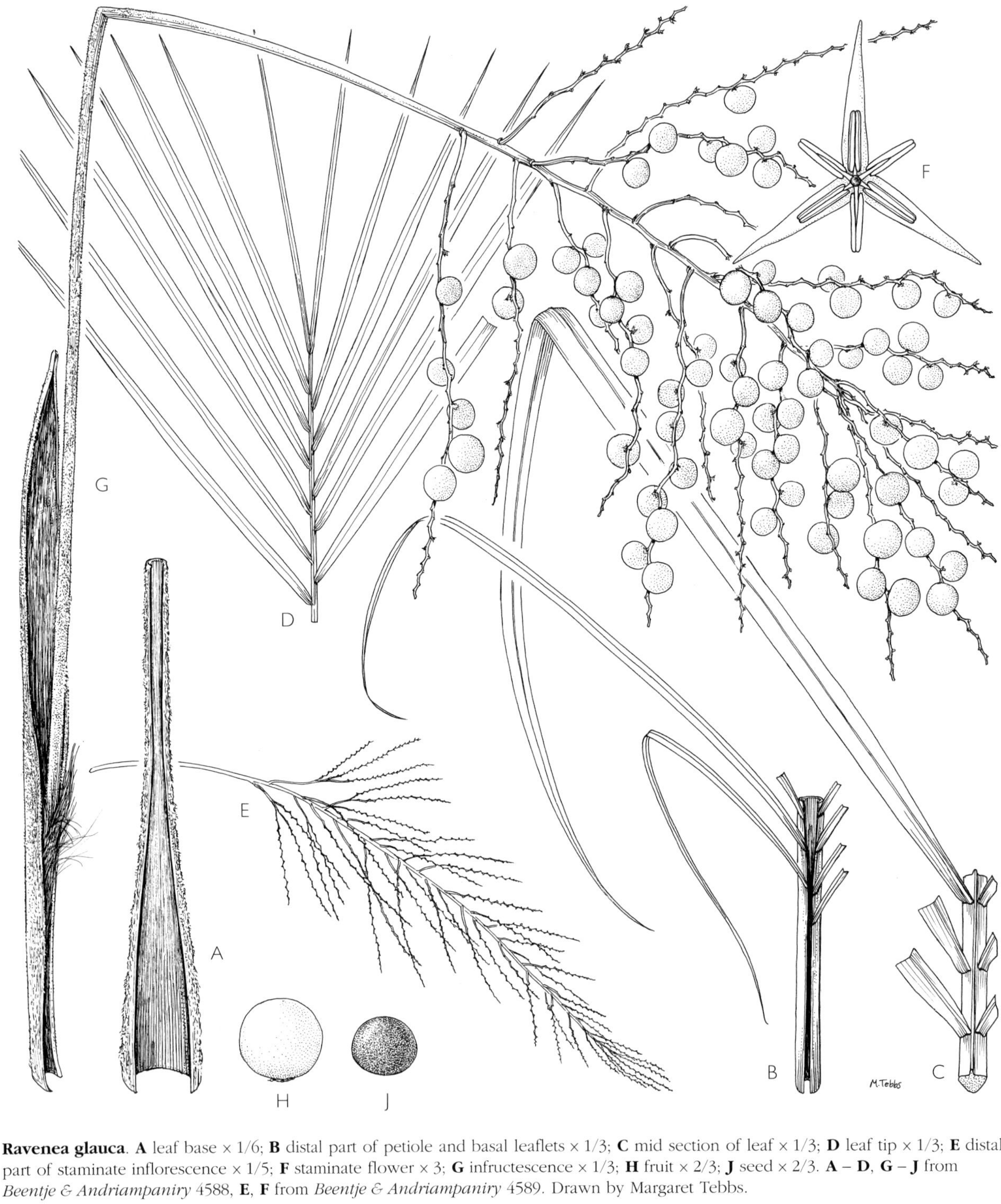

Ravenea glauca. **A** leaf base × 1/6; **B** distal part of petiole and basal leaflets × 1/3; **C** mid section of leaf × 1/3; **D** leaf tip × 1/3; **E** distal part of staminate inflorescence × 1/5; **F** staminate flower × 3; **G** infructescence × 1/3; **H** fruit × 2/3; **J** seed × 2/3. **A – D**, **G – J** from *Beentje & Andriampaniry* 4588, **E**, **F** from *Beentje & Andriampaniry* 4589. Drawn by Margaret Tebbs.

these branched rachillae not present), erect, 70–110 cm; peduncle 35–64 cm long, proximally 5–7 mm across, distally 3–7 mm across, densely white-puberulous; common prophyll 3.5–13.5 cm and membranous; peduncular bracts 9–24 cm (inserted at c. 3 cm from the base of the peduncle), 22–55 cm (inserted at c. 6 cm), 60–105 cm (inserted at c. 6 cm), 62–107 cm (inserted at c. 12 cm), densely pubescent abaxially, adaxially smooth and glabrous; rachis 21–46 cm long, ± glabrous, with c. 30 unbranched and 10 branched first order branches in Andringitra, 0–1 unbranched and 25–52 branched ones in Isalo, with rachis bracts c. 15 × 0.3 cm; rachillae close, porrect, 2–12 cm long, 1 mm across, zigzag, especially distally, with dense flowers; pedicel 0.2–1.5 mm. **Staminate flowers** with the calyx connate for 0.3–1 mm, free lobes 0.8–1 × 0.8 mm, triangular, acute; petals free, 7.5–9 × 1.5 mm, gradually tapering to acute apex; anthers 2.5–3.2 × 1.3–1.4 mm, filaments 0.6–0.7 mm, the 3 epipetalous stamens adnate for 0.7 mm; pistillode 1 × 0.4 mm. **Pistillate inflorescence** solitary, erect in bud, spreading at anthesis, pendulous in fruit, 66–158 cm long, branching to one order; peduncle 45–105 cm, slightly flattened, proximally 6–10 × 4–10 mm across, distally 4–10 mm across, pale brown- or grey-white tomentose, distally glabrescent; prophyll 6–18 cm, membranous; peduncular bracts respectively 13–34 (inserted at 0–3 cm from the base of the peduncle), 27–48 × 2.2–2.6 (inserted at 0–6 cm), 60–103 (inserted at 6–7 cm) and 59–158 × 1.6–2 cm (inserted at 10–14 cm), the distal two with apices at the same level; all bracts grey-puberulous to white floccose-tomentose abaxially, adaxially glabrous, smooth and rich brown when dry; non-tubular peduncular bract 4–9 × 0.3–0.4 cm; rachis 17–40 cm long, proximally white-tomentose or with grey bloom, otherwise glabrous; most proximal rachis bracts 38 × 5 mm; rachillae spreading or porrect, slightly zigzag, 24–35 in number, 4–26 cm long, 1.2–2 mm across, sinuous, with spaced flowers, with slightly bulbous base 3.5–4 × 2 mm, white-waxy; bracteole 2.3 × 1.8 mm; pedicels 0.5–5 mm. **Pistillate flowers** with the calyx connate for 0.8 mm, free lobes triangular, acute, 1–3 × 1.0–2 mm; petals 6.5–8 × 1.2–1.7 mm, ovate, acute; staminodes 1.8 × 0.6 mm; ovary 2.5 × 1.3 mm. **Fruit** yellow, 20–22 × 22–23 mm, 1-seeded; stigmatic remains lateral or subbasal. **Seed** 16–18 × 18–19 mm, dark brown; seedcoat brown, 0.2 mm thick; embryo 4 mm. **Eophyll** pinnate.

Ravenea glauca

Note. The protologue describes populations from the Andringitra, but all collections we have seen are from a later date.

The illustration in the Flore de Madagascar (1945) has a habit sketch which is based on a photograph of *Humbert* 13698, which is a *R. xerophila*.

Ravenea glauca, growing in a canyon at Isalo (Photo: B. Du Puy & D. Du Puy).

Ravenea glauca. Tree in fruit, Isalo (*Beentje* 4588).

SPECIMENS SEEN. Ihosy: Isalo, W of Ranohira, anno 1955 (fr.), *Humbert* 28705 (K, P); idem, July 1992 (pist.), *Beentje & Andriampaniry* 4711 (BH, K, MO, P, TAN) and (stam.), *Beentje & Andriampaniry* 4712 (K, MO, TAN); Isalo, Canyon des Singes, Jan. 1964 (fr.) , *Bosser* 18867 (P, TAN); idem, Feb. 1992 (fr.), *Beentje* 4588 (BH, K, MO, P, TAN) and (old stam.), *Beentje* 4589 (BH, K, MO, P, TAN); Isalo, Angabolava, Nov. 1960 (yfr.), *Leandri* 3923 (P); Isalo, Canyon des Rats, Nov. 1989 (fr.), *B. Du Puy et al.* MB 450 (BH, K, MO, P, TAN); Isalo, without precise locality, Oct. 1924 (stam., pist.), *Perrier* 16550 (P). Ambalavao: Andringitra, W slopes, 1200–1800 m, Sept. 1916 (pist.), *Perrier* 12061 (P); idem, 1500–2000 m, April 1921 (stam., fr.), *Perrier* 13649 (Holotype P); idem, Canton Sendrisoa, July 1951 (stam.), *Razafindrakoto RN* 3601 (P); idem, inselberg 11 km S of Ambalavao, Aug. 1972 (pist.), *Cremers* 2292 (TAN).

11. RAVENEA HILDEBRANDTII

Ravenea hildebrandtii

One of the smallest *Ravenea* species. In Victorian times this was a popular species for indoor cultivation in Europe. The epithet honours the collector of the type, J.M. Hildebrandt (1847–1881).

DISTRIBUTION. Comoro Islands: Grand Comore, Moheli, Anjouan.
HABITAT. Moist forest at altitudes of 600–900 m. Once reported from riverine forest at this altitude.
LOCAL NAME. *Inazi* (fide St John) which sounds suspiciously like *Mnazi*, the Swahili name for coconut.
USES. Not recorded.
CONSERVATION STATUS. ? Endangered. All forest in the Comoro Islands is under severe threat. This species has not been collected for a long time.

Ravenea hildebrandtii *C.D. Bouché*
Monatsschrift Vereines Beförd. Gartenbaues Königl. Preuss. Staaten 1878: 197, 324 (1878); H. Wendl. in J.D. Hook., Bot. Mag. 10: t.6776 (1884); Baill. in Bull. Mens. Soc. Linn. Paris (1895): 1186 (1895); Becc., Palme del Madagascar 49, fig. 39, pl. 47 (1914); Jum. & H. Perrier, Fl. Madagascar 30: 169, fig. 47 (1945); J. Dransf. & N.W. Uhl, Principes 30 (4): 157 (1986); Beentje, Kew Bull. 49: 654 (1994). Holotype: illustration of young plant in Monatsschrift Vereines Beförd. Gartenbaues Königl. Preuss. Staaten 1878: 323; protype: cult. in Hort. Bot. Reg. Kew, 21 May 1884 (K).

Slender undergrowth palm, sometimes flowering when almost acaulescent. **TRUNK** 1–6 (–12) m, usually less than 4 m, 5–6 cm diam. Trunk swollen at base, to 10 cm diam.; internodes short, hardly visible in older stems; bark grey-brown, smooth, in older trees somewhat fissured longitudinally. **LEAVES** 12–25 in the crown, arching; sheath c. 14.5 x 5 cm, the margins sparsely or hardly fibrous, densely tomentous abaxially, the tomentum white, later turning pale brown; petiole 28–50 cm long, 0.7–1 cm wide proximally, 6–11 mm distally, adaxially deeply channelled, glabrous, abaxially rounded and densely brown-tomentose or short-pubescent, with sharp edges; rachis 48–99 cm long, in mid-leaf 5–7 mm wide, adaxially shallowly grooved proximally, changing to keeled (keel to 5 mm wide) to angled distally, laterally grooved, sparsely tomentose or pubescent; leaflets on opposite sides of the rachis probably at an angle of 90° with each other, 20–43 on each side of the rachis, dark olive green on both surfaces, linear, acuminate, the basal ones 8–28 x 0.2–1.2 cm, the median ones 27–45.5 x 1–2 cm (interval 2–3.2 cm), the distal ones 7.5–37 x 0.7–1.9 cm, the distal pair sometimes slightly connate at the base, midrib with few to many small, pale, tattered, rounded ramenta abaxially along its whole length (always?), main veins faint, 2–4. **STAMINATE INFLORESCENCES** solitary (though *St.John* 26543 states "2 per axil"), branched to 1 order (at the base to 2 orders in *St.John* 26543); peduncle very slender, 17–35 cm long, 2–4 mm diam. both proximally and distally, rounded in cross-section, tomentose, glabrescent; prophyll 5–7.5

Ravenea hildebrandtii, in cultivation, Fairchild Tropical Garden (Photo: Scott Zona).

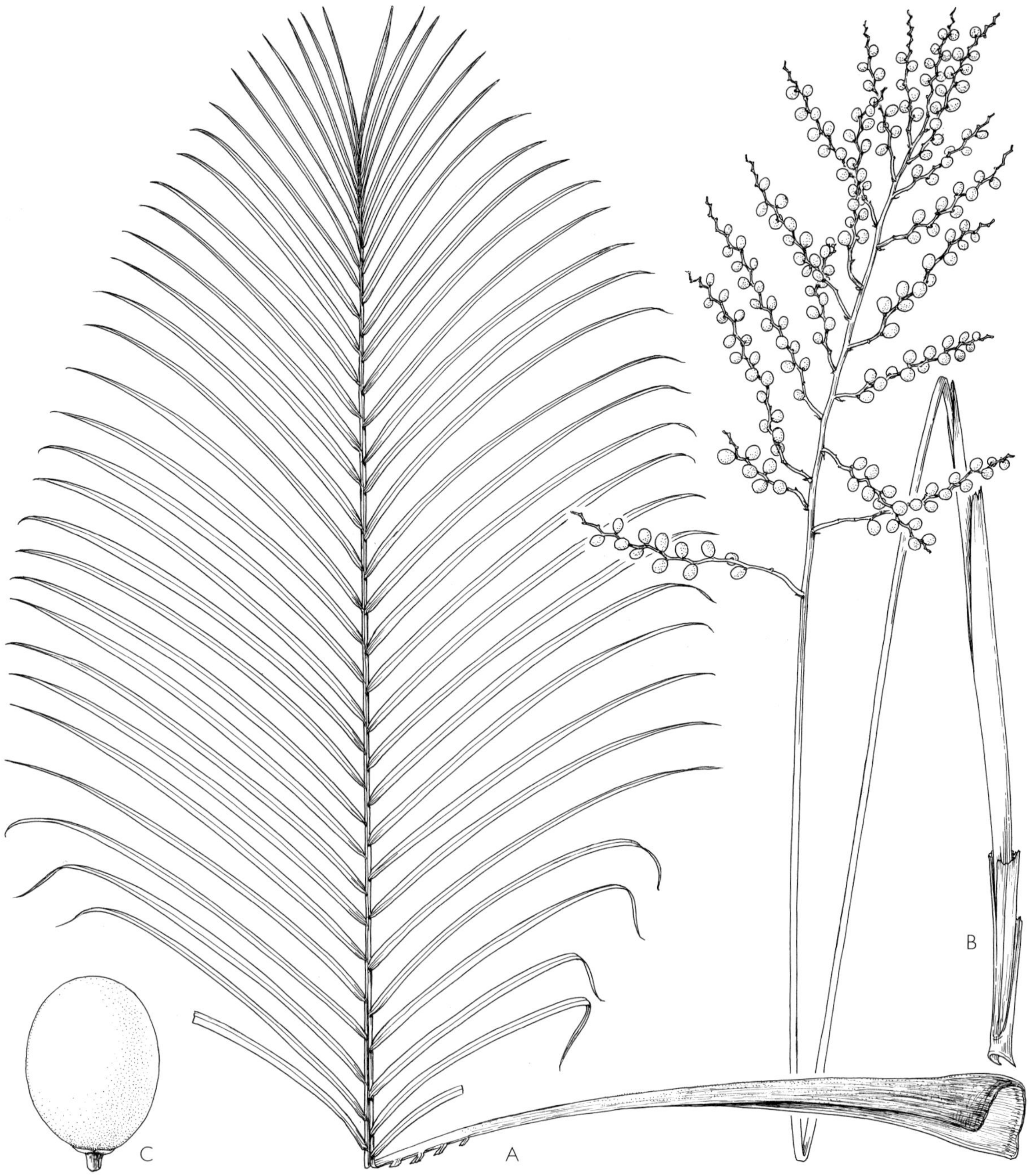

Ravenea hildebrandtii. **A** leaf × 1/5; **B** infructescence × 2/3; **C** fruit × 3. All from *St. John* 26544. Drawn by Margaret Tebbs.

× 1–1.7 cm, rounded at apex, membranous, tending to tatter irregularly, abaxially with scattered soft brown tomentum; peduncular bracts c. 10.5 cm, 20.5 cm, c.57 cm, 57–60 × 2 cm (all measured from base of peduncle), the proximal pair membranous; distal pair distally ± keeled, thick, coriaceous to almost woody, adaxially glabrous, abaxially densely pale grey to pale brown tomentose; non-tubular peduncular bracts 0.9–4 × 0.15–0.3 cm; rachis 10–34 cm, with 33–45 rather crowded, spirally arranged, slender rachillae; rachis bracts 10–45 × 2–3 mm; rachillae 2.4–10 cm long, 0.7–1.2 mm diam., bearing spiral or subdistichous solitary flowers; pedicels 0.25–0.75 mm long. **STAMINATE FLOWERS** 3–5 mm long, dull white, slightly fragrant; calyx connate for 0.5 mm, with 3 triangular, acute or mucronate lobes to 1–1.3 × 1 mm; petals 3.5–4.6 × 1.5–1.7 mm, joined only at the very base, acute or acuminate, 3-veined; stamens 6, with short filaments, these fleshy, c. 0.5 mm long, anthers elongate, 2.0–2.25 × 0.3–0.7 mm; pistillode conspicuous, narrowly pyramidal, 1.5 × 0.4 mm. **PISTILLATE INFLORESCENCE** solitary (though *St.John* 26544 states "2 per axil"), branched to 1 order; peduncle 65–136 cm (*Humblot* 1395 states 300 cm) long, slightly flattened, 4–6 × 3.5 mm across proximally, distally 3–4 mm, tomentose but glabrescent; prophyll c. 9 cm long; peduncular bracts like those of the staminate inflorescence, c. 12 cm, 26–33 cm, 32–47 cm and 47 cm (all measured from base of peduncle), and more coriaceous or woody; rachis 17–22 cm, glabrous, with 19–31 branches; rachis bracts 5–25 × 1–2.5 mm; rachillae 2.5–13.5 cm long, 1–1.2 mm across, with spaced flowers; pedicel 0.5–1.5 mm. **PISTILLATE FLOWERS** with the calyx connate for c. 0.5 mm, with 3 short to long, acuminate, triangular lobes to 1.5 mm; petals 3, free, 2.5–2.6 × 0.5–1 mm, 3-veined, eventually disintegrating leaving the veins free; staminodes 6, filaments very slender, flattened, the anthers sagittate, flattened, empty; ovary flask-shaped, 3 × 1 mm; stigmas 3, short, recurved. **FRUIT** ellipsoid or broadly ovoid, 9–10 × 5–8 mm, one-seeded, stigmatic remains lateral near the base; epicarp yellow, drying wrinkled, mesocarp apparently thin. **SEED** rounded, c. 5 mm diam.; seed-coat brown, very thin.

SPECIMENS SEEN. Grand Comore: near Maoueni, La Grille, Dec. 1967 (stam.), *Bernardi* 11646 (K, P); Charboni-La Grille, Aug.1961 (stam.), *St. John* 26543 (K); and August 1961 (fr.), *St. John* 26544 (BH, K); Combani forest, Nov. 1884 (stam.), *Humblot* 1395 (P); "on high mountains", April 1884 (stam.), *Humblot* 1608 (P). Moheli: Mironjani, Nov. 1968 (fr.), *Schlieben* 11214 (P). Anjouan: March 1877 (stam.), *Bewsher* 34 (P); and April 1861 (fr.), *Kirk* s.n. (K) Without locality: (stam.), *Humblot* 395 (K, P).

CULTIVATED: Royal Botanic Gardens, Kew, 21 May 1884, (stam., protype K); and 20 July 1886, (fr., K); and 17 April 1891, (stam., pist., K); and 18 June 1901, (stam., K); Hannover, Herrenhausen Botanic Garden, Sept. 1884 (stam., yfr.), comm. *Wendland* s.n. (K); and May 1885 (stam.), *Wendland* s.n. (K).

12. RAVENEA NANA

A mystery. Until more is known about its habit and inflorescences (multiple? solitary?), we are unable to include this in the key or discuss its affinities, though it resembles *R. hildebrandtii* from the Comoros. It differs from this species in the much denser tomentum on the leaf-sheath, the smaller leaf rachis, and the larger fruit. It is curious that the intensive collecting on Marojejy during the late 1980s did not result in a re-collecting of this species.

DISTRIBUTION. E Madagascar, between Marojejy and Andohahela.
HABITAT. Ericoid bush or ?low forest on rocky sites, gneiss & quartzite; (?400–) 1100–1900 m; the altitude information on the printed label of the type might be wrong; we believe this is a high-altitude species.
LOCAL NAME. Not recorded.
USES. Not recorded.
CONSERVATION STATUS. ?Endangered. Only known from three areas, and not collected in the last thirty years.

Ravenea nana *Beentje*
Kew Bull. 49: 665 (1994). Type: Madagascar, near Andapa, *Humbert & Capuron* 21947 (Holotype P; isotype K).

Small palm. **TRUNK** 3–4 m tall. **LEAF** sheath 12.9–16 × 1.5 cm, with thin downward-pointing marginal fibres, densely brown-tomentose; petiole 18–67 cm, proximally 0.5–1.6 × 0.4–1 cm, distally 0.7–1.2 × 0.5–0.8 cm, densely pubescent/tomentose on both surfaces, slightly to deeply channelled adaxially, convex abaxially, with sharp edges; rachis (25–) 33–40 cm, in mid-leaf 6–7 mm wide, with dense grey-brown tomentum on both surfaces; leaflets on opposite sides of the rachis at an angle of 90° with each other, 18–29 on each side of the rachis, the proximal 12–26 × 0.1–1.3 cm (interval 0.8–2 cm), median 16–35 × 0.9–2 cm, without visible ramenta or with dense minute ramenta, acuminate, main vein 1, with scales on margins only, distal 7.5–23 × 0.3–1.5 cm. **STAMINATE INFLORESCENCE** branched to 1 order; peduncle 28 (or more) × 0.4 × 0.25 cm, densely brown-pubescent; prophyll unknown; peduncular bracts c. 10 cm, 18–25 cm, 29–40 (or more) cm, 30–41 cm; rachis bract c. 11 × 2 mm; rachis 9–13 cm, with 25–40 branches, scaly proximally, glabrous distally; rachillae 2–5 cm, 0.5 mm across, straight; pedicel 0.5–1 mm. **STAMINATE FLOWERS** with the calyx with very short connate part and lobes 0.8–1.1 mm long; petals 4–4.4 × 1.8 mm, connate for 0.3–0.4 mm

Ravenea nana

by the filament callus of the antesepalous stamens; filaments in 2 series, the antesepalous 0.4–0.5 mm, the antepetalous adnate to the petals for 0.8–1.3 mm, free for 0.2–0.3 mm; anthers 1.5–2 × 0.3–0.4 mm; pistillode 0.6–0.9 mm. **PISTILLATE INFLORESCENCE** 59–92 cm, branched to 1 order; peduncle 35 (or more?)–77 cm, proximally 3–6 mm across, distally 2–3.5 mm across, with dense red-brown scales, glabrescent; prophyll 6–15 cm, tattering; peduncular bracts 10–13 cm, 19–38 cm, 57–82 cm, 44–92 cm, abaxially densely scaly, adaxially glabrous, chestnut brown; non-tubular peduncular bract 4–18 × 0.3–0.5 cm; rachis 6.3–13 cm, scaly, glabrescent; rachillae 14–27 in number, 1–7.5 cm, 1.2–2 mm across, sinuous; pedicel 0.5–1.5 mm, bracteole 1–2 mm. **PISTILLATE FLOWERS** with the sepals 1–1.5 mm; petals (fibres only) 2.5–?5 × 1.5–1.6 mm. **FRUIT** 1.3–2.1 × 1.1–1.7 cm, one-seeded, stigmatic remains lateral to subbasal.

SPECIMENS SEEN. Andapa: Marojejy, W of Manantenina R., 1700–1900 m, March 1949 (fr.), *Humbert* 23695 (K, P); near Andapa, 400–600 m, Nov./Dec. 1948 (old stam., old pist., yfr.), *Humbert & Capuron* 21947 (Holotype P; isotype K); Mt Beondroka, N of Maroambihy, March 1949 (fr.), *Humbert* 23490 (K, P). Fianarantsoa: Andrambovato, Nov. 1963 (fr.), *Bosser* 18297 (P, TAN). Tolanaro: Mt. Papango, Nov. 1928 (yfr.), *Humbert* 6354 (P).

13. RAVENEA DRANSFIELDII

This rare species was one of the 'links' between *Ravenea* and *Louvelia*. It is a rather short palm of the middle story of rain forest. The basal leaflets hang down limply, as shown in our photograph. The flowers are very fleshy and have a musty or sickly smell.

DISTRIBUTION. E Madagascar: between Marojejy Mts. and Ifanadiana.
HABITAT. Moist forest, on slight to steep slope or hill crest; 425–1700 m.
POLLINATION. Flowers visited by beetles (fide Dransfield).

Ravenea nana. **A** median portion of rachis with leaflets × 1/2; **B** portion of staminate inflorescence × 1/2; **C** staminate flower × 5; **D** portion of pistillate inflorescence × 2/3; **E** seed × 1/2. **A** – **C** from *Humbert & Capuron* 21947, **D** – **E** from *Humbert* 23695. Drawn by Rosemary Wise.

Ravenea dransfieldii. Tree in lower story of forest on Masoala Peninsula.

Ravenea dransfieldii. Detail of pistillate inflorescence (*Dransfield et al.* JD6374).

LOCAL NAME. *Anivo* (Betsimisaraka–Masoala); *Ovotsarorona* (Betsimisaraka–Betampona); *Lakatra* (for stam.) and *Lakabolavo* (for pist.) (Tanala).
USES. Apex said to be poisonous by some, but also reports of palm-heart being eaten; used in hat-making.
CONSERVATION STATUS. Vulnerable. Population numbers are low at every site. The Ifanadiana population is under threat of destruction.

Ravenea dransfieldii *Beentje*
Kew Bull. 49: 656 (1994). Type: Madagascar, Hiaraka, *Dransfield et al.* JD6372 (Holotype K; isotypes BH, MO, P, TAN).

Medium-sized palm. **TRUNK** 4–7 m, 10–18 cm diam. (near crown 9.5–15 cm); bark pale brown, scars 2.5–6 cm, internodes 4–7 cm; wood black, extremely hard; wounds in bark bleed watery fluid. Base of crown bulbous, 24 cm across. **LEAVES** 11–17 in crown, porrect, straight, held on edge in distal half; sheaths open, 45–63 × 15–20 cm, pale or bright green, covered in white or brown tomentum, glabrescent, the margins with reflexed fibre spines; petiole absent; rachis 3.3–4.4 m long, 4.3–13 × 1.2–6 cm proximally, 2–2.8 × 1.1–2 cm in the middle part, here either deeply channelled or with a keel c. 5 mm wide, green with scattered brown scales; leaflets dark green, 70–84 on each side of the rachis, the lowermost reflexed or pendulous (and grouped in 3s), the rein margin with almost woody dentiform excrescences; other leaflets in one plane or slightly pendulous, pale or dark green, leathery, ramenta lacking, the proximal (65) 95–120 × 1.3–4.7 cm (and once with the proximal 2 pairs distant by 30 cm from the more distal pairs), median 72–100 × 3–5 cm (interval 3–4 cm), distal 11–37 × 0.5–3 cm, the top pair connate for 1/5, the rachis continuing past the distal pair for > 1 cm, ramenta very few or none, small sinuous transverse veinlets visible in dried material. **STAMINATE INFLORESCENCE** solitary, erect, c. 126 cm, branched to 2 orders with erect rachillae; peduncle 44–66 cm long, distally 0.5–2.5 × 0.5–1.25 cm; prophyll unknown; peduncular bracts about 42 cm, 52–66 × 9 cm, 97–140 cm, 100–139 × 8 cm; non-tubular peduncular bract 4.7–7.9 × 0.1–0.8 cm; rachis 30–50 cm with 1–6 branched and c. 50 unbranched first order branches; rachillae 18–35 cm long, 1.3 mm across; pedicels 1.5–3 mm; bracteole

Ravenea dransfieldii. **A** distal part of sheath and proximal leaflets × 1; **B** mid-portion of leaf, adaxial view × 1; **C** leaf tip × 1. **A** from *Dransfield et al.* JD6372, **B – C** from *Beentje & Andriampaniry* 4732. Drawn by Rosemary Wise.

Ravenea dransfieldii. **A** portion of staminate inflorescence × 2/3; **B** staminate flower × 5; **C** stamen × 5; **D** part of pistillate inflorescence × 2/3; **E** pistillate flower × 5; **F** part of infrutescence × 2/3. **A – C** from *Dransfield et al.* JD6732, **D – E** from *Dransfield et al.* JD6374, **F** from *Beentje & Andriampaniry* 4732. Drawn by Rosemary Wise.

Ravenea dransfieldii. Detail of staminate inflorescence (*Dransfield et al.* JD6372).

2.3–2.8 mm, connate for 0–1 mm; flowers foetid or sickly scented, cream-coloured becoming grey-brown, becoming hot at anthesis, not abscising; petals very quickly deliquescing, leaving included fibres. **Staminate flowers** with the calyx connate for 1.8–3 mm, 2.3–2.6 mm across, free lobes triangular, acute, 0.8–1.7 × 0.8–1.2 mm; petals fleshy, free, narrowly ovate, acute, 4.5–6.5 × 1.5–2.5 mm; anthers 3–3.3 × 1–1.3 mm, the filaments 0.2 mm (anther inserted 0.5 mm up the petal when antepetalous) to 0.6–1.6 mm (antesepalous); pistillode 0.3–0.7 × 0.3–0.5 mm. **Pistillate inflorescence** solitary, interfoliar, branching to one order, erect, largely obscured by the leaf sheaths; peduncle c. 26 cm long, proximally 3 × 1.4 cm, distally 2.5–2.8 × 1.3 cm across; prophyll 28 × 9 cm, white, membranous; peduncular bracts almost fleshy, pale cream-coloured when fresh, becoming brown, the proximal small (36.5–49 × 3.7–6.5 cm), pubescent externally, glabrous internally, the upper two large, up to 74 cm long; rachis c. 18 cm long, the rachillae many, tightly packed, appressed-erect, 9.5–27 cm long, 1.5–2 mm across; pedicels 1.5–8 mm, with an 80% connate bracteole 1.5–3.5 mm long. **Pistillate flowers** foetid or sickly-scented, cream-coloured becoming grey-brown, becoming hot at anthesis, not abscising; petals very quickly deliquescing, leaving included fibres; calyx connate for 1 mm, 1.5 mm across, free lobes 1–1.5 × 1.3 mm; petals 4.2–8.5 × 2.3–3.2 mm, c. 1 mm thick; staminodes 6, 1.5 mm long; ovary 3.5–4.2 × 2.3–3 mm. **Fruit** orange (fide Humbert), ovoid or ellipsoid, 15–20 × 12–15 mm, one-seeded; stigmatic remains subbasal. **Seed** ellipsoid, 15 × 10 mm, seedcoat black, 0.2 mm thick, with outer sclerified pericarp-like black layer.

Note. Distinguished from other species with solitary staminate inflorescences branched to two orders by the absence of a petiole, the sinuous veins connecting the longitudinal veins, the three epipetalous stamens and the short pistillate rachis.

Specimens seen. Andapa: Marojejy East, 1500–1700 m, March 1949 (fr.), *Humbert* 23682 (K, P); Maroantsetra: Hiaraka, Oct. 1986 (stam.), *Dransfield et al.* JD6372 (Holotype K; isotypes BH, MO, P, TAN), idem (pist.), *Dransfield et al.* JD6374 (K, TAN); idem, Ampanga River, Oct. 1986 (stam.), *Dransfield et al.* JD6384 (K, P, TAN) and (pist.), *Dransfield et al.* JD6385 (K, TAN); Toamasina: Betampona, Oct. 1991 (stam.), *Beentje* 4486 (BH, K, MO, P, TAN); Ifanadiana: Ambinanindrano, Nov. 1991 (old stam.), *Beentje* 4532 (BH, K, MO, P, TAN); idem, July 1992 (fr.), *Beentje & Andriampaniry* 4732 (BH, K, MO, P, TAN).

14. Ravenea xerophila

One of the rarest *Ravenea* species. This would probably be an excellent species for cultivation, as it grows in very dry sites, and is an extraordinarily beautiful palm. The Latin name means 'dry-loving'.

Distribution. S Madagascar, between Ampanihy and the Ampingaratra Mts.
Habitat. Dry spiny *Didiereaceae/Euphorbia* bush or dry low forest (Humbert), on laterite or gneiss; 200–700 m. May grow in groups.
Local Name. *Ahaza*, *Anivo* (Mahafaly); *Anivona* (Antandroy).
Uses. Leaves used to weave winnowing baskets and hats.

Ravenea xerophila. Tree growing in dry thorn forest, Antanimora.

CONSERVATION STATUS. Endangered. 65 trees with trunks seen, and some 80 seedlings in the two populations known in recent times. The Tranoroa site is being destroyed by overgrazing.

Ravenea xerophila *Jum.*
Ann. Inst. Bot.-Géol. Colon. Marseille sér. 5, 1 (1): 28 (1933); Cat. Pl. Madagascar, Palmae: 25 (1938); Jum. & H. Perrier, Fl. Madagascar 30: 176 (1945); Beentje, Kew Bull. 49: 659 (1994). Type: Madagascar, between Menarandra (Tranasoa) and Manambovo (Ankoba) Rivers, *Perrier* 18653 (Holotype P).

Solitary medium-sized palm. **TRUNK** 1.5–8 m, 13–30 cm diam., widened at the very base; distal half of trunk (or in smaller trees the entire trunk) covered in densely packed sheath remnants; near crown c. 10 cm diam.; bark brown; internodes 2.5–8 cm; nodal scars 0.5–1 cm, occasionally with tattered sheath remnants appearing as threads; wood hard; trunk sometimes producing a yellow latex-like substance when slashed. **LEAVES** (11–) 18–22 per crown, glaucous, gracefully arching, with stiff segments, the old leaves marcescent; sheath 36–55 × 13–20 cm, white- to dark brown tomentose, the tomentum quickly deciduous, the distal part waxy-white; petiole 22–60 cm long, proximally 2.8–6 × 2.0–3.5 cm, distally 1.4–3.0 × 1.4–2 cm, slightly channelled adaxially, with rather sharp edges, proximally red-brown tomentose, distally glabrous and yellow or with whitish bloom when dry, abaxially strongly convex; rachis 1–2.1 m long, in mid-leaf 0.9–2 × 0.8–1.6 cm, pale green with white bloom; leaflets pale green, stiff, coriaceous, the leaflets on opposite sides of the rachis at an angle of 90° with each other,

Ravenea xerophila. **A** leaf base × 1/3; **B** proximal portion of rachis and tip of petiole × 1/2; **C** median portion of rachis × 1/2; **D** leaf tip × 1/2; **E** part of infructescence × 1/3; **F** fruit × 2/3; **G** seed in vertical section × 2/3. **A** from *Beentje & Andriampaniry* 4753, **B** – **D** from *Beentje & Andriampaniry* 4751, **E** – **G** from *Beentje & Andriampaniry* 4754. Drawn by Rosemary Wise.

slightly glaucous, acuminate, 47–55 on each side of the rachis, regular, coriaceous, glaucous, the proximal ones 33–110 × 0.3–1.5 cm, median 43–64 (94) × 1.3–2.4 cm (interval 3.5–4 cm), distal 11–44 × 0.3–1.3 cm, secondary veins very visible, tertiary veins indistinct, ramenta red-brown or grey, elongate, 1–2 (–3) near the base of lower leaflets, to large patches of dense ramenta in the proximal third of the midrib, upper 12 pairs without ramenta. **STAMINATE INFLORESCENCE** solitary, 55–58 cm, branched to 1 order; peduncle 15–19 (or more) cm, distally 5 × 3 mm across; prophyll not seen; peduncular bracts seen 31 cm, 41–62 cm, 44–62 cm; rachis 29–40 cm, with c. 100 branches; rachillae 1.5–7.5 cm, 0.6–1 mm across; pedicels c. 0.2 mm long, 0.7 mm across, with a 1–1.3 mm bracteole. **STAMINATE FLOWERS** with the calyx connate for 0.3 mm, free lobes 1.9 × 0.3 mm; petals 6–7 × 2.4–2.5 mm, connate by the filamental callus of the antesepalous stamens; stamens in 2 series, the antesepalous ones with a filamentous callus 0.6 × 0.4 mm, anthers 2 0.9 mm; antepetalous ones adnate to the petals for c. 1.1 mm, with no free filament, anthers 1.9–2 × 0.9 mm. **PISTILLATE INFLORESCENCE** solitary, 60–80 (or more) cm, branched to 1 order; peduncle 36–40 cm, proximally 8–13 mm across, 6–13 × 3–9 mm across distally, usually with whitish bloom, occasionally puberulous, glabrescent; prophyll not seen; peduncular bracts white to brown- pubescent, (in bud 15, 21, 20–32 × 1 cm, 31–40 × 1 cm), densely white-tomentose, more brown distally, the most distal one inserted at 21 cm from the base of the peduncle; rachis 20–35 cm long; rachillae porrect, 22–41 in number, 5–33 cm long, 1–2.2 mm across, with slightly bulbous base, proximally 4–7 × 2–5 mm, sinuous, more distally zigzag, with spaced flowers; pedicels 0.5–8 mm long. **PISTILLATE FLOWERS** with the calyx connate for 0.9–1.2 mm, with free lobes 1–1.3 mm; petals 4–6 × 1.5 mm, narrowly triangular, acute; staminodes 1.7 mm; ovary broadly ovoid to conical, 2 mm. **FRUIT** 15–22 × 17–27 mm, yellowish, 1-, 2- or 3-seeded, 20–26 mm across in 1-seeded fruit, lobed in the more-seeded ones and then 22–23 × 22–34 mm; stigmatic remains subapical to lateral; endocarp slightly sclerified, loose and rather thin. **SEED** 16–18 × 12–18 mm, globose, hemispherical or in the shape of a third of a sphere, yellow-brown, with white endosperm with small central lacuna.

NOTE. The Humbert collections are from a different habitat than the other collections (i.e. dry forest) and, in some instances, have larger leaflets, longer rachillae and smaller fruits. The protologue mentions an epiphytic orchid found associated with this palm; HB believes the same orchid grows in the population he visited, where it seemed restricted to the leaf sheaths of this palm. It is probably a species of *Cymbidiella* (D. Du Puy, pers. comm.)

SPECIMENS SEEN. Beloha?: between Menandra R. and Manambara R., June 1932 (pist.), *Perrier* 18653 (Holotype P). Amboasary Atsimo: Mandrare valley, Anadabolava, Dec. 1933 (fr.), *Humbert* 12491 (P); Mananara basin, W slopes of mountains between Andohahela and Elakelaka R., between Ampahiso and Mahamaro, early 1934 (fr.), *Humbert* 13698 (P, TAN); Manambolo valley, near Isomono, Dec. 1933 (fr.), *Humbert* 13017 (P). Ampanihy: 3 km W of Saboko near Tsimilofo, March 1988 (ster.), *Willing* 64 (K, MO); idem, Jan. 1990 (ster.), *B. & D. Du Puy* MB 575 (K, MO, P, TAN); idem, Feb. 1992 (ster.), *Beentje* 4590 (K, MO, TAN). Ambovombe: Andalatanosy, Dec. 1992 (dead stam.), *Beentje & Andriampaniry* 4751 (K, MO, P, TAN), (stam.) *Beentje & Andriampaniry* 4753 (K, TAN), (fr.) *Beentje & Andriampaniry* 4754 (K, TAN); 17 km N of Antanimora, March 1992 (dry fr.), *Phillipson* 3991 (TAN). Without locality, 1986 (seed only), *Razafindratsira* s.n. (K).

15. RAVENEA LAKATRA

This species can be recognized immediately by the woody 'steps' on its trunk – actually remains of the leaf sheaths. All woody parts of this species are rock-hard. The scientific name is taken from the local name.

Ravenea lakatra. Mature tree at Manombo (*Beentje* 4523).

Ravenea lakatra. Infructescence (*Beentje* 4523).

DISTRIBUTION. E Madagascar: between Andasibe and Vangaindrano.

HABITAT. Moist lowland forest on slight mid slope or ridge crest; 90–850 m

LOCAL NAME. *Lakatra*; *Tsilanitafika* (cannot be cut down by an army).

USES. At Andasibe extensively used as a source of fibre for weaving high-quality hats, the young leaves being harvested; this prevents the trees from growing, and most populations are pruned to a perpetually juvenile rosette stage.

CONSERVATION STATUS. Endangered. Occurs in a very limited number of sites, several of which are being destroyed rapidly. The population at Mantady consists of pruned rosettes; the two southern populations have less than 20 trunked trees.

Ravenea lakatra *(Jum.) Beentje*
Kew Bull. 49: 662 (1994).

SYNONYM:
Louvelia lakatra Jum., Ann. Inst. Bot.-Géol. Colon. Marseille sér. 4, 5 (1): 50 (1927); Cat. Pl. Madagascar, Palmae: 15 (1938); Jum. & H. Perrier, Fl. Madagascar 30: 164 (1945). Type: Madagascar, Analamazaotra, Anevoka, *Perrier* 12022 (Holotype P).

Moderate-sized palm. **TRUNK** 4.5–14 m, 13–17 cm diam., to 11 cm across near the crown; nodal scars c. 1 cm (to 3.5 cm near crown), internodes 10–16 cm (c. 2 cm near crown), light brown; wood very hard, with black fibre layer; sheath remnants clothing distal part of trunk; short, erect woody sheath base remnants set at nodes over the whole length of trunk, very hard, to 7 cm long; base of crown c. 22 cm across. **LEAVES** 8–10 in the crown, porrect, slightly arching, held on edge in distal part of leaf; sheath 75–80 × 16 cm, with down-pointing fibres on margins, white-tomentose, with black internal fibres; petiole 80–160 cm, proximally 5–6 × 3–6 cm across, distally 4 × 3 cm across, channelled, with very sharp (almost sawtooth) edges, proximally closely denticulate, with white or grey tomentum but glabrescent; rachis 2.3–3.5 m, in mid-leaf 1.2–2.2 × 1.3 cm, with white to grey tomentum but quickly glabrescent; leaflets stiff, in one plane, mid-green, 87–98 on each side of the rachis, the proximal 42–102 × 0.8–3.8 cm, the median 66–77 × 2.7–4.7 cm (interval 3–4 cm), distal 23–49 × 0.7–2.3 cm, few small ramenta near the rachis, but these deciduous, sinuous transverse veinlets clearly visible. **STAMINATE INFLORESCENCE** interfoliar, solitary, branched to 1 order (only seen as dead); peduncle not seen; prophyll not seen; peduncular bracts seen 89, 103 cm; rachis c. 92 cm; rachillae 6–30 cm, 1–1.5 mm across; pedicel 1.2–1.6 mm. **STAMINATE FLOWERS** with the calyx connate for 1 mm, with free lobes 0.8 × 0.2–0.3 mm; petals probably connate for 1.8 mm, free lobes 5–5.5 × 1.3 mm; stamens ? in 2 series, the epipetalous ones adnate for/inserted at 2.2–2.5 mm, the filaments absent, the anthers 1.5 × 0.5 mm. **PISTILLATE INFLORESCENCE** solitary, erect, interfoliar among old leaf bases in fruit, branched to 1 order; peduncle 82–130 cm, proximally 3.5–4 × 1.5–2 cm, distally 2.5 × 1.5 cm, densely white-brown pubescent; prophyll 10–25 × 5 cm, tattering, white, membranous; peduncular bracts 20–43 cm, 30–40 cm (inserted at 22–24 cm from the base of the peduncle), 140 cm (inserted at 48 cm), 88–119 cm (inserted at 60–87 cm), all densely white- to brown-pubescent abaxially, glabrous adaxially; rachis 44–86 cm, white-tomentose, with 50–70 porrect rachillae; rachillae 7–65 cm, 2 mm across, sinuous, proximally 11 × 5 mm; pedicels 3–6 mm; bracteole 1.5 × 1.5 mm, connate for up to 1 mm. **PISTILLATE FLOWERS** with the calyx connate for 2–2.5 mm, 3.5 mm across, the free lobes 0.8–2 × 0.9–1.2 mm; petals 5–8 × 2.5–5.5 mm. **FRUIT** blackish, slightly depressed

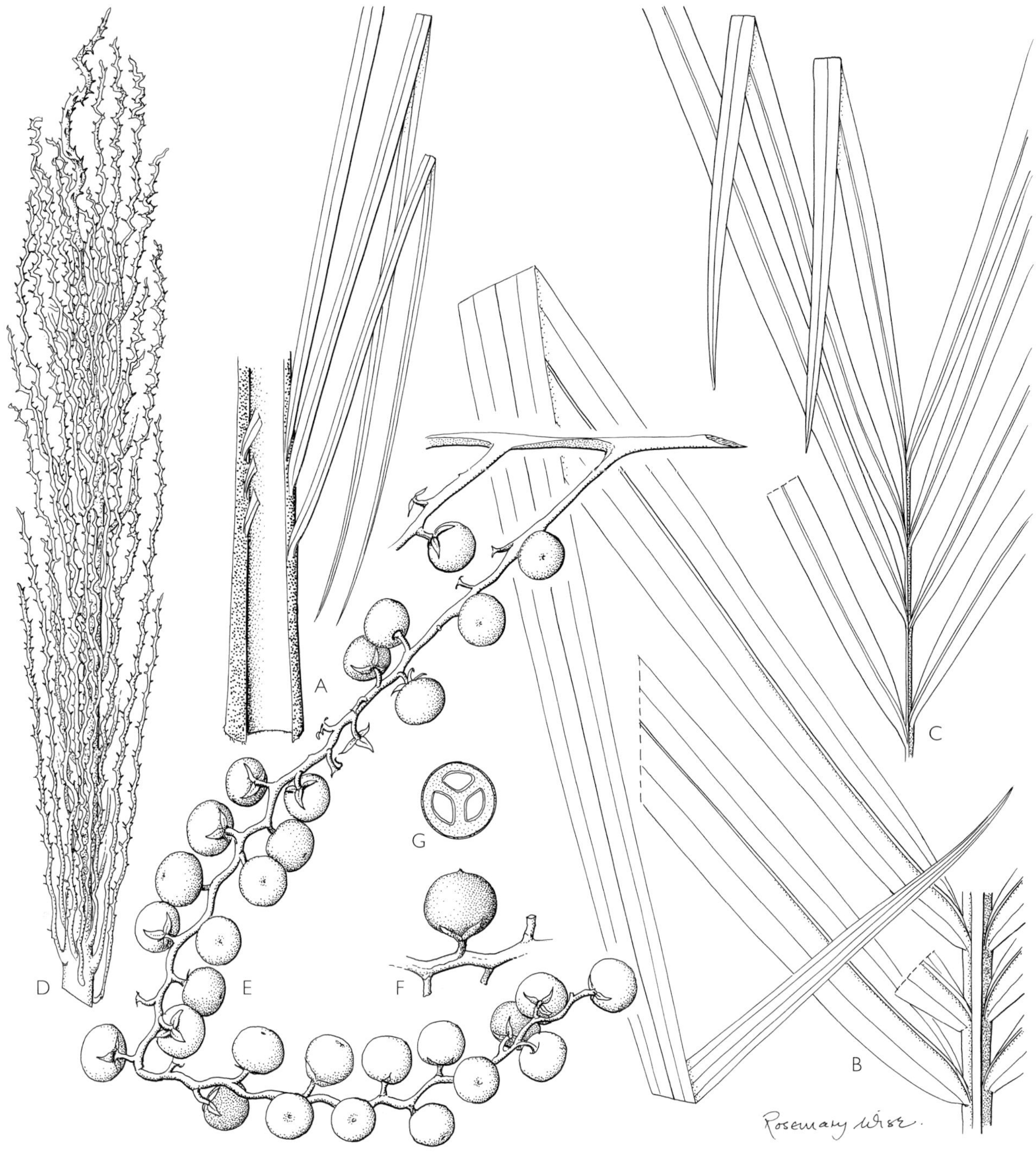

Ravenea lakatra. **A** proximal portion of rachis and tip of petiole × 1/4; **B** median portion of rachis × 1/2; **C** leaf tip × 1/2; **D** tip of staminate inflorescence × 1/2; **E** part of infructescence × 1/2; **F** part of rachilla and one fruit × 2/3; **G** fruit in cross section × 2/3. **A**, **B**, **D** from *Beentje & Andriampaniry* 4793, **C**, **E**, **F** from *Beentje* 4523. Drawn by Rosemary Wise.

globose, 15–20 × 18–21 mm, with terminal stigmatic remains, 1-/2-/3-seeded; each seed with sclerified layer 8–11 mm and 6–10.6 mm across with sharp distal acumen 2.5–3 mm, the layer 0.3–0.35 mm thick. **Seed** 9–10 mm, 5–10 mm across, with distal acumen of 1.5 mm; seedcoat black, 0.2 mm thick.

Note. The type and the material found much further south agree perfectly as for the leaf and inflorescence details. The Flora (Jumelle & Perrier 1945), but not the protologue, describes the tree as being 15–30 m high, and with a diameter of 15–30 cm, while the wood is described as white and soft. We treat these later additions with suspicion.

The seed is unique within the genus in having a sharp pointed apex.

Specimens seen. Moramanga: Andasibe, Nov. 1986 (juv.), *Dransfield et al.* JD6424 (K, P, TAN); Anevoka, (prob. Dec. 1914) (fr.), *Perrier* 12022 (Holotype P). Manakara: 39 km NNE of Manakara, July 1992 (fr.), *Beentje & Andriampaniry* 4726 (BH, K, MO, P, TAN). Farafangana: Manombo, Nov. 1991 (yfr.), *Beentje* 4523 (BH, K, MO, P, TAN); idem, Jan. 1993 (old stam.), *Beentje & Andriampaniry* 4793 (K, TAN).

16. RAVENEA ALBICANS

This tree is immediately recognizable by the white undersides of the leaflets, and the 'zebra-striping' on the leaf rachis. Despite these outstanding characters, it remained uncollected for about seventy years after the first collection by Perrier. The Latin name refers to the white leaflet undersides.

Ravenea albicans
(**x**: sight records)

DISTRIBUTION. NE Madagascar; the protologue says common between Fenerive (Fenoarivo) and Antalaha, but does not cite any specimens for this distribution. The palm is certainly not common nowadays.
HABITAT. Moist forest at 100–400 m alt.; steep mid slope on ultramafic soil.
LOCAL NAME. *Hoza-tsiketra* (Betsimisaraka)
USES. Heart edible.
CONSERVATION STATUS. Endangered. Only known from two recent localities, in one of which where we counted 40 trees with trunks, about a hundred young ones without trunks, and several seedlings. The distribution area was very limited. The second locality only came to light when the book was going to press.

Ravenea albicans. Detail of crown, showing whitish undersides of leaflets and staminate inflorescence, Makiraha (Photo L. Holloway).

Ravenea albicans. Close-up of petiole to show pattern of scales (*Beentje et al.* 4643).

Ravenea albicans *(Jum.) Beentje*
Kew Bull. 49: 663 (1994).

SYNONYM:
Louvelia albicans Jum., Ann. Inst. Bot.-Géol. Colon. Marseille sér. 5, 1 (1): 5 (1933); Cat. Pl. Madagascar, Palmae: 15 (1938); Jum. & H. Perrier, Fl. Madagascar 30: 162 (1945). Pollen illustrated in Genera Palmarum (1987): 27, Plate 1.7: H. Type: Madagascar, Masoala, *Perrier* 11939 (Holotype P).

Moderate-sized undergrowth palm. **TRUNK** 3–9 m, up to 11 cm diam., covered in marcescent, litter-trapping sheath bases. **LEAVES** 8–14 in the crown, porrect, 3–4 m long; sheath c. 72 cm, with brown-velvety tomentum over green; petiole 0–34 cm, proximally about 5 × 4 cm across, distally 3.7 × 3.3 cm across, channelled; rachis c. 3.7 m long, in mid-leaf 2–2.2 cm wide with 8 mm wide keel, with transverse (zebra) striping especially abaxially; leaflets stiff, in one plane, 45–48 on each side of the rachis, the proximal 62–65 × 2.2–2.5 cm, median 75–93 × 6.6–8 cm (the protologue gives 50–60 × 2–2.5, but in the type the apices are broken off and the leaflets are folded) (interval 4.5–5 cm), distal 20–25 × 0.8–1.5 cm, abaxially with continuous white tomentum. **STAMINATE INFLORESCENCES** solitary, interfoliar, branching to 1 order, hardly exserted, mostly hidden among sheaths and leaf remnants; peduncle distally 12 × 6 mm, dense white- puberulous; with 3(–4?) bracts, the first being destroyed at anthesis, the ones seen 25 cm, 30 cm, 32–35 cm; rachillae 5–8 cm long, 2.5 mm across. **STAMINATE FLOWERS** with the calyx reduced to 3 small triangular teeth < 2 mm; petals elongate, triangular and acuminate, 5.5–6.5 mm, free or hardly connate at the base; episepalous stamens with large filament, appearing free among the spread

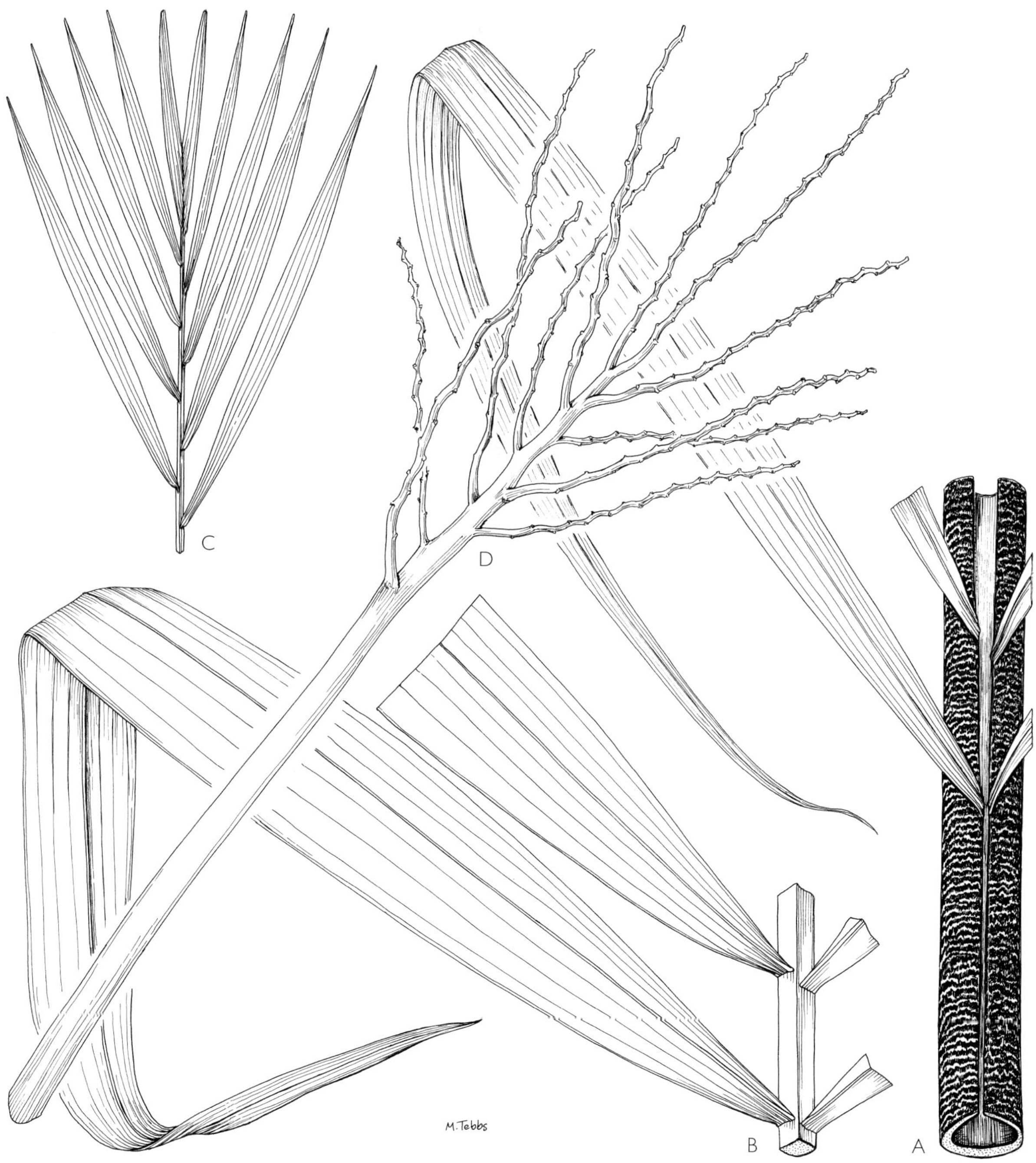

Ravenea albicans. **A** distal part of petiole, showing scales in bands and proximal leaflets × 1/2; **B** mid-portion of leaf, adaxial view × 1/3; **C** leaf tip × 1/5; **D** old pistillate inflorescence × 2/5. **A** – **C** from *Beentje & Dransfield* 4812, **D** from *Beentje et al.* 4643. Drawn by Margaret Tebbs.

petals; epipetalous stamens with filament connate to inner side of petals, the anthers therefore appearing sessile. **PISTILLATE INFLORESCENCES** solitary, interfoliar, erect, 126–150 cm, branched to 1 order; peduncle 94–120 cm, proximally 2.5–3.4 × 1.2–1.7 cm in diam., distally 1.5–2.7 × 0.8–1.3 cm in diam.; prophyll c. 18 × 7.5 cm; peduncular bracts seen 27 cm, 34 cm, 46 cm; rachis 8–13 cm; rachillae 8–15 in number, 16–29 cm long; pedicels 1–2 mm. **PISTILLATE FLOWERS** unknown. **FRUIT** unknown. **EOPHYLL** bifid, with the characteristic white abaxial surface of the leaflets.

NOTE. We believe that this species only flowers and fruits at intervals, i.e. not every year, since all the trees we checked in Mananara Biosphere Reserve over a period of two years were sterile, with only ancient inflorescences.

SPECIMENS SEEN. Masoala, presumably near Antalaha, no date (stam.), *Perrier* 11939 (Holotype P); Mananara: W of Antanambe, April 1992 (old pist.), *Beentje et al.* 4630 (K), 4643 (K, MO, P, TAN); idem (seedling), *Beentje et al.* 4644 (K); idem, Oct. 1994 (dead pist.), *Beentje & Dransfield* 4812 (K, TAN), and (seedl.), 4814 (K, TAN).
SIGHT RECORD. Maroantsetra: Makira(ha), 1955, *L. Holloway* (see photo).

17. RAVENEA LOUVELII

This is one of the strangest *Ravenea* species, with an archaic appearance. The inflorescences are almost hidden among the leaf bases, and are probably beetle-pollinated. The species is named for M. Louvel, head of the Forestry Department at Analamazaotra in the 1920s.

DISTRIBUTION. E Madagascar, only known from Andasibe.

Ravenea louvelii

Ravenea louvelii, showing litter-trapping habit and persistent leaf bases (*Dransfield et al.* JD6485).

HABITAT. Moist forest, steep slope near crest; 800–1000 m.
LOCAL NAME. *Lakamarefo, Siraboto.*
USES. Not recorded.
CONSERVATION STATUS. Endangered. Only known from a single site, where it grows outside the protected area. Population numbers are low; we have counted fewer than 25 trees with trunks, and rejuvenation seems minimal.

Ravenea louvelii *Beentje*
Kew Bull. 49: 664 (1994).

SYNONYM:
Louvelia madagascariensis Jum. & H. Perrier, Compt. Rend. Acad. Paris 155: 411 (1912); Ann. Inst. Bot.-Géol. Colon. Marseille sér. 3, 1 (1): 58, t. 33 (1913); Becc., Palme del Madagascar 53, fig. 43, tav. 49 (1914); Jum., Ann. Inst. Bot.-Géol. Colon. Marseille sér. 4, 5 (1): 47 (1927); Cat. Pl. Madagascar, Palmae: 16 (1938); Jum. &

Ravenea louvelii, growing on a ridge top near Andasibe.

H. Perrier, Fl. Madagascar 30: 165, fig. 46 (1945). Type: none indicated, locality Analamazaotra, "Lakamarefo", alt. 800 m (probably *Perrier* 12021 (Holotype P) which has "type" on the label, written by Jumelle).

Short stocky palm of the forest undergrowth. **TRUNK** 1.5–3 m, 15–20 cm diam. without leaf sheaths, c. 50 cm with the sheath bases; distal 1.5 m of trunk with sheath remnants, proximal part only with sheath bases; wood hard; internodes c. 3 cm. **LEAVES** 9–14 in crown, robust, porrect, stiff, with up to 15 marcescent leaves present; sheath 46–60 × 6.5–8 cm, channelled adaxially, rounded abaxially, swollen at base, with dense soft pale brown tomentum, turning fibrous with age, with long stiff brown marginal fibres; petiole 50–130 cm long, proximally 3–5.5 × 2–3.5 cm across, distally 2–3 × 1.2–1.9 cm across, especially abaxially covered in rich brown scales, with very sharp margins, adaxially channelled with the central part flat or with central ridge, abaxially rounded; rachis 2.2–3 m long, in mid-leaf c. 19 × 11 mm across, adaxially keeled with up to 9 mm wide flat keel top and abaxially flat, with sharp upper edges, especially abaxially covered in brown scales, continuing past the distal pair of leaflets for 10–21.5 cm; leaflets in one plane, slightly pendulous in distal part of leaflet, dark green, attenuate, 80–104 on each side of the rachis, the proximal ones 28–42 × 0.3–1 cm, median 47–67 × 1.4–2.6 cm (interval 1.5–2.5 cm), distal 10–40 × 0.4–1.8 cm, with some ramenta near the very base of the leaflet, or in young leaves

Ravenea louvelii. Infructescence (*Beentje* 4536).

Ravenea louvelii. Staminate inflorescence (*Dransfield et al.* JD6485).

with brown scales along the margins and apex, the midrib and the base, venation faint but visible. **STAMINATE INFLORESCENCE** hidden among the leaf bases, solitary, 30–40 cm long, branched to 1 order; peduncle 34–36.5 cm, proximally c. 8 × 6 mm across, scaly; prophyll c. 10 cm long, cream-coloured; peduncular bracts purplish, 12–15 cm, c. 20 cm, 36–38.5 cm, c. 40 cm, the most distal beaked, completely sheathing the inflorescence, covered in brown stellate scales; rachis 21–25 cm, with c. 80 crowded rachillae; proximal rachis bracts 4–12 × 1–3 mm; rachillae 1–7 cm long and 1–2.5 mm across, glabrous; pedicels 0.2–1 mm, 0.8–1 mm across; bracteoles 1.3–1.5 × 0.8 mm, connate for 0.4–0.7 mm. **STAMINATE FLOWERS** closely set, musty-scented, cream-coloured; calyx connate for 0.6–1 mm, 2.8 mm across, the free lobes 1–1.2 × 0.7–0.8 mm; petals 6.5–7 × 2.4–2.7 mm, ovate, acute, connate for 1 mm at the base by the fleshy base of the antesepalous filaments, deliquescent after anthesis; anthers 3.2–4 × 1–1.3 mm, the antesepalous basifixed with a fleshy filament of 1–2 × 1.5–2 mm, the antesepalous ones dorsifixed and inserted 2 mm up the petal; pollen white; pistillode 1.3 mm, 0.9–1 mm across, white. **PISTILLATE INFLORESCENCE** interfoliar, hidden among the leafbases, solitary, 30–40 cm long, branching to one order; peduncle c. 15 cm; prophyll c. 10 × 5 cm, cream-coloured; peduncular bracts purplish, 12–15 cm, 20 × 5 cm, 29–38 × 5.5 cm, 40 cm, the most distal beaked, completely sheathing the inflorescence, covered in brown scales; rachis c. 18 cm, with c. 30 spreading to erect rachillae, 2.5–7 cm long, 3.5–5.5 mm across, with lines of red-brown scales; rachis bracts 4 × 1 mm; pedicels 0.5–2 mm, 2.5–4 mm across; bracteoles 1.5–4 × 1–2 mm, connate with the pedicel for 0.5–1 mm. **PISTILLATE FLOWERS** slightly musty-scented, cream-coloured; calyx connate for 0.6–1 mm, 2.8–5 mm across, the free lobes 0.8–2 × 2.6–3 mm; petals 6.5–9.5 × 1.7–2.2 mm, fleshy, 1.5 mm thick, connate by 1 mm through the fleshy base of the antesepalous filaments; staminodes 6, 2.8–3.2 × 1.6–1.7 mm; ovary ovoid, 3–3.8 × 3 mm. **FRUIT** dark purple (protologue says yellowish at maturity), somewhat depressed globose, 13–14 × 15–20 mm (25 mm according to Beccari), 1–2-seeded; stigmatic remains subapical or terminal. **SEED** ovoid, sometimes flat on 1 face, 9–13 × 6–9 mm; seed coat black, 0.2 mm thick. **GERMINATION** remote, eophyll pinnate.

SPECIMENS SEEN. Moramanga: Analamazaotra, Feb. 1912 (stam., fr.), *Perrier* 12021 (Holotype P); idem, Nov. 1986, (stam.), *Dransfield et al.* JD6418 (K, P, TAN) and (pist.) *Dransfield et al.* JD6419 (K, TAN) and (seedling) *Dransfield et al.* JD6420 (K, TAN); idem, March 1988 (stam.), *Dransfield et al.* JD6485 (K, TAN) and (pist.) *Dransfield et al.* JD6486 (K, TAN); idem, Dec. 1991 (fr.), *Beentje* 4536 (BH, K, MO, P, TAN); idem, Aug. 1992 (ster.), *Beentje & Andriampaniry* 4745 (K, BH); idem, Dec. 1992 (dead stam.), *Beentje & Andriampaniry* 4769 (K, TAN).

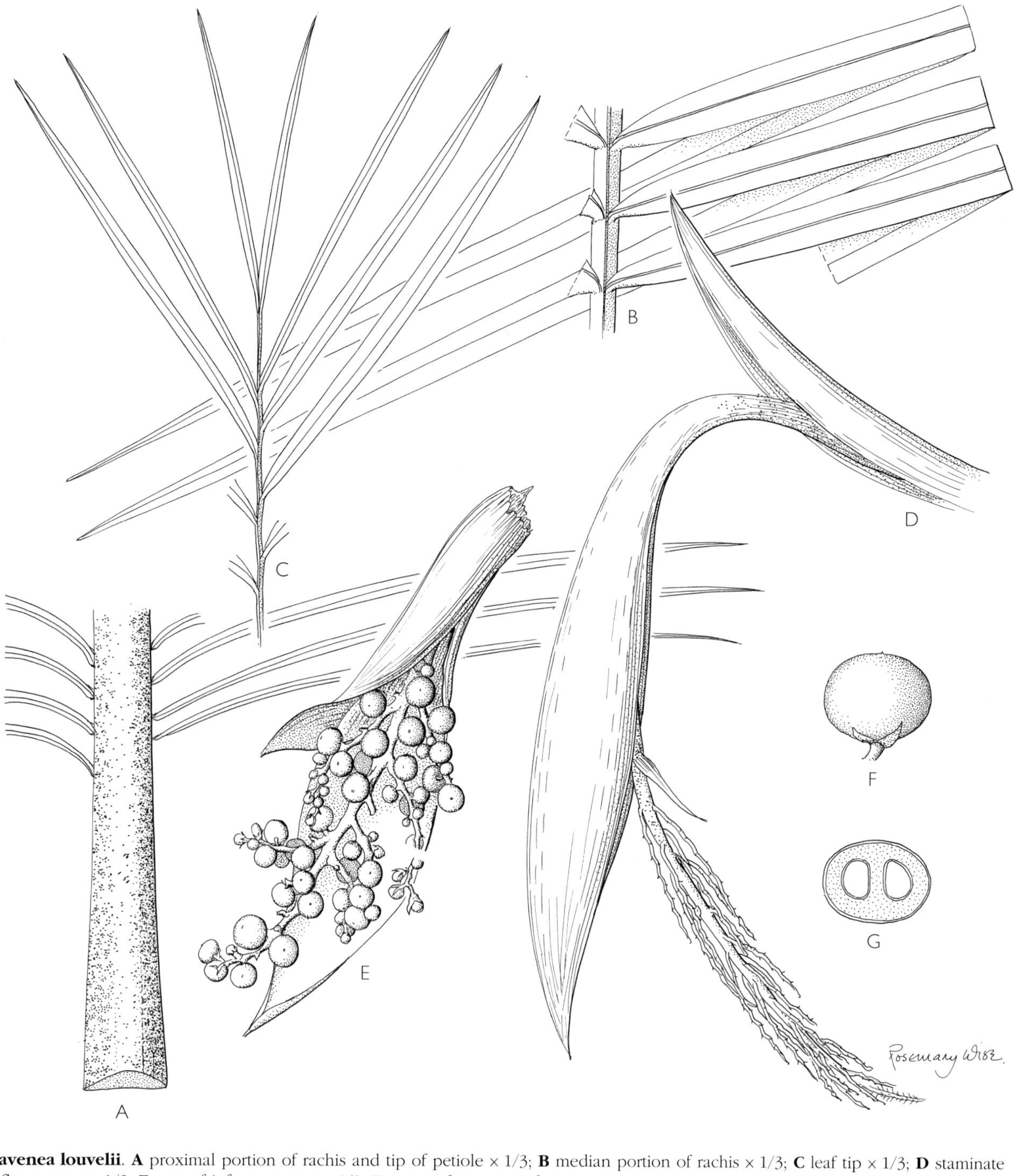

Ravenea louvelii. **A** proximal portion of rachis and tip of petiole × 1/3; **B** median portion of rachis × 1/3; **C** leaf tip × 1/3; **D** staminate inflorescence × 1/3; **E** part of infructescence × 1/3; **F** mature fruit × 1; **G** fruit in cross section × 1. **A – C**, **E – G** from *Beentje* 4536, **D** from *Dransfield et al.* JD6485. Drawn by Rosemary Wise.

SUBFAMILY ARECOIDEAE
TRIBE ARECEAE
SUBTRIBE ORANIINAE

VIII. ORANIA

Zipp. in Blume, Algemene Konst- en Letter-bode 1: 297 (1829). Type: *O. regalis* Zipp.

SYNONYMS:
Sindroa Jum., Ann. Inst. Bot.-Géol. Colon. Marseille sér. 5, 1: 11 (1933). Type: *S. longisquama* Jum. [= *O. longisquama* (Jum.) J. Dransf. & N.W. Uhl]

Halmoorea J. Dransf. & N.W. Uhl, Principes 28, 4: 164 (1984). Type: *H. trispatha* J. Dransf. & N.W. Uhl [= *O. trispatha* (J. Dransf. & N.W. Uhl) Beentje & J. Dransf.]

Solitary, pleonanthic monoecious palms. **STEM** often conspicuously ringed with leaf scars. **LEAVES** spiral or less often distichous, pinnate, deciduous under their own weight; crownshaft not present; sheath well-developed, thick and woody, splitting longitudinally opposite the petiole, usually densely tomentose, fibrous along the margins; petiole usually short, channelled adaxially, rounded abaxially; rachis much longer than the petiole, curved, abaxially rounded, adaxially angled but channelled near base; leaflets single-fold, regularly arranged and held in one plane, apices praemorse, abaxial surface with dense white indumentum and occasionally with ramenta along the main vein, midrib prominent adaxially, marginal nerves occasionally prominent, transverse veinlets obscure or occasionally conspicuous. **INFLORESCENCES** axillary, solitary, interfoliar, branching to 1–3 orders, apparently protandrous; peduncle stout, circular in cross-section, covered in caducous brown scales; prophyll persistent, short, tubular, 2-keeled, included within the subtending leaf, usually becoming frayed distally; peduncular bract one, rarely (but often in Madagascar species) two, borne just above the prophyll, very large and conspicuous, almost woody, tubular, often inflated, completely enclosing the inflorescence in bud, splitting along its length to expose the inflorescence, with a solid, flattened, lanceolate beak, tomentose, eventually deciduous; rachis longer than the peduncle, brown-tomentose, bearing spirally arranged low, glabrous, coriaceous, collar-like bracts subtending first-order branches; first-order branches bearing few, spirally arranged, similar bracts, each subtending a 2nd order branch; first-order branches often with a basal pulvinus; further branches subtended by an inconspicuous triangular bract; rachillae spreading, flexuous, bearing rather distant triads proximally and solitary or paired staminate flowers distally; triads superficial, subtended by a minute triangular rachilla bract; floral bracteoles minute or not visible. Staminate and pistillate **FLOWERS** superficially rather similar, cream-coloured. **STAMINATE FLOWERS** asymmetrical, narrower than the pistillate; calyx very short, flattened, partially connate with 3 low triangular lobes; petals 3, fleshy, distinct, valvate, thickened at the apex; stamens 3, 6 or (in Madagascar always, in Malesia sometimes) 9–30 , filaments distinct, short to moderate, rather fleshy, anthers elongate, basifixed, erect, with large connective, extrorse or latrorse; pollen elliptic or circular, monosulcate, with finely reticulate or finely rugulate, tectate or semi-tectate exine; pistillode usually lacking. **PISTILLATE FLOWERS** conical or pyramidal; calyx flattened, short, with 3 low, triangular lobes; petals 3, valvate, distinct, triangular; staminodes 3–12 (in Madagascar 9–12), very short, awl-shaped or well-developed; gynoecium trilocular, triovulate, pyramidal, stigmas 3, short, recurved at anthesis, ovule laterally attached, pendulous, probably hemianatropous. **FRUIT** developing from 1, 2 or 3 carpels, spherical or nearly so (but if developing from more than 1 carpel, bi- or tri-lobed), stigmatic remains sub-basal; epicarp smooth, mesocarp thin or thick, fleshy, traversed by numerous short radial fibers, endocarp rather thin, with a heart-shaped button. **SEED** spherical, basally attached with a circular hilum, the surface of the seed somewhat grooved by a sparse network of fibers, endosperm homogeneous, sometimes with a very small central hollow; embryo subapical or lateral. **GERMINATION** remote-tubular; eophyll bifid with praemorse apices, or pinnate (in *O. trispatha*).

Orania longisquama, growing at Amby near Manakara (*Beentje & Andriampaniry* 4678).

DISTRIBUTION. About nineteen species, three endemic to Madagascar, the others found in Malesia, with most species in New Guinea.

Seeds of all three species of *Orania* in Madagascar are difficult to distinguish, differing only in size (which itself is variable). Young plants should be distinguishable by their leaf arrangement and coloration. However, juveniles of *O. longisquama* can appear to be distichous at an early stage.

KEY TO ORANIA

1. Leaves distichous **2**
   Leaves spiral, inflorescence branching to 3 orders .... 1. **O. longisquama**
2. Inflorescence branching to 2 orders; leaflets with red-brown ramenta .... 2. **O. ravaka**
   Inflorescence branched to 3 (–4) orders; leaflets with pale grey ramenta .... 3. **O. trispatha**

1. ORANIA LONGISQUAMA

From a distance this palm looks slightly like *Ravenea madagascariensis*, but on closer inspection can immediately be distinguished by the praemorse leaflet tips. This species can be quite variable in size, from a compact, small palm to a canopy tree.

DISTRIBUTION. NW and E Madagascar.

HABITAT. Lowland rain forest; slope base, mid slope or ridge-crest; 40–550 m.

LOCAL NAMES. *Sindro, Anivona, Ovobolafotsy* (Betsimisaraka), *Vakapasy* (Antaimoro);

USES. Palm heart not edible or even said to be poisonous; HB has drunk some fruit sap without any ill effects. Wood used for house walls.

CONSERVATION STATUS. Rare. Fairly widespread, though nowhere really common.

Orania longisquama (Jum.) J. Dransf. & N.W. Uhl
Principes 28 (4): 164.

SYNONYM:
Sindroa longisquama Jum., Ann. Mus. Colon. Marseille sér. 5, 1 (1): 11 (1933); Cat. Pl. Madagascar, Palmae: 25 (1938); Jum. & H. Perrier, Fl. Madagascar 30: 160, fig. 45 (1945). Type: Madagascar, Masoala Peninsula, Marambo, November 1912, *Perrier* 11937 (Holotype P).

Solitary palm. **TRUNK** 4–20 m high, 10–25 cm diam., often with surface roots and a basal boss to 40 cm diam.; internodes 5–10 cm long, near the crown 1–2 cm long, brown, often with a reddish-green indumentum; nodal scars obvious or obscure, 1–6 cm high. Crownshaft absent. Wood yellowish, hard through a dense ring of reddish fibres just inside the bark. **LEAVES** spirally inserted, (6–7 fide Perrier) 8–15 (rarely with a few marcescent leaves), arching, slightly held on edge in their distal part; base of crown bulbous, 16–30 cm across; sheath 26–40 cm long, green with white indumentum turning pale brown to rusty on exposure, interface with petiole continuous, with a few marginal fibres; sheath and petiole together 80–103 cm; apparent petiole 33–120 cm long, 3–5.5 × 2.5–4 cm diam., channelled, (pale) red-brown, with a pale dense

Orania longisquama. **A** petiole and basal leaflets × 1/2; **B** mid section of leaf × 1/2; **C** leaf tip × 1/2; **D** first order branch of inflorescence × 1/5; **E** staminate flower, one petal removed × 4; **F** part of infructescence in young fruit × 1/3. **A** from *Dransfield et al.* JD6375, **B** – **F** from *Beentje & Andriampaniry* 4678. Drawn by Rosemary Wise.

indumentum and fibrous margins; rachis 1.3–2 m long, channelled in the proximal part, in mid-leaf 2–3 x 1–1.3 cm and keeled, white- to reddish-brown-pubescent or with scattered pale scales or with white bloom; leaflets regular, 47–65 on each side of the rachis, stiff or not very stiff, those on opposite sides of the rachis at an angle of 60–120 with each other, the proximal 38–84 x 0.4–4.5 cm, median 56–88 x 1.7–5 cm (interval 1.5–4 cm), distal 3.5–32 x 0.5–2.7 cm, mid-green, with yellow veins adaxially, with a thin whitish covering abaxially, main vein 1, with large (15–25 mm long) red-brown to whitish ramenta in patches to almost continuous along the midrib (rarely absent, HB 4730), with lines of scattered scales along the minor veins, bases sometimes thickened (HB 4616), apices praemorse, distal pair connate for 0.5–5 cm. **INFLORESCENCE** interfoliar, branching to 3 orders, erect to arching, with spreading branches (to almost pendulous in fruit); peduncle 25–50 cm long, proximally 3.5–10 x 2–6 cm diam., green with much red to brown pubescence; prophyll 26–45 cm long, 6–12 cm wide, borne 4–5 cm above the base of the peduncle, split abaxially, with toothed margins, green, brown-lepidote; peduncular bracts one or two (one in HB 4706, 4730, two in HB 4616, JD 6479), quickly deciduous, inserted at respectively 7–16 and 13–21 cm from the base of the peduncle, each 55–72 cm long and 6–14 cm wide, beaked for 6–12 cm, thin red-pubescent; rachis 58–69 cm long, zigzag, with scattered brown scales, with 12–22 branched and 6–12 unbranched first order branches, the proximal of these with an axis of up to 35 cm long and 1.8–4 x 0.6–2.2 cm diam., with up to 20 second order branches; rachillae pale green, turning blushing pink in fruit, 8–36 cm long (possibly lengthening in fruit stage), 3–7 mm diam., glabrous or with scattered red scales; triads distant. **STAMINATE FLOWERS** with the sepals connate for 0.6–0.8 mm, free for 1.6–2.2 mm, 1.3–1.8 mm wide; petals 5.5–9.5 x 3–4.5 mm; stamens 12–20 (possibly in three antepetalous bundles–sometimes falling off in equal groups, with the removal of the petal), filaments 0.8–1.5 mm long with slightly bulbous base, anthers almost basifixed, 3–6 x 0.7–1 mm, some unequally sagittate, others equally sagittate; pistillode not seen. **PISTILLATE FLOWERS** with the sepals 1.3–3.2 x 2.2–5.2 mm; petals 3.5–7 x 3.6–6.7 mm, fleshy; staminodes 8–10, 1–2.5 mm long, narrowly triangular; ovary 3–9 mm high, 3.2–14 mm diam., with three styles 0.5–1 mm long and papillose or pubescent. **FRUIT** green, globose or obovoid, occasionally (as in JD 6375) 2- or 3-lobed, or with two abortive lobes basally, 4–5.5 x 3–4.5 cm. **SEED** globose with a basal conical bump, 3.1–4.4 x 3.2–4.2 cm, with large central lacuna when almost mature, this filled with sweet to slightly bitter fluid, later turning to homogeneous endosperm. **GERMINATING SEEDS** with elongate white cotyledonary petiole.

Orania longisquama

NOTE. The leaf sheath sometimes splits to leave a tongue-like structure, reaching up to the proximal leaflets, producing an apparent petiole.

SPECIMENS SEEN. Analalava: NNE of Maromandia, July 1992 (fl., y.fr.), *Beentje & Andriampaniry* 4706 (BH, K, MO, P, TAN). Antalaha: Marambo, Nov. 1912 (fl.), *Perrier* 11937 (type; P); Mahavinitra near Ambatobe, Dec. 1972 (fr.), *Moore* 10118 (P, TAN). Maroantsetra: Hiaraka, Oct. 1986 (y.fr.), *Dransfield et al.* JD6375 (K, TAN); idem, April 1989 (fr.), *Du Puy & Du Puy* MB 152 (K, TAN); Antalavia, Feb. 1988 (bud), *Dransfield et al.* JD6479 (K, P, TAN). Mananara Avaratra: Antanambe, Oct. 1991 (fr.), *Beentje* 4455 (BH, K, MO, P, TAN); idem, April 1992 (bud), *Beentje et al.* 4616 (K, TAN). Toamasina: Betampona, Oct. 1991 (fr.), *Beentje* 4493 (BH, K, MO, P, TAN). Ifanadiana: Ambinanindrano, July 1992 (fl., fr.), *Beentje & Andriampaniry* 4730 (BH, K, MO, P, TAN). Manakara: Amby, May 1992 (fl., fr.), *Beentje & Andriampaniry* 4678 (BH, K, MO, P, TAN).

Orania longisquama. Crown, showing inflorescence and fruit (*Beentje & Andriampaniry* 4678).

2. ORANIA TRISPATHA

A majestic distichous palm of great beauty, and the largest member of the genus in Madagascar. The trunk is frequently swollen at the base. Because of its striking distichous leaves this species has been much sought after by enthusiasts.

DISTRIBUTION. E Madagascar.

HABITAT. Lowland rain forest; flat ground near streams, swamp edges; 50–400 m.

LOCAL NAMES. *Sindro* or *Sindroa* (Betsimisaraka); *Anivo* (Antaisaka).

USES. Palm heart not edible. Wood used in hut construction (Manombo).

CONSERVATION STATUS. Critical. Known only from

three sites, all of which are under threat of destruction, with very low population numbers (less than fifty trees are known).

Orania trispatha

Orania trispatha *(J. Dransf. & N.W. Uhl) Beentje & J. Dransf.* **comb. nov.**

SYNONYM:
Halmoorea trispatha J. Dransf. & N.W. Uhl, Principes 28 (4): 166 (1984). Type: Madagascar, Masoala, across Onive R. from Ambatobe and upriver from Ambohitralanana, 8 April 1971, *Moore* 9921 (Holotype BH, not seen; isotype P).

Solitary palm. **TRUNK** 20–22 m, 20–35 cm diam.; root boss 30 cm high, 40–60 cm diam.; internodes 12–15 cm, light brown or grey; nodal scars 5 cm; wood hard. Crownshaft absent. **LEAVES** distichous, forming a fan, 10–12 in the crown, held on edge in their distal part; sheath c. 60 cm long, without ligules or markings, green with brown margins, grading into petiole, with rusty brown indumentum and white wax; petiole 75–200 cm long, channelled with sharp edges, 2.3–7 x 2.9–6 cm diam., densely grey-brown pubescent; rachis 2–2.3 m long, in mid-leaf 2.6–3 cm wide, keeled; leaflets regular, 60–65 on each side of the rachis, those on opposite sides of the rachis at an angle of c. 150°, the proximal 58–93 x 0.6–4 cm, median 65–99 x 3–9.5 cm (interval 4–5 cm), distal 10–53 x 0.5–2 cm, distal pair hardly connate, discolorous, abaxially with a thin white covering, with minute brown scales, the midrib with many pale grey ramenta to 5 cm long, main veins 3–5, also with large pale ramenta, apices obliquely praemorse. **INFLORESCENCE** interfoliar, branched to 3–4 orders, spreading; peduncle 16–25 cm long, 5.5–6 cm diam., slightly flattened, densely red- or brown-pubescent; prophyll woody, 30–40 cm long, 6.5–10 cm wide, borne at c. 5 cm above the base of the peduncle, brown-pubescent; peduncular bracts two, the first inserted at 3–10 cm from the base of the peduncle, 60–80 cm long with a beak of 8 cm, the second inserted at 7–18 cm from the base of the peduncle, 58 cm long with a beak of 7 cm, both woody, inflated, terete, green with brown scales, inflated, deciduous in the fruiting stage; rachis 30–40 cm long, with 8–12 branched first order branches, these proximally 3 x 1 cm diam.; rachillae 15–46 cm long, 6–7 mm diam., zigzag, reddish-scaly to glabrous; triads distant. **STAMINATE FLOWERS** asymmetrical, fleshy; sepals 1.5–4 x 4 mm, petals 9 x 5–9 mm; stamens 27–30, basifixed, latrorse, the filaments very short, the anthers c. 6 x 1 mm; pistillode absent. **PISTILLATE FLOWERS** asymmetrical; sepals 5 x 5 mm, petals 17–33 x 12–15 mm (in fruit); staminodes 11–12, thread-like, c. 2 mm; ovary 3-lobed, c. 5 x 8 mm, with papillose stigmas to 1.5 mm. **FRUIT** green, when slightly immature semi-globose or obovoid and 3.9–4.5 cm diam., but more often 2-lobed or 3-lobed and 5–5.5 cm high and 5–8 cm diam., when immature containing a bitter-tasting fluid solidifying into a homogeneous endosperm; stigmatic remains sub-basal. **EOPHYLL** pinnate (*Beentje* 4522).

NOTE. The protologue states the pinnae are white-waxy on the adaxial surface; this should be the abaxial surface. The generic distinction could not be upheld, when we discovered that *Orania longisquama* may have either one or two peduncular bracts; with more material available, and with *O. ravaka* being described here, the number of stamens does not remain a generic character, but rather a specific one; the number of staminodes for *O. ravaka* encompasses the range of both *O. trispatha* and *O. longisquama*. The calyx in all taxa seems to be connate to varying degrees at its very base.

The specimen from Manombo (about 800 km S of the Masoala/ Mananara populations) has a different 'feel' to the leaf, with hardly any surface covering of a whitish substance, and fewer little red scales.

SPECIMENS SEEN. Antalaha: Ambatobe, April 1971 (fl.), *Moore* 9921 (type, P). Maroantsetra: Sahavary, Oct. 1986 (fr.), *Dransfield et al.* JD6400 (K, P, TAN). Farafangana: Manombo, Nov. 1991 (fr.), *Beentje* 4522 (BH, K, MO, P, TAN).

Orania trispatha, view into the distichous crown, Manombo (*Beentje* 4522).

3. ORANIA RAVAKA

The name 'ravaka' is the Malagasy for ornament, jewel; this is a particularly elegant tree, rather like *O. trispatha*, but with a more slender trunk.

DISTRIBUTION. NE Madagascar.
HABITAT. Lowland rain forest; ridge crests and valley bottoms; 200–550 m.
LOCAL NAMES. *Sindro* (Betsimisaraka); *Vapakafotsy* (Tsimihety).
USES. Not recorded.
CONSERVATION STATUS. Vulnerable. Known only from three sites, but with fairly large population numbers, though outside the protected area; total number estimated at 500.

Orania ravaka *Beentje* **sp. nov.**
Palma disticha foliolis ramentas castaneas infra ferentis a ceteris speciebus Madagascariensibus inflorescentia in 2 ordines ramificanti et a *O. trispatha* foliolis paucioribus minoribus et inflorescentia longiore graciliore differt. Typus: Madagascar, Sahavary, *Dransfield et al.* JD6401 (Holotypus K; isotypi P, TAN).

Solitary palm. **TRUNK** 10–15 m tall, 9–20 cm diam.; diameter just below crown < 10 cm; internodes 2.5–5 cm long, pale brown or grey-brown; nodal scars obscure; basal boss c. 35 cm diam., with surface roots. **LEAVES** 6–8 in the crown, distichous; sheath open, 35 cm long, with marginal fibres; apparent petiole 30 cm long, channelled; sheath

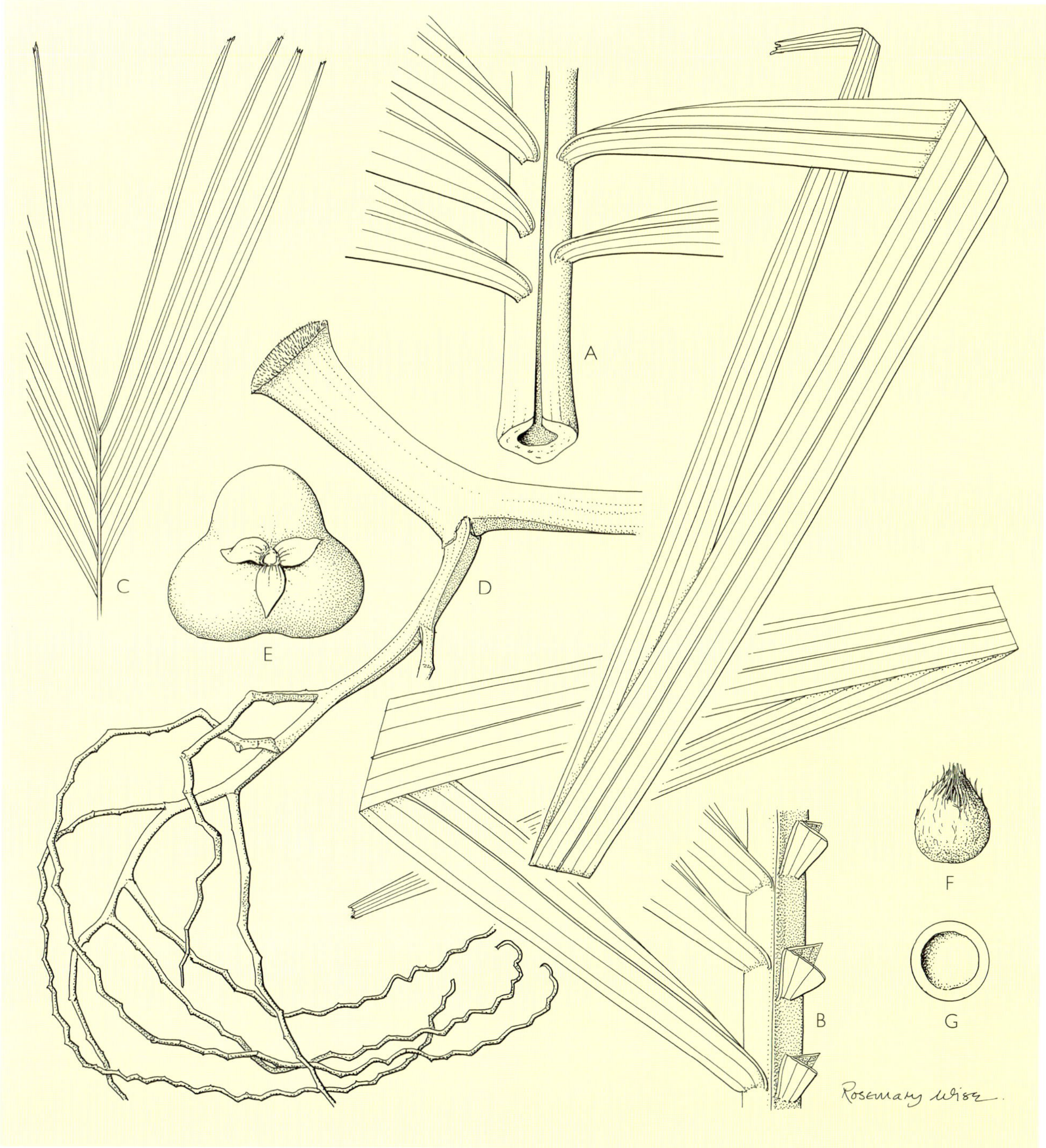

Orania trispatha. **A** distal part of petiole and basal leaflets × 1/3; **B** mid section of leaf × 1/3; **C** leaf tip × 1/3; **D** first order branch of inflorescence × 1/4; **E** fruit × 1/2; **F** endocarp × 1/2; **G** seed in section × 1/2. All from *Beentje* 4522. Drawn by Rosemary Wise.

Orania ravaka, growing in groups along a stream in the Manarara Avaratra Biosphere Reserve.

and apparent petiole combined 65–94 cm long, with patches of dense red-brown or grey scales and wax, distally c. 1.3 × 0.9–1 cm diam.; rachis 1.2–1.8 m long, in mid-leaf 1–2 cm wide, densely scaly, with scattered scales or glabrous; leaflets 33–44 on each side of the rachis, regular, stiff, those on opposite sides of the rachis at an angle of c. 90 with each other, strongly discolorous and much paler abaxially, the proximal 36–87 × 0.4–2.4 cm, the median 47–77.5 × 2.5–4.8 cm (interval 3.5–4.5 cm), the distal 18–45 × 0.4–2.8 cm, distal pair hardly connate, apices obliquely praemorse, main veins 3–5, midrib (and occasionally on other major veins as well) with large red-brown ramenta to 15 mm, with scattered red small scales, with visible sinuous transverse veinlets. **INFLORESCENCE** interfoliar, branching to 2 orders, spreading, protandrous; peduncle curved, 42–60 cm long, distally 12–18 × 5–12 mm diam., rusty-tomentose; prophyll borne 2.5–13 cm above the base of the peduncle, 16–32 cm long, 3.5–6 cm wide, with patches of red-brown scales; peduncular bract(s) one or two (in the same population), deciduous, inserted at 6.5–14

Orania ravaka. Detail of crown with (left) inflorescence at pistillate anthesis, and (right) inflorescence at staminate anthesis, Hiaraka.

cm and 16 cm from the base of the peduncle, 43–63.5 cm long, 7–9 cm diam., inflated, beaked; rachis with c. 13 first order branches, these proximally to 17 x 8 mm diam.; rachillae cream at anthesis, 7–40 cm long, 2–4 mm diam., zigzag, with rusty scales; triads distant. **STAMINATE FLOWERS** cream, not scented; sepals 1.5–2.4 x 2.2–2.8 mm; petals 8–13 x 4–6 mm; stamens 22–27, the filaments 0.7–2.5 mm (connate for up to 0.5 mm) with a bulbous base to 1 mm diam., anthers basifixed, slightly sagittate, equal or slightly unequal, 6.3–7.6 x 1.2–1.3 mm; pistillode not seen. **PISTILLATE FLOWERS** greenish, with sepals 1–4.5 x 4–5.5 mm; petals 10–15 x 9–11 mm; staminodes 9–12, thread-like; ovary globose, 10–11 mm diam., with c. 1 mm long papillose stigmas. **FRUIT** yellow or pale brown, slightly obovoid or globose, 4–6 cm diam., with sub-basal stigmatic remains. **SEED** depressed globose, 2.5–3.8 cm high, 3.2–4.2 cm diam., with fibrous surface; endosperm homogeneous; cotyledonary stalk up to 50 cm long and 6 mm diam.

NOTE. Differs from the other Madagascar species in the inflorescence branched to 2 orders, and from the other distichous species by the fewer leaflets which are smaller (with red-brown rather than grey ramenta), and the longer, more slender inflorescence.

SPECIMENS SEEN. Maroantsetra: Hiaraka, Oct. 1986 (fl.), *Dransfield et al.* JD6376 (K, P, TAN); Sahavary, Oct. 1986 (fl., fr.), *Dransfield et al.* JD6401 (Holotype K; isotypes P, TAN); idem, Feb. 1988 (y.fr.), *Dransfield et al.* JD6458 (K, TAN); idem, April 1989 (fr.), *Du Puy & Du Puy* MB 161 (K, TAN). Mananara Avaratra: Antanambe, Oct. 1991 (fl., fr.), *Beentje & Andriampaniry* 4466 (BH, K, MO, P, TAN).

Orania ravaka. **A** mid section of leaf x 1/3; **B** leaf tip x 1/2; **C** base of inflorescence x 1/3; **D** first order branch of inflorescence and tip of peduncle x 1/3; **E** detail of rachilla x 1; **F** staminate flower, one petal removed x 3; **G** fruit x 2/3; **H** seed in vertical section x 2/3. **A**, **B**, **G**, **H** from *Dransfield et al.* JD6458, **C – F** from *Dransfield et al.* JD6401. Drawn by Rosemary Wise.

SUBTRIBE DYPSIDINAE

IX. DYPSIS

Noronha ex Mart. Hist. Nat. Palm. 3: 180 (1838); Benth. & Hook. f., Gen. Plant. 3: 909 (1883); Becc., Bot. Jahrb. Syst. 38, Beibl. 87: 14 (1906), Becc., Palme del Madagascar 9 (1912); Jum., Ann. Inst. Bot.-Géol. Colon. Marseille sér. 3, 6 (1): 21 (1918); Jum., Bull. Acad. Malgache 6: 1 (1923); Jum. & H. Perrier, Fl. Madagascar 30: 92 (1945); Becc. & Pic.-Serm., Webbia 11: 146 (1956); N.W. Uhl & J. Dransf., Genera Palmarum 353 (1987). Lectotype: *Dypsis forficifolia* Mart.

SYNONYMS:
Chrysalidocarpus H.Wendl., Bot. Zeit. 36: 117 (1878); Benth. & Hook. f., Gen. Plant. 3: 822 (1883); Becc., Bot. Jahrb. Syst. 38, Beibl. 87: 30 (1906); Palme del Madagascar 37 (1914); Jum., Ann. Inst. Bot.-Géol. Colon. Marseille sér. 3, 10: 5 (1922); Jum. & H. Perrier, Fl. Madagascar 30: 92 (1945); Becc. & Pic.-Serm., Webbia 11: 150 (1956); N.W. Uhl & J. Dransf., Genera Palmarum 346 (1987). Type: *C. lutescens* H. Wendl. (= *Dypsis lutescens* (H. Wendl.) Beentje & J. Dransf.), **synon. nov.**

Phloga Noronha ex Hook. f., Benth. & Hook. f., Gen. Plant. 3: 877, 909 (1883); Becc., Bot. Jahrb. Syst. 38, Beibl. 87: 10 (1906); Palme del Madagascar 3 (1912); Jum. & H. Perrier., Fl. Madagascar 30: 122 (1945); Becc. & Pic.-Serm., Webbia 11: 141 (1956); N.W. Uhl & J. Dransf., Genera Palmarum 351 (1987). Type: *Phloga nodifera* (Mart.) Noronha ex Salomon (= *Dypsis nodifera* Mart.), **synon. nov.**

Neodypsis Baill., Bull. Mens. Soc. Linn. Paris 148: 1172 (1894); Becc., Bot. Jahrb. Syst. 38, Beibl. 87: 38 (1906); Palme del Madagascar 5 (1912); Jum., Compt. Rend. Acad. Sc. Paris 179: 448 (1924); Ann. Inst. Bot.-Géol. Colon. Marseille sér. 4, 2: 5 (1924); Jum. & H. Perrier, Fl. Madagascar 30: 136 (1945); Becc. & Pic.-Serm., Webbia 11: 145 (1956); N.W. Uhl & J. Dransf., Genera Palmarum 349 (1987). Type: *Neodypsis lastelliana* Baill. (= *Dypsis lastelliana* (Baill.) Beentje & J. Dransf.), **synon. nov.**

Neophloga Baill., Bull. Mens. Soc. Linn. Paris 2: 1173 (1894); Hist. Plantes 13: 372 (1895); Becc., Bot. Jahrb. Syst. 38, Beibl. 87: 19 (1906); Palme del Madagascar 19 (1912); Jum., Ann. Inst. Bot.-Géol. Colon. Marseille sér. 4, 6 (3): 7 (1929); Ann. Inst. Bot.-Géol. Colon. Marseille sér. 5, 1 (3): 7 (1933); Jum. & H. Perrier, Fl. Madagascar 30: 53 (1945); Becc. & Pic.-Serm., Webbia 11: 148 (1956); N.W. Uhl & J. Dransf., Genera Palmarum 348 (1987). Type: *Neophloga commersoniana* Baill. (= *Dypsis commersoniana* (Baill.) Beentje & J. Dransf.), **synon. nov.**

Haplodypsis Baill., Bull. Mens. Soc. Linn. Paris 2: 1167 (1894). Type: *Haplodypsis pervillei* Baill. (= *Dypsis pervillei* (Baill.) Beentje & J. Dransf.).

Haplophloga Baill., Bull. Mens. Soc. Linn. Paris 2: 1168 (1894). Type: *Haplophloga poivreana* Baill. (= *Dypsis poivreana* (Baill.) Beentje & J. Dransf.).

Dypsidium Baill., Bull. Mens. Soc. Linn. Paris 2: 1172 (1894). Type: *Dypsidium catatianum* Baill. (see Becc.& Pic.-Serm. 1955) (= *Dypsis catatiana* (Baill.) Beentje & J. Dransf.).

Phlogella Baill., Bull. Mens. Soc. Linn. Paris 2: 1174 (1894). Type: *Phlogella humblotiana* Baill. (= *Dypsis humblotiana* (Baill.) Becc.).

Trichodypsis Baill., Bull. Mens. Soc. Linn. Paris 2: 1174 (1894). Type: *Trichodypsis hildebrandtii* Baill. (= *Dypsis hildebrandtii* (Baill.) Becc.).

Vonitra Becc., Bot. Jahrb. Syst. 38, Beibl. 87: 18 (1906); Agric. Colon. 5: 320 (1911); Palme del Madagascar 6 (1912); Jum., Ann. Inst. Bot.-Géol. Colon. Marseille sér. 4, 5: 5 (1927); Jum. & H. Perrier, Fl. Madagascar 30: 128 (1945); Becc. & Pic.-Serm., Webbia 11: 142 (1956); N.W. Uhl & J. Dransf., Genera Palmarum 344 (1987), Type: *Vonitra fibrosa* (C.H. Wright) Becc. (*Dictyosperma fibrosum* C.H. Wright) (= *Dypsis fibrosa* (C.H. Wright) Beentje & J. Dransf.), **synon. nov.**
NOTE. The protologue of the genus is based on the specimen *Baron* 2319, which is the type of *V. thouarsiana* Becc. non *Dypsis thouarsiana* Baillon. *V. thouarsiana* sensu Becc. is synonymous with, and predated by, *V. fibrosa* (C.H. Wright) Becc.

Adelodypsis Becc, Bot. Jahrb. Syst. 38 Beibl. 87: 18 (1906). Type: *Adelodypsis gracilis* (Bory) Becc. (*Dypsis gracilis Bory*) (= *D. pinnatifrons* Mart.).

Macrophloga Becc., Palme del Madagascar 47 (1914); Becc. & Pic.-Serm., Webbia 11: 146 (1956). Type: *Macrophloga decipiens* (Becc.) Becc. (= *Dypsis decipiens* Becc.).

Antongilia Jum., Ann. Inst. Bot.-Géol. Colon. Marseille sér. 4, 6 (2): 19 (1928); Jum. & H. Perrier., Fl. Madagascar 30: 126 (1945); Becc. & Pic.-Serm., Webbia 11: 144 (1956). Type: *Antongilia perrieri* Jum. (= *Dypsis perrieri* (Jum.) Beentje & J. Dransf.), **synon. nov.**

Very small to very large unarmed pleonanthic monoecious palms. **STEMS** solitary or clustered, very short to massive, rarely subterranean, creeping-rhizomatous, erect, in one species climbing, sometimes branched aerially by apparent dichotomy. **LEAVES** pinnate or pinnately ribbed, neatly abscising or marcescent; sheath tubular, rarely almost open, usually forming a well-defined crownshaft, sometimes fibrous, in a few species with abundant pendulous piassava, sheath surface variously scaly and/or waxy or glabrous, auricles sometimes present; petiole absent or short to long, variously glabrous, scaly or hairy; blade entire, entire-bifid, or divided into single or multi-fold reduplicate leaflets, regularly or irregularly arranged, sometimes fanned within groups to produce a plumose appearance, leaflets usually entire, rarely praemorse, very rarely discolorous, often with abundant minute punctiform scales on both surfaces and ramenta along the main rib abaxially. **INFLORESCENCES** mostly interfoliar, more rarely infrafoliar, spicate or branched to 1–4 orders, apparently protrandrous (?always); peduncle usually elongate, basal branches not sharply divaricate; prophyll often borne above the base of the peduncle; peduncular bract usually conspicuous, exserted and caducous; rachillae variously glabrous or scaly and hairy; rachilla bracts low, generally inconspicuous, sometimes conspicuous; flowers borne in triads of a central pistillate flower and two lateral staminate flowers, triads superficial or slightly sunken in shallow pits. **STAMINATE FLOWERS** symmetrical, ± rounded to bullet-shaped, sometimes very small; sepals imbricate; petals valvate, basally briefly connate; stamens 3 (antesepalous or antepetalous) or 6 (very rarely 1, 2, 4 or 5 as monstrosities), 3 staminodes sometimes present, these either antesepalous or antepetalous, very rarely adnate to the pistillode; pollen elliptic, monosulcate or trichotomosulcate, with variously ornamented exine; pistillode present or absent. **PISTILLATE FLOWERS** about the same size as the staminate; sepals rounded, broadly imbricate; petals imbricate with triangular valvate tips; staminodes usually present, minute, toothlike, usually 3 or 6 at one side of the ovary; pistil pseudomonomerous, often strongly asymmetrical (especially in smaller species), stigmas 3,

Dypsis madagascariensis, in the forest at Lokobe.

apical, sometimes eccentric, ovule form unknown. **FRUIT** borne with persistent calyx and corolla, spherical, ellipsoid, fusiform or rarely curved, stigmatic remains basal, often obscured by perianth; epicarp often brightly coloured or jet-black or rarely dull green or brown; mesocarp thin, fleshy or fibrous; endocarp usually thin, fibrous. **SEED** closely adhering to the endocarp, endosperm homogeneous, sometimes deeply penetrated by regular grooves, or weakly to strongly ruminate; embryo subbasal. **GERMINATION** adjacent-ligular; eophyll bifid or pinnate.

DISTRIBUTION. 140 species; Madagascar, also two on Comoro Islands and one on Pemba Island.

NOTES. At the start of the present project, as Genera Palmarum (Uhl & Dransfield 1987) went to press, there was already uncertainty about the reliability of generic distinctions in the *Dypsidinae*. Until the present work, subtribe *Dypsidinae* consisted of seven accepted genera and another seven genera have at one time or another been recognised in the subtribe. This reflects the uncertainty about characters that arose from very incomplete material; this resulted in the few original taxa receiving generic status, with taxonomists in later years contorting themselves to wrestle new taxa within narrowly circumscribed genera. A good example is the distinction between *Chrysalidocarpus* and *Neophloga*; the most solid character was size and number of pinnae ("Palms large with many pinnae: *Chrysalidocarpus* / Palms small with few pinnae: *Neophloga*") while the type species of each of these genera were of almost exactly the same size, with roughly the same number of pinnae. These two genera form an almost seamless chain of species.

Neodypsis and *Chrysalidocarpus* formed a different problem; they had always been very close. The only consistent difference between them was a single character (endosperm ruminate or homogeneous), which in other groups within tribe *Areceae* could be variable in otherwise apparently natural genera. The ruminations can vary enormously, not only in size but in character; there are species in which the seed coat is sculptured, and the ruminations are caused by the undulations in the outer seed coat; in other species the seed coat is smooth on the outside but the intrusions appear to derive from the inner coat. In a taxon such as *Neodypsis baronii* (which bears a strong resemblance in gross morphology to *Chrysalidocarpus lutescens*) the ruminations can be less than half a millimeter deep and difficult to spot; in other taxa the ruminations take up as much space as the endosperm itself. This character might not be a single one, since the ruminations might arise in two (or even three) different ways. In thirteen other genera (out of 58) in the tribe (e.g. from the New World: *Oenocarpus/Jessenia, Euterpe, Prestoea;* Old World: *Iguanura, Hydriastele, Veitchia, Gronophyllum*) this character may vary within a genus. We suggest here that this is the case in *Dypsidinae*, supported by a host of other characters which are constant within the tribe. What had divided the two genera was not so much a real difference, as the weight of history. The monotypic genus *Antongilia* was based on a single species, distinct from *Neodypsis* by what surely were wrongly observed characters; specimens of the same taxon were also described in two different species in *Chrysalidocarpus*!

Both *Phloga* and *Vonitra* seemed to form good genera within the subtribe. *Phloga* has didymous anthers and very distinct ovate and acuminate leaflets, grouped along the rachis; *Vonitra* has branched trunks, many inflorescences in various stages of development, and the curious 'piassava', or leaf sheath fibres hanging all around the upper trunk, as well as didymous anthers, but strictly regular pinnae. However, *Antongilia*, as well as several taxa found by us in recent years bridge the gap between *Vonitra* and the main group: we can present a series (figure on next page) of species with strictly didymous anthers, slowly grading into the normal main group type of anther; the piassava character is variable, e.g. in *V. nossibensis*, refound in 1992 after eighty years of obscurity; several of the new taxa have unbranched trunks (as has *V. nossibensis*) or trunks branched underground; and some of the new taxa have only a single inflorescence per plant. This line of argument also causes the *Phloga* character of didymous anthers to lose its generic importance, which leaves only the distinctive leaflets; and these occur also in *Neophloga linearis*, refound in 1992, with characters perfectly matching the bulk of *Neophloga* in all other respects.

Dypsis with three stamens in the staminate flower seemed unambiguously defined – a species has either six stamens, or three, and if it has three it must be a *Dypsis*. So defined, the genus consists of mostly undergrowth palmlets, with inflorescences with (usually) many rachillae – though in the absence of staminate flowers it was very difficult to distinguish *Dypsis* from many *Neophloga* or even *Phloga*. However, there is more to the staminate flower of *Dypsis* than the number of stamens. The position and shape of the stamens and the presence or absence of staminodes and pistillode are of significance and variations in these parameters provided the basis for the recognition of further genera: *Trichodypsis* (Baillon 1894) and *Adelodypsis* (Beccari 1906). Later these genera were subsumed once again into *Dypsis*, now as subgenera (Beccari 1912). *Dypsis* in this enlarged sense has members with some of the smallest of all palm flowers, rarely exceeding 1 mm in diameter, a factor that makes the crucial examination of the flowers very difficult.

Given that the base number of stamens is probably six, the three-stamen condition could have arisen by the loss of one of the whorls of stamens. In the subgenus *Dypsis* there are three antesepalous, short stamens, no staminodes and a minute pistillode; in *Adelodypsis* the stamens are also antesepalous but the anthers are sagittate and there is a columnar pistillode in the flower centre; in *Trichodypsis* three antepetalous fertile stamens alternate with three antesepalous staminodes, the anthers are didymous and there is a very small pistillode. The structure and positioning of the stamens gives quite a different appearance to the flower at anthesis. In members of *Trichodypsis* the fertile anthers are closely appressed to each other at the tip of the flower, while the fact that the staminodes are minute and antesepalous means that there is a minute aperture into the flower just above the staminodes next to the gape in the corolla. The natural history of the flower form is not understood and there have been no observations of the pollination of these minute flowers, but we are impressed by the differences between the flowers and suggest that there is probably some functional significance that relates to access by pollinators. That the three stamens can be either antepetalous or antesepalous suggests to us that reduction from six to three has taken place at least twice in evolutionary history. If this is true then in a cladistic sense the genus *Dypsis* (sensu stricto) is polyphyletic and unnatural. If there are at least two evolutionary lines, as we suggest here, then there are some really remarkable parallels in overall form between taxa with antesepalous stamens and taxa with antepetalous stamens. The most extraordinary example is displayed by *Dypsis zahamenae* and the closely related *D. angustifolia* that have antesepalous stamens, while *D. pachyramea* with almost identical, unusually condensed hairy rachillae has antepetalous stamens. Yet, despite all this variation in flower form among the dypsids with three stamens the homogeneity of the endosperm remains consistent (as it does with the other group of small understorey palms within the dypsids, the former *Neophloga*). However, we need to stress that fruits are unknown for many of these palms. Further new combinations of floral characters have been discovered which cloud the rather neat distinctions between

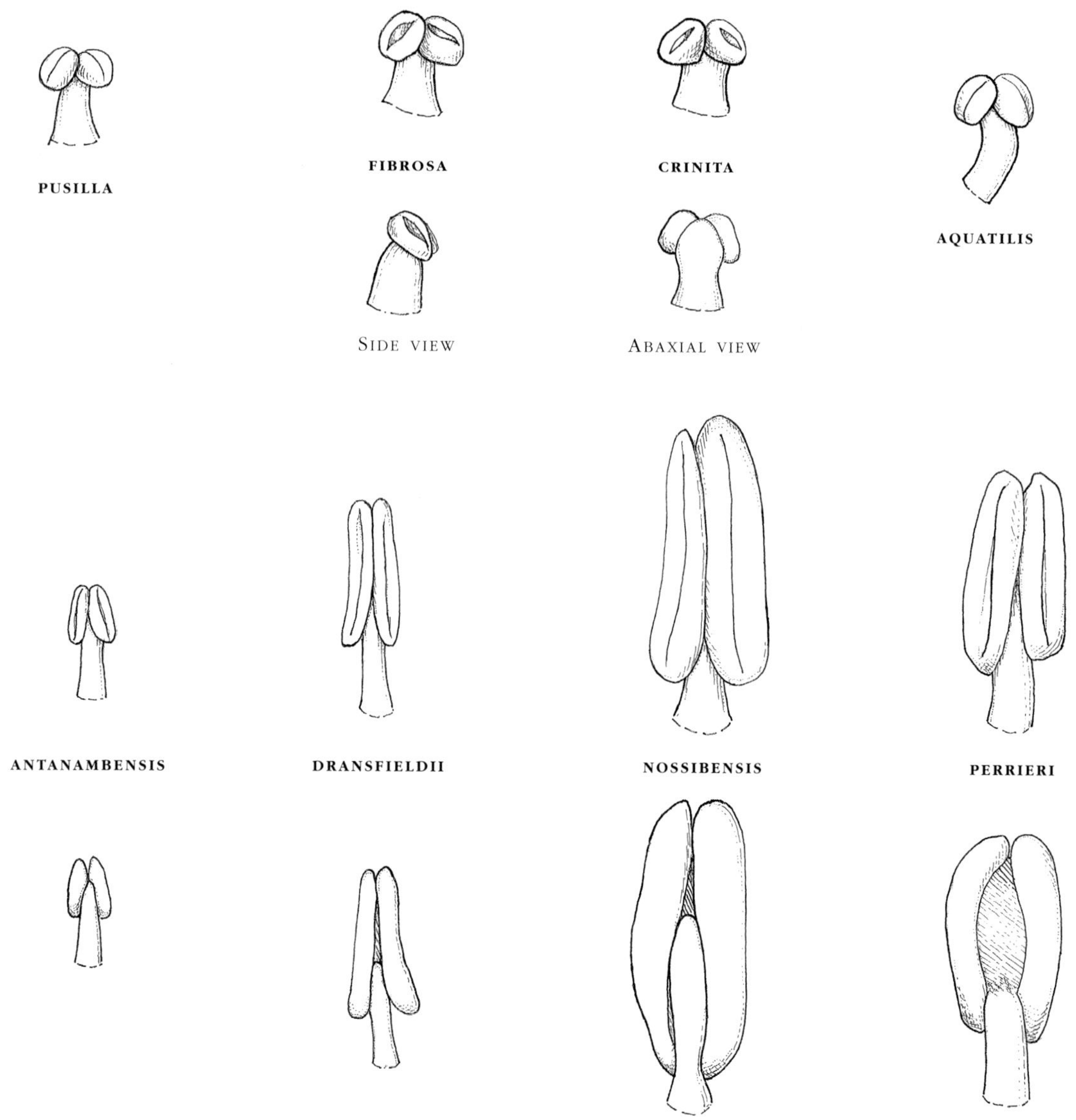

the three subgenera (or genera) of *Dypsis*. An extraordinary new taxon from the Marojejy Massif (*Dypsis mirabilis)* has three antesepalous stamens and three antepetalous staminodes that are adnate to the central columnar pistillode. One collection of *Dypsis lantzeana* from the Masoala Peninsula (a species that normally has three antesepalous stamens) has consistently a single stamen (still antesepalous), while another collection of the same species has some flowers with two stamens, and some with three. Collections of *Dypsis ambilaensis* display some staminate flowers with four, five or even six stamens next to the normal flower with three. *Dypsis thermarum* has characters such as dwarf habit, short inflorescence, densely hairy rachillae, minute globular rounded flowers, reminiscent of a member of *Trichodypsis*, yet it has six stamens.

There is a further development of the anthers, as yet unrecorded. In a few species, e.g. *Dypsis lokohensis* from the Marojejy massif, the stamens are of a bizarre form. There is a terete filament topped by a long reflexed bifid connective, the two arms of which are divergent and bear, at their tips, elongate anthers. This form of stamen is somewhat reminiscent of stamen form in the tribe *Geonomeae* (*Arecoideae*), an entirely New World tribe, and we have dubbed the stamen form "geonomoid". *D. fasciculata* and *D. thouarsiana* both have three geonomoid stamens, while *D. plurisecta* has six geonomoid stamens.

All this suggests to us that the three-staminate condition has arisen several times by reduction from six. Clearly our hypothesis demands a more rigorous testing using cladistic methods, but as yet our data are incomplete. It would seem that within the *Dypsidinae* there has been extraordinary radiation with little of the extinction that allows the recognition of disjunctions in variation and the demarcation of genera. It might be noted that the same condition, with either three or six stamens occurring within a single genus, is known to occur in *Geonoma*, *Orania*, *Areca* and several other genera.

In the new broad sense, *Dypsis* is a large genus with nearly 140 species, ranging from massive palms of the forest canopy to minute undergrowth palmlets, among the smallest of all palms. We wish that this now cumbersome genus could be divided up satisfactorily into smaller units, but there are so many intermediates between the historic genera and even between some other groupings that have become obvious during our present study, that we have had to abandon any attempt at a formal division, based on our current knowledge. There is one great practical advantage to the inclusion of all genera of the *Dypsidinae* in the single genus *Dypsis*. This relates to the description and identification of incomplete material. In the past, it was essential to have male flowers and fruit in order to be able to assign a specimen to a genus. Despite this Jumelle described new species of *Chrysalidocarpus* without knowing whether the seed had homogeneous endosperm and new species of *Neodypsis* without knowing whether the endosperm was ruminate. Now it is possible at least to apply the generic name with certainty, as they are all species of the newly broadened genus *Dypsis*. Many species can be identified as long as leaves and inflorescences are available, without the necessity of counting stamens, but for some of the smaller species it is still necessary to count stamens or to examine them to see of what form they are, or to check whether fruit has ruminate or homogeneous endosperm. However, for the most part, the keys will allow identification without recourse to a dissecting microscope, and should thus be much easier to use.

Within the enlarged genus we have nevertheless made some informal groupings and it is by these groupings that we have arranged the order of species in the book. In this way, species which appear somewhat similar in the field should be near to each other in the text, and thus will be easy to compare.

GROUP 1.

This group consists of large tree palms that reach into the canopy and includes species formerly described in both *Chrysalidocarpus* and *Neodypsis*. The species within this group have perhaps given us most taxonomic trouble. Historic collections of these major tree palms tend to be inadequate; even with a strong mental image or photographs of the living palm, and good modern herbarium collections, we have had difficulty in matching the new collections with the historic types. The group ranges from species such as *D. bejofo* with fascicled leaflets to species such as *D. lastelliana* with regularly arranged leaflets. All members of the group have staminate flowers with six equal stamens and the endosperm may be homogeneous or ruminate.

1. **Dypsis bejofo** *Beentje*
2. **Dypsis canaliculata** *(Jum.) Beentje & J. Dransf.*
   Neodypsis canaliculatus Jum.
3. **Dypsis hovomantsina** *Beentje*
4. **Dypsis ceracea** *(Jum.) Beentje & J. Dransf.*
   Neodypsis ceraceus Jum.
5. **Dypsis ampasindavae** *Beentje*
   Neodypsis loucoubensis Jum.
6. **Dypsis tsaravoasira** *Beentje*
7. **Dypsis nauseosa** *(Jum. & H. Perrier) Beentje & J. Dransf.*
   Neodypsis nauseosus Jum. & H. Perrier
8. **Dypsis oropedionis** *Beentje*
9. **Dypsis pilulifera** *(Becc.) Beentje & J. Dransf.*
   Chrysalidocarpus piluliferus Becc.,
   Chrysalidocarpus paucifolius Jum.
10. **Dypsis mananjarensis** *(Jum. & H. Perrier) Beentje & J. Dransf.*
    Chrysalidocarpus mananjarensis Jum. & H. Perrier
    Chrysalidocarpus fibrosus Jum.
11. **Dypsis malcomberi** *Beentje*
12. **Dypsis prestoniana** *Beentje*
13. **Dypsis tokoravina** *Beentje*
14. **Dypsis ifanadianae** *Beentje*
15. **Dypsis lastelliana** *(Baill.) Beentje & J. Dransf.*
    Neodypsis lastelliana Baill.
16. **Dypsis leptocheilos** *(Hodel) Beentje & J. Dransf.*
    Neodypsis leptocheilos Hodel
17. **Dypsis ligulata** *(Jum.) Beentje & J. Dransf.*
    Neodypsis ligulatus Jum.
18. **Dypsis saintelucei** *Beentje*
19. **Dypsis ovobontsira** *Beentje*

20. Dypsis tanalensis *(Jum. & H. Perrier) Beentje & J. Dransf.*
Neodypsis tanalensis Jum. & H. Perrier
21. Dypsis ankaizinensis *(Jum.) Beentje & J. Dransf.*
Chrysalidocarpus ankaizinensis Jum.
Neodypsis lobatus Jum.
22. Dypsis madagascariensis *(Becc.) Beentje & J. Dransf.*
Chrysalidocarpus madagascariensis Becc.
Chrysalidocarpus madagascariensis var. *oleraceus* (Jum. & H. Perrier) Jum.
Chrysalidocarpus madagascariensis forma *oleraceus* (Jum. & H. Perrier) Jum.
Chrysalidocarpus oleraceus Jum. & H. Perrier
Chrysalidocarpus madagascariensis var. *lucubensis* (Becc.) Jum.
Chrysalidocarpus lucubensis Becc.
23. Dypsis decaryi *(Jum.) Beentje & J. Dransf.*
Neodypsis decaryi Jum.

GROUP 2.

This group consists of a single species, *D. decipiens*. Its habit of dichotomously branching underground and its swollen trunk, combined with the massive leaves with fascicled leaflets, give it an appearance unlike any other member of the enlarged genus. The staminate flowers have six equal stamens; endosperm is homogeneous.

24. Dypsis decipiens *(Becc.) Beentje & J. Dransf.*
Chrysalidocarpus decipiens Becc.
Macrophloga decipiens (Becc.) Becc.

GROUP 3.

This group is linked to Group 1, there being large members of group 3 that could almost be included within Group 1. The group consists of solitary or clustering subcanopy palms with staminate flowers with six equal stamens; endosperm may be homogeneous or ruminate, and there are intermediate conditions, such as that found in the fruit of *D. baronii*, where the endosperm is only slightly penetrated by seed coast tissue. The group includes familiar palms such as *D. lutescens*, *D. baronii* and *D. cabadae*.

25. Dypsis basilonga *(Jum. & H. Perrier) Beentje & J. Dransf.*
Neodypsis basilongus Jum. & H. Perrier
26. Dypsis ambositrae *Beentje*
27. Dypsis heteromorpha *(Jum.) Beentje & J. Dransf.*
Neodypsis heteromorphus Jum.
28. Dypsis baronii *(Becc.) Beentje & J. Dransf.*
Chrysalidocarpus baronii Becc.
Neodypsis baronii (Becc.) Jum.
Chrysalidocarpus propinquus Jum.
Neodypsis compactus Jum.
29. Dypsis andrianatonga *Beentje*
30. Dypsis serpentina *Beentje*
31. Dypsis onilahensis *(Jum. & H. Perrier) Beentje & J. Dransf.*
Chrysalidocarpus onilahensis Jum. & H. Perrier
Chrysalidocarpus midongensis Jum.
Chrysalidocarpus brevinodis H. Perrier
32. Dypsis acuminum *(Jum.) Beentje & J. Dransf.*
Chrysalidocarpus acuminum Jum.
33. Dypsis lutescens *(H. Wendl.) Beentje & J. Dransf.*
Chrysalidocarpus lutescens H. Wendl.
Chrysalidocarpus baronii var. *littoralis* Jum. & H. Perrier
?Chrysalidocarpus glaucescens Waby
34. Dypsis arenarum *(Jum.) Beentje & J. Dransf.*
Chrysalidocarpus arenarum Jum.
35. Dypsis psammophila *Beentje*
36. Dypsis pembana *(H.E. Moore) Beentje & J. Dransf.*
Chrysalidocarpus pembanus H.E. Moore
37. Dypsis cabadae *(H.E. Moore) Beentje & J. Dransf.*
Chrysalidocarpus cabadae H.E. Moore
38. Dypsis humblotiana *(Baill.) Beentje & J. Dransf.*
Chrysalidocarpus humblotianus (Baill.) Becc.
Phlogella humblotiana Baill.
39. Dypsis lanceolata *(Becc.) Beentje & J. Dransf.*
Chrysalidocarpus lanceolatus Becc.
40. Dypsis pumila *Beentje*
41. Dypsis tsaratananensis *(Jum.) Beentje & J. Dransf.*
Neodypsis tsaratananensis Jum.
42. Dypsis oreophila *Beentje*
Neodypsis gracilis Jum.
Phloga gracilis (Jum.) H. Perrier
43. Dypsis ambanjae *Beentje*
Phloga sambiranensis Jum.
44. Dypsis coursii *Beentje*
45. Dypsis rivularis *(Jum. & H. Perrier) Beentje & J. Dransf.*
Chrysalidocarpus rivularis Jum. & H. Perrier
46. Dypsis marojejyi *Beentje*

GROUP 4.

The members of this group were formerly included in the genus *Neophloga*, but being relatively large, generally multistemmed undergrowth palms with staminate flowers with six equal stamens they form a link between the old genera *Neophloga* and *Chrysalidocarpus*. Endosperm is homogeneous. These are beautiful palms that would make fine ornamentals.

47. Dypsis commersoniana *(Baill.) Beentje & J. Dransf.*
Neophloga commersoniana Baill.
Neophloga pygmaea Pic.-Serm.
48. Dypsis humbertii *(Jum.) Beentje & J. Dransf.*
Neophloga humbertii Jum.
49. Dypsis scottiana *(Becc.) Beentje & J. Dransf.*
Neophloga scottiana (Becc.) Becc.
Phloga scottiana Becc.
Neophloga affinis Becc.
50. Dypsis singularis *Beentje*
51. Dypsis intermedia *Beentje*
52. Dypsis mcdonaldiana *Beentje*

GROUP 5.

This group is linked to group 4. It consists of generally slender forest undergrowth palms formerly included in *Neophloga*, with numerous lanceolate leaflets and lax inflorescences with relatively few branches and staminate flowers with six equal stamens and homogeneous endosperm. It also includes *D. scandens*, the only climbing palm in Madagascar, very distinctive in its habit, but by its leaflets and inflorescence it seems to be linked to *D. jumelleana*.

53. Dypsis jumelleana *Beentje & J. Dransf.*
Neophloga lanceolata Jum.
54. Dypsis linearis *(Becc.) Beentje & J. Dransf.*
Neophloga linearis Becc.
Neophloga procumbens Jum. & H. Perrier
Chrysalidocarpus ambolo Jum.
Neophloga mananjarensis Jum. & H. Perrier
55. Dypsis bonsai *Beentje*
56. Dypsis caudata *Beentje*
57. Dypsis scandens *J. Dransf.*

GROUP 6.

This group consists of two species only, in inflorescence and flower structure similar to the last group, but the leaves are usually entire bifid. The two species are generally clustering palms of moderate size of the forest undergrowth.

58. Dypsis faneva *Beentje*
59. Dypsis fanjana *Beentje*

GROUP 7.

The members of this group, formerly included in *Neophloga*, have inflorescences with very distinctive rachillae. They are long and relatively thick, almost catkin-like, and the flowers may be partially sunken in pits within the rachillae, or the rachilla bracts are conspicuous and partially obscure the flowers buds. Inflorescences may be spicate or branched to one or two orders, and leaves may be entire-bifid or pinnate. Staminate flowers have six equal stamens, and the endosperm is homogeneous. All members of this group are moderate-sized undergrowth palms from the northeast of the island.

60. Dypsis boiviniana *Baill.*
Chrysalidocarpus oligostachya Becc.
Neophloga oligostachya (Becc.) H. Perrier
61. Dypsis mangorensis *(Jum.) Beentje & J. Dransf.*
Neophloga mangorensis Jum.
Neophloga littoralis Jum.
62. Dypsis sanctaemariae *J. Dransf.*
63. Dypsis soanieranae *Beentje*
64. Dypsis curtisii *Baker*
Neophloga curtisii (Baker) Becc.
65. Dypsis pervillei *(Jum.) Beentje & J. Dransf.*
Neophloga pervillei (Baill.) Becc.
Haplodypsis pervillei Baill.

GROUP 8.

This group, that includes some of the prettiest palms of Madagascar, consists of dwarf undergrowth palms, formerly included in *Neophloga*, with sparsely branched inflorescences and staminate flowers with six equal stamens; the endosperm is homogeneous. The leaves are generally pinnate, though the group does include a few species with entire, bifid leaves.

66. Dypsis elegans *Beentje*
67. Dypsis heterophylla *Baker*
Neophloga heterophylla (Baker) Becc.
Dypsidium vilersianum Baill.
Dypsidium emirnense Baill.
Neophloga emirnensis (Baill.) Becc.
Neophloga majorana Becc.
Dypsis rhodotricha Baker
Neophloga rhodotricha (Baker) Becc.
Neophloga linearis var. *distachya* Jum.
68. Dypsis concinna *Baker*
Neophloga concinna (Baker) Becc.
Neophloga triangularis Jum. & H. Perrier
Neophloga concinna forma *triangularis* (Jum. & H. Perrier) Jum.
Neophloga tenuisecta Jum. & H. Perrier
Neophloga microphylla Becc.
69. Dypsis schatzii *Beentje*
70. Dypsis corniculata *(Becc.) Beentje & J. Dransf.*
Neophloga corniculata Becc.
71. Dypsis thiryana *(Becc.) Beentje & J. Dransf.*
Neophloga thiryana Becc.
72. Dypsis trapezoidea *J. Dransf.*

GROUP 9.

The members of this rather poorly defined grouping have fascicled leaflets and branched inflorescences and staminate flowers with six equal stamens; where the fruit is known, the endosperm is homogeneous.

73. Dypsis hiarakae *Beentje*
74. Dypsis confusa *Beentje*
75. Dypsis sahanofensis *(Jum. & H. Perrier) Beentje & J. Dransf.*
Chrysalidocarpus sahanofensis (Jum. & H. Perrier) Jum.
Neophloga sahanofensis Jum. & H. Perrier

GROUP 10.

This group consists of three species of undergrowth palmlet with entire, bifid leaves and inflorescences with numerous branches. The staminate flowers have six equal stamens and the endosperm is homogeneous. Two of the members of this group were formerly included in *Neophloga*.

76. Dypsis eriostachys *J. Dransf.*
77. Dypsis lutea *(Jum.) Beentje & J. Dransf.*
Neophloga lutea Jum.
Neophloga lutea Jum. var. *transiens* Jum. & H. Perrier
78. Dypsis betamponensis *(Jum.) Beentje & J. Dransf.*
Neophloga betamponensis Jum.

GROUP 11.

This is a large group of very small palms, formerly included in the genus *Neophloga*, with spicate inflorescences with staminate flowers with six equal stamens. The endosperm is homogeneous. Leaves are usually entire bifid, but may occasionally be divided into a few irregular broad leaflets or, in one species, *D. integra*, may be entire and with a minute apical cleft.

79. Dypsis mahia *Beentje*
80. Dypsis pulchella *J. Dransf.*
81. Dypsis plurisecta *Jum.*
82. Dypsis andapae *Beentje*
83. Dypsis occidentalis *(Jum.) Beentje & J. Dransf.*
Neophloga occidentalis Jum.
84. Dypsis montana *(Jum.) Beentje & J. Dransf.*
Neophloga montana Jum.
85. Dypsis bernierana *(Baill.) Beentje & J. Dransf.*
Neophloga bernierana (Baill.) Becc.
Haplophloga bernierana Baill.
86. Dypsis poivreana *(Baill.) Beentje & J. Dransf.*
Neophloga poivreana (Baill.) Becc.
Haplophloga poivreana Baill.
87. Dypsis catatiana *(Baill.) Beentje & J. Dransf.*
Neophloga catatiana (Baill.) Becc.
Dypsidium catatianum Baill.
Neophloga indivisa Jum. & H. Perrier
88. Dypsis coriacea *Beentje*
89. Dypsis minuta *Beentje*
90. Dypsis tenuissima *Beentje*
91. Dypsis lucens *(Jum.) Beentje & J. Dransf.*
Neophloga lucens Jum.
92. Dypsis simianensis *(Jum.) Beentje & J. Dransf.*
Neophloga simianensis Jum.
93. Dypsis integra *(Jum.) Beentje & J. Dransf.*
Neophloga integra Jum.
94. Dypsis digitata *(Becc.) Beentje & J. Dransf.*
Neophloga digitata Becc.
95. Dypsis brevicaulis *(Guillaumet) Beentje & J. Dransf.*
Neophloga brevicaulis Guillaumet

GROUP 12.

The palms in this grouping were formerly included in *Dypsis* in the strict sense; they are small palms of the forest undergrowth. There are only three stamens in the staminate flower, they have didymous anthers and they are opposite the sepals; the endosperm is homogeneous. This combination of characters defined § *Dypsis*. This group includes the type of the whole genus, *D. forficifolia*.

96. Dypsis forficifolia *Noronha ex Mart.*
Dypsis hirtula Mart.
Dypsis littoralis Jum.
Dypsis masoalensis Jum.
97. Dypsis interrupta *J. Dransf.*
98. Dypsis lantzeana *Baill.*
Dypsis lantzeana var. *simplicifrons* Becc.
99. Dypsis remotiflora *J. Dransf.*
100. Dypsis louvelii *Jum. & H. Perrier*
101. Dypsis zahamenae *J. Dransf.*
Dypsis humbertii H. Perrier
102. Dypsis angustifolia *(H. Perrier) Beentje & J. Dransf.*
Dypsis humbertii H. Perrier var. *angustifolia* H. Perrier

Group 13.

Members of this group were formerly included in *Dypsis* in the strict sense; they are moderate-sized palms of the forest undergrowth. The three stamens are opposite the sepals, the anthers are sagittate rather than didymous and there is a conical pistillode; the endosperm is homogeneous. These features were used to define § *Adelodypsis*.

103. Dypsis pinnatifrons *Mart.*
Areca gracilis Thouars.
Dypsis gracilis Bory ex Mart.
Adelodypsis gracilis (Bory ex Mart.) Becc.
Dypsis gracilis var. *sambiranensis* Jum. & H. Perrier
Dypsis sambiranensis (Jum. & H. Perrier) Jum.
Adelodypsis sambiranensis (Jum. & H. Perrier) H.P. Guérin
Chrysalidocarpus sambiranensis (Jum. & H. Perrier) Jum.
Phloga polystachya Becc. (non Baker)
104. Dypsis procera *Jum.*
Dypsis longipes Jum.
105. Dypsis paludosa *J. Dransf.*
Dypsis boiviniana sensu Becc. non Baill.
Adelodypsis boiviniana Becc. non Baill.

Group 14.

There is only one species in this group; in habit it resembles members of group 13, but the staminate flowers are of most unusual form. The three fertile stamens are opposite the sepals and have large fleshy filaments and small rounded anthers, and there are three staminodes, opposite the petals but fused to the rather massive pyramidal pistillode in the centre of the flower.

106. Dypsis mirabilis *J. Dransf.*

Group 15.

The three members of this group have only three stamens that are borne opposite the sepals; the anthers are pendulous and divergent and of a form reminiscent of those of geonomoid palms. One species, *D. thouarsiana*, is known only from fragments and nothing is known of its habit. The other two species are small or moderate palms of the forest undergrowth.

107. Dypsis fasciculata *Jum.*
108. Dypsis lokohoensis *J. Dransf.*
109. Dypsis thouarsiana *Baill.*

Group 16.

This is a group of large to moderate palms, one of which was previously included in the genus *Antongilia*, and several others formed the genus *Vonitra*. Most of the members of this group have staminate flowers with six usually biseriate stamens, usually but not always with didymous anthers. The leaf sheaths are generally fibrous and the leaves themselves do not absciss neatly from the stem. The endosperm is always ruminate. This group is linked to Group 3 via *D. perrieri*, *D. moorei* and *D. dransfieldii* (that are included here) and to Group 7.

110. Dypsis perrieri *(Jum.) Beentje & J. Dransf.*
Antongilia perrieri Jum.
Chrysalidocarpus auriculatus Jum.
Chrysalidocarpus ruber Jum.
111. Dypsis moorei *Beentje*
112. Dypsis dransfieldii *Beentje*
113. Dypsis nossibensis *(Becc.) Beentje & J. Dransf.*
Chrysalidocarpus nossibensis Becc.
Vonitra nossibensis (Becc.) H. Perrier
Vonitra loucoubensis Jum.
114. Dypsis crinita *(Jum. & H. Perrier) Beentje & J. Dransf.*
Vonitra crinita Jum. & H. Perrier
115. Dypsis utilis *(Jum.) Beentje & J. Dransf.*
Vonitra utilis Jum.
116. Dypsis fibrosa *(Wright) Beentje & J. Dransf.*
Vonitra fibrosa (Wright) Becc.
Dictyosperma fibrosum Wright
Vonitra thouarsiana Becc. non *Dypsis thouarsiana* Baillon
117. Dypsis antanambensis *Beentje*
118. Dypsis pusilla *Beentje*
119. Dypsis aquatilis *Beentje*

Group 17.

This consists of a single species, *D. nodifera*, previously included in a separate genus, *Phloga*. There are six biseriate stamens with didymous anthers and the fruit has ruminate endosperm. Vegetatively this palm is indistinguishable from *D. pinnatifrons* in Group 13.

120. Dypsis nodifera *Mart.*
Phloga nodifera (Mart.) Salomon
Phloga polystachya Noronha ex Baker
Phloga polystachya Noronha ex Baker var. *stenophylla* Becc.
Dypsis polystachya Baker

Group 18.

Members of this group are small undergrowth palms which have staminate flowers with three stamens with didymous anthers that are borne opposite the petals; there are usually three staminodes opposite the sepals and there is only a minute pistillode. Endosperm is homogeneous. These characters were used to define § *Trichodypsis* of *Dypsis* in the strict sense. Occasionally there may be six stamens (as rarely in *D. ambilaensis* and perhaps constantly in *D. thermarum*).

121. Dypsis thermarum *J. Dransf.*
122. Dypsis angusta *Jum.*
123. Dypsis anovensis *J. Dransf.*
Dypsis linearis Jum.
124. Dypsis ambilaensis *J. Dransf.*
125. Dypsis glabrescens *(Becc.) Becc.*
Trichodypsis glabrescens Becc.
126. Dypsis laevis *J. Dransf.*
127. Dypsis viridis *Jum.*
128. Dypsis hildebrandtii *(Baill.) Becc.*
Trichodypsis hildebrandtii Baill.
129. Dypsis bosseri *J. Dransf.*
130. Dypsis furcata *J. Dransf.*
131. Dypsis lanuginosa *J. Dransf.*
132. Dypsis mocquerysiana *(Becc.) Becc.*
Trichodypsis mocquerysiana Becc.
Dypsis mananarensis Jum.
133. Dypsis cookei *J. Dransf.*
134. Dypsis beentjei *J. Dransf.*
135. Dypsis pachyramea *J. Dransf.*
136. Dypsis monostachya *Jum.*
137. Dypsis spicata *J. Dransf.*
138. Dypsis acaulis *J. Dransf.*

Taxa for which material is insufficient for determining affinities.

139. Dypsis canescens *(Jum. & H. Perrier) Beentje & J. Dransf.*
Chrysalidocarpus canescens Jum. & H. Perrier
140. Dypsis ramentacea *J. Dransf.*

Key to Dypsis

How to use this key

This key is based on herbarium material, because that was the only way we could compare like with like. If this key is used for fresh material, one has to take into account shrinkage of some organs on drying. Fruits, for example, have shrunk on drying, from 16 × 15 to 11 × 8 mm!

Most people know their palms instantly and 'instinctively', their minds quickly checking several characteristics, the 'look' of the palm. These keys are a formalized expression of such instinctive actions. The problem with a key to 140 species is that one has to take into account the variation which species show; a taxon like *D. paludosa* shows up in seven places in two different key parts; *D. heterophylla* is even more variable and shows up in nine different places.

Some commonly used terms or characters in this key:

abaxial/adaxial: abaxial is the lower surface (away from the axis), adaxial the upper surface.

stems/trunk: we generally use trunk for larger palms, but to some extent these terms are interchangeable.

Inflorescence interfoliar/infrafoliar: interfoliar is among the leaves, infrafoliar is below the leaves.

Endosperm homogeneous/ruminate: seed with a homogeneous endosperm is totally white on the inside, while ruminate seed has dark intrusions into the white.

Leaflets abaxially with minute scattered scales: this is a useful character, but it is difficult to see without a hand lens. The scales appear as tiny red or darkish dots, usually along the minor veins.

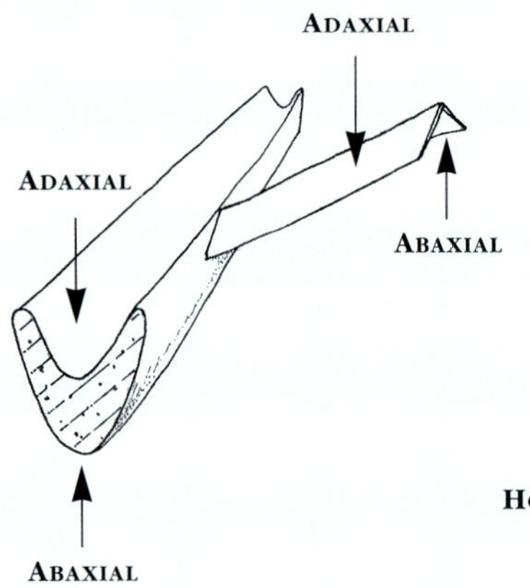

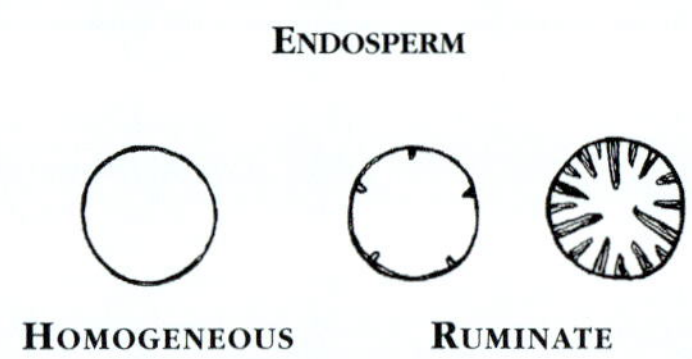

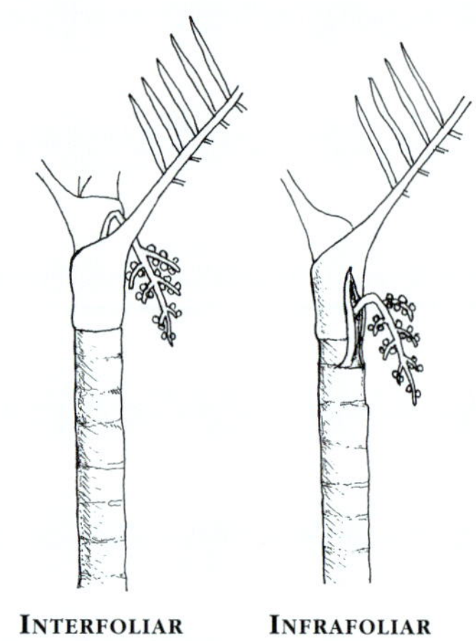

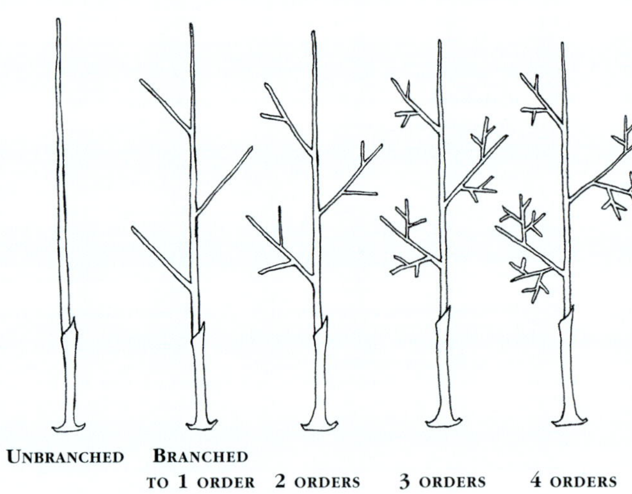

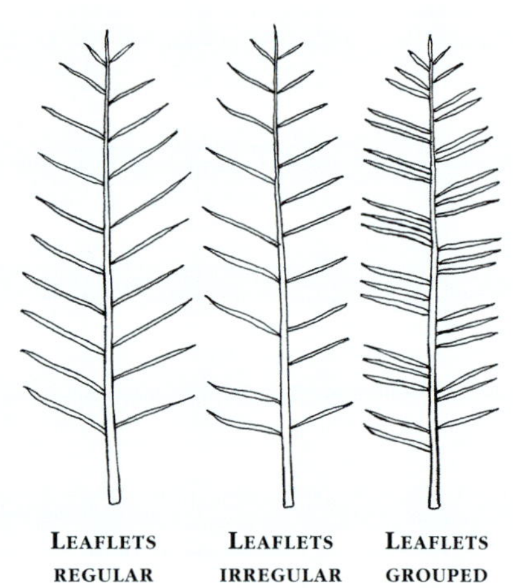

k1 Inflorescence unbranched (or with two rachillae, i.e. bifurcate) . . . . **KEY 1–a**
Inflorescence branched, with three or more rachillae **k2**

k2 Inflorescence branched to 1 order (rarely with a single bifurcate rachilla) . . . . **KEY 2–b**
Inflorescence branched to 2–4 orders **k3**

k3 Leaves entire . . . . **KEY 3–c**
Leaves pinnate **k4**

k4 Leaflets < 10 on each side of the rachis . . . . **KEY 4–d**
Leaflets ≥ 10 on each side of the rachis **k5**

k5 Leaflets grouped or (slightly) irregular . . . . **KEY 5–e**
Leaflets regular . . . . **KEY 6–f**

Note that at the end of the keys (page 145) there are lists of diagnostic characters for non-flowering or -fruiting palms.

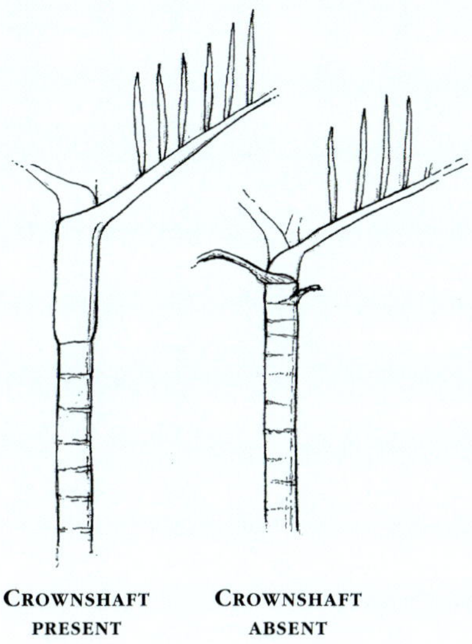

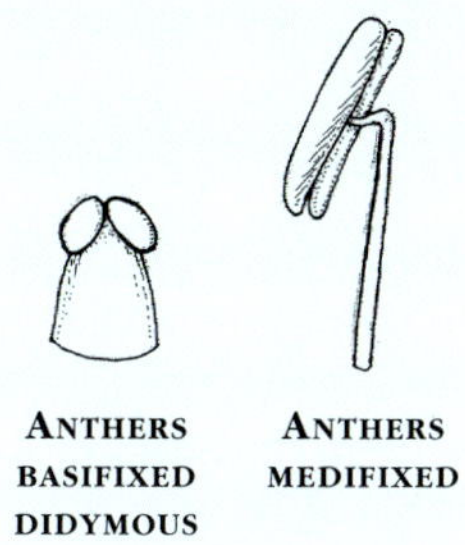

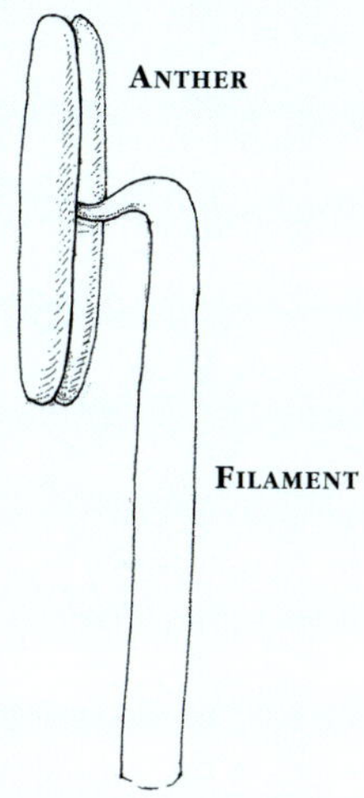

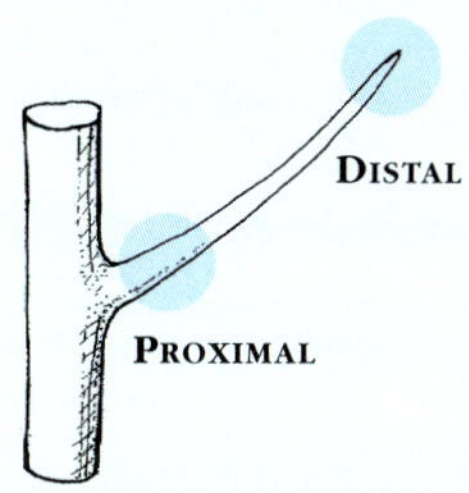

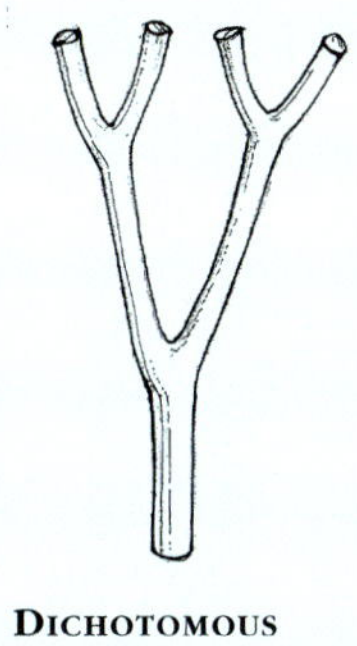

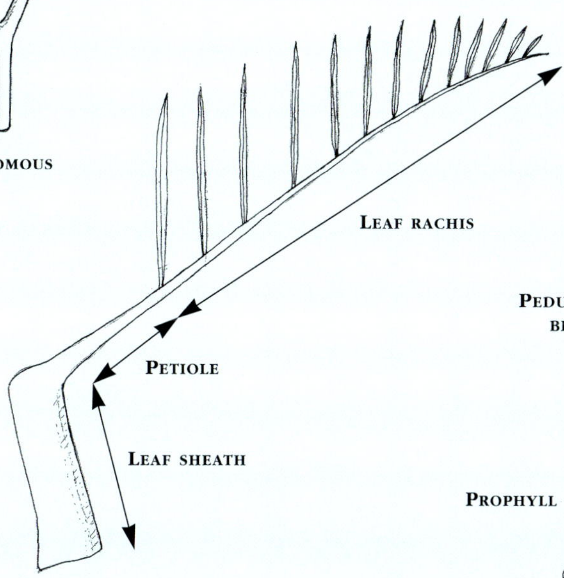

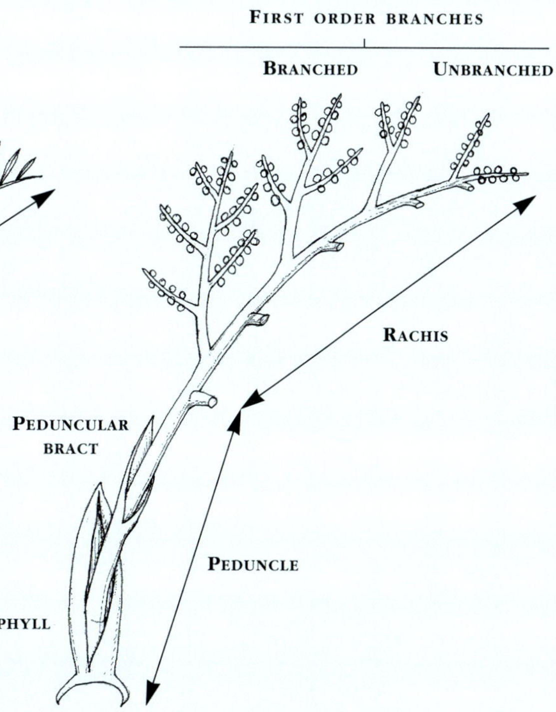

KEY 1 – INFLORESCENCE UNBRANCHED

a1 Leaves entire and bifid (rarely with two pairs of leaflets, the proximal very small) **a2**
Leaves on flowering stems pinnate with two or more leaflets on each side of the rachis **a17**

a2 Leaves lobed for more than half their length, or about half their length **a3**
Leaves lobed for less than half their length **a10**

a3 Leaf lobes less than 2 × as long as the costa (leaf midrib) **a4**
Leaf lobes more than 3 × as long as the costa **a7**

a4 Leaf blade 29–48 cm long **a5**
Leaf blade 17–26 cm long **a6**

a5 Peduncle c. 13 cm long; rachilla/spike c. 9 cm long . . . . **138. D. acaulis**
Peduncle 22–33 cm long; rachilla/spike 13–22 cm long . . . . **82. D. andapae**

a6 Leaf lobes as long as the costa; leaf sheath closed; fruit 5–6.5 mm long . . . . **67. D. heterophylla**
Leaf lobes 2 × as long as the costa; leaf sheath open; fruit 6–13 mm long . . . . **85. D. bernierana**

a7 Leaf blade 11–19 cm long, the lobes 8–15 × 0.7–1.3 cm; leaf sheath 2–6 cm long
. . . . **90. D. tenuissima**
Leaf blade 19–42 cm long, the lobes 16–37 × 1.2–4.5 cm; leaf sheath 6–13 cm long **a8**

a8 Usually an almost stemless palm; leaf sheaths open almost to their base; triads dense . **94. D. digitata**
Palms with stems 15–80 cm tall; leaf sheaths closed for more than half; triads distant **a9**

a9 Leaf sheath 6–9 cm long; rachilla/spike 4–12 cm long, minutely puberulous or glabrous
. . . . **89. D. minuta**
Leaf sheath 11–13 cm long; rachilla/spike 11–26 cm long, densely scaly . . . . . . **86. D. poivreana**

a10 Plant almost stemless (the stem underground); leaf blade 25–150 cm long . . . **95. D. brevicaulis**
Palms with stems > 20 cm tall; leaf < 41 cm long **a11**

a11 Leaf blade notched, the notch < 4 cm deep; clustering palms **a12**
Leaf blade lobed, the lobes > 4 cm long; clustering or solitary palms **a13**

a12 Leaf blade narrowly triangular, 2.5–5 cm wide, glabrous; triads distant . . . . . **92. D. simianensis**
Leaf blade obovate, 4–7 cm wide, with minute scales scattered over minor veins; triads dense . . . . **93. D. integra**

a13 Rachilla/spike glabrous, 23–24 cm long . . . . **91. D. lucens**
Rachilla/spike puberulous or scaly **or** if glabrous, then < 15 cm long **a14**

a14 Rachilla/spike glabrous **a15**
Rachilla/spike puberulous or scaly **a16**

a15 Stamens 6; blade obtriangular, about as long as the inflorescence . . . . **87. D. catatiana**
Stamens 3, with 3 staminodes; blade obovate, much longer than the inflorescence . **137. D. spicata**

a16 Anthers didymous; leaf chartaceous . . . . **87. D. catatiana**
Anthers versatile; leaf leathery . . . . **88. D. coriacea**

a17 Leaf sheath 16–18 cm, leaf rachis 70–95 cm; inflorescence > 1 m; solitary palm 4–5 m tall
. . . . **63. D. soanieranae**
Leaf sheath < 14 cm, leaf rachis < 40 cm; inflorescence < 75 cm; palms < 2.5 m tall **a18**

a18 Some of the proximal and/or median leaflets multifold, flat and attached to the rachis for more than 1 cm **a19**
All leaflets (except the top pair) composed of a single fold, the attachment less than 1 cm . . . **a22**

a19 Leaf rachis 29–40 cm long; leaf sheath 8–14 cm long . . . . **83. D. occidentalis**
Leaf rachis < 28 cm long; leaf sheath < 9 cm long (except **67. D. heterophylla**) **a20**

a20 Stamens 3; leaflets 2 on each side of the rachis; solitary palm from Marojejy area . . **137. D. spicata**
Stamens 6; leaflets 2 or more on each side of the rachis; solitary or clustering palms **a21**

a21 Leaflets grouped, the terminal pair at a narrow angle; stamens versatile . . . . **67. D. heterophylla**
Leaflets regular, the terminal pair at a wide angle; stamens didymous . . . . **87. D. catatiana**

a22 Leaflets 9–12 on each side of the rachis **a23**
Leaflets ≤ 6 on each side of the rachis (NOTE: **67. D. heterophylla** may be intermediate and comes up in both leads) **a25**

a23 Rachilla/spike 4–10 cm long . . . . **67. D. heterophylla**
Rachilla/spike 12–24 cm long **a24**

a24 Leaf sheath 12–13 cm long; prophyll 16–35 cm long . . . . **64. D. curtisii**
Leaf sheath 8–9 cm long; prophyll 8–14 cm long . . . . **65. D. pervillei**

a25 Leaflets without punctiform scales on the minor veins **a26**
Leaflets with small punctiform scales on the minor veins **a27**

a26 Solitary palm; leaflets 21–29 cm long; spike/rachilla 6–17 cm, puberulous . . . . . . **94. D. digitata**
Clustering palm; leaflets 9–14 cm long; spike/rachilla 3–4 cm, glabrous . . . . . **90. D. tenuissima**

a27 Terminal leaflets joined for 7–8 cm . . . . **84. D. montana**
Terminal leaflets joined for < 5 cm **a28**

a28 All leaf sheaths open for > 50%; solitary palm, stem 5–40 cm high . . . . **94. D. digitata**
Leaf sheaths closed, except for near the very apex; clustering palm, stem > 50 cm high **a29**

a29 Rachis/spike 9–21 cm long; stamens 3, with 3 triangular staminodes . . . . . **136. D. monostachya**
Rachis/spike 4–10 cm long; stamens 6 . . . . **67. D. heterophylla**

Key 2 – Inflorescence branched to 1 order

b1 Leaves entire or with up to 5 leaflets on each side of the rachis **b2**
Leaves pinnate with 6 or more leaflets on each side of the rachis **b31**

b2 Rachillae more than 10 cm long **b3**
Rachillae less than 10 cm long **b9**

b3 Rachillae 30–40 cm long . . . . **62. D. sanctaemariae**
Rachillae less than 28 cm long **b4**

b4 Petiole 10–25 cm long . . . . **61. D. mangorensis**
Petiole less than 10 cm long **b5**

b5 Rachillae glabrous **b6**
Rachillae puberulous or scaly, at least around the triads **b7**

b6 Leaf 13–30 cm long, leaf sheath 7–11 cm; peduncle glabrous . . . . **68. D. concinna**
Leaf 30–55 cm long, leaf sheath 10–17 cm; peduncle densely scaly . . . . **77. D. lutea**

b7 Leaf sheath 12–15 cm; stem 2.5–5 m tall; stamens 6, didymous . . . . **59. D. fanjana**
Leaf sheath 6–12 cm; stem 2–3 m tall; stamens 3, or when 6, then versatile **b8**

b8 Stamens 3, geonomoid; distal leaflet or leaf lobed for more than 10 cm . . . **108. D. lokohoensis**
Stamens 6, versatile; distal leaflet or leaf notched for up to 5 cm . . . . **69. D. schatzii**

b9 Rachillae 1–3 cm long with continuous triads and /or bracts and hairs, giving a fat, furry spike-like appearance; leaflets with large ramenta; anthers 3 didymous **b10**

Rachillae with distant triads, always some rachilla surface visible **b12**

b10 Stamens antesepalous; peduncle 11–32 cm long; leaf sheath 7–12 cm **b11**

Stamens antepetalous; peduncle < 10 cm long; leaf sheath < 6 cm . . . . **135. D. pachyramea**

b11 Leaf lobed for less than a third, the lobes 3–9 cm long . . . . **101. D. zahamenae**

Leaf lobed to three quarters, the lobes 20–40 cm long. . . . . **102. D. angustifolia**

b12 Inflorescence with very many (c. 90) short rachillae . . . . **132. D. mocquerysiana**

Inflorescence with fewer than 30 rachillae **b13**

b13 Rachillae glabrous or papillose, but without scales or hairs **b14**

Rachillae scaly or puberulous **b22**

b14 Rachillae 14–25 in number **b15**

Rachillae 2–12 in number **b16**

b15 Leaf sheath c. 11 cm; peduncle c. 41 cm; inflorescence rachis c. 16 cm; stamens 6 . . . . **78. D. betamponensis**

Leaf sheath 6–8 cm; peduncle 18–23 cm; inflorescence rachis 6–11 cm; stamens 3 . . . . **125. D. glabrescens**

b16 Leaf pinnate with 2–5 leaflets **b17**

Leaf entire **b18**

b17 Leaf sheath 6–12 cm; inflorescence rachis 0–5 cm, with 2–6 rachillae; stamens 6 . . . . **67. D. heterophylla**

Leaf sheath 5–6 cm; inflorescence rachis 5–8 cm, with 7–11 rachillae; stamens 3 . . **127. D. viridis**

b18 Leaf sheath 5–6 cm; stems < 1 m tall; rachillae 2–5 cm; leaf lobed for more than three quarters **b19**

Leaf sheath 6–17 cm; stems 1–2.5 m tall; rachillae 4–10 cm; leaf lobed for less than three quarters **b20**

b19 Petiole 5–7 cm; leaf without ramenta . . . . **79. D. mahia**

Petiole 9–15 cm; leaf with ramenta . . . . **140. D. ramentacea**

b20 Peduncle 33–77 cm; leaf sheath 10–17 cm . . . . **77. D. lutea**

Peduncle 7–28 cm; leaf sheath 6–12 cm **b21**

b21 Peduncle densely puberulous . . . . **67. D. heterophylla**

Peduncle glabrous. . . . . **68. D. concinna**

b22 Rachillae 3–9 in number **b23**

Rachillae 10–28 in number **b27**

b23 Stamens 6 . . . . **b24**

Stamens 3 . . . . **b27**

b24 Rachillae 4–13 cm long; leaf sheath 6–12 cm **b25**

Rachillae 2–4 cm long; leaf sheath 3–7 cm **b26**

b25 Peduncle 7–19 cm long; entire leaf split for about half, the lobes 7–12 cm . . **67. D. heterophylla**

Peduncle 28–31 cm long; entire leaf split for about a quarter, the lobes 2–7 cm . . . **69. D. schatzii**

b26 Stamens didymous; leaf entire . . . . **80. D. pulchella**

Stamens geonomoid; leaf pinnate with 5 leaflets . . . . **81. D. plurisecta**

b27 Stamens 6 (Ranomafana) . . . . **121. D. thermarum**
Stamens 3 . . . . **b28**

b28 Stamens antesepalous; rachillae reflexed; Moramanga area or Marojejy **b59**
Stamens antepetalous; rachillae porrect or spreading; coastal **b29**

b29 Rachillae short and stocky, c. 1.5–2.5 mm diam. . . . . **122. D. angusta**
Rachillae more slender, scarcely exceeding 1 mm diam. **b30**

b30 Leaf blade usually entire bifid, sometimes irregularly divided with close leaflets with 2–5 folds . . . . **125. D. glabrescens**
Leaf blade regularly divided into narrow, rather distant leaflets with 1–2 folds . . **123. D. anovensis**

(leaves pinnate with 6 or more leaflets on each side of the rachis)

b31 Leaflets regular **b32**
Leaflets grouped or irregular **b41**

b32 Leaflets > 40 on each side of the rachis; rachillae > 34 cm long **b33**
Leaflets < 35 on each side of the rachis; rachillae < 32 cm long **b34**

b33 Solitary palm; petiole c. 3.5 m long; median leaflets 80–81 cm long . . . . **111. D. moorei**
Clustering palm; petiole < 1 m long; median leaflets 52–53 cm long . . . . **117. D. antanambensis**

b34 Median leaflets 3–11 cm, wedge-shaped with praemorse apices . . . . **71. D. thiryana**
Median leaflets linear and with an acute apex **b35**

b35 Leaf sheath 20–40 cm long; petiole usually > 12 cm **b36**
Leaf sheath 5–19 cm long; petiole usually < 10 cm **b38**

b36 Petiole 75–100 cm; median leaflets 38–51 cm long . . . . **118. D. pusilla**
Petiole 6–32 cm; median leaflets 16–40 cm long **b37**

b37 Clustering, branching palm; rachillae 2–10 cm long; endosperm ruminate . . . **29. D. andrianatonga**
Solitary, unbranched palm; rachillae 10–25 cm long; endosperm homogeneous . . **32. D. acuminum**

b38 Leaflets 19–21 on each side of the rachis; stem 2–7 cm diam.; fruit 12–26 mm, endosperm ruminate . . . . **40. D. pumila**
Leaflets 6–9 on each side of the rachis; stem < 1.5 cm diam.; fruit unknown **b39**

b39 Rachillae > 7 cm long; stamens 3, geonomoid . . . . **108. D. lokohoensis**
Rachillae 2–7 cm long **b40**

b40 Petiole up to 0.5 cm long; stamens 6, geonomoid . . . . **81. D. plurisecta**
Petiole 3–12 cm long; stamens 3, didymous . . . . **122. D. angusta**

b41 Median leaflets < 12 cm long, wedge-shaped with praemorse (ragged) apices **b42**
Median leaflets with acute or caudate apices **b43**

b42 Clustering palm; rachillae glabrous; fruit 9–11 × 3–5 mm . . . . **71. D. thiryana**
Solitary palm; rachillae scaly; fruit c. 18 × 6 mm . . . . **72. D. trapezoidea**

b43 Median leaflets ovate, with 2–4 cm long caudate driptips; Masoala Peninsula . . . . **56. D. caudata**
Median leaflets with attenuate apices **b44**

b44 Leaf sheath > 15 cm **b45**
Leaf sheath < 15 cm (Note: some species come up in both leads) **b51**

b45 Branching, floppy palm; petiole 40–75 cm . . . . **30. D. serpentina**
Erect, unbranched palms; petiole < 25 cm (except **42. D. oreophila**, 2–50 cm) **b46**

b46 Rachillae glabrous **b47**
Rachillae scaly or hairy **b48**

b47 Leaflets 25–45 on each side of the rachis; rachillae 3–15 cm long; endosperm ruminate . . . . **42. D. oreophila**
Leaflets 10–18 on each side of the rachis; rachillae 13–30 cm long; endosperm homogeneous . . . . **53. D. jumelleana**

b48 Rachillae 25–70 cm long **b49**
Rachillae 3–24 cm long **b50**

b49 Leaf sheath 21–30 cm long; petiole 5–24 cm . . . . **60. D. boiviniana**
Leaf sheath 17–20 cm long; petiole absent . . . . **62. D. sanctaemariae**

b50 Leaflets 25–45 on each side of the rachis; endosperm ruminate . . . . **42. D. oreophila**
Leaflets 11–23 on each side of the rachis; endosperm homogeneous . . . . **54. D. linearis**

b51 Rachillae glabrous **b52**
Rachillae scaly or hairy **b55**

b52 Rachillae 13–30 cm long . . . . **53. D. jumelleana**
Rachillae 1–12 cm long **b53**

b53 Stamens 3, didymous; leaf sheath 5–6 cm long; leaflets 6–7 on each side of the rachis . . . . **127. D. viridis**
Stamens 6, versatile; leaf sheath 6–12 cm long; leaflets 6–25 on each side of the rachis **b54**

b54 Leaflets 6–13 on each side of the rachis; peduncle densely puberulous . . . . **67. D. heterophylla**
Leaflets 11–25 on each side of the rachis; peduncle glabrous . . . . **68. D. concinna**

b55 Stamens 3, geonomoid; leaflets 6–9 on each side of the rachis; Marojejy Mountains . . . . **108. D. lokohoensis**
Stamens 6, versatile; leaflets 6–23 on each side of the rachis **b56**

b56 Leaf sheath > 13 cm long; leaflets fanned within the groups . . . . **54. D. linearis**
Leaf sheath 6–12 cm long (occasionally 13–15 cm in **70. D. corniculata**, with leaflets in one plane) **b57**

b57 Inflorescence erect, orange; leaflets fanned within the groups . . . . **55. D. bonsai**
Inflorescence arching, green; leaflets more or less in one plane **b58**

b58 Median leaflets 0.4–1.3 cm wide; prophyll 4–15 cm long, borne at 3–5 cm above the base of the peduncle; fruit 5–7 × 4–5 mm . . . . **67. D. heterophylla**
Median leaflets 1.2–2.6 cm wide; prophyll 16–27 cm long, borne at c. 6 cm above the base of the peduncle; fruit 10–12 × 4–6 mm . . . . **70. D. corniculata**

b59 Petiole absent or up to 2 cm long; Moramanga area . . . . **100. D. louvelii**
Petiole 2–9 cm long; Marojejy . . . . **108. D. lokohoensis**

KEY 3 - INFLORESCENCE BRANCHED TO 2 OR MORE ORDERS, LEAF ENTIRE

c1 Palm stemless or nearly so (stem underground); leaf with yellow stripe down the middle . . . . **134. D. beentjei**

Palms with distinct stems; leaf uniformly green **c2**

c2 Rachillae 15–50 cm long; leaf sheath 17–31 cm long **c3**

Rachillae < 15 cm long; leaf sheath < 15 cm long (except **77. D. lutea**; see either lead) **c5**

c3 Leaf lobed for less than half **c4**

Leaf lobed for more than three-quarters . . . . **105. D. paludosa**

c4 Leaf sheath 10–17 cm long; petiole up to 5 cm long; stamens 6 . . . . **77. D. lutea**

Leaf sheath 17–31 cm long; petiole (3–) 10–24 cm long; stamens 3 . . . . **104. D. procera**

c5 Leaf lobed for more than half **c6**

Leaf lobed for less than half (for leaves lobed about half, see either lead) **c10**

c6 Rachillae glabrous; stamens 6 . . . . **c7**

Rachillae scaly or hairy; stamens 3 . . . . **c8**

c7 Leaf 13–30 cm long; inflorescence rachis up to 10 cm long . . . . **68. D. concinna**

Leaf 45–55 cm long; inflorescence rachis c. 16 cm long . . . . **78. D. betamponensis**

c8 Inflorescence with 1–2 branched first order branches; stamens antesepalous . . . . **100. D. louvelii**

Inflorescence with > 37 branched first order branches; stamens antepetalous **c9**

c9 Petiole 7–13 cm long; costa c. 2 cm long . . . . **130. D. furcata**

Petiole up to 6 cm long; costa 8–19 cm long **c17**

c10 Rachillae glabrous **c11**

Rachillae scaly or hairy **c14**

c11 Inflorescence with 1–5 branched first order branches; rachillae 3–17 cm; stamens 6 . . . . **c12**

Inflorescence with > 5 branched first order branches; rachillae 2–6 cm; stamens 3 . . . . **c13**

c12 Leaf sheath 7–11 cm long; leaf blade 13–30 cm long . . . . **68. D. concinna**

Leaf sheath 10–17 cm long; leaf blade 30–55 cm long . . . . **77. D. lutea**

c13 Stamens antesepalous . . . . **96. D. forficifolia**

Stamens antepetalous . . . . **124. D. ambilaensis**

c14 Leaf sheath c. 15 cm long; leaf blade c. 48 cm long, without ramenta . . . . . . **131. D. lanuginosa**

Leaf sheath < 10 cm long; leaf blade < 40 cm long, usually with ramenta **c15**

c15 Leaf blade 12–21 cm long . . . . **128. D. hildebrandtii**

Leaf blade > 30 cm long **c16**

c16 Rachillae c. 30; stamens 6; fruit 16 mm long . . . . **76. D. eriostachys**

Rachillae 50– 100; stamens 3; fruit 10 mm long . . . . **98. D. lantzeana**

c17 Rachillae glabrous; coastal white sand forest around Toamasina . . . . **124. D. ambilaensis**

Rachillae scaly; lowland rain forest around Bay of Antongil . . . . **132. D. mocquerysiana**

KEY 4 – INFLORESCENCES BRANCHED TO 2–4 ORDERS, LEAFLETS 2–9 ON EACH SIDE OF THE RACHIS

NOTE: in this group, but not keyed due to lack of information, is **109. D. thouarsiana**

d1 Inflorescence branched to 4 orders; triads at distant (c. 7 mm) intervals . . . . . **99. D. remotiflora**
Inflorescence branched to 2–3 orders; triads at intervals of less than 4 mm **d2**

d2 Inflorescence rachis 1–11 cm long **d3**
Inflorescence rachis > 11 cm long (two species come up in both leads) **d14**

d3 Leaflets abaxially with scattered minute scales **d4**
Leaflets abaxially glabrous; SE Madagascar . . . . **48. D. humbertii**

d4 Rachillae glabrous **d5**
Rachillae scaly or hairy. **d7**

d5 Leaflets unequal, some multifold and broad **d6**
Leaflets all narrow, single-fold and < 1 cm wide, leathery and metallic green . . . . **133. D. cookei**

d6 Leaf rachis 15–22 cm long; inflorescence rachis 9–11 cm long . . . . **124. D. ambilaensis**
Leaf rachis c. 11 cm long; inflorescence rachis 5–8 cm long . . . . **127. D. viridis**

d7 Rachillae 6–23 cm long, the most proximal usually > 9 cm **d8**
Rachillae ≤ 7 cm long. **d9**

d8 All leaflets ± equal; stamens 6, versatile . . . . **70. D. corniculata**
Leaflets unequal, some multifold and broad; stamens 3, geonomoid. . . . . . . **108. D. lokohoensis**

d9 Rachillae < 50 in number **d10**
Rachillae c. 90 in number . . . . **132. D. mocquerysiana**

d10 Leaf rachis 37–40 cm long. . . . . **129. D. bosseri**
Leaf rachis < 30 cm long. **d11**

d11 Inflorescence with < 3 branched first order branches; petiole 3–14 cm long. **d12**
Inflorescence with > 5 branched first order branches; petiole up to 3 cm long
. . . . **128. D. hildebrandtii**

d12 Stamens 6; Ranomafana . . . . **121. D. thermarum**
Stamens 3 . . . . **d13**

d13 Leaflets irregular; Bay of Antongil . . . . **127. D. viridis**
Leaflets regular; Mananjara/Farafangana area . . . . **122. D. angusta**

d14 Rachillae glabrous **d15**
Rachillae scaly or hairy **d20**

d15 Leaf sheath 20–30 cm long; petiole > (11–) 40 cm long **d16**
Leaf sheath < 15 cm long; petiole < 15 cm long **d17**

d16 Leaflets abaxially with scattered scales; inflorescence rachis 12–25 cm long . . . **105. D. paludosa**
Leaflets abaxially without scattered scales; inflorescence rachis c. 70 cm long . . **106. D. mirabilis**

d17 Leaf rachis 40–60 cm long; stamens 6 . . . . **50. D. singularis**
Leaf rachis 15–38 cm long; stamens 3 . . . . **d18**

d18 Rachillae 6–7.5 cm long . . . . **126. D. laevis**
Rachillae 2–5.5 cm long **d19**

d19 Stamens 3 antesepalous; Masoala Peninsula to Île Sainte Marie . . . . **96. D. forficifolia**
Stamens 3 antepetalous; Tampolo to Ambila-Lemaitso . . . . **124. D. ambilaensis**

d20 Leaf sheath 17–32 cm long; rachillae 10–50 cm long; leaflets 25–72 cm long **d21**
Leaf sheath 5–16 cm long; rachillae 1–14 cm long; leaflets 4–30 cm long **d23**

d21 Leaflets abaxially with ramenta present; stamens 6 . . . . **58. D. faneva**
Leaflets abaxially without ramenta; stamens 3 . . . . **d22**

d22 Petiole (3–) 10–25 cm long; rachillae 10–18 in number . . . . **104. D. procera**
Petiole up to 7.5 cm long; rachillae c. 50 in number . . . . **105. D. paludosa**

d23 Leaflets abaxially without either scales or ramenta; petiole 5–30 cm long; stamens 6 . . . . **d24**
Leaflets abaxially with scales, and often with ramenta; petiole up to 12 cm long; stamens 3 . . **d25**

d24 Inflorescence branched to 2 orders, with a rachis 20–33 cm long; rachillae 3–14 cm long . . . . **47. D. commersoniana**
Inflorescence branched to 3 orders, with a rachis 15–16 cm long; rachillae 0.5–2.5 cm long . . . . **51. D. intermedia**

d25 Stems usually 2–4 m tall; stamens 3, antesepalous **d26**
Stems usually < 2 m tall; stamens 3, antepetalous **d27**

d26 Rachillae sparsely puberulous . . . . **96. D. forficifolia**
Rachillae densely hairy and scaly, triads almost obscured . . . . **98. D. lantzeana**

d27 Inflorescence arching; uplands S of Analamazaotra . . . . **128. D. hildebrandtii**
Inflorescence erect; lowlands around the Bay of Antongil . . . . **132. D. mocquerysiana**

KEY 5 – INFLORESCENCE BRANCHED TO 2–4 ORDERS, LEAFLETS MANY AND GROUPED/IRREGULAR

e1 Median leaflets > 50 cm long **e2**
Median leaflets < 50 cm long (Note: four taxa are included in both leads) **e25**

e2 Rachillae densely scaly or densely puberulous **e3**
Rachillae glabrous or with very few scales **e6**

e3 Median leaflets 97–112 cm long; leaf sheath > 1 m long . . . . **12. D. prestoniana**
Median leaflets < 70 cm long; leaf sheath < 50 cm long **e4**

e4 Litter-accumulating palm; leaf rachis 3–4 m long; inflorescence rachis > 1 m long. . **46. D. marojejyi**
Palms not accumulating litter; leaf rachis < 1.5 m long; inflorescence rachis < 50 cm long. **e5**

e5 Leaflets c. 32 on each side of the rachis; inflorescence branched to 3 orders; stamens 6 . . . . **45. D. rivularis**
Leaflets 10–13 on each side of the rachis; inflorescence branched to 2 orders; stamens 3 . . . . **105. D. paludosa**

e6 Leaf sheath < 30 cm long; stem < 6 cm diam.; clustering palm **e7**
Leaf sheath > 30 cm long; stem > 7 cm diam. **e8**

e7 Petiole 12–46 cm long; leaflets 55–60 on each side of the rachis . . . . **41. D. tsaratananensis**
Petiole 0–8 cm long; leaflets 10–13 on each side of the rachis . . . . **105. D. paludosa**

e8 Inflorescence branched to 2 orders **e9**
Inflorescence branched to 3 orders **e15**

e9 Clustering palm with ventricose trunks, > 50 cm diam.; central Madagascar . . . . **24. D. decipiens**
Solitary palm, or trunk < 15 cm diam. **e10**

e10 Stem 2–7 m tall, 10–15 cm diam.; inflorescence interfoliar, peduncle > 40 cm **e11**
Stem > 6 m tall, > 15 cm diam.; inflorescence infrafoliar, peduncle < 30 cm **e12**

e11 Leaf sheath c. 40 cm long; ramenta absent from abaxial leaflet surface . . . . **25. D. basilonga**
Leaf sheath 60–100 cm long; ramenta present on abaxial leaflet surface . . . . . **26. D. ambositrae**

e12 Leaflets without scattered minute scales on the abaxial surface (ramenta may be present) . . . **e13**
Leaflets with scattered minute scales on the abaxial surface **e14**

e13 Palm 15–25 m tall; fruit c. 2 cm, with ruminate endosperm . . . . **1. D. bejofo**
Palm 6–15 m tall; fruit c. 1 cm, with homogeneous endosperm . . . . **3. D. hovomantsina**

e14 Leaf sheath glabrous and waxy; leaflets with continuous ramenta on the abaxial surface
. . . . **2. D. canaliculata**
Leaf sheath distally densely scaly; leaflets without any ramenta on the abaxial surface
. . . . **21. D. ankaizinensis**

e15 Leaf rachis < 3 m long; inflorescence interfoliar, peduncle 52–200 cm **e16**
Leaf rachis > 3 m long; inflorescence infrafoliar, peduncle 13–40 cm (unknown for **4. D. canaliculata**) **e18**

e16 Solitary palm, trunk 45–50 cm diam.; peduncle 2 m long . . . . **13. D. tokoravina**
Clustering palm, the stems 7–20 cm diam.; peduncle < 1.3 m long **e17**

e17 Leaves tristichous; leaflets strongly grouped; endosperm homogeneous
. . . . **22. D. madagascariensis**
Leaves spiral; leaflets irregular; endosperm ruminate (?) . . . . **26. D. ambositrae**

e18 Leaflets with scattered minute scales on the abaxial surface **e19**
Leaflets without scattered minute scales on the abaxial surface (ramenta may be present) . . . **e20**

e19 Pistillate petals 5–6 mm long; fruit 16–20 x 8–13 mm, with ruminate endosperm . . . **4. D. ceracea**
Pistillate petals 2–3 mm long; fruit 5–7 mm, with homogeneous endosperm . . . . **9. D. pilulifera**

e20 Ramenta absent from the abaxial surface of the leaflets . . . . **10. D. mananjarensis**
Ramenta present on the abaxial surface of the leaflets **e21**

e21 Central Plateau palm; fruit c. 1 cm, with ruminate endosperm . . . . **8. D. oropedionis**
East Coast palm; fruit, if with ruminate endosperm, > 2 cm **e22**

e22 Leaf sheath waxy, near the petiole with a dense patch of furry hair; fruit c. 1 cm, with homogeneous endosperm . . . . **3. D. hovomantsina**
Leaf sheath waxy, near the petiole glabrous or with scattered scales **e23**

e23 Peduncle 13–20 cm long; prophyll 20–30 cm long; fruit > 2 cm, with ruminate endosperm . . . . **1. D. bejofo**
Peduncle 20–30 cm long; prophyll 40–75 cm long; fruit < 1 cm, with homogeneous endosperm **e24**

e24 Leaf sheath with scales; peduncular bract 37–70 cm; Marojejy to Moramanga . . . . **9. D. pilulifera**
Leaf sheath glabrous; peduncular bract 72–117 cm; Andohahela . . . . **11. D. malcomberi**

e25 Inflorescence branched to 3–4 orders **e26**
Inflorescence branched to 2 orders **e34**

e26 Short squat palm, trunk 20–30 cm diam., litter-accumulating, with 18–20 leaves . **46. D. marojejyi**
Trunk/stem < 20 cm diam.; not accumulating; < 15 leaves in the crown **e27**

e27 Clustering palm **e28**
Solitary palm **e30**

e28 Comoro Islands; little known, median leaflets 3.5–7 cm wide . . . . **39. D. lanceolata**
Madagascar; median leaflets 1–3 cm wide **e29**

e29 Trunk 7–20 cm diam.; median leaflets > 45 cm long . . . . **22. D. madagascariensis**
Stem < 2 cm diam.; median leaflets 8–24 cm long . . . . **49. D. scottiana**

e30 Tristichous palm of dry and riverine forest; leaf rachis 1.6–3 m long . . . **22. D. madagascariensis**
Palm with spiral leaves of wet forests; leaf rachis < 1.4 m long (–2.2 in **103. D. pinnatifrons**) **e31**

e31 Leaflets ± in one plane; rachillae densely puberulous; next to streams . . . . **45. D. rivularis**
Leaflets fanned within the groups; rachillae glabrous or nearly so **e32**

e32 Leaflets 12–21 on each side of the rachis; inflorescence erect; fruit < 1 cm, with homogeneous endosperm . . . . **73. D. hiarakae**
Leaflets (12) 23–59 on each side of the rachis; inflorescence arching **e33**

e33 Leaf sheath 25–48 cm long, rachis 70–220 cm; stamens 3; endosperm homogeneous . . . . **103. D. pinnatifrons**
Leaf sheath 12–31 cm long, rachis 20–80 cm; stamens 6; endosperm ruminate . . **120. D. nodifera**

e34 Climbing palm; internodes > 20 cm long; distal leaflets reflexed, with basal swelling . . . . **57. D. scandens**
Erect or leaning palm; internodes < 15 cm long (similar leaflets only in **120. D. nodifera**) . . **e35**

e35 Rachillae > 15 cm long **e36**
Rachillae < 15 cm long (Note: 4 taxa are included in both leads) **e45**

e36 Leaflets with scattered minute scales on the abaxial surface **e37**
Leaflets without scattered minute scales on the abaxial surface (ramenta may be present) **e42**

e37 Trunk 8–18 cm diam.; fruit > 2 cm long, with ruminate endosperm . . . . **44. D. coursii**
Trunk < 6 cm diam.; fruit < 1.8 cm long, with homogeneous endosperm (ruminate in **120. D. nodifera**) **e38**

e38 Leaflets in tight groups, fanned, often with a swollen base; endosperm ruminate **120. D. nodifera**
Leaflets in looser groups, not or only slightly fanned; endosperm homogeneous **e39**

e39 Leaflets 24–29 on each side of the rachis, in a few groups within which the leaflets are regularly spaced . . . . **97. D. interrupta**
Leaflets 10–23 on each side of the rachis, grouped but not interrupted regular **e40**

e40 Stamens 6; leaflets 11–23 on each side of the rachis, the median 7–31 cm long; leaflets slightly fanned **e41**
Stamens 3; leaflets 10–13 on each side of the rachis, the median 30–50 cm long; leaflets in one plane . . . . **105. D. paludosa**

e41 Clustering palms, with scaly distal internodes; leaf rachis densely pubescent . . . . . **54. D. linearis**
Solitary palms, with glabrous internodes; leaf rachis with scattered scales . . . . . . **73. D. hiarakae**

e42 Leaf rachis c. 1 m long; stem 3–8 cm diam. . . . . **75. D. sahanofensis**
Leaf rachis < 0.9 m long; stem 1–4 cm diam. **e43**

e43 Rachillae up to 24 cm long; stamens 6 (unknown for **52. D. mcdonaldiana**); fruit < 1.4 cm . . **e44**
Rachillae 20–50 cm long; stamens 3; fruit c. 1.4 cm . . . . **107. D. fasciculata**

e44 Inflorescence with 4–16 branched first order branches; leaflets glabrous . . **52. D. mcdonaldiana**
Inflorescence with 1–3 branched first order branches; leaflets with scales on distal margins . . . . **54. D. linearis**

e45 Leaflets ovate and acuminate, often with swollen base; endosperm ruminate . . . **120. D. nodifera**
Leaflets linear-lanceolate, without swollen base **e46**

e46 Median leaflets 8–15 × 0.4–0.6 cm; leaflets 41–65 on each side of the rachis, regular within the distant groups . . . . **66. D. elegans**
Median leaflets usually wider; leaflets irregular or in tight groups **e47**

e47 Leaf sheath 18–55 cm long, waxy; leaflets 25–47 on each side of the rachis; endosperm ruminate **e48**
Leaf sheath often smaller, not waxy; leaflets 10–28 on each side of the rachis; endosperm homogeneous **e50**

e48 Solitary palm, stem 8–18 cm diam.; fruit > 2 cm . . . . **44. D. coursii**
Clustering palm, stems 3–6 cm diam.; fruit < 1.5 cm **e49**

e49 Peduncle 6–12 cm long; inflorescence rachis 1.5–9 cm long; seed 6–8 × 3–7 mm . **42. D. oreophila**
Peduncle c. 42 cm long; inflorescence rachis c. 18 cm long; seed 11–13 × 9–10 mm . . . . **43. D. ambanjae**

e50 Inflorescence with 1–3 branched first order branches **e51**
Inflorescence with 4 or more branched first order branches (one taxon occurs in both leads) . . . **e54**

e51 Rachillae scaly or hairy **e52**
Rachillae glabrous . . . . **68. D. concinna**

e52 Inflorescence orange, erect; fruit < 1 cm **e53**
Inflorescence green, arching; fruit 1–1.2 cm . . . . **70. D. corniculata**

e53 Leaf sheath 13–35 cm, scaly; leaf rachis 20–70 cm long . . . . **54. D. linearis**
Leaf sheath 6–9 cm long, densely pubescent; leaf rachis 10–20 cm long . . . . **55. D. bonsai**

e54 Leaflets with scattered minute scales on the abaxial surface **e54**
Leaflets without scattered minute scales on the abaxial surface; inflorescence rachis 7–40 cm . . **e55**

e55 Leaflets grouped, 8–24 cm long in mid-leaf; rachillae 0.7–6.5 cm long . . . . **49. D. scottiana**
Leaflets irregular, 21–43 cm long in mid-leaf; rachillae 4–20 cm long . . . . . **52. D. mcdonaldiana**

e54 Inflorescence rachis 1–9 cm long . . . . **70. D. corniculata**
Inflorescence rachis 10–36 cm long . . . . **74. D. confusa**

KEY 6 – INFLORESCENCE BRANCHED TO 2–4 ORDERS, LEAFLETS MANY AND REGULAR

f1 Inflorescence branched to 2 orders **f2**
Inflorescence branched to 3–4 orders **f19**

f2 Leaflets with scattered minute scales on the abaxial surface **f3**
Leaflets without scattered minute scales on the abaxial surface (ramenta may be present) **f11**

f3 Almost stemless palm in or near water; leaflets 21–24 x 1–1.5 cm . . . . **119. D. aquatilis**
Palms with distinct stems **f4**

f4 Solitary palm, trunk 20–30 cm diam.; leaf sheath c. 1 m, red-tomentose . . . . **110. D. perrieri**
Clustering palm (rarely appearing solitary), stems < 12 cm diam.; leaf sheath < 0.6 m **f5**

f5 Distal leaflets forming a flabellum, joined for > 5 cm and multifold **f6**
Distal leaflets similar to others, single-fold **f7**

f6 Leaflets with ramenta; stamens 6 . . . . **58. D. faneva**
Leaflets without ramenta; stamens 3 . . . . **105. D. paludosa**

f7 Coastal palms, growing near the sea; endosperm homogeneous **f8**
Inland palms; endosperm ruminate. **f9**

f8 Petiole 19–37 cm long; leaflets 44–59 on each side of the rachis . . . . **33. D. lutescens**
Petiole 60–72 cm long; leaflets 28–30 on each side of the rachis . . . . **34. D. arenarum**

f9 Branching palm with floppy stems, 1.5–2.5 cm diam.; inflorescence rachis 4–14 cm
. . . . **29. D. andrianatonga**
Erect, unbranched palms, stems 2–12 cm diam.; inflorescence rachis 10–35 cm. **f10**

f10 Endosperm deeply ruminate (up to 7 mm) . . . . **27. D. heteromorpha**
Endosperm shallowly ruminate (< 3 mm) . . . . **28. D. baronii**

f11 Median leaflets c. 120 cm long. . . . . **17. D. ligulata**
Median leaflets < 80 cm long **f12**

f12 Inflorescence with swollen bases to first order branches; upper trunk with fibrous mass **f13**
Inflorescence without such bases; trunks clean **f15**

f13 Petiole up to 30 cm long; leaf rachis < 1.7 m long; leaflets without ramenta . . . . **112. D. dransfieldii**
Petiole 39–100 cm long; leaf rachis > 2 m long; leaflets with ramenta on abaxial surface **f14**

f14 Palm solitary, not branched; petiole 39–40 cm long. . . . . **113. D. nossibensis**
Palm clustering, usually branched; petiole 90–100 cm long . . . . **114. D. crinita**

f15 Petiole absent; median leaflets 75–80 x 1.5 cm (insufficiently known species). . **139. D. canescens**
Petiole usually present; leaflets 35–75 cm long. **f16**

f16 Leaf sheath green, distally pubescent; leaflets 12–21 on each side of the rachis; SE Madagascar
. . . . **52. D. mcdonaldiana**
Leaf sheath waxy-white, distally sparsely scaly; leaflets 30–65 on each side of the rachis **f17**

f17 Solitary palm; leaflets c. 30 on each side of the rachis. . . . . **32. D. acuminum**
Clustering palm; leaflets 40–65 on each side of the rachis. **f18**

f18 Inland palm from N and W Madagascar; leaf rachis > 1 m long . . . . **31. D. onilahensis**
Coastal palm from Toamasina area. . . . . **35. D. psammophila**

f19 Branched palms, with dense fibrous mass around upper trunks; rachillae with swollen base . . . **f20**
Unbranched palms with clean trunks; rachillae bases not swollen **f21**

f20 Trunk 25–37 cm diam.; median leaflets 77–92 cm long; rachillae 69–91 cm long . . . **115. D. utilis**
Trunk 5–18 cm diam.; median leaflets 46–71 cm long; rachillae 17–53 cm long . . **116. D. fibrosa**

f21 Clustering palm, stems < 15 cm diam.; leaf rachis < 2 m; endosperm homogeneous **f22**
Solitary palm (except **18. D. saintelucei**), stems > 13 cm diam.; leaf rachis > 2 m; endosperm ruminate (except **14. D. ifanadianae**; unknown for **6. D. tsaravoasira**) **f24**

f22 Leaflets with scattered minute scales on the abaxial surface **f23**
Leaflets without scattered minute scales on the abaxial surface . . . . **37. D. cabadae**

f23 Leaflets 1.3–3 cm wide; Madagascar, and widely cultivated . . . . **33. D. lutescens**
Leaflets 3–4 cm wide; Pemba Island . . . . **36. D. pembana**

f24 Leaf sheath 30–65 cm long, with dense red or brown tomentum; inflorescence interfoliar **f25**
Leaf sheath c. 1 m long, densely red-tomentose . . . . **110. D. perrieri**
Leaf sheath 70–150 cm long, waxy or green with a few distal scales **f28**

f25 Tristichous palm; leaf sheath open; leaflets with scattered minute scales on the abaxial surface . . . . **23. D. decaryi**
Leaves spirally inserted; leaf sheath closed; leaflets without scattered minute scales on the abaxial surface (ramenta may be present) **f26**

f26 Petiole c. 47 cm long; leaflets without ramenta; rachillae 10–20 cm long (Mananara only) . . . . **19. D. ovobontsira**
Petiole up to 20 cm long; leaflets with ramenta abaxially; rachillae 20–50 cm long **f27**

f27 Petiole up to 10 cm long; fruit 18–24 x 12–17 mm . . . . **15. D. lastelliana**
Petiole c. 17 cm long; fruit 10–12 mm . . . . **16. D. leptocheilos**

f28 Leaflets with scattered minute scales and ramenta on the abaxial surface **f29**
Leaflets completely glabrous on the abaxial surface **f31**

f29 Trunk 6–10 m tall, c. 14 cm diam.; leaflets 59–61 on each side of the rachis; SE Madagascar . . . . **18. D. saintelucei**
Trunk usually > 10 m tall, > 18 cm diam.; leaflets > 100 on each side of the rachis (unknown for **20. D. tanalensis**) **f30**

f30 Tristichous palm; peduncle 22–26 cm long . . . . **6. D. tsaravoasira**
Palm with spirally inserted leaves; peduncle 83–96 cm long . . . . **7. D. nauseosa**
NOTE: unknown for **20. D. tanalensis** which cannot be keyed out)

f31 Tristichous palm; leaf sheath 110–150 cm long; endosperm ruminate . . . . . . **5. D. ampasindavae**
Palm with spirally inserted leaves; leaf sheath c. 70 cm long; endosperm homogeneous . . . . **14. D. ifanadianae**

Diagnostic characters for palms when not in flower or fruit

Comoro Islands:
39. D. lanceolata
38. D. humblotiana

Pemba Island:
36. D. pembana

Palm stemless:
119. D. aquatilis
134. D. beentjei
95. D. brevicaulis
138. D. acaulis

Trunk ventricose, cigar-shaped:
24. D. decipiens

Branching palms:
29. D. andrianatonga
30. D. serpentina
33. D. lutescens
57. D. scandens
114. D. crinita
115. D. utilis
116. D. fibrosa
118. D. pusilla

Climbing palm:
57. D. scandens

Litter-collecting palms:
46. D. marojejyi
110. D. perrieri

Palm with fibrous mass obscuring the upper trunk:
112. D. dransfieldii
113. D. nossibensis
114. D. crinita
115. D. utilis
116. D. fibrosa
118. D. pusilla

Tristichous palms:
1. D. bejofo
3. D. hovomantsina
5. D. ampasindavae
6. D. tsaravoasira
8. D. oropedionis
9. D. pilulifera
10. D. mananjarensis
11. D. malcomberi
13. D. tokoravina
18. D. saintelucei
22. D. madagascariensis
23. D. decaryi
27. D. heteromorpha
28. D. baronii
29. D. andrianatonga
33. D. lutescens
36. D. pembana

Leaf sheath densely reddish tomentose
15. D. lastelliana
16. D. leptocheilos
19. D. ovobontsira
23. D. decaryi (tristichous)
3. D. hovomantsina has distal tuft of red brown tomentum, but proximal part glabrous

Leaflets praemorse:
71. D. thiryana
72. D. trapezoidea

Leaflets long-caudate:
56. D. caudata

Leaf sheath > 60 cm long
1. D. bejofo
2. D. canaliculata
3. D. hovomantsina
4. D. ceracea
5. D. ampasindavae
6. D. tsaravoasira
7. D. nauseosa
8. D. oropedionis
9. D. pilulifera
10. D. mananjarensis
11. D. malcomberi
12. D. prestoniana
13. D. tokoravina
14. D. ifanadianae
18. D. saintelucei
20. D. tanalensis
21. D. ankaizinensis
22. D. madagascariensis
24. D. decipiens
26. D. ambositrae

1. DYPSIS BEJOFO

One of the most impressive palms of Madagascar. The massive trunk and enormous leaves, sometimes arranged in three rows in an arching crown, give this tree a majestic appearance. This would be a spectacular ornamental tree for grand landscaping in the tropics. Seeds have been widely distributed (but see note below). The local name has supplied us with the specific epithet.

DISTRIBUTION. Only known from Maroantsetra and Mananara, around the Bay of Antongil.
HABITAT. Moist lowland forest; steep slope; 200–400 m.
LOCAL NAMES. *Bejofo*, *Hovotraomby* (Betsimisaraka).
USES. Not recorded.
CONSERVATION STATUS. Endangered. Only known from two sites, where numbers are low.

Dypsis bejofo

Dypsis bejofo *Beentje* **sp. nov.**
Palma excelsa foliis maximis foliolis aggregatis inflorescentia curta multo ramosa seminibus profunde canaliculatis distincta. Typus: Madagascar, Maroantsetra: Andranofotsy R., Sahavary, Oct. 1986, *Dransfield et al.* JD6405 (Holotypus K; isotypi P, TAN).

Solitary canopy palm. **TRUNK** 15–25 m high, 25–40 cm diam., near the crown c. 20 cm diam.; internodes 20–35 cm, pale reddish brown, near the crown 2–10 cm, dark green to ashy grey; crownshaft white, waxy. **LEAVES** 7–10 in the crown, sub-tristichous, porrect and arching; sheath 1–2 m, 25–40 cm diam., abaxially white waxy and distally dense red-brown scaly, adaxially peach-coloured; petiole 12–34 cm long, 4.5–15 cm wide, densely pubescent; rachis 3–6 m long, brown-scaly, in mid-leaf 2.5–5 x 1.5–4 cm, channelled in the proximal 2 m; leaflets 80–100 on each side of the rachis, in groups of 5–7 and fanned within the groups, group interval 8–10 cm, rich green, stiff or arched, the distal part often bent down, the proxi-

Dypsis bejofo, a massive palm, reaching above the forest canopy, Sahavary.

Dypsis bejofo. David Cooke holds the massive waxy leaf sheath, Sahavary (*Dransfield et al.* JD6405)

mal 90–200 × 0.4–1.7 cm and often with the most proximal pendulous, median 72–144 × 1.3–4 cm (interval 0.5–2 cm), distal 30–69 × 0.6–1.8 cm, main vein 1, with few ramenta 5–6 mm long on the midrib, otherwise glabrous and waxy, apices unequally bifid, attenuate. **INFLORESCENCE** infrafoliar, branched to 2 (– 3) orders, 0.5 × 0.5–1 m, yellow-green with sub-pendulous rachillae; peduncle 13–20 cm long, 5–10 × 3–7 cm diam., green with scattered scales; prophyll 20–30 cm long, to 19 cm wide, borne on the peduncle at 2–4 cm, with scattered scales; peduncular bract inserted at 5–9 cm from the base of the peduncle, 54–91 cm long and 10–17 cm wide, with scattered scales, flushed pink or pale lavender, split except for the beak; non-tubular peduncular bract rarely present, once seen 18 cm long; rachis 20–52 cm long, glabrous or with scattered scales, with 15–17 first order branches, these proximally 1.3–6 × 0.5–2.2 cm diam.; rachillae 20–44 cm long (to 73 cm in fruit), green, stiff, 3.5–10 mm diam., glabrous, with dense, slightly sunken triads; rachilla bract 2 mm long. **STAMINATE FLOWERS** scented like *Sambucus nigra*; sepals 2.3–2.8 × 2.1–2.4 mm; petals connate for c. 1 mm, free parts 4.1–4.5 × 2.3–2.5 mm; stamens 6, slightly biseriate (offset 0.2 mm), filaments 1.5–1.8 mm, cylindrical, anthers 2.2–2.7 × 0.7–1 mm; pistillode 4 × 1 mm. **PISTILLATE FLOWERS** in young fruit with sepals 2.8–3.2 × 2.5–3.7 mm; petals 3.7–4.3 × 4–5.3 mm. **FRUIT** with very fibrous endocarp, 20–25 × 18–21 mm. **SEED** ellipsoid, black, deeply grooved, 17–23 × 15–20.5 mm; endosperm deeply penetrated, the grooves regular and dense, 2–9 mm deep.

NOTE. When this tree is sterile it resembles *D. pilulifera*, which is less massive and has less strikingly

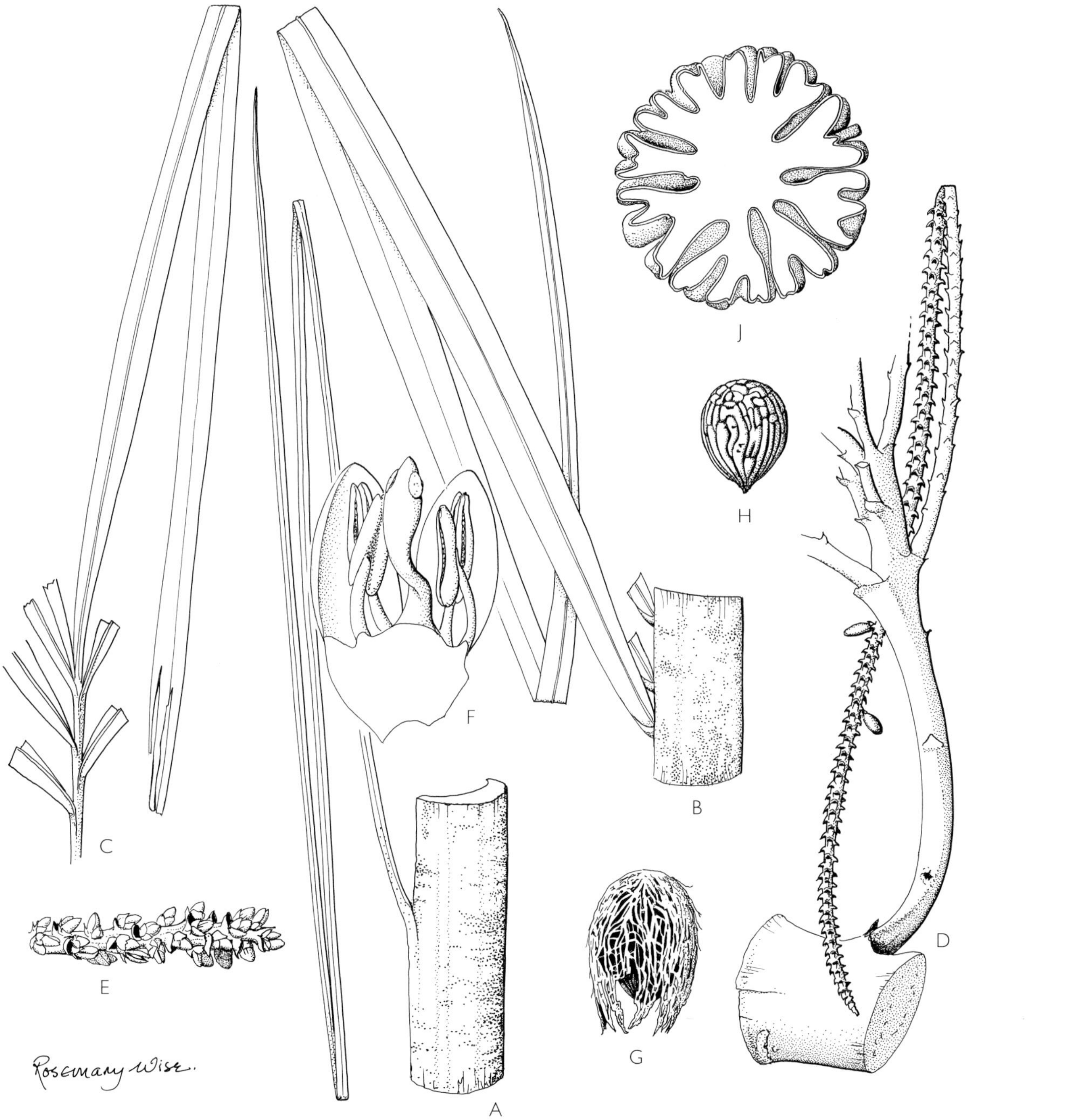

Dypsis bejofo. **A** Proximal part of leaf with lowermost leaflet × 1/3; **B** mid section of leaf × 1/3; **C** leaf tip × 1/2; **D** part of inflorescence showing tip of peduncle and basal first order branch × 1/2; **E** detail of rachilla × 2/3; **F** staminate flower in section × 6; **G** seed enveloped by fibrous endocarp × 1; **H** seed × 2/3; **J** seed in cross section × 2.5. **A**, **C** – **F** from *Dransfield et al.* JD6461, **B**, **G** – **J** from *Dransfield et al.* JD6405. Drawn by Rosemary Wise.

plumose leaves. It is probably close to *D. hovomantsina* and *D. canaliculata*. The seed is most unusual and distinctive and until recently we thought of a form unique within the genus. However, in 1994 JD saw similar, but smaller, seed with entirely different seedlings in cultivation in Australia, seed source unknown, but 'from Madagascar'.

Specimens seen. Maroantsetra: Andranofotsy R, Sahavary, Oct. 1986 (bud, fr.), *Dransfield et al.* JD6405 (Holotype K; isotypi P, TAN), and Feb. 1988 (fl., y.fr.), *Dransfield et al.* JD6461 (K, P, TAN). Mananara Avaratra: Manambato forest, Feb. 1987 (y.fr.), *M. Nicoll* 439 (K, TAN); Antanambe, Oct. 1991 (bud, fr.), *Beentje* 4485 (BH, K, MO, P, TAN).

2. Dypsis canaliculata

This species is something of a mystery. It has not been found since 1951, and the flowers are unknown to science. The two collections made so far are far apart geographically, but seem to belong to the same taxon. The name was given to indicate that the leaf rachis is channelled in its lower half, but this is not so exceptional as Jumelle seemed to think!

Distribution. Only known from the Manongarivo area and from near Ampasimanolotra.
Habitat. Forest on sandstone; c. 200 m.

Dypsis canaliculata. **A** mid section of leaf × 1/4; **B** subapical portion of leaf × 1/4; **C** detail of undersurface of leaflet × 2; **D** first order branch of inflorescence × 1/2. All from *Perrier* 15413. Drawn by Margaret Tebbs.

LOCAL NAMES. *Lopaka* (Antankarana); *Monimony* (Betsimisaraka).
USES. Palm-heart bitter, said to be poisonous by some.
CONSERVATION STATUS. Possibly extinct; not seen since 1951.

Dypsis canaliculata *(Jum.) Beentje & J. Dransf.* **comb. nov.**

SYNONYM:
Neodypsis canaliculatus Jum., Ann. Inst. Bot.-Géol. Colon. Marseille sér. 4, 2 (2): 9 (1924); Jum., Cat. Pl. Madagascar, Palmae: 17 (1938); Jum. & H. Perrier, Fl. Madagascar 30: 141 (1945). Type: Madagascar, Bas-Sambirano, W base of Mt Kalabenono, Perrier 15413 (Holotype P).

Solitary palm. **TRUNK** 10–15 m high, 30–40 cm diam.; nodal scars very visible. **LEAVES** with sheath c. 1 m long, glabrous, whitish-green and waxy; petiole absent; rachis 3–4 m long (fide Perrier) or c. 6 m (fide Cours), in mid-leaf 2.5–3.5 cm wide, waxy, channelled adaxially at least proximally, rounded abaxially; leaflets bright green, grouped and fanned within the groups, the group interval 2–3 cm, very many, not stiff, interval 0.2–1.3 cm, proximal not seen, median 75–116 x 2.6–3.1 cm, distal to 13–30 x 0.3–0.8 cm, glaucous, with many ramenta proximally, the ramenta bifid, to 2 mm long, 2–3 mm high, and with minute reddish scales over the entire abaxial surface, main vein 1, apex unequally attenuate. **INFLORESCENCE** infrafoliar, branched to 2 orders, short and pendulous; parts seen include first (?) order branches with a glabrous rachis of up to 20 cm, proximally up to 1.7 x 0.8 cm diam., with up to 8 rachillae; rachillae 35–48 cm long, 5–6 mm diam., glabrous, waxy, with distant triads in pits, the rachilla bracts proud, rounded or obtuse. **STAMINATE FLOWERS** unknown. **PISTILLATE FLOWERS** unknown. **FRUIT** ellipsoid or rounded, 25–30 x 17–21 mm when dry. **SEED** subglobose, 15–21 x 12–16 mm; endosperm deeply and densely ruminate.

NOTE. The protologue states the type is *Perrier* 1541, but this is changed in the Flora to 15413 (which is correct). The type at P is currently without fruit or flowers.

In the protologue there are the following data not on the type label: trunk green, 10–15 m high, 30–40 cm diam. Leaves 3–4 m long; sheath green, glabrous; petiole absent, rachis 2.5 cm wide; leaflets irregularly grouped (group interval c. 2 cm), distal 30 x 0.7–0.8 cm. Inflorescence branched to 2 orders; rachillae c. 50 cm.

This taxon resembles *D. bejofo*, but is distinct in the almost continuous ramenta on the midrib and the scattered scales, and by the absence of a petiole.

The *Cours* specimen is rather poor, and is included with some doubt. The fruit in this specimen has densely and irregularly ruminate endosperm, distinct from that of *D. bejofo*.

SPECIMENS SEEN. Ambilobe: W base of Mt Kalabenono, no date (infl. only), *Perrier* 15413 (P, type). Ampasimanolotra: near Anivoranokely and Andrambolahikely, April 1951 (fr.), *Cours* 4494 (K, P, TAN).

3. DYPSIS HOVOMANTSINA

An impressive tree; we found a second site only three months before the book went to press. The leaf-sheaths are distinct from any other, with their lower parts white-waxy and a golden-brown furry part at the very top. The name comes directly from the local name, which means stinking and refers to the smelly palm-heart.

DISTRIBUTION. Only known from Maroantsetra and Mananara.

Dypsis hovomantsina. View of crown, showing golden-brown hairs at petiole base, Mananara Avaratra (*Beentje & Dransfield* 4819).

HABITAT. Rain forest, steep slopes either near valley bottom or near ridge crest; 50–300 m.

LOCAL NAMES. *Hovomantsina* (Betsimisaraka; 'mantsina' means stinking)

USES. Palm-heart smelly but still eaten; no other uses.

CONSERVATION STATUS. Critical. Only known from two sites, both of which are under pressure from agriculture; populations consist of few individuals.

Dypsis hovomantsina *Beentje* **sp. nov.**
inter species maximas foliolis aggregatis vagina folii versus basim glabra ceracea apice brunneo-tomentosa staminibus 6 endospermio homogeneo distincta. Typus: Madagascar, Antalavia, *Dransfield et al.* JD6744 (Holotypus K; isotypi P, TAN).

Solitary palm. **TRUNK** 6–15 m tall, 20–30 cm, diam., near the crown 15–20 cm diam.; internodes 10–35 cm, pale to reddish-brown basally, grey-green distally, nodal scars 1–1.5 cm, pale brown; crownshaft well-developed, 1–1.2 m, c. 25 cm diam., grey-green and covered in white wax except distally, where red-brown furry. **LEAVES** spirally inserted, occasionally tristichous, 6–7 in the crown, curved, with an untidy look, the leaflets held in all directions; sheath pale green to whitish, proximally waxy and glabrous, with distal dense soft pale reddish brown persistent tomentum, adaxially peach-coloured; petiole 10–56 cm, 7 × 5.5 cm diam., channelled with sharp edges, with dense red-brown indument; rachis 3–3.5 m long, in mid-leaf 3.5 cm wide, with keel; leaflets 80–96 on each side of the

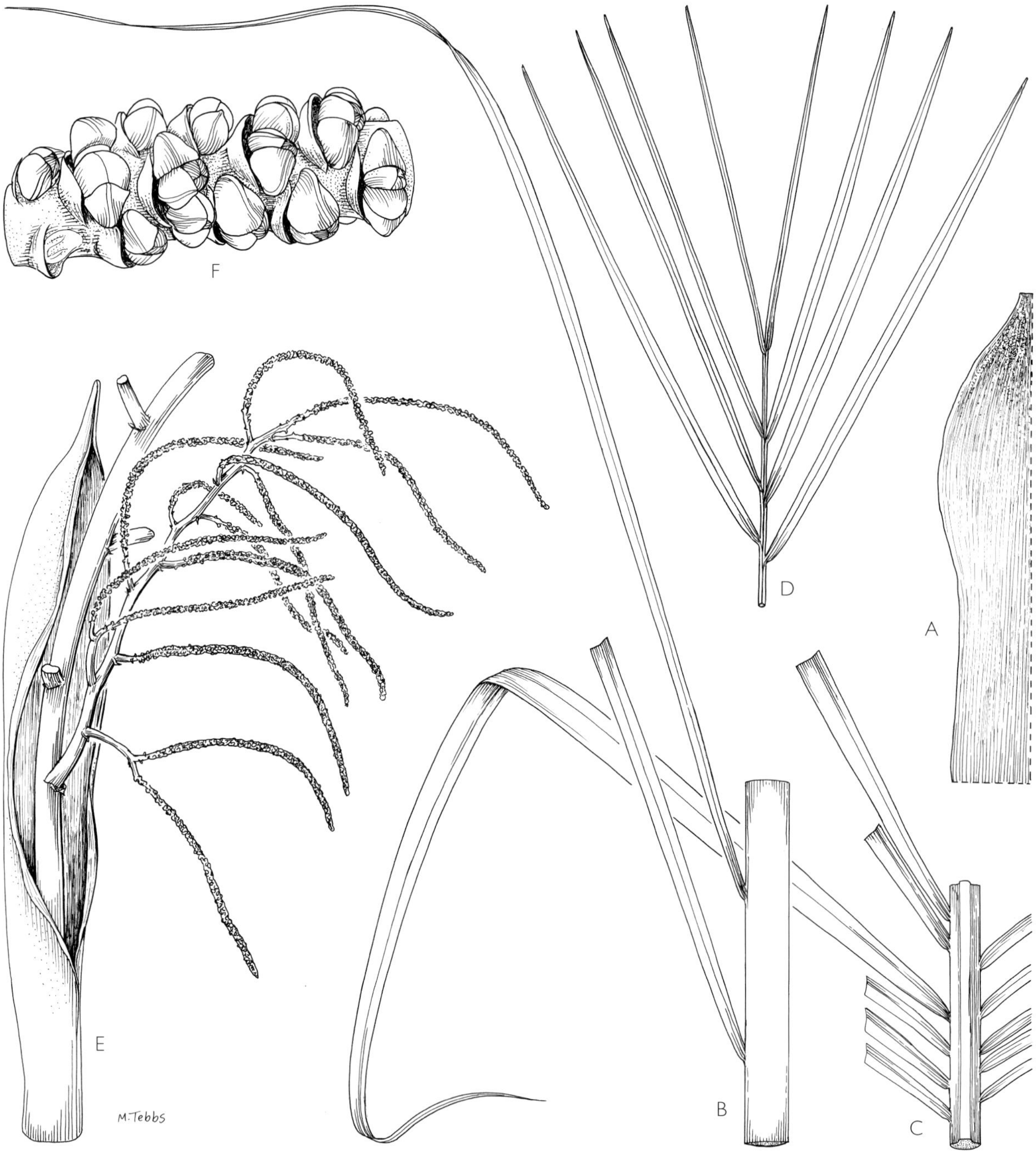

Dypsis hovomantsina. **A** leaf sheath × 1/10; **B** distal part of petiole with lowermost leaflets × 1/6; **C** mid section of leaf × 1/6; **D** leaf tip × 1/6; **E** basal portion of inflorescence rachis × 1/4; **F** detail of rachilla × 3. All from *Dransfield et al.* JD6744. Drawn by Margaret Tebbs.

rachis, grouped and fanned in 3s–6s, the group interval 3–7 cm, bent downwards at two-thirds of their length, the proximal 120–242 × 0.7–2.4 cm, median 123–135 × 3.3–4 cm, distal 22–50 × 0.6–2 cm, connate for 1.5–4 cm, with a few large laciniate ramenta, without scattered scales. **INFLORESCENCE** infrafoliar, c. 0.5–1 × 0.5–1 m, branched to 2–3 orders; peduncle 18–26 cm, diam. 7 × 3 cm; prophyll 21–30 cm, borne at 3–4 cm above the base of the peduncle; peduncular bract deciduous, inserted at 9–11 cm from the base of the peduncle, 45–66 × 10–12 cm, not beaked or briefly beaked with triangular beak, split for 90%; non-tubular peduncular bract occasionally present on upper peduncle, to 20 cm; rachis 48 cm, with 14 branched and 10 unbranched first order branches; rachillae stiff or sub-pendulous, green to yellow-green, 16–40 cm long, 4–8 mm diam., glabrous, with dense flowers. **STAMINATE FLOWERS** only known in rather young bud, with sepals 2.1–3 × 2.5–2.8 mm, very hooded; petals < 1.5 mm; stamens 6, < 1.3 mm, probably versatile. **PISTILLATE FLOWERS** only known from the young fruit stage, with sepals 2.7–3.1 × 3–3.3 mm; petals 3.5–3.8 × 3–4 mm; staminodes 6, flat, tooth-shaped, 0.5 mm. **FRUIT** not known, except for the fibrous endocarp. **SEED** obovoid with pointed base, 9–10 × 7–8 mm, with homogeneous endosperm. **EOPHYLL** entire, deeply bifid.

NOTE. Possibly closest to *D. tokoravina*, but with a totally different leaf sheath.

SPECIMENS SEEN. Maroantsetra: Antalavia, Feb. 1988 (y.fr.), *Dransfield et al.* JD6473 (K, P, TAN); idem, Nov. 1989 (bud), *Dransfield et al.* JD6744 (Holotype K; isotypes P, TAN). Mananara Avaratra: Antanambe, Oct. 1994 (fl., fr.), *Beentje & Dransfield* 4819 (K, TAN); idem, Oct. 1994 (bud), *Beentje & Dransfield* 4827 (K, TAN).

4. DYPSIS CERACEA

This species has not been collected for almost fifty years, and the material is rather poor. This is another mystery, resembling *D. bejofo* and *D. canaliculata*.

DISTRIBUTION. Only known from the Marojejy area and Betampona.

HABITAT. Moist lowland forest; 450 m (fide Cours).
LOCAL NAMES. *Lafaza* (Betsimisaraka).
USES. Leaves used in thatching.
CONSERVATION STATUS. ?Extinct; not seen since 1949.

Dypsis ceracea *(Jum.) Beentje & J. Dransf.* **comb. nov.**

SYNONYM:
Neodypsis ceraceus Jum., Ann. Inst. Bot.-Géol. Colon. Marseille sér. 5, 1 (1): 18 (1933); Jum., Cat. Pl. Madagascar, Palmae: 17 (1938); Jum. & H. Perrier, Fl. Madagascar 30: 144 (1945). Type: Madagascar, Betampona near Ambodiriana, *Perrier* 17474 (Holotype P).

Solitary palm. **TRUNK** 8–15 m high, cylindrical, 8–15 cm diam. **LEAVES:** sheath medium brown with a dense layer of wax, the young sheath bright pink, without ligules or auricles, distally with medium-sized laciniate scales; petiole c. 35 cm long, waxy, with scattered scales; rachis in mid-leaf 1.4–1.8 cm wide, keeled, with scattered scales, slightly waxy; leaflets in groups of 4–6, the group interval > 6 cm, the leaflet interval 0.5–1 cm, "hanging towards the middle" (Perrier); proximal 82–86 × 2.5–3.2 cm, median 55–92 × 2.3–2.9 cm, distal 13–39 × 0.3–2 cm, adaxially dark green, abaxially glaucous, with sparse large ramenta (6–10 mm long) on the proximal midrib, with minute reddish scales scattered over the minor veins, main vein 1, with slightly thickened margins, apex unequally attenuate. **INFLORESCENCE** interfoliar, branched to 3 orders (or more); prophyll coriaceous; peduncular bract coriaceous, deciduous, with a rather long beak; first order branch seen with secondary rachis of 6–20 cm, proximally 1–3 × 0.5–1.3 cm, glabrous, with 7–12 second order branches; rachis bracts up to 12 mm long; rachillae 16–30 cm long, 3–4.5 mm diam., with distant triads in shallow pits; rachilla bracts proud, rounded to acute. **STAMINATE FLOWERS** not seen. **PISTILLATE FLOWERS** not seen at anthesis; in the fruit the sepals are 3.7–4.1 × 3.9–4.7 mm (the outermost largest), obtuse, with membranous margins; petals 5–6 × 3.9–4.7 mm, the innermost largest; staminodes 0.5–1.1 mm, thin and flat. **FRUIT** ellipsoid, 16–20 × 8.5–12.5 mm, rounded at base and apex, with sub-basal stigmatic remains (with a ridge to the stalk); endocarp fibrous, with slightly anasto-

mosing fibres. **SEED** ellipsoid, 12–13 × 5–6 mm, pointed at the base, rounded at the apex, attached to the spot corresponding to the stigmatic remnant on the outside; endosperm deeply ruminate, the ruminations reaching the middle.

NOTE. The following data from the protologue do not appear on the type label: found at 100 m altitude. This species resembles both *D. bejofo* and *D. canaliculata* but is distinct in the smaller and thinner trunk, the more branched inflorescence and the much smaller fruit.

SPECIMENS SEEN. Andapa: Mt Mainampango, Amtalavanio, Jan. 1949 (fr.), *Cours* 3234 (K, P, TAN). Toamasina: Betampona, Dec. 1925 (fr.), *Perrier* 17474 (P, type).

5. DYPSIS AMPASINDAVAE

A rare species, with the leaves in three ranks. The new name refers to the village nearest to the forest where the type was found, and where HB stayed when he refound the species ('the sand across the bay' in Malagasy).

DISTRIBUTION. Restricted to Nosy Be and Manongarivo Mountains.
HABITAT. Moist lowland forest, on steep mid slope; 10–200 m.
LOCAL NAMES. *Lavaboka* (Tsimihety).

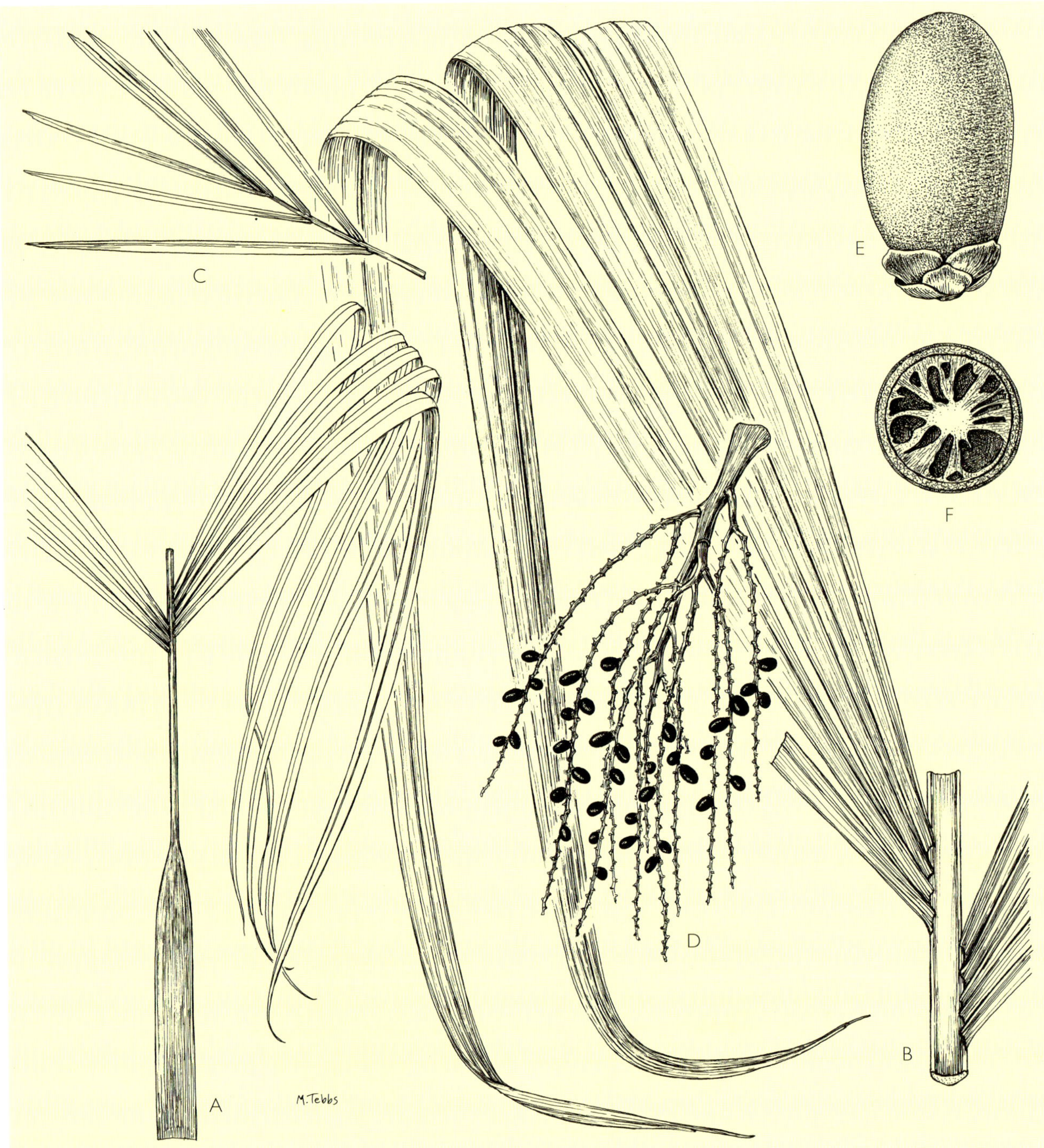

Dypsis ceracea. **A** leaf sheath, petiole and basal leaflets × 1/5; **B** mid section of leaf × 1/2; **C** leaf tip × 1/5; **D** part of infructescence × 1/5; **E** fruit × 2; **F** fruit in cross section × 2. All from *Perrier* 17474. Drawn by Margaret Tebbs.

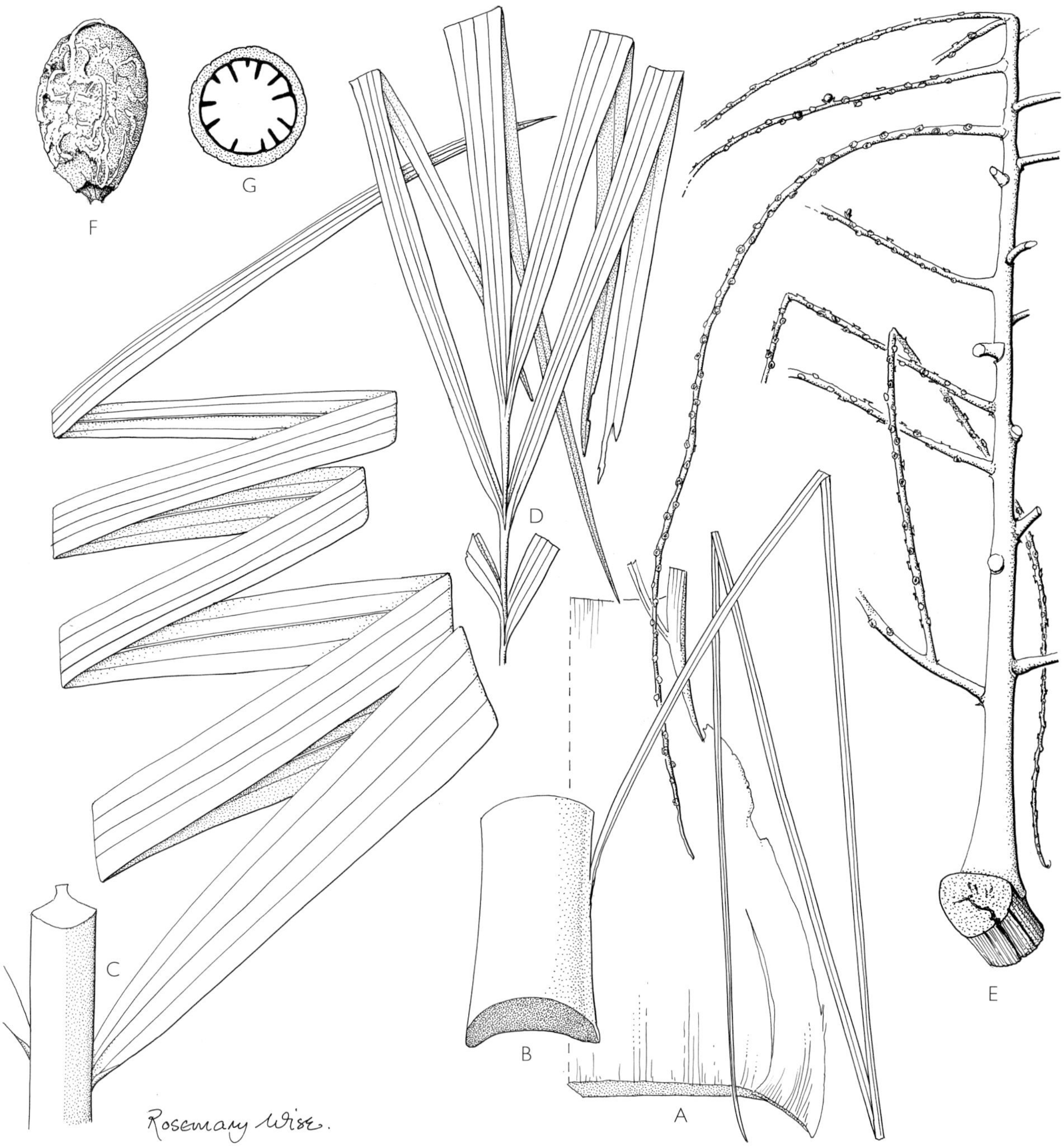

Dypsis ampasindavae. **A** distal part of leaf sheath × 1/3; **B** lowermost leaflet × 1/3; **C** mid section of leaf × 1/3; **D** leaf tip × 1/3; **E** first order branch of inflorescence × 1/3; **F** mature fruit × 2; **G** mature fruit in cross section × 2. **A** – **E** from *Beentje & Andriampaniry* 4700, **F**, **G** from *Beentje & Andriampaniry* 4583. Drawn by Rosemary Wise.

USES. Used in house construction; palm-heart eaten. **CONSERVATION STATUS.** Endangered. Known from only two sites, both of which are protected; but tree poaching is a severe threat in Lokobe Special Reserve (see Adany *et al.* 1994). HB has seen less than 25 trees in the two sites.

Dypsis ampasindavae *Beentje* **nomen nov.**

SYNONYM:
Neodypsis loucoubensis Jum., Ann. Inst. Bot.-Géol. Colon. Marseille sér. 5, 1 (1): 17 (1933); Jum., Cat. Pl. Madagascar, Palmae: 18 (1938); Jum. & H. Perrier, Fl. Madagascar 30: 156 (1945), **synon. nov.** Type: Nosy Be, Lokobe, *Perrier* 18730 (Holotype P).

NOTE. The epithet 'loucoubensis' is too close to that of *Chrysalidocarpus lucubensis* Becc. of 1906, a synonym of *D. madagascariensis*.

Large solitary palm. **TRUNK** 12–15 m, 18–25 cm diam., with basal swelling 20 cm high and 40–55 cm diam., with surface roots; internodes 10–17 cm long, pale brown; nodal scars 1–3 cm, grey; near the crown 17 cm diam., and the internodes here c. 17 cm long, nodal scars 2 cm. **LEAVES** 9–11 in the crown, tristichous, porrect to arching with pendulous leaflets; sheath 110–146 cm long, 20–23 cm diam., pale green with some wax, half open, without auricles or with slight auricles to 1 cm high; petiole absent or up to 18 cm long, 5–8.5 × 3–4.5 cm diam., channelled, green and with scattered scales or glabrous; rachis 3.6–5 m long, in mid-leaf 2–3 × 1.8–2.5 cm diam. and keeled, with scattered scales; leaflets 84–103 on each side of the rachis, regular, pale or mid-green, in one plane but with the distal part pendulous, the proximal 105–137 × 0.8–2.1 cm, median 105–170 × 2.7–5.1 cm (interval 2.5–4 cm), distal 18–30 × 0.4–2.3 cm, the distal pair joined for c. 10 cm, main veins 1, quite glabrous, apices acute and bilobed. **INFLORESCENCE** infrafoliar, erect and spreading, 1.3–1.5 × 1.2–1.4 m, branched to 3 orders; peduncle 25–34 cm long, proximally 6.5–12 × 3.5–7 cm diam., with scattered scales; prophyll 50–60 × 10–13 × 4 cm, borne at 7–9 cm above the base of the peduncle; peduncular bract inserted at 15–16 cm from the base of the peduncle, c. 90 cm long, 14 × 9 cm diam., green with a white bloom, beaked for 6 cm; rachis 90–120 cm long, pale whitish green, with 18–22 branched and 15–21 non-branched first order branches, the most proximal of these with a rachis of up to 32 cm, at its base to 2.5–5 × 1.2–1.5 cm diam., with up to 21 second order branches and rachillae; rachillae whitish green, 24–58 cm long, 3.5–5 mm diam., glabrous; triads distant, in slight pits; rachilla bract small, acute. **STAMINATE FLOWERS** cream; sepals 1.5–2.1 × 1.4–1.8 mm, broadly ovate, gibbous proximally, keeled, acute to truncate, with ciliolate membranous margins; petals 3–3.4 × 1.7–2 mm, ovate, acute; stamens 6, uniseriate, filaments 1.2–1.5 mm long, thin, anthers 2–2.3 × 0.5–0.6 mm, dorsifixed, versatile, the locules parallel and obtuse; pistillode c. 1.5 mm high, columnar, 0.5 mm diam. **PISTILLATE FLOWERS** cream; sepals 1.8–2.4 × 2.2–2.6 mm (– 3.6 mm in fruit), broadly ellipsoid, rounded, concave, ciliolate; petals 2.7–3.4 × 2.3–2.5 mm (–4 × 4.5 mm in fruit), concave, striate, with membranous margins, ciliolate; staminodes 6, flat, obtuse, 0.2–0.8 mm long; gynoecium 3–3.4 × 2.2–2.7 mm. **FRUIT** ovoid, 10–13 × 7.5–9 mm, apex obtuse with an asymmetrical point; endosperm fibrous, the fibres slightly anastomizing. **SEED** ellipsoid, 9–11 × 7–8 mm, pointed at the base, rounded at the apex, with a subaequatorial depression; endosperm ruminate, the ruminations distant and 1–2 mm deep.

Dypsis ampasindavae

CONSERVATION STATUS. Endangered. Only known from three sites, two of which are under agricultural pressure; numbers within the populations are low, and we have seen less than thirty altogether.

NOTE. Closest to the other large tristichous species, *D. tsaravoasira* and *D. pilulifera.* Distinct in its longer rachillae and homogeneous endosperm.

SPECIMENS SEEN. Nosy Be: Lokobe, Sept. 1932 (fr.), *Perrier* 18730 (Holotype P); idem, July 1992 (fl.), *Beentje & Andriampaniry* 4697 (K, MO, TAN), and (bud), *Beentje & Andriampaniry* 4700 (BH, K, MO, P, TAN). Ambanja: Manongarivo, S of Ambalafary, Feb. 1992 (fr.), *Beentje et al.* 4583 (BH, K, MO, P, TAN).

6. DYPSIS TSARAVOASIRA

A majestic, tristichous palm. The leaves are in three ranks.Although the material is incomplete, this is clearly a distinct taxon. The name derives from the local name.

DISTRIBUTION. Only known from Marojejy, Maroantsetra and Mananara.
HABITAT. Open primary forest, steep upper slopes or ridgetop hollows; 275–1050 m.
LOCAL NAMES. *Tsaravoasira, Hovotravavy, Lavaboko* (Betsimisaraka).
USES. Palm-heart edible.

Dypsis tsaravoasira *Beentje* **sp. nov.** inter species maximas tristichas foliolis regulariter dispositis inflorescentiis multo ramosis numero foliolorum infra glandibus dispersis tectorum distincta. Typus: Madagascar, Sahavary, hills E of village, Andilampananina, *Dransfield et al.* JD6463 (Holotypus K; isotypi P, TAN)

Solitary palm. **TRUNK** 10–25 m, 18–40 cm diam, prominently ringed above, scarcely below, 7.5–25 cm across near crown, internodes 5–15 cm, pale brown, distally green. Crownshaft green, swollen, 1–1.5 m. Wood pink. **LEAVES** 5–9, tristichous, porrect, stiff to arcuate; sheath 69–150 cm long, 12 cm across, green, distally densely scaly, with or without irregular ligule c. 22 mm; petiole 0–13 cm, distally 4.5 × 4 cm, densely scaly to glabrous; rachis strongly arcuate, 2–3.5 m long, glabrous or scaly, in mid-leaf 2–3 cm wide; leaflets 102–120 on each side of the rachis, regular, stiff to arcuate, dull dark green, the ones on opposite sides of the rachis in one plane or at a slight angle, less conspicuously so near the tip, mid-green, proximal leaflets 68–135 × 0.5–3.1 cm with conspicuous pendulous reins, median 81–127 × 2.2–3 cm (interval 1.5–3 cm), distal 10–48 × 0.2–2 cm, apex single or bifid, unequal, main vein 1, thickened margins, rest faint, scattered tufts of ramenta, and with a few scattered scales on the minor veins. **INFLORESCENCE** infrafoliar, branching to 3 orders, arching with pendulous rachillae; peduncle 22–26 cm, strongly curved, distally c. 4 × 3 cm, with

Dypsis tsaravoasira, the crowns of curving leaves visible in the distance, Antalavia.

scattered scales; prophyll 41–54 cm, borne at 6–9.5 cm above the base of the peduncle, c. 14 cm wide, persistent, pale brown abaxially, chestnut-red adaxially; peduncular bract inserted at c. 14 cm above the base of the peduncle, deciduous; rachis 47–50 cm, first order branches 14–20, 15 × 9 mm across, glabrous or with minute scattered scales, all axes green; rachillae cream-coloured, pendulous, 13–53 cm, 3–4 mm across, glabrous; triads spaced to dense, sunken; rachilla bract obtuse. **Staminate flowers** only known in young bud. **Pistillate flowers** in young fruit with sepals 2.8–3.7 × 4–5.2 mm, ciliolate; petals 4.5–5 × 5.6–6 mm; staminodes c. 1.2 mm long. Young **Fruit** 4–5 × 5–5.5 mm.

Note. This taxon resembles *D. pilulifera* but is distinct by the regular leaflets and the presence of scattered scales on the leaflets.

Specimens seen. Andapa: Marojejy E, N of Mandena, Nov. 1989 (y.fr.), *Dransfield et al.* JD6760 (K, P, TAN). Maroantsetra, Sahavary, hills E of village, Andilampananina, Feb. 1988 (y.fr.), *Dransfield et al.* JD6463 (Holotype K; isotypes P, TAN); Antanambe, Oct. 1991, (dead infl.), *Beentje & Andriampaniry* 4467 (K, TAN).

Allied specimen: close, but with clear differences are two specimens from Andapa: Marojejy, Nov. 1984 (y.fr.), *Dransfield et al.* JD6762 (K, P, TAN); idem, S base of Mt. Beondroka, Oct. 1989 (dead infl.), *Miller & Randrianasolo* 4493 (K, P, TAN); local names: *Voanioala, Lavaboko, Tsaravoasira*. A solitary palm 15–20 m high, 15–20 cm diam., with 8–10 tristichous leaves; leaf sheath 62 cm, hardly scaly; petiole 0–5 cm; rachis c. 4 m with c. 70 leaflets on each side, the median 85–104 × 3.3–3.6 cm. Inflorescence branched to 2 orders; rachillae pink, 11–26 cm long, 3–4.5 mm diam., glabrous,

Dypsis tsaravoasira. **A** proximal part of leaf with distal part of sheath, petiole and lowermost leaflets × 1/3; **B** mid section of leaf × 1/3; **C** leaf tip × 1/3; **D** first order inflorescence branch × 1/3; **E** young staminate flower, one petal removed × 15; **F** first order branch of infructescence × 1/3. All from *Dransfield et al.* JD6463. Drawn by Rosemary Wise.

with distant triads in slight pits. No flowers have been seen; fruit only seen when young.

Similar is possibly *Moore* 9919 from Masoala: Ambohitralanana, April 1971 (fr.) with the leaf sheath c. 90 cm, petiole 45 cm, rachis 3.2 m, but this has interfoliar inflorescences with rachillae 19–29 cm and fruits 16–17 × 12–13 mm, with homogeneous endosperm; the leaves are said to be in five ranks. The local name is *Buresy*.

7. DYPSIS NAUSEOSA

We found this species some eighty years after the type (and up to that time, only) collection by Perrier, though we failed to find it on Vatovavy. The name indicates the supposed poisonous palm-heart, which is also reflected in local names.

DISTRIBUTION. Known only from the coastal parts of Fianarantsoa Province.
HABITAT. Eastern forest, possibly rather dry; 50-200 m.
LOCAL NAMES. *Rahoma*, *Mangidibe* (Antaisaka, means very bitter), *Laafa* (general palm name).
USES. Wood used for roofing beams, outer wood used for floorplanks. Palm-heart bitter, said to be poisonous.
CONSERVATION STATUS. Critical. Only known from two recent collections, both from areas where the forest is being cut down at a rapid rate. Less than twenty trees are known to exist.

Dypsis nauseosa *(Jum. & H. Perrier) Beentje & J. Dransf.* **comb. nov.**

SYNONYM:
Neodypsis nauseosus Jum. & H. Perrier, Ann. Inst. Bot.-Géol. Colon. Marseille sér. 3, 1 (1): 19, pl. VI (1913); Jum., Ann. Inst. Bot.-Géol. Colon. Marseille sér. 4, 2 (2): 31 (1924); Jum., Cat. Pl. Madagascar, Palmae: 18 (1938); Jum. & H. Perrier, Fl. Madagascar 30: 156 (1945). Type: Madagascar, Mt Vatovavy, *Perrier* 12087 (Holotype P).

Solitary palm. **TRUNK** 7.5–15 m, 20–25 cm diam., with a basal swelling 47–58 cm diam.; internodes 10–23 cm, pale brown, nodal scars c. 3 cm; wood medium soft, white or pink. **LEAVES**

Dypsis nauseosa. Inflorescence (*Beentje & Andriampaniry* 4510).

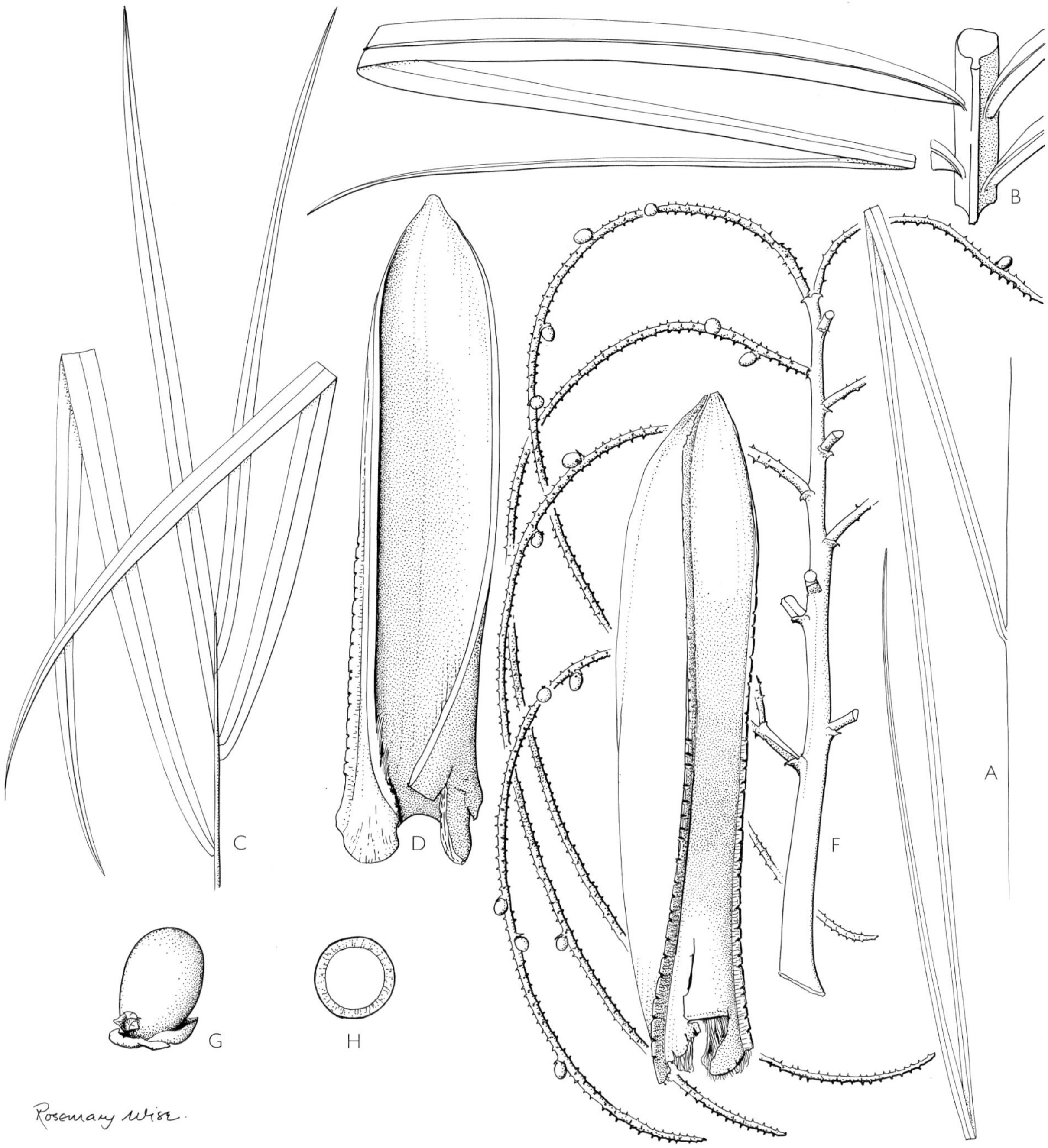

Dypsis nauseosa. **A** lowermost leaflet × 1/4; **B** mid section of leaf × 1/4; **C** leaf tip × 1/4; **D** prophyll, adaxial view × 1/5; **E** prophyll, abaxial view × 1/5; **F** first order branch of infructescence × 1/3; **G** fruit × 1; **H** fruit in cross section × 1. All from *Beentje & Andriampaniry* 4510. Drawn by Rosemary Wise.

12–13 in the crown, spirally arranged, stiff, porrect-spreading; sheath (open?) 91–105 cm and 26–28 cm diam., without ligules or with ligules c. 2 × 2 cm, glabrous, green (turning pale brown) with white waxy covering and scattered scales, adaxially cherry-red; petiole absent or to 27 cm long, 6.5–10 cm wide, c. 3.5 cm thick; rachis 3.4–3.9 m long, in mid-leaf c. 3 × 2 cm diam. and here channelled or keeled; leaflets 108–131 on each side of the rachis, regular, in one plane, spreading to almost pendulous, the proximal 78–123 × 1.3–2.2 cm, median 90–133 × 3.3–4.3 cm (interval 2–4 cm), distal 30–50 × 1.2–2.5 cm, top pair not connate, sometimes abaxially with a few small black scales proximally and few ramenta, apices unequally acute, bifid, main vein 1 (– 5). **Inflorescence** infrafoliar, erect and spreading, 150–200 × 80 cm, branched to 3 orders; peduncle 83–96 cm long, proximally c. 8 × 5 cm diam., distally c. 3.5 × 3 cm, curved; prophyll 43–50 × 10–12 cm; peduncular bract deciduous, 70–100 × 14 cm (inserted at 25–35 cm from the base of the peduncle), pale brown with dense but scattered reddish scales, beaked and closed for 9–30 cm; non-tubular peduncular bract c. 1.5 × 4 cm; rachis 50–90 cm long, puberulous, with 13–17 branched and 9–10 unbranched first order branches, the proximal of these with a rachis of up to 18 cm, at base 2.5–3 × 0.4–1 cm diam., and with up to 12 rachillae or second order branches; rachillae 14–47 cm, 3–5 (in fruit up to 6) mm diam., glabrous; rachilla bract c. 1 mm, broadly triangular, acute; triads distant, in shallow pits. **Staminate flowers** in bud ovoid, 2.5 mm high; sepals as large as wide, truncate, keeled. **Pistillate flowers** in fruit with sepals 1.5–2 × 2.3 mm; petals 2.8–4 × 3.3–4.4 mm, unequal, imbricate and broadly ovate proximally, near the apex with a

small triangular part, ciliolate; staminodes six, flat, up to 0.7 mm high. Young **fruit** fleshy; mature fruit not seen. **Seed** ellipsoid with slightly asymmetrical pointed base and rounded apex, 15–16 × 12–14 mm, with a very adherent endocarp, close-fitting, appearing fibrous at base but more distally so anastomosing that it seems a smooth coat with a few pits; sub-basal depression of 3 mm diam.; endosperm ruminate, the intrusions many, wavy and irregular, 2–6 mm deep.

Specimens seen. Mananjary: Mt Vatovavy, Oct. 1911 (fl.), *Perrier* 12087 (Holotype P). Manakara: Amby, May 1992 (dead infl.), *Beentje & Andriampaniry* 4679 (K, TAN). Farafangana: Manombo, Nov. 1991 (y. fr.), *Beentje & Andriampaniry* 4510 (BH, K, MO, P, TAN); idem, Nov. 1991 (seed) *Beentje & Andriampaniry* 4516 (K).

8. Dypsis oropedionis

Although the material of this palm is incomplete, it needs a name; it is clearly distinct, and severely threatened. The specific name is Greek for 'of the plateau', referring to the habitat.

Distribution. Known from two sites on the western side of the central high plateaux.
Habitat. Relict dry evergreen plateau forest in steep-sided valleys; 1100–1450 m.
Local names. Not recorded.
Uses. Not recorded.
Conservation Status. Critical. Numbers within the two sites were low; both populations are without real protection, with their habitat under serious threat from annual fires as well as tree-cutting.

Dypsis nauseosa, a canopy palm of the forest at Manombo.

Dypsis oropedionis *Beentje* **sp. nov.**
inter species maximas foliolis fasciculatis inflorescentiis multo ramosis pedunculo breve rachillis glabris endospermio ruminato distincta. Type: Madagascar: Ambohitsaratelo, *Dransfield et al.* JD6446 (Holotypus K; isotypus TAN).

Solitary palm. **TRUNK** 8–20 m tall, 27–35 cm diam., near the crown c. 20 cm diam.; internodes 10–20 cm, grey-brown but distally green, nodal scars pronounced, 4–5 cm; crownshaft grey-green, covered in white wax. **LEAVES** tristichous, 6–11 in the crown, porrect to spreading; sheath grey-green, white-waxy, 80–157 cm, without auricles; petiole 25–35 cm long, channelled, densely scaly, 5.5–6 × 4.5–5 cm diam.; rachis c. 3.1 m, in mid-leaf c. 3.5 cm wide; leaflets 80–172 on each side of the rachis, in groups of 3–9 and fanned within the groups, stiff with the distal 1/3 pendulous, the proximal 80–93 × 1.7–2.1 cm, median 102–110 × 2–3.5 cm, distal 15–43 × 0.8–1.8 cm, main vein 1, plus thickened margins, with few ramenta to 8 mm long, otherwise glabrous, apices unequally bifid, attenuate. **INFLORESCENCE** infrafoliar, branched to 3 orders, spreading, 85–130 × 90–100 cm; peduncle 18–40 cm, c. 10 × 6 cm diam., glabrous; prophyll c. 72 × 20 cm, borne at c. 11 cm above the base of the peduncle, erect, split over its length; peduncular bract deciduous, inserted at 18 cm from the base of the peduncle, 87 cm long, 24 cm wide, closed for its distal 7 cm beak; rachis with c. 27 branched and 13 unbranched first order branches, these proximally 3.4–4 × 1.4–1.5 cm diam.; rachis bracts to 7 mm long; rachillae spreading-pendulous, cream, 10–37 cm long, 3.5–5 mm diam., glabrous, with distant

Dypsis oropedionis, surviving in a forest fragment in the midst of degraded man-made grassland.

triads. **STAMINATE FLOWERS** with sepals 1.4–1.5 × 1–1.2 mm; petals connate for c. 0.5 mm, the free parts 1.8–2.1 × 1.2–1.3 mm; stamens 6, uniseriate, filaments c. 1 mm and curved, anthers 0.8–0.9 × 0.4–0.5 mm; pistillode c. 1.3 × 0.3–0.4 mm. **PISTILLATE FLOWERS** unknown. **FRUIT** only known from endocarps, ellipsoid, 7.5–10 × 6–7.5 mm, with almost free fibres. **SEED** subglobose to ellipsoid, 7–8 × 5.5–6 mm; endosperm ruminate, with few columnar intrusions to 1.5 mm deep and 1 mm diam.

NOTE. Probably closest to *D. pilulifera*.

SPECIMENS SEEN. Miandrivazo: Ambohitsaratelo, Nov. 1986 (fl., fr.), *Dransfield et al.* JD6446 (Holotype K; isotype TAN). Ankazobe: Manerinerina, April 1992 (fr.), *Beentje et al.* 4659 (K, TAN).

9. DYPSIS PILULIFERA

A large and beautiful tristichous palm, virtually unknown until a few years ago, but found recently in several places.

DISTRIBUTION. Sambirano region, Marojejy, and Mantady.

HABITAT. Moist submontane forest; slight or steep mid slopes; alt. 750–950 m.

LOCAL NAMES. *Ovomamy* (Tsimihety), *Lavaboko* (Marojejy), *Hozatanana* (Mantady).

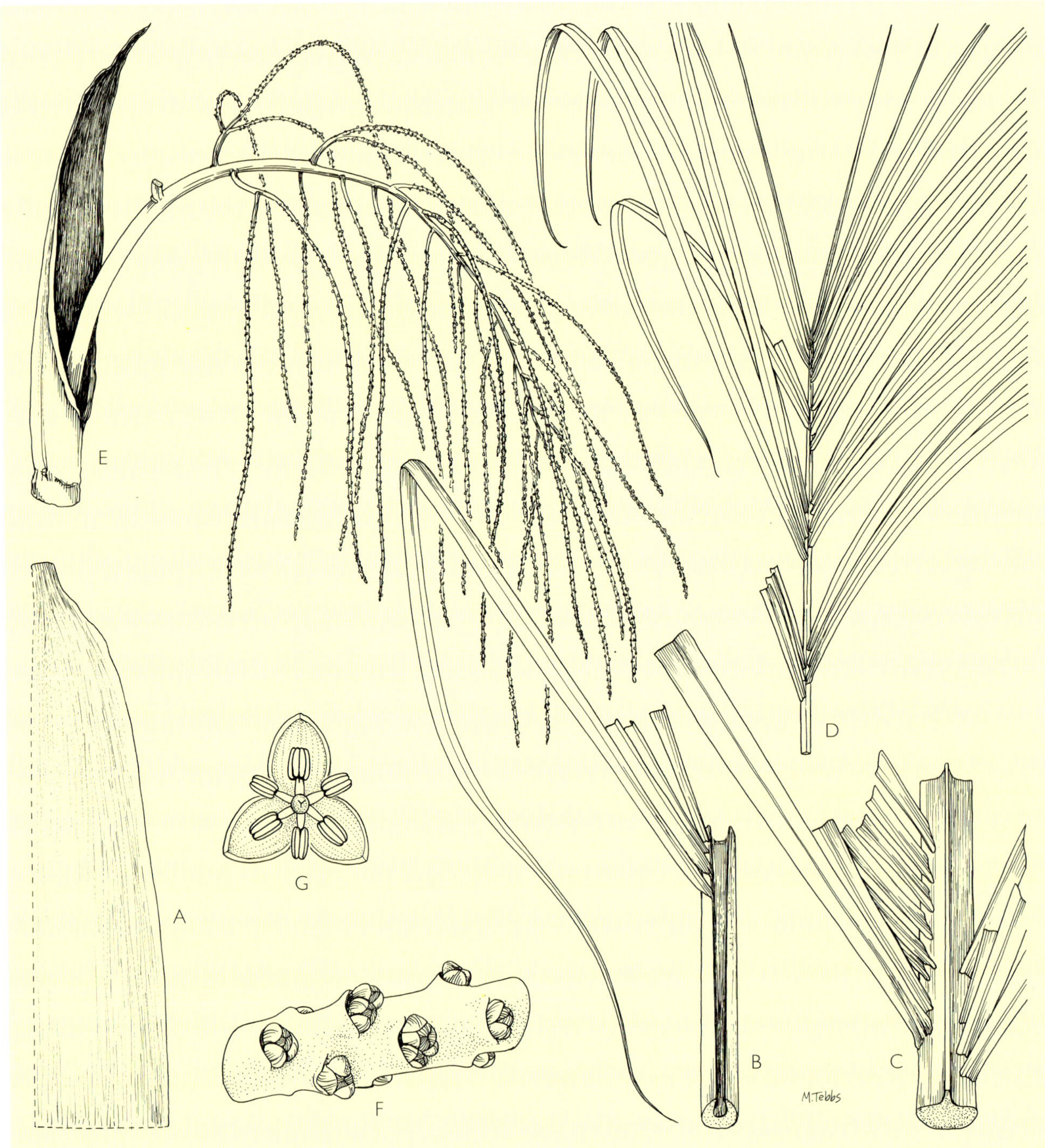

Dypsis oropedionis. **A** leaf sheath × 1/4; **B** distal part of petiole with lowermost leaflets × 1/4; **C** mid section of leaf × 1/4; **D** leaf tip × 1/4; **E** inflorescence × 1/4; **F** detail of rachilla × 3; **G** staminate flower × 6. All from *Dransfield et al.* JD6446. Drawn by Margaret Tebbs.

USES. Good palm heart, and cut down for this.
CONSERVATION STATUS. Vulnerable. Though the populations in the Marojejy and Mantady occur within protected areas, the harvest of the palm-heart continues, with fatal consequences for the trees.

Dypsis pilulifera *(Becc.) Beentje & J. Dransf.* **comb. nov.**

SYNONYMS:
Chrysalidocarpus piluliferus Becc. in Bot. Jahrb. Syst. 38, Beibl. 87: 37 (1906); Jum. & H. Perrier in Ann. Inst. Bot.-Géol. Colon. Marseille sér. 3, 1 (1): 45 (1913); Becc. in Palme del Madag.: 46, t. 45, fig. 37 (1914); Jum. in Ann. Inst. Bot.-Géol. Colon. Marseille sér. 10, 3: 21 (1922); Jum. in Cat. Pl. Madagascar, Palmae: 11 (1938); Jum. & H. Perrier in Fl. Madagascar 30: 118 (1945). Type: Madagascar, without locality, Lastelle s.n. anno 1841 (Holotype P).

Chrysalidocarpus paucifolius Jum., Ann. Inst. Bot.-Géol. Colon. Marseille sér. 10, 3: 9 (1922); Jum., Cat. Pl. Madagascar, Palmae: 11 (1938); Jum. & H. Perrier, Fl. Madagascar 30: 102 (1945). Type: Madagascar, Analamazaotra, *Perrier* 12004 (Holotype P).

Dypsis pilulifera

Solitary palm. **TRUNK** 8–30 m, 10–40 cm diam., 10–12 cm diam. near crownshaft, internodes 20–60 cm proximally, 2–3 cm near the crownshaft, green becoming greyish or dark brown, conspicuously ringed, nodal scars c. 1.5 cm; wood medium hard. **LEAVES** 4–9, tristichous, slightly arching; sheath 1.1–1.7 m long, 3 mm thick, abaxially green, pale brown to peach-coloured with some wax and with slightly sunken large (3–8 x 1–2 mm) reddish-tomentose scales, the continuous cover flaking and glabrescent, with disintegrating auricles of c. 4 x 6 cm; petiole absent or up to 40 cm long, with the same indument as the sheath but with only remnants of the scales present, 7 x 4 cm in diam., canaliculate; rachis 2.9–5 m long, in mid-leaf c. 2 x 1.5 cm, with small scattered scales, proximally canaliculate, distally keeled; leaflets grouped and fanned or slightly irregular, 70–144 on each side of the rachis, closely set, the proximal 56–160 x 0.5–1.8 cm, median 88–124 x 2.4–3.7 cm, distal 10–44 x 0.4–2 cm, glabrous or with minute scattered scales, with distant red ramenta, main vein 1, with unequally acute apices. **INFLORESCENCE** infrafoliar, branched to 3–4 orders, erect in bud, pendulous in fruit; peduncle c. 23 cm long, 8 x 3.5 cm diam.; prophyll 40–50 x 9–15 cm, borne at c. 3 cm above the base of the peduncle, beaked for up to 10 cm, pale brown abaxially, chestnut brown adaxially; peduncular bract inserted at c. 7 cm above the base of the peduncle, 37–70 x 6.5–13 cm, split over its length but closed in the distal part which forms a beak 10 cm long; rachis with up to 20 first order branches; rachillae cream, (7–) 20–40 cm long, 2.5–3 mm diam., with quite dense triads in pits. **STAMINATE FLOWERS** (from detached material) with sepals 1–1.2 x 1.5–2 mm; petals 1.8–2 x 1.8–2 mm; stamens 6, filaments 1.7–1.9 mm, connate at the base for up to 0.2 mm, uniseriate, anthers c. 1.4 x 0.7 mm, versatile; pistillode columnar, 1.8 x 0.7 mm. **PISTILLATE FLOWERS** with sepals 1.3–2.1 x 1.4–2.2 mm, hooded and rounded; petals 2–2.9 x 1.8–2.5 mm; staminodes six, 0.2–0.6 mm high; gynoecium 2–2.8 mm high, 1.5–2.3 mm diam., slightly asymmetrical with spreading terminal stigmas. **FRUIT** globose, 5–7 mm, with fibrous endocarp, the fibres anastomosing. **SEED** globose or nearly so, 4–5 mm, rounded at base and apex with homogeneous endosperm.

Dypsis pilulifera on the slopes of Marojejy (Photo: D.N. Cooke).

NOTE. Despite the fact that the type of *D. pilulifera* consists of an inflorescence with fruits and lacks any vegetative material, it is one and the same as the other populations from the North, rather than *D. mananjarensis*. Another collection by de Lastelle (the type of *D. lastelliana*) is also without locality, but judging by the distribution of that species, must have been from NE Madagascar.

Beccari and Jumelle state that the type of *D. pilulifera* was called *Neodypsis lastelliana* by Baillon, but

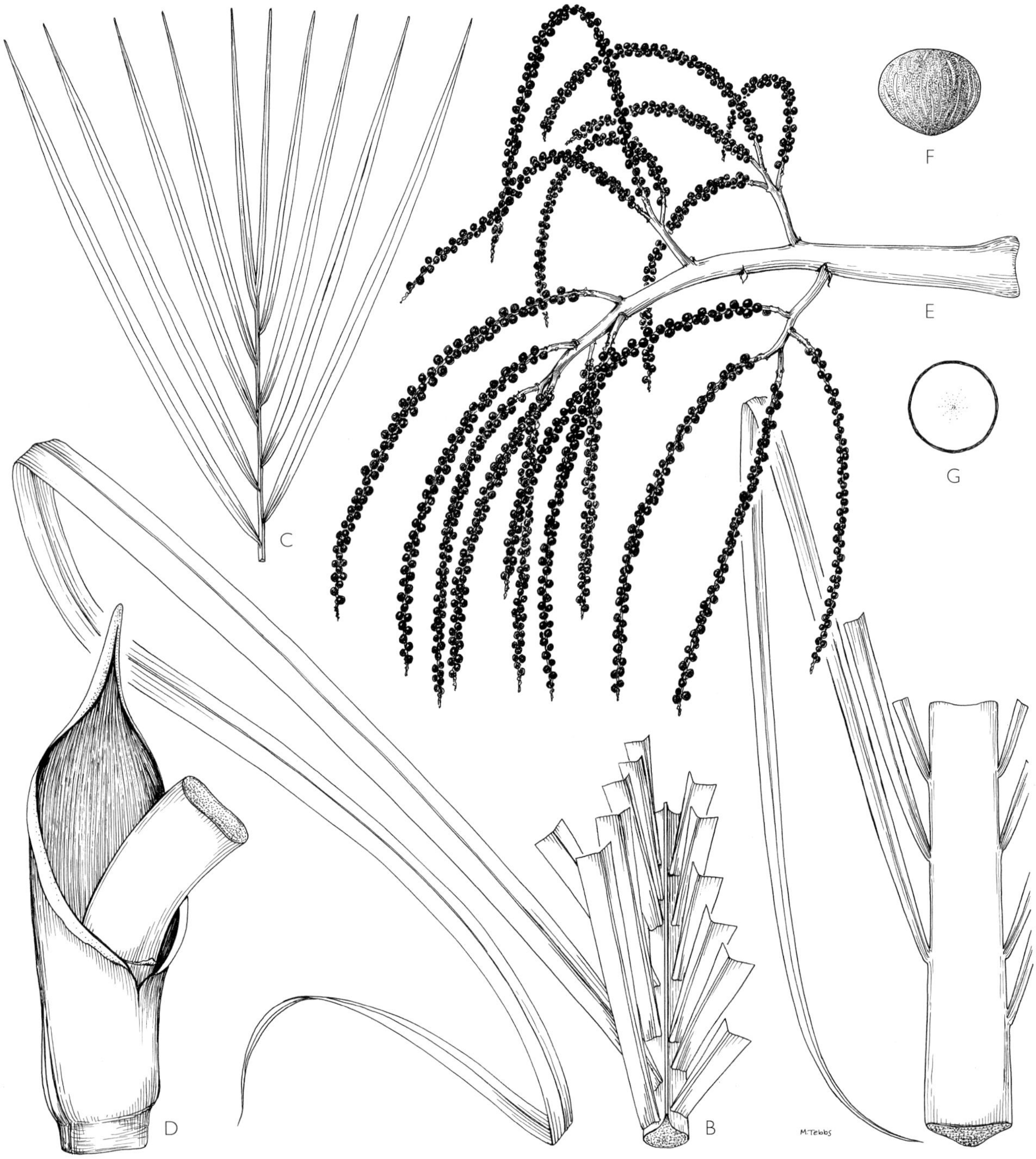

Dypsis pilulifera. **A** basal part of leaf rachis × 1/4; **B** mid section of leaf × 1/4; **C** leaf tip × 1/4; **D** part of peduncle and peduncular bract × 1/5; **E** first order branch of infructescence × 1/5; **F** endocarp × 3; **G** seed in cross section × 3. All from *Beentje et al.* 4574. Drawn by Margaret Tebbs.

in fact Baillon compared it to that taxon, and wrote that it was quite likely a different species.

The material of *Chrysalidocarpus paucifolius* is scanty, but a second collection made in the type area some seventy-five years later is conspecific, and allows us to state with almost complete certainty that this taxon is identical to *D. pilulifera*.

Close to *D. mananjarensis*, from which it can be distinguished by the indument: reddish tomentose scales in *D. pilulifera*, against large white waxy scales in *D. mananjarensis*.

SPECIMENS SEEN. Ambanja: Manongarivo, May 1909 (fl., fr.), *Perrier* 11961 (P); idem, Bekolosi, Feb. 1992 (fl., sd.), *Beentje et al.* 4574 (BH, K, MO, P, TAN); Antsatrotro, March 1993 (fl.), *Malcomber et al.* 2309 (K, P). Andapa: Marojejy Est, Nov. 1989 (bd., fr.), *Dransfield et al.* JD6766 (K, TAN). Moramanga: Analamazaotra, sine die (fl.), *Perrier* 12004 (P; type of *C. paucifolius*); Mantady, April 1992 (dead infl.), *Beentje et al.* 4654 (K).

LOCALITY UNKNOWN: anno 1841 (fr.). *De Lastelle* s.n. (P, type of *D. pilulifera*).

10. DYPSIS MANANJARENSIS

A large and beautiful tristichous palm, in the horticultural world known as 'mealy bug', because of the prominent scales on the leaf sheath, petiole and rachis. These scales are white and waxy, with scalloped edges, and are unique within the genus. The scientific name comes from the fivondronana or 'county' of Mananjary, where the type was collected.

Dypsis mananjarensis

DISTRIBUTION. East coast between Vatomandry and Tolanaro.
HABITAT. Moist or dry forest (remnants), also ?rejuvenating in savoka; slight or steep mid slopes; alt. 30–200 m.
LOCAL NAMES. *Laafa*, *Lakatra* (general palm names on East coast), *Ovodaafa* (Antaisaka).
USES. Good palm heart, and cut down for this; the rachis produces a fibre formerly much used by the Betsimisaraka.
CONSERVATION STATUS. Vulnerable. Though this species may be locally common, the total number HB has seen are only a few hundred. Over the whole distribution area the vegetation is being destroyed for agricultural land or by burning; the cutting for palm-heart targets this palm specifically.

Dypsis mananjarensis *(Becc.) Beentje & J. Dransf.* **comb. nov.**

SYNONYMS:
Chrysalidocarpus mananjarensis Jum. & H. Perrier in Ann. Inst. Bot.-Géol. Colon. Marseille sér. 3, 1 (1): 33, t. 17 (1913); Jum. in Ann. Inst. Bot.-Géol. Colon. Marseille sér. 10, 3: 25 (1922); Jum. in Cat. Pl. Madagascar, Palmae: 10 (1938); Jum. & H. Perrier in Fl. Madagascar 30: 117 (1945). Type: none mentioned, but loc.: Madagascar, Vatovavy forest, *Perrier* 12073 (P).

Chrysalidocarpus fibrosus Jum. in Ann. Inst. Bot.-Géol. Colon. Marseille sér. 10, 3: 10 (1922); Jum. in Cat. Pl. Madagascar, Palmae: 9 (1938); Jum. & H. Perrier in Fl. Madagascar 30: 103 (1945), **synon. nov.** Type: Madagascar, forests in Vatomandry region, *Perrier* 14158 (P). NOTE: within the enlarged genus *Dypsis*, the epithet fibrosa/um has been used earlier for *Dictyosperma fibrosum* = *Vonitra fibrosa* = *Dypsis fibrosa*.

Solitary palm. **TRUNK** 6–25 m, cylindrical, 14–29 cm diam. (to 12 cm diam. near the crown), internodes 10–50 cm (near the crown to 5 cm), green, ringed, nodal scars white, 2–4 cm; wood very hard with red fibres. **LEAVES** tristichous, porrect to arching, 6–10 per crown; sheath 0.6–1.6 m long, a quarter to two-thirds open, whitish green to pale yellow-green to bright orange-yellow in recently exposed sheaths, with characteristic slighly sunken, white, waxy, large (4–12 x 2–3 mm) scales with scalloped edges, auricles c. 1 cm; petiole absent or up to 12 cm long, with scales as the sheath, 5–10 x 3–7 cm diam., deeply canaliculate with sharp edges; rachis 3–3.5 m long, in mid-leaf 2.5–4 cm wide, proximally canaliculate, more distally keeled, pale brown, with scales as the sheath; leaflets slightly irregular or in groups of 3–7, but always in several planes, arching, 121–149 on each side of the rachis, the group interval 1.5–17 cm, basal leaflets 35–150 (–300) x 0.8–2.8 cm, median 100–135 x 2.2–4.6 cm, distal 18–50 x 0.5–2.5 cm, acuminate, glabrous, waxy, main vein 1 with thickened margins. **INFLORESCENCE** infrafoliar, branched to 3 orders, arching and with pendulous rachillae; peduncle 18–40 cm long, 6–12 x 3.5–7 cm diam., pale green, glabrous; prophyll erect, 45–65 x 11–17 cm, clearly 2-winged, borne at 5–8 cm above the base of the peduncle; peduncular bract inserted at 15–18 cm above the base of the peduncle, deciduous, 80–120 x 10–20 cm, beaked for

Dypsis mananjarensis, a magnificent tree palm at Amby *(Beentje & Beentje 4727)*.

6–10 cm, closed in the distal 15 cm, coriaceous, with a waxy bloom; non-tubular peduncular 8–29 × 4–7 cm; rachis 70–106 cm long, pale pastel green with 14–23 branched and 14–20 unbranched first order branches; rachillae whitish to pale pastel green, 17–58 cm long, 3–5 mm diam. **STAMINATE FLOWERS** with sepals 1.3–1.6 × 1.2–1.5 mm, red-spotted; petals 2–2.2 × 1.4–1.5 mm; stamens 6, uniseriate, filaments 1.7–2 mm, anthers 1–1.3 × 0.6–0.7 mm, versatile; pistillode 1.5–1.7 × 0.4–0.7 mm. **PISTILLATE FLOWERS** with sepals 1.5–2 × 1.6–2.3 mm, petals 2.2–2.4 × 1.8–2.3 mm; staminodes 0.2–0.4 mm; gynoecium 2–3.5 × 2–4 mm. **FRUIT** globose or nearly so, 4–6 mm, with fibrous endocarp, the fibres slightly anastomosing. **SEED** globose, 3.5–4.5 mm, with sub-basal depression, rounded at the base, apiculate or rounded at the apex, with homogeneous endosperm (occasionally with undulate edges).

NOTE. Collections from several large populations in the Manakara area show that the variation within populations may range from almost regular leaflets in almost one plane to grouped and fanned leaflets (Perrier describes the leaflets of *C. fibrosus* as "non groupés mais non tous inserés sur un même plan"). The group interval of 11–17 cm in the type of *C. mananjarensis* is rather extreme, the norm being 1–6 cm; it is possible that the leaf parts in this type come from a leaf from a young plant. The protologue of *C. mananjarensis* states that the type is a

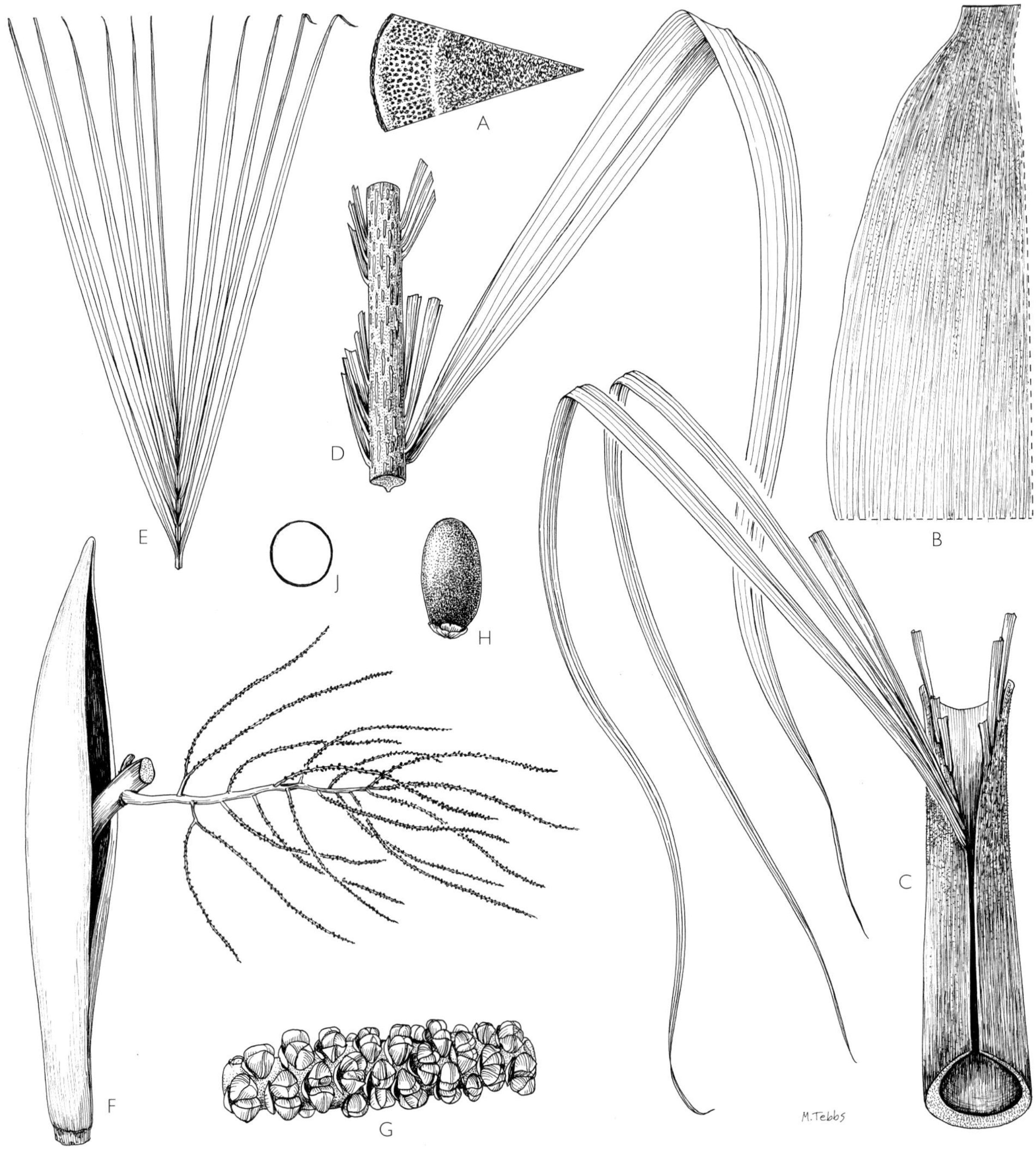

Dypsis mananjarensis. **A** section of stem × 1/6; **B** leaf sheath × 1/6; **C** distal part of petiole with lowermost leaflets × 1/4; **D** mid section of leaf, viewed from beneath, showing patches of white scales × 1/4; **E** leaf tip × 1/4; **F** basal portion of inflorescence rachis and part of peduncle and peduncular bract × 1/4; **G** detail of rachilla × 2; **H** young fruit × 3; **J** fruit in cross section × 3. All from *Beentje & Beentje* 4727. Drawn by Margaret Tebbs.

robust palm with a short trunk 4–6 m high, and this is certainly the size of an young tree; all flowering or fruiting specimens we have observed had trunks over 6 m tall, and usually over 10 m.

Specimens seen. Vatomandry: 6 km S of Ambalatenina, Oct. 1991 (y. fr.), *Beentje* 4506 (BH, K, MO, P, TAN); Levazy, SW of Vatomandry, Dec. 1921 (bud, fr.), *Perrier* 14158 (P, type of *D. fibrosa*). Ifanadiana: Ambinanindrano, Jan. 1993 (stam.fl.), *Beentje & Andriampaniry* 4798 (K). Mananjary: Mt Vatovavy, Oct. 1911 (bud, fr.), *Perrier* 12073 (P, type of *D. mananjarensis*). Manakara: Amby, July 1992 (y.fr.), *Beentje & Beentje* 4727 (BH, K, MO, P, TAN); idem, July 1992 (stam.fl.), *Beentje* 4728 (K); idem, Jan. 1993 (fr.), *Beentje & Andriampaniry* 4796 (K, TAN). Farafangana: Manombo, Nov. 1991 (bud), *Beentje* 4521 (BH, K, MO, P, TAN). Tolanaro: 6km N of Mandromodromotra, March 1992 (y.fr.), *Beentje & Andriampaniry* 4613 (K, TAN).

Sight record: Ifanadiana: Ranomafana, 2 km N of town, 1991, fide Beentje (said to be over 50 years old); HB has also seen seedlings with the typical 'mealy-bug' scales in Ranomafana National Park forest.

Dypsis malcomberi

11. Dypsis malcomberi

An impressive and massive tristichous palm. The species is named for Simon Malcomber, enthusiastic plant collector and tree climber, who took a great deal of trouble to obtain more material of this rare species.

Distribution. Only known from the Andohahela.
Habitat. Moist forest; slight or steep mid slopes, occasionally near forest streams; alt. 400–800 m.
Local names. *Rahosy, Vakaka* (Antanosy).
Uses. The outer wood is used to make planks for walls.
Conservation Status. Vulnerable. Only known from a single small area. Though this species occurs in a protected area (Andohahela is a Reserve Spéciale) the felling for construction purposes seems to continue.

Dypsis malcomberi *Beentje* **sp. nov.**
D. piluliferae et *D. mananjarensi* similis sed vagina folii glabra forma rachillarum bracteas triadium conspicuas et flores confertim tectarum recedit. Typus: Madagascar, Andohahela, Col Tanatana, *Dransfield et al.* JD6779 (Holotypus K; isotypi P, TAN).

Solitary palm. **Trunk** 15–25 m, 25–35 cm diam., distally 14–15 cm diam., at the base slightly bulbous with a few aerial roots, internodes 10–28 cm (distally 5–6 cm), slightly stepped, finely fissured and brown, ringed, nodal scars c. 2 cm. **Leaves** tristichous, 6–8 per crown, plumose; sheath bright green to yellow-green, waxy, swollen, 1.5–2 m, 17–25 cm diam., closed or open for up to one quarter of its length, glabrous, auricles absent or up to 4 × 10 cm; petiole 20–50 cm, 4–8 cm diam., bright green, glabrous, canaliculate with sharp edges; rachis 3–4 m long, bright green, in mid-leaf 4–5 cm diam., proximally canaliculate, distally keeled; leaflets in groups of 2–8 or only slightly irregular, the group interval 1–13 cm, the leaflets fanned to almost in one plane, drooping in their distal part, 135–188 on each side of the rachis, basal 80–110 (–220) × 0.6–3.2 cm, median 93–135 × 2.8–4.6 cm, distal 10–50 × 1–2 cm, attenuate, main vein 1, bright green, not waxy, glabrous but for the ramenta. **Inflorescence** infrafoliar, branched to 3–4 orders, arching and with pendulous rachillae in fruit; peduncle green, 20–30 cm long, 6–16 × 3–5 cm diam., glabrous; prophyll 42–73 × 12–20 cm, borne at 8–9 cm above the base of the peduncle, glabrous, splitting abaxially; peduncular bract borne at 15 cm above the base of the peduncle, deciduous, 72–117 × 16 cm, beaked for 8–21 cm, green, glabrous with a waxy bloom; rachis 60–124 cm, green, glabrous, with 18–21 branched and 12–17 unbranched first order branches; rachillae 15–48

Dypsis malcomberi, a common tree palm at Andohahela.

cm long, 3.5–8 mm diam., glabrous, triads dense with proud, slightly reflexed bracts. **STAMINATE FLOWERS** pale green, with sepals 1.8–2.3 × 2.2–2.8 mm; petals 2.3–2.6 × 1.8–1.9 mm; stamens 6, cream, uniseriate, filaments c. 1 mm in bud, anthers c. 1.3 × 0.6 mm, versatile; pistillode c. 1.5 × 0.5 mm. **PISTILLATE FLOWERS** (in fruit) with sepals 2.6–3.2 × 2.4–2.8 mm; petals 3–3.5 × 3–4.2 mm; staminodes 0.5–1.3 mm, flat; gynoecium not seen. **FRUIT** pale orange, globose to ellipsoid, 8–10 × 4–7 mm, with hardly fibrous endocarp; **SEED** c. 5.5 × 4 mm, with homogeneous endosperm.

NOTE. Clearly allied to *D. mananjarensis* and *D. pilulifera*; though very similar, in the field these three taxa have a different appearance.

Probably the same is: Midongy: 24km S of Midongy, May 1992 (fr.), *Beentje & Andriampaniry* 4670 (K, TAN), but the material is too scanty to be certain; it comes from wet forest at 800 m altitude, but was collected from a tree felled for palm-heart some time before.

SPECIMENS SEEN. Tolanaro: upper Manampanihy valley between Eminiminy and the Saindro Col, Feb. 1934 (fr.), *Humbert* 14032 (P); Col Tanatana, Andohahela, Nov. 1989 (fl., fr.), *Dransfield et al.* JD6779 (K, P, TAN); 13km N of Ezoambo, March 1992 (bud), *Beentje & Andriampaniry* 4596 (BH, K, MO, P, TAN); idem, March 1992 (fr.), *Beentje & Andriampaniry* 4603 (BH, K, MO, P, TAN); Andohahela R, May 1992 (bud), *Malcomber et al.* 1542 (K, MO, TAN).

SIMILAR SPECIMENS: two collections from the Andohahela [Tolanaro: Andohahela, Tsilotsilo Col, Jan./Feb. 1934 (bud), *Humbert* 13977 (K, P); idem, (fl.), *Humbert* 14136 (K, P); alt.

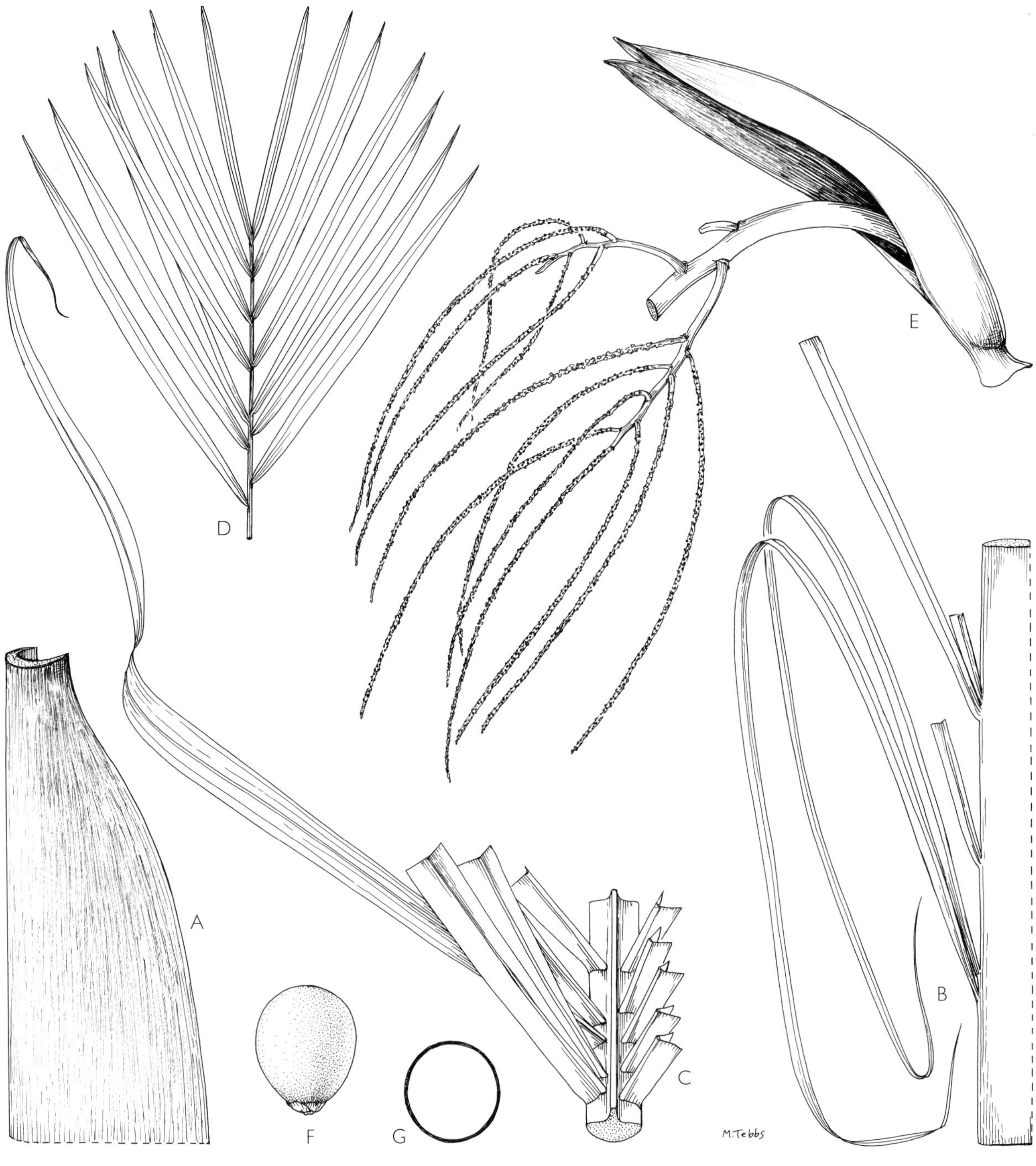

Dypsis malcomberi. **A** leaf sheath × 1/6; **B** distal part of petiole with lowermost leaflets × 1/6; **C** mid section of leaf × 1/6; **D** leaf tip × 1/6; **E** part of inflorescence × 1/6; **F** fruit × 2; **G** fruit in cross section × 2. All from *Dransfield et al.* JD6779. Drawn by Margaret Tebbs.

1000–1200 m, from the natural limits of the moist forest] are similar, but with a different 'feel' to them. They represent a large solitary palm, 10–12 m, and according to Humbert 'very ornamental'. The leaves have a large, waxy sheath; the leaflets are grouped, with lines of small scattered scales, and of similar size to *malcomberi*. The branching of the inflorescence is unclear, but the rachillae are 6–20 cm long and c. 5 mm diam. The staminate flower parts are slightly larger than *malcomberi*; so are the pistillate flower parts, with petals 5–5.2 mm long. The fruit of this taxon is not known.

12. DYPSIS PRESTONIANA

A species which was discovered through serendipity: HB was trying to refind *Chrysalidocarpus midongensis* (now a synonym of *D. onilahensis*), unaware that that taxon came from a totally different Midongy (a small hamlet in the Itremo Mountains), failed to find it, but spotted this elegant tree on the skyline of a hill. The specific epithet honours Mr. Paul Preston, President of McDonald's Restaurants Limited (UK), who sponsored the four-year Palms of Madagascar fellowship.

DISTRIBUTION. Only known from the Midongy area; an old collection from Mahanoro.

Dypsis prestoniana. A massive canopy palm in the forest at Midongy.

HABITAT. Moist forest; slight slope; clay soils derived from laterite; alt. 50–550 m.

LOCAL NAMES. *Tavilo* (Betsimisaraka); *Babovavy, Tavilo* (Antaisaka).

USES. Palm-heart edible (fide *Perrier*).

CONSERVATION STATUS. Vulnerable. At present only known from the Midongy area, where numbers do not exceed two hundred. None occurs in a protected area. HB has visited the Mahanoro area, which is nowadays devoid of tree palms.

Dypsis prestoniana *Beentje* **sp. nov.**
Palma excelsa inter species maximas foliolis aggregatis rachillis dense puberulis endospermio homogeneo distincta. Typus: Madagascar, 6km ENE of Midongy, *Beentje & Andriampaniry* 4672 (Holotypus K; isotypi BH, MO, P, TAN).

Solitary palm. **TRUNK** 4–12 m, 24–40 cm diam.; basal boss about 15 cm high, 40 cm diam.; internodes 10–15 cm, pale grey-brown, distally green and smooth; nodal scars c. 3.5 cm, pale grey. **LEAVES** 8–10 in the crown, spiral, porrect to spreading, arched gracefully in the distal part; sheath 90% open, green to pale brown or grey, with waxy covering, with flaking brown tomentum of peltate scales, c. 114 × 28 cm, with distinct rounded shoulders, the sheath in cross-section slightly triangular with thick middle section and thin flat margins; petiole 0–17 cm, c. 7 × 6 cm diam., deeply channelled; rachis c. 4.4 m, channelled in its proximal part (channel 6 cm wide near petiole) but in midleaf c. 3.5 × 3 cm diam., with slightly sunken circular waxy scales; leaflets c. 164 on each side of the rachis, grouped and fanned within the groups, the groups 3–9 in number, dense and irregular, c. 4 cm apart, the individual leaflets 0.3–2 cm apart, the proximal 76–123 × 1–2.5 cm, the most proximal often with long pendulous reins, median 97–112 × 2.7–4.7 cm, distal 17–50 × 1.3–2.6 cm, main veins 3, plus very thickened margins, abaxially glaucous with white wax, as well as with dense patches of large (– 8 mm) red-brown ramenta on the midrib near the base, leaflet apices unequally bifid and acute. **INFLORESCENCE** interfoliar, erect with spreading or recurved branches, branched to 3 orders, c. 226 × 90 cm; peduncle c. 118 cm long, 12 × 4 cm diam. proximally, 8 × 4.5 cm diam. distally, yellow with brown scales, turning green with red-brown scales;

Dypsis prestoniana. **A** leaf sheath, petiole and basal leaflets × 1/2; **B**, **C** mid section of leaf × 1/2; **D** leaf tip × 1/2; **E** infructescence in young fruit × 1/2 . All from *Beentje & Andriampaniry* 4672. Drawn by Rosemary Wise.

prophyll c. 60 × 20 cm (to 1.2 m, fide Perrier), woody, densely scaly, slightly beaked, only opening near the apex; peduncular bract deciduous (inserted at 55 cm), the distal portion seen with white wax and minute peltate scales, probably beaked; non-sheathing peduncular bract at 91 cm (4.5–5 cm high, base around peduncle), at 105 cm (3.5 cm high, 9 cm wide), at 110 cm (1 × 9 cm); rachis c. 106 cm, densely puberulous, with c. 26 branched and 8 unbranched first order branches; main first order branches with a rachis of up to 42 cm, proximally up to 4 × 1.5 cm diam., with up to 12 branched and 9 unbranched secondary branches; rachillae 9–42 cm, densely puberulous, 2.5–4 mm diam., whitish or pale yellow; triads dense, set in slight pits subtended by entire, rounded bracts. **STAMINATE FLOWERS** in bud with sepals 1.2–1.3 × 1.2–1.3 mm; petals 1.5–1.8 × 1.3–1.4 mm; stamens 6, 1-seriate, with filaments c. 1 mm and narrowly cylindrical, anthers 1–1.2 × 0.4 mm with parallel locules, versatile and dorsifixed; pistillode c. 0.8 × 0.3 mm. **PISTILLATE FLOWERS** with sepals 1.5–2.2 × 1.6–2.4 mm, concave, orbicular, rounded, minutely ciliolate; petals 2–2.6 × 2.1–2.8 mm, broadly obovate with a small apiculus; staminodes 6, 0.4–1 mm high, thin and flat; ovary c. 2.5 × 1.8 mm, asymmetrical. **FRUIT** ellipsoid with rounded apex, 12–15 × 6–8 mm, orange; endocarp fibrous. **SEED** narrowly ellipsoid, 11–12 × 5–5.5 mm, acute at base, rounded at apex, with homogeneous endosperm with slight marginal undulations.

NOTE. Quite distinct from other large solitary species with grouped leaflets by its long and interfoliar inflorescence with densely puberulous rachillae, and homogeneous endosperm.

SPECIMENS SEEN. Mahanoro: Masomeloka, Jan. 1922 (fl.), *Perrier* 14202 (P). Midongy: 6 km ENE of Midongy, May 1992 (fr.), *Beentje & Andriampaniry* 4672 (Holotype K; isotypes BH, MO, P, TAN; type); 8 km ENE of Midongy, May 1992 (fr.), *Beentje & Andriampaniry* 4673 (K, TAN).

13. DYPSIS TOKORAVINA

An immense and very beautiful palm of primary forest. We have not seen the flowers, but the tree is so distinctive, especially with its enormous leaf sheaths which are open and swollen, that it deserves to be named.

DISTRIBUTION. Maroantsetra and Mananara.
HABITAT. Lowland rain forest; edge of swamp in valley bottom and ridge tops; c. 420 m.
LOCAL NAMES. *Tokoravina* (Betsimisaraka; toko means group, ravina leaf).
USES. Not recorded.
CONSERVATION STATUS. Endangered. Only known from two populations, with few (< 20) individual trees known; the population at Antanambe is in an area under agricultural pressure.

Dypsis prestoniana. View of crown with infructescence (*Beentje & Andriampaniry* 4672).

Dypsis tokoravina, growing in the Mananara Biosphere Reserve (Photo: B. Rogers).

Dypsis tokoravina *Beentje* **sp. nov.**
Palma ingens inter species maximas foliolis aggregatis vagina folii reclusa tumida geniculata rachillis longis gracilibus endospermio homogeneo distinctissima. Type: Madagascar, Antalavia, *Dransfield et al.* JD6739 (Holotypus K; isotypi P, TAN).

Solitary palm. **TRUNK** c. 20 m, c. 60 cm diam. at the base, 45–50 cm diam. at breast height, 30 cm diam. near the crown; internodes c. 1.2 cm, dull reddish brown, vertically fissured. **LEAVES** 10–14 in the crown, subtristichous, arcuate; sheath scarcely forming crownshaft, 0.7–1 m long, very swollen, almost kneed, open for much of its length, greyish brown, bright red-brown within, with lateral auricles; petiole 6–34 cm long, c. 6 × 8 cm diam., deeply grooved; rachis grey-brown; rachis c. 2.7 m long, in mid-leaf c. 5 cm wide and 4 cm deep, keeled; leaflets 80–110 on each side of the rachis, stiff, held irregularly in groups of 3–8 in different planes, curled, grey-waxy beneath, interval between the groups 4–5 cm, between the leaflets within the groups 1–2 cm, the proximal c. 115 × 1.3–2.6 cm, median 100–128 × 3.3–4 cm, distal 25–37 × 1–2.3 cm, main vein 1, with thickened leaf margins, with tufts of ramenta and scattered scales on the minor veins, distal pair joined for c. 2 cm, dentate. **INFLORESCENCE** interfoliar, branched to 3 orders, huge, c. 3 m, with rather spreading rachillae; peduncle c. 2 m long, c. 8 cm wide near the base; peduncular bract c. 1 m long, curved, reddish brown tomentose; first order branches with an axis of up to 42 cm, proximally 15 × 9 mm diam., with up to 12 rachillae; rachillae numerous, very slender (c. 2.5 mm diam.), ?glabrous, with shallow pits. **STAMINATE FLOWERS** unknown. **PISTILLATE FLOWERS** unknown. **FRUIT** obovoid, with pointed base, 15–20 × 11–13 mm; endocarp fibrous. **SEED** not seen entire, but with homogeneous endosperm. **EOPHYLL** entire, bifid.

Dypsis tokoravina. **A** proximal part of leaf with lowermost leaflets × 1/3; **B** mid section of leaf × 1/3; **C** leaf tip × 1/3; **D** first order inflorescence branch × 1/3; **E** empty endocarp × 2; **F** seedling × 1/2. **A – D** from *Dransfield et al.* JD6739, **E**, **F** from *Dransfield & Beentje*. JD7507. Drawn by Rosemary Wise.

NOTE. Though this species slightly resembles giants such as *D. bejofo* and *D. pilulifera*, the open leaf sheaths make it immediately recognizable. The open sheath and large size are reminiscent of *D. prestoniana*, but that taxon has a more slender trunk.

SPECIMENS SEEN. Maroantsetra: Antalavia, Nov. 1989 (dead infl.), *Dransfield et al.* JD6739 (K, P, TAN, type). Mananara Avaratra: Antanambe, Oct. 1994 (fr.), *Dransfield & Beentje* 7507 (K, TAN).

14. DYPSIS IFANADIANAE

A rather slender tree palm reminiscent of *D. lastelliana* through the pendulous leaflets, but immediately distinct from that species by its green leaf-sheath. The name comes from the nearest large town and administrative unit.

DISTRIBUTION. Only known from the Ifanadiana area.
HABITAT. Lowland rain forest; steep mid slopes; 200–450 m.
LOCAL NAMES. Not recorded.
USES. Not recorded.
CONSERVATION STATUS. Critical. Restricted to a small area, where clearing of forest is continuing apace; less than fifty trees known.

Dypsis ifanadianae *Beentje* **sp. nov.**
Palma insignis habitu *D. lastellianae* superficialiter similis sed vagina folii viride non brunneo-tomentosa, *D. nauseosae* affinis sed

Dypsis ifanadianae, a tree palm growing near Ifanadiana (*Beentje & Beentje* 4729).

fructu parvo endospermio homogeneo differt. Typus: Madagascar, Ifanadiana: 34km E of Ifanadiana, *Beentje & Beentje* 4729 (Holotypus K; isotypi MO, TAN).

Slender solitary palm. **TRUNK** 15–24 m tall, conspicuously stepped and ringed, 18 cm diam.; internodes 12 cm long, orange-brown with white nodal scars 1.5 cm broad, with age the whole trunk becoming pale grey to almost white; upper internodes green with white scars. **LEAVES** 7 in the crown, spirally inserted, porrect with leaflets in one plane but curving downwards; sheath 72 cm long, at the base 20 cm wide (when flattened), green, 50–75% open, with minute scattered scales, with brown triangular ligules 5 × 2.3 cm; petiole 30 cm long, proximally 6 × 4 cm, distally 4 × 3 cm, channelled with sharp edges, waxy and with minute scattered scales; rachis c. 3 m long, proximally channelled, in mid-leaf 1.4–2.5 cm diam., sharply keeled, waxy and ? with minute scattered scales; leaflets regular, c. 55 on each side of the rachis, the proximal 90–108 × 1–3 cm, median 104–110 × 4.3–5 cm (interval 4 cm), distal 15–37 × 0.4–2.2 cm, glaucous and glabrous, with attenuate apices, main veins 5, prominent. **INFLORESCENCE** infrafoliar, with curved peduncle, branched to 3 orders; peduncle 57 cm long, proximally 4 × 2 cm, distally 3 × 2 cm diam.; prophyll c. 89 cm, borne at c. 5 cm above the base of the peduncle, 10 cm wide, 2-winged; peduncular bract inserted at 26 cm from the base of the peduncle, 65 cm long, the distal 18 cm closed, beaked for 4 cm; rachis 60 cm long, with c. 21 branched and c. 16 unbranched first order branches, these proximally 2.8 × 1.6 cm; rachillae 12–33 cm long, 3–4 mm diam., with distant triads in slight pits, with triangular, entire bracts. **STAMINATE FLOWERS** unknown. **PISTILLATE FLOWERS** with imbricate sepals 2.5–3 × 2.5–3.5 mm, orbicular, rounded, slightly spurred at the base; petals

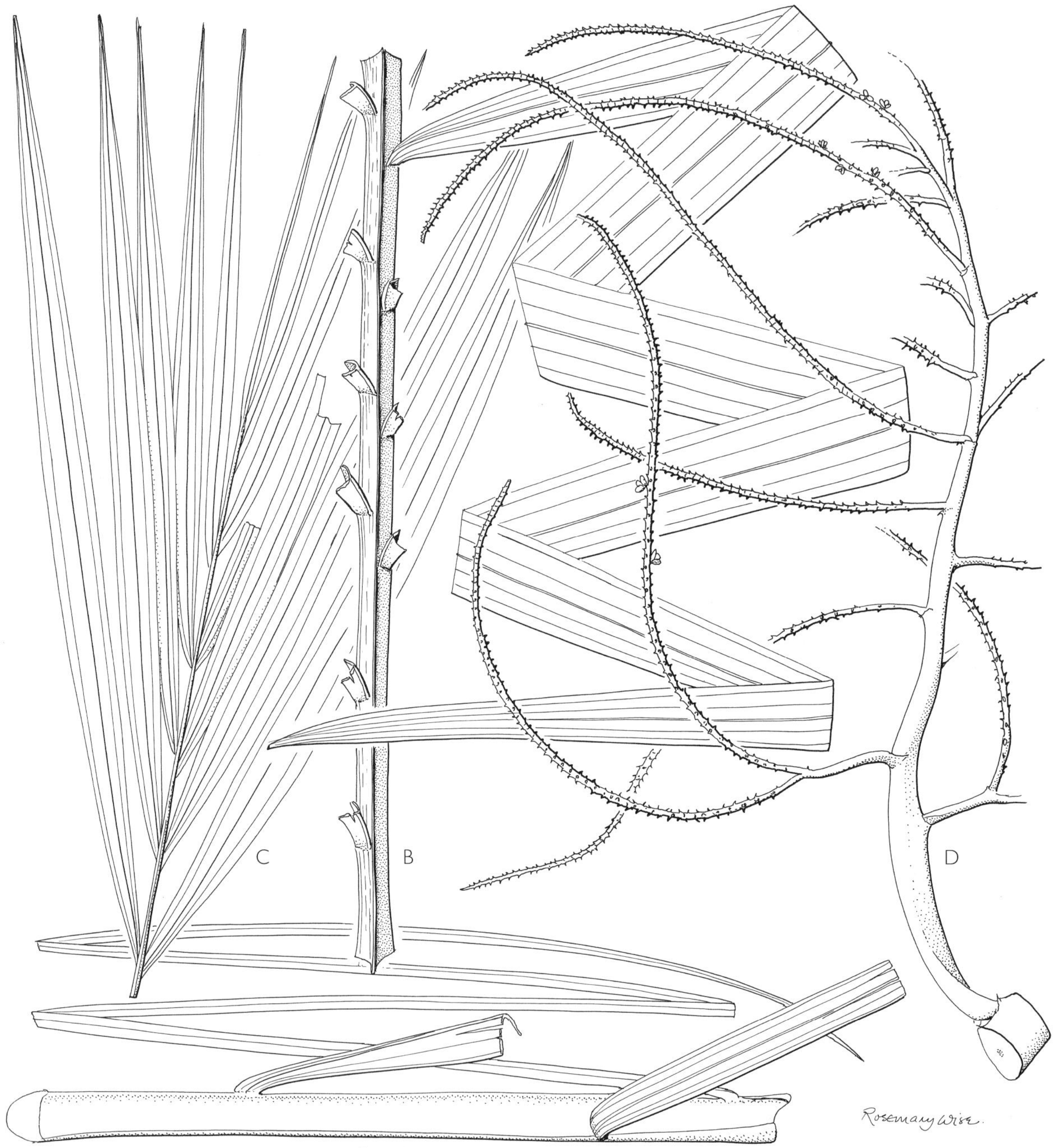

Dypsis ifanadianae. **A** distal part of petiole with lowermost leaflets × 1/3; **B** mid section of leaf × 1/3; **C** leaf tip × 1/3; **D** first order inflorescence branch × 1/3. All from *Beentje & Beentje* 4729. Drawn by Rosemary Wise.

4–4.5 x 3–5 mm, hooded, orbicular with minute acute apex; staminodes 0.8–1.3 mm, flat and narrow; ovary c. 3.5 x 4.5 mm, asymmetrical, the stigmas slightly lateral and 1.3 mm high. **FRUIT** 8 x 7–10 mm; endocarp with anastomosing fibres. **SEED** transversely ellipsoid, broader in one plane than in the other, 6.5 x 5.5 x 8–9 mm, with homogeneous endosperm. **EOPHYLL** bifid.

NOTE. This species is rather close to *D. nauseosa* but distinct by its small fruit with homogeneous endosperm.

SPECIMENS SEEN. Ifanadiana: 34km E of Ifanadiana, July 1992 (fl., fr.) *Beentje & Beentje* 4729 (Holotype K; isotypes MO, TAN).

SIGHT RECORD. This species has been observed by HB at Ambinanindrano and Andranomaitso.

15. DYPSIS LASTELLIANA

This is a very elegant species, now widespread in cultivation. The densely red-brown hairy crownshaft is particularly ornamental. In cultivation it seems to require warmer and moister conditions than *D. leptocheilos*. The name refers to the collector of the type, de Lastellé.

DISTRIBUTION. NW, NE and E Madagascar.
HABITAT. Moist lowland forest on slopes (gneiss, quartz, granite), often rather open or near forest mar-

Dypsis lastelliana. **A** portion of leaf sheath x 1/3; **B** petiole and basal leaflets x 1/3; **C** mid section of leaf x 1/3; **D** leaf tip x 1/3; **E** peduncular bract x 1/12; **F** part of infructescence x 1/5; **G** fruit x 2/3; **H** fruit in cross section x 2/3. All from *Dransfield et al.* JD6359. Drawn by Margaret Tebbs.

gins, or near water, in ravines, also in coastal forest on white sand; 1–450 m.

LOCAL NAMES. *Menavozona* (Betsimisaraka, meaning Red Neck, referring to the leaf sheath), *Sira* or *Ravintsira* (Betsimisaraka, meaning Salt or Salt Leaf, referring to the former practice of making salt from the pith).

USES. Pith formerly used to make salt; palm-heart bitter, inedible, said to be poisonous by the Sakalava and Tsimihety.

CONSERVATION STATUS. Not threatened. Seems to be fairly widespread, over a wide altitude range.

Dypsis lastelliana *(Baill.) Beentje & J. Dransf.* **comb. nov.**

SYNONYM:

Neodypsis lastelliana Baill., Bull. Soc. Linn. Paris 148: 1172 (1894); Becc., Bot. Jahrb. Syst. 38, Beibl. 87: 38 (1906), as *lastelleana*; Becc., Palme del Madagascar 5, fig. 2, t. 3 (1912), as *lastelleana*; Jum. & H. Perrier, Ann. Inst. Bot.-Géol. Colon. Marseille sér. 3, 1 (1): 12, fig. 5, t. 2–3 (1913), as *lastelleana*; Jum., Ann. Inst. Bot.-Géol. Colon. Marseille sér. 4, 2 (2): 16 (1924); Jum., Cat. Pl. Madagascar, Palmae: 18 (1938); Jum. & H. Perrier, Fl. Madagascar 30: 150, fig. 42 (1945). Type: Madagascar, without precise location, *de Lastellé* anno 1841 (Holotype P).

Solitary palm. **TRUNK** 5–15 m tall, 18–25 cm diam., with swollen base; wood hard; internodes 8–10 cm, pale green to grey waxy green, nodal scars c. 4 cm, pale brown; crownshaft 70–75 cm tall, rich velvet red-brown. **LEAVES** 9–15 in the crown, spirally inserted; sheath 40–60 cm long, partially open, adaxially brilliant cherry-red, abaxially densely red-brown pubescent; petiole 0–10 cm long, proximally to 11 x 3.5 cm, distally c. 6 x 3.7 cm, channelled; rachis yellowish, to 3.8 m long, in mid-leaf 1.2–2.2 x 1.5–1.6 cm diam. and here either channelled or keeled, glabrous, sometimes waxy; leaflets (50 -) 94–102 on each side of the rachis, regular, somewhat pendulous, the proximal 39–66 x 0.7–2 cm, median 56–89 x 2.4–4.3 cm, distal 18–54 x 0.4–2.3 cm, with a few tufts of laciniate ramenta on the proximal part of the abaxial midrib, but otherwise glabrous, main veins 1, prominent adaxially, as well as margins thickened, apices bifid and acute. **INFLORESCENCE** interfoliar, branched to 3 orders, spreading, 1.3–2.2 x 1.2 m; peduncle 60–96 cm long, distally 2.5–3.2 x 1.7–1.5 cm; prophyll 30–52 x 6.5–11 cm, borne at 3–13 cm above the base of the peduncle, rusty pubescent; peduncular bract deciduous, inserted at c. 30 cm from the base of the peduncle, 80–110 cm long, c. 12 cm wide when flat, splitting along its length but for the upper c. 28 cm, with a beak of up to 7 cm long, densely reddish pubescent; rachis c. 97 cm long, yellow-green, sometimes tinged red, with 13–17 branched and c. 10 unbranched first order branches, the proximal of these with a rachis 2.5–3.5 x 1–1.3 cm diam. at the base, to 45 cm long, and with up to 15 second order branches; rachis bracts up to 3 cm long, triangular and acute; rachillae pale or yellow-green, 27–47 cm long, 3–6 mm diam., glabrous, with dense, slightly sunken triads. **STAMINATE FLOWERS** only known from rather young buds, with sepals 2–2.3 x 1.4–2.3 mm, keeled, slightly gibbous, concave, with membranous wings, truncate or nearly so; receptacle 0.2–0.6 mm high; petals 1.7–2.7 x 1.3–1.6 mm, ovate, acute; stamens 6, very slightly biseriate, the antepetalous stamens inserted slightly higher up, filaments 0.5–0.6 mm and thin-cylindrical, anthers 1.3–1.6 x 0.5–0.7 mm, the locules parallel and obtuse; pistillode c. 2.2 x 0.6 mm. **PISTILLATE FLOWERS** with sepals 1.7–2.4 x 2.2–2.6 mm; petals 2.4–2.8 x 1.8–2.4 mm; staminodes c. 0.5–0.6 mm high; gynoecium c. 2.5 x 1 mm. **FRUIT** obovoid with a narrowed base, slightly asymmetrical, 18–24 x 12–17 mm, with obtuse apex; endocarp fibrous, with anastomosing fibres. **SEED** 12–21 x 10.5–16 mm, the base apiculate with a small depression just above the apiculus, this corresponding to the position of the embryo, the apex rounded; endosperm deeply ruminate, the intrusions many and 2–4 mm deep. **SEEDLING** with two scale leaves, the distal one densely scaly; eophyll pinnate, with petiole scaly with reddish and silvery scales; leaflets of eophyll 4–6 on each side of the rachis.

Dypsis lastelliana, remaining after forest clearance, Sahavary.

NOTE. The type consists of a partial inflorescence with staminate and pistillate buds, but there is no doubt that it is conspecific with the other cited material; *Miller & Randrianasolo* 4491 is at the same stage of development and is an almost exact match.

Baillon mentions another de Lastellé specimen in the protologue, this one with fruits, and states this probably belongs to another species. The fruit has homogeneous endosperm. Beccari (1906, l.c.) describes the proper fruit of *D. lastelliana* from a Baron collection, and identified the fruiting specimen of de Lastellé as *Chrysalidocarpus pilulifferus* (= *D. pilulifera*).

SPECIMENS SEEN. Ambanja: Manongarivo Mts, Ambalafary, Feb. 1992 (fr., seedling), *Beentje & Zo Lalaina* 4580 (K, MO, TAN). Sambava: near Maroambihy, March 1949 (bud, old seed), *Humbert* 23391 (K, P). Andapa: Mt Beondroka, (fl.), *Miller & Randrianasolo* 4491 (K, MO, TAN). Maroantsetra: Maroantsetra, Oct. 1986 (ster.), *Dransfield et al.* JD6358 (K, TAN) and (bud) 6359 (K, P, TAN). Mananara Avaratra: Antanambe, Oct 1991 (dead infl.), *Beentje* 4452 (K, TAN). Soanierana-Ivongo: Soanierana, Sept. 1912 (old infl.), *Perrier* 12023 (P); E of Antasibe (Andasibe), Dec. 1938 (bud, fr.), *Lam & Meeuse* 5863 (K, L).

WITHOUT PRECISE LOCALITY: NE Madagascar, no date (fr. only), *Razafindratsira* s.n. (K). Central Madagascar, circa 1883 (fr.), *Baron* 2322 (K); without locality, anno 1841 (bud), *de Lastellé* s.n. (P, type); idem, circa 1912 (fl., y.fr.), *Perrier* 12080 (P).

SIGHT RECORD. Sainte-Marie: Kalalao forest (*Dransfield*, 1994).

SIMILAR SPECIMENS. *Beentje et al.* 4575 from the Manongarivo (altitude 785 m) is similar in most respects, but has grouped leaflets and rachillae 46–55 cm long. As there are no flowers or fruits we cannot describe it as a new taxon.

16. DYPSIS LEPTOCHEILOS

A species described from a cultivated tree in Tahiti; it resembles *D. lastelliana*, but the situation about its origin is confused. In cultivation this beautiful palm is much more tolerant of cooler and drier conditions than is *D. lastelliana*, making it more suitable for cultivation in southern California. The name is derived from the Greek words for slender lip.

DISTRIBUTION. Said to occur in W Madagascar, but no voucher specimens from the wild are known.
HABITAT. Said to be at low elevations along and in rocky, sandy, seasonally dry streambeds (no vouchers).
LOCAL NAMES. Not recorded.
USES. Not recorded.
CONSERVATION STATUS. Unknown, until the status in the wild has become more clear.

Dypsis leptocheilos
(o: *Perrier* 15802)

Dypsis leptocheilos *(Hodel) Beentje & J. Dransf.* **comb. nov.**

SYNONYM:
Neodypsis leptocheilos Hodel, The Palm Journal Jan. 1993: 9 (1993). Type: cultivated material originating from Madagascar and grown in Tahiti, Papeari, *Hodel* 1144 (Holotype BH; isotype K).

Solitary palm. **TRUNK** to 10 m, 25 cm diam., slightly flared at the base; internodes 12 cm, nodal scars 2 cm, white. **LEAVES** c. 15 in the crown, porrect; sheath partially open in outer leaves, c. 62 cm long, 5 mm thick, covered in dense rusty brown tomentum, with auricles 3.5 cm high; petiole 17 cm long, 7–9 cm wide, channelled adaxially, densely tomentose but glabrescent; rachis to 4 m long, proximally channelled and tomentose, more distally becoming flattened and eventually keeled distally, glabrous distally; leaflets up to 103 on each side of the rachis, regular, slightly drooping, green, linear, straight, acuminate, the proximal 55 × 0.5–0.8 cm, median 85 × 4 cm, distal 55 × 0.5–0.8 cm, glabrous but for some ramenta on the proximal part of the midrib abaxially, main veins 3. **INFLORESCENCE** interfoliar becoming infrafoliar in fruit, branched to 3 orders, to 1.7 × 1.4 m; peduncle to 65 cm long, 11 cm diam. proximally, 3–6 cm diam. distally, flattened, reddish tomentose over green; prophyll ± woody, c. 43 cm, borne at 12 cm above the base of the peduncle, c. 12 cm wide, tomentose; peduncular bract deciduous, woody, inserted at 26 cm from the base of the peduncle, to 70 × 10–15 cm, reddish tomentose; two smaller peduncular bracts present near the apex of the peduncle, to 17 cm long, tomentose; rachis 100 cm, flattened, grey-tomentellous, with 24 first order branches; rachis bracts to 3 cm long; rachillae spreading, to 30 cm long, 3 mm diam., green; with distant triads in pits; rachilla bract 0.5–0.75 mm high. **STAMINATE FLOWERS** with sepals 2–2.5 mm, keeled, truncate; petals on a 1 mm high receptacle, 2 × 1.5–1.75 mm, ovate, acute; stamens said to be connate for 1.75 mm, filaments 0.8–1 mm long, anthers 1 mm long, dorsifixed?; pistillode 1.75–2 mm, columnar. **PISTILLATE FLOWERS** with sepals 2 mm; petals 3 × 3 mm; gynoecium 4 mm high, 2–2.5 mm diam., with 1 mm long stigma lobes. **FRUIT** globose, dark brown, 10–12 mm; endocarp fibrous, the fibres anastomosing rather intensely. **SEED** 8.5–10 × 8.5–9 mm, globose, pointed at the base, rounded at the apex, with a subbasal depression corresponding to the embryo; ruminations of the endosperm irregular, to 3.5 mm deep and up to half the diameter of the seed wide. **EOPHYLL** pinnate.

NOTE. Close to *D. lastelliana* but differs in rachilla bract (thinner in *leptocheilos*), and fruit/seed size. Differences enumerated in the protologue with another species of *Neodypsis* (it is unclear which one) in the same garden in Tahiti, or with the description of *N.*

Dypsis leptocheilos, view of type plant in cultivation in Tahiti (Photo: D. Hodel).

lastelliana in the Flore de Madagascar (notoriously unreliable) are not pertinent.

A problem is *Perrier* 15802 (P), from Ambilobe: upper Mananjeba R., c. 400 m alt., Aug. 1903 (fl.) with very red-cottony leaf sheaths, which is quite like *leptocheilos*; it has rachillae almost without any lip on the rachilla bract, and so is unlike either *leptocheilos* or *lastelliana*.

SPECIMENS SEEN. Cultivated material originating from Madagascar and grown in Tahiti, Papeari, April 1992 (fr., seedling), *D.R. & M. Hodel* 1144 (type, K). And probably: without locality, anno 1986 (seed), *Razafindratsira* s.n. (K).

17. DYPSIS LIGULATA

INSUFFICIENTLY KNOWN SPECIES

The name refers to the 'ligule' or auricle on the leaf sheath, a character much used by Jumelle, but of doubtful value in separating species.

DISTRIBUTION. NW Madagascar.
HABITAT. Moist forest on sandstone, at low elevation.
LOCAL NAMES. Not recorded.
USES. Palm-heart edible.

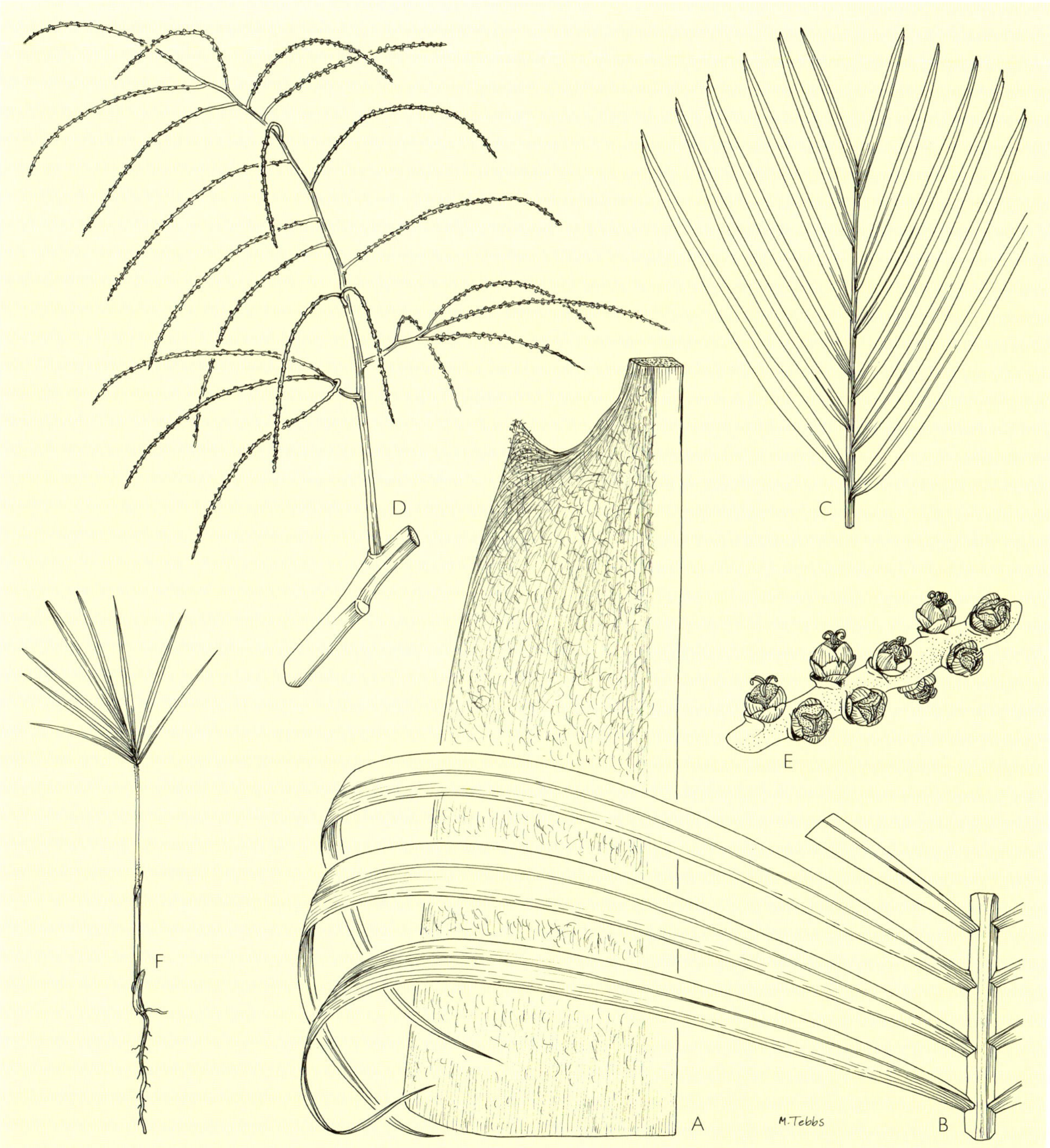

Dypsis leptocheilos. **A** leaf sheath × 1/5; **B** mid section of leaf × 1/5; **C** leaf tip × 1/5; **D** first order inflorescence branch × 1/5; **E** detail of rachilla × 2; **F** seedling × c. 1/2. All from *Hodel & Hodel* 1144. Drawn by Margaret Tebbs.

Conservation Status. Possibly extinct; not seen for over 70 years.

Dypsis ligulata *(Jum.) Beentje & J. Dransf.* **comb. nov.**

Synonym:
Neodypsis ligulatus Jum., Ann. Inst. Bot.-Géol. Colon. Marseille sér. 4, 2 (2): 19 (1924); Jum., Cat. Pl. Madagascar, Palmae: 18 (1938); Jum. & H. Perrier, Fl. Madagascar 30: 152 (1945). Type: Madagascar, Sambirano, Belinta, W base of Mt Kalabenona, *Perrier* 15414 (Holotype P).

Solitary palm. **Trunk** 4–6 m high, c. 20 cm diam., smooth, ringed. **Leaves** with the sheath (fide Perrier) whitish, glabrous, with distinct, unequal, obtuse auricles; petiole not seen; rachis in mid-leaf keeled, c. 1.8 cm wide and 2.2 cm high, with scattered scales; leaflets presumably regular, the proximal not seen, median c. 120 × 2.3–2.5 cm (interval 3.5–4 cm), distal 23–42 × 0.5–1.4 cm, the terminal pair joined for c. 6 cm, main veins 1 (–3), with thickened margins, with scattered red ramenta 2–4 mm long on the proximal part of the midrib, otherwise glabrous, apices unequally attenuate, bifid. **Inflorescence** branched to 2 orders; the only first order branch seen with a secondary rachis 27 cm long, proximally 1.2 × 0.6 cm, glabrous, with 17 rachillae; rachillae 24–34 cm long, 2–2.5 mm diam., with distant slightly sunken triads, the rachilla bracts small and rounded. **Staminate flowers** not seen. **Pistillate flowers** not seen; in fruit the persistent sepals 1.8–2.2 × 2.5–2.8 mm, petals 3.2–3.6 × 3.8–4.1 mm; staminodes 0.6–0.7 mm, flat. **Fruit** yellowish, ellipsoid, 12–16 × 7–8.5 mm, with an obtuse apex; endocarp fibrous, the fibres anastomosing. **Seed** ellipsoid, c. 13 × 7.5–8 mm, pointed at the base, obtuse at the apex, with a sub-aequatorial depression; endosperm ruminate, the ruminations few, distant, 1–2 mm deep.

Dypsis ligulata

Note. The protologue describes the auricles as 0.8–1.5 cm long; the petiole as 12 cm long; the proximal segments as 75–100 × 0.4–0.8 cm; inflorescence branched to 2 orders; prophyll 40 × 6 cm, with scattered scales. It also has the Sakalava name *Kindro.* None of these details is apparent from the Paris specimen. We are unable to comment on the affinities of this species.

Specimens seen. Ambilobe: Belinta, Feb. 1923 (fr.), *Perrier* 15414 (Holotype P).

18. Dypsis sainteluce i

A medium-sized, elegant palm which would probably do well in cultivation. It occurs on white sand, and so would probably do well in the drier tropics; the habitat is that of *D. lutescens*, one of the most successful ornamental palms.

Distribution. Only known from the Sainte-Luce forest in the extreme South-East of the island.
Habitat. Coastal forest on white sand; 10–20 m.
Local names. Not recorded.
Uses. Not recorded.

Conservation Status. Critical. Only known from a single forest, with numbers less than fifty; this is right in the middle of the area where mining operations for mineral sand (ilmenite ore) are proposed.

Dypsis sainteluce i *Beentje* **sp. nov.**
inter species arborescentes tristichas foliolis regulariter dispositis endospermio homogeneo *D. ampasindavae* affinis sed vagina folii clausa et inflorescentia interfoliacea longa differt. Typus: Madagascar, St. Luce, *Beentje* 4760 (Holotypus K; isotypi BH, MO, P, TAN).

Solitary palm (occasionally in clumps of 2-3). **Trunk** 6–10 m, c. 14 cm diam.; internodes 3–5 cm long, grey, green more distally; nodal scars 1.6 cm high, whitish; wood very hard, red; crownshaft waxy green. **Leaves** tristichous, 7–11 per crown, porrect, slightly arching distally; sheath c. 80 cm, 12 cm diam., closed, waxy pale green, smooth, with brown-tattering shoulders; petiole absent or up to 13 cm long, 2–2.5 × 2–2.5 cm diam., channelled with sharp edges, green, glabrous; rachis 2.3–2.4 m long, proximally deeply channelled and 4 × 3 cm, more distally keeled, green, in mid-leaf 1.5–2.3 cm wide, proximally with thin grey wax or scales, distally glabrous; leaflets 59–61 on each side of the

Dypsis sainteluce i, in coastal forest, Sainte Luce (*Beentje* 4760).

Dypsis saintelucei. **A** tip of leaf sheath, petiole and basal leaflets × 2/3; **B** mid section of leaf × 2/3; **C** leaf tip × 2/3; **D** first order branch of inflorescence × 2/3; **E** staminate flower in vertical section × 10; **G** fruit in vertical section × 2. **A** – **E** from *Beentje* 4760, **F**, **G** from *Beentje & Andriampaniry* 4609. Drawn by Rosemary Wise.

rachis, regular, stiff, pale waxy green abaxially, the proximal 116–134 (–188) × 2.2–3.5 cm (16 cm apart, the most proximal often pendulous), median 90–104 × 3–3.7 cm (interval 2.5–4 cm), distal 15–40 × 1–2.2 cm, main veins 1, only the midrib prominent on both surfaces with large distant brown ramenta, the other veins faint and with densely set small red-brown glands, apices attenuate, unequally bifid. **Inflorescence** solitary, interfoliar at anthesis, erect within sheath, 175–225 cm, branched to 3 orders; peduncle 89–138 cm, proximally 6–10 × 2–2.5 cm and convex abaxially, concave adaxially, distally c. 7.5 × 4 cm diam., glabrous; prophyll 100–130 cm, borne at 30–56 cm above the base of the peduncle, c. 9 cm wide, glabrous; peduncular bract 68–73 cm, closed in the distal 6 cm, beaked for c. 2 cm, inserted at 61–97 cm, abaxially pale waxy green, adaxially chestnut-brown; rachis 70–87 cm, green, glabrous, with 17 branched and 11–20 unbranched first order branches; rachillae 16–27 cm, green, glabrous, c. 5 mm diam.; triads distant, superficial, with low, rounded bract. **Staminate flowers** slightly scented; sepals 1.8–2.5 × 2.2–2.9 mm, keeled, gibbous, broadly ovate, concave, the broad margins membranous; petals yellow, connate for 1–1.6 mm to the receptacle, free lobes 2.9–3.5 × 3–3.2 mm, ovate, acute, with adaxially and proximally with 2 swellings on each side of the base of the antepetalous stamen; stamens 6, uniseriate, filaments 3.8–4.4 mm long, connate at the very base for c. 0.3 mm, anthers 2.4–2.5 × 1 mm; pistillode columnar, 1.8–2.3 × 1.4–1.6 mm. **Pistillate flowers** only known in bud, with sepals 2.6–2.8 × 3.3–3.4 mm, ciliolate; petals c. 2.2 × 2.5 mm; staminodes 6, 0.4–0.6 mm; ovary asymmetrical, c. 2.3 × 2.3 mm. **Fruit** not known but for seed and endocarp found under *Beentje & Andriampaniry* 4609; endocarp fibrous. **Seed** ellipsoid, 11.5–13 × 7 mm, rounded at one end, pointed at the other; endosperm deeply ruminate.

Note. In its group of large tristichous palms with regular leaflets and ruminate endosperm closest to *D. ampasindavae* but distinct in the closed leaf sheath and the interfoliar inflorescence, resulting in long peduncles and very long bracts.

Specimens seen. Tolanaro: 5km W of Manafiafy (Ste. Luce), March 1992 (ster.), *Beentje & Andriampaniry* 4609 (K); idem, Dec. 1992 (fl.), *Beentje* 4760 (Holotype K; isotypes BH, MO, P, TAN).

19. Dypsis ovobontsira

This is a beautiful palm of restricted distribution. The name is taken from the Betsimisaraka name for this species.

Distribution. Only known from the Mananara Biosphere Reserve.

Habitat. Moist forest, steep mid slope, ultramafic soils with deep humus layer; c. 265 m.

Local names. *Ovobontsira* (Betsimisaraka).

Uses. Not recorded.

Conservation Status. Critical. Only known from a single site, where less than ten individuals are known.

Dypsis ovobontsira *Beentje* **sp. nov.**
inter species arborescentes foliolis regulariter dispositis endospermio ruminato inflorescentia interfoliacea pedunculo longo foliis spiraliter dispositis vagina folii hirsutissima distincta. Typus: Madagascar, Antanambe, *Beentje et al.* 4645 (Holotypus K; isotypi BH, MO, P, TAN)

Solitary palm. **Trunk** to 8–10 m, 13 cm diam., near crown 9 cm diam.; internodes 14 cm, near crown 2–2.5 cm; wood pink, with dense fibre-layer below bark. **Leaves** spiral, 6, arching; sheath c. 62 cm, at crown 14 cm diam., green with dense brown and white scales; petiole c. 47 cm, green with dense white scales, proximally 6 × 5 cm, distally 5 × 4.5 cm, channelled with sharp edges; rachis 2.5–2.6 m, in midleaf 2.5 × 2.3 cm, green, waxy with sparse scattered scales; leaflets 68–69 on each side of the rachis, very regular, stiff, in 1 plane, the proximal 68–71 × 3.3–4.2 cm (sometimes with long pendulous reins), median 86–90 × 4.6–5.2 cm, dark green, glabrous. **Inflorescence** interfoliar, arching

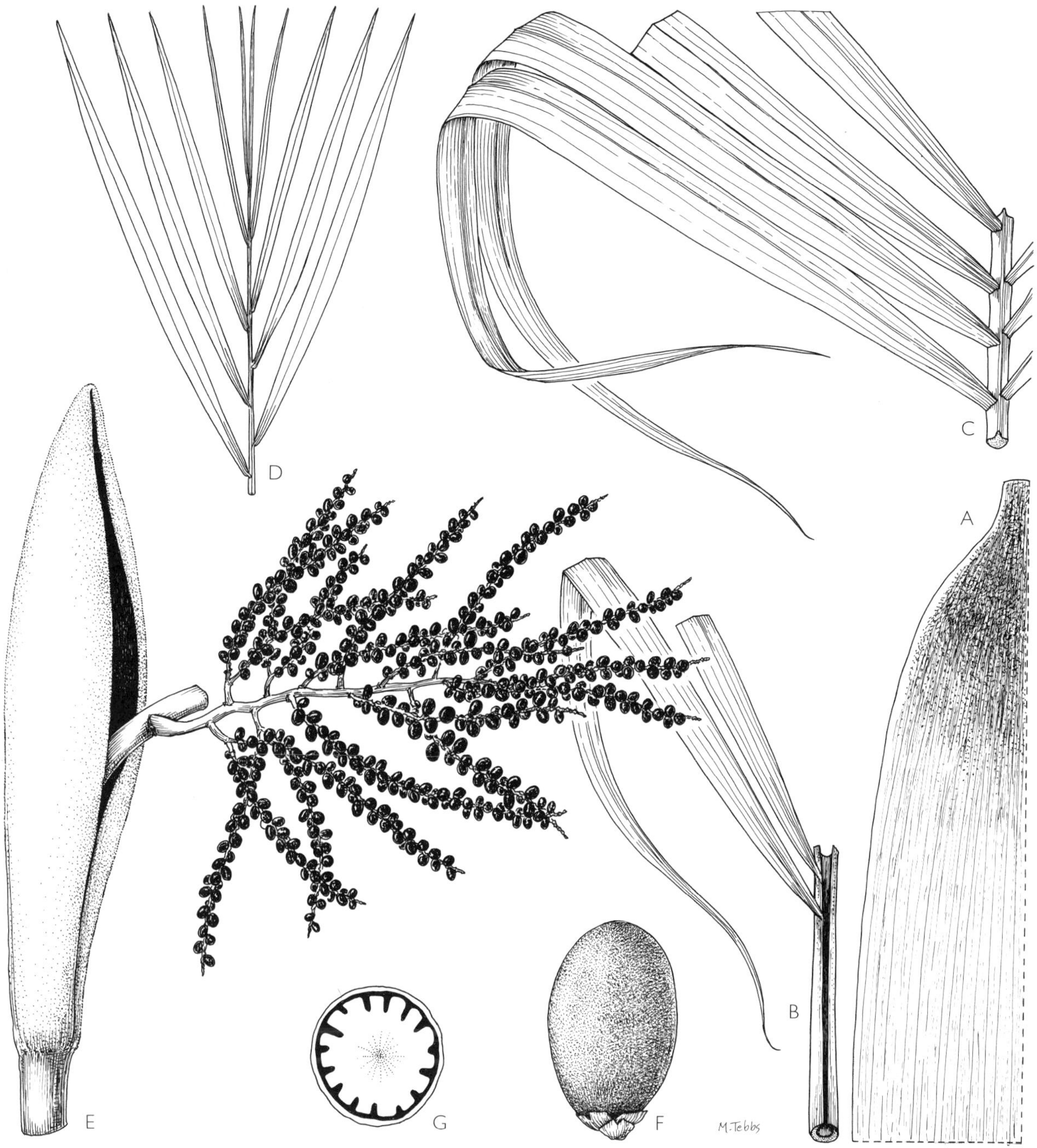

Dypsis ovobontsira. **A** leaf sheath × 1/4; **B** distal part of petiole with lowermost leaflets × 1/4; **C** mid section of leaf × 1/4; **D** leaf tip × 1/4; **E** part of infructescence × 1/4; **F** fruit × 3; **G** fruit in cross section × 3. All from *Beentje et al.* 4645. Drawn by Margaret Tebbs.

with spreading rachillae, 170 × 90 cm, branched to 3 orders; peduncle 79 × 7 × 2 cm proximally, distally 7 × 4 cm; prophyll borne at 34 cm above the base of the peduncle, 56 cm × 14 cm, split very much on 1 side; peduncular bract deciduous, borne at 53 cm above the base of the peduncle; rachis with 22 branched and 19 unbranched first order branches, these proximally flattened, 4 × 1.7 cm; rachillae 10–18.5 cm, c. 6 mm across. **Flowers** unknown. **Fruit** green, 15–17 × 13–15 mm when fresh, 10 × 7 mm when dried. **Seed** 13–15 × 11–13 mm, rounded at the apex, pointed at the base, broader than wide, with sub-basal depression; endosperm slightly ruminate.

Note. In its group of large palms with regular leaflets and ruminate endosperm distinct by the interfoliar inflorescence with its long peduncle and bracts, the very hairy leaf sheath, and the spirally inserted leaves. It does not really resemble any other species closely.

Specimens seen. Mananara Avaratra: 10 km W of Antanambe, April 1992 (fr.), *Beentje et al.* 4645 (type; BH, K, MO, P, TAN).

20. DYPSIS TANALENSIS

INSUFFICIENTLY KNOWN SPECIES

DISTRIBUTION. Known from a single collection from the Vohipeno area, south of Manakara. It was named for the people living in this area, the Tanala.
HABITAT. Lowland rain forest; 100 m.
LOCAL NAMES. *Matitanana*, *Matitana* (Tanala, meaning dead hand, supposedly after the dead inflorescences when fallen on the ground).
USES. Not recorded.
CONSERVATION STATUS. Possibly extinct. Not seen for over eighty years; HB visited the area, which now seems devoid of tree palms.

Dypsis tanalensis *(Jum. & H. Perrier) Beentje & J. Dransf.* **comb. nov.**

SYNONYM:
Neodypsis tanalensis Jum. & H. Perrier, Ann. Inst. Bot.-Géol. Colon. Marseille sér. 3, 1 (1): 18 (1913); Jum., Ann. Inst. Bot.-Géol. Colon. Marseille sér. 4, 2 (2): 29 (1924); Jum., Cat. Pl. Madagascar, Palmae: 19 (1938); Jum. & H. Perrier, Fl. Madagascar 30: 155, fig. 44 (1945). Type: Madagascar, Matitanana basin, *Perrier* 12072 (Holotype P; not mentioned in protologue, but the only specimen mentioned in Jumelle 1924).

LEAVES with the sheath with sloping shoulders without any sign of auricles, tomentose with peltate scales with white-laciniate edges; petiole 12 cm long (cut lengthwise in the type), densely tomentose; rachis in mid-leaf white-tomentose, distally keeled; leaflets regular or grouped?, in mid-leaf the interval 1–2 cm; proximal leaflets c. 115 x 1.2 cm, median 129 x 2.4–2.6 cm, near-distal 63 x 2.4 cm, main vein 1, abaxial midrib with a few basal reddish ramenta to 6 mm, with scattered minute reddish glands on the minor veins, apices bifid, unequally attenuate. **INFLORESCENCE** branched to 1 order or more?; rachillae 26–33 cm long, 2.5 mm diam. in flower, 4 mm diam. in fruit, with minute bundles of bristles but glabrescent, the triads distant, slightly sunken, with small acute rachilla bracts. **STAMINATE FLOWERS** with sepals 1.8–2 x 1.6–2.5 mm; petals 3.8–4.4 x 2.1–2.5 mm; stamens 6, ?slightly 2-seriate, filaments 1.4–1.8 mm in bud, to 2.6 mm at anthesis, anthers 1.8–2.5 x 0.6–0.8 mm, versatile; pistillode 2–2.8 x 0.4–0.5 mm. **PISTILLATE FLOWERS** not seen at anthesis, in fruit with sepals 2–2.8 x 3–3.6 mm; petals 3–3.4 x 3.8–4.5 mm; staminodes 0.7–0.8 mm. **FRUIT** ellipsoid, 9–14 x 5.5–7 mm. **SEED** 8–13 x 4.5–6 mm; endosperm ruminate, with quite dense ruminations, 1–2 mm deep.

NOTE. Data given in the protologue, but not apparent from the type or its label: **TRUNK** 15–20 m high, diam. 20–50 cm, internodes greyish, nodal scars very evident. **LEAVES** (incl. sheath) 4–6 m long; sheath 1–1.5 by 0.4–0.5 m (but then the authors go on and say it is channelled adaxially–so it is uncertain whether they refer to the petiole or the rachis), smooth, green, distally with a thin whitish puberulous indument; rachis 3–5 m; leaflets regular, median 130 x 3 cm. Every trunk with 3–4 interfoliar to infrafoliar **INFLORESCENCES;** these pendulous, branched to 3 or 4 orders, 1.5–2 by 1.2–1.6 m; prophyll 50 x 15 cm, opening near the apex only; peduncular bract 80 x 13 cm.

The specimen looks like *D. pilulifera*, but the ruminate endosperm separates it from that taxon. Among ruminate endosperm taxa it resembles *D. hovomantsina* and *D. tsaravoasira* - but in every case the material of *D. tanalensis* does not have quite the same look or feel to it.

According to the protologue, the epiphytic orchid *Oeonia* often grows on this palm.

SPECIMENS SEEN. Vohipeno: Matatana (Matitana) basin, no date (fl., fr.), *Perrier* 12072 (P, type).

21. DYPSIS ANKAIZINENSIS

Yet another species of which various specimens were housed in different genera. It is still somewhat of a mystery, and it combines characters from several other species, without being quite the same as any of them. The name derives from Ankaizina, the southern foothills of Mt. Tsaratanana.

DISTRIBUTION. Only known from Mt Tsaratanana.
HABITAT. Montane forest, in moist depressions or on ridge crests; 1400–2000 m. Said to be very common.
LOCAL NAMES. *Laboka*, *Hovatra*, *Lavaboka*.
USES. Palm-heart slightly bitter, but edible.
CONSERVATION STATUS. Unknown. Not seen for over seventy years, but the Tsaratanana area has not been visited by botanists for a long time.

Dypsis ankaizinensis *(Jum.) Beentje & J. Dransf.* **comb. nov.**

SYNONYM:
Chrysalidocarpus ankaizinensis Jum., Ann. Inst. Bot.-Géol. Marseille sér. 10, 3: 23 (1922); Jum., Cat. Pl. Madagascar, Palmae: 8

(1938); Jum. & H. Perrier, Fl. Madagascar 30: 119 (1945). Type: Madagascar, Tsaratanana, *Perrier* 11936 (Holotype P)

Neodypsis lobatus Jum., Ann. Inst. Bot.-Géol. Colon. Marseille sér. 4, 2 (2): 13 (1924); Jum., Cat. Pl. Madagascar, Palmae: 18 (1938); Jum. & H. Perrier, Fl. Madagascar 30: 142 (1945), **synon. nov**. Syntypes: Tsaratanana, *Perrier* 16227 (P), 16227bis (P).

Solitary palm. **TRUNK** 10–15 m high, 15–40 cm diam. **LEAVES** with sheath very large, green (Perrier), distally with dense dark brown laciniate scales, with square shoulders, not very large, to c. 8.5 cm wide when flat; petiole 2–13 cm long, up to 4 cm diam., flat with sharp margins and with dense dark scales; rachis in mid-leaf 1.3–2.2 cm wide, proximally channelled, distally keeled, with scattered white hairs and reddish glands; leaflets not very stiff, the distal half pendulous, in groups of 2–4 in mid-leaf, the group interval 2.5–5 cm, the leaflet interval 0.2–0.6 cm; proximal 47–132 × 0.6–1.2 cm, median 50–72 × 1.9–2.7 cm, distal 13–33 × 0.5–2.5 cm, main vein 1, margins thickened, sometimes slightly glaucous, pale faint glands scattered on the minor veins, apices bifid, unequally attenuate or acute. **INFLORESCENCE** interfoliar or infrafoliar, branched to 2 orders, erect or arching; prophyll 25–30 cm long, c. 6.6 cm wide, with scattered scales, borne at 14 cm above the base of the peduncle; peduncular bract 42–70 cm long, c. 10 cm wide, with scattered scales, beaked for 2 cm, opening over its entire length; first order branches with a secondary rachis of 6–18 cm, proximally 1.2–1.8 × 0.4–0.7 cm diam., glabrous, with 6–14 rachillae; rachillae 14–25 cm long, 2.5–4 mm diam., with distant to close slightly sunk triads, rachilla bracts proud, acute. **STAMINATE FLOWERS** seen as densely packed buds, with sepals 2.7–3.1 × 2.5–3.3 mm; petals 2.5–2.8 × 2.2 mm; stamens 6, uniseriate, filaments c. 1 mm and thin, anthers c. 1.3 × 0.6 mm; pistillode c. 1.8 × 0.5 mm. **PISTILLATE FLOWERS** not seen, but fruit with persistent sepals 3–3.4 × 3.2–4 mm; petals 4.5–5.2 × 3.5–6 mm; staminodes 0.8–1 mm. **FRUIT** ellipsoid, 13–17 × 8.5–12

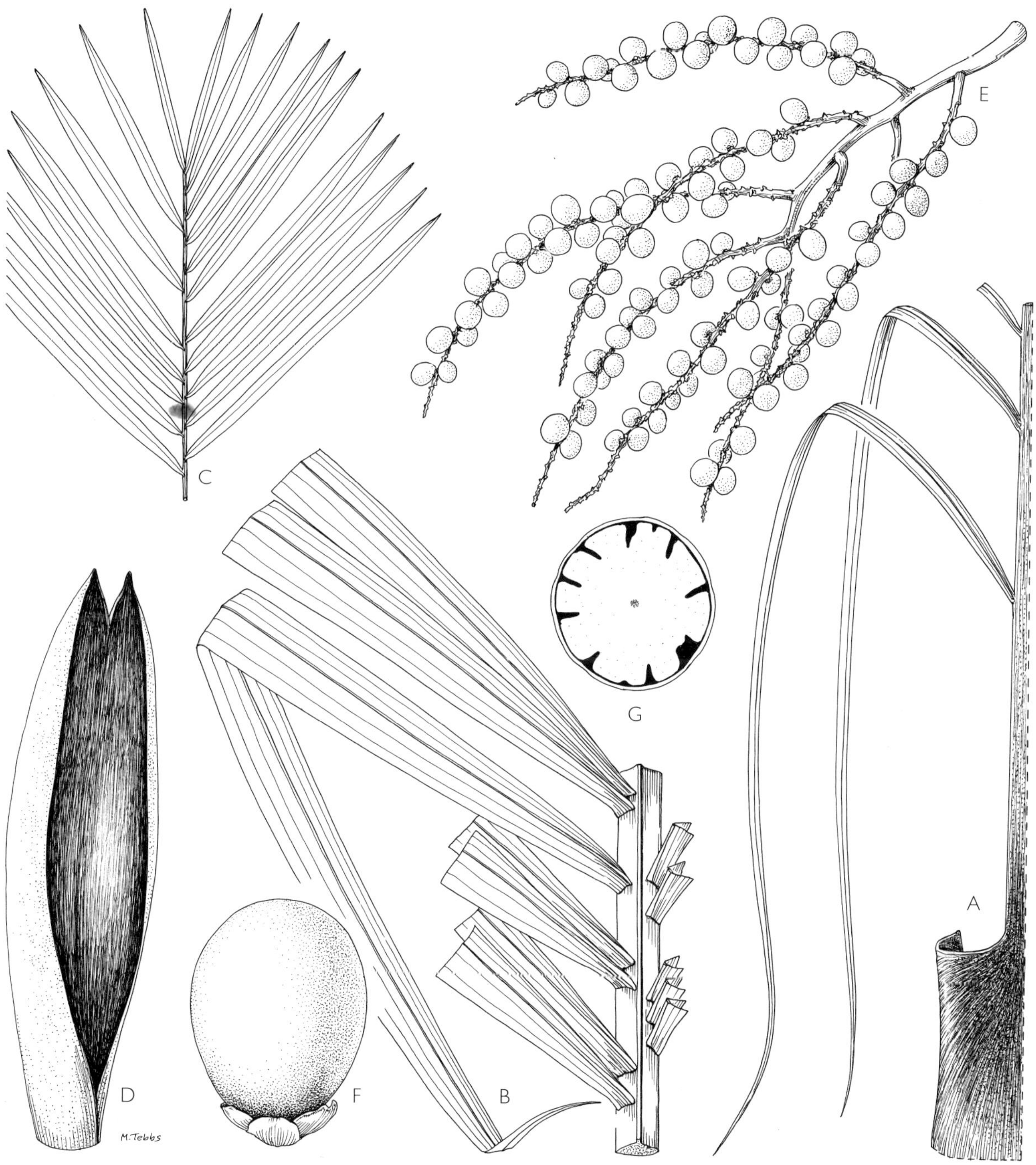

Dypsis ankaizinensis. **A** sheath, petiole and basal leaflets × 1/3; **B** mid section of leaf × 1/3; **C** leaf tip × 1/5; **D** peduncular bract × 1/4; **E** first order branch of infructescence × 1/3; **F** fruit × 2.5; **G** fruit in cross section × 2.5. All from *Perrier* 16227. Drawn by Margaret Tebbs.

mm (to 25 × 18 mm when fresh), with a rounded apex. **SEED** ellipsoid, 9.5–15 × 7.5–10 mm, with pointed base and rounded apex; endosperm ruminate, the ruminations distant, 1–2 mm deep.

NOTE. *Neodypsis lobatus* is clearly the same taxon, and comes from the same habitat from the same mountain. Jumelle states that in the palms from ridge crests (16227bis) the stem is twice as short as in the ones from slope forest; so is the sheath.

We are uncertain about the affinities of this species. The inflorescence has the look of *D. tsaravoasira*, but the leaflets are grouped and, in their turn, look more like *D. hovomantsina* - but the leaf sheath is glabrous. This is one of those species which seem to combine features of several others.

SPECIMENS SEEN. Ambanja/Ambilobe/Bealanana: Mt Tsaratanana area, anno 1912 (y.fr.), *Perrier* 11936 (P, type); idem, April 1924 (fr.), *Perrier* 16227 (syntype; P); idem, April 1924 (fr.), *Perrier* 16227bis (syntype; P).

22. DYPSIS MADAGASCARIENSIS

This species is widespread in the northwest and west of the country, occurring in drier forest than most other *Dypsis* species; it even survives in moister localities in dry bushland. This is a handsome palm which does well in cultivation, and is widespread throughout the tropics; in its native country it even grows in forest margins next to the beach.

DISTRIBUTION. NW and W Madagascar.
HABITAT. Moist rain forest or semi-deciduous dry or plateau forest; rare in palm grassland; in dry bushland usually in gullies or ravines; may grow very close to the sea shore; alt. 1–650 m.
POLLINATION & DISPERSAL. Flowers visited by bees and/or other insects (fide *Moore* 9026); fruit eaten by

Dypsis madagascariensis. View of crown, Lokobe.

black lemur and *Lepilemur dorsalis*, and seed dispersed by black lemur (Birkinshaw, pers. comm.).

LOCAL NAMES. *Hiribiry*, *Kizobazo*, *Faribazo* (all Sakalava, fide Jumelle & Perrier), *Madiovozona* (Sakalava, fide Rakotovao), *Kindro* (Tsimihety).

USES. Excellent palm-heart; fruit eaten by children; outer wood used for floorboards of houses, and much cut for this purpose in Lokobe Special Reserve (Adanay *et al.*, 1994).

CONSERVATION STATUS. Rare. Although widespread, habitat destruction is reducing the number of populations; harvesting for palm-heart and floorboards (the last particularly within the Special Reserve of Lokobe) might put the species into the Vulnerable category in the near future. Current population estimated at several thousands.

Dypsis madagascariensis *(Becc.) Beentje & J. Dransf.* **comb. nov.**

SYNONYMS:

Chrysalidocarpus madagascariensis Becc., Bot. Jahrb. Syst. 38, Beibl. 87: 35 (1906). Jum., Ann. Inst. Bot.-Géol. Colon. Marseille, 5, 1 (3): 12 (1933); Jum., Cat. Pl. Madagascar, Palmae: 9 (1938); Jum. & H. Perrier, Fl. Madagascar 30: 112 (1945). Types: cultivated specimens from the Botanical Gardens at Calcutta and Buitenzorg (nowadays Bogor), and Mauritius (quoad fructus) (syntypus, K!).

Dypsis madagascariensis. **A** distal part of petiole with lowermost leaflet × 1/3; **B** mid section of leaf × 1/3; **C** leaf tip × 1/3; **D** first order inflorescence branch × 1/3; **E** staminate flower × 8; **F** fruiting rachillae × 1/3; **G** mature fruit × 3; **H** mature fruit in section × 3. All from *Beentje* 4561. Drawn by Rosemary Wise.

Chrysalidocarpus madagascariensis var. ***lucubensis*** (Becc.) Jum., Ann. Inst. Bot.-Géol. Colon. Marseille sér. 5, 1: 15 (1933); Jum., Fl. Madagascar 30: 116 (1945), **synon. nov.**
Chrysalidocarpus lucubensis Becc., Bot. Jahrb. Syst. 38, Beibl. 87: 35 (1906), & Palme del Madagascar 44, t. 43 (1914). Type: Madagascar, Nosy Be, Lokobe forest, *Hildebrandt* 2918 (Holotype P; isotypes K, WU).

Chrysalidocarpus madagascariensis var. ***oleraceus*** (Jum. & H. Perrier) Jum., Fl. Madagascar 30: 114 (1945) **synon. nov.**
Chrysalidocarpus madagascariensis forma ***oleraceus*** (Jum. & H. Perrier) Jum., Ann. Inst. Bot.-Géol. Colon. Marseille sér. 5, 1: 15 (1933).
Chrysalidocarpus oleraceus Jum. & H. Perrier, Ann. Inst. Bot.-Géol. Colon. Marseille 3, 1 (1): 42, t. 21–22 (1913); Jum., Ann. Inst. Bot.-Géol. Colon. Marseille sér. 10, 3: 29 (1922). Syntypes: not indicated in the protologue, but according to localities and citation, Jumelle (1929) probably Madagascar, Boina, *Perrier* 145 (syntype P), 149 (we believe this was a reading error for 145), 11982 (not seen).

Dypsis madagascariensis
(**x**: sight records)

Elegant solitary or clustering palm in tufts of 2–4. **TRUNK** 2–18 m tall, 7–20 cm diam.; internodes (0.5–) 5–12 cm long, green weathering to grey, nodal scars (0.2–) 1.5–2.5 cm; wood very hard because of an outer layer of tough fibres; crownshaft green, white-waxy, 8–20 cm diam. **LEAVES** 7–12 in the crown, tristichous, porrect and arching distally; sheath 75% to completely open in the outermost leaves, (27–) 40–63 cm (to 70 cm in young trees), adaxially orange, abaxially green with white waxy overlay, proximally glabrous, distally with scattered reddish scales, with irregular ligules varying from rounded shoulders to small but distinct triangular processes to 3.5 mm high, and sometimes ciliate with brown hairs (fide Jumelle, on *Perrier* 17570); petiole 12–40 cm long (to 50 cm in young trees), 1.3–5 × 1.8–3.5 cm diam., channelled with sharp edges, abaxially with red-brown tomentum but glabrescent, waxy; rachis 1.6–3.1 m long, proximally channelled, waxy and with vestiges of reddish pubescence, in mid-leaf keeled, 0.7–2 × 0.4–1.5 cm and glabrous or nearly so; leaflets (30–) 88–126 (–177) on each side of the rachis, in groups of 2–6, fanned within the groups, the groups 0.5–4.5 cm distant, the leaflets with their distal part drooping, the proximal 59–118 × 0.5–1.5 cm, the median 47–95 × 1.2–2.2 (–3.2) cm (interval within the groups 0.3–0.5 cm), the distal 5.5–39 × 0.5–1.8 (–2.6) cm, mid-green, abaxially waxy-pale green, main veins 1, the others very faint, proximally on the abaxial midrib with a few tufts of laciniate red–brown ramenta to 14 mm long, caducous in old leaves, otherwise glabrous or with scattered red glands on the minor veins, apices acute, unequally bifid. **INFLORESCENCE** interfoliar, branched to 3 orders (rarely to 4 orders), 1–1.5 m long, c. 1 m wide, arching with spreading rachillae; peduncle 52–60 cm long, proximally 5–7 × 3–3.5 cm, distally 1.5–5 × 1–3 cm diam., reddish-tomentose to tomentellous but glabrescent and then green; prophyll 35–61 cm, borne at 3–6 cm above the base of the peduncle, 4–8 cm wide, coriaceous, densely pubescent to glabrous; peduncular bract inserted at 14–23 cm from the base of the peduncle, 53–81 cm long, closed for 3–9 cm distally, beaked for 3–5.5 cm, white-waxy, with large laciniate reddish scales but glabrescent, abscising and carried upwards by the lengthening inflorescence; open peduncular bract c. 1 cm high, 3.5 cm wide; rachis 62–96 cm long, with flaking reddish pubescence, with 15–22 branched and 8–9 unbranched first order branches, the proximal of these proximally 0.9–3 × 0.4–1.2 cm diam., subtended by a bract 0.5–1.8 (6.2) cm high; rachillae 10–40 cm, 2–4 mm diam., green with some traces of reddish scales, with distant to quite densely set triads in shallow pits. **STAMINATE FLOWERS** yellow-green with sepals 1.5–1.6 × 1.3–1.8 mm, keeled, basally gibbous, membranous apart from the keel, hooded distally; petals connate to the receptacle for 0.6–1.2 mm, the free lobes 1.8–2.3 × 2–2.4 mm, ovate, acute, proximally on the adaxial side with a swelling on each side of the filament insertion; stamens 6, in 1 series, though the antepetalous slightly wider at the base, filaments connate at the base for 0.3–0.4 mm, 2.2–2.7 mm long, anthers 1.5–1.7 × 0.6–0.7 mm, dorsifixed, versatile; pistillode columnar, 1.3–2.3 × 0.9–1 mm. **PISTILLATE FLOWERS** with sepals 1.9–2.3 × 1.8–2.4 mm, orbicular, concave, apiculate, not membranous; petals connate with the receptacle for 0.4 mm, 2.3–2.5 × 1.8–2.8 mm, orbicular, concave, apiculate; staminodes 0.3–0.4 mm, flat, dentiform; ovary 2.3–2.8 mm high, asymmetrical, 1.9–2.2 mm diam. **FRUIT** green turning purplish with waxy bloom, slightly obovoid or ellipsoid, 10–16 × 5–10 mm, pointed at the base or rounded at both base and apex. **SEED** narrowly ellipsoid, pointed at base, rounded at apex, 9–12 × 5–6 mm; endosperm homogeneous. **EOPHYLL** bifid.

NOTE. We consider the varieties to be forms not worthy of official status. The wetter forest form is slightly more robust with longer median leaflets than the dry forest form. The palm pictured in Palme del Madagascar for *C. lucubensis* growing in Buitenzorg (= Bogor) probably provided the material for the type of *D. madagascariensis*! We do not believe the distinction between clustering and solitary stems is of significance here. We have seen specimens in a small population in the wild (Mahajanga area) which could be either of the two, and they seemed identical in all other respects.

This species is remarkably like *Dypsis onilahensis*, but the endosperm and habit differ, as well as the habitat.

The altitude given for Hildebrandt's type of *C. lucubensis* (500–800 m) seems excessive, given the fact that the highest point on Nosy Be is 430 m.

SPECIMENS SEEN. Ambilobe: Mt Ambohipiraka, Feb. 1960 (fr.), *Humbert & Cours* 32891 (K, P) & 32920 (K, P); (differs in short sheath, short nodes, short leaflets). Nosy Be: Lokobe, April 1879 (fl., fr.), *Hildebrandt* 2918 (K, P, WU, type of *C. lucubensis*); idem, c. 1924 (fl., fr.), *Perrier* 16100 (P); idem, c. 1926 (bud, fr.), *Perrier* 17554 (P); idem, Oct. 1963 (fl.), *Moore* 9026 (TAN); Ampasindava, Jan. 1992 (fl., y.fr.) *Birkinshaw & Adany* 105 (K). Ambanja: Ankaramibe, Jan. 1992 (fl., fr.), *Beentje* 4561 (K, MO, TAN); Ambalafary, Feb. 1992 (fl., fr.), *Beentje et al.* 4579 (K, MO, P, TAN). Iharana: 6km E of Daraina, June 1992 (ster.), *Beentje & Andriampaniry* 4693 (K, TAN). Mahajanga: 46km SE of Mahajanga, Dec. 1991 (dead infl.), *Beentje* 4553 (K, MO, P, TAN). Mitsinjo: 7km W of Katsepy, Dec. 1991 (fl., y.fr.), *Beentje* 4554 (BH, K, MO, P, TAN); Mitsinjo outskirts, cult., Sept. 1989 (fl., fr.), *B. Du Puy et al.* MB 410 (K); (probably) Mahabo, Jan. 1926 (fl., y.fr.), *Perrier* 17570 (K, P). Soalala: Andranomavo to Soalala, Jan. 1954 (bud), *Rakotovao RN* 6402 (K, P). Maevatanana: 66km N of Andriba, Nov. 1989 (fl.), *Cheek & Rakotozafy* 2380 (K, TAN). Antsalova: Ambodiriana, Jan. 1960 (fl., fr.), *Leandri* 2717 (K, P); Tsiandro area, Feb. 1960 (fl.), *Leandri* 2988 (K) & (fl., y.fr.), *Leandri* 2997 (K, P); SE of Antsalova, Aug. 1987 (fl.), *Phillipson* 2235 (K, P, TAN); Tsingy de Bemaraha, Feb. 1975 (fr.), *Cremers* 3788 (K, P, TAN); idem, Beraketa, Oct. 1932 (fl.), *Leandri* 386 (K, P, TAN); idem, Salapango gorge, March 1933 (fr.), *Leandri* 1104 (P). Miandrivazo: 25km W of Belobaka, March 1990 (fr.), *Du Puy et al.* MB714 (K, TAN). Mahabo: near Janjina, Dec. 1971 (bud), *Guillaumet* s.n. (K, P).

Locality uncertain: W Madagascar, (fr.), *Perrier* s.n. (P); Boina, Ambodisoha?, Nov. 1896 (y.fr.), *Perrier* 145 (P, syntype *C. oleraceus*).

Sight records. Antsiranana: Analamerana, Ampondrakely R. valley, 1994 (*D. Du Puy*). Nosy Be: Nosy Komba, 1992 (*Beentje*). Ambilobe, Ambanja: common in the Ambanja–Ambilobe area. 1992 (*Beentje*); Ankify peninsula, 1992 (*Beentje*). Analalava: 11km N of Maromandia, 1992 (*Beentje*). Antsohihy: 20–30km N of Antsohihy, 1992 (*Beentje*); Bora Réserve Speciale, 1992 (*Beentje*). Maevatanana: occasional along main road for c. 130 km southwards, 1992 (*Beentje*).

Cultivated. Mozambique: Hort. Maputo, May 1971 (fr.), *Balsinhas* 1849 (K). Mauritius: Hort. Mauritius, anno 1880 (seed), *anonymous* s.n. (K, syntype). India: Hort. Madras, anno 1900, *Bourne* s.n. (K). Sri Lanka: Hort. Peradeniya, July 1986 (y.fr.), *Rutherford & Bandara* R-100 (K). Thailand: Khao Chong, Nov. 1990 (fl.), *Barfod & Ueachirakan* 41.432 (K). Singapore: Hort. Singapore, Oct. 1929 (fl.), *Furtado* s.n. (K). Indonesia: Hort. Bogor, May 1936 (fr.), *Furtado* s.n. (K).

23. Dypsis decaryi

The famous 'Triangle Palm' is widespread in cultivation, and is grown in a wide variety of climates; this is unlike its native status, where it is restricted to a very narrow habitat 'niche', and as a result has a very restricted distribution. The name refers to the collector of the type, Raymond Decary.

Distribution. S Madagascar, confined to a small area.

Habitat. Dry forest or bush on stony soil, mid slope; 80–600 m.

Natural history. J. Ratsirarson (pers. comm.) has observed both Black Parrot and *Lemur catta* feeding on the fruit mesocarp; he also found pig droppings full of seed of *D. decaryi*; he observed bees and flies visiting the flowers at anthesis.

Local names. *Laafa* (Ranopiso).

Uses. Leaf used for thatching; fruits eaten by children, and formerly used to prepare a fermented drink; seeds exported for horticultural use, as the species is a prized ornamental.

Conservation Status. Vulnerable. Only known from a small area, where nearly all seed is harvested for export; fires are a threat. Population estimated at a thousand. This species is listed on CITES Annexe II.

Dypsis decaryi *(Jum.) Beentje & J. Dransf.* **comb. nov.**

Synonym:
Neodypsis decaryi Jum., Ann. Inst. Bot.-Géol. Colon. Marseille sér. 5, 1 (1): 15 (1933); Jum., Cat. Pl. Madagascar, Palmae: 17 (1938); Jum. & H. Perrier, Fl. Madagascar 30: 148, fig. 41 (1945); Read, Principes 5: 71, fig. 42, 43 (1961). Syntypes: Madagascar, Fort Dauphin–Bevilany, *Humbert & Swingle* 5715 (syntype P; isosyntype TAN); Bevilany, *Decary* 10294 (Syntype P) & 10299 (Syntype P); without loc., *Perrier* 18654 (not seen).

Dypsis decaryi, growing in transitional forest at Andohahela.

Solitary palm. **TRUNK** (0.5–) 3–6 (–10) m high, 30–40 cm diam.; internodes 3–10 cm, grey. **LEAVES** tristichous, 18–24, 1.5–3.25 m, porrect, arched, twisted around the rachis, marcescent; sheath open, 30–45 cm long, 40–65 cm wide when flattened, yellow-green with dense white wax usually overlaid by thick reddish pubescence, with ligules 7 cm high, 10 cm wide, orange turning grey-brown; petiole 33–50 cm long, proximally 6–7.5 × 9–10 cm diam., distally 5 × 5 cm, channelled with sharp margins, abaxially with reddish pubescence but glabrescent; rachis 2.2–3 m long, densely pubescent but glabrescent, channelled proximally, in mid-leaf 2–3 × 1.3–2.3 cm and keeled; leaflets 55–97 on each side of the rachis, regular, glaucous, at an angle of 90° with the leaflets on the opposite side of the rachis, the proximal 80–140 × 0.5–1 cm, (the most proximal often with reins and pendulous), median 58–89 × 2–3.5 cm (interval 2–5 cm), distal 10–60 × 0.4–1.6 cm, the top pair not or hardly connate, one or two tuft(s) of long brown-red ramenta present on the proximal part of the abaxial midrib, and lines of scattered minute reddish scales present on the fainter veins, main vein 1, very prominent adaxially, as well as thickened margins, apex unequal, bifid in median leaflets. **INFLORESCENCE** interfoliar, 125–178 × 120 cm, widely spreading, branched to 3 orders; peduncle 50–58 cm long, with scattered scales, proximally c. 5 × 3 cm diam., distally c. 3.5 × 2 cm diam.; prophyll 25–63 cm long, borne at 8–12 cm above the base of the peduncle, erect, hooded, split for 90% with only the proximal part closed, with scattered scales; peduncular bract inserted at 18–20 cm from the base of the peduncle, 40–55 cm long, open over most of its length but closed and beaked for the distal 5–19 cm, with scattered scales; rachis c. 118 cm, with all axes flaking and densely scaly, with 20–26 branched and c. 18 unbranched first order branches, these proximally flattened, their base up to 7 × 5 mm; most proximal first order branches with their rachis up to 50 cm long and up to 22 second order branches (8 of these branched again); rachillae pale yellow-green, 12–26 cm long, 1–4 mm diam.; triads, distant, slightly sunken, with small yellowish flowers. **STAMINATE FLOWERS** with sepals 1.7–2.1 × 1.6–2 mm, concave, keeled and proximally gibbous, elliptic, rounded, entire; petals on a 1.2–1.5 mm high receptacle, 3.2–3.5 × 1.8–2.3 mm, elliptic, fleshy, acute; stamens 6, very slightly biseriate, with the antepetalous stamens inserted slightly higher than the antesepalous ones, filaments 2.4–3 mm long, thin, anthers 1.7–2 × 1 mm, dorsifixed, versatile, held horizontally at anthesis, the locules parallel and obtuse; pistillode cylindrical, c. 1.6 mm high and 1 mm diam. **PISTILLATE FLOWERS** unknown, but from fruiting material sepals 2.4–2.8 × 2.6–3.3 mm, broadly ovate with a small apiculum; petals c. 3.3 × 4.3 mm, with broad membranous wings and a small fleshy triangular apex; staminodes c. 1 mm high. **FRUIT** ovoid, later subglobose, 15–22 × 12–19 mm, with rounded apex;

Dypsis decaryi

Dypsis decaryi. Close-up of crown, Andohahela (Photo: B. Du Puy & D. Du Puy).

Dypsis decaryi. Detail of infructescence (Photo: B. Du Puy & D. Du Puy).

Dypsis decaryi. **A** mid section of leaf × 1/3; **B** leaf tip × 1/3; **C** undersurface of leaflet base × 1; **D** first order inflorescence branch × 1/3; **E** part of infructescence × 1/2; **F** fruit × 1; **G** seed in cross section × 1. **A – D** from *Dransfield et al.* JD6772, **E – G** from *Beentje & Andriampaniry* 4614. Drawn by Rosemary Wise.

mesocarp fleshy-fibrous, endocarp fibrous with anastomosing fibres. **Seed** subglobose to ellipsoid, 17–19 × 15–17 mm, slightly asymmetric with an apiculate base (1 mm), with shallow anastomosing grooves over its surface, with rounded apex, and an equatorial depression corresponding to the embryo; endosperm ruminate, with ruminations up to 6 mm deep.

Note. Quite similar to *D. madagascariensis* but altogether neater in appearance, with its more compact habit and the densely three-ranked leaves.

Specimens seen. Tolanaro: Bevilany, Sept. 1928 (fl.), *Humbert & Swingle* 5715 (syntype; P, TAN); idem, Aug. 1932 (fl.), *Decary* 10294 (syntype; P) and (fl.), *Decary* 10296 (P), and (fl., fr.), *Decary* 10299 (syntype; P); Andohahela parc. 3, Dec. 1989 (fr.), *Dransfield et al.* JD6772 (K, TAN); Ranopiso, March 1992 (fl., fr.), *Beentje & Andriampaniry* 4614 (K, MO, TAN).

Locality unclear (but most likely from the same small area as the above): Col d'Ambato and E slopes of Vohipaly, 400–600 m alt., Feb. 1934 (fl., fr.), *Humbert* 14155 (P).

Cultivated: Ambovombe, Oct. 1931 (fl.), *Decary* 9303 (K, P); near Lutheran Hospital at Manambaro, Feb. 1975 (fl., y.fr.), *Croat* 31918 (TAN).

24. DYPSIS DECIPIENS

One of the most imposing palms of the island, but that may have something to do with its occurring in the austere surroundings of the Central Plateau, where it stands out dramatically. This makes a wonderful ornamental, able to withstand some cold (though not freezing) and periods of dry weather. The species name means 'deceiving', indicating that it closely resembles something else; in this case, it is a particularly inappropriate name!

DISTRIBUTION. Central Madagascar, between Ankazobe and Fianarantsoa.
HABITAT. Plateau forest (remnants), either near streams or in rocky sites; alt. 1400–2000 m.
LOCAL NAMES. *Betefaka, Manambe* (Imerina), *Sihara leibe* (Betsileo).
USES. Good palm heart; leaves used for erosion control (Betsileo).
CONSERVATION STATUS. Endangered. The number of individuals in the wild is estimated at around two hundred; their distribution area is prone to destruction and fire. This species is listed in CITES Annexe II.

Dypsis decipiens
(**x**: sight records)

Dypsis decipiens *(Becc.) Beentje & J. Dransf.* **comb. nov.**

SYNONYMS:
Chrysalidocarpus decipiens Beccari, Bot. Jahrb. Syst. 38, Beibl. 87: 36 (1906); Jum., Ann. Inst. Bot.-Géol. Colon. Marseille sér. 10, 3: 26 (1922); & Cat. Pl. Madagascar, Palmae: 8 (1938); Jum. & H. Perrier, Fl. Madagascar 30: 111, fig. 30 (1945). Lectotype: Madagascar, central Madagascar, *Baron* 3271 (K, lectotype, indicated here; isotype P).
Macrophloga decipiens (Beccari) Beccari, Palme del Madag.: 47, t. 46, fig. 38 (1914), pro parte (excl. *Perrier* 12088).

Imposing and handsome clustering palm, sometimes appearing solitary but more often in pairs or with younger shoots at the base. **TRUNK** 6–20 m high, ventricose; 50–70 cm diam., at the very base 30–40 cm diam., near the crown 25–30 cm diam.; internodes 4–5 cm long, grey, more distally shiny green, nodal scars c. 2 cm high, grey-brown; crownshaft pale waxy-grey-green. **LEAVES** 9–12 in the crown, spirally inserted, porrect; sheath c. 70 cm long, pale green with waxy white bloom, adaxially dark chestnut red-brown, one-third to half open in the oldest leaf, with brown ligules where the sheath margin makes a right angle towards the base of the petiole, with waxy scales or glabrous; petiole 10–25 cm long, proximally c. 11 × 5 cm, distally c. 7 × 5 cm diam., deeply channelled with sharp edges; rachis c. 2.2 m long, channelled proximally, in mid-leaf 3–4 × 2–2.7 cm diam. and keeled, with whitish tomentum or glabrous; leaflets c. 90 on each side of the rachis, in groups of 2–6, fanned within the groups giving the leaf a plumose appearance (though almost in one plane in very young trees), stiff with only the apices bending over, the proximal 70–94 × 1.3–1.7 cm (the most proximal often very long, narrow and pendulous), median 73–101 × 2.8–4.3 cm, distal 26–42 × 0.8–1.2 cm, in mid-leaf interval < 0.5–1 cm, the interval between the groups 3–8 cm, main veins 1, the other veins faint, apices attenuate and unequally bifid, with sparse large (1 cm) red-brown ramenta on the proximal midrib, and many small scattered reddish scales on the fainter veins. **INFLORESCENCE** infrafoliar, strongly curved, branched to 2 orders, c. 110 × 65 cm; peduncle 15–17 cm long, 9–10 × 4–6 cm diam.; prophyll 39–41 × c. 10 cm, borne at 6–6.5 cm above the base of the peduncle, split abaxially,

Dypsis decipiens, growing on a rocky hill, Itremo Massif (Photo: D. Du Puy).

Dypsis decipiens. Form with rather glaucous leaves, Itremo Massif (Photo D. Du Puy).

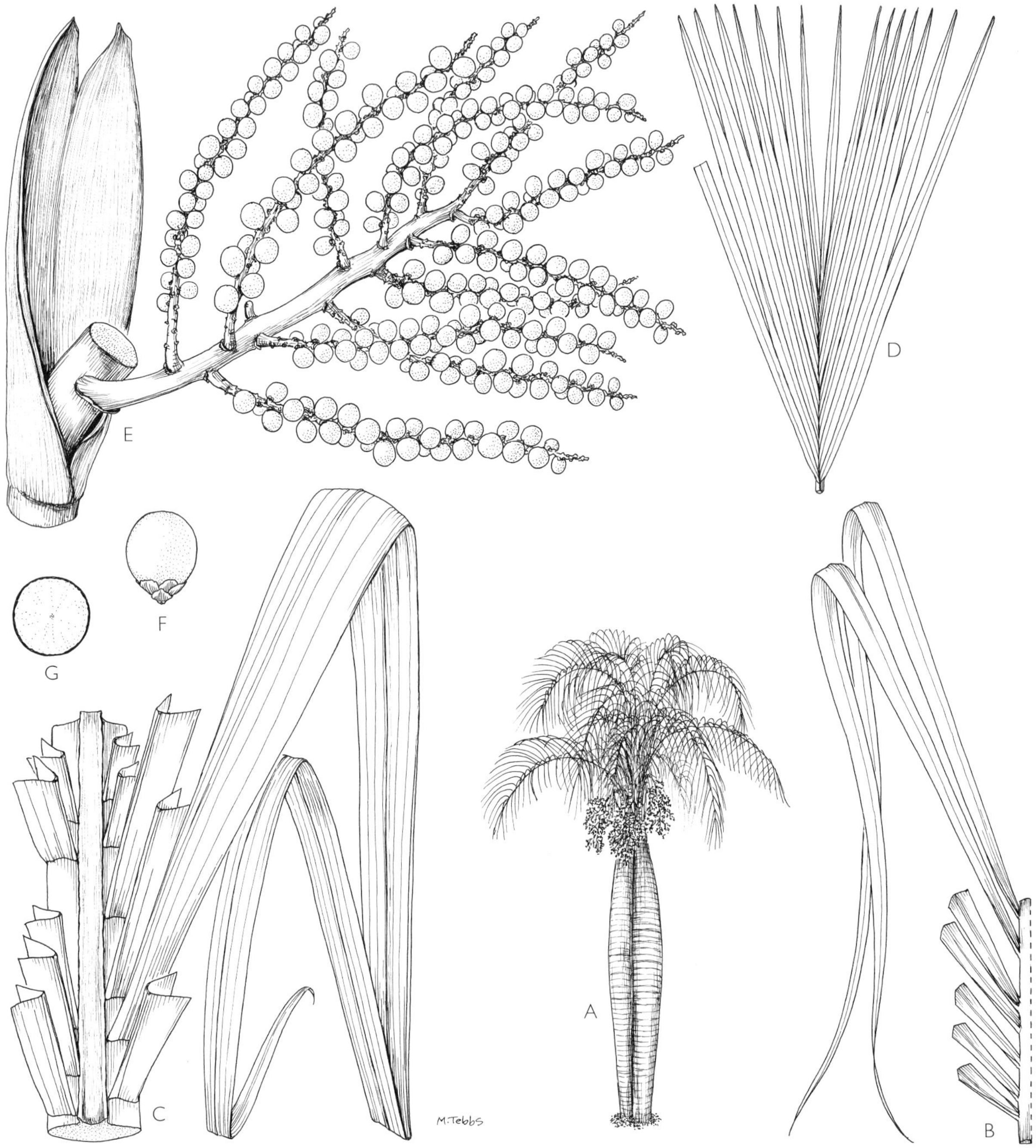

Dypsis decipiens. **A** habit, greatly reduced; **B** proximal part of leaf with lowermost leaflets × 1/4; **C** mid section of leaf × 1/4; **D** leaf tip × 1/4; **E** part of infructescence with first order branch × 1/4; **F** fruit × 1/2; **G** fruit in cross section × 1/2. **A**, from a photograph, **B – G** from *Beentje et al.* 4658. Drawn by Margaret Tebbs.

but distally with a small horizontal adaxial split, dark brown, ± glabrous; peduncular bract insertion point uncertain, quickly deciduous, 42–58 × 12–16 cm, opening over its whole length except for the beaked apex 5–6 cm long, pale waxy brown with scattered scales; rachis c. 60 cm, greyish white, glabrous, with c. 13 branched and 18 unbranched first order branches, the more proximal of these with flattened bases to 4.7 × 1.3 cm diam.; rachis bracts c. 9 mm high; rachillae 7–40 cm long, 3.5–7 mm diam., with distant triads in pits; rachilla bracts 2.5–3 × 3.5 mm, acute. **STAMINATE FLOWERS** with sepals 2–4 × 1.8–2.5 mm, concave, keeled, gibbous, with membranous margins, apiculate; petals connate for 1.3–1.6 mm, the free lobes 3.8–5.6 × 2.3–3.7 mm, ovate, acute to obtuse; stamens 6, in 1 series, the filaments (2–) 3.2–5 mm long and connate at base for 0.3 mm, anthers 2.3–2.8 × 0.8–1.4 mm, dorsifixed, versatile; pistillode columnar,1.9–2.7 × 0.7–1.5 mm. **PISTILLATE FLOWERS** with sepals 5.1–6.3 × 3.5–5 mm, concave, orbicular, with small hooded tip; petals resembling the sepals, 4–5.3 × 3–3.8 mm, concave, ovate, with small apiculus, staminodes 0.3–0.5 mm, dentiform; ovary 3.5–4.3 mm high, 1.5–2 mm diam. **FRUIT** broadly ellipsoid or almost globose, colour unknown, 22–25 × 20–22 mm; endocarp very fibrous with long anastomosing fibres. **SEED** 10–20 × 11.5–18mm, with rounded base and apex; endosperm homogeneous.

NOTE. Beccari established the new genus *Macrophloga* based on two different taxa: his own *Chrysalidocarpus decipiens*, and *Neodypsis basilongus* based on a Perrier specimen (no. 12088) with seed showing ruminate endosperm. True *Chrysalidocarpus decipiens* has homogeneous endosperm. The combination *Neodypsis decipiens* (Becc.) Jumelle & Perrier existed only in Beccari's imagination; Jumelle & Perrier (1913) mentioned the possibility but specifically refrained from making it.

HB has seen bees visiting the flowers of a cultivated specimen in Antananarivo.

Specimens seen. Ankazobe: Manankazo, sine die (bud), *Perrier* 12099 (P). Manjakandriana: (probably all from Andrangolaoka), *Baron* 502 (K, P; syntype), *Baron* 3271 (K, syntype); & *Baron* s.n. (K); Ambatoloana to Mandraka, ?1936 (fr.), *Boiteau* s.n.(P). Ambositra: 24km S of Ambositra, April 1992 (seed), *Beentje et al.* 4658 (BH, K, MO, P, TAN). Ambohimahasoa: Ankafina forest, March 1881 (fl.), *Hildebrandt* 3974a (K, syntype).

Cultivated. Antananarivo Botanic Garden, Jan. 1938 (fl.), *Herb. Jard. Bot. Tan.* 3091 (K, P); idem, Oct. 1940 (fl.), *Decary* 15889 (P); idem, April 1971 (y.fr.), *Moore* 9923 (P).

Sight records. Manerinerina (*Dransfield & Beentje*). Manjakandriana: Ikopa R. sources (*Perrier*). Antsirabe: Vavato (*Perrier*). Ambositra: Ranomena (*Beentje*); Ilaka (*Dransfield & Beentje*). Ambohimahasoa: NW of Ranomafana (*Beentje*).

Dypsis decipiens, view of crown in flower, Itremo Massif (Photo: D. Du Puy).

25. Dypsis basilonga

A rare and elegant palm, apparently confined to a single hill which is now surrounded by a sea of cultivation and secondary vegetation. We have seen this species *in situ* in the type locality, but the inflorescences were too rotten to make a proper specimen. It is a compact, rather graceful palm, common in the low-canopy thin-stemmed small-crown forest just below the summit of Mt Vatovavy at 450–500 m altitude, growing on the edges of cliffs in what is probably a wind-swept habitat. The species name refers to the 'basal' leaflets nearest the petiole, which are sometimes very long.

Distribution. Only known from Vatovavy.
Habitat. Small-crown, submontane forest, on gneiss; 300–500 m.
Local names. *Madiovozona* (Tanala; meaning 'clean neck').
Uses. Excellent palm-heart.
Conservation Status. Endangered. Single-site status; the only protection of the forest derives from local fady (taboos).

Dypsis basilonga *(Jum. & H. Perrier) Beentje & J. Dransf.* **comb. nov.**

Synonyms:
Neodypsis basilongus Jum. & H. Perrier, Ann. Inst. Bot.-Géol. Colon. Marseille sér. 2, 3, 1 (1): 16, pl. 4 (1913); Jum., Ann. Inst. Bot.-Géol. Colon. Marseille sér. 4, 2 (2): 11 (1924); Jum., Cat. Pl.

Madagascar, Palmae: 17 (1938); Jum. & H. Perrier, Fl. Madagascar 30: 144, fig. 39 (1945). Type not indicated (protologue) but from Jumelle (1924): Madagascar, Mt. Vatovavy, *Perrier* 12088 (Holotype P).

Solitary palm. **STEM** 2–5 m tall, 10–15 cm diam.; internodes short; crownshaft well-developed, whitish, c. 40 cm long. **LEAVES** strongly curved, c. 6–7 in the crown, 1–1.5 m long; sheath white and waxy, c. 40 cm long, glabrous, without ligules; petiole 14–16 cm long, c. 1.5 cm wide, channelled, with patches of dense tomentum; rachis c. 1 m long; leaflets more than 30 on each side of the rachis, in groups of 2–3, the proximal leaflets with a very long gap between the basal pair and the next pair, the most proximal to 117 × 3 cm, the next 75 × 1.9 cm, median 64–68 × 2.6–3.1 cm, the group interval 4.5–5.5 cm, the leaflet interval 0.2–0.3 cm, distal 16–40 × 1–1.7 cm, glaucous abaxially, with lines of minute reddish scales on the minor veins, main vein 1, with thickened margins, apices unequally attenuate. **INFLORESCENCE** interfoliar, c. 80 cm long, branched to 2 orders; peduncle 40–60 cm long, proximally 5 × 0.5 cm, straight within the sheath, then curved through 180 ° so the branched part hanging; prophyll borne at c. 40 cm above the base of the peduncle, waxy; peduncular bract inserted at c. 50 cm from the base of the peduncle; first order branches with a secondary rachis of up to 11 cm, proximally 2 × 0.8 cm, with up to 8 rachillae, glabrous; rachillae 15–19 cm long, c. 4 mm diam., with distant triads in pits; rachilla bracts proud and rounded. **STAMINATE FLOWERS** not seen. **PISTILLATE FLOWERS** not seen; sepals in fruit rounded; petals twice as long as the sepals (fide Beccari). **FRUIT** (see note) ellipsoid, c. 20 × 9–10 mm, with rounded base and apex; endocarp fibrous, with anastomosing fibres. **SEED** oblong, with pointed base and rounded apex; endosperm ruminate with shallow distant intrusions.

NOTE. The confusion with *D. decipiens* was caused by Beccari, who thought the two species were the same

Dypsis basilonga. **A** leaf base × 1/5; **B** mid section of leaf × 1/2; **C** leaf tip × 1/5; **D** first order inflorescence branch × 1/2. All from *Perrier* 12088. Drawn by Margaret Tebbs.

and united them in *Macrophloga*; the new genus was necessary because of the ruminate endosperm of the fruit of *Perrier* 12088. The leaves in the genus description were based on *D. decipiens*.

The fruit is now missing from the type; there is, however, a photo which shows the fruit attached to a loose rachilla. The description of the fruit and seed is taken from Beccari (1914), who is accurate in his descriptions.

The protologue adds the following data, not apparent from the type or its label: Solitary, stem 4–6 m, diam. 12 cm; leaves gracefully curved; sheath whitish waxy, adaxially pinkish, 3 cm wide; inflorescence interfoliar, branched to 2 orders; Staminate flowers in bud 4 mm long. Pistillate flowers with sepals 1.5 × 1.5 mm, petals 3 × 2.5 mm; ovary cylindrical.

SPECIMENS SEEN. Mananjary: Mt Vatovavy, Oct. 1911 (fr.), *Perrier* 12088 (P, type).

Dypsis ambositrae

26. DYPSIS AMBOSITRAE

A graceful palm of the high plateaux, which would probably do quite well in cultivation. We have looked for seeds but not found any, so far. This species will probably become extinct in the near future, unless some rapid action is taken to safeguard the remaining trees, now numbering less than ten. The name comes from the town which lies between the known populations.

DISTRIBUTION. Central Madagascar, near Ambositra.
HABITAT. ?Forest; among rocks or in riverine forest remnants, medium or steep mid slope; alt. 1300–1500 m.
LOCAL NAMES. Not recorded.
USES. Not recorded.
CONSERVATION STATUS. Critical. In 1992 twelve trees of this species were known, all growing in or next to agricultural areas; in 1994, at least five of these had been cut down or burnt.

Dypsis ambositrae *Beentje* **sp. nov.**
D. oreophilae et *D. tsaratananensi* caulibus caespitosis foliolis aggregatis staminibus 6 affinis sed inflorescentia glabra vagina folii magna ceracea, foliis spiraliter dispositis foliolis magnis differt. Typus: Madagascar, Ilaka Afovoany, *Beentje & Andriampaniry* 4742 (Holotypus K; isotypi BH, MO, P, TAN).

Clustering palm in tufts of 2–3, sometimes (*Beentje & Andriampaniry* 4615) appearing solitary when in regularly burnt terrain. **STEM** 3–7 m tall, c. 12 cm diam.; internodes 10–20 cm, pale brown to grey (green and ringed when young), nodal scars 0.5 cm, grey; wood hard; base of stem slightly wider, with some surface roots; slight bulge in upper trunk in one older tree; crownshaft pale waxy grey-green. **LEAVES** 7–11, spiral, gracefully arching, with stiff leaflets; sheath 64–103 cm, pale green with a white bloom, ligules 2 cm; petiole 9–30 cm long, 3–6 × 2.2–4.5 cm diam., channelled with soft edges; rachis 2.1–2.8 m, in mid-leaf 2.2–3.5 cm wide, green; leaflets 74–84 on each side of the rachis, grouped only very slightly in 2s–5s, in one plane, the leaflets on opposite sides of the rachis at an angle of 90°, stiff with only the apices pendulous, apices attenuate, unequally bifid, the proximal 69–144 × 0.3–1.8 cm, (first

Dypsis ambositrae. View of crown of mature tree, near Ambositra (*Beentje & Andriampaniry* 4615).

interval c. 29 cm, more distal 3.5–9 cm), median 89–114 × 2.3–3 cm (leaflet interval 0.2–2 cm, group interval 2–3.5 cm), distal 18–58 × 0.8–2.5 cm, abaxially with distant tufts of pale grey ramenta over almost whole length of midrib, with scattered scales very faint to invisible, main veins faint, with only the midrib very prominent on the adaxial surface. **INFLORESCENCE** interfoliar, branched to 2 (3 in a few cases) orders, with the basal part within the closed sheath, the prophyll hidden and the peduncular bract spreading from the top of the sheath; peduncle 68–123 cm long, distally 9 × 5 cm diam., green, glabrous, curved outside the sheath; prophyll c. 91 cm, borne at c. 32 cm above the base of the peduncle, 11.5 cm wide, narrowly 2-winged; peduncular bract deciduous, about 80 cm, beaked (c. 5 cm) and closed distally, pale waxy grey, inserted c. 48 cm from the base of the peduncle; open peduncular bract 14 × 7 cm; rachis 84–102 cm, with 23–24 branched and 14–17 unbranched first order branches, in a few cases some of the proximal branches branched twice more, but not more than 3 in the entire inflorescence; all axes green with white bloom; first order branches proximally 2–3 × 0.5–1 cm; rachillae 14–32 cm, 3–4 mm diam., with flattish base and distant to rather dense triads, hardly sunken in slight pits with entire, obtuse or acute bracts. **STAMINATE FLOWERS** with sepals 2.2–2.5 × 1.6–2.2 mm, keeled, gibbous at the base, broadly ovate, obtuse, the margins membranous; petals connate for 0.2–0.5 mm, the free lobes 2.8–3 × 2.8–3.2 mm, ovate or elliptic, acute, sometimes with hooded apex; stamens 6, uniseriate, the filaments connate for 0.2–0.5 mm, 2.8–3.2 mm long, anthers 2.1–2.3 × 1 mm; pistillode 2.2–2.3 mm, columnar, 0.8–1 mm diam. **PISTILLATE FLOWERS** with sepals 2.4–3 × 3–4.1 mm, broadly ovate, rounded; petals hardly connate at the base, 3.5–4.1 × 4–5 mm, imbricate but for the apiculate apex, broadly ovate, concave; staminodes 6, 0.3–1.6 mm, narrow and flat; ovary asymmetrical, 2.7–4.8 × 2.8–4 mm, with indistinct pyramidal stigmas.

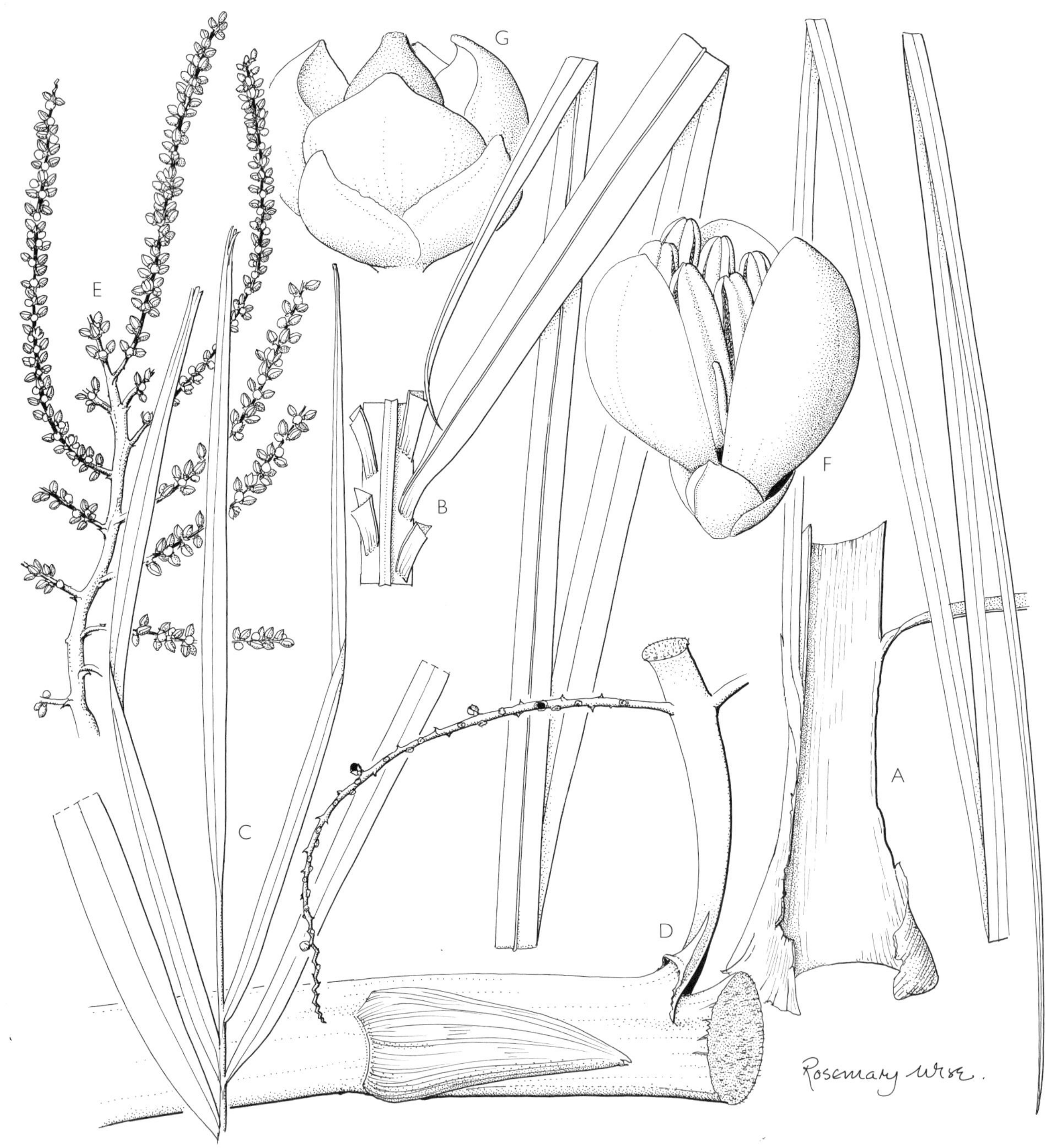

Dypsis ambositrae. **A** Proximal part of leaf showing top of sheath and lowermost leaflets × 1/3; **B** mid section of leaf × 1/3; **C** leaf tip × 1/3; **D** part of inflorescence showing tip of peduncle and basal first order branch × 1/3; **E** inflorescence tip × 1/3; **F** staminate flower × 10; **G** pistillate flower × 6. All from *Beentje & Andriampaniry* 4742. Drawn by Rosemary Wise.

Fruit only known from carbonized remnants, c. 14 x 10.5 mm, possibly with fibrous endocarp, possibly with ruminate endosperm.

Note. With its grouped leaflets and glabrous inflorescences branched to 2 orders, this species is allied to *D. oreophila* and *D. tsaratananensis*, from which it is easily distinguishable by its larger leaf sheaths, longer leaves with larger leaflets, the much longer inflorescences, and the larger fruit.

Specimens seen. Ambositra: 8km NNW of Ilaka Afovoany, July 1992 (fl.), *Beentje & Andriampaniry* 4742 (BH, K, MO, P, TAN; type); 45km S of Ambositra, March 1992 (old fr.), *Beentje & Andriampaniry* 4615 (K).

27. Dypsis heteromorpha

This palm has not been collected since 1959, and the material available to us was fairly fragmentary. A high altitude species, with the name indicating the variation in the leaf division: from regularly pinnate with many leaflets, to entire-leaved on young shoots.

Distribution. N Madagascar: Tsaratanana, Marojejy and Anjanaharibe.
Habitat. Moist montane forest; 1300–2200 m.
Local names. Not recorded.

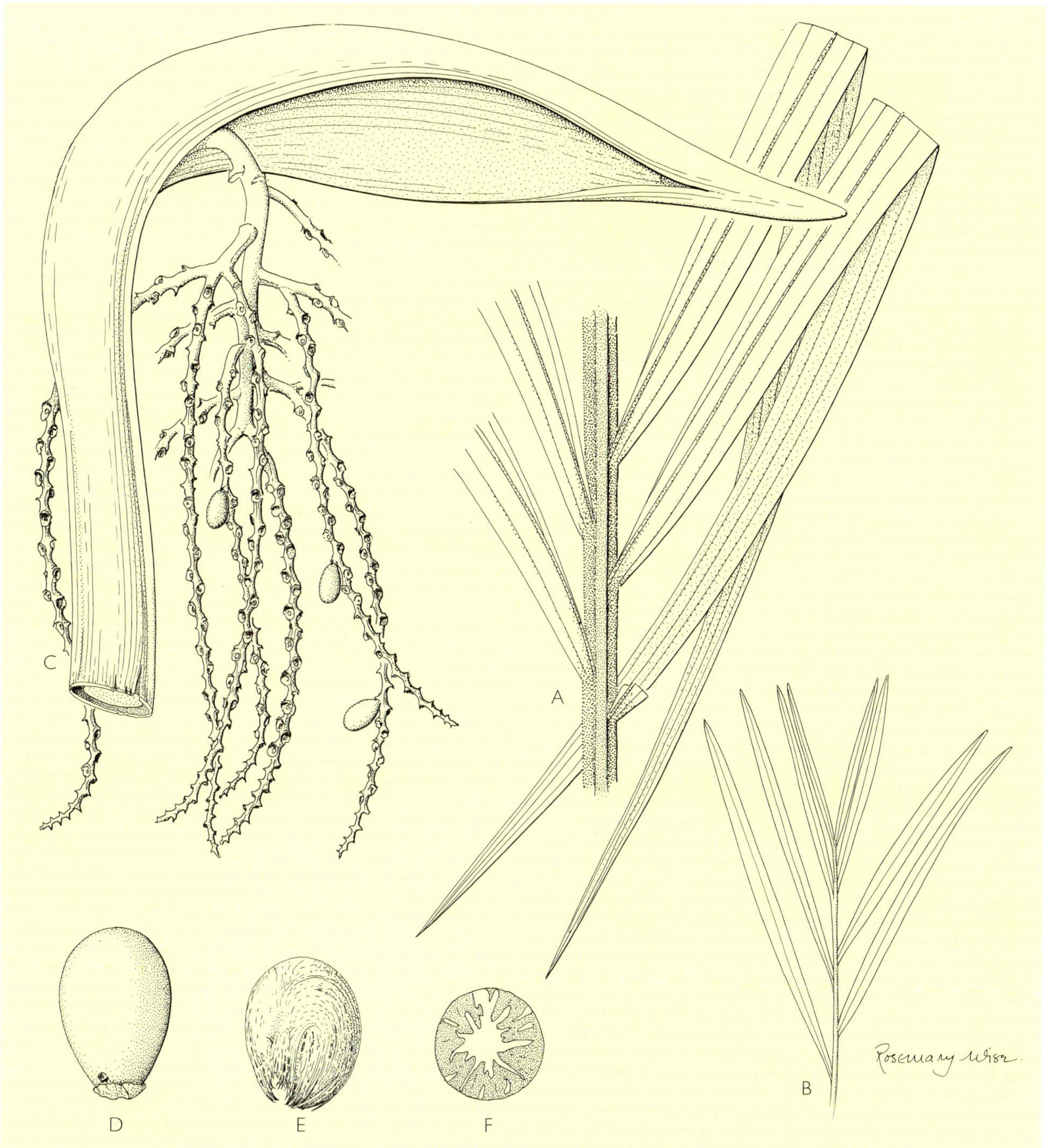

Dypsis heteromorpha. **A** mid section of leaf x 1/2; **B** leaf tip x 1/4; **C** infructescence x 1/2; **D** fruit x 1; **E** fibrous endocarp x 1; **F** seed in cross section x 1. **A**, **B** from *Humbert & Saboureau* 31725, **C – F** from *Humbert et al.* 24766. Drawn by Rosemary Wise.

Uses. Not recorded.
Conservation Status. Uncertain, but probably rare; the distribution area is not well known botanically.

Dypsis heteromorpha *(Jum.) Beentje & J. Dransf.* **comb. nov.**

Synonym:
Neodypsis heteromorphus Jum., Ann. Inst. Bot.-Géol. Colon. Marseille sér. 4, 2 (2): 20 (1924); Jum., Cat. Pl. Madagascar, Palmae: 18 (1938); Jum. & H. Perrier, Fl. Madagascar 30: 145 (1945). Syntypes: Madagascar, Tsaratanana, *Perrier* 11935 (not seen); 15266 (Syntype P); 15266bis (Syntype P).

Clustering palm in tufts of 3–6, but sometimes appearing solitary. **Stems** 3–12 m high, 8–12 cm diam.; internodes green, nodal scars obvious. **Leaves** c. 10 in the crown, tristichous (fide *Humbert & Sabourea*u 31725); sheath 27–50 cm long, to 7 cm wide when flattened, with sloping shoulders, adaxially red-brown, abaxially very waxy, in the distal part with dense reddish laciniate scales; petiole absent or up to 35 cm long, 1–1.8 cm diam., densely pubescent or with scattered scales; rachis c. 1.5 m long, in mid-leaf 0.9–1.3 cm wide, slightly keeled, pubescent on both surfaces; leaflets (number unknown) regular, the proximal 20–62 x 0.2–2.5 cm, median 41–67 x 1–3.5 cm (interval 1.5–3.5 cm), distal 7–35 x 0.6–2.5 cm, main veins 1–5, and with thickened margins, abaxially with scattered reddish glands on the minor veins and continuous or scattered laciniate ramenta 3–20 mm long on the midrib, apex bifid, unequally attenuate, young shoots with entire, deeply bilobed leaves 26–50 cm long with a costa 2–4 cm and lobes 24–46 cm long, or leaves with 2–4 leaflets (not seen), on a 64–75 cm long petiole. **Inflorescence** infrafoliar, branched to 2 orders, pendulous; peduncle 20–40 (+) cm long, proximally 2–3.5 x 0.8–1.3 cm, distally 1–2 x 0.7–1.2 cm, glabrous; prophyll 38–52 x 3–8.5 cm, borne at 9–20 cm above the base of the peduncle, open for a third or over its whole length, erect, with scattered scales; peduncular bract inserted at 17–29 cm from the base of the peduncle, 27–55 (+) cm long, open over its whole length, beaked for 0.5–4 cm, with scattered scales; non-tubular peduncular bracts one or two, 2–5 cm long; rachis 10–35 cm long, glabrous, with 8–10 branched and 7–9 unbranched first order branches, first order branches with a rachis 1.5–13 cm long and 1.1–1.7 x 0.5–0.8 cm diam. proximally with 3–9 rachillae; rachis bracts conspicuous, to 2.5 cm, pale brown; rachillae 5–23 cm long, 2–4 mm diam., with dense triads (more distant in fruit), superficial or slightly sunk, with triangular acute rachilla bracts and very pronounced bracteoles. **Staminate flowers** with sepals 2.7–3.2 x 3.5–4.4 mm; petals 3.5–3.8 x 2.5–3 mm; stamens 6, slightly 2-seriate (offset 0.2 mm), filaments in ripe buds 1.8–2.1 mm, cylindrical, anthers 2.2–2.4 x 0.9–1.2 mm; pistillode c. 2.4 mm high, 1.2 mm diam. **Pistillate flowers** with sepals 3–4 x 3.3–5mm; petals (at young bud stage) 3.6–3.8 x c. 3.5 mm, very concave (4–5.5 x 5–7.2 mm in young fruit); staminodes not seen; gynoecium when young c. 3 mm high, 1.8 mm diam. **Fruit** ellipsoid, 17–23 x 14–22 mm, rounded at the apex; endocarp fibrous, the fibres anastomosing. **Seed** ellipsoid, 16–22 x 13–21 mm, obtuse at the base with a sub-aequatorial depression, rounded at apex, with slight surface grooving; endosperm ruminate, the ruminations distant and 1–7 mm deep. **Eophyll** bifid.

Note. Related to *D. baronii* and *D. onilahensis*, and possibly a high-altitude variant of the former; distinct by deep-ruminate endosperm.

Specimens seen. Andapa / Ambilobe / Bealanana: Mt Tsaratanana, Oct. 1912 (y.fr.), *Perrier* 11935 (P); idem, Jan. 1923 (fr.), *Perrier* 15266 (P, syntype); idem, sine die (fl., fr.), *Perrier* 15266 bis (P, syntype). Bealanana: Ambohimirahavavy Mts, Jan./Feb. 1951 (fr.), *Humbert & Capuron* 25281 (K, P). Andapa: N Anjanaharibe, Dec. 1950/Jan. 1951 (fr.), *Humbert et al.* 24766 (K, P); Ambatoharanana valley to upper Antsahaberoka, Nov./Dec. 1959 (y.fr.), *Humbert & Sabourea*u 31725 (K, P).

28. Dypsis baronii

A common species of the rain forest of the eastern escarpments. It closely resembles *D. lutescens*, but that is strictly a littoral species, confined to a narrow strip of vegetation close to the sea; and *D. onilahensis*, from drier localities on the Western side of the island. *D. baronii* is a graceful palm, and is often seen in the gardens of central Madagascar, particularly in Antananarivo. This is such a fine ornamental that it should be much more widely grown outside Madagascar than it is at present, particularly in view of its occurrence in upland areas. The species was named after the Reverend Richard Baron (1847–1907) who collected the type and many other plants in 1880–1897.

Distribution. North, Central and E Madagascar.
Habitat. Moist montane forests, bamboo-dominated forests; usually on steep mid-slopes, less often on ridge crests; survives in half-shade or full sun; 850–1470 m.
Local names. *Faribazo* (Imerina, "sugarcane tree"); *Tongalo* (Betsimisaraka).
Uses. Excellent palm-heart; fruit edible and sweet. Very elegant palm, cultivated in Antananarivo and on the plateau as an ornamental.
Conservation Status. Not threatened. The species occurs over a large area.

Dypsis baronii *(Becc.) Beentje & J. Dransf.* **comb. nov.**

Synonyms:
Chrysalidocarpus baronii Becc., Bot. Jahrb. Syst. 38, Beibl. 287: 33 (1906); Becc., Palme del Madagascar 39, fig. 30, t. 37 (1914); Jum.,

Dypsis baronii, perched on cliff edge, Marojejy (*Dransfield et al.* JD6769).

Ann. Inst. Bot.-Géol. Colon. Marseille sér. 2, 10 (3): 11 (1922). Lectotype (chosen here): Central Madagascar, *Baron* 3270 (lectotype K; isotype P).
Neodypsis baronii (Becc.) Jum., Compt. Rend. Acad. Paris 179: 249 (1924); Jum., Ann. Inst. Bot.-Géol. Colon. Marseille sér. 4, 2 (2): 23, fig. 1 (1924); Jum., Cat. Pl. Madagascar, Palmae: 16 (1938); Jum. & H. Perrier, Fl. Madagascar 30: 146, fig. 40 (1945).

Chrysalidocarpus propinquus Jum., Ann. Inst. Bot.-Géol. Colon. Marseille sér. 2, 10 (3): 19 (1922); Jum., Cat. Pl. Madagascar, Palmae: 11 (1938). Type: Madagascar, Analamazaotra, *Perrier* 12018 (Not seen).

Neodypsis compactus Jum., Ann. Inst. Bot.-Géol. Colon. Marseille sér. 5, 1 (1): 13 (1933); Jum., Cat. Pl. Madagascar, Palmae: 17 (1938); Jum. & H. Perrier, Fl. Madagascar 30: 154, fig. 43 (1945), **synon. nov.** Type: Madagascar, Mt. d'Ambre, *Perrier* 18870 (Holotype P).

Dypsis baronii

Clustering palm in clumps of 3–5, rarely appearing solitary. **STEMS** 2–8 m high, rarely with a single branching point, 2.5–12 [–22] cm diam., near the crown 2.5–5.5 cm diam.; internodes 4–35 cm, near the crown as short as 1.3 cm, grey, grey-green or blackish; nodal scars c. 0.5 cm, faint, white; wood hard, with a dense layer of hard red fibres just underneath the bark; crownshaft to 10 cm diam., pale green to pale yellow, waxy, the unexposed sheaths peach-coloured; occasionally with the remnants of sheaths remaining on the distal part of the trunk, but usually the leaves abscising neatly. **LEAVES** 4–8 in the crown, spiral or tristichous, arching, the young leaves sometimes held on edge in their distal half; sheath 28–60 cm long, pale green to pale yellow, waxy, the unexposed sheaths peach-coloured, only distally densely scaly, with auricles to 2 cm high; petiole 0–37 (–53) cm long, proximally 1–2.3 x 1.2–2.5 cm, distally 0.8–1.3 x 0.8–1 cm diam., with dense but flaking red to dark scales, therefore appearing crimson when young, later with scattered scales, slightly channelled; rachis 0.5–1.2 m long, abaxially densely scaly but glabrescent, in mid-leaf 0.8–1.6 cm wide and keeled; leaflets 35–60 on each side of the rachis, regular, in one plane, dark green, stiff with the distal part pendulous, the proximal 19–100 x 0.3–1.1 cm, median 25–49 [–77] x 0.9–2.7 cm, distal 7–36 x 0.3–1.8 cm, sometimes glabrous but more often with quite a few ramenta (2–5 mm long) proximally, on the minor veins often with scattered reddish bumps, these being the remnants of the quite dense bases of stellate-laciniate reddish scales on the minor veins in young leaves, main vein 1, plus rather thickened margins, apices bifid, unequally attenuate. **INFLORESCENCE** interfoliar or infrafoliar, branched to 2 orders, arching; peduncle 24–62 cm long, proximally 0.8–2.2 x 0.5–0.8 cm, distally 1.8–3.5 x 0.9–2 cm diam., glabrous, proximally red, distally green; prophyll 25–73 x 2.2–4 [–5.5] cm, borne at 2.5–22 (–45) cm above the base of the

Dypsis baronii. View of crown in fruit, Mt. d'Ambre (Photo: B. Du Puy & D. Du Puy).

Dypsis baronii. Close up of staminate flowers, Ambohitsaratelo (*Dransfield et al.* JD6447).

peduncle, erect, often hooded, dark crimson to pale brown, distally with scattered scales; peduncular bract inserted at 22–54 from the base of the peduncle, deciduous, 20–44 (–65) cm long, erect and hooded, deep crimson or brown, carried upwards by the lengthening inflorescence; non-tubular peduncular bracts usually 2, 1–6 (–17.5) × 0.6–1.4 cm; rachis 13–33 cm long, glabrous, with 5–21 branched and 7–10 unbranched first order branches, the proximal with a secondary rachis of up to 4.5 [–11.5] cm long and 4–13 × 2–6 mm diam. and with 4–9 [–15] rachillae, rachis bracts 4–20 mm long; rachillae 3–24) cm long, [1.5–] 2.5–4.5 mm diam., glabrous, pinkish to pale green, with distant or dense slightly sunken triads; rachilla bracts 1–1.5 mm, acute. **Staminate flowers** faintly scented; sepals cream with brown tips and 1.9–3 × 2–3.1 mm; petals proximally white, distally red, connate for 0.5–1.8 mm, free parts 2.5–4 × [1.8–] 2.5–3.5 mm, often with swellings proximally on each side of the filament insertion; stamens 6, biseriate (offset 0.2–0.4 mm), filaments white, (0.8 in bud–) 2.4–4.5 mm long, 0.4–1.3 mm wide with the antepetalous ones flatter than the cylindrical antesepalous ones, anthers 1.1–2 × 0.5–1.4 mm; pistillode 1.4–3.3 × 0.6–1.5 mm. **Pistillate flowers** with sepals 2.2–2.8 × [1.7–2.4] 2.5–3 mm, slightly apiculate; petals 2.8–3.5 × [2.3–2.6] 2.8–4.2 mm; staminodes 6, 0.4–1 mm; ovary 2.8–3.3 × 2.8 mm. **Fruit** yellow, ellipsoid or subglobose, 10–20 × 8–16 mm; endocarp fibrous. **Seed** ellipsoid or slightly obovoid, 9.5–12 × 7.5–11 mm, pointed at the base, rounded at the apex, with a sub-basal depression, the outside slightly grooved; endosperm ruminate, the intrusions corresponding to the grooves, 1–1.5 [–3] mm deep and medium dense.

Note. *Chrysalidocarpus propinquus* is said by Jumelle to be a form on denuded rocks with a short petiole and short, narrow leaflets, a very short peduncle and

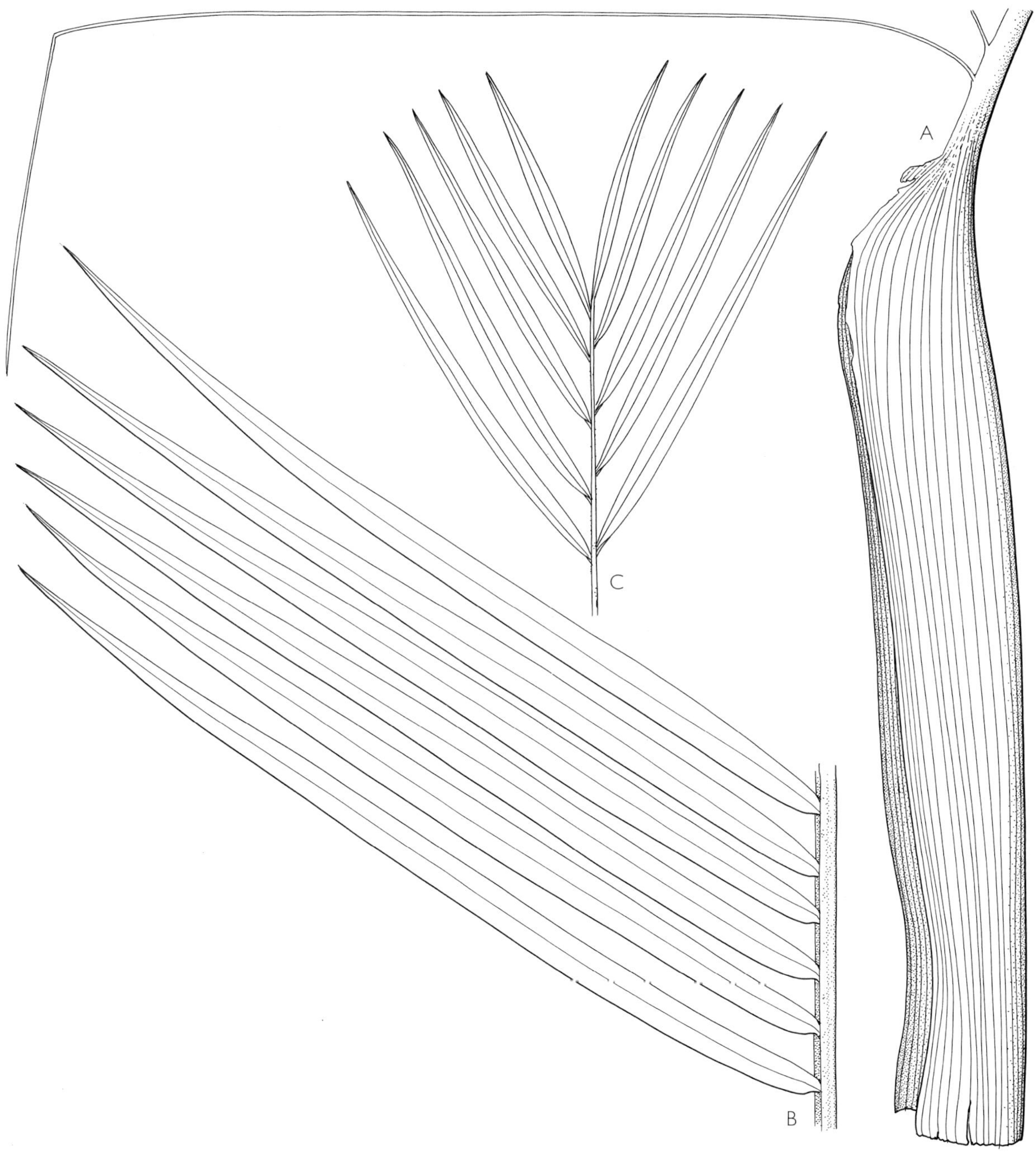

Dypsis baronii. **A** leaf sheath and basal leaflet × 1/2; **B** mid section of leaf × 1/2; **C** leaf tip × 1/2. All from *Beentje & Andriampaniry* 4544. Drawn by Rosemary Wise.

rachillae, and small fruit. It was put into synonymy by Jumelle in 1924.

We have included *N. compactus* in the synonymy of *D. baronii*; the differences with the bulk of *D. baronii* are *N. compactus* being called solitary and having a diameter of 10–22 cm; in *N. compactus* the median leaflets are generally longer, 49–77 cm long (bulk of population 25–49 cm); the first order branches have a generally longer secondary rachis, 4.5–11.5 cm, and the petals are narrower. Except for the trunk characters all these differences are differences of degree rather than absolute; several collections of *D. baronii* s.s. have been called solitary; we believe the thicker trunk is the only real character. With a bit more we would have kept them apart, but they look so similar that *N. compactus* has to be included in synonymy; the extremes for *N. compactus* are indicated by [square brackets] in the description.

HB found many small red ants in a ripe infructescence (*Beentje & Rabarilala* 4412). In Mantady and at Maromiza HB has seen two stems which had a single branch at c. 1 m above the base.

This species is extraordinarily close to *D. onilahensis*, and only differs in the habitat and the ruminate endosperm; the ruminations in *D. baronii* are difficult to see at times, being very small, and *D. baronii*

Dypsis baronii. **A** inflorescence × 1/2; **B** staminate flower × 5; **C** fruiting rachillae × 1/2; **D** mature fruit × 2; **E** fruit in cross section × 2. **A**, **C – E** from *Beentje & Andriampaniry* 4544, **B** from *Beentje & Andriampaniry* 4771. Drawn by Rosemary Wise.

has been found in sites which are really in *D. onilahensis* territory, such as the forest of Ambohitsaratelo. In the absence of fruit, several collections could not be identified as belonging to one or the other [Miandrivazo: NW of Ambohitsaratelo-Bebao, July 1974 (fl), *Morat* 4590 (P, TAN); idem, Nov. 1986 (fl., y.fr.), *Dransfield et al.* JD6447 (K, P, TAN)] but since *Dorr et al.* 3532 from the same locality has ruminate endosperm, they are more likely *D. baronii.*

SPECIMENS SEEN. Antsiranana: Mt d'Ambre, Nov. 1932 (fl., fr.), *Perrier* 18870 (P, type of *N. compactus*); idem, June 1970 (fl.), *Bosser* 20367; idem, near the summit, Oct. 1991 (fr.), *Malcomber et al.* 976; idem, Bianamalo, June 1989 (ster.), *B. Du Puy et al.* MB 217 (K) & (fl.) MB 222 (K). Andapa: Mt Ambodilaitra, March 1949 (y.fr.), *Humbert* 23287 (K, P;differs in longer rachillae than rest, 19–24 cm); Ambatosoratra, Jan. 1949 (y.fr.), *Cours* 3342 (P); Marojejy E, N of Mandena, Nov. 1989 (bud), *Dransfield et al.* JD6769 (K, TAN). Ambatondrazaka: Manaka Est, Jan. 1959 (dead infl), *Rakotovao RN* 11019 (K, P); idem, April 1961 (fr.), *Rakotovao* RN 11859 (K); Ambatoharanana near Antsevabe, March 1951 (fr.), *Cours* 4061 (K, P, TAN). Manjakandriana: Angavokely, Oct. ?1924 (fl.), *Perrier* 15883 (P); Mandraka, Feb. 1985 (y.fr.) *Barnett et al.* 455 (K, MO, P); idem, Feb. 1985 (fr.) *Dorr et al.* 3729 (K, MO). Moramanga: Andasibe, Nov. 1986 (bud), *Dransfield et al.* JD6426 (K, P, TAN); idem, Dec. 1991 (fl), *Beentje & Andriampaniry* 4534 (BH, K, MO, P, TAN); idem, Dec. 1991 (fl., y.fr.) *Beentje & Andriampaniry* 4544 (BH, K, MO, P, TAN); Mantady, Dec. 1992 (fl.), *Beentje & Andriampaniry* 4771 (K, TAN); Maromi(ha)za, Feb. 1926 (fr.), *Perrier* 15990 (P); idem, March 1991 (fr.), *Beentje & Rabarilala* 4412 (BH, K, MO, P, TAN), 4414 (K, TAN); Lakato Road, Nov. 1972 (fl.), *Guillaumet* 4030 (P, TAN); Rahobevava, March 1951 (fr.), *Cours* 4297 (K, P, TAN). Miandrivazo: NW of Ambohitsaratelo-Bebao, Jan. 1985 (fr.), *Dorr et al.* 3532 (K, MO, P). Fianarantsoa: Vohiparara, July 1992 (dead infl.), *Beentje & Andriampaniry* 4716 (BH, K, MO, P, TAN). **WITHOUT PRECISE LOCALITY:** Central Madagascar, (bud), *Baron* 3270 (K, type of *N. baronii*), (bud) 4509 (K, P), (bud) 6068 (K, syntype); anno 1847–1852 (leaf only), *Boivin* s.n. (P); without any data, *Perrier* 12082 (P).
CULTIVATED: Analamazaotra (fr.), *Perrier* 15989 (P); Antananarivo, Antanimena, 1924 (fr.), *Perrier* 16061 (P).

Dypsis andrianatonga, showing branching and infructescence, Bekolosi (*Beentje et al.* 4571).

29. DYPSIS ANDRIANATONGA

A rather small branching palm which is restricted to the high mountain massifs of northern Madagascar. All individuals we have seen displayed branching. The species seems closest to *D. baronii.* The epithet means 'the nobleman has arrived' and comes from the local name of the species.

DISTRIBUTION. Manongarivo and Marojejy Massif.
HABITAT. Open moist montane forest or heath vegetation, occasionally on rocks in denser forest; 700–1800 m.
LOCAL NAMES. *Tsiriki andrianatonga* (Tsimihety, *tsiriki* being a general palm name).
USES. Leaf decoction used in drink for convalescence, highly prized.
CONSERVATION STATUS. Rare. The distribution of this species is limited. In Manongarivo it is not uncommon in a rather narrow vegetation belt on Bekolosi Mountain.

Dypsis andrianatonga *Beentje* **sp. nov.**
D. baronii et *D. serpentinae* similis, a *D. baronii* habitu ramificanti foliolis paucioribus inflorescentia minore grana majore et a *D. serpentina* petiolo breviore foliolis regulariter dispositis inflorescentia majore rachillis pluribus differt. Typus: Madagascar, Manongarivo, Bekolosi, Jan. 1992, *Beentje & Quansah* 4559 (Holotypus K; isotypi MO, P, TAN).

Clustering palm in tufts of 8–14. **STEMS** 2–9 m tall, 1.5–2.5 cm diam., snaking and procumbent but with the distal part erect, proximally and/or distally branching at the nodes, often rooting at the branching points; internodes proximally 11–18 cm, distally 1–2.5 cm, dark green, glabrous, nodal scars 0.3–0.7 cm, grey-brown, slightly stepped. **LEAVES** c. 5 per crown, spiral to almost tristichous, arching-porrect, 80–190 cm; sheath 20–39 cm, closed but occasionally split proximally, without obvious auricles or with minute ones to 5 mm high, green, slightly waxy, proximally glabrous or with laciniate reddish scales, distally with some scattered scales; petiole 6–32 cm, proximally 5.5–10 x 3–5 mm diam. and with a triangular fleshy extension of the sheath lining, channelled proximally, distally 3–6 x 3–5 mm diam., densely pubescent or with scattered scales; rachis 42–128 cm, in mid-leaf to 7 mm wide, keeled, densely pubescent or with few scattered scales; leaflets regular, attenuate, 12–35 on each side of the rachis, the proximal 21–43 x 0.6–1.8 cm, median 16–35 x 2–3.2 cm (interval 3–5.3 cm), distal 3–20 x 0.3–1.9 cm, the distal pair joined for up to 0.5 cm, main veins 1–3, and with thickened margins, with occasional ramenta to 6 mm, with scattered scales on the veins and margins. **INFLORESCENCE** infrafoliar, branched to 1–2 orders, erect proximally, curved in the distal part of the peduncle through some 140°; peduncle 7–40 cm, proximally 4–12 x 3–3.5 mm diam., distally 4–6 x 2.5–4 mm diam., waxy; prophyll 12–59 cm, borne at 1.5–24 cm above the base of the peduncle, pale brown with scattered scales; peduncular bract persistent or deciduous, inserted at 4–28 cm from the base of the peduncle, 20–29 cm long, hooded, open all the way except the distal 4.5 cm, with a minute beak, with scattered scales; non-tubular peduncular bract 0.2–2 × 1 cm; rachis 4–14 cm, with 9–17 first order branches, sometimes a few (up to 5) of these with a secondary rachis of up to 1.2 cm and 2–3 (–4) rachillae; rachillae 2–10 cm long, 2–3.5 mm diam., glabrous or minutely puberulous, zigzag; triads dense to distant, sunken, with acuminate rachilla bract to 2 mm. **STAMINATE FLOWERS** with sepals 2.2–2.7 × 2.3–2.6 mm, red-spotted on keel and apex; petals connate for 1.2–1.6 mm, free parts 3–3.4 × 2.4–3.2 mm; stamens 6, uniseriate, the filaments connate for c. 0.6 mm, free for 2.8–3 mm, flattened in their proximal half and cylindrical distally, anthers 1.4–1.7 × 0.5–0.8 mm, versatile, obtuse; pistillode 2.5–2.6 × 1.2–1.3 mm. **PISTILLATE FLOWERS** with sepals 2–3 × 2.8–3.6 mm; petals in young bud 2.5–3.3 × 2.4–2.8 mm, concave, striate, (in fruit) 3–4 × 3.3–5.2 mm, ciliolate; staminodes 6, flat, 0.3–0.8 mm high; ovary 1.5–3.7 × 0.9–2 mm. **FRUIT** green, colour unknown when ripe, ellipsoid, 9–20 × 7–15 mm, rounded at the base and apex; mesocarp very thin; endocarp fibrous. **SEED** ellipsoid, 15–18 × 11–14 mm, with ruminate endosperm, the intrusions few, distant, regular and 1–2.5 mm deep.

NOTE. The branching of this taxon seems to conform to the branching pattern described for *D. lutescens* (Fisher 1973) but in *D. andrianatonga* seems the rule rather than the exception. Side branches are smaller than the main stem, and distal side branches are smaller than proximal ones. Flowering and fruiting does take place on the side branches; it is not certain whether this occurs only on the side branches, or also on the main stem.

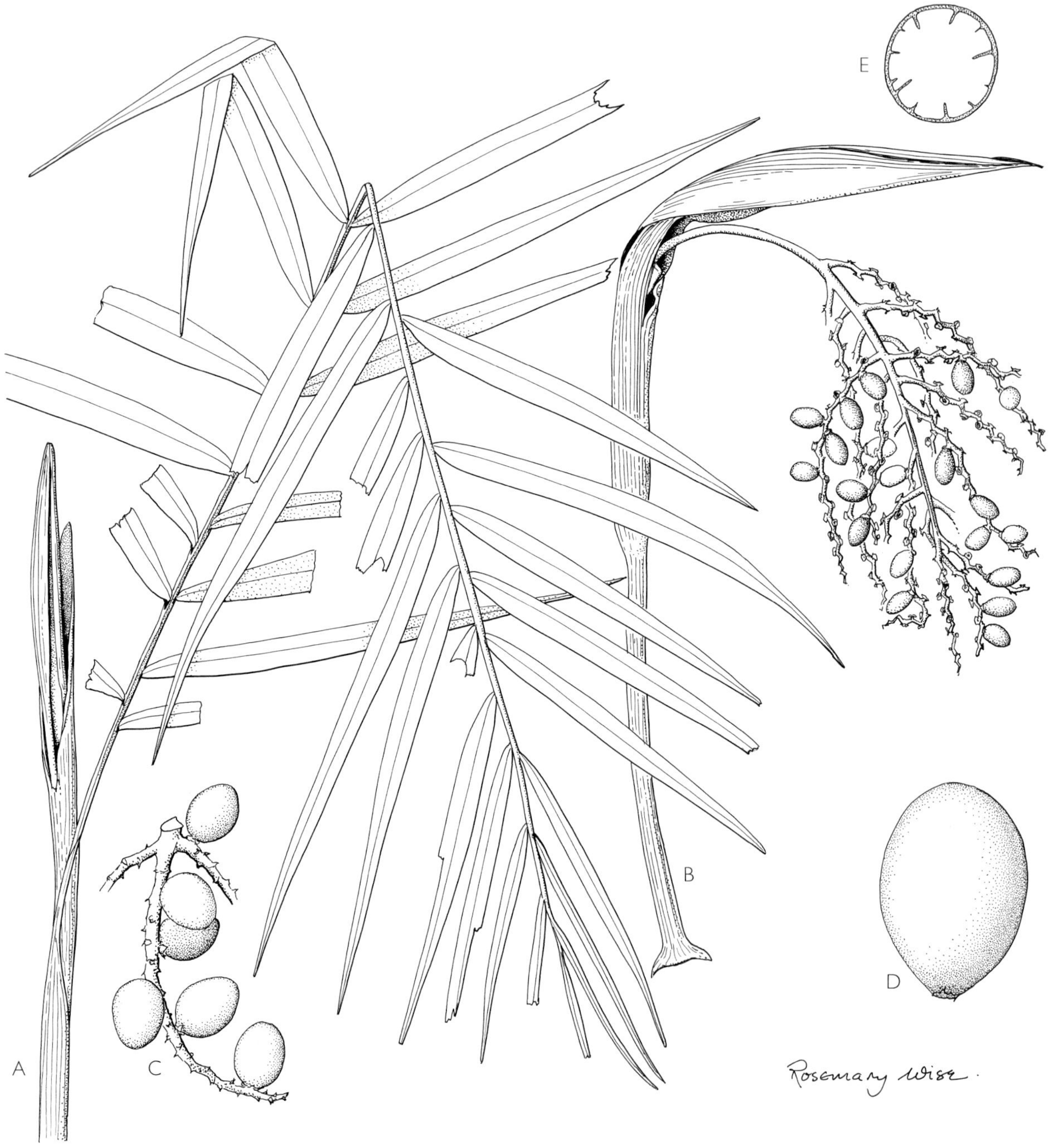

Dypsis andrianatonga. **A** leaf and base of inflorescence × 1/2; **B** infructescence × 1/2; **C** fruiting rachillae × 1/2; **D** fruit × 2; **E** seed in cross section × 1. All from *Beentje et al.* 4571. Drawn by Rosemary Wise.

SPECIMENS SEEN. Ambanja: Manongarivo, Bekolosi, Jan. 1992 (old infl.), *Beentje & Quansah* 4559 (Holotype K; isotypi MO, P, TAN); idem, Feb. 1992 (fr.), *Beentje et al.* 4571 (K, MO, TAN); Antsatrotro, Sept. 1991 (fl.), *Malcomber & Razafimandimbison* 885 (K, P). Sambava: Betsomanga massif, Nov. 1950 (y.fr.), *Humbert & Capuron* 24305 (K, P); Mt Beondroka, March 1949 (fr.), *Humbert* 23559 (K, P). Andapa: Marojejy, W slopes, Nov./Dec. 1948 (dead infl.), *Humbert & Capuron* 22287 (K, P); Ambatoharanana valley to upper Antsahaberoka, Nov. 1959 (ster.), *Humbert & Saboureau* 31883 (K, P); Marojejy, E peak area, Oct. 1988 (y.fr.), *Miller et al.* 3512 (MO, TAN); Marojejy, Dec. 1972 (fl., y.fr.), *Guillaumet* 4108 (TAN).

Dypsis serpentina. Detail of infructescences (*Dransfield & Beentje* 7502).

30. DYPSIS SERPENTINA

In its habit this is a most unusual and curious, rather than beautiful species.The stems appear to flop over under their own weight, and branch, the branches being of smaller diameter than the axes below the branching point. In this way the plant develops into a thicket of aerial stems that flop about the surrounding vegetation. Not strictly a climber, this palm is nevertheless scarcely self-supporting. The name reflects the habit of the palm: snaking through the undergrowth, and also alludes to the soil type, although it grows on ultramafic soils rather than true serpentine ones.

DISTRIBUTION. Only known from the Mananara Biosphere Reserve.
HABITAT. Lowland rain forest; may form thickets on steep mid slopes or in heath-like forest on ridgetops, with *Satranala* and *Pandanus* on very thick humus layer on ultramafic soil; 240–280 m.
LOCAL NAMES. Not recorded.

USES. Not recorded.

CONSERVATION STATUS. Vulnerable. Single-site status, but is fairly abundant in this site, which is protected.

Dypsis serpentina *Beentje* **sp. nov.**

D. baronii et *D. andrianatongae* similis sed foliolis aggregatis et petiolo longo, a *D. baronii* vagina breve et caule ramificanti, a *D. andrianatonga* grana minore bractea pedunculi breviore differt. Typus: Madagascar, Mananara Avaratra Biosphere Reserve, *Beentje et al.* 4646 (Holotypus K; isotypi BH, MO, P, TAN).

Solitary or clustering rather floppy palm. **STEMS** 5–6 m, 3–4 cm diam., starting vertical, then leaning over with almost horizontal part, the apex again vertical; nearly all stems seen branched twice or more, with the branches closely parallel; internodes distally 1–8.5 cm, glabrous, proximally corky, cracked vertically with lenticels, dull grey-brown, the upper branches green; nodal scars 0.8–1.2 cm, brown. **LEAVES** c. 6 in the crown, spirally inserted; sheath 21–26 cm long, green with a white bloom, with a few scattered scales distally; petiole 43–75 cm long, distally 2–3 mm wide, with scattered scales; rachis 50–120 cm long, in mid-leaf 0.2–0.3 cm wide; leaflets 7–17 on each side of the rachis, irregular or in groups, the group interval 4–9 cm, proximal leaflets 40–70 × 2.4–3.5 cm, median 25–46 × 1–1.6 cm, distal 13–20 × 0.7–1.7 cm, main veins 1–2 (distal 2–3), but in multifold leaflets 3–5, few ramenta, otherwise glabrous, some sinuous transverse veins, distal leaflets connate for c. 2 cm, dentate over a width of c. 2 mm, distal pair joined for c. 0.6 cm. **INFLORESCENCE** infrafoliar, branched to 1 order, recurved; peduncle 8–10 cm long, distally 0.4–0.5 × 0.3–0.5 cm diam., stellate-scaly; prophyll 12–18 cm long, 1.5–2 cm wide, borne at 2–3 cm above the base of the peduncle; peduncular bract inserted at 5–7 cm from the base of the peduncle, 7–12.5 cm long; rachis 2.5–3 cm long, with scattered stellate scales, with 5–7 branches; rachillae

Dypsis serpentina. **A** stem with inflorescence × 1/3; **B** stem with infructescence × 1/3; **C, D** basal leaflets × 1/3; **E** mid section of leaf × 1/3; **F** leaf tip × 1/3; **G** fruit × 2.5; **H** fruit in cross section × 2.5. All from *Dransfield & Beentje* JD7502. Drawn by Rosemary Wise.

4–8 cm long, 1–2 mm diam., with scattered scales; triads distally distichous. **STAMINATE FLOWERS** with sepals 1.9–2 x 1.5–1.8 mm, the innermost the widest; petals in young bud c. 1.8 x 1.3 mm; stamens 6, uniseriate?, filaments in bud 0.3 mm, anthers 0.6–0.8 x 0.3 mm; pistillode c. 0.8 x 0.2 mm. **PISTILLATE FLOWERS** with sepals 3–3.2 x 2.6–3.3 mm; petals 3.5–4 x 2.8–3.2 mm; staminodes 0.8–1 mm; gynoecium c. 4.2 x 2 mm. **FRUIT** purplish-tinged, ovoid, 12–14 x 9–10 mm. **SEED** c. 11 x 8.5 mm, endosperm ruminate, the ruminations 1–2.5 mm deep, distant, embryo lateral near the base.

NOTE. Related to *D. andrianatonga* and *D. baronii.*

SPECIMENS SEEN. Mananara Avaratra: Antanambe, April 1992 (fl.), *Beentje et al.* 4646 (type; BH, K, MO, P, TAN); idem, Oct. 1994 (fr.), *Dransfield & Beentje* JD7502 (K, TAN); idem, Oct. 1994 (fl, fr.), *Beentje & Dransfield* 4813 (K, TAN).

31. DYPSIS ONILAHENSIS

A handsome, medium-sized clustering palm which would do well in cultivation in the drier tropics. The species name comes from the Onilahy River, south of Toliara.

DISTRIBUTION. NW and W Madagascar and South-Central Madagascar.
HABITAT. Riverine forest, alt. 50–300 m (in the North) or riverine to evergreen forest (remnants) at 750–2400 m (Mountains of the Centre-North; Centre-South). Humbert states the palm is abundant on the rocky banks at flood level of the Analabe River in the north.
LOCAL NAMES. *Kindro* (Antankarana); *Sihara* (Bara).
USES. Not recorded.
CONSERVATION STATUS. Vulnerable. Though the distribution area is fairly large, the habitat is prone to destruction by fire. Numbers are estimated at less than a thousand. The population in the Isalo National Park is well-protected, but numbers less than a few hundred individuals.

Dypsis onilahensis *(Jum. & H. Perrier) Beentje & J. Dransf.* **comb. nov.**

SYNONYMS:
Chrysalidocarpus onilahensis Jum. & H. Perrier, Ann. Inst. Bot.-Géol. Colon. Marseille sér. 3, 1 (1): 37. t. 18 (1913); Jum., Ann. Inst. Bot.-Géol. Colon. Marseille sér. 10, 3: 15 (1922); Cat. Pl. Madagascar, Palmae: 10 (1938); Fl. Madagascar 30: 108 (1945). Type: Madagascar, Onilahy basin, Mt. Votaka near Benenitra, *Perrier* 12074 (Holotype P).

Chrysalidocarpus midongensis Jum., Ann. Inst. Bot. Géol. Colon. Marseille sér. 10, 3: 17 (1922); Cat. Pl. Madagascar, Palmae: 10 (1938); Fl. Madagascar 30: 109 (1945). Type: Madagascar, E of Midongy, *Perrier* 12499 (Holotypus P), **synon. nov.**

Chrysalidocarpus brevinodis Perrier, Not. Syst. 8: 47 (1939); Jum. & H. Perrier, Fl. Madagascar 30: 104, fig. 29: 4–6 (1945). Syntypes: Madagascar, upper Mananjeba R basin, *Perrier* 15800 (P); Analamera, Analabe R. banks, *Humbert* 19245 (P, see NOTE), **synon. nov.**

Clustering palm in tufts of 3–10 (occasionally appearing as solitary, fide Humbert). **STEMS** 2–20 m high, 2.5–15 cm diam. (–30 cm, fide Perrier), distally stepped and ringed; internodes 0.3–20 cm, proximally grey, distally dark green, nodal scars 0.2–0.5 cm, white. **LEAVES** 5–7 in the crown, gracefully arching to half-pendulous; sheath 18–60 cm, 4.5–6 cm diam., open for about one third, abaxially dark grey-green, pale green to pale brown but nearly always with a white waxy bloom, distally reddish-pubescent but glabrescent, adaxially rich red-brown, without or with only slight ligules to 1.5 x 4 mm; petiole absent or up to 40 cm long, proximally 1.2–2.3 x 1–1.5 cm with an adaxial triangular swelling as a sheath lining extension, distally 1.3–1.7 x 0.6–1.6 cm, slightly channelled with sharp edges, glabrous or with reddish scales on both surfaces, dark green with flecks of dark brown; rachis 1–1.8 m long, proximally channelled, in mid-leaf keeled and 0.7–1.7 cm wide, with reddish scales or glabrous; leaflets 42–64 on each side of the rachis, regular, in one plane, opposite leaflets also in one plane (in Isalo) or at an angle of 90° (elsewhere), slightly pendulous, curving or stiff with only the distal part arching, dark green and hardly shiny adaxially, mat and glaucous abaxially, the proximal (22–) 38–106 x 0.2–2 cm (interval up to 19 cm), median 35–75 x 0.8–2.5 cm (interval 0.8–4 cm), distal 8–44 x 0.2–1.9 cm (the distal pair joined for up to 2 cm), main veins 1–5, rather faint with only the midrib adaxially prominent, adaxially with proximal or scattered red-brown rather large (6–30 mm) ramenta on the midrib, otherwise glabrous, apices long-attenuate, bifid. **INFLORESCENCE** interfoliar or infrafoliar at anthesis, infrafoliar at the fruiting stage, curved with spreading rachillae, branched to 2 orders (once to 3 orders in *Humbert* 7025); peduncle 15–78 cm long, usually curving through 180°, proximally 0.6–4 x 0.4–1.2 cm diam., distally 0.9–4 x 0.6–1.5 cm, with scattered scales or glabrous; prophyll 11–87 cm, borne at 3.5–45 cm above the base of the peduncle, 2–6 cm wide with narrow wings to 3 mm wide, split only at the apex for some 3 cm or open for up to 50%, yellow turning pale brown with scattered reddish scales distally; peduncular bract often quickly deciduous (in trees with long peduncles), inserted at 6–42 cm from the base of the peduncle, 18–55 cm long, beaked for 0.5–4 cm, splitting completely (except for the beak), adaxially glabrous and red-brown, abaxially pale green with scattered red-brown laciniate scales; non-tubular peduncular bracts usually two near the peduncle apex, 0.2–2.5 x 0.5 cm; rachis 16–40 cm long, waxy pale green, with 5–15 branched and 8–12 unbranched first order branches, the most proximal with a rachis of up to 7.5 (–26.5 in

Dypsis onilahensis, growing at Zazafotsy.

Humbert 18432) cm and 0.8–1.7 × 0.3–0.5 cm proximally, with 4–12 rachillae; rachis bracts up to 1 cm; rachillae 6–30 cm long, sinuous distally, 1.5–4.5 mm diam., glabrous, with distant triads in slight pits with an acute triangular rachilla bract 1–1.5 mm. **STAMINATE FLOWERS** creamy green or yellowish, with a sweet or unpleasant smell, with sepals 1.4–2.9 × 1.5–2.8 mm, concave, acute, keeled and gibbous; petals free or connate up to 1 mm, 2.2–4.2 × 1.5–3.4 mm, ovate and concave, acute; stamens 6, uniseriate or very vaguely biseriate (then with 0.1 mm difference in insertion and a callus between the filaments), filaments 1.2–3 mm long and thinly cylindrical, anthers 1.3–2.8 × 0.6–1.3 mm, dorsifixed, versatile, the locules parallel or proximally slightly divergent; pistillode 1–2.3 × 0.4–1.3 mm. **PISTILLATE FLOWERS** with sepals 2.2–3.1 × 2.2–3.2 mm, rounded; petals 2.5–3.5 × 2.5–4 mm; staminodes 0.3–1 mm; gynoecium 2.8–3.5 × 1.5–2.9 mm. **FRUIT** yellow, waxy?, ovoid to subglobose, 10–18 × 7–15 mm, the apex rounded; endocarp very fibrous with almost free fibres. **SEED** black, 9–17 × 6.5–12 mm, ellipsoid with pointed base and rounded apex, and a small subbasal depression; endosperm homogeneous. **EOPHYLL** bifid.

NOTE. We are sinking *C. brevinodis* and *C. midongensis* as there are no distinguishing characters to separate these taxa from the bulk of *D. onilahensis*. Even the habitats seem to be identical. The protologue of *C. brevinodis* mentions as one of the syntypes *Humbert* 19032: Analamera, Analabe R banks. We believe this is an error; in the Flore the number (with the same data) is cited as 19245; on the sheets of *Humbert* 19245 (with the same data) Perrier has written *C. brevinodis sp. nov.* Perrier states that the stems are sometimes branched.

In the southern populations there is a gradual cline from north to south as regards leaflet length (decreasing) and number (increasing), and inflorescence

Dypsis onilahensis, in a gorge at Isalo (Photo: B. Du Puy & D. Du Puy).

parts lengths such as peduncle, prophyll, peduncular bract (all increasing). The type of *D. onilahensis* is from the middle of the range, and nicely intermediate.

The name *midongensis* refers to an obscure hamlet in the Itremo Massif, rather than to the large town in the south-east; before HB had realized this, he went to the large town and searched for this taxon. Not surprisingly, he did not find it, but in a nice case of serendipity, he found *D. prestoniana.*

D. acuminum is very close, only distinguished by being solitary; by its fewer leaflets (30) which are on the short side, and by the inflorescence which is branched to one order only (though the type has one bifurcate rachilla).

This species is extraordinaly close to *D. baronii,* and only differs in the habitat, the absence of minute reddish glands on the leaflets (not always present in *D. baronii*), and the homogeneous endosperm; the ruminations in *D. baronii* are difficult to see at times, being very small, and *D. baronii* has been found in sites which are really in *D. onilahensis* territory, such as the forest of Ambohitsaratelo. In the absence of fruit, several collections could not be identified as belonging to one or the other [Miandrivazo: NW of Ambohitsaratelo-Bebao, July 1974 (fl), *Morat* 4590 (P, TAN); idem, Nov. 1986 (fl., y.fr.), *Dransfield et al.* 6447 (K, P, TAN)] but since *Dorr et al.* 3532 from

Dypsis onilahensis, near Ranohira (Photo: B. Du Puy & D. Du Puy).

the same locality has ruminate endosperm, they are more likely to be *D. baronii*. This whole complex deserves further study.

SPECIMENS SEEN. Antsiranana: Analamera, Analabe R banks, Jan. 1938 (fr.), *Humbert* 19245 (P, syntype of *C. brevinodis*). Ambilobe: upper Mananjeba basin, Aug. 1913 (fl., fr.), *Perrier* 15800 (P, syntype of *C. brevinodis*). Ambanja: Manongarivo, 1200 m, May 1909 (fl.), *Perrier* 12083 (P); Tsaratanana, upper Sambirano valley, Nov./Dec. 1937 (fr.), *Humbert* 18432 (P). Andapa: Mt Mainborondro, March 1949 (fr.), *Humbert* 23377 (K, P). ?Maevatanana: Tampoketsa d'Antongodrahoza, upper Betsiboka R, Sept. 1922 (fl.), *Perrier* 14839 (P). Ankazobe: Ikopa R, March 1920 (fr.), *Decary* 7556 (P). Ambatofinandrohana: E of Midongy de l'Ouest, Feb. 1919 (fr.), *Perrier* 12499 (P, type of *C. midongensis*). Ihosy: Isalo, Oct. 1924 (fl.), *Perrier* 16548 (P); idem, Oct. 1924 (bud), *Humbert* 2822 (P); idem, NW of Ranohira, July 1992 (buds), *Beentje & Andriampaniry* 4709 (BH, K, MO, P, TAN); idem (seedling), *Beentje & Andriampaniry* 4713 (K); idem, Feb. 1990 (fr.), *B. Du Puy et al.* MB 673 (K); SW of Ranohira, Feb. 1990 (y.fr.), *Phillipson* 3406 (K, MO, TAN); 22km NNE of Zazafotsy, Feb. 1992 (seed), *Beentje & Andriampaniry* 4586 (K); idem, Dec. 1992 (fl.), *Beentje* 4768 (K, MO, P, TAN). Benenitra: Mt Votaka near Benenitra, July 1910 (fl.), *Perrier* 12074 (P, type of *C. onilahensis*). Betroka: Vohipolaka, Nov. 1933 (fl.), *Humbert* 11653 bis (P). Betroka/Befotaka: Ivohibe Peak, Sept. 1921 (bud), *Perrier* 12079 (P). Amboasary Atsimo: Ivakoany Massif, W slopes, Dec. 1928 (fl.), *Humbert* 7025 (K, P) and Nov.-Dec. 1933 (fl., y.fr.), *Humbert* 12248 (P). Tolanaro: Beampingaratra Mts, Maloto R valley Oct./Nov. 1928 (fl.), *Humbert* 6283 (P); upstream from Mahamavo, Jan./Feb. 1934 (fr.), *Humbert* 13905 (K, P); Col du Vavara, Nov. 1928 (fl.), *Humbert* 6539 (K, P).

Dypsis onilahensis. **A** habit, greatly reduced; **B** leaf sheath × 1/3; **C** distal part of petiole with lowermost leaflets × 1/3; **D** mid section of leaf × 1/3; **E** leaf tip × 1/3; **F** basal portion of inflorescence rachis and part of peduncle and peduncular bract × 1/3; **G** fruit × 2; **H** fibrous endocarp × 2; **J** endocarp and seed in cross section × 2. **A** – **F** from *Beentje* 4768, **G** – **J** from *Humbert* 18432. Drawn by Rosemary Wise.

32. DYPSIS ACUMINUM

A rare species, which may well be the same as *D. onilahensis*. The Latin name means 'of the peaks'.

DISTRIBUTION. N Madagascar: Manongarivo and Marojejy Mountains.
HABITAT. Montane forest; alt. 700–1900 m.
LOCAL NAMES. *Lafaza* (Manongarivo).
USES. Not recorded.
CONSERVATION STATUS. Unknown. Only known from a single recent collection, but it is possible this taxon occurs on more high mountains of the north, most of which are not well known botanically.

Dypsis acuminum *(Jum.) Beentje & J. Dransf.* **comb. nov.**

SYNONYM:
Chrysalidocarpus acuminum Jum., Ann. Inst. Bot.-Géol. Colon. Marseille sér. 3, 10, 3: 16 (1922); Jum., Cat. Pl. Madagascar, Palmae: 7 (1938); Jum. & H. Perrier, Fl. Madagascar 30: 109 (1945). Type: Madagascar, Manongarivo, *Perrier* 15801 (Holotype P).

Solitary, moderate palm. **STEMS** 4–6 m tall, 8 cm diam.; internodes dark green to greyish, nodal scars forming obvious rings. **LEAVES** 5–6, arching, 1–1.5 m long; sheath 30–32 cm long, when flattened 8–10 cm wide, adaxially dark reddish purple, abaxially

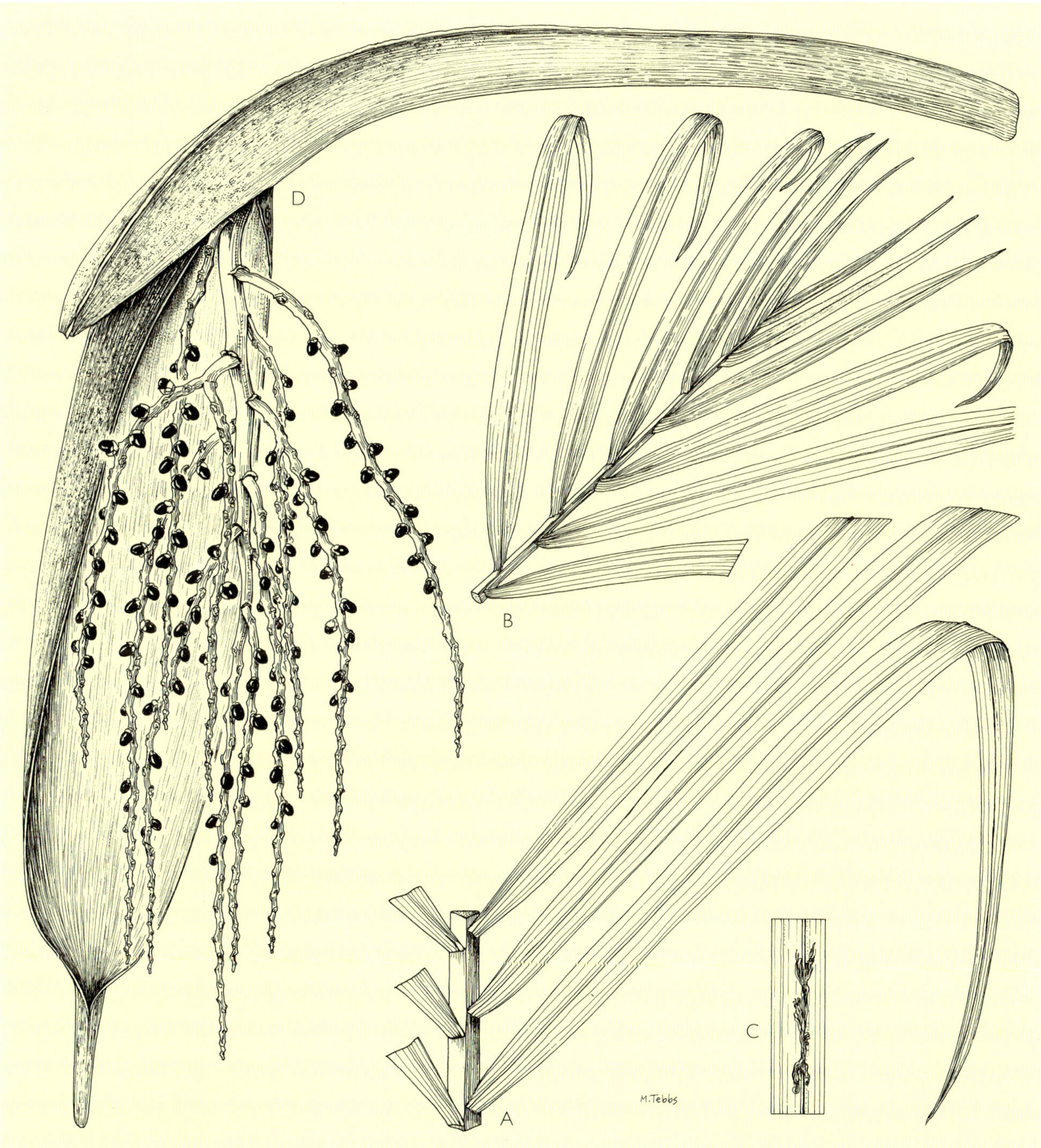

Dypsis acuminum. **A** mid section of leaf × 1/2; **B** leaf tip × 1/2; **C** detail of ramenta × 4; **D** young infructescence × 1/2. All from *Perrier* 15801. Drawn by Margaret Tebbs.

pale brown distally with scattered black scales, waxy, without ligules but with rounded shoulders; petiole 11–16 cm, proximally c. 1.2 × 0.6 cm, distally 0.8–1.3 × 0.6–0.7 cm diam., channelled with sharp edges, red-brown with blackish scales; rachis with dense to scattered blackish scales, in mid-leaf 0.6–1.2 × 0.6–0.7 cm diam.; leaflets regular, c. 30 on each side of the rachis, acuminate, proximal 35–36 × 0.3–1 cm, median 27–39 × 1.2–1.8 cm (interval 1.5–3 cm), distal 6–30 × 0.3–1.4 cm, main veins 1, glabrous but for 3–4 scattered ramenta. **INFLORESCENCE** interfoliar at anthesis, infrafoliar in fruit, branched to 1 or 2 orders; peduncle 28–41 cm long, proximally c. 2 × 0.4 cm, distally 1.2 × 0.6 cm, glabrous; prophyll 30–35 cm long, borne at 6–9 cm above the base of the peduncle, with scattered scales but glabrescent, open for the distal 10 cm; peduncular bract inserted at 18–20 cm from the base of the peduncle, 36–37 cm long, split over its whole length or except for the distal 3–4 cm, beaked for 3–4 cm, with dense but scattered scales; non-tubular peduncular bracts 1–1.3 cm, near the apex of the peduncle; rachis 12–34 cm, with 15–18 unbranched first order branches, in the type with one of the branches bifurcate; rachillae 10–25 cm long, 2.5–4 mm diam., glabrous, with quite dense triads, slightly sunken in pits; rachilla bract c. 2 mm long, acute to acuminate. **STAMINATE FLOWERS** unknown. **PISTILLATE FLOWERS** only known from the young fruiting stage, with the sepals 2.8–3.6 × 3–4 mm; petals 3.5–5.2 × 4–6 mm; staminodes 0.5–1 mm; gynoecium probably c. 4 mm high. **FRUIT** ellipsoid, 9–10 × 6–7 mm., rounded at apex; endocarp fibrous, with few anastomations. **SEED** ellipsoid, 8–8.5 × 5.5–6 mm, the base with a slight bump and a sub-basal depression corresponding to the embryo, the apex rounded; endosperm homogeneous.

NOTE. The altitude on the type is indicated as 2000 m, but the highest point in the Manongarivo Mts., Antsatrotro, is 1876 m. This taxon is probably the same as *D. onilahensis*; only the branching pattern of the inflorescence is distinct, being much less branched in *D. acuminum*.

SPECIMENS SEEN. Ambanja: Manongarivo, probably Antsatrotro, May 1909 (y.fr.), *Perrier* 15801 (P, type). Andapa: Marojejy, Feb. 1989 (y.fr.), *Miller & Lowry* 3941 (P, TAN).

33. DYPSIS LUTESCENS

One of the best-known Madagascar palms (at least under its old name): this species is very widespread in cultivation all over the tropics, doing well under a wide range of conditions. This is in strange contrast to its native country, where it is restricted to a special habitat: white sand forest in a narrow strip close to the sea. The name means 'becoming yellow' and refers (probably) to the leaf sheath, petiole and rachis.

DISTRIBUTION. E Madagascar.
HABITAT. Littoral forest or heath vegetation on white sand, also on rock; persists in secondary growth and may be locally common. Alt. 5–35 m.
LOCAL NAMES. *Rehazo*, *Lafahazo*, *Lafaza* (Betsimisaraka).
USES. Local uses not recorded, but this is one of the most important ornamental palms in commerce.
CONSERVATION STATUS. Not threatened. Total numbers of this palm in the wild are estimated at more than a thousand.

Dypsis lutescens *(H. Wendl.) Beentje & J. Dransf.* **comb. nov.**

SYNONYMS:
Chrysalidocarpus lutescens H. Wendl., Bot. Zeitg. 36, 8: 117 (1878); Beccari, Bot. Jahrb. 38, Beibl. 87: 32 (1906); Beccari, Palme del Madagascar 38 (1914); Jum., Ann. Inst. Bot.-Géol. Colon. Marseille

Dypsis lutescens, in coastal forest on white sand, Ambila-Lemaitso (*Dransfield et al.* JD6441).

sér. 10, 3: 13 (1922); Jum., Cat. Pl. Madagascar, Palmae: 9 (1938); Jum. & H. Perrier, Fl. Madagascar 30: 103, fig. 29 (1945). Type: none indicated. Lectotype (selected here): Madagascar, Ambila-Lemaitso, *Dransfield et al.* JD6441 (Holotype K; isotypes BH, P, TAN).

Chrysalidocarpus baronii var. ***littoralis*** Jum. & H. Perrier, Ann. Inst. Bot.-Géol. Colon. Marseille sér. 3, 1 (1): 35 (1913). Type: none mentioned, but *Perrier* 12058 corresponds with locality in protologue; Madagascar, near Tampolo R mouth, *Perrier* 12058 (Holotype P).

Chrysalidocarpus glaucescens Waby, Kew Bull. 1923:376 (1923). Type: cult. Victoria Gardens, Trinidad, *Waby* 9849 (Holotype K).

Graceful clustering palm in tufts of 4–20 plants. **STEMS** 1–7 m high, occasionally with 1–2 small branches, 5–12 cm diam., the base occasionally with a swelling to 30 cm high, 40 cm diam.; internodes 2–12 cm, yellowish or pale grey-brown, more distally green to grey with waxy white bloom. **LEAVES** 5–11, spiral or more often tristichous, porrect and strongly arching; sheath yellowish with white waxy bloom, (28–) 39–60 cm, 11–15 cm diam., abaxially with dense scattered scales distally, adaxially orange and glabrous, with slight ligules (to 3 mm) or with rounded shoulders; petiole 19–37 cm long, proximally 1.7–3.5 × 2.5 cm, distally 0.7–2 × 0.8–2 cm, channelled with sharp edges, yellow or yellowish orange, with few abaxial scales, proximally on the adaxial side with a triangular swollen extension to the sheath lining; rachis 1.1–1.9 m long, proximally channelled, in mid-leaf keeled and 1.5–1.8 × 0.9 cm diam., yellow or yellow-orange, with small scattered scales; leaflets 44–59 on each side of the rachis, regular, stiff, in one plane on each side of the rachis but the leaflets on opposite sides at an angle of 90–120°, adaxially green, abaxially slightly waxy and grey, the proximal 35–66 × 0.6–1.8 cm, median 44–70 × 1.3–3 cm (interval 1.7–4 cm), distal

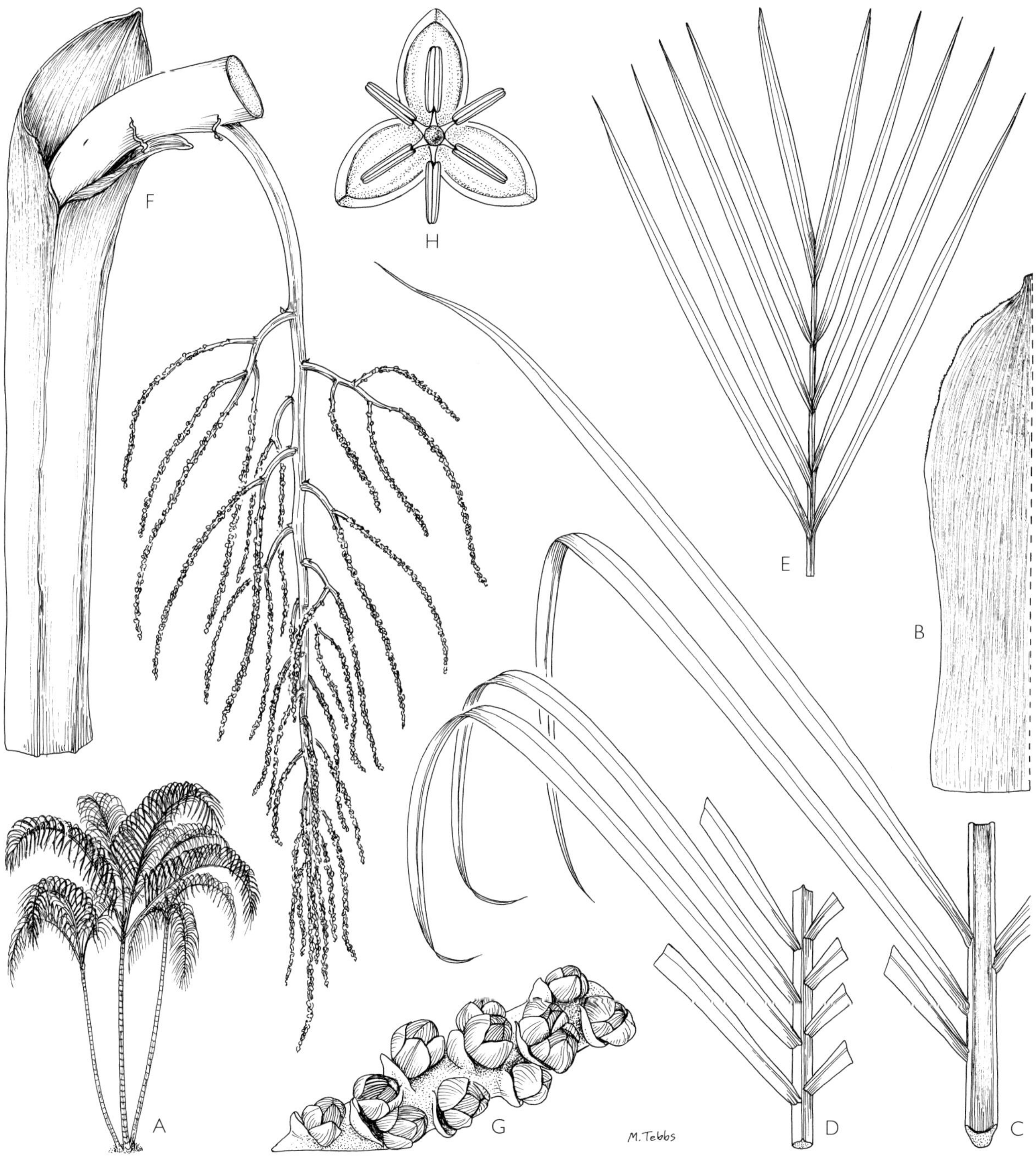

Dypsis lutescens. **A** habit, greatly reduced; **B** leaf sheath × 1/4; **C** distal part of petiole with lowermost leaflets × 1/4; **D** mid section of leaf × 1/4; **E** leaf tip × 1/4; **F** basal portion of inflorescence rachis and part of peduncle and prophyll × 1/3; **G** detail of rachilla × 3; **H** staminate flower × 7. All from *Dransfield et al.* JD6441. Drawn by Margaret Tebbs.

7–37 × 0.6–1.5 cm (terminal pair joined for up to 3.5 cm), the apices attenuate, with 5–9 faint veins but with only the midrib very prominent on both surfaces, with a few tufts of large ramenta on the abaxial midrib, and with many small scattered glands in long lines on the minor veins (these sometimes absent?). **INFLORESCENCE** interfoliar, sometimes the fruiting stage infrafoliar, spreading, with spreading rachillae, branched to 3 orders (rarely to 2 or 4 orders); peduncle 34–88 cm long, flattened, proximally 2.7–6 × 1–3 cm, distally c. 1.3–2.7 × 0.6 cm diam., glabrous; prophyll 31–102 cm, borne at 5–47 cm above the base of the peduncle, 3.3–4.2 cm wide, with narrow wings, split only at the apex for 7–16 cm, with scattered scales; peduncular bract inserted at 38–68 cm from the base of the peduncle, 48–60 cm long, 4.5 cm wide, closed for the distal 10 cm, with a beak of 2–5 cm, pale brown, waxy and glabrous or with a few scattered scales, abscising and carried up by the lengthening inflorescence and rather quickly deciduous; non-tubular peduncular bracts few, 3–5 × 10–12 mm; rachis 20–110 cm, glabrous, with 5–13 branched and 8–14 unbranched first order branches, the proximal of these with flattened bases 1–1.5 × 0.6–0.7 cm and subtended by bracts of 6–7 mm high; rachillae 6–30 cm long, 2–5 mm diam.; triads distant proximally, more dense distally, inserted in shallow pits; rachilla bracts proud, acute. **STAMINATE FLOWERS** with sepals 1.6–1.8 × 2–2.4 mm, hooded, rounded, gibbous, keeled; petals connate for 1–1.3 mm to the receptacle, free for 1.3–2.2 × 2–2.3 mm, ovate, acute; stamens 6, uniseriate, the filaments 2.8–3.2 mm long, anthers 1.6–1.8 × 0.8 mm, dorsifixed; pistillode 2.5–2.8 × 0.4–0.5 mm. **PISTILLATE FLOWERS** with sepals 2–2.2 × 2.4–2.5 mm, hooded, neither gibbous nor keeled; petals free, 2.8–3.2 × 2.3–3 mm, orbicular or broadly ovate, with small apicula; staminodes 0.3–0.4 mm, thin, empty; ovary c. 3 × 1.4 mm, with short stigmas. **FRUIT** yellow, ellipsoid to obovoid, 12–18 × 7–10 mm, with a pointed apex; endocarp long-fibrous, the fibres almost free. **SEED** ovoid with an obtuse apex and a pointed base, 11–16 × 6–9.5 mm, with homogeneous endosperm. Germination adjacent-ligular; eophyll bifid.

NOTE. Wendland described the genus *Chrysalidocarpus* and the species *Chrysalidocarpus lutescens* at the same time; he also mentioned that this was a common indoors palm in Europe, known as *Areca* or *Hyophorbe indica* or *lutescens*, and also as *Areca borbonica* or *A. dicksoni* (= *Dictyospermum album*); none of these names refers to the Madagascar material. [*Chrysalidocarpus lutescens* is not based on the Bory name *Hyophorbe lutescens*, since Wendland states categorically that he is describing the species as distinct from *Hyophorbe*. *Areca lutescens* Bory was described from specimens from Réunion, and is a synonym of *Hyophorbe indica* Gaertn. *Areca borbonica* is an old garden name for *Dictyosperma album* (Bory) H. Wendl. & Drude. None of these palms conforms to the description Wendland gave for his new species, with its green leaf-sheaths with a waxy bloom combined with a clustering habit].

With *D. arenarum* and *D. psammophila* it forms a complex that requires further study, particularly since all three taxa occur in the same area and almost in the same habitat. This complex seems close to *D. baronii* and *D. onilahensis*, to which it bears an uncanny resemblance.

We are tentatively including *Chrysalidocarpus glaucescens* Waby in synonymy. This species was based on a particularly glaucous and robust plant cultivated in Trinidad. It is certainly larger in all its parts than *D. lutescens* but we do not think it can be anything else.

SPECIMENS SEEN. Sambava: 7km NE of Anjangoveratra, June 1992 (sd.), *Beentje & Andriampaniry* 4690 (K, TAN, MO). Maroantsetra: Maroantsetra For. Sta., Oct. 1963 (ster.), *Moore* 9018 (TAN); Manambia, Oct. 1986 (y. infl.), *Dransfield et al.* JD6407 (K, P, TAN). Mananara Avaratra: Tapolo (Tampolo) R mouth, Oct. 1911 (bud), *Perrier* 12058 (P); Antanambe, Oct. 1991 (fl.), *Beentje* 4460 (BH, K, MO, P, TAN). Toamasina: Mahatsara For. Res., Feb. 1992 (fr.), *Noyes et al.* 922 (K. P); Foulpointe (Mahavelona), Dec. 1962 (fl.), *Bosser* 16826 (P); 45km N of Toamasina, Feb. 1975 (fl.), *Croat* 32483 (P, TAN). Ampasimanolotra: Ambila, May 1928 (fr.), *Decary* 6303 (K, P, TAN); 5km S of Ambila-Lemaitso, Nov. 1986 (fl.), *Dransfield et al.* JD6441 (K, P, TAN). Manakara: 5km S of Manakara, May 1992 (old infl.), *Beentje & Andriampaniry* 4676 (K, TAN).

SIGHT RECORDS. Sambava: occasional near Antsirabe Avaratra (*Beentje* 1992). Soanierana-Ivongo: common along the coastal road (*Beentje* 1992). Mahanoro: S of Mahanoro (*Beentje* 1991). Sainte-Marie: Ile Sainte-Marie (*Dransfield* 1994).

CULTIVATED. England: Hort. Kew, May 1901 (fl.), *anon*. s.n. (K); idem, June 1979 (fl) & Sept. 1984 (fr.), 000-73.12576 (K). Côte d'Ivoire: 17 km W of Abidjan, June 1970 (fl.), *Leeuwenberg* 8033 (K). Mozambique: Maputo, Feb. 1982 (fr.), *de Koning* 8645 (K). Singapore: Hort. Singapore, anno 1902, *Ridley* s.n. (K); and anno 1920, *anon* s.n. (K); and June 1929 (fl., fr.), *Furtado* s.n. (K). U.S.A.: Oahu, Hort. Honolulu, Nov. 1930 (fr.), *Wilder* s.n. (K). Trinidad & Tobago: Hort. Tobago, March 1909 (fl.), *Broadway* 2971 (K); Port of Spain, Feb. 1920 (fl.), *Waby* 98749 (K).

34. DYPSIS ARENARUM

This species can easily be confused with *D. lutescens*, which grows in the same area, but has a longer petiole and fewer leaflets; the seed is also much smaller. The name means 'of the sands', since the species occurs on sand near the sea.

DISTRIBUTION. Between Soanierana-Ivongo and Vatomandry.

HABITAT. Littoral forest near fresh water; alt. 1–15 m.

Local names. *Hiribiry* (fide Perrier).
Uses. Not recorded.
Conservation Status. Critical. The distribution area is small, the numbers of this species are thought to be very low, and the vegetation type is threatened by development and fires.

Dypsis arenarum *(Jum.) Beentje & J. Dransf.* **comb. nov.**

Synonym:
Chrysalidocarpus arenarum Jum., Ann. Inst. Bot.-Géol. Colon. Marseille sér. 10, 3: 17 (1922); Jum., Cat. Pl. Madagascar, Palmae: 8 (1938); Jum. & H. Perrier, Fl. Madagascar 30: 98 (1945). Type: Madagascar, Tampina, *Perrier* 13292 (Holotype P).

Clustering palm in tufts of c. 5. **Stems** 5–6 m tall, 6.5 cm diam.; internodes 6–8 cm; bark dark green, on older trunks brown; wood quite hard, slightly pinkish, moist. **Leaves** c. 10 in the crown, porrect, slightly arched distally, with opposite leaflets at an angle of 90° with each other; sheath 20–47 cm, whitish green to pale yellow-brown abaxially and distally with wax and scattered reddish scales, reddish brown and glabrous adaxially, turning into the petiole after a small sharp bend but without obvious ligules; petiole 60–72 cm, 1.8–2 x 1.5–1.6 cm proximally, 1.5 x 1.2 cm distally, channelled with sharp edges, pale brown with minute scattered scales; rachis 1.5–1.7 m, in mid-leaf 0.9–1.6 x 0.7–1.2 cm, keeled, pale brown with minute scattered scales to glabrescent; leaflets 28–30 on each side of the rachis, regular, stiff and straight, pendulous in their most distal part, the proximal 81–129 x 0.9–2.5 cm, median 67–80 x 2.2–3.6 cm (interval 4–6.5 cm), distal 16–36 x 0.9–2.1 cm, the terminal pair not or hardly (up to 0.5 cm) joined and with briefly

Dypsis arenarum. **A** basal leaflets × 1/3; **B** mid leaf leaflet × 1/3; **C** leaf tip × 1/3; **D** base of inflorescence × 1/3; **E** distal part of inflorescence showing bracts and rachillae × 1/3; **F** first order branch of inflorescence × 1. **A – E** from *Beentje* 4445, **F** from *Perrier* 13292. Drawn by Rosemary Wise.

(c. 0.3 cm) truncate and dentate apices, apices of median leaflets unequally bifid for 2–3 cm and attenuate, glaucous, with tufts of large red-brown ramenta to 4 mm long proximally and abaxially on the midrib, and with many minute scattered scales on the veins (invisible in *Guillaumet* 2527), main veins 1–5, faint; young leaves reddish. **INFLORESCENCE** interfoliar, the part outside the leaf sheath arching through 180°, branched to 2 orders; peduncle 34–57 cm long, proximally c. 2 × 0.5 cm, distally 1.5–2.3 × 0.6–1.2 cm, distally with scattered scales, glabrescent; prophyll 42–75 cm long and 4–5.5 cm wide, borne at 5–17 cm above the base of the peduncle, open for 16–25 cm and hooded distally with an acute apex, coriaceous, ?slightly waxy, abaxially pale brown with scattered scales, adaxially red-brown and glabrous; peduncular bract inserted at 27–34 cm from the base of the peduncle, 39–53 cm long, not or hardly beaked, split over its entire length or closed in the distal part, distally hooded and 12 cm wide, pale brown with scattered scales, not deciduous but remaining contiguous with the prophyll for the most part and similar to the prophyll; non-tubular peduncular bracts near the apex of the peduncle, 1.3–2 cm long, triangular, acute; rachis 11–28 cm long, glabrous, with 5–14 branched and 12–20 unbranched first order branches; first order branches subtended by rachis bracts to 1 × 0.5 cm, with a secondary rachis of up to 6 cm long and with flattened base 9–17 × 3–8 mm and with 2–8 rachillae; rachillae porrect, 10–31 cm, 2–4 mm diam., with rather dense triads in slight pits and with distinct, entire, triangular, acute rachilla bracts. **STAMINATE FLOWERS** with sepals 2–2.2 × 1.8–2 mm; petals connate for c. 0.5 mm, 3.1–3.4 × 1.9–2 mm; stamens 6, slightly unequal with the antepetalous ones with filaments slightly wider at the base, filaments 1.6–2 mm long and cylindrical, anthers 1.8–2 × 0.6 mm, the locules parallel and obtuse, dorsifixed and versatile; pistillode c. 2.4 mm high, 0.6 mm diam. **PISTILLATE FLOWERS** with sepals 2.7–3 × 2.4–3.4 mm; petals 3.5–4.2 × 3–4 mm; staminodes 0.6–0.8 mm, flat; gynoecium 3.2–3.6 × 2.9–3 mm. **FRUIT** ovoid to ellipsoid with a rounded apex, 10–12 × 8–9 mm, with anastomosing fibrous endocarp. **SEED** slightly obovoid with rounded apex and apiculate base, 8–9.5 × 5.5–6 mm, with subaequatorial depression; endosperm homogeneous.

NOTE. Differs from *D. lutescens* (which occurs in the same localities, in the same habitat) in the longer petiole, the smaller number of leaflets, the longer petals and the more robust rachillae; the inflorescence branches to 2 orders, while in *D. lutescens* it usually branches to 3 orders (but occasionally is branched to 2 orders). From *D. psammophila* it differs in the generally larger leaves (petiole, rachis, leaflets) with many scattered scales; the much longer prophyll and stouter rachillae.

SPECIMENS SEEN. Fenoarivo Atn.: Tampolo, Feb. 1970 (fr., young leaf), *Guillaumet* 2527 (K, P). Ampasimanolotra: Tampina, Nov. 1920 (fl.), *Perrier* 13292 (P, type) and (fr.), *Perrier* 15988 (P); 8 km S of Ambila-Lemaitso, Sept. 1991 (dead infl.), *Beentje* 4445 (BH, K, MO, P, TAN).

35. DYPSIS PSAMMOPHILA

Another species close to *D. lutescens*, but much more slender with spindly stems towering above the coastal bush. The name means 'sand-loving', since this taxon is restricted to coastal white sands.

DISTRIBUTION. E Madagascar, between Soanierana-Ivongo and Ambila-Lemaitso.
HABITAT. Coastal forest on white sand; alt. 5 m.
LOCAL NAMES. Not recorded.
USES. Not recorded.
CONSERVATION STATUS. Critical. The distribution area is small, and over the whole area the habitat is being destroyed. Numbers are estimated at less than a hundred.

Dypsis psammophila *Beentje* **sp. nov.**
D. lutescens affinissima sed caulis gracilioribus nigris, vagina folii breviore, foliolis brevioribus angustioribus, inflorescentia in 2 ordines ramificanti differt. Typus: Madagascar, Ambila-Lemaitso, *Dransfield* JD6495 (Holotypus K; isotypus TAN).

Clustering palm. **Stems** to 4 (–6) m, 3 cm diam., 1.7 cm diam. near crown; bark tending to be black, internodes c. 15 cm near base, 1.2–1.8 cm near crown; crownshaft 40 cm, green with white wax. **Leaves** arcuate, c. 2 m; sheath 27–30 cm, with dense scattered scales, no ligule; petiole 20 cm, proximally 0.5 × 0.45 cm, distally 0.5 × 0.4 cm diam., with minute scales; rachis c. 76 cm long, with densely scattered scales; leaflets c. 40 on each side of the rachis, regular, strongly curved, proximal 27–41 × 0.3–0.8 cm, median 36–37 × 0.7–0.9 cm, distal 8.5–37 × 0.2–0.9 cm, adaxially glaucous, main vein 1, ramenta few or sometimes 0–1, basal, otherwise glabrous, apex attenuate, unequally bifid. **Inflorescence** branching to 2 orders; peduncle 31 cm, with non-sheathing peduncular bracts (10–12 mm) at 26 and 29 cm above the base of the peduncle; prophyll 21–2 cm; peduncular bract 41 cm (from base of peduncle, not insertion point), glabrous; rachis c. 13 cm, 7 × 4 mm diam. proximally; narrowly triangular, 7 mm; rachillae 13–18 cm, with bare basal part 1.5–2 mm diam., glabrous, with triads distant and in slight pits; rachilla bracts broadly ovate, acute. **Staminate flowers** unknown. **Pistillate flowers** known only as remnant parts on fruit; sepals 1.4–1.5 × 1.3–1.7 mm, broadly ovate, rounded, minutely ciliolate; petals c. 2 × 1.8–2.4 mm, imbricate with ?valvate apiculate tips, broadly ovate, minutely ciliolate. **Fruit** ellipsoid, 9–10 × 5–6 mm; endocarp fibrous. **Seed** 7–8.5 × 4–5 mm, rounded at both base and apex; endosperm homogeneous.

Note. Differs from *D. lutescens* in the smaller dimensions of the leaf (sheath, leaflets) with no scattered scales abaxially; the inflorescence always branched to 2 orders only; the shorter rachis; the shorter pistillate sepals; and the smaller fruit and seed. The differences with *D. arenarum* are enumerated under the latter.

Moore 9920 from the Masoala Peninsula (Ambatobe

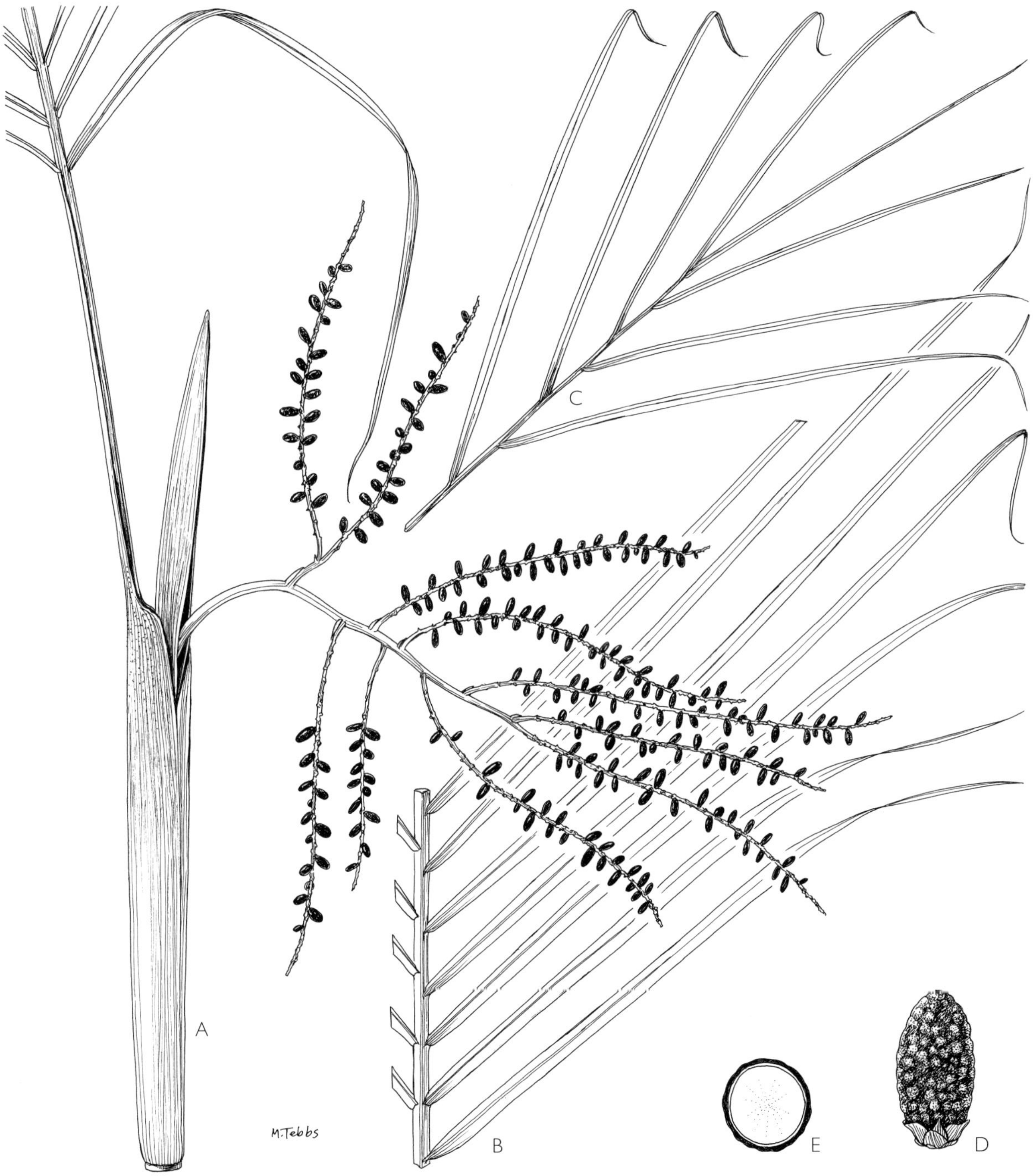

Dypsis psammophila. **A** leaf sheath, petiole and basal leaflets, and infructescence × 1/3; **B** mid section of leaf × 1/3; **C** leaf tip × 1/3; **D** fruit × 2; **E** fruit in cross section × 2.5. All from *Dransfield et al.* JD6495. Drawn by Margaret Tebbs.

on the Onive R., April 1971, fr., *Moore* 9920 at P) seems intermediate between this taxon and *D. onilahensis*. It occurs in low canopy rain forest, has the inflorescence and fruit of the former, but the leaves resemble those of the latter – although admittedly on the short side. It has no petiole. The habit is also more like *onilahensis*, with a solitary trunk c. 10 m high and 7.5 cm in diameter. The local name was *lafazovombona*.

Specimens seen. Ampasimanolotra: 5km S of Ambila-Lemaitso, Nov. 1986 (bud), *Dransfield* JD6437 (K, TAN); idem, March 1988 (fr.), *Dransfield et al.* JD6495 (K, TAN; type).

Sight record. Soanierana-Ivongo: S of Manompana (*Beentje*). Sainte-Marie: Île Sainte-Marie (*Dransfield*).

36. Dypsis pembana

A curious 'outlier' of the genus, only known from Pemba Island, just off the African mainland, from which it takes its name. Pemba has several more links with Madagascar: there is the fruit bat *Pteropus voeltzkowi*, with the other members of its genus in Madagascar, the Comoro Islands and through to southeast Asia and the Pacific; and the Aroid *Typhonodorum lindleyanum* occurs in Pemba, Zanzibar/Unguja and Madagascar.

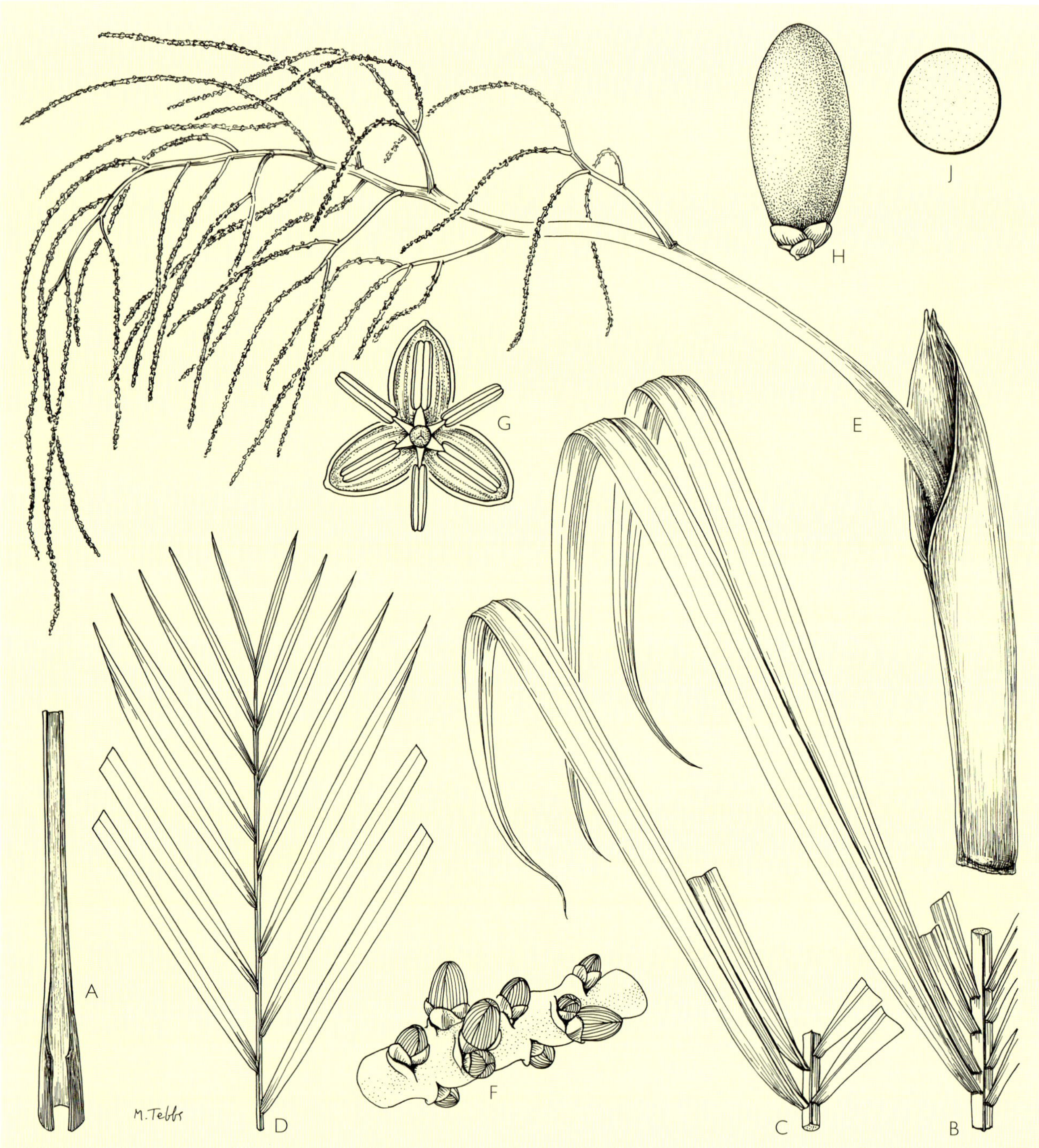

Dypsis pembana. **A** basal part of petiole × 1/6; **B** basal part of leaf rachis × 1/4; **C** mid section of leaf × 1/4; **D** leaf tip × 1/4; **E** inflorescence × 1/4; **F** detail of rachilla × 5; **G** staminate flower × 8; **H** fruit × 2.5; **J** seed in section × 2.5. **A – C**, **E – H**, **J** from *Williams* H177/56, **D** from *Greenway* 1488. Drawn by Margaret Tebbs.

Dypsis pembana, growing in the Ngezi Forest, Pemba.

DISTRIBUTION. Tanzania: Pemba.
HABITAT. Moist lowland forest, littoral forest; alt. 1–50 m.
LOCAL NAMES. *Mpapindi*, less often *Mpopo wa mwitu* (Swahili).
USES. HB has seen trunks of this species used to construct a football-goal near Ngezi forest; no other uses known.
CONSERVATION STATUS. Vulnerable. Numbers are estimated at 3000, but these are all within a single forest.

Dypsis pembana *(Moore) Beentje & J. Dransf.* **comb. nov.**

SYNONYMS:
Chrysalidocarpus pembanus Moore, Principes 6: 109 (1962), as *C. pembana*; Dransfield, Fl. Trop. E. Africa: Palmae: 46 (1986). Type: Tanzania, Pemba, Ngezi forest, *Williams* H 177/56-7 (Holotype BH; isotype EA, K).

Chrysalidocarpus sp. of Williams, U.O.P.Z.: 190 (1949).

Clustering palm. **STEMS** 4–12 m high, 6–15 cm diam.; internodes to 24 cm long, pale brown to green, strongly ringed with leaf-scars. **LEAVES** c. 10 in the crown, arching, arranged in ± 3 ranks; sheath 50–60 cm long, waxy green, sparsely tomentose when young; rachis to 2.4 m long, in mid-leaf to 1.5 cm wide; leaflets 40–50 on each side of the rachis, regular, arching, the leaflets on opposite sides of the rachis at an angle of c. 90° with each other, dark green adaxially, abaxially waxy, the proximal 70–76 × 1.3–2.8 cm, median 46–74 × 3–3.9 cm (interval 4–5 cm), distal 14–45 × 1.4–2.4 cm, main vein 1, with 2 marginal veins, midrib prominent adaxially, apices acuminate or attenuate, abaxially with a dense cover of minute waxy scales, with small shining brown scales on all the veins, with one or a few irregularly spaced large bifid ramenta with brown centres and laciniate margins on the midrib. **INFLORESCENCE** interfoliar, branched to 3–4 orders with spreading branches, lengthening in fruit by some 40%; peduncle c. 60 cm long, stout, ± flattened, densely reddish tomentose, distally curving through 90°; prophyll > 30 cm, c. 5 cm wide, glabrous, dull waxy; peduncular bract 30–55 cm long, splitting over its length, rusty-pubescent or glabrous and waxy, beaked for 2–3 cm, deciduous; first order branches slightly reddish-pubescent but glabrescent, with up to 15 second order branches; rachillae glabrous, 11–19 cm long, 1–2 mm diam.; triads distant; rachilla bract 0.5–0.7 mm, obtuse to acute. **STAMINATE FLOWERS** only known from buds, with sepals 1.3–1.6 × 1.4–1.8 mm, concave, proximally gibbous, keeled, ciliolate; petals 2.3–2.8 × 1.5–1.8 mm; stamens 6, slightly biseriate, offset 0.2 mm, the filaments 1.4–1.5 mm and thin-cylindrical, the anthers 1.3–1.5 × 0.4–0.7 mm; pistillode columnar, 1.8–2.8 × 0.6 mm. **PISTILLATE FLOWERS** unknown at anthesis, the petals in fruit 2–2.6 mm long. **FRUIT** dark red, oblong-ovoid, 12–15 × 5–7 mm; endocarp fibrous, the fibres anastomosing. **SEED** 10.5–11 × 5–5.5 mm; endosperm homogeneous.

SPECIMENS SEEN. Pemba: Ngezi Forest, July 1901 (ster.), *Lyne* 100 (K); idem, Feb. 1929 (y.fr.), *Greenway* 1488 (K); idem, anno 1956 (fr.), *Williams* 177/56-7 (type, EA, K)

SIGHT RECORD. Pemba: 5 km N of Wete (*Beentje*). Cultivated at Zanzibar town.

37. DYPSIS CABADAE

This species is widespread in cultivation, but unknown from the wild: it might be from Madagascar or the Comoro Islands.

DISTRIBUTION. Only known from cultivated plants; origin unknown.
HABITAT. Unknown.
LOCAL NAMES. Not recorded.
USES. Not recorded.
CONSERVATION STATUS. Unknown.

Dypsis cabadae *(Moore) Beentje & J. Dransf.* **comb. nov.**

SYNONYM:
Chrysalidocarpus cabadae Moore, Principes 6: 108 (1962). Type: Cuba, Soledad, cultivated at Atkins Garden, *Moore* 6517 (Holotype BH).

Clustering palm in tufts of up to 14 stems. **STEMS** to 10 m tall, to 9 cm diam.; internodes 9–12.7 cm long, smooth, glossy, green, nodal scars prominent, pale. **LEAVES** c. 10 in the crown, arching distally; sheath 50–76 cm long, bright green with a glaucous bloom, sparsely dotted with reddish fimbriate scales but glabrescent, with square shoulders; petiole 25–33 cm long (–60 cm in young shoots) with a prominent callus pad at the base, glabrous, channelled, 1.7–2 × 2–2.2 cm diam. proximally, 2–2.3 × 1.7–1.9 cm distally; rachis to 1.7 m long, green adaxially, yellowish abaxially, glabrous, in mid-leaf 1–1.3 cm wide, slightly keeled; leaflets

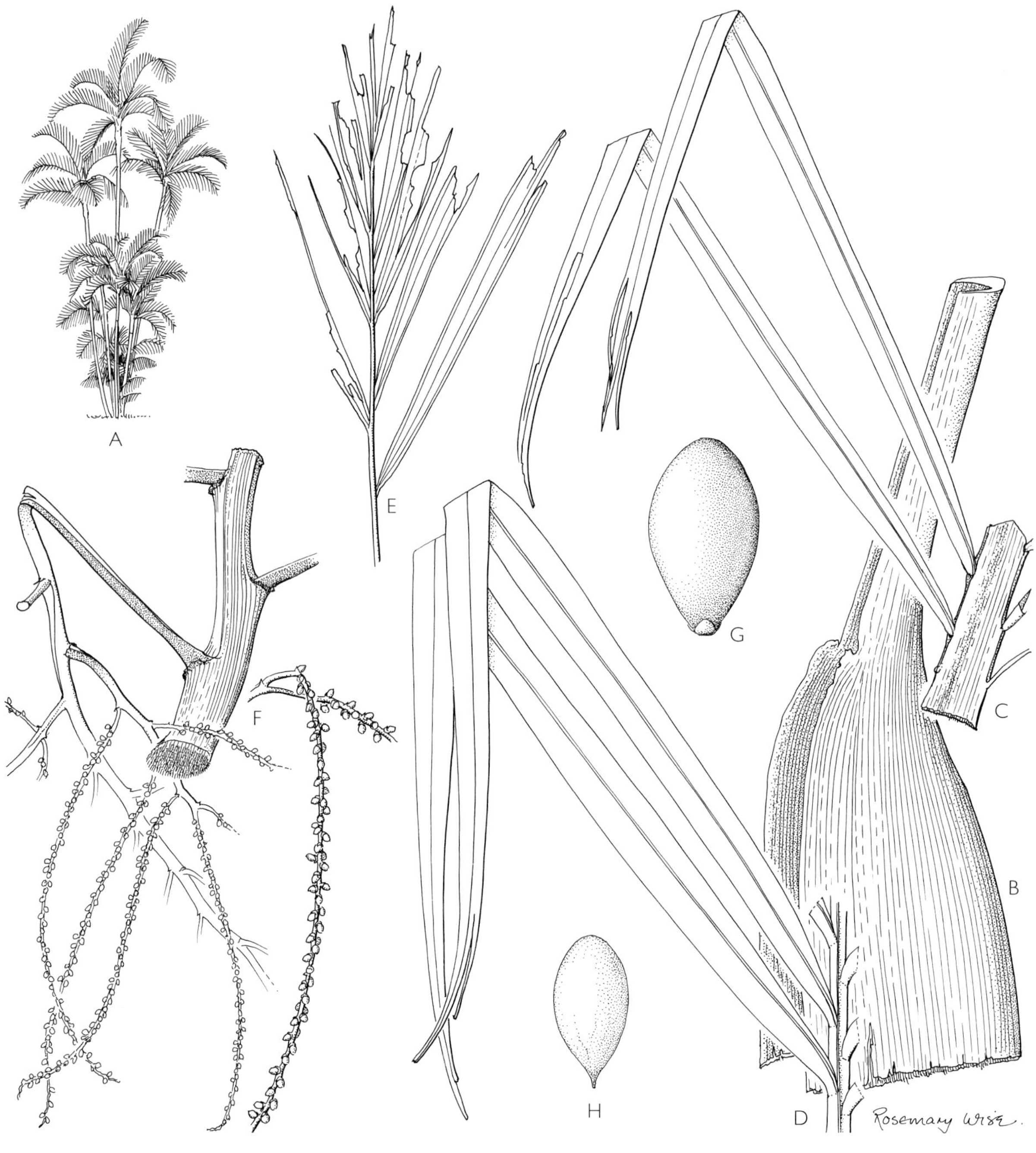

Dypsis cabadae. **A** habit, greatly reduced; **B** leaf sheath and petiole × 1/3; **C** basal part of leaf rachis × 1/3; **D** mid section of leaf × 1/3; **E** leaf tip × 1/3; **F** part of inflorescence × 1/3; **G** fruit × 2.5; **H** seed × 2.5. **A** from a photograph in *Moore* 6101, **B** – **H** from *Moore* 6517. Drawn by Rosemary Wise.

regular, 24 (in young plants)–60 on each side of the rachis, those on opposite sides of the rachis at an angle of c. 45° with each other, dark green, shiny, the proximal 42–57 × 1.6–2.1 cm, median 51–59 × 1.8–2.3 cm, distal 6–27 × 0.5–1.6 cm, midrib yellowish abaxially, with one or a few ramenta (1–4 mm long) proximally, otherwise glabrous, waxy, midrib and marginal veins prominent, apices acute. **INFLORESCENCE** interfoliar, branched to 3 orders, c. 1.5 m long, erect, green; peduncle 59–69 cm long, with dense to sparse minute rusty scales, proximally 4–5 × 1.2–2 cm, distally 1.5–2 × 1–1.2 cm diam.; prophyll green or glaucous, 44–52 cm long, 3.7–5 cm wide, borne at 11–13.5 cm above the base of the peduncle, splitting obliquely at the apex for about 50%, with dense to sparse minute rusty scales; peduncular bract green or glaucous, inserted at 29–34 cm from the base of the peduncle, 40–54 cm long, 4 cm wide, splitting on one side, closed for a 3 cm beak, with dense to sparse minute rusty scales, eventually deciduous; rachis c. 70 cm long, glabrous, with 20–22 branched and 8–10 unbranched first order branches, the proximal of these with a rachis of up to 40 cm; rachillae 8–18 cm long, c. 1 mm diam., glabrous; triads distant. **STAMINATE FLOWERS** with sepals green, 1.6–2.1 × 1.8–2.5 mm; petals yellow-green, connate for c. 0.8 mm, free for 2.2–2.6 × 1.6–2.1 mm; stamens 6, biseriate (offset c. 0.4 mm), the filaments 2.2–2.6 mm long, thin, the anthers 1.6–1.7 × 0.8–1 mm, versatile, the locules parallel; pistillode white, columnar, c. 3 mm long. **PISTILLATE FLOWERS** with sepals 1.4–1.7 × 1.8–2.2 mm; petals 2.2–2.7 × 2–2.8 mm; staminodes 0.3–0.4 mm; ovary not seen. **FRUIT** ellipsoid, scarlet, 9–12 × 4–6 mm, apex rounded; exocarp smooth,

Dypsis cabadae, cultivated in Caracas Botanic Garden.

mesocarp thin and fleshy and fibrous, endocarp ± adherent to the seed, fibrous, anastomosing. **SEED** ellipsoid, 8–9 x 4–5 mm, pointed at the base, rounded at the apex; endosperm homogeneous, embryo lateral; raphe branches ascending from the base and loosely anastomosing. **SEEDLING** with eophyll bifid, occasionally with a few scales abaxially.

NOTE. The affinities of this species seem to be with *D. pembana* from Pemba, but it is certainly not the same.

SPECIMENS SEEN. Cuba: Soledad, Cienfuegos, Atkins Garden, Feb. 1952 (old infl., fr.). *Moore* 6101 (BH, paratype); idem, Feb. 1953 (fl., fr.), *Moore* 6517 (BH, holotype).

38. DYPSIS HUMBLOTIANA

INSUFFICIENTLY KNOWN SPECIES

The material of this species is too scanty for us to say much about it, or to include it in the key. The name refers to the collector of the type, Léon Humblot.

DISTRIBUTION. Comoro Islands: Grand Comore.
HABITAT. Forest; no further data.
LOCAL NAMES. Not recorded.
USES. Not recorded.
CONSERVATION STATUS. Unknown.

Dypsis humblotiana *(Baill.) Beentje & J. Dransf.* **comb. nov.**

SYNONYMS:
Chrysalidocarpus humblotianus (Baill.) Beccari, Bot. Jahrb. Syst. 38, Beibl. 87: 33 (1906); Becc., Palme del Madagascar 40 (1914); Jum. & H. Perrier, Fl. Madagascar 30: 100 (1945).
Phlogella humblotiana Baill., Bull. Soc. Linn. Paris: 1174 (1894). Type: Comoros, Grand Comore, Combani forest, *Humblot* 1338 or 338 (Holotype P, isotype K).

TRUNK to 15 m tall, to 15 cm diam. **LEAVES** with the rachis in mid-leaf 1.5–2 cm wide, keeled, glabrous; leaflets probably regular with an interval of 1.7–2.3 cm, median 64–70 x 2.7–3 cm, occasionally with a single ramentum to 1 cm long proximally, otherwise glabrous, main vein 1, and thickened margins, apices unequally acute. **INFLORESCENCE** branched to 2 or more orders, probably to 3 orders; rachillae 9–12 cm long, c. 1.5 mm diam., glabrous, with distant superficial triads, with proud rachilla bracts with rounded apices. **STAMINATE FLOWER** buds (very young) with sepals 1.3–1.5 x 1.2–1.5 mm; petals < 0.8 x 0.8 mm; stamens 6, uniseriate (at least at this stage), < 0.7 mm high. **PISTILLATE FLOWERS** not seen. **FRUIT** not seen.

NOTE. Labels on the type sheets bear the number

338 in profusion, but the single label with date and locality has the number 1338.

Specimens seen. Grande Comore: Combani forest, Oct. 1884 (bud), *Humblot* 1338 or 338 (K, P; type).

Similar specimens. Comoro Islands: without further locality, anno 1847–1852 (ster.), *Boivin* s.n. (P) is a palm only known from the leaves: sheath 20.5 cm long, c. 2 cm diam., waxy with reddish scales, with ragged sloping shoulders; petiole c. 1.5 cm long and 0.9 × 10.5 cm diam.; rachis c. 58.5 cm long, in mid-leaf 3–4 mm wide, slightly keeled and glabrous; leaflets c. 23 on each side of the rachis, irregular or in groups of 2–3, the more distal regular, ?in one plane, the proximal 25–32 × 0.2–0.6 cm, median 43–47 × 1.7–1.8 cm, distal 9–16 × 0.5–1 cm; main vein 1 (– 3); glabrous?; acuminate; the distal pair not connate. Because it is sterile it is impossible to decide on its status.

39. Dypsis lanceolata

This would be a wonderful ornamental. Hull describes it as 'even more beautiful than *D. cabadae*'. The name refers to the shape of the leaflets, which are unusually broad for the genus *Dypsis*.

Distribution. Comoro Islands: Grande Comore, Moheli.
Habitat. Mid-altitude rain forest; 500–1000 m.
Local names. *Ivovowo* (fide Hull).
Uses. Not recorded.
Conservation Status. Unknown, but at least Vulnerable.

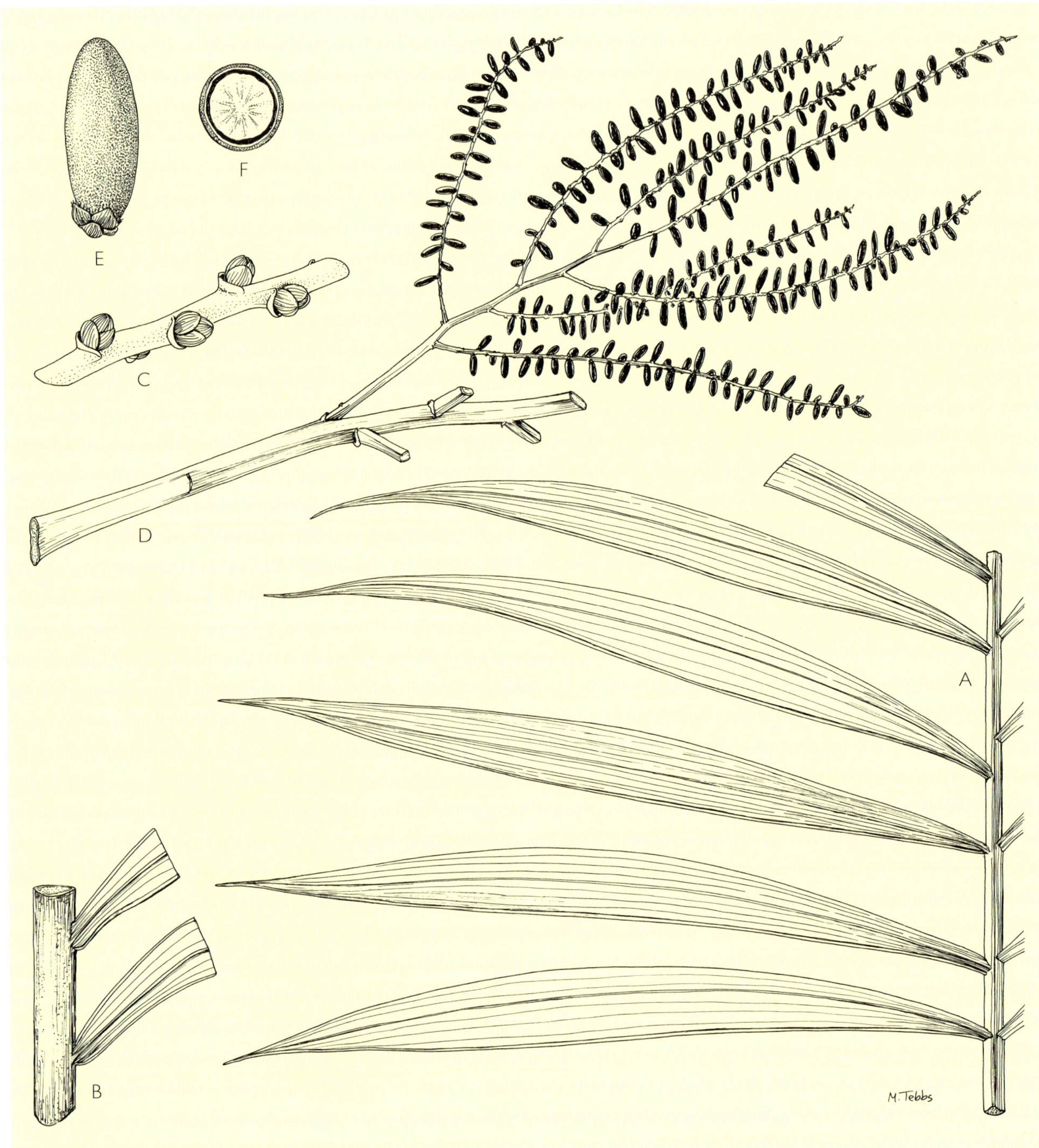

Dypsis lanceolata. **A** mid portion of leaf × 1/3; **B** detail of leaflet bases × 1; **C** detail of rachilla × 3; **D** part of infructescence × 1/3; **E** fruit × 2; **F** fruit in cross section × 2. **A** – **C** from *Humblot* 1347, **D** – **F** from *Bernardi* 11750. Drawn by Margaret Tebbs.

Dypsis lanceolata
(**x**: sight records)

Dypsis lanceolata *(Becc.) Beentje & J. Dransf.* **comb. nov.**

Synonym:
Chrysalidocarpus lanceolatus Beccari, Bot. Jahrb. Syst. 38, Beibl. 87: 34 (1906); Jum. & H. Perrier, Fl. Madagascar 30: 97 (1945). Type: Comoros, Combani Forest, *Humblot* 1347 (Holotype P, isotype K).

Clustering palm. **Stems** 5–6 m high; nodal scars pronounced. **Leaves** "somewhat plumose" (*Hull*); petiole distally 1.5–1.7 cm diam., reddish pubescent on both surfaces, channelled; rachis 1.8–1.9 m long, in mid-leaf 1–1.6 cm wide, keeled, densely scaly or with scattered pale scales; leaflets slightly irregular (interval in mid-leaf 1.5–5 cm), proximal 38–43 × 1–2.7 cm, median 30–48 × 3.5–7 cm, distal 4–24 × 0.7–3.8 cm, main veins 3–5, with very conspicuous thickened margins, with several large (0.5–1 cm long) pale-coloured laciniate ramenta on midrib and main veins proximally, and faint minute reddish scales in longitudinal lines on the main and minor veins on the type, but absent in modern collections, acuminate. **Inflorescence** branched to 3 orders or more, c. 60 cm long; rachis bract (one seen) 4.5 cm long, narrowly triangular; rachillae 13–24 cm long, 1.5–3 mm diam., glabrous, with distant superficial triads. **Staminate flowers** with sepals 1.2–1.4 × 1.4–1.6 mm; petals connate for 0.5 mm, free for 2.6–2.9 × 1.4–1.6 mm, spreading at full anthesis; stamens 6, uniseriate, filaments 1 (in closed flowers)–3 (in fully open flowers) mm long, narrowly cylindrical, anthers 1.4 × 0.5–0.6 mm; pistillode 1.6–1.7 mm high, 0.4–0.6 mm diam. **Pistillate flowers** not seen. **Fruit** ellipsoid, 13–17 × 6–10 mm, with rounded apex; endocarp fibrous, with anastomizing fibres. **Seed** slightly obovoid with obtuse apex, (10–) 13–16 × 5–7 mm, with homogeneous endosperm.

Note. Affinities of this taxon are unclear.

Specimens seen. Grande Comore: Combani forest, Oct. 1884 (fl.), *Humblot* 1347 (K, P; type); Mt Kartala, above Boboni, Oct. 1993 (fr.), *Hull* s.n. (K). Moheli: Mt St. Antonio, Dec. 1967 (fr.), *Bernardi* 11750 (K, P).
Sight records. Anjouan: Mt Ntingui, around Lac Dzialandzei (*Hull*).
Cultivated. Hull reports that this species is cultivated at the Galawa Beach hotel at the N end of Grande Comore.

40. Dypsis pumila

A species from high mountains in the north; the specific name means 'dwarf', and refers to the habit: in this case short and stocky rather than the 'tiny' it often means. This species holds the altitude record for Madagascar palms as it occurs at a higher elevation than any other.

Distribution. NE Madagascar, only known from the Marojejy Mts.
Habitat. On gneiss and quartzite, in ericoid vegetation or montane forest, also in high altitude swamp; 1500–2100 m.
Local names. Not recorded.
Uses. Not recorded.
Conservation Status. Vulnerable. Only known from a single locality, which is protected. Not collected in recent years.

Dypsis pumila *Beentje* **sp. nov.**
Palma nana monticola si caespitosa *D. heteromorphae* affinis sed multo breviore vagina folii minima foliis minoribus inflorescentiis semel ramificantibus, si solitaria *D. acuminum* affinis sed endospermio ruminato vagina folii minima differt. Typus: Madagascar, Marojejy summit, *Cours* 3576 (Holotypus P; isotypi K, TAN).

Dwarf palm 0.3–1 m, apparently solitary. **Stem** erect or procumbent, 2.2–7 cm diam., internodes 2–5 mm distally. **Leaves** 3–4 per crown; sheath 11–19 cm long, c. 2 cm diam., waxy and distally densely scaly, without auricles but with square shoulders; petiole absent or up to 4 cm, 7–11 × 4–8 mm diam., densely scaly; rachis

Dypsis pumila

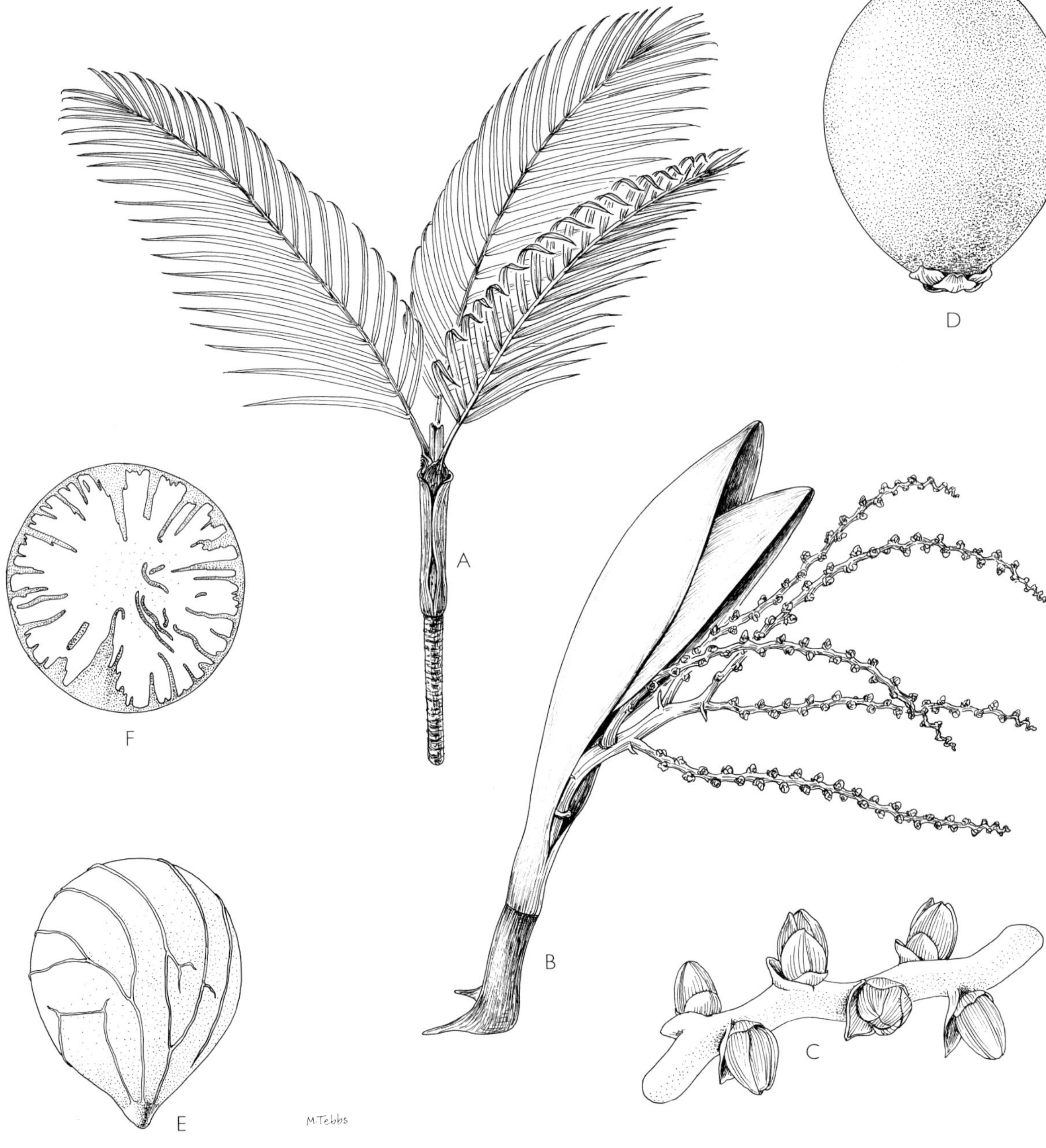

Dypsis pumila. **A** habit × 1/5; **B** inflorescence × 1/2; **C** detail of rachilla × 3; **D** fruit × 2; **E** seed × 2; **F** seed in cross section × 2. **A**, **B**, **D** – **F** from *Humbert & Cours* 23799, **C** from *Cours* 3576. Drawn by Margaret Tebbs.

26–47 cm, densely scaly; leaflets regular, 19–21 on each side of the rachis, the proximal 11–18 × 0.3–1.2 cm, median 13–20 × 1–2.1 cm (interval 0.8–2 cm), distal 2.5–9 × 0.2–1 cm, apices attenuate, not bifid, main vein 1–3, as well as thickened margins, few to many small scales on midrib and margins abaxially, with a few scales on the proximal midrib adaxially. **INFLORESCENCE** infrafoliar, branching to 1 order; peduncle 7–16 cm, 4–8 mm diam.; prophyll dark green, 17–24 cm long, borne at 2.8–8 cm above the base of the peduncle, 3–3.5 cm wide, the distal half open; peduncular bract inserted at 4.5–12.2 cm, 10–15 × 3.6 cm, open, with scattered small scales; rachis 3–5 cm long, glabrous, with 9 rachillae (always?); rachillae 6–10 cm long, 1.5–3 mm diam., glabrous, the triads distant, sunk in pits; rachilla bracts acute, entire. **STAMINATE FLOWERS** with sepals 2.3–3.9 × 1.8–3.2 mm, ovate or elliptic, proximally slightly gibbous, keeled, with membranous margins, acute or obtuse, entire; receptacle 1.8–2 mm high; petals 3–5.8 × 2.6–3.6 mm, ovate or elliptic, striate, acute; stamens 6, uni- or biseriate, filaments 2.8–5 × 0.7–1 mm, flat, anthers 1.4–1.8 × 0.8–1 mm, dorsifixed, versatile, the locules slightly diverging proximally, obtuse; pistillode c. 1.8–3 × 0.7–1 mm. **PISTILLATE FLOWERS** with sepals 2.6–3.2 × 1.7–3.3 mm (to 4.7 mm wide in fruit), (broadly) ovate, concave, acute; petals (in bud) imbricate with a minute valvate apex, broadly ovate, 2.2–2.8 × 1.8–2.2 mm, or (in fruit) white and reddish, 4.2–4.8 × 5–5.5 mm; staminodes (four seen) flat, up to 0.6 mm high; gynoecium c. 2.2 × 1.5 mm. **FRUIT** subglobose to slightly obovoid, 17–26 × 12–20 mm, pointed at the base, rounded at the apex; endocarp fibrous, but not adhering to the seed. **SEED** slightly obovoid, 16–17 × 13–14 mm, pointed at the base, rounded at the apex, with a faint subequatorial depression; endosperm ruminate with many intrusions, 2–6 mm deep.

NOTE. This species is near *D. heteromorphus* but is solitary, much shorter, has a shorter sheath, smaller leaves, and the inflorescence branches to only one order; it is also near *D. acuminum* but is distinct by its ruminate endosperm and shorter sheath.

SPECIMENS SEEN. Marojejy E peak, Dec. 1948 (dead infl.), *Humbert* 22756 (K, P); idem, March 1949 (fl., fr.), *Humbert* 23679 (K, P); idem, March/April 1949 (fr.), *Humbert & Cours* 23799 (K, P); Marojejy summit, April 1949 (fl.), *Cours* 3576 (K, P, TAN); Marojejy, without further loc., 1900 m, Nov. 1972 (??), *Guillaumet* 4059 (TAN).

41. DYPSIS TSARATANANENSIS

A rare clustering palm from a montane habitat. The name refers to the type locality, Tsaratanana.

DISTRIBUTION. Only known from Mt Tsaratanana.
HABITAT. 1000–1700 m.
LOCAL NAMES. *Kindro* (Antankarana).
USES. Palm-heart edible and highly esteemed.
CONSERVATION STATUS. Uncertain. Not seen for more than seventy years, but this mountain has not been

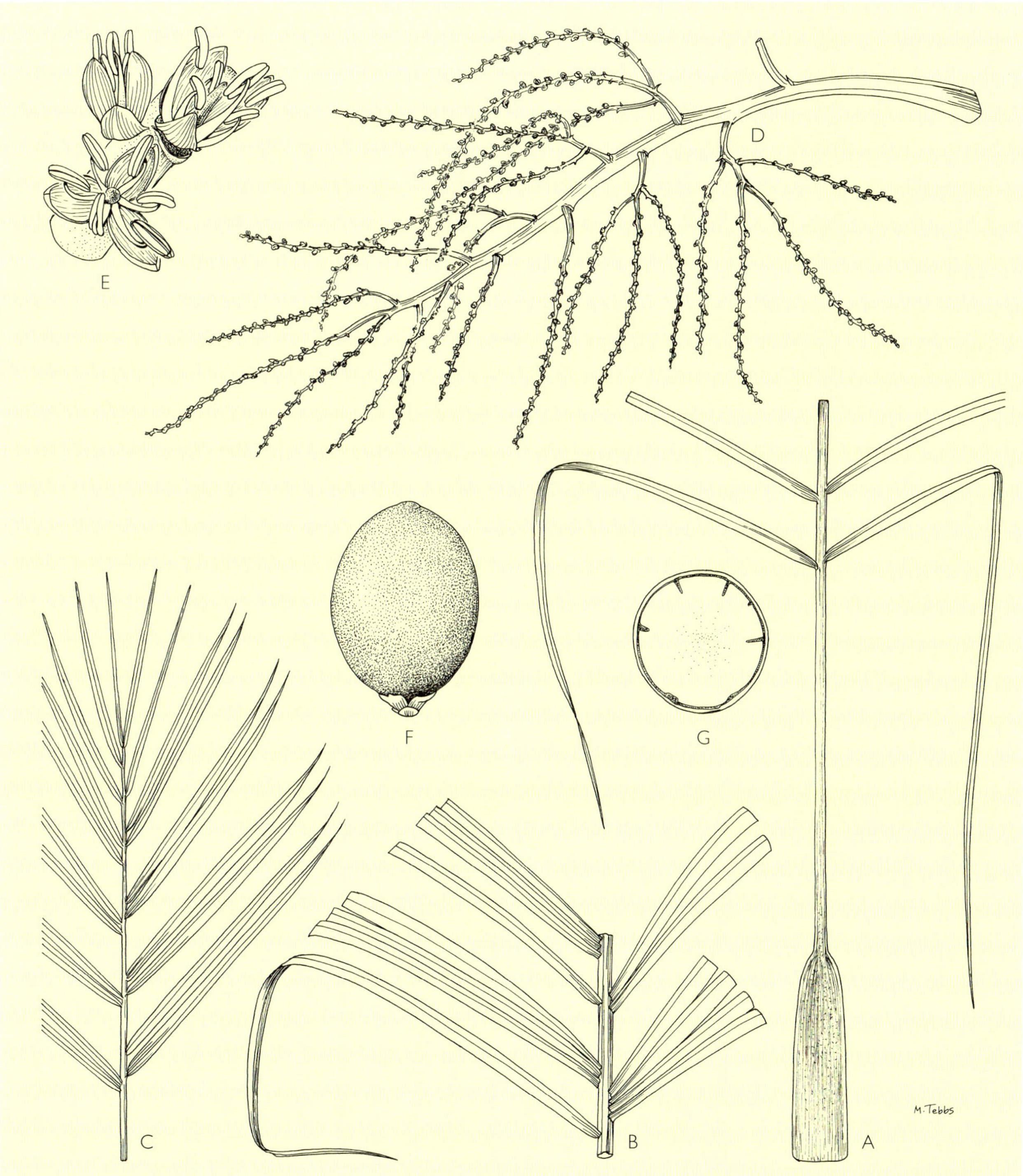

Dypsis tsaratananensis. **A** leaf sheath, petiole and basal leaflets × 1/5; **B** mid section of leaf × 1/5; **C** leaf tip × 1/5; **D** part of inflorescence × 1/4; **E** detail of rachilla × 5; **F** fruit × 2.5; **G** fruit in cross section × 2.5. All from *Perrier* 15265. Drawn by Margaret Tebbs.

visited by botanists for a long time. Numbers are unknown. The fact that the palm-heart was so popular bodes ill for this species, but at least it is a clustering one which makes harvesting less final.

Dypsis tsaratananensis *(Jum.) Beentje & J. Dransf.* **comb. nov.**

SYNONYM:
Neodypsis tsaratananensis Jum., Ann. Inst. Bot.-Géol. Colon. Marseille sér. 4, 2 (2): 15 (1924); Jum., Cat. Pl. Madagascar, Palmae: 19 (1938); Jum. & H. Perrier, Fl. Madagascar 30: 140, fig. 38 (1945). Lectotype (selected here): Madagascar, Tsaratanana, *Perrier* 15265bis (Holotype P).

Clustering palm, intermediate in size between *D. lutescens* and *D. baronii* (Perrier). **LEAVES:** sheath 18 cm or more long, pale brown, distally with small scales, 5.5 cm wide when flat; petiole 12–46 cm long, proximally 1.5 × 0.4 cm, distally 1.1 × 0.3 cm, with sparse scales; rachis in mid-leaf 1.3–1.5 cm wide, keeled, sub-glabrous or with some white waxy scales; leaflets 55–60 on each side of the rachis, in groups of 1–4, group interval 3.5–8 cm, the proximal 40–70 × 0.7–0.9 cm, median 55–60 × 1.5–1.7 cm (interval < 0.5 cm), distal 13–30 × 0.5–1.7 cm, main vein 1, margins thickened, glabrous, apices unequally bifid, attenuate. **INFLORESCENCE** interfoliar, branched to 2 orders (more?), recurved; peduncular bract (fide Perrier) beaked, deciduous; rachis c. 24 cm long, with 9 branched and 16 unbranched first order branches; first order branches seen (distal?) with a rachis to 4 cm long, 8 × 3 mm diam., with 5 rachillae, glabrous; rachillae 9–19 cm long, c. 2 mm diam., glabrous, with distant superficial triads; rachilla bract rounded. **STAMINATE FLOWERS** with sepals 2.3–2.5 × 1.8–2 mm; petals connate to the receptacle for 1.8–2.2 mm, free parts 4.3–4.8 × 2–3 mm; stamens 6, uniseriate, filaments 3.6–4.5 mm, flattened, anthers 1.5–1.6 × 0.6–0.8 mm, versatile, obtuse; pistillode 2–2.3 × 0.8–1 mm. **PISTILLATE FLOWERS** with sepals 2.5–2.7 × 2.2–3.5 mm; petals 3–3.8 × 3 mm; staminodes c. 0.4 mm; ovary 2 × 1.2 mm. **FRUIT** ellipsoid, 12–15 × 9–11 mm, rounded at the apex; endocarp fibrous, the fibres much anastomosing. **SEED** 13–14 × 9–10.5 mm, with few ruminations to 1 mm deep, corresponding to slight grooves on the outside.

NOTE. Data in protologue, not apparent in type or its label: inflorescence branched to 2 or 3 orders. The flowers (present in one of the syntypes) are not described in the protologue, which is most unusual; Jumelle only mentions buds.

The leaves, and local name, are reminiscent of *D. madagascariensis*, but the inflorescence and fruit look more like *D. baronii*. This species is close to *D. oreophila*.

SPECIMENS SEEN. Ambanja/Bealanana: Tsaratanana massif, Dec. 1922 (fl., fr.), *Perrier* 15265 (P, syntype); idem, May 1924 (fr.), *Perrier* 15265bis (P, lectotype).

42. DYPSIS OREOPHILA

Another clustering, montane species, with distinctive small inflorescences. The epithet *oreophila* means mountain-loving, a reference to the habitat. This is a very attractive species.

DISTRIBUTION. Tsaratanana, Marojejy, high ground near Maroantsetra and Mandritsara.
HABITAT. Moist montane forest, on steep slopes; 500–1700 m. Said to be frequent in the forests of the Tsaratanana peaks, between 1000–1600 m (protologue).

Dypsis oreophila, in mountain forest, Marojejy.

Local names. *Kindro, Lafaza* (Antankarana); *Fitsiriky* (Sambirano Sakalava).

Uses. Palm-heart edible, highly esteemed. The hollowed out stems are used for blowpipes.

Conservation Status. Vulnerable. The distribution area is limited. The fact that the palm-heart is highly esteemed bodes ill for the future. Numbers unknown.

Dypsis oreophila. Detail of inflorescence (*Dransfield et al.* JD6756).

Dypsis oreophila *Beentje* **nom. nov.**

Synonyms:
Neodypsis gracilis Jum., Ann. Inst. Bot.-Géol. Colon. Marseille 1, 1: 19 (1933); Jum., Cat. Pl. Madagascar, Palmae: 17 (1938). Type: Madagascar, Tsaratanana, *Perrier* 16076 (P, Lectotype, chosen here).
Phloga gracilis (Jum.) H. Perrier, Fl. Madagascar 30: 126 (1945), **synon. nov**.
Note: the specific epithet was predated in *Dypsis*; *Adelodypsis gracilis* Becc. from 1906 is a synonym of *D. pinnatifrons*.

Clustering palm, occasionally appearing solitary. **Trunk** 2–8 m high; diam. 3–4 cm, stepped and ringed, with some of the bark corky (diseased?); internodes distally dark green, 1–3.5 cm, scars 0.2–1.2 cm wide, pale brown; crownshaft conspicuous, pale green with wax. **Leaves** 6–8 in the crown; young leaves tinged pink; sheath pale brown, 18–55 cm long, with rounded shoulders, pale green or brown, waxy, distally with few to many reddish scales; petiole 2–50 cm long, proximally 0.5–1.5 × 0.3–0.9 cm, distally 0.5–0.9 × 0.4–0.5 cm diam., with many scattered reddish scales, proximally with a triangular adaxial extension to the sheath lining; rachis 0.5–1.5 m long, in mid-leaf 0.3–1.6 cm wide, slightly keeled, with many reddish scales or only with some white waxy scale remnants; leaflets 25–45 on each side of the rachis, in groups of 2–5 in mid-leaf, group interval 2.5–12 cm, fanned within the groups, the proximal 8–33 × 0.4–1.8 cm, median 14–46 × 1.1–2.7 cm (interval 0.3–2 cm), distal 5.5–30 × 0.3–2.6 cm, main vein 1, plus thickened margins, with scattered scales to glabrous, or with a few ramenta 3–5 mm long, apices bifid, unequally attenuate. **Inflorescence** infrafoliar, recurved, branched to 1–2 orders; peduncle 6–11.5 cm, proximally 9–10 × 4–5 mm diam., distally c. 5 × 5 mm, glabrous; prophyll 8–34 cm × 1.2–4 cm, borne at 2–4.5 cm above the base of the peduncle, split over most of its length, pale brown, waxy, with scattered scales to glabrous; peduncular bract inserted at 3.5–7 cm from the base of the peduncle, 10–24 cm long, splitting over its entire length except for the 1–5 cm long beak, sometimes deciduous, pale brown, waxy, with scattered scales or glabrous; non-tubular peduncular bract c. 4.5 cm long; rachis 1.5–9 cm long, glabrous, with 0–9 branched and 5–19 unbranched first order branches, the proximal of these with a rachis up to 2 cm long, 5–8 × 3–4 mm diam., with up to 4 rachillae; rachillae 3.5–14.5 cm, green to cream at anthesis, sinuous, 2–3 mm diam., with scattered reddish scales or glabrous, the triads spirally arranged, quite dense, slightly sunken; rachilla bract rounded or acute, 1.5–2 mm; flowers cream or reddish, without scent. **Staminate flowers** with sepals 2.8–3 × 2.6–3.2 mm, broadly ovate, proximally gibbous, keeled, acute or obtuse, ciliolate; petals connate to a receptacle 1.8 mm high, free for 2.8–3 × 2.6–3 mm, triangular and acute; stamens 6, equal, the filaments connate at their base for 0.3–0.5 mm, 2.8–3.3 × 0.8 × 0.3 mm, flattened with reddish specks, the anthers 1.7–2 × 0.8–0.9 mm, dorsifixed, versatile, the locules parallel and obtuse; pistillode conical-columnar, 2.4–2.8 mm high, 0.6–1.2 mm diam. **Pistillate flowers** with sepals 1.8–3.5 × 2.3–3.8 mm, keeled, rounded, ciliolate; petals 2.8–4.5 × 2.6–4.6 mm, ciliolate; staminodes six, flat, elongate, 0.6–1.3 mm long; ovary c. 3.8 × 3.2 mm. **Fruit** subglobose to slightly obovoid, rounded at base and apex, 5–11 × 3–8 mm; mesocarp 0.5 mm thick; endocarp fibrous, densely anastomosing. **Seed** ellipsoid, 6–7.5 × 3.5–7 mm, pointed at the base, rounded at the apex, with a slight subbasal depression, the surface with anastomosing channels; endosperm ruminate, the intrusions somewhat dense, irregular, 1.5–2.5 mm deep.

Note. Close to *D. tsaratananensis*, but differs in the much shorter inflorescence rachis (6–9 rather than 24+ cm), fewer leaflets (25–45 on each side of the rachis,

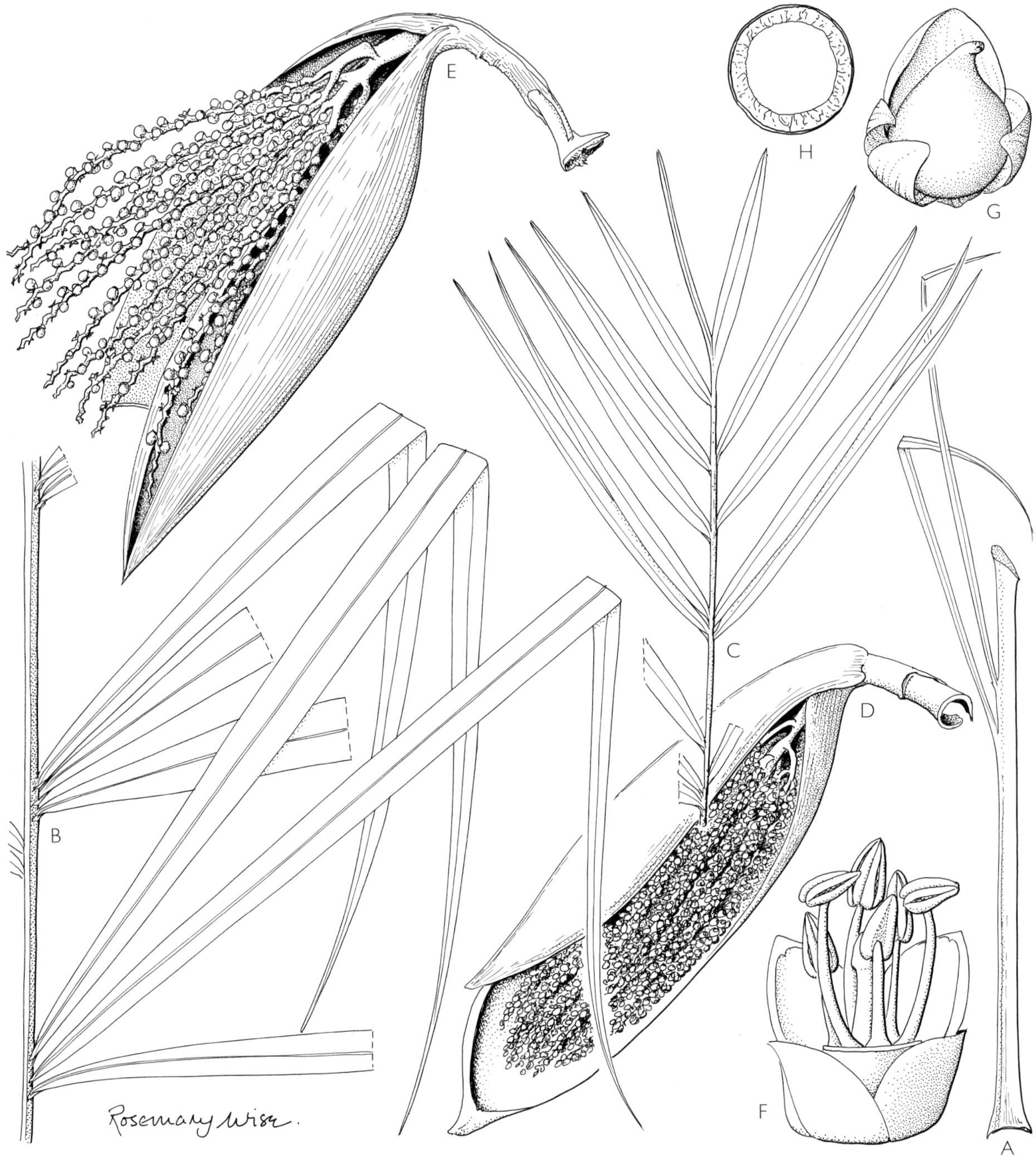

Dypsis oreophila. **A** petiole with lowermost leaflets × 1/3; **B** mid section of leaf × 1/3; **C** leaf tip × 1/3; **D** inflorescence at staminate anthesis × 1/3; **E** young infructescence × 1/3; **F** staminate flower × 7; **G** very young fruit × 4; **H** fruit in cross section × 3. All from *Dransfield et al.* JD6756. Drawn by Rosemary Wise.

rather than 55–60 on each side of the rachis, smaller leaflets (median 14–46, not 55–60 cm), fruit (5–11 × 3–8, not 12–15 × 9–11) and seed (6–8 × 3–7, not 13–14 × 9–11). Most of these characters are not very strong, except for the fruit and seed ones, which in our opinion are important. Maybe further collections will blur the distinctions.

It is rather unclear why this was included in *Phloga* in the Flora (Jumelle and Perrier, 1945). Staminate flowers were unknown, so the only reason must have been its size. Jumelle and Perrier (1945) included *Phloga sambiranensis* as a synonym, but this has a much longer inflorescence and we believe this is wrong.

SPECIMENS SEEN. Ambanja/Ambilobe/Bealanana: Massif de Tsaratanana, Jan. 1923 (y.fr.), *Perrier* 15647 (P, syntype); idem, April 1924 (bud, fr.), *Perrier* 16076 (P, syntype). Andapa: Marojejy East Peak, NW of Mandena, Oct. 1988 (fl., y.fr.), *Miller* 3510 (P); idem, Feb. 1989 (fr.), *Miller & Lowry* 3975 (K, MO); idem, Nov. 1989 (fl.), *Dransfield et al.* JD6756 (K, P, TAN); W slopes of Mt Beondroka, Oct. 1989 (fr.), *Miller & Randrianasolo* 4378 (K). Maroantsetra: Mafaipoza, Oct. 1912 (fr.), *Perrier* 12055 (P). ?Mandritsara: between Andalimena and Mandritsara, Nov. 1922 (bud, y.fr.), *Perrier* 14980 (P).

43. DYPSIS AMBANJAE

One of the mysteries of the genus, only known from the type, and not seen for over eighty years. The species name comes from the town of Ambanja.

DISTRIBUTION. Only known from the type, from the upper Sambirano River.
HABITAT. Rain forest; c. 500 m.
LOCAL NAMES. Not recorded.
USES. Not recorded.
CONSERVATION STATUS. ?Extinct. Not collected for over 80 years.

Dypsis ambanjae *Beentje* **nomen nov.**

SYNONYM:
Phloga sambiranensis Jum., Ann. Inst. Bot.-Géol. Colon. Marseille sér. 5, 1 (3): 18 (1933); Jum., Cat. Pl. Madagascar, Palmae: 23 (1938); **synon. nov.** Type: Madagascar, Sambirano, *Perrier* 12070 (Holotype P).

The epithet *sambiranensis* is predated in *Chrysalidocarpus* in the same publication (Jumelle, 1933), but in an earlier fascicle (later combined as *Dypsis sambiranensis*, a synonym of *D. pinnatifrons*). The new epithet derives from the nearest large town and administrative unit in which the type was found.

Clustering palm. Stems 4–5 m tall, 5–6 cm diam.; internodes distally 2.5–3.5 cm long, nodal scars c. 0.2 cm, distal diameter of stems c. 2.3 cm. **LEAVES** pinnate; sheath c. 32 cm, proximally glabrous and waxy, distally with scattered scales, with rounded opening, auri-

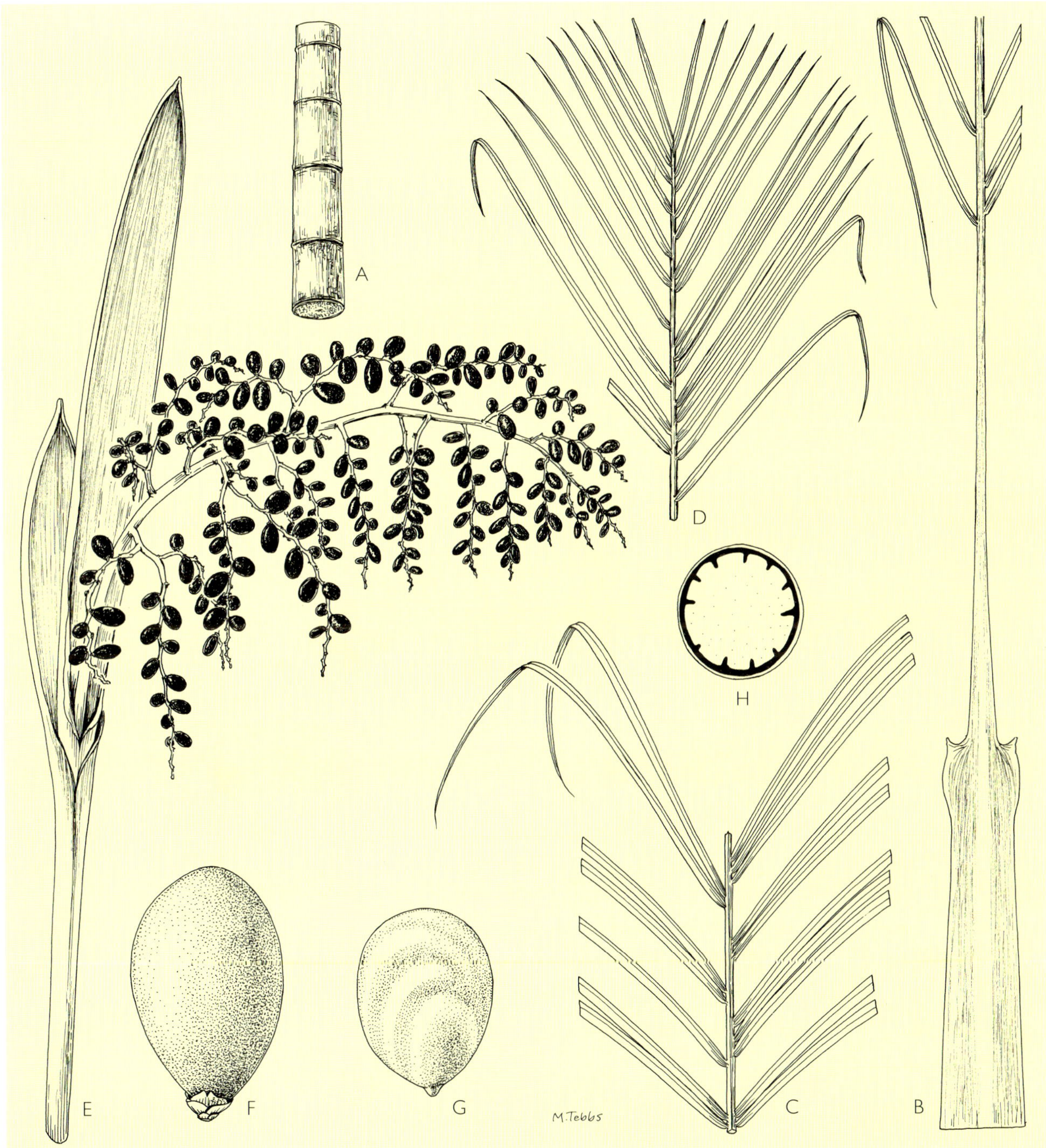

Dypsis ambanjae. **A** portion of stem × 1/3; **B** sheath, petiole and basal leaflets × 1/3; **C** mid section of leaf × 1/3; **D** leaf tip × 1/3; **E** infructescence × 1/3; **F** fruit × 2.5; **G** seed × 2.5; **H** seed in cross section × 2.5. All from *Perrier* 12070. Drawn by Margaret Tebbs.

cles absent; petiole 13–27 cm long, proximally (in a young leaf) 8 × 7 mm, distally 4 × 6 mm, channelled adaxially; rachis slender, in a young leaf 1 m long, in mid-leaf 3–5 mm wide, slightly keeled, almost glabrous; leaflets c. 47 on each side of the rachis, in groups of 2–6, the group interval 4–5 cm, rather stiff, linear, in a young leaf the proximal 38–40 × 0.3–0.5 cm, median 34–36 × 0.6–0.8 cm, distal 8–23 × 0.3–0.6 cm, main vein 1, ramenta few, 1–3 mm long, or absent, rest of leaf glabrous, apices unequally bifid, attenuate. **INFLORESCENCE** branched to 2 orders; peduncle c. 42.5 cm long, distally curved, proximally c. 8 × 3 mm diam., distally c. 7 × 4 mm diam.; prophyll c. 52 cm long, 3 cm wide, borne at c. 23 cm above the base of the peduncle, waxy, glabrous, open in its upper 20 cm; peduncular bract inserted at c. 35 cm from the base of the peduncle, 29–40 cm long, opening over its length except for the beak of 5 mm; rachis c. 18 cm long, glabrous with c. 8 branched and 13 unbranched first order branches, the proximal of these with a secondary rachis up to 2.5 cm long and proximally 4 × 2 mm diam., with up to 4 rachillae; rachillae 4–11 cm long, c. 1.5 mm diam., glabrous, sinuous, with distant superficial triads; rachilla bracts c. 1 mm, acute. **STAMINATE FLOWERS** unknown. **PISTILLATE FLOWERS** with sepals 1.8–2.2 × 2.8–3 mm; petals 3–3.3 × 3.3–3.6 mm; staminodes 0.6–0.7 mm; ovary 2.5–3.3 × 1.8–2 mm. **FRUIT** ellipsoid or slightly obovoid, 11–14 × 8–11 mm, the apex rounded; endocarp very fibrous, with interwoven fibres. **SEED** ellipsoid, 11.5–12.5 × 9.5–10 mm, pointed at the base with a sub-basal depression, rounded at the apex; endosperm ruminate, the ruminations distant and c. 1 mm deep.

NOTE. According to the Flore de Madagascar, this is synonymous with *P. gracilis* (now *D. oreophila*), and the description is a hotchpotch – the small fruit of *D. oreophila* is not mentioned, nor is the type of *D. sambiranensis*. The look of the plant is different, and the inflorescence is much larger; we believe this is a taxon in its own right.

SPECIMENS SEEN. Ambanja: upper Sambirano, Dec. 1912 (fl., fr.), *Perrier* 12070 (P, type).

44. DYPSIS COURSII

A curious montane species, with a short, wide trunk. The name refers to one of the collectors, G. Cours, who made many good palm collections between 1939 and 1952.

DISTRIBUTION. Only known from the Marojejy area.
HABITAT. Moist montane forest or dense sclerophyll forest on ridges; on gneiss and quartzite, (400–) 900–1850 m.
LOCAL NAMES. Not recorded.
USES. Not recorded.
CONSERVATION STATUS. Vulnerable. Distribution area small, but protected. Numbers unknown, but thought to be low.

Dypsis coursii *Beentje* **sp. nov.**
Palma distinctissima caule breve lato foliolis in greges distantes dense aggregatis fructo grande profunde ruminato. Typus: Madagascar, Marojejy W, near Doanyanala col, *Humbert* 23159 (Holotypus P; isotypus K).

Solitary palm. **STEM** 2–8 m, 8–18 cm diam., near the apex c. 2.2 cm diam. **LEAVES** c. 4 in the crown; sheath 18–36 × 9 cm, with dense reddish tomentum and wax, and with triangular auricles to 5 × 3 cm; petiole 4–27 cm long, 1.1–1.3 × 0.3–0.7 cm diam., with few scattered scales; rachis 0.4–1 m long (*Humbert* 23159: leaves 2 m long), in mid-leaf 0.6–1 × 0.4 cm diam., with scattered scales; leaflets 35–39 on each side of the rachis, in groups of pairs proximally, in 2–7 medially (group interval 2.2–7 cm), the proximal 10–29 × 0.2–1.2 cm, median 12–34 × 1.6–3.5 cm (interval 0.7–1.2 cm), distal 6–17 × 0.4–2.2 cm, main veins 1–5, and thickened margins, ramenta few or none, plus red scattered scales on midrib and veins, apex acute, unequally bifid. **INFLORESCENCE** infrafoliar, pendulous,

branching to (1–) 2 orders; peduncle 23–70 cm, 5–10 × 2–4 mm diam., with dense to few scattered scales; prophyll 12–26 cm long, borne at 1–6 cm above the base of the peduncle, 1.6–2 cm wide; peduncular bract inserted at 2–17 cm, deciduous, 24 cm long; sterile bract inserted at 19–25 cm, 0.2–3 × 0.6–3 cm; rachis 19–36 cm, glabrous or nearly so, with 6 branched (first order rachis to 20 cm, proximally to 8 × 4 mm diam., with up to 9 rachillae) and 8–22 unbranched first order branches; rachis bracts to 3 mm; rachillae 1–27 cm, 1.2–3 mm diam., glabrous; triads distant, superficial; rachilla bract c. 1 mm, obtuse. **Staminate flowers** yellowish; sepals 1.5–1.6 × 1.3–1.8 mm; petals 2.7–3 × 1.6–2 mm; stamens 6, uniseriate, filaments 1.6–2.8 mm long with triangular base and cylindrical distal half, anthers 1.2–1.6 × 0.6–0.9 mm; pistillode 0.6–1 mm high, 0.4–0.6 mm diam. **Pistillate flowers** with sepals 1.7–2.8 × 1.5–3.9 mm; petals 4–5 × 2.5–5.5 mm; staminodes 0.3–0.8 mm long; ovary 2.5–3 mm high, 0.9–1.5 mm diam. **Fruit** ellipsoid or slightly obovoid, 20–35 × 15–25 mm, pointed at the base, rounded at the apex; endocarp flaky. **Seed** obovoid, c. 25 × 13–17 mm, pointed at the base, rounded at the apex, the surface slightly channelled, with a sub-basal or lateral depression; endosperm ruminate, the ruminations many, 2–7 mm deep.

Note. Distinct by its short, wide trunk, the distant groups of densely set leaflets, and large ruminate fruits.

Specimens seen. Mt Beondroka, N of Maroambihy, March 1949 (fr.), *Humbert* 23463 (K, P); Marojejy W, near Doanyanala col, Jan./Feb. 1949 (fl.), *Humbert* 23159 (K, P); Marojejy E, W of Manantenina R, Dec. 1948 (bud), *Humbert* 22558 (K, P); Marojejy E trail, Oct. 1988 (fl., y.fr.), *Miller at al.* 3566 (TAN); Marojejy, 1700-1800 m, Nov. 1972 (fl.), *Guillaumet* 4105 (P, TAN); idem (bud), *Guillaumet* 4106 (P, TAN). Sambava: Mt. d'Ambatosoratra, Jan. 1949 (fr.), *Cours* 3299 (K, P, TAN).

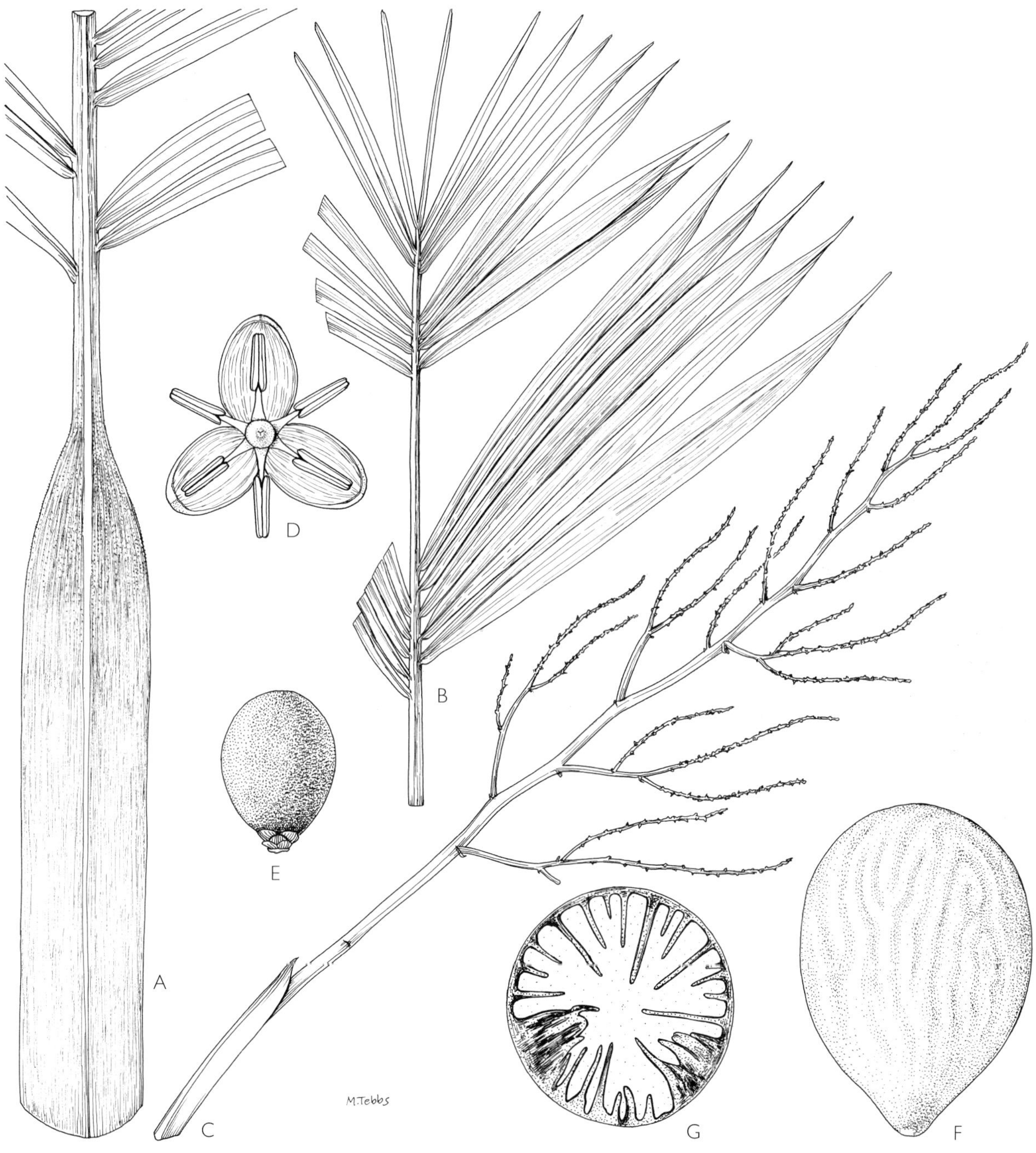

Dypsis coursii. **A** leaf sheath, petiole and basal leaflets × 1/3; **B** mid section of leaf and leaf tip × 1/3; **C** inflorescence × 1/3; **D** staminate flower × 6; **E** fruit × 1; **F** seed × 2; **G** seed in cross section × 2. All from *Humbert* 23463. Drawn by Margaret Tebbs.

45. DYPSIS RIVULARIS

A charming palm, but with a rather untidy look to it, due to the irregular grouping of the leaflets. The name indicates its habit of growing along rivers. This species may be in cultivation in Queensland, where JD has seen juveniles of a palm that seem to match the herbarium specimens from the wild.

DISTRIBUTION. Mananjeba River, Manongarivo, Ankarafantsika.
HABITAT. Moist forest stream edge, among boulders; alt. 130–300 m.
LOCAL NAMES. *Sarimadiovozona* (Sakalava).
USES. Not recorded.
CONSERVATION STATUS. Vulnerable. Only known from two recent sites; total population estimated at less than a hundred.

Dypsis rivularis *(Jum. & H. Perrier) Beentje & J. Dransf.* **comb. nov.**

SYNONYM:
Chrysalidocarpus rivularis Jum. & H. Perrier, Ann. Inst. Bot.-Géol. Colon. Marseille sér. 3, 1 (1): 40, t. 20 (1913); Jum., Ann. Inst. Bot.-Géol. Colon. Marseille sér. 10, 3: 20 (1922); Jum., Cat. Pl. Madagascar, Palmae: 11 (1938); Jum. & H. Perrier, Fl. Madagascar 30: 116 (1945). Type: Madagascar, Boina, *Perrier* 11970 (Holotype P).

Solitary palm, with untidy crown and stilt roots. **TRUNK** 4–5.5 m, 4.5–15 cm diam., internodes 3.5–6 cm long, near the crown c. 2 cm. **LEAVES** spirally inserted, arching, 7–14 in the crown; sheath 3/4 open, 26–43 cm, yellow with few scattered reddish scales and erose ligules c. 1 cm long; petiole absent or up to 2 cm; rachis c. 1.4 m long, proximally 1.5–2 cm wide, in mid-leaf 0.8–1.3 cm wide, reddish with dense pale flaking indument, glabrescent; leaflets c. 32 on each side of the rachis, in groups of 2–5 in the proximal and middle part of the leaf to almost over the entire length of the leaf, the groups 8–19 cm apart, but often with the leaflets in the distal half or third regular; proximal leaflets 12–64 x 0.3–3.7 cm, median 42–68 x 3.3–6 cm (interval within groups 1.5–3.5 cm), distal 13–36 x 0.6–2.8 cm, main veins 5–7, veins with many minute reddish scales abaxially and sometimes adaxially, few or no ramenta, apex (unequally) attenuate. **INFLORESCENCE** interfoliar at anthesis to infrafoliar in fruit, c. 90 cm, branching to 3 orders; peduncle 43–46 cm, densely puberulous, proximally c. 8 x 3 mm, distally c. 7 x 4 mm diam.; prophyll 39–56 cm, borne at 8–14 cm above the base of the peduncle, 1.6–4.5 cm wide, with rather dense reddish scales and sometimes with a waxy bloom; peduncular bract inserted at c. 26 cm, 20–32 cm long, beaked for c. 1 cm, with scattered scales, deciduous; rachis 30–50 cm, densely reddish puberulous, with 17–20 branched and 3–7 unbranched first order branches; proximal first order branches to 50 cm long, with 7–10 branched and 10 unbranched second order branches, proximally to 1.5 x 0.7 cm diam.; rachillae 3–19 cm, densely puberulous, c. 1 mm diam.; triads distant, superficial, with entire rounded bract. **STAMINATE FLOWERS** unknown at anthesis, but in young bud with sepals 0.7–1 x 1–1.3 mm, ciliolate; petals 1.2–1.6 x 1.2–1.4 mm; stamens 6, biseriate, with filaments 0.4–0.5 mm long, cylindrical, with didymous anthers 0.5–0.6 x 0.5 mm; pistillode < 0.5 mm. **PISTILLATE FLOWERS** yellow, without discernible scent, with sepals 0.8–1.1 x 1.4–1.8 mm, broadly elliptic, rounded, minutely ciliate; petals 1.8–2.6 mm long, with the lower part membranous, concave, 1–1.4 x 1.3–2 mm, the upper part fleshy, triangular, flat, 0.8–1.2 x 1.2–1.4 mm; staminodes 6, 0.2–0.3 mm; ovary 1.5–1.8 mm high, asymmetrical, 1.8–2.2 mm diam., with very divergent stigmas 0.5–0.6 mm long. **FRUIT** ellipsoid, slightly curved, 12–14 x 5–7 mm, asymmetrical; base and apex rounded; endocarp fibrous, the fibres hardly anastomosing. **SEED** 10–11 x 4.5–5 mm, rounded at base, slightly pointed at apex, with homogeneous endosperm.

Dypsis rivularis, detail of crown, Ambalafary (*Beentje* 4562).

NOTE. There are some differences between the type (which is from the Ankarafantsika) and the Manongarivo populations. These consist of the width of the leaflets (much wider in the type: 6 cm as opposed to 3–4 cm) and the length of the rachillae (shorter in the type: 3–7 cm as opposed to 8.5–19 cm). However, we feel these differences can be attributed to variation between populations, and we do not believe they warrant taxonomic distinction, since the similarities between the type and the other population are large.

Jumelle describes the palm as 5–10 m tall. This information does not appear on any of the Perrier specimen labels, and we have not included it in the description. There are recent but unsubstantiated reports of a population of large palms growing along a river in the Ankarafantsika, which might confirm Jumelle's data. He also describes the male flowers as having sepals 1.35 × 1.5 mm; petals 2 × 2 mm; anthers with short, wide filaments and a connective projecting beyond the locules, which are said to be divergent at the base. We have only been able to find young staminate buds on *Perrier* 15803; these differ from Jumelle's description.

SPECIMENS SEEN. Ambilobe: upper basin of Mananjeba R., July 1913 (fl., y.fr.), *Perrier* 15803 (P). Ambanja: SE of Beraty, May 1989 (y. bud, fr.), *B. Du Puy et al.* MB 176a (K, TAN); Ambalafary, Jan. 1992 (fl.), *Beentje et al.* 4562 (BH, K, MO, P, TAN); idem, Feb. 1992 (y.fr.), *Beentje et al.* 4581 (K, MO, TAN). Marovoay: Ankarafantsika, without date, (y.fr.), *Perrier* 11970 (Holotype, P; the label only says Boina, but publications give the above site). Ambato-Boeni: Tsaramandroso, April 1952 (fl.), *Ramamongisana* 4185 RN (P).

Dypsis rivularis. **A** leaf sheath, petiole and basal leaflets × 1/5; **B** mid section of leaf × 1/5; **C** leaf tip × 1/5; **D** inflorescence × 1/3; **E** detail of rachilla × 3; **F** fruit × 2.5; **G** fruit in vertical section × 2.5. All from *Beentje et al.* 4562. Drawn by Margaret Tebbs.

46. Dypsis marojejyi

This a most distinctive squat robust palm of the undergrowth, abundant on the broad ridges in submontane forest on Marojejy. It has a short stem and leaves that do not fall off neatly, the crown tending to trap litter. In this respect it resembles *D. perrieri* but the leaves, while being about the same size, have grouped leaflets, and the inflorescence is quite different, being much more diffusely branched. Roots from neighbouring trees tend to grow into the litter that accumulates in the crown, and as the palm grows, these zigzag tree roots continue to grow from sheath to sheath, eventually being exposed. Ferns are also abundant in the crown of this palm. It somewhat resembles *D. coursii* but has a much more robust stem, with much larger leaves and longer, narrower leaflets. The species name is derived from the type locality, Marojejy. As far as we know, this species is not in cultivation.

Distribution. Only known from the Marojejy massif.
Habitat. Submontane rain forest; 700–1100 m. May be locally abundant.
Local names. *Menamosona beratyraty* (Betsimisaraka).
Uses. Not recorded.
Conservation Status. Vulnerable. Distribution restricted to a single, albeit protected, area.

Dypsis marojejyi *Beentje* **sp. nov.**
Palma solitaria detritifera foliis marcescentibus caule breve lato, foliolis longis aggregatis inflorescentia in 3 ordines ramificanti, endospermio ruminato distincta. Typus: Madagascar, Marojejy, NW of Mandena, *Dransfield et al.* JD6755 (Holotypus K; isotypus TAN).

Dypsis marojejyi. Inflorescences (*Dransfield et al.* JD6755).

Dypsis marojejyi, a squat litter-trapping palm of mountain forest, Marojejy.

Squat solitary palm. **Trunk** 3–6 m tall, 20–30 cm diam., near the crown c. 20 cm diam.; stilt roots present near the base (always?); internodes c. 2 cm. **Leaves** 18–20 in the crown, the upper part of the trunk with marcescent leaves and sheath remnants, litter-accumulating; sheath 20 cm long, open, not forming a crownshaft, with dense rusty-brown pubescence, c. 40 cm wide; petiole 0–10 cm, densely reddish-pubescent, 3–3.5 × 2 cm diam.; rachis 3–4 m long, in mid-leaf 1.1–2.3 cm wide, reddish-tomentose; leaflets c. 60 on each side of the rachis, in groups of 3–6 and fanned within the groups, stiff, the group interval 4–9 cm, bright to dark green, the proximal 29–41 × 0.3–0.4 cm, median 45–70 × 2.5–5 cm (interval 0.5–2 cm), distal 15–35 × 0.7–2.7 cm, the apical pair joined for up to 3 cm and multi-fold (always?), main veins 3–5, with ramenta on the main veins, and with reddish scattered scales on the minor veins, apices unequally acute, the distal ones praemorse, young leaf liver-coloured. **Inflorescence** interfoliar, porrect with spreading branches, branched to 3 orders; peduncle c. 80 cm (estimate from a slide), distally 2.2–3 × 1.2–1.5 cm diam., green with dense reddish scales; prophyll 50–54 cm long, proximally bright crimson, distally brown, with scattered scales; peduncular bract 46–70 cm long, dark green with brown scales; rachis c. 110 cm long, green with dense brown stellate scales, the rachis bracts 8 mm long, with c. 21 branched and 12–17 unbranched first order branches, the proximal of these with a secondary rachis of 30–60 cm, proximally c. 1 × 0.4 cm diam., with up to 16 rachillae; rachillae yellow-green, 9–32 cm long, 1.5–3 mm diam., stellate-puberulous, with distant superficial or slightly sunken triads; rachilla bract 0.5–1 mm, rounded. **Staminate flowers** with sepals 1.5–2 × 1.2–2.3 mm, ciliolate; petals on a 0.5–0.6 mm high receptacle, 2.2–3.2 × 1.6–2.5 mm; stamens 6, biseriate (offset c. 0.3 mm, the antepetalous stamens inserted higher up and with slightly wider filaments), filaments 1.2–1.6 mm in

Dypsis marojejyi. **A** basal part of leaf rachis with lowermost leaflets × 1/3; **B** mid section of leaf × 1/3; **C** leaf tip × 1/3; **D** tip of peduncle and basal first order inflorescence branch × 1/3; **E** fruit × 1.2; **F** seed in cross section × 2. **A**, **C**, **D** from *Dransfield et al.* JD6755, **B**, **E**, **F** from *Miller & Lowry* 4185. Drawn by Rosemary Wise.

bud, cylindrical and fat (0.6 mm diam.), anthers 1.8–2 × 0.8–1 mm, versatile; pistillode c. 1.3 × 0.6–0.9 mm. **PISTILLATE FLOWERS** with sepals 1.3–2.5 × 1.7–3.5 mm; petals 2.7–4.6 × 2.5–5 mm; staminodes six, 0.2–0.6 mm; gynoecium (in bud) c. 2.5 × 0.8 mm. **FRUIT** pale yellow-green, ellipsoid to slightly obovoid, 22–25 × 14–18 mm, the apex rounded, pustulate; endocarp with densely anastomosing fibres. **SEED** slightly obovoid, 18–20 × 13–16 mm, with pointed base and rounded apex; endosperm ruminate, the intrusions dense, 2–5 mm deep.

NOTE. At first we thought this litter-accumulating palm belonged in a group with the other litter-collectors, such as *D. perrieri*. But the structure of the inflorescence indicates it is closer to taxa such as *D. madagascariensis, D. coursii* and *D. rivularis*.

SPECIMENS SEEN. Andapa: Marojejy W, Ambatoharanana valley to upper Antsahaberoka R, Nov./Dec. 1959 (y.fr.), *Humbert & Saboureau* 31702 (K, P); Marojejy E, NW of Mandena, Feb. 1989 (fr.), *Miller & Lowry* 4185 (K, MO); idem, N of Mandena, Oct. 1988 (bud), *Miller* 3509 (K, MO, P); idem, Nov. 1989 (fl., y.fr.), *Dransfield et al.* JD6755 (Holotype K; isotype TAN).

47. DYPSIS COMMERSONIANA

A little known species from the southern lowland rain forests. The name refers to the collector of the type, Philibert Commerson (1727–1773), who collected in Madagascar in 1770–1771.

DISTRIBUTION. SE Madagascar.
HABITAT. Lowland rain forest; c. 120 m.
LOCAL NAMES. Not recorded.
USES. Not recorded.
CONSERVATION STATUS. Critical. The distribution area is small, and under severe pressure by an expanding population. Nearly all lowland rain forest in the area has now been cleared.

Dypsis commersoniana *(Baill.) Beentje & J. Dransf.* **comb. nov.**

SYNONYMS:
Neophloga commersoniana Baill., Hist. Plantes 13: 372 (1895); Becc., Bot. Jahrb. Syst. 38, Beibl. 87: 22 (1906); Becc., Palme del Madagascar 20, fig. 13, t. 18 (1912); Jum., Cat. Pl. Madagascar, Palmae: 19 (1938); Jum. & H. Perrier, Fl. Madagascar 30: 90 (1945). Type: Madagascar, without locality, *Commerson* s.n. (Holotype P).

Neophloga pygmaea Pic.-Serm., Webbia 11: 149 (1956), **synon. nov.** See **NOTE.**

Clustering palm. **STEMS** 1.8–5 m tall. **LEAVES** irregularly pinnate; sheaths only known from their distal part, with rounded shoulders and a few scattered scales; petiole 5–21 cm long, 2.5–3 mm diam., flat adaxially, with dense minute reddish scales; rachis 21–38 cm long, in mid-leaf 2–2.5 mm wide, with dense to scattered scales; leaflets in groups of 2 or irregular, 4–7 on each side of the rachis (interval 2–9 cm), distally sigmoid, proximal 5–23 × 0.4–1.6 cm, median 11–28 × 1.3–2.5 cm, cuneate at the base, acuminate, the distal leaflets often praemorse-denticulate on the distal lower margin, glabrous, top pair forming a deeply lobed flabellum 15–33 cm long, connate for 6–11 cm, the lobes 12–23 × 3.4–5 cm, with denticulate-praemorse apices 2–4 cm wide and with the teeth continuing along the distal margin, and 6–7 main veins, leaflets glabrous except for the very base. **INFLORESCENCE** interfoliar, branched to 2 orders; peduncle c. 24 cm long outside the sheath, compressed, 2.5–6 mm wide distally, with rather dense minute reddish scales; prophyll c. 15 cm long outside the sheath, opening only in the distal 1–3 cm, with scattered scales; peduncular bract inserted at 12 cm above the sheath apex, c. 13 × 0.6 cm, with scattered scales, split over its length, deciduous; rachis 20–33 cm long, with 6–11 branched and 10–13 unbranched first order branches, the proximal with a rachis to 8 cm and up to 7 rachillae; rachis bracts up to 4 × 2.5 mm; rachillae 3–14 cm long, 0.5–1 mm diam., minutely puberulous; triads rather distant, superficial; rachilla bract concave, apiculate. **STAMINATE FLOWERS** with sepals 0.5–0.7 × 0.7–1 mm, keeled and gibbous, orbicular, rounded; petals 1.2–1.8 × 0.9–1.2 mm, elliptic, acute, striate; stamens 6, slightly biseriate (0.2 mm offset, the inner higher), the filaments 0.4–1 mm long, thin, the anthers 0.8–1.3 × 0.3–0.5 mm, dorsifixed, versatile, with parallel acute locules; ovary rudiment with wide base, distally subtrigonous-pyramidal, 0.6–0.8 × 0.2–0.3 mm. **PISTILLATE FLOWERS** with sepals 0.5–0.6 × 0.5–1 mm; petals 2–2.4 × 1.8–2.3 mm; staminodes 6, minute; gynoecium when young to 1 mm high. **FRUIT** only known when young, up to 9 × 3 mm, with rather pointed apex.

NOTE. Baillon (1894b) described only the genus *Neophloga*, without giving a specific name; he cites the specimen *Commerson* s.n., "palmula microcarpa caudice sesquipedala" from Madagascar, and his genus description leaves no doubt that he is describing a true member of the *Dypsidinae*. Baillon states that Martius saw this plant, and thought it identical to *Hyophorbe indica* Gaertn. Baillon also states that it is distinct from *Hyophorbe*, closer to, but different from, *Dypsis* and distinct from *Areca lutescens* of Bory (which was, again, a *Hyophorbe*). We have seen the Commerson type in Paris (which is annotated by both Baillon and Beccari), and we can confirm that the Baillon description was drawn up based on it.

In Baillon (1895) the genus description is repeated

Dypsis commersoniana. **A** distal part of leaf × 1/2; **B** part of inflorescence × 1/2; **C** detail of rachilla × 3. All from *Commerson* s.n. Drawn by Margaret Tebbs.

with a footnote mentioning the binomial *Neophloga commersoniana* H.Bn.; in its synonymy are mentioned *Hyophorbe commersoniana* Mart. and *H. indica* Mart.

The inclusion of these synonyms gave Pichi-Sermolli (1956) reason to believe that the name *Neophloga commersoniana* should be cited as "(Mart.) Baill." and that though the genus description was truly about *Neophloga*, the species mentioned was in reality a *Hyophorbe*. We believe this is erroneous. If Baillon based his genus description on the Commerson collection from Madagascar ("palmula microcarpa caudice sesquipedala"), then his naming this species *Neophloga commersoniana* in his 1895 work is logical. The fact that Baillon thought that *Hyophorbe commersoniana* of Martius was identical (which it was not), does not matter; the *Neophloga* description is based on true Dypsid material, and so is the name *Neophloga commersoniana* Baill.

Curiously enough, the species is not treated in Jumelle's 1929 revision of *Neophloga*; he mentions the species (p. 12) but fails to include it in his key, and there is no description; probably the lack of bracts in the type made Jumelle hesitate about it being a *Neophloga* as he saw the genus. It is, however, treated in the Flora (Jumelle & Perrier 1945), but it might have been included by Perrier, who edited this volume after Jumelle's death.

Close to *D. humbertii* (which might be the same as this species) and *D. scottiana*, which is distinct by much shorter rachillae.

SPECIMENS SEEN. Tolanaro: Manantenina–Soavola, Nov. 1971 (fl.), *Guillaumet* 3901 (K, P); Lakandava, Jan. 1990 (y.fr.), *Rabevohitra* 2208 (P). Madagascar, without locality, without date, (fl.), *Commerson* s.n. (Holotype P); also without locality or date (bud), *Goudot* s.n. (P).

48. DYPSIS HUMBERTII

This palm is probably identical to *D. commersoniana*, but more material needs to be collected in southern Madagascar to make certain. The name refers to the collector of the type (and many other excellent palm collections), later Director of the Musée National d'Histoire Naturelle in Paris, Henri Humbert (1887–1967).

DISTRIBUTION. SE Madagascar.
HABITAT. Forest; 60–300 m.
LOCAL NAMES. Not recorded.

Dypsis humbertii. **A** crown × 1/2; **B** open staminate flower × 8. All from *Humbert* 5817. Drawn by Margaret Tebbs.

USES. Not recorded.
CONSERVATION STATUS. Unknown. Not seen for over sixty years.

Dypsis humbertii *(Jum.) Beentje & J. Dransf.* **comb. nov.**

SYNONYM:
Neophloga humbertii Jum., Ann. Inst. Bot.-Géol. Colon. Marseille sér. 5, 1 (1): 20 (1933); Jum., Cat. Pl. Madagascar, Palmae: 21 (1938); Jum. & H. Perrier, Fl. Madagascar 30: 90 (1945). Type: Madagascar, Manantantely forest, *Humbert* 5817 (Holotype P).

Palm 2–3 m, very slender. **LEAVES** 4 or more in the crown; sheath c. 6.6 cm, with oblique opening, with minute auricles to 2 mm high, red-brown, with scattered scales; petiole 5–6 cm long, c. 2 mm diam., with scattered scales; rachis 17–18 cm long, c. 2 mm diam., with scattered scales; leaflets 2–5 on each side of the rachis, in 2 sub-opposite groups distant for 2.8–5 cm, proximal 8–9 × 0.7–1 cm, median 10–13 × 0.9–1.2 cm, the top pair forming a flabellum 12.5–14 × 9 cm, connate for 4.5–5.5 cm, the lobes 8–9 × 2.5–2.6 cm, proximal leaflets with 1 main vein, acuminate, distal leaflets with 6 main veins, the apices truncate and dentate, also the distal outer margins dentate, all leaflets glabrous or with a few scattered scales proximally. **INFLORESCENCE** c. 34 cm, branched to 2 orders; peduncle c. 19 cm long, distally 1.5–2 mm diam., with scattered scales; prophyll 13–16 × 0.3 cm, with scattered scales, open for the distal 5 cm or near the apex only; peduncular bract deciduous, inserted at 9 cm from the base of the peduncle; rachis c. 10 cm, with scattered scales, with c. 2 branched and 10 unbranched first order branches, the proximal with 2 rachillae each; rachillae 5.5–8 cm long, 1 mm diam., minutely puberulous but glabrescent, with distant superficial triads. **STAMINATE FLOWERS** with sepals 0.5–0.6 × 0.7–1 mm; petals 1.8–2.1 × 1.2–1.3 mm, striate; stamens 6, slightly biseriate (offset 0.25 mm), filaments 1–1.3 mm long, the base slightly triangular, the distal part thin, anthers 1–1.3 × 0.4–0.5 mm, dorsifixed, versatile, the locules parallel and acute; pistillode 0.8–0.9 × 0.3–0.4 mm. **PISTILLATE FLOWERS** unknown. **FRUIT** unknown.

NOTE. Very close to *D. commersoniana* and possibly not distinct.

SPECIMENS SEEN. Tolanaro: Manantantely forest, Sept. 1928 (fl.), *Humbert* 5817 (Holotype P).

49. DYPSIS SCOTTIANA

A very distinct species from the white sand forest of southern Madagascar, with elegant, slender inflorescences with short rachillae. This species would probably do very well as an ornamental; not only is it beautiful, but its habitat indicates it would tolerate fairly dry climates, probably along the lines of *D. lutescens*. The name refers to the collector of the type, George Francis Scott Elliot (1862–1934) who collected in Madagascar between 1888 and 1890.

DISTRIBUTION. SE Madagascar.
HABITAT. Forest on white sand, heath scrub on white sand, once found in lowland rain forest; 10–515 m.
LOCAL NAMES. *Raosy* (Antanosy).
USES. Not recorded.
CONSERVATION STATUS. Vulnerable. Distribution area small, in a specialized habitat which is under threat of mining operations and fire. Numbers are estimated at less than five hundred.

Dypsis scottiana *(Becc.) Beentje & J. Dransf.* **comb. nov.**

Dypsis scottiana, growing in coastal forest north of Tolanaro.

Synonyms:

Neophloga scottiana (Becc.) Becc., Bot. Jahrb. Syst. 38, Beibl. 87: 22 (1906); Becc., Palme del Madagascar 22, fig. 15, t. 20 (1912); Jum., Cat. Pl. Madagascar, Palmae: 22 (1938); Jum. & H. Perrier, Fl. Madagascar 30: 91 (1945).

Phloga scottiana Becc., J. Linn. Soc. 29: 61 (1891); Baill., Bull. Soc. Lin. Paris 150: 1195 (1894). Type: Madagascar, forest near Fort Dauphin (Tolanaro), *Scott Elliot* 2615 (Holotype K).

Neophloga affinis Becc., Bot. Jahrb. Syst. 38, Beibl. 87: 22 (1906); Becc., Palme del Madagascar 21 (1912); Jum., Cat. Pl. Madagascar, Palmae: 19 (1938); Jum. & H. Perrier, Fl. Madagascar 30: 91 (1945). Type: Madagascar, Fort Dauphin (Tolanaro), *Cloisel* 108 (Holotype P), **synon. nov**.

Clustering palm in tufts of 3–16. **Stems** 2–4 m tall, 0.6–2 cm diam.; internodes 1–5 cm, pale grey, nodal scars 2–3 mm, slightly stepped. **Leaves** 4–7 in the crown, porrect to spreading; sheath 8–31 cm long, 0.8–2 cm diam., 2/3 closed in outermost leaves, light brown with dense red scales, glabrescent but distally always with some patches of scales, with small triangular auricles up to 2 mm high; petiole 4–30 cm long, 0.2–0.5 × 0.2–0.3 mm diam., adaxially flat or slightly channelled, abaxially convex, with scattered scales, proximally often with a raised triangular continuation of the inner sheath lining; rachis 15–66 cm long, in mid-leaf 1–4 mm wide, with scattered scales on both surfaces; leaflets 11–27 on each side of the rachis, in groups of 2–8 (regular or nearly so in *Decary* 10729), interval between the groups 2.5–8 cm, interval between the leaflets 0.5–2 cm, the proximal 7–22 × 0.3–0.8 cm, median 8–24 × 0.6–2 cm, distal 6–15 × 0.4–2.7 cm, proximal and median with cuneate base, 1–3 main veins (prominent adaxially), and acuminate apex, glabrous or with a few scattered scales near the base and on the margins, the distal pair joined for 0.4–3.5 cm, with 3 main veins, narrow and dentate at the apex. **Inflorescence** interfoliar to infrafoliar,

Dypsis scottiana. **A** leaf and inflorescence × 1/2; **B** part of infructescence × 1/3; **C** fruit × 2.5; **D** fruit in cross section × 2. All from *Beentje & Andriampaniry* 4599. Drawn by Rosemary Wise.

erect to spreading to semi-pendulous, branched to (2) 3 (4) orders, 25–55 cm long; peduncle 12–25 cm long, proximally 3–6 mm diam. and flattened, distally 1.6–5 mm diam. and cylindrical, densely scaly but glabrescent; prophyll 14–26 cm long, borne at 2–3.5 cm above the base of the peduncle, up to 1.3 cm wide, pale brown with scattered scales, opening near the apex only; peduncular bract quickly deciduous, inserted at 7–14 cm from the base of the peduncle, 15–19 cm long, split in its distal half except for the distal 1–2 cm, with a narrow beak of 0.5–1 cm, pale brown with scattered scales; rarely with a small (1.5 mm) non-tubular peduncular bract in the distal part of the peduncle; rachis 7–40 cm long, scaly, with 7–27 branched and 0–10 unbranched first order branches, the proximal ones near their base 1.5–3 mm diam. and hardly flattened; rachillae 0.7–6.5 cm long, c. 1 mm diam., sparsely scaly, with distant superficial triads, with green to cream flowers. **STAMINATE FLOWERS** with sepals 0.5–0.8 × 0.8–1 mm, keeled, gibbous at the base, broadly ovate, rounded; petals on a 0.4 mm high receptacle, 1.4–1.8 × 1.4–1.6 mm, ovate to elliptic, acute, opening only slightly; stamens 6, at anthesis poking through the slits in the corolla, with the anthers held vertically, their openings upward, slightly 2-seriate, the antepetalous filaments with small swellings at the base and inserted slightly higher than the antesepalous ones, 1.2–1.5 mm long, thin, anthers 1.2–1.3 × 0.4–0.5 mm, dorsifixed, versatile, the locules parallel and obtuse; pistillode 0.7–0.8 mm high, ellipsoid, 0.5–0.6 mm diam. **PISTILLATE FLOWERS** with sepals 0.8–1.1 × 1.2–1.8 mm, broadly ovate, rounded, concave; petals proximally imbricate and broadly ovate, distally triangular, fleshy and acute, 2–2.3 × 1.6–2.2 mm; staminodes 6, 0.3–0.5 mm, dentiform, flat; gynoecium 2.4–2.6 × 2–2.2 mm, asymmetrically gibbous. **FRUIT** red, ellipsoid, 6–11 × 3.5–6.5 mm, with slightly pointed apex; mesocarp fleshy, c. 1 mm thick; endocarp fibrous with free fibres. **SEED** ellipsoid, 6.5–9 × 3–5 mm, obtuse to pointed at both ends; endosperm homogeneous. **EOPHYLL** bifid.

NOTE. *N. affinis* was described by Beccari based on an inflorescence collected by Cloisel (called Choisel by Beccari), which resembles that of *N. scottiana* closely. Beccari distinguished the species by the branching of the inflorescence: *N. scottiana* branched to two orders, *N. affinis* branched to three orders. However, the type of *N. scottiana* is clearly branched to three orders, and we hereby put *N. affinis* into the synonymy of *D. scottiana*.

Jumelle mentions the two species only briefly in his 1929 revision of *Neophloga*, since he did not believe the two taxa belonged to the genus. Not hampered by the fact that he had not seen the type of *N. scottiana*, he based his opinion on the description by Beccari (to whom he usually refers to as 'the Italian botanist'). In the Flora (Jumelle & Perrier 1945) the species are treated under *Neophloga*, with the brief remark that they are intermediate between *Neophloga* and *Chrysalidocarpus*.

A collection from lowland rain forest (*Beentje* 4599) matches the collections from white sand forest and heath scrub at much lower altitudes; the main differences are the higher number of leaflets (up to 27 on each side of the rachis, rather than the 11–19 of the white sand populations) and the slightly longer petiole. The inflorescence agrees perfectly, however, and we see no reason to distinguish the forms formally. A collection from Manombo Forest, much further north, looks similar to *D. scottiana*, but with the following differences: leaf sheath 6.5–9.5 cm long; **LEAFLETS** *7–11* on each side of the rachis; inflorescence branched to 2 orders; prophyll and peduncular bract slightly shorter; rachis with *1–5 branched* and 8–9 unbranched first order branches; **STAMINATE FLOWERS** with sepals 0.7–1.2 mm. Other characters overlap [Farafangana: Manombo, Jan. 1993 (fl.), *Beentje & Andriampaniry* 4782 (K, MO, P, TAN]. It is not the same, but is verging towards it!

SPECIMENS SEEN. Tolanaro: forest near Fort Dauphin (Tolanaro), May 1890 (fl.), *Scott Elliot* 2615 (Holotype K); between Pic St. Louis and the sea, 1928 (fl.), *Humbert* 5958 (P); 13 km N of Ezoambo, March 1992 (fr.), *Beentje & Andriampaniry* 4599 (K, MO, TAN); 24 km N of Tolanaro, April 1989 (ster.), *Rabevohitra et al.* 1932 (K, P); idem, March 1992 (fl.), *Beentje & Andriampaniry* 4607 (BH, K, MO, P, TAN) and (fl., fr.), *Beentje & Andriampaniry* 4608 (BH, K, MO, P, TAN); St. Luce forest, March 1992 (fl., fr.), *Phillipson et al.* 3961 (K); idem, Dec. 1992 (fl.), *Beentje* 4759 (K, MO, P, TAN); L. Mananivo–Andriamanga, March 1989 (fl.), *Dumetz et al.* 616 (K, P); NW Marokoky, March 1989 (y.fr.), *Dumetz et al.* 618 (K, P, TAN); Mandena, March 1989 (fl., fr.), *Dumetz et al.* 503 (MO); idem, (fl.), *Dumetz et al.* 527 (K, P); idem, April 1989 (bud, fr.), *Rabevohitra et al.* 1888 (K, P). Fort Dauphin (Tolanaro), without further locality, without date (fl.), *Cloisel* 108 (P, type of *N. affinis*); idem, Oct. 1932 (fl., fr.), *Decary* 10729 (K, P, TAN).

50. DYPSIS SINGULARIS

A handsome small palm, sadly on its way to extinction. The name indicates that it is a peculiar taxon (due to the didymous stamens on a palm of this habit), as well as the fact that it is known from a single collection from only one site. Nevertheless, it is a distinct species.

DISTRIBUTION. Only known from Manombo forest.
HABITAT. Lowland rain forest; slope base; c. 45 m.
LOCAL NAMES. Not recorded.

USES. Not recorded.

CONSERVATION STATUS. Critical. Single-site status, in a forest being gradually destroyed by fire, shifting cultivation and logging; numbers low, estimated at less than a hundred.

Dypsis singularis *Beentje* **sp. nov.**

Palma concinna foliolis paucis connatis inflorescentia in 2 ordines ramificanti *D. commersonianae* affinis sed antheris didymis distincta. Typus: Madagascar, Manombo, *Beentje* 4513 (Holotypus K; isotypus TAN).

Solitary palm to 6 m. **STEM** distally c. 1 cm diam; internodes distally c. 0.5 cm, nodal scars c. 1 mm. **LEAVES** c. 6 in the crown, porrect, pinnate; sheath 9–13 cm long, with scattered reddish scales which are rather dense distally, and auricles to 1 cm long; petiole 0–1.5 cm long, 3–7 × 2–3 mm diam., with scattered scales; rachis 40–59 cm long, in mid-leaf 2.5–4 mm wide, with tiny scattered scales; leaflets 3–7 on each side of the rachis, solitary or in groups of 2, group interval 6–16 cm, leaflet interval 0.5–0.8 cm, proximal 8–16 × 0.6–2.2 cm, median 16–25 × 1.5–5 cm, distal 14–19 × 3.5–6 cm, main veins 3–5 (but faint), up to 7 in the distal pair, with scattered scales on the major and minor veins, distal pair joined for 4–8 cm, dentate over a width of 0.5–1.7 cm. **INFLORESCENCE** interfoliar, branched to 2 orders; peduncle 51–61 cm long, 2.5–3 mm diam., glabrous; prophyll 29–37 cm long, 8–18 mm wide, borne at 4–10 cm above the base of the peduncle, open in the distal 1.5–4 cm with scattered scales; peduncular bract inserted at 18–20 cm from the base of the peduncle, 36–41 cm long, open in the distal 3.5–5 cm, with scattered scales; rachis 37–69 cm long, glabrous, with 13–23 branched and 15–23 unbranched first order branches, the proximal to 12.5 cm long and with up to 7 rachillae; rachillae 5–19 cm long, 1 mm diam., glabrous. **STAMINATE FLOWERS** in young bud with sepals 0.5–0.7 ×

Dypsis singularis. **A** crown with inflorescence × 1/5; **B** detail of rachilla × 4. All from *Beentje* 4513. Drawn by Margaret Tebbs.

Dypsis singularis, at Manombo (*Beentje* 4513).

0.5–0.8 mm; petals 0.8–1.3 × 0.7–0.8 mm; stamens 6, uni– or biseriate, didymous, filaments 0.3 mm long (and possibly connate for 0.5 mm?), anthers 0.2 × 0.3 mm; pistillode c. 0.3–0.4 × 0.4 mm. **PISTILLATE FLOWERS** with sepals 0.6–0.7 × 0.6–1 mm; petals 1.2–1.5 × 0.9–1.6 mm; staminodes invisible; pistil 1.5–1.6 mm high, c. 1.8–1.9 mm diam. **FRUIT** unknown.

NOTE. With its few broad, connate leaflets and inflorescence branched to 2 orders it resembles *D. commersoniana* and *D. humbertii* but is distinct in the didymous anthers.

SPECIMENS SEEN. Farafangana: Manombo, Nov. 1991 (bud), *Beentje* 4513 (Holotype K; isotype TAN).

51. DYPSIS INTERMEDIA

A strange little palm, only known from a single forest, with leaves resembling those of *D. commersoniana*, and the inflorescence that of *D. scottiana*; the name refers to this.

DISTRIBUTION. Only known from Manombo.
HABITAT. Lowland rain forest; gentle slope; 30–60 m.
LOCAL NAMES. Not recorded.
USES. Not recorded.
CONSERVATION STATUS. Critical. Only known from a single forest, which is disappearing steadily. Numbers are estimated at less than fifty.

Dypsis intermedia *Beentje* **sp. nov.**
Palma concinna foliolis paucis connatis inflorescentia in 2 ordines ramificanti *D. commersonianae* affinis sed antheris didymis distincta. Typus: Madagascar, Manombo, *Beentje & Andriampaniry* 4794 (Holotypus K; isotypus TAN).

Clustering palm in tufts of 5–7. **STEMS** 2–10 m tall, 1.5–3 cm diam.; internodes 6–12 cm, green, glabrous; one sterile stem seen branched once. **LEAVES** 7–8 in the crown, pinnate, spreading; sheath 12–15 cm long, with flaking patches of red-brown scales; petiole 14–30 cm long, 3–5 × 2–3 mm diam., with scattered scales; rachis 27–37 cm long, in mid-leaf c. 2–3 mm wide, with scattered scales; leaflets 3–7 on each side of the rachis, in groups of 2, group interval 7–20 cm, the proximal 9–21 × 0.7–2.8 cm, median 13–25 × 0.9–3.3 cm (interval 1–2 cm), with unequal, slightly truncate (proximal half of leaflet) and acuminate (distal half of leaflet) apices, distal 14–32 × 5.4–8 cm, connate for 9–16 cm, main veins 1–5 (to 9 in the distal pair), glabrous except for the very base, apices unequally acuminate, the lower margin of the tip with several teeth, distal pair dentate over a width of c. 4 cm, with teeth on the outside margins as well. **INFLORESCENCE** interfoliar, branched to 3 orders; peduncle 30–34 cm long, distally 4–6 × 3–4 mm diam., densely scaly; prophyll 24–28 cm long, 8–9 mm wide, borne at c. 6 cm above the base of the peduncle, opening in the distal 1 cm; peduncular bract not seen, inserted at c. 19 cm from the base of the peduncle; rachis 15–16 cm long, scaly, with c. 17 branched and 8 unbranched first order branches, the proximal of these with an axis of up to 7.5 cm and 5 branched and 10 unbranched second order branches; rachillae 0.5–2.5 cm long, c. 1 mm diam., minutely puberulous, with dense triads. **STAMINATE FLOWERS** with sepals 0.6–0.7 × 0.6–0.8 mm; petals in bud 1.3–1.5 × 0.8–1 mm; stamens 6, uniseriate, filaments 0.4–0.7 mm, anthers 0.7–0.8 × 0.3–0.4 mm, versatile; pistillode c. 0.3–0.4 × 0.3 mm. **PISTILLATE FLOWERS** unknown. **FRUIT** unknown.

NOTE. Close to both *D. commersoniana* and *D. scottiana*, but probably closer to the latter.

SPECIMENS SEEN. Farafangana: Manombo, Nov. 1991 (dead infl.), *Beentje* 4517 (K); idem, Jan. 1993 (bud), *Beentje & Andriampaniry* 4794 (K, TAN).

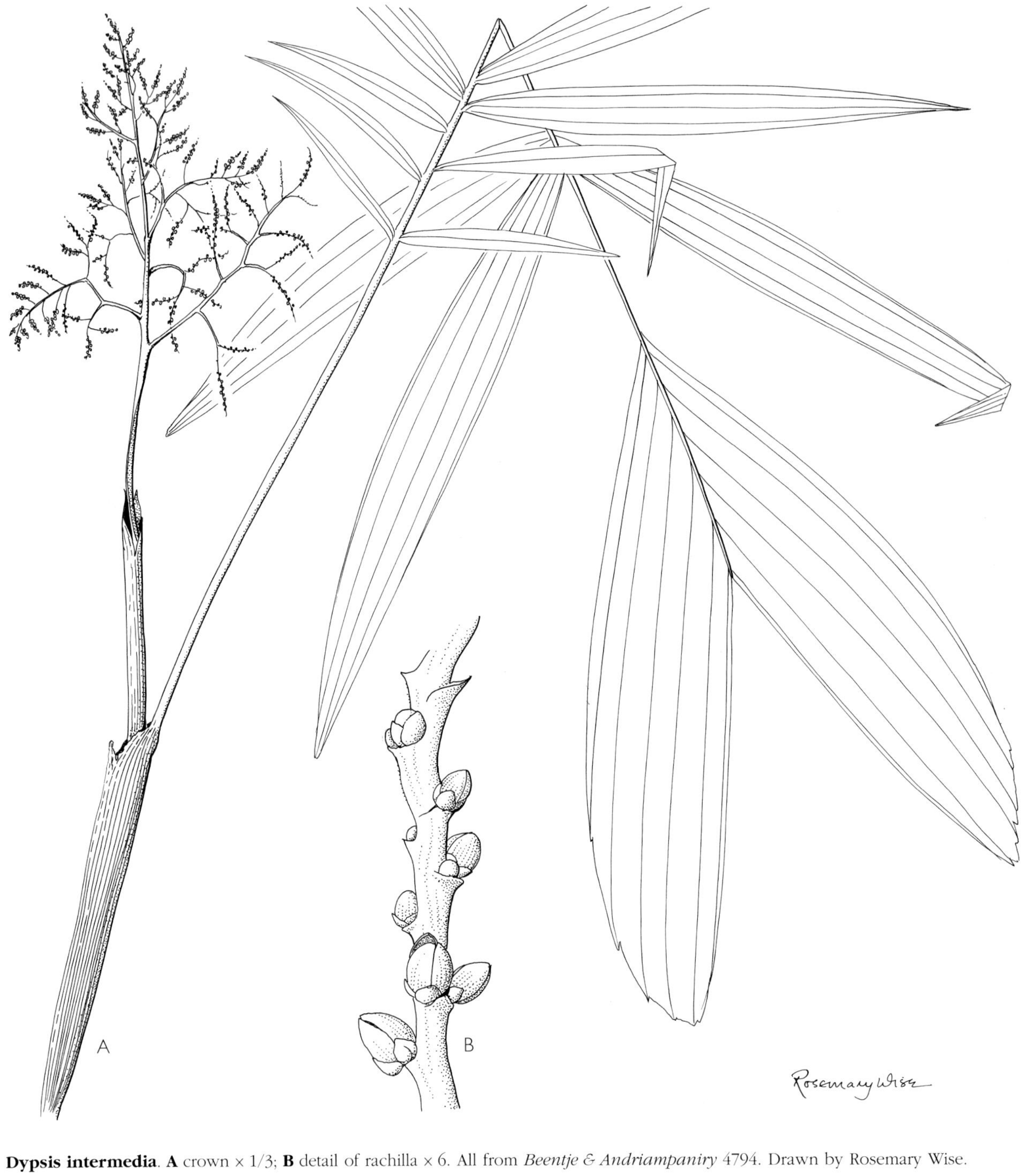

Dypsis intermedia. **A** crown × 1/3; **B** detail of rachilla × 6. All from *Beentje & Andriampaniry* 4794. Drawn by Rosemary Wise.

52. Dypsis mcdonaldiana

Named after the company that funded HB for four years of the project on Madagascar palms: McDonald's Restaurants (UK). This is in the tradition of Linnaeus, the father of Botany, who named taxa after his benefactors (and nasty weeds after his enemies, so we have to add that this is a beautiful palm!).

Distribution. Only known from the mountains of SE Madagascar.
Habitat. Lowland or submontane rain forest edge; slight midslope or wet valley bottom among boulders; 300–1100 m.
Local names. Not recorded.
Uses. Not recorded.
Conservation Status. Vulnerable. Restricted to a small area, which is only partially protected.

Dypsis mcdonaldiana *Beentje* **sp. nov.**
D. scottianae versosimiliter affinis sed inflorescentia robustiore rachillis longioribus distincta. Type: Madagascar, Andohahela, *Beentje & Andriampaniry* 4591 (Holotype K; isotypi MO, P, TAN).

Clustering or subcolonial palm in groups of 8–20. **Stems** 3–5 m, 1–3 cm diam.; internodes 1–5 cm, brown, more distally green. **Leaves** 7–10, spiral, flushed reddish when young; sheath 12–20 cm, pale green with distal red-brown pubescence, with tiny rounded ligules, the outer sheaths half open; petiole 5–10 cm long, 4–9 × 2–4 mm diam., abaxially with scattered scales, adaxially flat with sharp edges; rachis 37–90 cm, abaxially with scattered scales, in mid-leaf keeled, 3–5 mm diam.; leaflets 12–21 on each side of the rachis, regular or slightly irregular (in mid-leaf 2–4 cm apart), with top flabellum, the proximal 14–28 × 0.3–1.5 cm, median 21–43 × 1.6–2.4 cm, distal 0.8–29 × 0.8–2.9 (–7) cm, the distal pair joined for up to 7.5 cm and at the apex dentate over a width of 3–15 mm, main veins 1–3 (to 6 in the distal ones), apices attenuate, unequally bifid, no ramenta or scales. **Inflorescence** (inter- and) infra-foliar, 45–90 cm, branched to 2 orders; peduncle 18–33 cm, proximally 5–8 × 2–3 mm, distally 4–7 × 2–6 mm diam., with sparse red-brown scales; prophyll 16–27 × 0.9–1 cm, borne at 4–8 cm above the base of the peduncle, with scattered scales, split in the distal 1–4 cm only; peduncular bract deciduous, not seen; rachis 16–27 cm, glabrous or nearly so, with 4–6 (–16) branched and 11–13 unbranched first order branches, the proximal of these to 8 cm long with up to 8 rachillae; rachillae 4–20 cm, 1–1.5 mm diam., puberulous to glabrous. **Staminate flowers** unknown. **Pistillate flowers** with sepals 0.8–1 × 1–1.2 cm; petals 1.6–2.5 × 1.5–1.8 cm; staminodes not seen; ovary subglobose, c. 1.5 mm high, c. 1.3 mm diam., with a trifid stigma. **Fruit** red, 6–9 × 3–4 mm; remnant sepals 1 × 1 mm, petals 2.2–2.4 × 1.8 mm, with lower part imbricate and minutely ciliate, upper part valvate and triangular; endocarp fibrous, fibres not anastomosing. **Seed** with homogenous endosperm. **Eophyll** bifid.

Note. Probably closest to *D. scottiana*, but with more robust inflorescences with longer rachillae.

The (Guillaumet) specimen from the highest ele-

Dypsis mcdonaldiana. Crown and infructescence (*Beentje & Andriampaniry* 4591).

vation has much shorter and scaly rachillae while the lowland ones have longer glabrous ones.

Close to this species is *Bosser* 18599 from Vondrozo: W of Vondrozo, alt. 700 m, Dec. 1963 (bud); the sheath is longer (29 cm), the petiole shorter (5 mm), the rachis is 28 cm with only 6 leaflets on each side; the peduncle is 68 cm, but the rest of the inflorescence is similar.

SPECIMENS SEEN. Tolanaro: E slopes of N part of Anosy chain, Nov. 1971 (old infl.), *Guillaumet* 3980 (P); Andohahela, col Tanatana, Dec. 1989 (y.fr.), *Dransfield et al.* JD6782 (K, TAN); 13km N of Ezoambo, March 1992 (fr.), *Beentje & Andriampaniry* 4591 (K, MO, P, TAN, type).

53. DYPSIS JUMELLEANA

One of the commonest palms of the well-known forest at Analamazaotra, but very rare elsewhere. The name refers to Henri Lucien Jumelle (1866–1935), who collaborated with Perrier de la Bâthie on the publication of the palm flora of Madagascar and who described many species.

DISTRIBUTION. Known from Analamazaotra, and from old collections in Zahamena and Anjozorobe.

HABITAT. Submontane rain forest on gentle to steep slopes; 800–1300 m.

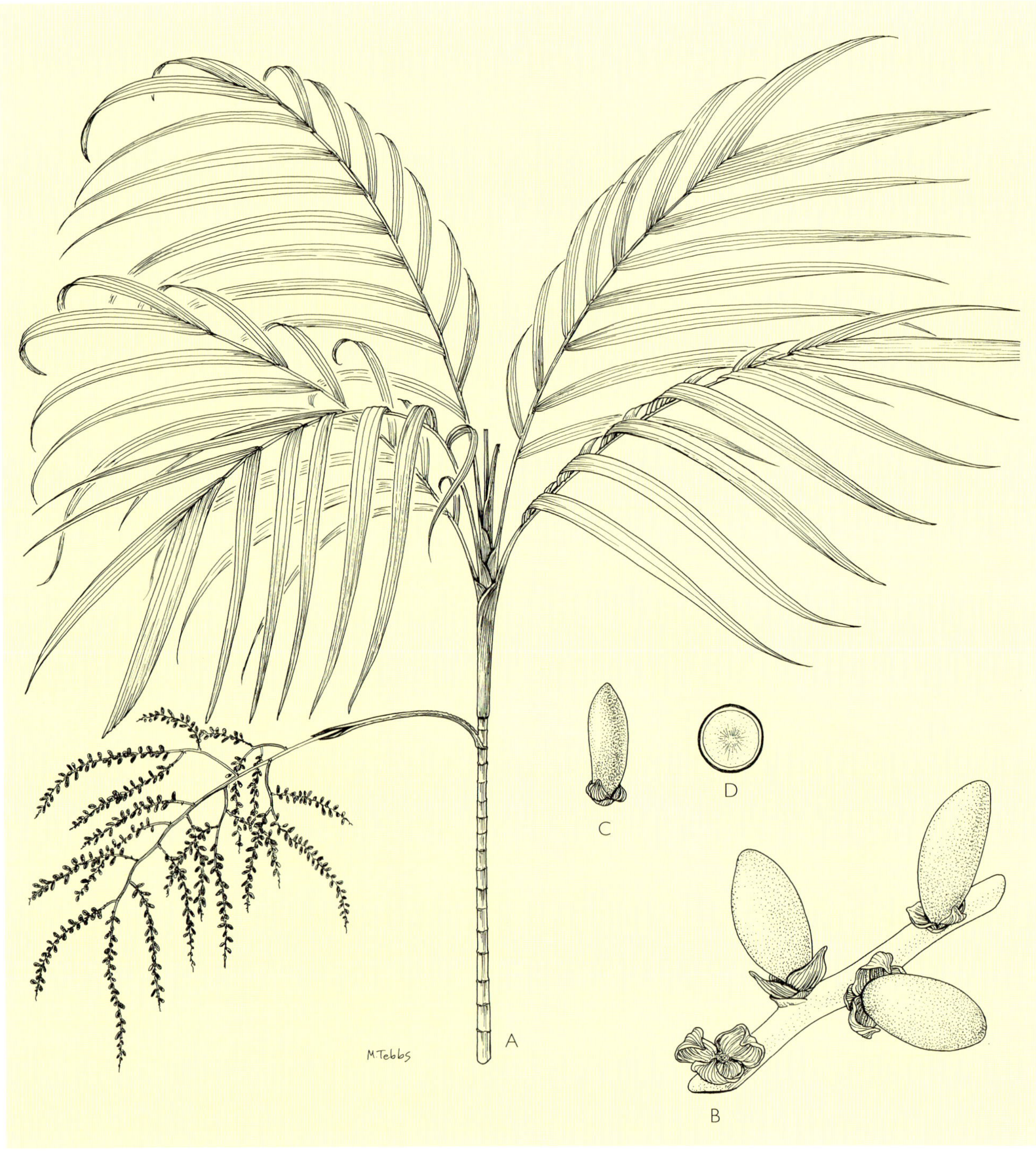

Dypsis mcdonaldiana. **A** crown × 1/8; **B** detail of fruiting rachilla × 3; **C** fruit × 2.5; **D** fruit in cross section × 3. All from *Beentje & Andriampaniry* 4591. Drawn by Margaret Tebbs.

Dypsis jumelleana, in rather open forest at Analamazoatra.

Local names. Not recorded.
Uses. Not recorded.
Conservation Status. Vulnerable. Only known from a single recent site, which is protected.

Dypsis jumelleana *Beentje & J. Dransf.* **nomen nov.**
[basionym of the specific epithet was predated in *Chrysalidocarpus lanceolatus* Becc.]

Synonym:
Neophloga lanceolata Jum., Ann. Inst. Bot.-Géol. Colon. Marseille sér. 4, 6 (3): 45 (1929); Jum., Cat. Pl. Madagascar, Palmae: 21 (1938); Jum., Fl. Madagascar 30: 88, fig. 25 (1945), **synon. nov**. Syntypes: Madagascar, Ambatovola, *Perrier* 11985 (syntype, not seen); Analamazaotra, *Perrier* 11986 (syntype, not seen), 11987 (syntype, not seen), 11988 (Lectotype P, chosen here); excluding *Perrier* 18043 (see *D. linearis*).

Clustering palm in tufts of 4–6. **Stems** 2–4 m high, 0.5–1.5 cm diam.; internodes 6–13 cm long, near the crown 1–4 cm, green with reddish scales, later turning brown-grey. **Leaves** 3–8, spirally inserted, porrect; sheath 8–20 cm long, pale green with scattered scales, closed, sometimes with a black pre-split line edged in 4–5 mm long black fibres, without auricles or (rarely) with triangular auricles to 2.5 mm high; petiole 1.5–15 cm long, 1.5–4 × 2–2.5 mm diam., flat to slightly channelled, with scattered scales; rachis 24–56 cm long, in mid-leaf 2–3 mm wide, with scattered scales; leaflets 10–18 on each side of the rachis, in irregular groups of 2–4 (group interval 2–12 cm) or just irregular, in one plane (not fanned within the groups), folded, the proximal 8–23 × 0.2–0.7 cm, median 14–27 × 0.5–2.2 cm, distal 10–18 × 1.3–2 cm, the distal pair joined for 1–2 cm and with 2–8 mm wide dentate apices, other leaflets acuminate and with 1 main midrib and 2 fainter veins, distal pair with 3 main veins, all leaflets with scattered minute reddish scales on the main and minor veins. **Inflorescence** interfoliar, pendulous, branched to 1 order; peduncle 22–43 cm long, distally 2–3 mm diam., green with scattered scales but glabrescent; prophyll 14–31 cm long, 5–8 mm wide, borne at 1–13 cm above the base of the peduncle, pale brown with scattered scales, opening at the apex for 1–2.5 cm; peduncular bract inserted at 13–29 cm from the base of the peduncle, 13–14 cm long but rapidly deciduous, seen once and that time opening in the distal 5–6 cm; second peduncular bract seen once, inserted some 6 cm below the apex of the peduncle, c. 3 cm long; rachis absent or up to 6.5 cm long, glabrous, with 2–10 rachillae; rachillae 13–30 cm long, 1–1.5 mm diam., with spaced triads. **Staminate flowers** custard yellow, with sepals 0.7–1 × 0.6–1 mm; receptacle 0.5–0.6 mm high; petals 1.2–1.7 × 0.7–1.3 mm; stamens 6, equal, the filaments 0.3–0.8 mm long, triangular with a 0.3 mm wide base, the anthers 0.4–0.8 × 0.4–0.5 mm, dorsifixed and versatile; pistillode 0.2–0.4 × 0.2–0.3 mm. **Pistillate flowers** with sepals 0.5–1 × 0.6–1.4, unequal with one markedly asymmetrical; petals 1.5–2 × 2–2.2 mm; staminodes 0.2–0.3 mm; gynoecium 1.2–1.5 × 0.8–1.4 mm on a 0.2 mm stalk. **Fruit** red, subglobose, 9–12 mm diam., fleshy, with fibrous endocarp with free fibres. **Seed** 6.5–9 × 6–8 mm, with homogeneous endosperm.

Note. We have found two of the syntypes, *Perrier* 11988 from Analamazaotra, and *Perrier* 18043 from the lower Mangoro R. These are not the same taxon. We have chosen *Perrier* 11988 as the lectotype; the other number is not included in the description (see *D. linearis*). The protologue has characters drawn from both these sheets.

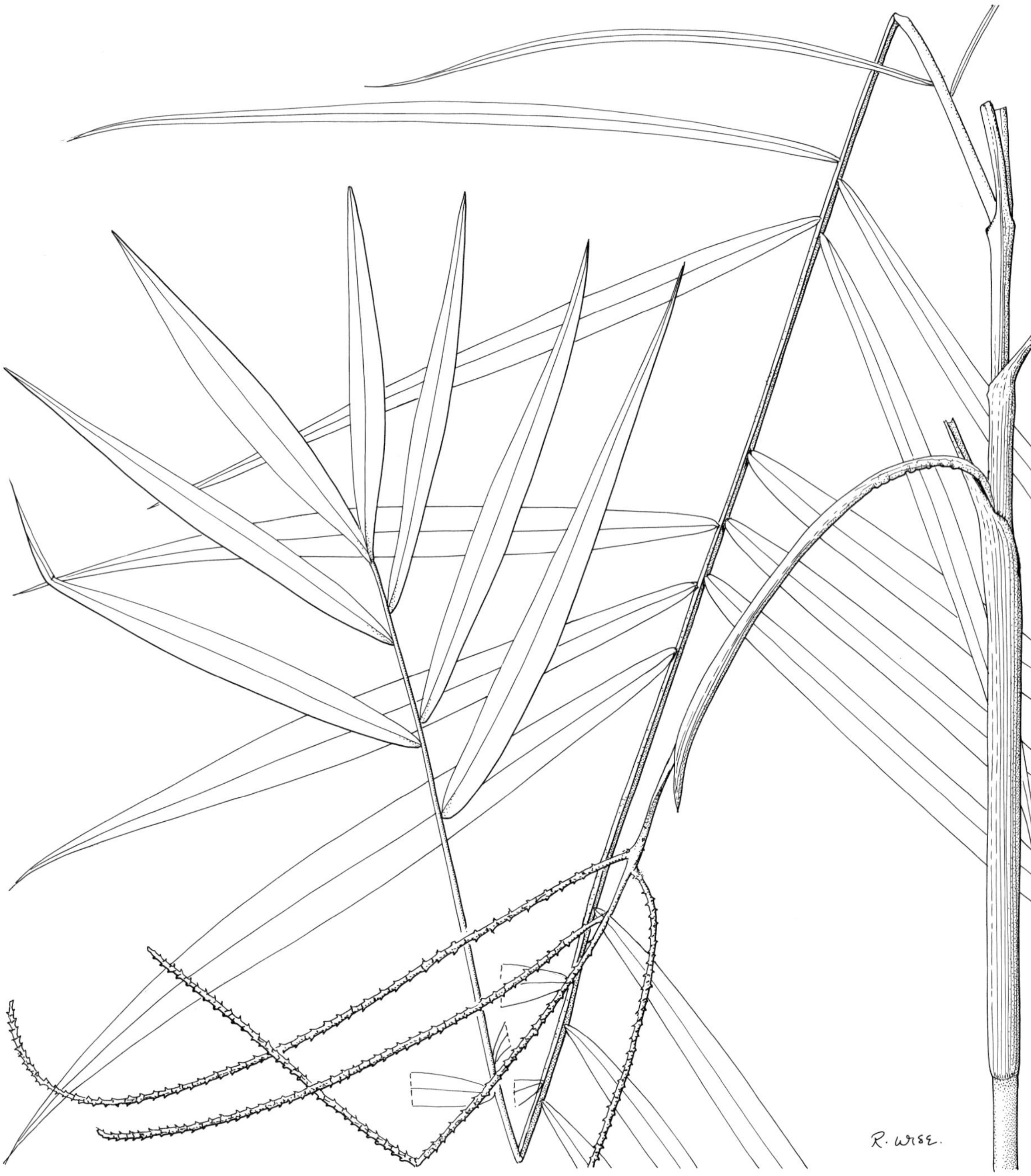

Dypsis jumelleana. Crown, showing inflorescence × 2/5. From *Dransfield et al.* JD6490. Drawn by Rosemary Wise.

A fairly nondescript species, whose relationships are probably with *D. linearis*.

Humbert & Cours 17908 differs slightly from the other collections in the narrower leaflets which are grouped much more tightly within the groups, but this character is fairly variable within this taxon.

Probably the same is: Mananara Avaratra, Andravolasoa, April 1992 (old infl.), *Beentje et al.* 4640 (K, TAN) from rain forest at 330 m; a solitary palm to 6 m, with 22 leaflets on each side of the rachis, of which the proximal are 6.5–11 cm long, and glabrous or nearly so; the inflorescence rachis is 15 cm long, with 15 rachillae. In other respects it conforms to the above description, and has the look of being the same taxon, but in some ways it is also close to *D. corniculata*.

SPECIMENS SEEN. Ambatondrazaka: Zahamena, Andrangovalo Massif, Oct. 1937 (fl.), *Humbert & Cours* 17908 (P); Ambatoharanana near Antsevabe, March 1951 (old infl.), *Cours* 4058 (P, TAN). Anjozorobe: N of Anjozorobe, Nov. 1922 (bud), *Perrier* 14969 (P). Moramanga: Analamazaotra, July 1913 (fl.), *Perrier* 11988 (P, lectotype); idem, Feb. 1926 (fr.), *Perrier* 15986 (P); idem, Dec. 1932 (fl.), *Leandri* 721 (K, P); idem, Nov. 1986 (fl.), *Dransfield et al.* JD6412 (K, TAN) and (fl.) JD6429 (K, TAN); idem, March 1988 (fl.), *Dransfield et al.* JD6490 (K, TAN); idem, March 1991 (y.fr.), *Beentje & Rabarilala* 4411 (K, TAN) and (fr.) 4418 (K, TAN); idem, Aug. 1992 (old infl.), *Beentje & Beentje* 4744 (K).

54. DYPSIS LINEARIS

A small to medium-sized, fairly nondescript palm, which can, at times, look very much like *Dypsis nodifera*. The specific name indicates the narrowness of the leaflets of the type, but this is a variable character. Some forms, particularly those on poor soils on quartzite ridges, can be very attractive and would make fine ornamentals; it is not known whether this species is currently in cultivation.

DISTRIBUTION. Quite widespread on the middle altitudes of the eastern escarpment between Zahamena and Andohahela.

HABITAT. Bamboo-rich montane forest on steep slopes, secondary bush on white gritty sand; hillcrest or mid-slope; 250–1450 m.

LOCAL NAMES. *Tsirikabidy*, *Sinkara* (Moramanga). *Ambolo* (fide Perrier), which means "to be smelled". *Ovana* (Tanala, fide Beentje, same name as for *Dypsis nodifera*), meaning to be changed (which we find funny, since we changed our minds several times about the status of this taxon); *Sirahazo* (Tanala, fide Beentje).

USES. Not recorded.

CONSERVATION STATUS. Not threatened. Widespread, with several populations in protected areas.

Dypsis linearis *(Becc.) Beentje & J. Dransf.* **comb. nov.**

SYNONYMS:

Neophloga linearis Becc., Bot. Jahrb. Syst. 38, Beibl. 87: 26 (1906); Becc., Palme del Madagascar 30, fig. 23, t. 29 (1912); Jum., Ann. Inst. Bot.-Géol. Colon. Marseille sér. 4, 6 (3): 43 (1929); Jum., Cat. Pl. Madagascar, Palmae: 21 (1938); Jum. & H. Perrier, Fl. Madagascar 30: 86 (1945). Lectotype (chosen here): Madagascar, Ambohimitombo forest, *Forsyth Major* 606 (K).

Neophloga procumbens Jum. & H. Perrier, Ann. Inst. Bot.-Géol. Colon. Marseille sér. 3, 1 (1): 27, fig. 8 (1913); Jum., Cat. Pl. Madagascar, Palmae: 22 (1938); Jum., Fl. Madagascar 30: 82, fig. 22, 1-3 (1945), **synon. nov**. Type: none indicated, but locality E flank Andringitra Massif, R. Ihovika, 1000 m (probably *Perrier* 11977 which has the same data, holotype P).

Chrysalidocarpus ambolo Jum., Ann. Bot.-Géol. Colon. Marseille sér. 4, 6 (3): 8 (1929); Jum., Cat. Pl. Madagascar, Palmae: 8 (1938), **synon. nov.** This was a nomen nov. because in *Chrysalidocarpus* there was already a *C. mananjarensis*.

Neophloga mananjarensis Jum. & H. Perrier, Ann. Inst. Bot.-Géol. Colon. Marseille sér. 3, 1 (1): 26, pl. 12 (1913); Jum. & H. Perrier, Fl. Madagascar 30: 81 (1945). Type: not indicated in the protologue, but the locality mentioned there conforms to the single specimen mentioned in the Flora: Madagascar, Mt. Vatovavy, *Perrier* 12071 (Holotype P).

Dypsis linearis, form with broad leaflets, growing on a quartzite ridge near Ifanadiana (*Beentje & Andriampaniry* 4731)

Dypsis linearis, form with rather narrow leaflets growing in submontane forest, Andohahela (*Dransfield et al.* JD6777).

Clustering in tufts of up to 4–40 stems or apparently solitary palm. **STEM(S)** 1–7 m high, sometimes leaning on other vegetation, 0.9–4 cm diam., internodes 2–15 cm, pale green, with dense red-brown to purple scales, turning glabrous and grey with age, nodal scars c. 5 mm; sometimes aerial roots present at some 25 cm from the ground, 6 mm across; crownshaft 13–16 cm long, 1.3–2 cm diam., yellow to pale green. **LEAVES** 4–8 in the crown, pinnate, porrect, and sometimes marcescent leaves present; sheath 13–35 cm long, pale green with dense red-brown scales, auricles absent or up to 1 cm, fringed with red scales; petiole absent or up to 16 cm long, 2–6 mm diam., pubescent-scaly, adaxially flat; rachis 22–70 cm long, in mid-leaf 1.5–3 mm wide, pubescent-scaly; leaflets 11–23 on each side of the rachis, in groups of 2–5, fanned within the groups, occasionally with slightly swollen nodes at the base, group interval 3–13 cm, proximal 3.5–22 × 0.1–1.3 cm, median 7–26 × 0.3–3 cm, distal 7–18 × 0.2–2.7 cm, main veins 1–3, apices attenuate, proximally scaly, with few minute scattered scales, to glabrous, with prominent scales on the distal margins, distal pair joined for 0.2–2.5 cm, with 1 main vein, with dentate apices 2–6 mm wide. **INFLORESCENCE** interfoliar to infrafoliar, branched to 1 (–2) orders, with orange axes, arching; peduncle 13–53 cm long, distally 2–7 mm diam., pubescent; prophyll 7–40 cm long, 7–10 mm wide, borne at 1–11 cm above the base of the peduncle, open for the distal 1–7 cm, patchily pubescent; peduncular bract inserted at 5–45 cm from the base of the peduncle, circumscissile and rapidly deciduous, 9–24 cm long, with scattered scales, with a beak to 3 cm long; non-tubular peduncular bract 0.2–3.8 cm long; rachis 1.5–17 cm long, densely pubescent, orange, with 0 (–3) branched and 4–12 unbranched first order branches; rachillae 5–24 cm long, 1.5–5 mm diam., with dense to scattered scales (rarely glabrous) and distant triads. **STAMINATE FLOWERS** with sepals 1.2–2 × 0.7–2.3 mm; petals orange, connate for 0.5–0.8 mm, free lobes 1.5–2.8 × 1–2.2 mm; stamens 6, uniseriate or rarely slightly biseriate (offset 0.1–0.3 mm), filaments 0.4–1.2

Dypsis linearis. **A** distal part of stem and leaf sheaths × 2/3; **B** leaf and infructescence × 2/3; **C** base of infructescence × 2/3; **D** fruit × 3; **E** fruit in cross section × 3. All from *Beentje & Andriampaniry* 4731. Drawn by Rosemary Wise.

mm, anthers 0.8–1.8 × 0.4–1 mm, versatile; pistillode 0.6–1.3 × 0.2–0.7 mm, conical with obtuse apex. **PISTILLATE FLOWERS** with sepals 1–2 × 1.3–3.3 mm; petals 2.5–3.6 × 2.3–3.5 mm; staminodes 0.2–0.6 mm; gynoecium 2.8–4 × 2–3.2 mm. **FRUIT** yellow to red, 6–9 × 4–6 mm. **SEED** 5–6.5 × 3–3.3 mm, with homogeneous endosperm.

NOTE. The protologue of *D. linearis* incorporates characters of both syntypes (*Forsyth Major* 604 and 606) but as the floral characters derive clearly from *Forsyth Major* 606, we choose that as the holotype. *Forsyth Major* 604 is *D. heterophylla*.

We have decided that the differences between *C. ambolo, N. procumbens* and *D. linearis* are spurious, and we hereby put *C. ambolo* and *N. procumbens* in the synonymy of *D. linearis*, which is the oldest name.

The type of *C. ambolo* has up to six petals, the innermost much smaller than the outer; there may be up to eight stamens. The peduncular bract of *C. ambolo* was considered to be that of a *Chrysalidocarpus*, and therefore Jumelle felt in 1928 this was a *Chrysalidocarpus*, but the specific epithet was already in use in that genus.

C. ambolo caused Jumelle problems. He had drawn up a list of differences between the genera *Neophloga* and *Chrysalidocarpus* based on the peduncular bract (deciduous, beaked and splitting all over from the apex downwards in *Chrysalidocarpus*, remaining and splitting only at or near the apex in *Neophloga*), the leaf sheath (membranous in *Neophloga*, coriaceous in *Chrysalidocarpus*) and the habit, plus a few more characters of the "sometimes, but not always" kind. This taxon, first described by him in *Neophloga*, then had to be put in *Chrysalidocarpus* because of the peduncular bract, which was beaked; a problem was that the leaf sheath was of the *Neophloga* kind (membranous). In 1945 it had, without explanation, been returned to *Neophloga* with the note that it had all the characters of that genus, but the peduncular bract of a *Chrysalidocarpus*.

N. procumbens was mentioned by Jumelle (1929), but not treated; Jumelle wondered whether this was a *Neophloga* or a *Chrysalidocarpus*, because of the lack of a peduncular bract on the older inflorescence he saw.

N. linearis var. *distachya* is *D. heterophylla*.

Moramanga: Ambatovola, 1912 (fl.), *Perrier* 11994 (P) is probably this species, but has regular leaflets. It is described by Perrier as clustering and 1–2.5 m high, but has the typical yellow-orange inflorescence axes. One inflorescence has a bifurcate proximal rachilla.

SPECIMENS SEEN. Ambatondrazaka: Bemainty to Androndramanitra, March 1951 (fr.), *Cours* 4234 (P); Anonokambo (Nonokambo), Jan. 1967 (bud), *Bogner* 186 (K). Moramanga: Mantady, Dec. 1991 (bud), *Beentje & Andriampaniry* 4538 (K, MO, TAN); idem, Dec. 1992 (fl.), *Beentje & Andriampaniry* 4777 (K, TAN); Ambalafary, Anivomaro, Feb. 1968 (bud), *Service Forestier* SF 26668 (P). Ambatondrazaka: Manakambahiny, Oct. 1958 (bud), *Rakotovao* RN 9600 (K, P). Ambositra: Ambohimitombo forest, Jan. 1895 (fl.), *Forsyth Major* 606 (K, lectotype). Ampasimanolotra/Nosy Varika: Sahalampona-Ampita, Jan. 1945 (bud), *Cours* 2437 (K, P, TAN). Mahanoro: Bas-Mangoro, Oct. 1927 (bud), *Perrier* 18043 (P). Ifanadiana: Ambohirafia, March 1985 (fr.), *Dorr* 3891, 3892 (K); Ranomafana, March 1991 (fl., fr.), *Beentje* 4424 (K, TAN) and 4426 (K); idem, July 1992 (bud), *H. Beentje & J. Beentje* 4735 (K); idem, trail to Maharira, March 1992 (y.fr.), *Malcomber & Rakoto* 1333 (K, P); Ambinanindrano, July 1992 (fl., fr.), *Beentje & Andriampaniry* 4731 (BH, K, MO, P, TAN); idem, Jan. 1993 (fl.), *Beentje & Andriampaniry* 4797 (K, MO, TAN); 3 hours walk E of Tsaratanana, March 1991 (fl., y.fr.), *Beentje* 4436 (K, TAN). Mananjary: Mt Vatovavy, Oct. 1911 (fl.), *Perrier* 12071 (P, type of *C. ambolo/N. mananjarensis*). Ambalavao: Ambatomboay, Oct. 1954 (bud), *Rakotovao* 574 (P); E slopes of Andringitra Mts, near Ihovika R., Sept. 1911 (bud), *Perrier* 11977 (P, type of *N. procumbens*). Ambalavao:Betroka/Iakora: Mt. Kalambatitra, Nov. 1933 (fl.), *Humbert* 11859 (P). Iakora: between Kalambatitra Col and the Manambolo R., Nov. 1933 (fl.), *Humbert* 12109 (P). Tolanaro: Akaramy R. upstream from Mahamavo, Jan./Feb. 1934 (fl.), *Humbert* 13904 (P); Andohahela, col Tanatana, Dec. 1989 (fl.), *Dransfield et al.* JD6777 (K, TAN).
WITHOUT ANY LOCALITY: no date (before 1946) (bud), *Homolle* 2437 (P)–probably the Cours collection with a wrong label.

55. DYPSIS BONSAI

A beautiful little palm, with a name which is Japanese for 'dwarf tree' or 'dwarfed tree'; we believe this is a wind-dwarfed taxon, and it is one of the most beautiful of the smaller Madagascar palms; of course, the taxon has no connection with Japan, but it looks like a bonsai tree.

DISTRIBUTION. Marojejy area and Masoala Peninsula; possibly Zahamena.
HABITAT. Low forest or ericoid vegetation on ridge crests; 1000–1700 m.
LOCAL NAMES. Not recorded.
USES. Not recorded.

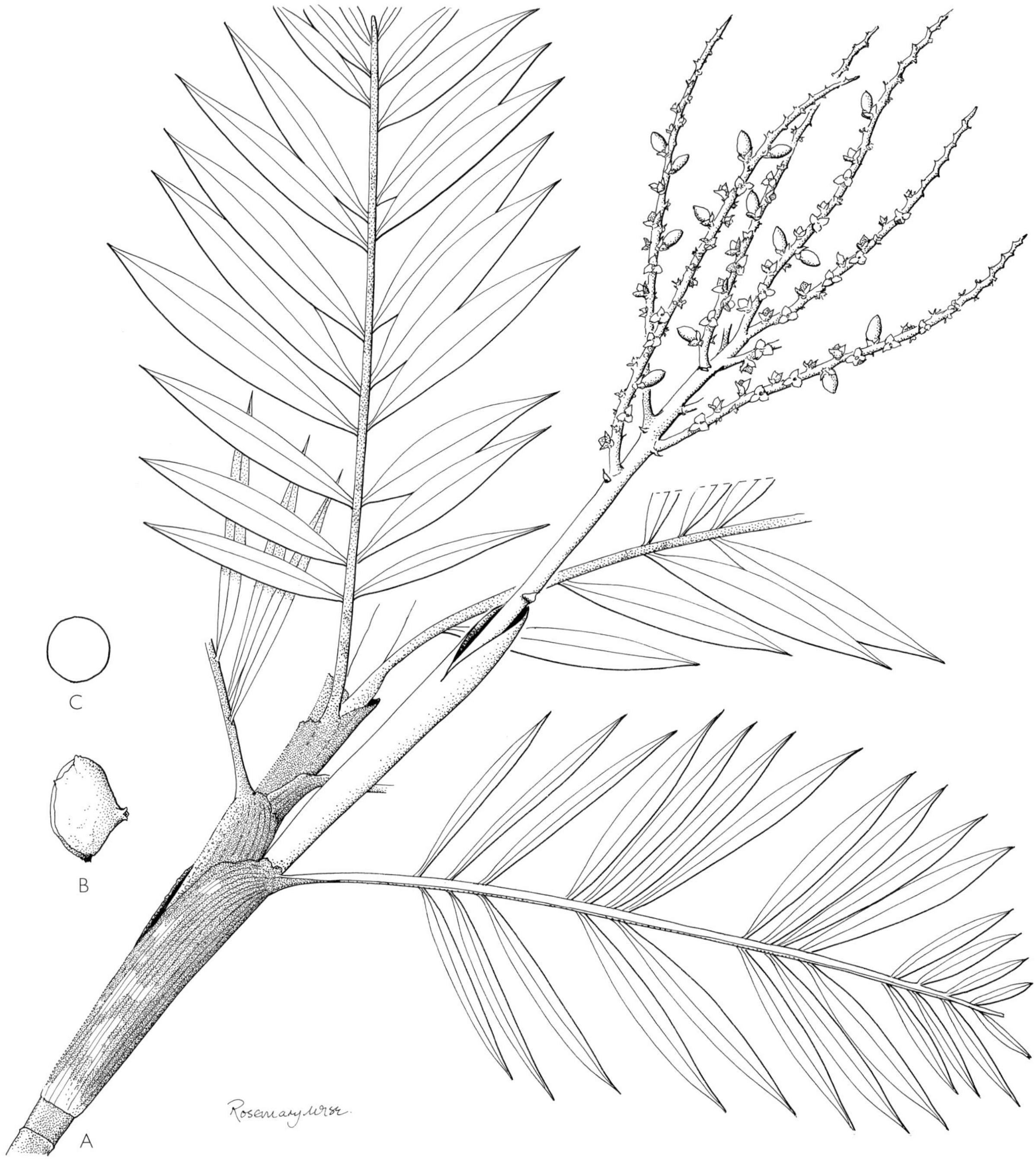

Dypsis bonsai. **A** crown and infructescence × 2/5; **B** fruit × 2; **C** seed in section × 2. All from *Zjhra & Hutcheon* 203.
Drawn by Rosemary Wise.

CONSERVATION STATUS. Vulnerable. Occurs in a fire-susceptible habitat, over a limited area. Numbers presumably low (possibly fewer than 300).

Dypsis bonsai *Beentje* **sp. nov.**

A ceteris speciebus *Dypsidis* solitariis foliolis parvis fasciculatis inflorescentia breve erecta aliquanto robusta axibus aurantiacus distincta. Typus: Madagascar, Ambanizana, Andohitsitondroina Pk, *Schatz & Modeste* 2897 (Holotypus K; isotypi MO, P, TAN).

Solitary palm. **STEMS** 1–2 m tall; distal internodes 0.7–1.3 cm, 6–8 mm diam, reddish pubescent; nodal scars 2 mm. **LEAVES** 4 in the crown (always?), pinnate; sheath 6.5–9 cm long, the outer open, densely reddish pubescent, with auricles 3 mm; petiole 1–4 cm long, 2–2.5 mm wide, densely to sparsely pubescent; rachis 10–18 cm long, in mid-leaf c. 2 mm wide, densely to sparsely pubescent; leaflets 10–14 on each side of the rachis, in groups of 2–5, the group interval 1.5–5 cm, the proximal 4–9 × 0.2–1 cm, median 5.5–11 × 0.6–2.2 cm, distal 3–5 × 0.2–1.2 cm, main veins faint, 1 (–5), with scattered scales or almost glabrous, but distally on the margin with some longer scales, apices acuminate, unequal, distal pair joined for 0.5–1.2 cm, narrowly dentate. **INFLORESCENCE** interfoliar, branched to 1–2 orders with orange axes; peduncle (13–) 22–28 cm long, 3–6 mm diam., densely pubescent; prophyll 13–20 cm long, 8–11 mm wide, with scattered scales, open in the distal few cm; peduncular bract inserted at c. 15 cm from the base of the peduncle, c. 11 cm long, open for two-thirds, with a 3 mm long beak, quickly deciduous; non-tubular peduncular bract 2–6 × 5 mm; rachis 3–5.5 cm long, puberulous, with up to 3 (but usually without) forked first order branches, and 5–13 unbranched first order branches; rachillae 4–12 cm long, 1.5–2.5 mm diam., densely pubescent or puberulous, or with sparse scales. **STAMINATE FLOWERS** with sepals orange, 1.3–1.8 x 1.2–1.7 mm; petals orange, 2–2.6 x 1.3–1.5 mm (on an up to 0.6 mm high receptacle); stamens 6, white, uniseriate, the filaments 0.6–0.9 mm long and thin, the anthers 1.2–1.3 x 0.5–0.7 mm, dorsifixed with parallel locules; pistillode 0.7–1 x 0.3 mm, conical. **PISTILLATE FLOWERS** with sepals 1–2 x 1.3–2.2 mm; petals 3–3.4 x 3–3.6 mm; staminodes 6, 0.2–1 mm; pistil c. 2.5 x 1.9 mm. **FRUIT** only seen young, then golden yellow and c. 8.5 x 4 mm, with subaequatorial stigmatic remains. **SEED** 8 x 3.5 mm, with homogeneous endosperm.

NOTE. Slightly similar to *D. linearis* in the stout, erect inflorescence with orange axes and very hairy, rather fat rachillae; but distinct in much slighter build, the build and size of the leaflets. Other relationships are probably with *D. concinna* and *D. heterophylla.*

Nosy Varika: Sakaleona valley, June 1939 (fl., y.fr.), *Decary* 14220 (P) is similar, but differs in the petiole (6–11 cm long), rachis (to 23 cm long), peduncle (14–16.5 cm), rachillae 14–18 cm long; the peduncular bract is inserted at 8 cm from the base of the peduncle, and is 10.5 cm long. *Pistillate flowers* were within the range given above; fruit was 6–6.5 x 4–5 mm, and seed 4 x 3 mm, with homogeneous endosperm.

SPECIMENS SEEN. Andapa: E Marojejy, W of Manantenina R, March 1949, (y.fr.), *Humbert & Cours* 23729 (P); idem, (bud), *Humbert* 23682b (K, P); Mt Beondroka, N of Maroambihy, March 1949 (bud), *Humbert* 23493 (K, P). Maroantsetra: Ambanizana, Andohitsitondroina Pk, Dec. 1989 (fl.), *Schatz & Modeste* 2897 (Holotype K; isotypes MO, P, TAN); idem, March 1992 (y.fr.), *Zjhra & Hutcheon* 203 (K).

Dypsis bonsai

SIMILAR SPECIMENS. Several collections from the Ambatondrazaka/Zahamena area are reminiscent of both *D. bonsai* and *D. concinna*: Ambatondrazaka: Sahamalaza, Aug. 1937 (bud), *Herb. Jard. Tananarive* 18.8.1937 (K, P); Zahamena, March 1941 (old infl., fr.), *Decary* 16680 (P); Manaka E, April 1961 (old infl.), *Serv. Eaux & Forêts RN* 11850b (K, P). Differences with above description: sheath 8–10 cm, petiole 0–0.2 cm, rachis 11–34 cm; leaflets 14–21 on each side of the rachis, the proximal 1–5 cm long; almost glabrous. Inflorescence with peduncle 14–21 cm; peduncular bract 4–8 cm long (inserted at 10–13 cm), rachis 1–6 cm with 4–8 branches; rachillae 3–6.5 cm; fruit 7–9.5 mm x 4–5 mm, seed c. 6.5 x 3.5 mm.

56. DYPSIS CAUDATA

A very distinctive species with its long-acuminate, even caudate leaflet tips (hence the name), which are pendulous at almost right angles to the main part of the leaf. The custard yellow flowers are particularly attractive.

DISTRIBUTION. Masoala Peninsula, known from a single site.

Dypsis caudata. A beautiful palm with neat foliage and custard yellow flowers, Antalavia (*Dransfield et al.* JD6478).

Dypsis caudata. **A** crown and infructescence × 1/2; **B** fruit × 3; **C** seed in section × 3. All from *Dransfield et al.* JD6478. Drawn by Rosemary Wise.

Habitat. Lowland rain forest; steep slope or valley bottom; 50–300 m.

Local names. Not recorded.

Uses. Not recorded.

Conservation Status. Critical. Only known from a single site, in an unprotected area. Numbers are thought to be low.

Dypsis caudata *Beentje* **sp. nov.**

A ceteris speciebus 6-staminatis caulibus caespitosis foliolis fasciculatis ovatis abrupte caudati distincta. Typus: Madagascar, Antalavia, *Dransfield et al.* JD6478 (Holotypus K; isotypus TAN).

Clustering palm. **Stems** 1–3 m, 0.5–1.5 cm diam.; internodes 1–6 cm, green when young, later blackish, distally with dark scales; crownshaft indistinct. **Leaves** 6–11 in the crown; sheath c. 10 cm long, pale green with dense dark scales or later with scattered scales; petiole 6–20 cm long, 2–3 mm diam., rusty-scaly, becoming more glabrous; rachis 25–46 cm long, in mid-leaf 2–2.5 mm wide, rusty-scaly; leaflets 11–15 on each side of the rachis, in groups of 2–4, group interval 4–7 cm, proximal 8–10 × 0.5–1.5 cm, median 8–14 × 1.5–2.2 cm, distal 6–8 × 1–2.2 cm, narrowly obovate, with abrupt and pendulous drip-tips 2–4 cm long, main vein 1, with scattered scales on the minor veins and when young with bands of scales on the adaxial midrib, and on the margins (especially on the distal margin), distal pair joined for < 1 cm, dentate over c. 10 mm. **Inflorescence** branched to 1 order, custard orange-yellow; peduncle c. 31 cm long, c. 1.5 mm diam., scaly; prophyll 19–29

cm long, 4–7 mm wide, opening in the distal 1 cm only, with scattered scales; peduncular bract inserted at c. 16 cm from the base of the peduncle, c. 11 cm long; non-tubular peduncular bract 2–4 mm long; rachis c. 7 cm long, with c. 11 branches; rachillae orange, 4–5.5 cm long, c. 1 mm diam., minutely puberulous and scaly. **Staminate flowers** with sepals 0.7–0.8 × 0.7–0.8 mm; petals orange, 1.5–1.7 × 1.4–1.5 mm; stamens 6, biseriate (offset 0.2 mm), filaments 0.6–0.8 mm, thin; anthers 1–1.2 × 0.7 mm; pistillode c. 0.8 × 0.3–0.4 mm. **Pistillate flowers** with sepals 0.7–0.8 × 0.6–1 mm; petals orange, 2.2–2.3 × 1.5–2.4 mm; staminodes 6, c. 0.4 mm; pistil c. 2.2 × 1.5 mm. **Fruit** orange, 8–11 × 3.5–5 mm. **Seed** c. 7 × 3.5 mm, with homogeneous endosperm.

Note. Probably related to *D. linearis* but very distinct in its leaflets with their long drip-tips.

Specimens seen. Maroantsetra: Antalavia, Feb. 1988 (fl., y.fr.), *Dransfield et al.* JD6478 (Holotype K; isotype TAN); idem, April 1988 (fr.), *Gentry & Schatz* 62179 (K); idem, Nov. 1989 (bud), *Dransfield et al.* JD6741 (K, TAN).

57. Dypsis scandens

This remarkable species is the first climbing palm to be recorded for Madagascar. In habit and texture, it bears an uncanny resemblance to the central American climbing palm, *Chamaedorea elatior* Mart., so much so that on first finding it in November 1994, we had to examine the inflorescences closely to convince ourselves that the plant was an Arecoid palm rather than a Ceroxyloid.

Its discovery, just before the manuscript of this book was completed, emphasises yet again the extraordinary richness of the Madagascar palm flora and how much there may yet be to discover and describe. The species name is Latin for 'climbing'.

Dypsis scandens, growing on a quartzite ridge near Ifanadiana.

Distribution. Ifanadiana area, only known from one site.

Habitat. Low canopy forest with small crowns on poor soils on quartzite ridge; 500 m.

Local names. *Olokoloka* (Tanala).

Uses. Stems harvested for splitting to make fish traps, bird cages and hats. Said to be widespread in the area, but much harvested.

Conservation Status. Probably endangered if not critical. The forests in this area are not protected, and are under pressure from shifting cultivation.

Dypsis scandens *J. Dransf.*, **sp. nov.**
habitu scandenti, caudicibus gracilibus vaginis foliorum glabris, foliolis divaricatis, basin pulvinatis, inflorescentia in 2 ordines ramificanti instanter distinguibilis. Typus: Madagascar, Ifanadiana, *Dransfield & Beentje* JD7515 (Holotypus K; isotypi BH, P, TAN).

Clustering, climbing palm. **Stems** to 8–10 m long, flexible, 7–12 mm diam., internodes 20–31 cm long, bright green, with scattered dark brown scales, nodal scars c. 2 mm wide; sheathed stem c. 1.5–1.8 mm diam. Stems carrying about 15 green leaves and several dead marcescent leaves. **Leaf**-sheaths 15–30 cm, pale green, smooth, with thin white wax, glabrous, turning dark

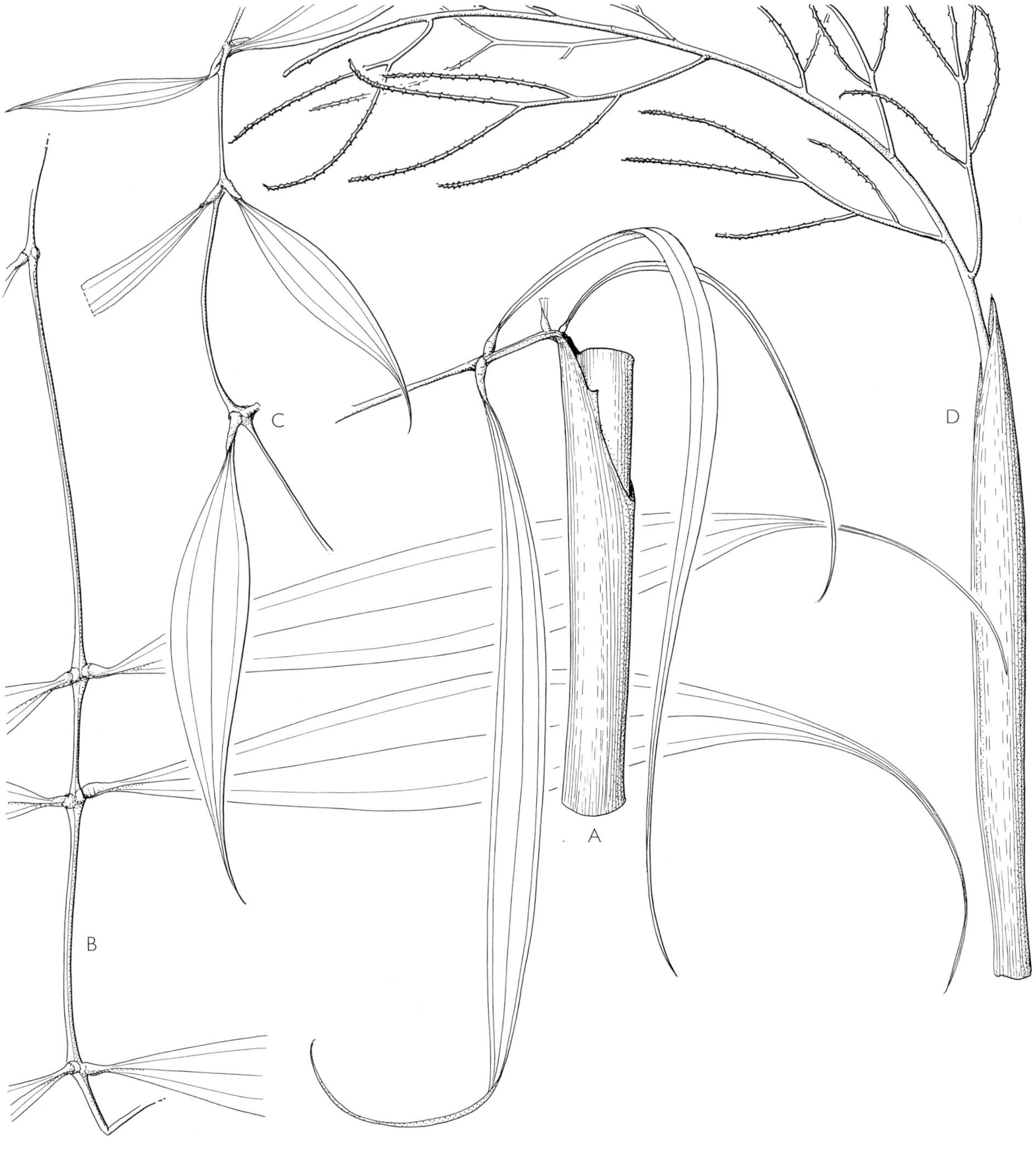

Dypsis scandens. **A** leaf base and stem enclosed by leaf sheaths × 2/5; **B** mid portion of leaf × 2/5; **C** leaf tip × 2/5; **D** inflorescence × 2/5. All from *Dransfield & Beentje* JD7515. Drawn by Rosemary Wise.

blackish brown on drying, auricles absent; petiole absent; rachis 1.1–1.45 m, 7 mm wide at the base, triangular in section, tapering to c. 1 mm diam. at the tip; leaflets c. 15–18 on each side of the rachis, grouped, lanceolate, long acuminate, mostly strongly reflexed and with a conspicuous basal woody pulvinus, basal leaflets arranged singly, then two distant groups of two leaflets, then mostly singly to the tip, the leaflets dull green, glabrous, turning dirty brown-black on drying, basal leaflets 20 × 0.2 cm, median 24–30 × 3–3.5 cm, distal to 6 × 1.5 cm. **Inflorescence,** only dead mummified material available, interfoliar, branched to 2 orders; peduncle mostly enclosed by the subtending leaf sheath, 16–45 cm, 7–20 mm wide at the base, 3–10 mm diam. near the tip; prophyll borne up to 10 cm above the peduncle base, to 15–34 × 1.6–3.1 cm, membranous, somewhat striate, glabrous; peduncular bract only fragments known, c. 15 × 1.5 cm, much tattered; rachis 27–40 cm long, bearing c. 13–15 first order branches; lowermost branches 12–18 cm, bearing 3–6 rachillae; rachillae c. 40–50 in total, 8–12 cm long, c. 1.2 mm diam., triads c. 1–3 mm distant. **Staminate flower** rounded; sepals ± rounded, gibbous, keeled, c. 0.8 mm diam., smooth; petals triangular, c. 1.5 × 1.5 mm, striate; stamens 6, somewhat biseriate, the antepetalous longer than the antesepalous, filaments 0.3–0.5 mm, anthers elongate c. 1 × 0.2 mm, joined to connective throughout their length; pistillode low, conical. **Pistillate flower** with sepals rounded, c. 0.8 mm diam., smooth; petals triangular, striate, 1.8 × 2 mm, enlarging to 2 × 2 mm in fruit, ovary c. 1 mm diam. **Fruit** ellipsoid, 8 × 4.5 mm; endosperm homogeneous.

Note. Unfortunately the type bears only dead inflorescences; however, we found one mummified fruit and several flowers still attached to the inflores-

cences and this has allowed a rather complete description to be prepared. The climbing habit makes this species instantly identifiable. The leaves have distinctive distant reflexed leaflets that are grossly swollen at the base in the manner of those of *D. pinnatifrons* and *D. nodifera*.

These reflexed leaflets presumably act, as in *Chamaedorea elatior*, as grapnels that help to support the long flexible stems. Leaf texture and inflorescence, flower and fruit structure suggest that the relationships of *D. scandens* are probably with *D. jumelleana* and related species.

SPECIMENS SEEN. Ifanadiana: c. 10 km east of Ifanadiana, Nov. 1994 (dead infl.), *Dransfield & Beentje* JD7515 (Holotype K; isotypes BH, P, TAN)

58. DYPSIS FANEVA

A fairly rare species from lowland rain forest, but highly distinctive. The name *faneva* (flag in Malagasy) refers to the leaves, which are remarkable for the large terminal flabellum.

DISTRIBUTION. Maroantsetra, Mananara and near Fenoarivo.
HABITAT. Littoral forest or moist lowland rain forest, on steep or level slope; 1–300 m.
LOCAL NAMES. *Tsinkiara mavinty* (Betsimisaraka).
USES. None recorded.
CONSERVATION STATUS. Endangered. The lowland rain forest in this area is not well protected, and numbers are low; we have seen less than fifty individuals.

Dypsis faneva *Beentje* **sp. nov.**
D. boivinianae affinis sed inflorescentia in plures ordines ramificanti rachillis brevibus *D. procerae* et *D. paludosae* superficialiter similis sed staminibus 6 non 3 differt. Typus: Antalavia, *Dransfield et al.* JD6465 (Holotypus K; isotypi P, TAN).

Clustering palm in tufts of 3–12 stems. **STEMS** 3–6 m, 3–5 cm diam.; somewhat stilt-rooted at base; internodes 3–10 cm, brown; wood soft; crownshaft moderately well-developed, with occasionally a yellowish stripe below the rachis. **LEAVES** 8–9 in the crown (sometimes with up to 4 marcescent leaves present), spirally inserted, porrect and arching; sheath pale creamy yellow or green, spotted with red, 24–30 cm long, with scattered brown scales (distally rather dense), with auricles to 3.2 cm high; petiole absent or up to 10 cm long, with scattered scales, channelled adaxially, 9–10 × 5–6 mm diam.; rachis 79–94 cm long, in mid-leaf 5–6 mm wide, with scattered scales; leaflets 8–21 on each side of the rachis, regular, stiff, straight, in one plane, somewhat arcuate, 1–4 folds wide, proximal 25–63 × 0.1–3.6 cm, median 62–72 × 1.7–11 cm (interval 6–7.5 cm), distal 42–55 × 7–10 cm, main veins 1–4, apices long-attenuate, main veins with ramenta, minor veins with scattered scales, distal pair joined for 11–17 cm, with 8 main veins, apices dentate over a width of 2–2.2 cm. **INFLORESCENCE** interfoliar, branched to 2 orders, spreading, c. 50 × 50 cm; peduncle 28–64 cm, distally 1.2–1.7 × 0.5–1 cm diam., rusty-pubescent; prophyll 29–54 cm long, borne at 13–17 cm above the base of the peduncle, 2.2–2.5 cm wide, with scattered scales, opening for the distal 4–6 cm; peduncular bract inserted at 21–30 cm from the base of the peduncle, 12–30 cm long, split over its distal 10–14 cm; rachis 14–19 cm, with 4–6 branching and 7–16 unbranched first order branches (the proximal with an axis of 9 cm, with 5 rachillae), all axes hairy; rachillae 10–28 cm long, with scattered scales, with distant triads. **STAMINATE FLOWERS** orange in bud; sepals 1.3–1.8 × 1.2–1.5 mm, the middle one sometimes very asymmetrical; petals 1.9–2.8 × 1.5–2.1 mm; stamens 6, didymous, 1- or biseriate (offset to 0.4 mm), filaments 0.8–1 mm, fat and ellipsoid, anthers 0.6–0.7 × 0.7–0.9 mm; pistillode c. 1 × 1.2 mm, pyramidal. **PISTILLATE FLOWERS** unknown. **FRUIT** unknown.

NOTE. Distinct from the very similar *D. mangorensis* in the more-branched inflorescence with shorter rachillae, and from the similar *D. procera* and *D. paludosa* in the six (not three) stamens.

SPECIMENS SEEN. Maroantsetra: Antalavia, Feb. 1988 (fl.), *Dransfield et al.* JD6465 (Holotype K; isotypes P, TAN). Mananara Avaratra: Antanambe, Oct. 1991 (dead infl.), *Beentje* 4454 (BH, K, MO, P, TAN); idem, April 1992 (old infl.), *Beentje et al.* 4618 (K, TAN). Fenoarivo Atn.: Tampolo–Fenoarivo, Feb. 1970 (bud), *Guillaumet* 2531 (K, P).

59. DYPSIS FANJANA

Rather like *D. faneva*, but with entire or hardly dissected leaves. The name comes from the local name for the species.

DISTRIBUTION. Only known from Mananara Biosphere Reserve.
HABITAT. Lowland rain forest, steep mid slope; 115–250 m.
LOCAL NAMES. *Fanjana* (Betsimisaraka).
USES. None recorded.
CONSERVATION STATUS. Endangered. Known from a

Dypsis faneva. **A** leaf sheath × 2/5; **B** distal part of petiole with lowermost leaflets × 2/5; **C** mid section of leaf × 2/5; **D** leaf tip × 2/5; **E** part of inflorescence × 2/5. All from *Beentje et al.* 4618. Drawn by Rosemary Wise.

single site, outside the protected area, where we have seen less than fifty individuals.

Dypsis fanjana *Beentje* **sp. nov.**
D. mangorensi simillima sed petiolo breve antheris didymis differt. Typus: Madagascar, Mananara Avaratra, Antanambe, *Beentje* 4453 (Holotypus K; isotypi BH, MO, TAN).

Clustering palm in tufts of 3–4, occasionally appearing solitary. **Stems** 2.5–5 m high, 1–1.5 cm diam.; internodes 2.5–6.5 cm long; stem occasionally with basal aerial roots to 20 cm long. **Leaves** 6–11 in the crown, spirally inserted, entire or rarely with a few leaflets, arching; sheath green, 12–15 cm long, pale reddish-brown when dry, with few scattered scales, with c. 5 mm high ligules; petiole absent or up to 8 cm long; lamina entire and 57–62 cm long, with scattered scales on the minor and major veins, the midrib 20–21 cm long, the lobes 34–39 x 4.3–8.5 cm, the apices narrowly dentate; or pinnate with rachis 18–24 cm long; 2 (–3) leaflets on each side of the rachis, the proximal 15–56 x 0.5–5.3 cm, the distal 40–44 x 3–4.1 cm, main veins 5, with scattered scales on minor and major veins, proximal leaflets acuminate, distal ones dentate over c. 5 mm, distal pair joined for c. 7 cm. **Inflorescence** interfoliar, branching to 1 order, porrect with pendulous rachillae; peduncle 24–47 cm long, with a few scattered scales; prophyll 22–27 cm long, opening in the distal 2–4 cm, with scattered scales; peduncular bract inserted at c. 20 cm from the base of the peduncle, c. 16 cm long, opening in the distal 3 cm, with scattered scales; rachis 2–13 cm, with 3–7 rachillae; rachillae 16–25 cm long, green, puberulous in the triad depressions, with distant yellow buds. **Staminate flowers** with sepals 0.7–0.8 x 0.6–0.7 mm; petals 1.2–1.6 x 0.9–1 mm; stamens 6, biseriate (offset c. 0.2 mm), didymous, connate for c. 0.3 mm, filaments c. 0.5 x 0.4–0.5 mm, anthers c. 0.3 x 0.5 mm; pistillode 0.3–0.4 x 0.4 mm. **Pistillate flowers** unknown. **Fruit** unknown.

Dypsis fanjana. Crown with inflorescences x 2/5. From *Beentje* 4453. Drawn by Rosemary Wise.

Dypsis fanjana

NOTE. Remarkably like *D. mangorensis* – but for the short petiole, the more slender rachillae with less conspicuous bracts and the didymous anthers.

Close but not the same is *Perrier* 12038 (P), Antalaha: Marambo, Oct 1912 (young bud), which has larger staminate sepals. The stamens are too young to study.

SPECIMENS SEEN. Mananara Avaratra: Antanambe, Oct. 1991 (bud), *Beentje* 4453 (Holotype K; isotypes BH, MO, TAN); idem, Oct. 1991 (old infl.), *Beentje* 4459 (BH, K, MO, P, TAN); idem, Oct. 1994 (bud), *Dransfield & Beentje* JD7506 (K, TAN).

60. DYPSIS BOIVINIANA

This is a palm of forest undergrowth in the lowlands of NE Madagascar, particularly around the Bay of Antongil and on Île Sainte-Marie. Based on a mixed gathering made by Boivin last century on Île Sainte-Marie, there has been a certain amount of confusion over this species. As interpreted here, it is an attractive palm with solitary or clustering erect stems and neat leaves with grouped leaflets and inflorescences branched to one order only. Together with *D. sanctaemariae, D. pervillei, D. soanieranae, D. curtisii* and *D. mangorensis* it forms a group of apparently closely related species; it is distinguished by the rather thick rachillae with crowded triads that are partially sunken in pits. We do not know whether *D. boiviniana* is in cultivation.

The name refers to the collector of the type, Louis Hyacinthe Boivin (1808–1852).

DISTRIBUTION. Mananara, Soanierana-Ivongo and Sainte-Marie; there is also an old collection from the Masoala Peninsula.
HABITAT. Open lowland rain forest or white sand forest next to a black water/ peat swamp, slight slope; 5–285 m.
LOCAL NAMES. *Talanoka, Tsingovatra* (Betsimisaraka).
USES. Not recorded.
CONSERVATION STATUS. Endangered. An estimated fifty individuals in three sites, only one of which is protected, and most of which are under threat.

Dypsis boiviniana *Baill.*
Bull. Soc. Linn. Paris 147: 1164 (1894), pro parte (see Note).

SYNONYMS:
Chrysalidocarpus oligostachya Becc., Bot. Jahrb. Syst. 38, Beibl. 87: 37 (1906); Becc., Palme del Madagascar 45 (1914); Jum., Cat. Pl. Madagascar, Palmae: 10 (1938). Type: Madagascar, Île Sainte-Marie, Ravine-tsara forest, *Boivin* s.n. anno 1850 (Holotype P, pro parte [see Note]).
Neophloga oligostachya (Becc.) H. Perrier, Fl. Madagascar 30: 81 (1945), **synon. nov.**

Solitary or clustering palm in tufts of 3–4. **STEM(S)** 2–8 m tall, 1.5–4.5 cm diam.; internodes 3–7 cm, brown to eau-de-nil, sometimes with woody, chestnut-brown lenticels, with dense red-brown indument when young; surface roots occasionally present. **LEAVES** 4–8 in the crown, porrect; sheath 21–30 cm long, 2–3 cm diam. or 6.5–7 cm wide when flattened, pale yellow or pale green tinged purple dis-

Dypsis fanjana, growing in the Mananara Avaratra Biosphere Reserve.

tally, with patches of dense red-brown tomentum, with triangular brown auricles 0.5–2 cm high; petiole 5–24 cm long, 5–8 × 2.5–6 mm diam., flat adaxially, with dense or scattered scales; rachis 29–75 cm long, pale ivory green, in mid-leaf 3–4 mm wide, with dense or scattered brown scales; leaflets 6–15 on each side of the rachis, dark green and stiff, in groups of 2–6, the interval between the groups (3–) 7–20 cm, the interval between leaflets 0.5–0.7 cm, the proximal 27–54 × 1.2–3.5 cm, the median 23–51 × 1.6–5 cm (increasing in size within the groups towards the distal side), the distal 19–44 × 2.1–4 cm, the distal pair joined for 2.5–4.5 cm, with 3–5 main veins and (0.3–1 cm wide) dentate apices, the other leaflets with 1–5 main veins and long-acuminate apices, all leaflets with minute scattered reddish scales on veins and larger scales on the (reflexed) margins. **INFLORESCENCE** interfoliar, branched to 1 order (2 in *Guillaumet* 2404), arching with semi-pendulous rachillae; peduncle 33–54 cm long, proximally 7–9 × 2–4 mm diam., distally 6–7 × 3.5–7 mm diam., densely reddish pubescent or puberulous; prophyll 31–49 cm long, 1.3–1.8 cm wide, borne at 4–16 cm above the base of the peduncle, open in the distal 4–8 cm, with scattered scales especially on the margins; peduncular bract deciduous at a very early stage, inserted at 7–31 cm from the base of the peduncle, 27–48 cm long; rachis 12–15 cm long, reddish-puberulous, with 5–8 rachillae; rachillae 25–70 cm long, 2–6 mm diam., densely puberulous, the triads dense to almost continuous, in pits. **STAMINATE FLOWERS** with sepals 1.5–2.2 × 1.2–1.8 mm, keeled; petals 2.2–2.6 × 1.5–1.8 mm, ovate or elliptic, acute; stamens 6, ± equal, filaments 1–1.2 mm long and narrow, anthers 1.6–1.7 × 0.8–0.9, the locules parallel and obtuse; pistillode 1.3 × 0.3–0.4 mm. **PISTILLATE FLOWERS** (only known in young fruit) with sepals 1.8–2.2 × 1.9–2.1 mm, orbicular, concave, slightly ciliolate, not keeled; petals with only a small gape, 2.5–2.8 × 2.2–2.3 mm; staminodes 5–6, 0.3–0.7 mm, thin. **FRUIT** c. 10 × 4 mm; endocarp fibrous, the fibres slightly anastomosing. **SEED** (very young) c. 5.3 × 2.3 mm, with homogeneous endosperm.

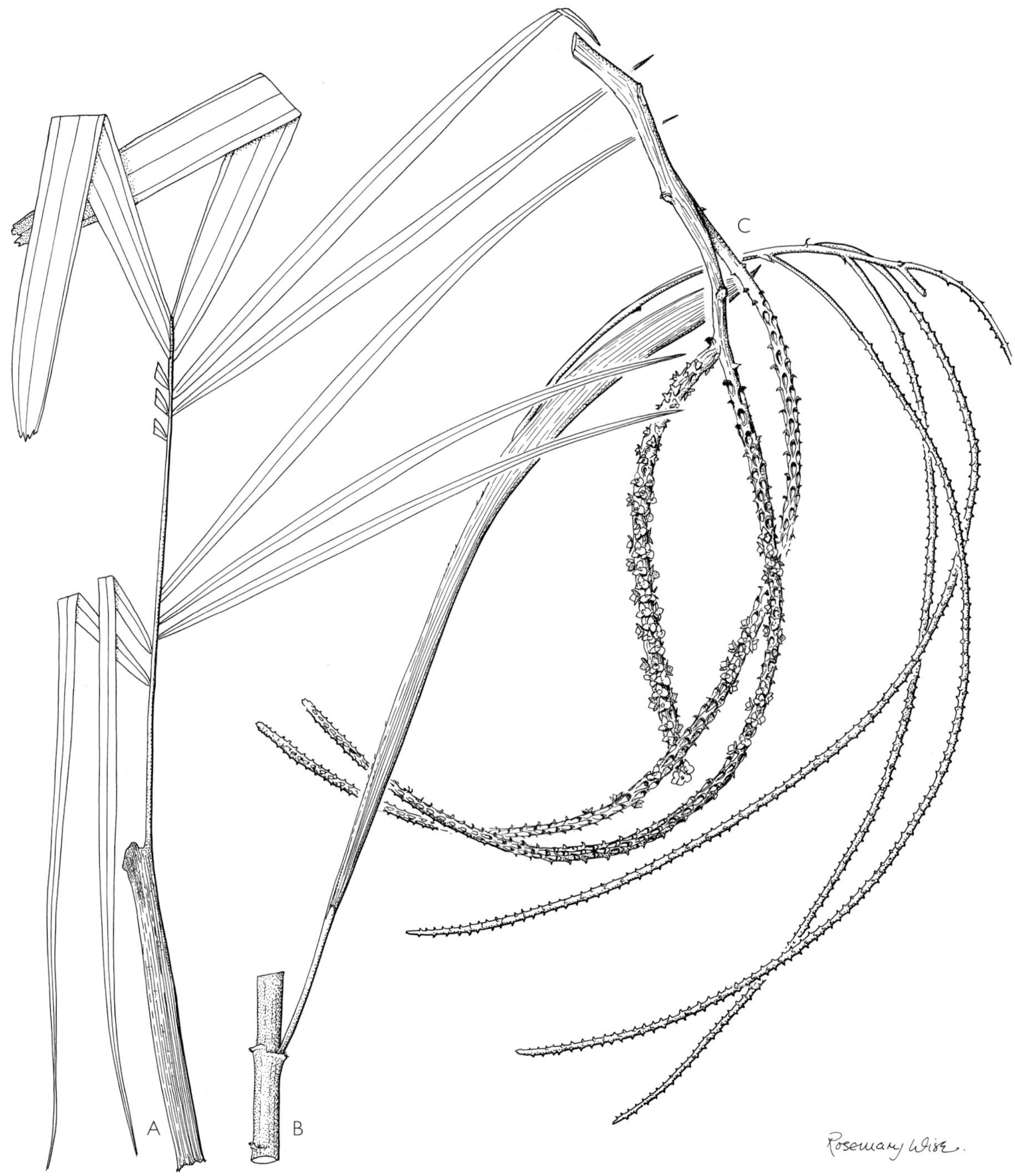

Dypsis boiviniana. **A** leaf × 2/5; **B** stem with slender rachillae × 2/5; **C** rachillae of robust form × 2/5. **A – B** from *Beentje et al.* 4622, **C** from *Beentje et al.* 4638. Drawn by Rosemary Wise.

Dypsis boiviniana, growing near the type locality on Île Sainte Marie (*Dransfield* JD7519).

NOTE. The application of the name *D. boiviniana* has caused considerable problems for all who have written about the palms of Madagascar since the time of Baillon. The name was published by Baillon (1894a) for specimens collected by Boivin on Île Sainte-Marie at Ravin-tsara in 1850. Specimens of two species were mixed together as a single gathering, either in the field, or, more likely, back in the herbarium. At any rate, Baillon describes *D. boiviniana* as a palm with irregularly arranged linear-lanceolate pinnae and an inflorescence branched to one order with about six rachillae, and with male flowers that contain three stamens. The type in fact consists of two different leaves and two different inflorescences which are, however, similar enough to explain Baillon's mistake. There is an entire inflorescence, branched to one order, with male flowers with six stamens; and there are three partial inflorescences closely resembling the former, but carrying flowers with three stamens, and bases showing that they are actually part of an inflorescence branched to more than one order. The leaves are also different: there are parts of a pinnate leaf with regular leaflets (see Beccari 1912, 1914, plate 17) and a complete pinnate leaf with grouped leaflets (see Beccari 1912, 1914, plate 44).

Beccari felt that the latter belonged with the entire inflorescence branched to one order, because the leaf sheath and lower part of the inflorescence fit; he thought that the parts of the regularly pinnate leaf might go with the partial inflorescences, but of course this is more difficult to prove. Beccari retained the name *Dypsis boiviniana* for the partial inflorescence with 3-staminate flowers, together with the regularly pinnate leaf; for the inflorescence branched to one order with 6-staminate flowers and the grouped leaflet leaf he established the name *Chrysalidocarpus oligostachya*. However, we believe that Beccari's interpretation of Baillon's name to be incorrect. Lectotypification should be based on the specimen that most nearly matches the protologue. Except for the number of stamens, Baillon's protologue of *D. boiviniana* describes in perfect detail the leaves and inflorescence of the specimen that Beccari later separated as the type of *Chrysalidocarpus oligostachya*. We therefore lectotypify *D. boiviniana* on the inflorescence with few thick rachillae of the first order and leaves with fascicled leaflets, and exclude the inflorescence fragments that represent slender rachillae of the second order that bear a covering of hairs (i.e. the fragments that Beccari interpreted as the type of *D. boiviniana*). Thus *Chrysalidocarpus oligostachya* and *Neophloga oligostachya* become synonyms of *Dypsis boiviniana*.

Dypsis boiviniana in Beccari's interpretation is an altogether different species which we have named *D. paludosa* (q.v.).

Beentje et al. 4638 differs from the other two specimens, that agree perfectly with the type, in the leaflets which are fewer in number, much longer (and the median ones much narrower as well) and in having clustered stems, while the other two collections seemed solitary.

Two specimens (*Perrier* 12051 and 12090) from the Mananara River (both at P) are reminiscent of this species, but the leaflets are regular. The inflorescences are in a very young stage and branch to two orders.

SPECIMENS SEEN. Antalaha: environs of Antalaha, Sept. 1912 (bud), *Perrier* 12033 (P). Mananara Avaratra: 10 km W of Antanambe, Oct. 1991 (bud), *Beentje* 4464 (BH, K, MO, P, TAN); idem, April 1992 (old infl.), *Beentje et al.* 4622 (BH, K, MO, TAN); idem, Oct. 1994 (bud), *Dransfield & Beentje* JD7509 (K, TAN); 5 km S of Antanambe, April 1992 (y. fr.), *Beentje et al.* 4638 (K, MO, P, TAN). Sainte-Marie: Ravine Tsara, 1850 (bud), *Boivin* s.n. (P, type); Kalalao forest, Nov. 1994 (fl.), *Dransfield* JD7519 (K, TAN). Soanierana-Ivongo: Soanierana, Feb. 1969 (fr.), *Guillaumet* 2404 (P, TAN).

61. DYPSIS MANGORENSIS

A medium-sized palm with entire or little-dissected leaves. The name comes from the Mangoro River, where the type was found; currently, the area is devoid of natural vegetation.

DISTRIBUTION. Mananara Biosphere Reserve, and an old collection from the lower Mangoro River.

HABITAT. Littoral or lowland rain forest; flat, or mid slope; 5–300 m.

LOCAL NAMES. Not recorded.

USES. Not recorded.

Dypsis mangorensis. **A** crown × 1/4; **B** leaf × 1/10; **C** detail of rachilla × 4. All from *Perrier* 18042. Drawn by Margaret Tebbs.

CONSERVATION STATUS. Critical. There is only one recent sighting; we saw less than twenty individuals in an area surrounded by agriculture, and probably under threat of agricultural conversion.

Dypsis mangorensis *(Jum.) Beentje & J. Dransf.* **comb. nov.**

SYNONYMS:
Neophloga mangorensis Jum., Ann. Inst. Bot.-Géol. Colon. Marseille sér. 4, 6 (3): 34 (1929); Jum., Fl. Madagascar 30: 78 (1945). Type: Madagascar, Mangoro Basin, *Perrier* 18042 (Holotype P).

Neophloga littoralis Jum., Ann. Inst. Bot.-Géol. Colon. Marseille sér. 4, 6 (3): 41 (1929); Jum., Fl. Madagascar 30: 80 (1945), **synon. nov.** Type: Madagascar, Mananara Basin, *Perrier* 12049 (Holotype P). NOTE: *Dypsis littoralis* Jum. of 1918 would have priority; this is now a synonym of *D. forficifolia* Mart..

Slender clustering palm in tufts of up to 7. **STEMS** 1–6 m high, 1.2–3 cm diam.; internodes c. 8 cm, near crown 2.5–3 cm, green, with flaking scales; crownshaft "eau-de-nil". **LEAVES** spirally inserted, c. 10 in the crown, entire or with up to 3 pairs of leaflets; sheath 9–13 cm long, pale green turning red-brown, with scattered to distally dense dark red scales, and sometimes with some wax, with narrowly triangular auricles to 15 mm; petiole 10–25 cm long, 2–3 mm diam., red-brown with scattered scales; rachis 17–35 cm long, with scattered scales, in mid-leaf c. 2 mm wide; lamina *entire* and then 35–51 cm long, connate for 8–14 cm, lobes 28–37 × 5.5–8.2 cm, the apices truncate over a width of 0.5–1 cm (also dentate on distal outer margins), main veins 7–11, with a scaly midrib and with (faint) scattered scales on minor veins; or *pinnate* with 2–3 leaflets on each side of the rachis, the proximal 35–40 × 2–4.7 cm and acuminate, interval to median c. 10 cm, the median 24–28 × 1.1–1.7 cm, long-acuminate, interval to distal 0.2–6 cm, distal 22–32 × 1.8–4.2 cm, connate for 2–6 cm, main veins 5–7, apices truncate, 10–15 mm wide, dentate, glabrous or with a few scattered scales proximally. **INFLORESCENCE** interfoliar, branched to 1 order, porrect to spreading; peduncle 14–53 cm long, proximally 6–8 × 2–4 mm, distally 2–5 mm diam., densely to sparsely pubescent; prophyll 15–35 × 0.7–1.3 cm, borne at 2–13 cm above the base of the peduncle, open in the distal 2–6 cm, glabrous or with scattered scales; peduncular bract 15–30 cm long, inserted at 13–22 cm from the base of the peduncle, open in the distal 2–4 cm or for more than halfway, with a few scattered scales, quickly deciduous; non-tubular peduncular bract occasionally present and c. 3 mm long; rachis 1.8–22 cm long, pubescent to glabrous, with 3–11 rachillae; rachillae 13–27 cm long, 1–2 mm diam., minutely puberulous to glabrous. **STAMINATE FLOWERS** with sepals 0.8–1.3 × 0.8–1.1 mm; petals 2–2.7 × 1.6–2.1 mm; stamens 6, in one series, the filaments 1–1.2 × 0.5 mm, cylindrical, the anthers 1.3–1.5 × 0.7–0.8 mm, dorsifixed, versatile, the locules parallel; pistillode 0.7–1.2 mm high, 0.4–0.6 mm diam. **PISTILLATE FLOWERS** unknown. **FRUIT** and **SEED** unknown.

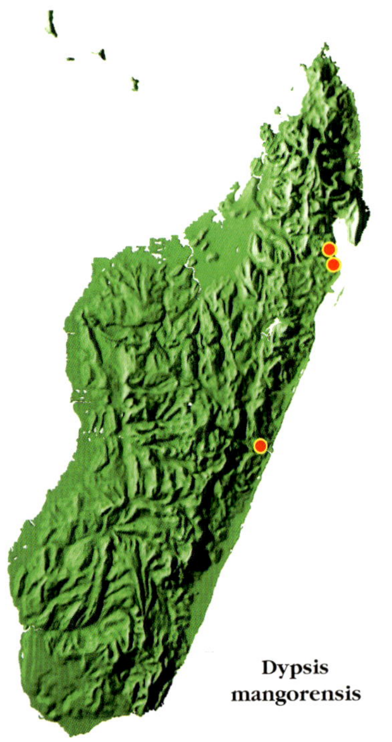

Dypsis mangorensis

NOTE. Remarkably like *D. faneva*, which however has didymous stamens.

SPECIMENS SEEN. Mananara Avaratra: Mananara, Oct. 1912 (fl.), *Perrier* 12049 (P, type of *N. littoralis*); Mananara, Oct. 1912 (y. bud), *Perrier* 12050 (P); Andravolasoa, April 1992 (old infl.), *Beentje et al.* 4639 (BH, K, MO, P, TAN). Mahanoro: Mangoro basin, Oct. 1927 (bud), *Perrier* 18042 (P, type of *N. mangorensis*).

SIMILAR SPECIMENS. *Guillaumet* 2528 (P) from Fenoarivo: Tampolo, probably belongs here. It differs in having an inflorescence branched to two orders, and in this respect somewhat resembles *D. faneva*. However, the rachillae and flowers are more similar to *D. mangorensis*. The two leaf fragments in the collection are probably not conspecific; the larger leaf seems to approach *D. paludosa*.

62. DYPSIS SANCTAEMARIAE

A handsome palm of the undergrowth of wind-pruned coastal forest on white sand raised-beaches on the Indian Ocean coast of Île Sainte Marie. It occurs in two forms that grow together, one with a pinnate leaf, the other with an entire leaf. In the field we thought that this species might be a form of *D. paludosa* with inflorescences branched to one instead two orders, but there are six stamens so the relationships are with *D. boiviniana*. The species epithet is derived from Sainte-Marie.

DISTRIBUTION. Known only from Île Sainte-Marie.
HABITAT. White sand forest; 20 m.
LOCAL NAMES. Not recorded.
USES. Not recorded.
CONSERVATION STATUS. Critical; known from a single population. The Forêt d'Ambohidena is under threat by the development of a new resort hotel.

Dypsis sanctaemariae *J. Dransf.* **sp nov.**
forma inflorescentiae floribus staminatis 6-staminibus *D. boivinianae* affinis sed foliis epetiolatis, integris-bifidis, vel foliolis approximatis non fasciculatis, rachillis gracilioribus bracteis inconspicuis distincta. Typus: Île Sainte Marie, Forêt d'Ambohidena, *Dransfield et al.* JD7526 (Holotypus K; isotypus TAN).

Clustering palm of the forest undergrowth. **STEMS** to 2.5 m tall, c. 2 cm diam.; internodes c. 3 cm long, basally grey-brown, distally green, with scattered brown scales. **LEAVES** c. 8 in crown, porrect, forming a "shuttlecock" and tending to trap litter; sheath yellow-

Dypsis sanctaemariae. **A** leaf × 2/5; **B** inflorescence × 2/5; **C** detail of rachilla with pistillate flowers × 2; **D** staminate flower × 10. All from *Dransfield et al.* JD7526. Drawn by Rosemary Wise.

green or crimson, 17–20 cm long, 2–3 cm diam., densely covered in evenly spaced red-brown scales, auricles triangular, 0.5–2.5 × 1–2 cm; petiole absent; blade to 120 cm long, entire, bifid, or with 2 very broad leaflets on one side of the rachis, the apical cleft to 48 cm deep, or irregularly divided into leaflets; costa or rachis 80–88 cm long, 7–9 mm wide at the base, abaxially densely covered with red-brown scales; leaflets very variable in width, to 10 on each side of the rachis, proximal leaflets 35 × 0.5 cm, then leaflets increasing in length and width towards the middle of the leaf, where 44 × 2.3 cm, then decreasing in length to the broad distal pair to 26.5 × 3.5 cm, adaxial surface with very sparse minute brown punctiform scales along major veins, abaxially with abundant brown punctiform scales and bands of looser brown scales. **INFLORESCENCE** interfoliar, branched to one order; peduncle 41–45 cm long, erect or strongly reflexed, c. 10 mm wide at the base, tapering to 4–6 mm wide distally, thinly covered with red-brown tomentum; prophyll inserted 8.5–10 cm above the base of the peduncle, 18–36 × 1.2–1.3 cm, striate, sparsely covered with red-brown scales; peduncular bract inserted at least 14 cm above the insertion of the prophyll, 25 × 2 cm when split and opened out, scaly as the prophyll; rachis 11–13 cm, 4–6 mm diam. at the base, red-brown tomentose; rachillae 5–6, pendulous, 30–40 cm long, c. 3.5 mm diam. when fresh, shrinking to c. 2 mm diam. in dried specimens, with a basal bare portion 1–6 cm long, surface with scattered short brown hairs; triads c. 2 mm distant, partially sunken in shallow pits, the rachilla bracts inconspicuous, c. 0.5 × 2.5 mm. **STAMINATE FLOWER** buds bluntly pointed, c. 2.5 × 1.5 mm; sepals rounded, 1.5 × 1.5 mm, irregularly keeled; petals 2.5 × 1.5 mm, triangular; stamens 6, filaments (in bud) 0.7 mm long, anthers 1.5 × 0.5 mm; pistillode conical, trifid, minute. **PISTILLATE FLOWER** 4 × 2.5 mm; sepals 1.8 × 1.8 mm,

broad imbricate; petals triangular 3.5 × 2 mm; staminodes 6, minute; ovary 3 × 2 mm, conical, stigmas eccentric. **FRUIT** unknown.

NOTES. The form of the inflorescence and stamen number suggest that this species is related to *D. boiviniana*. It may be distinguished from the latter species by the form of the leaf, which is entire and bifid, or with few irregular approximate leaflets as opposed to numerous fascicled and divergent leaflets, and the rachillae which are more slender and with inconspicuous rachilla bracts. It is similar to *D. mangorensis* but differs in the very much larger leaf, lacking a petiole and in the much longer rachillae.

Maroantsetra: Hiaraka, Oct. 1986 (bud), *Dransfield et al.* JD6367 (K, TAN) seems a very slender (clustering) plant of this species; the stem is 1 cm thick, the sheath 15 -16 cm long, the petiole is almost absent; the leaflets are regular and slightly smaller than in the above description; the inflorescence is also slender, with a 2 mm thick peduncle; there are four rachillae 21–23 cm long.

SPECIMENS SEEN. Île Sainte Marie: Lokintsy, Forêt d'Ambohidena, Nov. 1994 (fl.), *Dransfield et al.* JD7526 (Holotype K; isotype TAN), JD7527 (K, TAN).

63. DYPSIS SOANIERANAE

The description of this species is not complete, but the single long fat rachilla makes it so distinct that we feel obliged to give it a name, which refers to the locality.

DISTRIBUTION. Only known from Soanierana-Ivongo.
HABITAT. Lowland rain forest, 75 m.
LOCAL NAMES. *Tsinkary* (a rather general Betsimisaraka name for smaller palms).
USES. Not recorded.
CONSERVATION STATUS. Probably extinct. Not seen since 1938, in an area where the forest has been severely fragmented and is under continuing pressure.

Dypsis soanieranae *Beentje* **sp. nov.**
D. boivinianae affinis folio pinnato foliolis fasciculatis sed inflorescentia spicata rachilla longa distincta. Typus: Madagascar, Soanierana-Ivongo, Ambahoabe, *Lam & Meeuse* 5659 (Holotypus L).

Solitary palm. **STEM** 4–5 m tall, distally with internodes 4–6.5 cm long, c. 1.5 cm diam., distally with dense scales; nodal scars 2–5 mm. **LEAVES** pinnate; sheath 16–17.5 cm long, with scattered scales and ragged shoulders; petiole 13–13.5 cm long, 5–6 × 3–5 mm diam., flat adaxially, with scattered scales but glabrescent; rachis 71–95 cm long, with scattered scales, in mid-leaf 2–4 mm wide; leaflets 19–21 on each side of the rachis, in groups of 2–3, group interval 6–9 cm, the most proximal 17 cm below the next, proximal 17–25 × 0.4–0.8 cm, median 21–37 × 1.3–1.9 cm (interval 0.3–2 cm), distal 16–22 × 1.2–2.2 cm, lanceolate or slightly sigmoid, main veins 1–3, with faint reddish scattered scales on the minor veins, apices attenuate, distal pair joined for 1–3 cm, with 2–3 main veins and narrowly (< 4 mm) dentate apices. **INFLORESCENCE** interfoliar, unbranched; peduncle 40–47 cm long, dense red-pubescent, 3–4.5 mm in diam.; prophyll 20–28 × 0.8 cm, with scattered scales, open for the distal 1–2.5 cm, borne at c. 9 cm above the base of the peduncle; peduncular bract 22–27+ cm exserted from the sheath, opening in the distal 3–9 cm, densely scaly; rachilla (16 cm in Perrier, bud?) c. 72 cm long, 4–5 mm diam., densely puberulous, with rather densely packed triads. **STAMINATE FLOWERS** with sepals 1.7–2.2 × 1.2–1.5 mm; petals 2.7–3.2 × 1.6–1.8 mm; stamens 6, slightly biseriate (offset 0.1 mm), filaments 1–1.2 mm, thin, anthers

1.4–1.7 × 0.6–0.7 mm, versatile to slightly sagittate; pistillode 0.4–1.2 × 0.2–0.5 mm, trifid. **Pistillate flowers** not seen. **Fruit** not seen.

Note. Distinct among all *Dypsis* species by its unbranched inflorescence combined with the large pinnate leaf and long rachilla. The inflorescence reminds us of a long version of that of *D. pervillei*.

Specimens seen. Soanierana-Ivongo: Ambahoabe, Dec. 1938 (fl.), *Lam & Meeuse* 5659 (Holotype, L); Simianona (Simiane) R., Sept. 1912 (bud?), *Perrier* 12065 (FI, photo at K; absent from P).

64. Dypsis curtisii

Only represented by two ancient collections, this remains one of the lesser known palms of the island. The name refers to the collector of the type, Charles Curtis (1852–1928), who collected in Madagascar in 1881.

Distribution. Unfortunately, we are not sure where Curtis collected, and his number 117 (the only other specimen we have seen made by this collector) is an unidentifiable palm leaf, also without locality.

Dypsis soanieranae. **A** crown × 1/5; **B** detail of rachilla × 3; **C** staminate flower × 6. All from *Lam & Meeuse* 5659. Drawn by Margaret Tebbs.

Dypsis curtisii. **A** distal portion of leaf × 1/2; **B** inflorescence × 3/5. From *Curtis* 116. Drawn by Margaret Tebbs.

Jumelle & Perrier, in the Flore de Madagascar, state Imerina (= northern central Madagascar) for the type – which they also call *Curtis* s.n.

HABITAT. Montane forest; 1500–1800 m.

LOCAL NAMES. Not recorded.

USES. Not recorded.

CONSERVATION STATUS. Uncertain; Tsaratanana has not been visited by botanists in the recent past.

Dypsis curtisii *Baker*

J. Linn. Soc. 22: 526 (1887). Type: Madagascar, without locality, *Curtis* 116 (Holotype K).

SYNONYM:

Neophloga curtisii (Baker) Becc., Bot. Jahrb. Syst. 38, Beibl. 87: 30 (1906); Becc., Palme del Madagascar 36, fig. 28, pl. 35 (1914); Jum., Ann. Inst. Bot.-Géol. Colon. Marseille sér. 4, 6 (3): 30 (1929); Jum., Cat. Pl. Madagascar, Palmae: 20 (1938); Jum. & H. Perrier, Fl. Madagascar 30: 70, fig. 18 (1945).

STEM with internodes 3–3.5 cm long distally, 8–9 mm across, glabrous. **LEAVES:** sheath 12–13 cm long, with rounded shoulders and laciniate margins, with red-brown scattered scales; petiole 40–10 cm long, 2–4 mm diam., with scattered reddish scales; rachis 34–39 cm, in mid-leaf 2–3 mm diam., densely to sparsely dark reddish scaly; leaflets 9–12 on each side of the rachis, in groups of 2–3, the group interval 4.5–6.5 cm, sigmoid, narrowly ovate, folded or rarely slightly connate (for up to 1 cm) with the rachis, the proximal 8–18 × 0.3–2.3 cm, median leaflets 9–20

× 1.2–4 cm (interval 1–2 cm), distal 9–13 × 2–4.6 cm, the distal pair joined for 1–5 cm, dentate on the outer bend for 0.5–1.5 cm with quite long teeth, other leaflets with the apices long-acuminate, with 1–6 main veins, dark green adaxially with some red (almost spiny) scales on midrib, pale green abaxially with sinuous transverse veinlets and glabrous or with a few reddish scales on the midrib and the margins, but only rarely on the main veins. **INFLORESCENCE** interfoliar, unbranched or bifurcate; peduncle 22–34 cm, 1.5–3 × 1–2 mm diam., densely to sparsely scaly; prophyll 16–35 cm long, 0.5 cm wide, borne at 5–11 cm above the base of the peduncle, pale brown with few scales, opening near the apex for 1–2 cm; peduncular bract inserted at 14 -19 cm from the base of the peduncle, 12–15 cm long, opening near the apex for c. 2 cm but not at the beak which is up to 3 mm long, with scattered scales; rachilla 16–23 cm long, or (in *Perrier* 12039) with 2 rachillae c. 12.5 cm long (broken?), c. 2 mm diam., densely scaly with dark red to almost black small scales, with c. 147 distant triads in slight pits. **STAMINATE FLOWERS** in bud ovoid; sepals concave, keeled, slightly gibbous at base, only slightly ciliolate; receptacle c. 0.6 mm high; petals 2.2–2.3 × 1.4–1.5 mm, concave, ovate, slightly acute, striate; stamens 6, equal, filaments connate for 0.2 mm, 1.2 mm long, linear, anthers dorsifixed, versatile, 1.3–1.5 × 0.4 mm, linear, slightly apiculate, the locules parallel; pistillode 1.2 mm long, c. 0.3 mm diam. **PISTILLATE FLOWERS** ovoid-conical, slightly acute; sepals 0.6–1.3 × 1.2–2 mm, orbicular, hardly ciliolate, not really keeled; petals imbricate proximally, valvate distally, 1.8–2.9 × 1.4–2 mm, concave, ovate, obtuse, striate; staminodes 0.3–0.4 mm; ovary somewhat trigonous, 2.2–2.3 × 0.7–0.8 mm. **FRUIT** unknown.

NOTE. Close to *D. pervillei*, and possibly not distinct.

SPECIMENS SEEN. ? Bealanana: Tsaratanana, Oct. 1912 (fl.), *Perrier* 12039 (P); idem, Dec. 1912 (y.fr.), *Perrier* 12043 (P). Madagascar, Sept. 1881 (fl.), *Curtis* 116 (Holotype K).

65. DYPSIS PERVILLEI

Another of those rare palms with unbranched inflorescences, not seen for many years. The name refers to the collector of the type, Auguste Pervillé, who collected in Madagascar between 1837 and 1841.

DISTRIBUTION. The only certain site is Betampona.
HABITAT. Forest.
LOCAL NAMES. Not recorded.
USES. Not recorded.
CONSERVATION STATUS. Presumably extinct. Not seen for seventy years.

Dypsis pervillei *(Jum.) Beentje & J. Dransf.* **comb. nov.**

SYNONYMS:
Haplodypsis pervillei Baill., Bull. Soc. Linn. Paris 147: 1167 (1894). Type: Madagascar, NW region, *Pervillé* anno 1841 (Holotype P).
Neophloga pervillei (Baill.) Becc., Bot. Jahrb. Syst. 38, Beibl. 87: 26 (1906); Becc., Palme del Madagascar 29, fig. 22, t. 28 (1912); Jum., Ann. Inst. Bot.-Géol. Colon. Marseille sér. 4, 6 (3): 28 (1929); Jum., Cat. Pl. Madagascar, Palmae: 22 (1938); Jum. & H. Perrier, Fl. Madagascar 30: 71 (1945).

No data about habit. **LEAVES:** sheath 8–9 cm long, distally red-pubescent or with scattered scales, with auricles 3–15 mm high and triangular; petiole 2–9 cm long, 2–3.5 mm diam., with scattered reddish scales; rachis 19–35 cm long, in mid-leaf 1.5–2 mm wide, with scattered scales; leaflets 9–12 on each side of the rachis, in groups of 2–5, interval within the groups 0.5–0.7 cm, interval between the groups 4–9 cm, proximal leaflets 8–16 × 0.4–1.3 cm, median 11–23 × 0.8–1.5 cm, sigmoid, main vein 1, long acuminate, with scattered scales on minor veins on the abaxial side but glabrescent, with scattered scales on margins, distal pair 9–15 × 1–2 cm, connate for 2–2.5 cm, apices 5–10 mm wide and truncate and dentate, with 2–3 main veins. **INFLORESCENCE** unbranched, 25–60 cm; peduncle 12–31 cm

Dypsis curtisii

Dypsis pervillei

long, 2–4 mm diam., densely red-scaly; prophyll 8–14 cm long, 6–7 mm wide, borne at 1–4 cm above the base of the peduncle, with dense white or reddish scales, glabrescent, opening at the very apex; peduncular bract inserted at 4–11 cm from the base of the peduncle, 6–18 cm long, open in the distal 1–3.5 cm, with scattered scales; rachilla 14–24 cm long, 2–3 mm diam., densely puberulous; triads in slight pits. **STAMINATE FLOWERS** with sepals 1–1.6 × 0.9–1.7 mm; petals 2.6–2.8 × 1.3–1.5 mm; stamens 6, not seen, described by Baillon and Beccari as: filaments slender, biseriate according to Baillon but ± equal according to Beccari, anthers dorsifixed, versatile, oblong, obtuse, the locules parallel or slightly divergent at the base. **PISTILLATE BUDS** with sepals 1.2–1.4 × 1.3–1.9 mm; petals c. 3 × < 1 mm; staminodes at least four, 0.2–0.5 mm; gynoecium 2–2.5 × 1–1.7 mm. **FRUIT** unknown.

NOTE. On the type sheet labels (underneath the "Madagascar N. O. 1841" in ink) there is a note in pencil "plus probable Ste. Marie". In a little pocket on one of the type sheets there is some anther material (with some loose staminate petals 2–2.4 × 1.2 mm) with uniseriate filaments c. 1 mm long and thin, and anthers 1.4–1.5 × 0.7–0.8 mm, dorsifixed with parallel locules. Though difficult to key out from *D. heterophylla* (except for the rachilla), we feel this is nevertheless a distinct species, related to *D. curtisii* and *D. soanieranae*.

SPECIMENS SEEN. NW Madagascar, without further locality, 1841 (fl.), *Pervillé* s.n. (P, type). Toamasina: Betampona, Dec. 1925 (bud), *Perrier* 17470 (P).
WITHOUT LOCALITY: without date (fl.), *Goudot* s.n. (P).

Dypsis pervillei. **A** leaf × 3/5; **B** inflorescence × 3/5; **C** detail of rachilla × 3. All from *Pervillé* s.n. Drawn by Margaret Tebbs.

66. Dypsis elegans

The name has been chosen to reflect the great beauty of this small understorey palm. It would make an exceptionally beautiful ornamental, but might be too choosy in its habitat requirements to live anywhere but a greenhouse.

Distribution. Only known from Manombo forest.
Habitat. Lowland rain forest, on well-drained rocky soil; c. 65 m.
Local names. None recorded.
Uses. None recorded.

Conservation Status. Critical. HB has seen less than ten individuals in a forest which is being destroyed by fire and timber-cutting.

Dypsis elegans *Beentje* **sp. nov.**
foliis pinnis aggregatis sed in gregibus regulariter dispositis rachillis brevibus staminibus 6 distincta. Typus: Madagascar: Manombo, *Beentje & Andriampaniry* 4788 (Holotypus K; isotypus TAN).

Solitary or tufted.palm. **Stem(s)** to 2 m tall, 1–2 cm diam.; internodes 4–6 cm, green, with dense scales distally. **Leaves** c. 6 in the crown, pinnate, porrect and slightly arching; sheath 11–17 cm long, densely red-scaly, with sloping shoulders; petiole 9–14 cm long, 3–4 mm diam., densely scaly; rachis 50–68 cm long, in mid-leaf 2–3 mm wide, densely red-scaly; leaflets 41–65 on each side of the rachis, in groups of 4–5, group interval 3–4.5 cm, leaflet interval c.

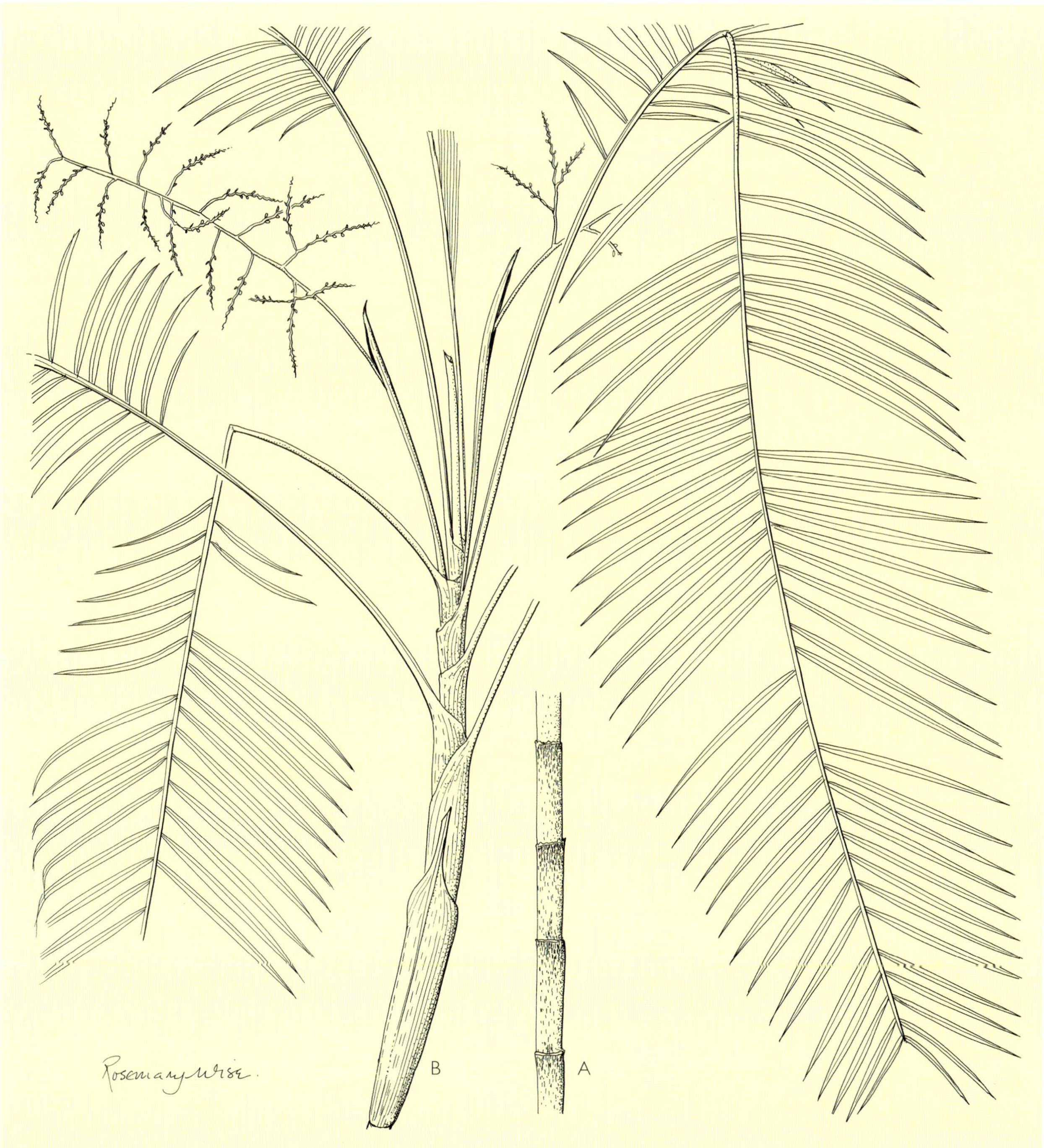

Dypsis elegans. **A** portion of stem × 1/2; **B** crown with leaves and inflorescences × 2/5. From *Beentje & Andriampaniry* 4788. Drawn by Rosemary Wise.

5 mm, the proximal 3–5 × 0.1–0.3 cm, median 8–15 × 0.4–0.6 cm, distal 5–9 × 0.3–0.9 cm and connate for 0.5–1.5 cm, main vein 1 (in the distal pair 3–4), with some scales on midrib and especially on the margins but mainly glabrous, apices attenuate, in the distal pair dentate over a width of 2–4 mm. **INFLORESCENCE** interfoliar, branched to 2 orders; peduncle 33–36 cm long, 1.5–2 mm diam., densely scaly; prophyll 10–14 cm long, 3–5 mm wide, with scattered scales, open in the distal 2 cm; peduncular bract with its apex at 7–8 cm above the apex of the prophyll, open in the distal 2 cm, with scattered scales; rachis 10–14 cm long, with scattered scales, with 5–6 branched and 11–12 unbranched first order branches, the most proximal of these with an axis of up to 2 cm and with up to 4 rachillae; rachillae 2.5–4.2 cm long, c. 1 mm diam., scaly, slightly recurved. **STAMINATE FLOWERS** yellow in bud, orange at anthesis; sepals 0.8–1 × 0.9–1.2 mm; petals 2–2.1 × 1.1–1.2 mm; stamens 6, biseriate (offset 0.2–0.3 mm), filaments 0.8–1 mm and thin, anthers 1.3–1.4 × 0.5–0.6 mm, versatile; pistillode c. 0.8 × 0.5 mm, columnar. **PISTILLATE FLOWERS** unknown. **FRUIT** unknown.

NOTE. Distinct in its leaves with groups of regularly spaced narrow leaflets, and the short rachillae. Relationships are vague (also because of the lack of pistillate and fruiting material) but possibly with *D. concinna*. Very similar to this is a collection from Mahanoro: lower Mangoro basin, alt. 300 m, Oct. 1927 (bud), *Perrier* 18049 (P); which has a shorter sheath (7–10 cm), shorter petiole (3–5 cm), shorter rachis (20–30 cm) with 28–32 leaflets which are smaller (4–7 × 0.6–0.9 cm); the inflorescence is branched to one order with, again, all parts shorter (peduncle 16 cm, prophyll 5.7 cm, rachis 5.8 cm, rachillae nine in number, 4.5–6.5 cm long); despite all these differences, we believe it is the same taxon; but the material is not enough to be certain. This locality appeared without natural vegetation when HB visited the area in 1991.

SPECIMENS SEEN. Farafangana: Manombo, Jan. 1993 (fl.), *Beentje & Andriampaniry* 4788 (Holotype K; isotype TAN).

67. DYPSIS HETEROPHYLLA

An aptly named species: *heterophylla* means with different or variable leaves, and that is certainly true. The type has entire leaves and pinnate leaves on the same stem. This is a small clustering palm of the submontane forest, with quite some variation, some forms being very decorative.

DISTRIBUTION. Reasonably widespread between Tsaratanana and Marojejy in the north to the Ambositra area.
HABITAT. Submontane rain forest; steep slopes or ridges; 550–1450 m.
LOCAL NAMES. Not recorded.
USES. Not recorded.
CONSERVATION STATUS. Rare; occurring in a wide area, but not common anywhere.

Dypsis heterophylla *Baker*, J. Linn. Soc. 22: 525 (1887). Type: Central Madagascar, *Baron* 486 (lectotype K, indicated here; see note).

SYNONYMS:
Neophloga heterophylla (Baker) Becc., Bot. Jahrb. Syst. 38, Beibl. 87: 28 (1906); Becc., Palme del Madagascar 33, fig. 26, t. 32 (1914); Jum., Ann. Inst. Bot.-Géol. Colon. Marseille sér. 4, 6 (3): 37 (1929); Jum., Cat. Pl. Madagascar, Palmae: 20 (1938); Jum. & H. Perrier, Fl. Madagascar 30: 74 (1945).

Dypsis rhodotricha Baker, J. Linn. Soc. 22: 525 (1887). Type: Madagascar, between Tamatave and Antananarivo, *Meller* 2 Aug. 1862 (Holotype K).
Neophloga rhodotricha (Baker) Becc., Bot. Jahrb. Syst. 38, Beibl. 87: 29 (1906); Becc., Palme del Madagascar 34, t. 27, t. 22, 33-34 (1914); Jum., Ann. Inst. Bot.-Géol. Colon. Marseille sér. 4, 6 (3): 25 (1929); Jum., Cat. Pl. Madagascar, Palmae: 22 (1938); Jum. & H. Perrier, Fl. Madagascar 30: 72, fig. 19 (1945), **synon. nov.**

Dypsidium vilersianum Baill., Bull. Soc. Linn. Soc. Paris 148: 1173 (1894). Type: Central Madagascar, *Le Myre de Vilers* s.n. (Holotype P).

Dypsidium emirnense Baill., Bul. Soc. Lin. Paris (1894): 1173. Type: Madagascar, Androngaloaka (Andrangolaoka), *Le Myre de Vilers* March 1889 (Holotype P).
Neophloga emirnensis (Baill.) Becc., Bot. Jahrb. Syst. 38, Beibl. 87: 28 (1906); Becc., Palme del Madagascar 32, fig. 25, pl. 31 (1912).

Neophloga majorana Becc., Bot. Jahrb. Syst. 38, Beibl. 87: 23 (1906); Becc., Palme del Madagascar 23, t. 21 (1912); Jum., Ann. Inst. Bot.-Géol. Colon. Marseille sér. 4, 6 (3): 36 (1929); Jum., Cat. Pl. Madagascar, Palmae: 21 (1938); Jum. & H. Perrier, Fl. Madagascar 30: 78, fig. 21: 1-2 (1945). Type: Madagascar, Ambohimitombo forest, *Forsyth Major* 605 (Holotype; see note), **synon. nov.**

Neophloga linearis Becc. var**. *distachya*** Jum., Ann. Inst. Bot.-Géol. Colon. Marseille sér. 4, 6 (3): 44 (1929); Jum. & H. Perrier, Fl. Madagascar 30: 88 (1945). Type: Madagascar, Vohitra R., Ambatovola, *Perrier* 18399 (Holotype P), **synon. nov**.

Clustering palm, often seeming solitary. **STEMS** 0.5–2.5 m tall, 7–8 mm diam.; internodes 3–6 cm long, dark green, distally 0.7–4 cm long, densely dull rusty-scaly. **LEAVES** 5–9 in the crown, spiral, entire or pinnate, porrect, sometimes marcescent; sheath 6–12 cm long, pale green and densely scaly with long-laciniate purple to red scales, with auricles absent or up to 1 cm long; petiole absent or up to 12 cm long, 1.5–3 mm wide, densely puberulous; rachis 9–28 cm long, densely puberulous to almost glabrous, in mid-leaf 1–2 mm wide, hardly keeled; lamina *entire* and 17–24 cm long, midrib

7.5–11.5 cm, lobes 7.5–14 x 1.2–3.5 cm, the apices dentate over a width of 5–12 mm and with some teeth on the outside margin as well, with 7–11 main veins, with reddish glands on midrib and minor veins; or *pinnate* with 2–13 leaflets on each side of the rachis, in groups of 2–5 or irregular, group interval 2–8 cm, the proximal leaflets 7–20 x 0.3–1.8 cm (occasionally much wider than the median and connate for up to 1.8 cm with the midrib, with 1–6 main veins), median 6–13 (– 19) x 0.4–1.3 cm, distal 5–15 x 0.6–3 cm, main veins 1–5, with a few to many scattered scales on the main and minor veins and distally with large marginal scales, apices attenuate to acuminate, distal pair joined for 0.5–5 cm, with 3–6 main veins, dentate over a width of 0.3–1 cm. **INFLORESCENCE** interfoliar, branched to 1 order or rarely (in the north, and the type of *N. rhodotricha*) unbranched, arching, 15–30 cm; peduncle 7–19 cm long, 1.5–2 mm diam. distally, densely puberulous; prophyll 4–15 cm long, 3–8 mm wide, borne at 3–5 cm above the base of the peduncle, dark brown, with scattered scales to almost glabrous, opening in the distal 1–3 cm; peduncular bract deciduous, inserted at 9–12 cm from the base of the peduncle, 3–9 cm long, split in the distal 2–3 cm; non-tubular peduncular bract occasionally present and 2–11 mm long; rachis absent or up to 5 cm long, with (1 -) 2–6 branches, with scattered scales; rachillae yellow-green, 4–10 cm long, 1–2.5 mm diam., with scattered stellate scales but glabrescent, with distant superficial triads. **STAMINATE FLOWERS** with sepals 0.7–1.6 mm x 0.8–1.5 mm; petals 1.6–3 x 1.1–1.6 mm, striate, occasionally on a 0.3–0.5 mm high receptacle; stamens 6, equal or biseriate, the antesepalous 0.2–0.4 mm lower than the antepetalous, connate for 0.2–0.3 mm, filaments 0.4–1.2 mm, white, thin, anthers 1–1.5 x 0.4–0.7 mm, dorsifixed versatile, slightly apiculate; pistillode trigonous-pyramidal, 0.6–1.2 x 0.2–0.6 mm, cylindrical. **PISTILLATE FLOWERS** with sepals 0.7–1.6 x 0.8–1.6 mm; petals 2–2.8 x 1.5–2.8 mm; staminodes 0.1–0.4 mm; pistil 2.1–2.2 mm high, 0.7–2.2 mm diam. **FRUIT** yellow turning orange and then red, ellipsoid, fleshy, 5–6.3 x 4–5 mm (9–11 x 4–5 mm in the type of *rhodotricha*). **SEED** c. 6 x 3.5–4 mm, with rounded to slightly pointed base and apex, with homogeneous endosperm.

NOTE. *Dypsis heterophylla* was described by Baker with *Baron* 486 and *Lyall* 323 as the only specimens mentioned. These syntypes are of different taxa. We believe *Baron* 486 represents Baker's idea of *D. heterophylla* best, not only because the other specimen is cited as "also, *Lyall* 323" but the protologue refers to the mixture of entire and pinnate leaves present in the *Baron* sheet; the *Lyall* sheet has only entire leaves. Therefore we have chosen *Baron* 486 as the lectotype of the name. *Lyall* 323, the lectoparatype of *D. heterophylla*, is *D. minuta.*

There are three sheets of the lectotype at Kew, all looking quite different from each other; however,

Dypsis heterophylla, on the slopes of Marojejy (*Dransfield et al.* JD6763).

one of the sheets is a perfect match for all the material cited here, including one of the syntypes of *N. concinna* – but not the lectotype of that taxon!

Beccari (1906) cites as type of *N. majorana* "in silva Amboimitombo... inter 1350–1440 m... *Forsyth Major*, 22. Dec. 1894 legit (n. 605 in herb. De Cand.)". Beccari (1912) elaborates on the protologue, and cites the same type, from the Herbarium De Candolle; the illustration has the legend "from the type sheet *Forsyth Major* n. 605 of the Paris Herbarium". This is no longer in Paris.

Jumelle (1929) bases his description of *N. majorana* on *Perrier* 11993 and 12007; he cites "*Forsyth Major*, Dec. 1894, 405 of the De Candolle Herbarium, according to Beccari" which sounds to us as if he had not seen the specimen; the Flore de Madagascar reiterates Jumelle (1929). There is a specimen labelled *Forsyth Major* 605 at Kew, with the correct locality and altitude, but the date is 25. I. 95; its leaves are entire, but with very short petioles; the ones photographed by Beccari (1912) have long petioles on the entire leaves, as is indeed usual on young plants. However, the Kew sheet also has an inflorescence associated with an entire leaf, and has the same general appearance.

We have put *N. majorana* into synonymy as the types of this taxon and *N. heterophylla* are very simi-

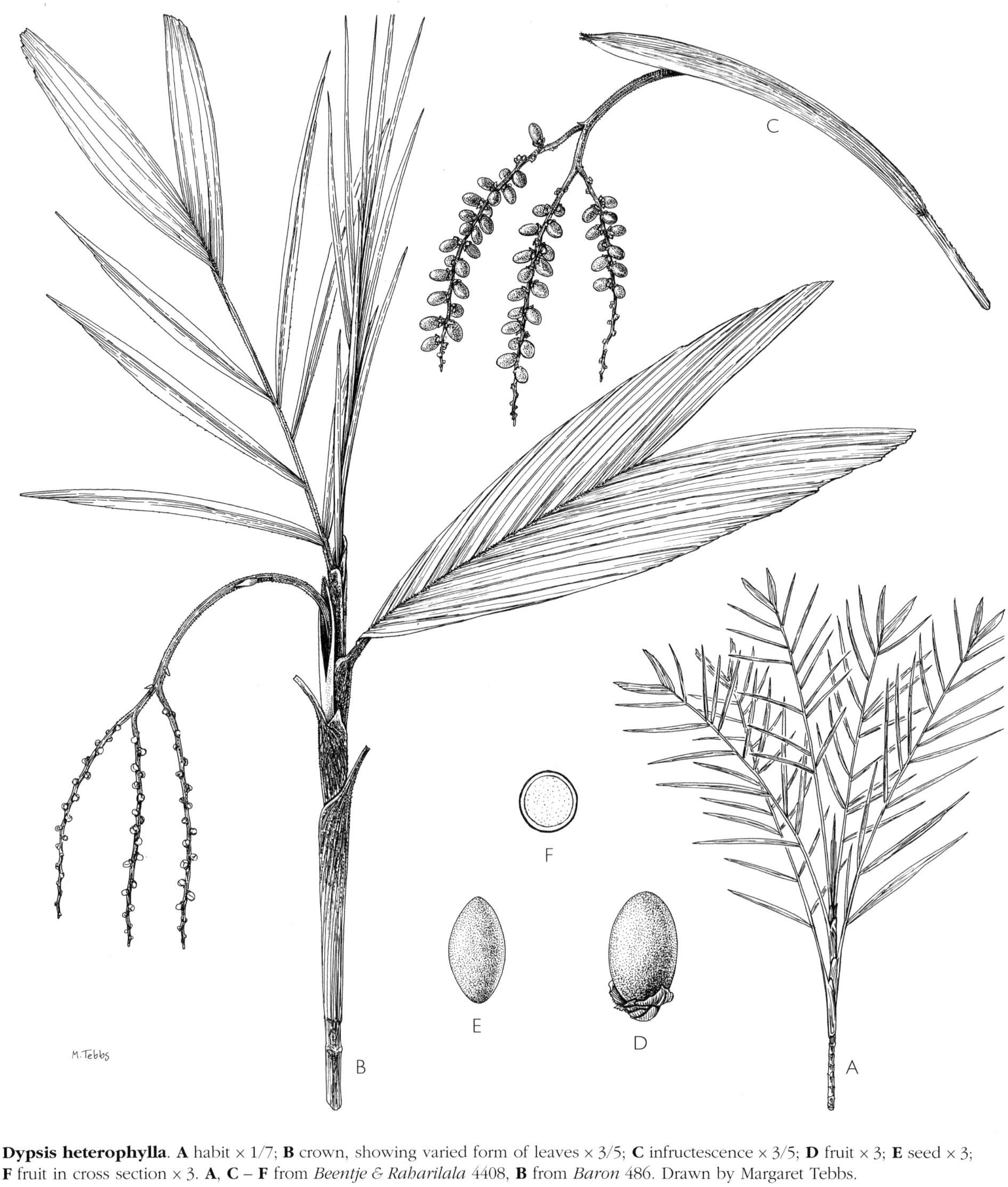

Dypsis heterophylla. **A** habit × 1/7; **B** crown, showing varied form of leaves × 3/5; **C** infructescence × 3/5; **D** fruit × 3; **E** seed × 3; **F** fruit in cross section × 3. **A**, **C** – **F** from *Beentje & Rabarilala* 4408, **B** from *Baron* 486. Drawn by Margaret Tebbs.

lar. *Heterophylla* seems a most appropriate name for this taxon. *N. rhodotricha* is distinct only by the fruit size, and therefore is brought into synonymy here. *Dypsidium vilersianum* was put into synonymy of *Neophloga rhodotricha* by Beccari in 1906; *Dypsidium emirnense / Neophloga emirnensis* was put into synonymy by Jumelle in 1929.

Specimens seen. Bealanana: Ankaizina near Bealana, Nov. 1929 (fl.), *Perrier* 15104 (P). Andapa: Marojejy East, Nov. 1989 (bud), *Dransfield et al.* JD6763 (K, P, TAN). Maroantsetra: Hiaraka, Oct. 1986 (bud), *Dransfield et al.* JD6370 (K, P, TAN). Amparafaravola: Andiamena road, Feb. 1970 (y. fr.), *Guillaumet* 2512 (K, P, TAN); idem, Andranobe forest, March 1970 (fl., y.fr.), *Bosser* 19853 (P). Ambatondrazaka: Zahamena, March 1941 (fl.), *Decary* 16503 (K, P); idem, *Decary* 16672, 16683 (P); Andrangovalo massif, Oct. 1937 (ster.), *Humbert & Cours* 17908bis (P). Anjozorobe: Analabe, Feb. 1928 (fl.), *Perrier* 18446 (P). Manjakandriana: Androngaloaka (Andrangolaoka), May 1889 (fr.), *Le Myre de Vilers* s.n. (P, type of *Dypsidium emirnense*). Moramanga: Ambatovola, Jan. 1928 (fl.), *Perrier* 18390 (P) and 18399 (P, type of *N. linearis* var. *distachya*); Mantady, March 1991 (fr.), *Beentje & Rabarilala* 4408 (K, TAN); Analamazaotra, Dec. ?? (fl.), *Perrier* 12003 (P); Perinet, June 1962 (fr.), *Bosser* 15946 (P). Ambositra: Ambohimitombo, Dec. 1894 (bud), *Forsyth Major* 604 (K, isotype of *N. linearis*); idem, Jan. 1895 (fl.), *Forsyth Major* 605 (K, type of *N. majorana*; see NOTE); near Ranomena, July 1992 (old infl.), *Beentje & Beentje* 4740 (BH, K, MO, P, TAN).
Without precise locality: "Central Madagascar", (fl.), *Baron* 486 (K, type of *Neophloga heterophylla*); idem, received Dec. 1883 (fl.), *Baron* 3248 (K); idem, no date, *Baron* 3002 (K). "Chiefly from NW Madagascar", received Sept. 1887 (fr.), *Baron* 5071 (K).
Without any locality: no date (old fl.), *Le Myre de Vilers* s.n. (P, type of *Dypsidium vilersianum*).
Locality uncertain: between Tamatave and Antananarivo, Aug. 1862 (fr.), *Meller* s.n. (K, type).

Similar specimen. "Central Madagascar", comm. Oct. 1882 (fl., fr.), *Baron* 1286 (K, syntype of *N. concinna*) but the proximal leaflets are too narrow (though with 2 main veins); pistillate sepals 0.8–1.2 × 1.6–2 mm, petals 2.5–3 × 2.3–2.5 mm, fruit 6.5–7 × 4–5 mm.

68. Dypsis concinna

A species with very narrow leaves which is locally common over a fairly small area. JD has seen beautiful plants of this species in cultivation in Queensland; it is most decorative and appears to be easily cultivated. The name *concinna* is Latin for neat or pretty.

Distribution. Zahamena and the Moramanga/ Mantady area.
Habitat. Submontane rain forest, often with much bamboo; gentle slopes or ridge tops; 800–1120 m.
Local names. Not recorded.
Uses. Not recorded.
Conservation Status. Vulnerable. Although the species is common at Analamazaotra, its distribution area is small.

Dypsis concinna *Baker*
J. Linn. Soc. 22: 526 (1887). Syntypes: Madagascar, without locality, *Baron* 1286 (K) & 3413 (K), *Pool* anno 1876 (K); lectotype (chosen here), Central Madagascar, *Baron* 3413 (lectotype K, isolectotype P).

Synonyms:
Neophloga concinna (Baker) Becc., Bot. Jahrb. Syst. 38, Beibl. 87: 27 (1906); Becc., Palme del Madagascar 31, fig. 24, t. 30 (1912); Jum., Ann. Inst. Bot.-Géol. Colon. Marseille sér. 4, 6 (3): 38 (1929); Jum., Cat. Pl. Madagascar, Palmae: 20 (1938); Jum. & H. Perrier, Fl. Madagascar 30: 84, fig. 24 (1945).

Neophloga triangularis Jum. & H. Perrier, Ann. Inst. Bot.-Géol. Colon. Marseille sér. 3, 1 (1): 32, pl. 14 (1913). Type: none indicated, locality Analamazaotra, probably *Perrier* 11971 or 11992 (not seen).
Neophloga concinna forma ***triangularis*** (Jum. & H. Perrier) Jum., Ann. Inst. Bot.-Géol. Colon. Marseille sér. 4, 6 (3): 40 (1929); Jum. & H. Perrier, Fl. Madagascar 30: 86 (1945), as var. *triangularis*, in error, **synon. nov.**

Neophloga tenuisecta Jum. & H. Perrier, Ann. Inst. Bot.-Géol. Colon. Marseille sér. 3, 1 (1): 30, pl. 13 (1913). Type: none indicated, but loc. Analamazaotra, 800 m, collines des bois, probably *Perrier* 11972 (Holotype P, isotype K) which has 'tenuisecta' written in Jumelle's hand.

Neophloga microphylla Becc., Palme del Madagascar 36 (1914). Type: Madagascar, prairies près de Mantawa (?), Emyrne, mid January 1889, *?Catat* (not seen).–synonymized by Jum. & H. Perrier.

Solitary or clustering palm. **Stems** to 2 m, 5–8 mm diam., erect or leaning, internodes 1–5 cm, densely to sparsely scaly with red-brown scales; nodal scars c. 1 mm. **Leaves** 5–12 in the crown, porrect, within one population pinnate or entire (in *Beentje* 4535), shiny rich green; sheath 7–11 cm long, closed, densely to sparsely scaly with

Dypsis concinna, in forest undergrowth at Analamazaotra.

long laciniate reddish scales, with small auricles to 5 mm high and with laciniate-scaly margins; petiole absent or up to 3 cm long (to 9 cm in some entire leaves), 2–3 mm diam., with scattered scales; rachis 11–29 cm long, in mid-leaf 1.5–2 mm wide, densely pubescent or with scattered scales on all surfaces; lamina when *entire* 13–30 × 3–5.3 cm, the lobes 4.5–21 × 1.5–2.4 cm, dentate over a width of 0.6–1 cm, the margins occasionally with a lacuna of 50–95% of the width between midrib and margin, with minute glands on minor veins, with 4–5 main veins; when *pinnate* with 11–25 leaflets on each side of the rachis, irregular or in groups of 2–7, in one plane, the group interval 1–6 cm, the leaflet interval 0.3–0.8 cm, the proximal leaflets 1–7 × 0.2–0.7 cm, median 3.5–10 × 0.4–1.3 cm, distal 2.5–6 × 0.3–1.7 cm, main vein 1, with faint or clear scattered scales on minor veins and distally with larger marginal scales, apices unequally acute to attenuate, terminal pair joined for 0.2–2.3 cm, with 2–3 main veins and dentate over a width of 0.4–1.5 cm, rarely with a large leaflet (10–12 × 1–1.5 cm) among the normal ones, either proximal or median, connate for up to 3.5 cm with the midrib. **INFLORESCENCE** interfoliar, branched to 1 order (rarely with up to three bifurcate rachillae), arching; peduncle 12–28 cm long, 1.5–2 mm diam. distally, glabrous or densely red-pubescent (only in JD6435); prophyll 7–22 cm long, 3–6 mm wide, with scattered scales, borne at 2–10 cm above the base of the peduncle, open for 1–3 cm at the apex; peduncular bract often quickly deciduous, inserted at 6–15 cm from the base of the peduncle, 4–9 cm long, 3–4 mm wide, with scattered scales, open for the distal 1–4 cm; rarely a non-tubular peduncular bract near the apex of the peduncle, 3–12 mm long; rachis absent or up to 10 cm long, with 2–9 rachillae (and occasionally with 1–3 more branched ones at the base of the rachis); rachillae 3–12 cm long, 1–1.5 mm diam., glabrous, with distant superficial triads. **STAMINATE FLOWERS** with sepals 0.6–1.1 × 0.7–1.3 mm, the middle one sometimes very asymmetrical; petals 1.5–2 × 1.1–1.4 mm; stamens 6, biseriate (offset 0.1–0.2 mm), the filaments 0.6–0.8 mm, thin, with anthers 0.9–1.3 × 0.3–0.6 mm, parallel and dorsifixed; pistil-

Dypsis concinna (form with pinnate leaves). **A** crown with pinnate leaves × 3/5; **B** basal entire bifid leaf × 3/5; **C** detail of petiole scales × 10; **D** infructescence × 3/4; **E** fruit × 2; **F** seed × 2; **G** fruit in section × 2. All from *Beentje* 4535. Drawn by Margaret Tebbs.

Dypsis concinna (form with entire bifid leaves). **A** habit × 1/4; **B** detail of rachilla × 3. All from *Beentje & Andriampaniry* 4550. Drawn by Margaret Tebbs.

lode 0.5–0.8 mm high, 0.2–0.4 mm diam., conical. **PISTILLATE FLOWERS** with sepals 0.8–1.3 × 0.8–1.8 mm; petals 2.3–3 × 1.7–3 mm; staminodes 6, 0.2–0.6 mm; pistil 2.2–2.3 × 1.2–1.8 mm. **FRUIT** red, ellipsoid, 6–16 × 3.5–7 mm, with fibrous endocarp, the fibres anastomosing little. **SEED** 5.5–8.5 × 3–4.5 mm, obtuse at both ends; endosperm homogeneous.

NOTE. One of the syntypes of *N. concinna*, *Baron* 1286, is *D. heterophylla*. These taxa are close, there is an overlap in variation, but still our feeling is that they are distinct species.

N. microphylla Becc. was distinguished from *N. concinna* by having only one inflorescence bract; it was put into synonymy by Jumelle in 1929. In the same publication *N. tenuisecta* was synonymized, and *N. triangularis* was reduced to a form of *N. concinna*. Perrier erroneously cited this taxon in the Flora (1945) as var. *triangularis* Jumelle. These entire-leaved forms are quite variable in their degree of lobing.

Dransfield et al. JD6435 has almost regular leaflets. Anosibe-an-Ala: confluence of Onive and Mangoro R., Feb. 1925 (y.fr.), *Perrier* 17217 (P) has the sheath 6.2 cm long; the lobes 3–4.5 × 3–3.5 cm; and the rachillae are puberulous rather than glabrous; it has the look of a slightly strange *D. concinna*.

Dequaire 27728 (piste de Besalana, possibly Ifanadiana/Ambohimanga area?) is probably *D. concinna*, but has the rachis 42.5 cm with 28 leaflets on each side, the median to 10.5 cm long; the peduncular bract is 10.3–10.5 cm long, the inflorescence rachis 4.5–7.5 cm, the rachillae to 12.5–14 cm long, or to 18 cm long in fruit.

Cours 4713 (Didy to Brickaville) has 3–9 branched rachillae, a longer rachis than normal, and more leaflets than normal. This is an extreme form of *D. concinna*; the few specimens with bifurcate proximal rachillae (*Rakotovao* RN 12115, *Cours* 4410) already have slightly longer rachis and slightly more leaflets than the stricly 1-order branched specimens. *Cours* 2312 (Andringitra massif, Ambodibaiso forest, Jan. 1945 in bud and fruit, P) looks like this taken to extremes: it is described as 7 m high and being used to make blowpipes; the leaf sheath is 17 cm, its inflorescence branches to 2 orders with a rachis of 5–20 cm and 3–5 branched + 6–10 unbranched first order branches. It certainly cannot be called *D. concinna* any more but does not match anything else either.

SPECIMENS SEEN. Ambatondrazaka: Ambatosoratra, Jan. 1962 (fr.), *Rakotovao* RN 12115 (P). Moramanga: Analamazaotra, anno 1912 (fl.), *Perrier* 11972 (P; type of *Neophloga tenuisecta*); idem, Feb. 1924 (fl.), *Perrier* 15983 (P); Andasibe, Feb. 1971 (fl.), *Cremers* 1419 (P); idem, April 1971 (ster.), *Moore & Mabberley* 9913 (P); idem, Nov. 1986 (bud), *Dransfield et al.* JD6411 (K, TAN) & JD6416 (K, TAN); idem, March 1988 (bud), *Dransfield et al.* JD6489 (K, TAN) & JD6488 (K, TAN) & JD6489 (K, P, TAN); idem, Dec. 1991 (fl.), *Beentje* 4535 (BH, K, MO, P, TAN); Anranumenabe, Nov. 1986 (fl.), *Dransfield et al.* JD6435 (K, TAN); Mantady, Dec. 1991 (bud), *Beentje & Andriampaniry* 4539 (K, TAN) & 4550 (K); idem, April 1992 (old fl., fr.), *Beentje & Andriampaniry* 4653 (BH, K, MO, P, TAN); idem, Dec. 1992 (fl.), *Beentje & Andriampaniry* 4550 (K, TAN); Rahobevava to Andasibe, March 1951 (fr.), *Cours* 4410 (P); Lakato, Dec. 1932 (fl.), *Leandri* 717 (P) and (fl.) 720 (P); idem, Sept. 1942 (fl.), *Decary* 18222 (K, P); Lakato road, June 1964 (ster.), *Bosser* 19739 (P); S of Moramanga, Feb. 1930 (bud), *Decary* 7082 (P) and (y.fr.) 7218 (P); idem, Nov. 1952 (old infl.), *Leandri* 1659 (P). Central Madagascar, anno 1885 (fr.), *Baron* 3413 (K, P, lectotype of *N. concinna*); idem, comm. Oct. 1882 (y.fr.), *Baron* s.n. (K). Without any locality, anno 1875? (y.fr.), *Pool* s.n. (K, syntype of *N. concinna*; identification not quite certain, a wretched specimen).

69. DYPSIS SCHATZII

This is a very handsome small undergrowth palm; it is named for the collector of the type, George Schatz of Missouri Botanical Garden, who has worked with us in the field, often helping greatly with logistic arrangements.

DISTRIBUTION. E Madagascar; only known from Betampona.
HABITAT. Lowland rain forest, steep mid slope; 500–565 m.
LOCAL NAMES. *Tsinkara* (Betsimisaraka).
USES. Stems used to make blowpipes.
CONSERVATION STATUS. Vulnerable. Only known from a single site; numbers estimated to be less than fifty.

Dypsis schatzii *Beentje* **sp. nov.**
D. integrae similis sed statura robustiore foliis majoribus inflorescentia ramosa triadibus confertim dispositis differt. Typus: Madagascar, Betampona, *Schatz et al.* 2687 (Holotypus K; isotypus MO).

Solitary or clustering palm. **STEM(S)** 2–3 m tall, to 1 cm diam.; internodes 2–6 cm long, distally with dense reddish scales, often with sheath remnants clothing the stem in its distal part. **LEAVES** 6–13 in the crown, entire or with 2 pinnae; sheath 8–12 cm long, dense reddish scaly, closed, with triangular auricles 0.6–1.2 cm long; petiole absent or up to 10 cm long, distally c. 2 mm wide, densely scaly; lamina dark green, when *entire* obovate, 17.5–41 cm long, 6.8–11 cm wide, with lobes 2.6–7 cm long; main veins 11–13, faint, only the midrib prominent; apices long-dentate, the teeth continuing along the outer margins to about halfway down the leaf, the teeth to 1 cm long, rarely a deep lacuna almost reaching the midrib; midrib densely scaly, minor veins with bands of scattered scales abaxially, as well as white-punctate abaxially which gives

the abaxial side a silvery appearance; when *pinnate* with a rachis 18–22 cm long, leaflets 2 on each side of the rachis, the proximal 9–10 × 1.1–1.7 cm, acuminate, leaflet interval 2–3 cm, distal leaflets obovate 17–24 cm long, c. 7 cm wide, with 3–5 main veins, connate for > 80%. **INFLORESCENCE** interfoliar, branched to 1 order; peduncle 28–31 cm long, distally 1.5–2 mm diam., densely reddish-scaly; prophyll 18–21 cm long, 6–7 mm wide, borne at c. 4 cm above the base of the peduncle, opening in the distal 1.5–2 cm, with scattered scales; peduncular bract inserted 14–18 cm from the base of the peduncle, 9–11 cm long, opening in the distal few cm, with scattered scales; non-tubular peduncular bracts 2–3 mm long, sometimes almost tubular; rachis 1–5 cm long, beige in fruit, with 3–6 first order branches; rachillae 6.5–13 cm long, minutely puberulous with reddish stellate scales and dense triads; rachilla bract 0.6–0.8 mm, rounded. **STAMINATE FLOWERS** with sepals 1.1–1.4 × 0.6–1.6 mm, the middle one very asymmetrical; petals 1.6–2 × 1.2–1.4 mm; stamens 6, slightly biseriate (offset < 0.1 mm), filaments 0.4–0.6 mm long, thin, anthers 0.8–1.3 × 0.4–0.5 mm, dorsifixed and with parallel locules; pistillode 0.6–1 × 0.3 mm. **PISTILLATE FLOWERS** not seen. **FRUIT** pinkish red, 13–14 × 4–6 mm. **SEED** c. 8 × 2.5 mm, with homogeneous endosperm.

NOTE. Relationships are probably with *D. integra* but distinct in the more robust stature, in all parts being larger, in the branched inflorescence with the triads more densely arranged.

A plant from Ifanadiana: upper Sahanary R., Nov. 1911 (bud), *Perrier* 11962 (P; type of *D. forficifolia* var. *reducta*) is very similar as for the leaves, but the inflorescence is much smaller, with a shorter peduncle but a longer rachis (10 cm) with twelve partially

Dypsis schatzii. **A** crown with infructescence × 1/2; **B** crown with bifurcate inflorescence × 1/2; **C** fruit in vertical section × 2. All from *Schatz* 2687. Drawn by Rosemary Wise.

reflexed rachilla 2.5–3.5 cm long. The sepals are c. 0.7 mm long, and the stamens seem to be didymous; the plant is in young bud, so we are unable to reach a conclusion as to its identity.

Specimens seen. Toamasina: Betampona, Dec. 1925 (bud), *Perrier* 17467 (P); idem, April 1989 (fr.), *Schatz et al.* 2687 (Holotype K; isotype MO, not seen); idem, Oct. 1991 (bud), *Beentje* 4487 (K, TAN).

70. Dypsis corniculata

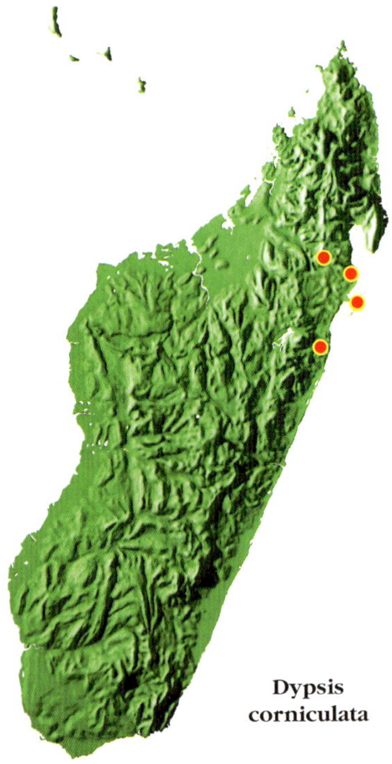

This is a very beautiful small and neat undergrowth palm with slender stems and dark green, shiny leaves. Based originally on a single inflorescence, this species remained for long a misunderstood taxon. The specific name is Latin for 'curved in the shape of a horn', presumably a reference to the young fruit.

Distribution. Mananara, Mandritsara, Sainte-Marie, Betampona.
Habitat. Rain forest, flat to steep mid slopes or ridgetops; 70–850 m.
Local names. Not recorded.
Uses. Not recorded.
Conservation Status. Vulnerable. Numbers are thought to be low (less than two hundred).

Dypsis corniculata *(Becc.) Beentje & J. Dransf.* **comb. nov.**

Synonym:
Neophloga corniculata Becc., Bot. Jahrb. Syst. 38, Beibl. 87: 24 (1906); Becc., Palme del Madagascar 25, fig. 18, t. 23 (1912); Jum., Cat. Pl. Madagascar, Palmae: 20 (1938). Type: Madagascar, Ile Sainte Marie, Tafondru forest, March 1847, *Boivin*, lectotype P, indicated here.

Clustering palm in groups of 5, or solitary but in subcolonial groups. **Stems** 1.5–6 m tall, occasionally with remnants of sheath bases; distal internodes 0.8–2 cm long, 5–20 mm diam., grey-brown, distally reddish pubescent; crownshaft pale green with dark brown scales. **Leaves** 6–10 in the crown; sheath 6–15 cm, with patches of reddish scales or with scattered scales, with laciniate auricles to 15 mm; petiole absent or up to 7.5 cm long, 2–2.5 mm diam., with scattered scales; rachis 13–40 cm long, in mid-leaf c. 2 mm wide, with dense to scattered scales; leaflets 9–18 on each side of the rachis, in groups of 2–4 but sometimes the distal ones regular, the group interval 3–6 cm, the proximal 4–10 × 0.4–1.8 cm, median 6–14 × 1.2–2.6 cm, distal 5–10 × 1–2.5 cm, main veins 1–3, with marginal bands of scattered scales, sometimes with scattered scales on the major and minor veins, and distally along the margin with some larger scales, apices unequally acuminate to almost praemorse, bases often flat and connate with the rachis for c. 5 mm, distal pair joined for 1.5–4 cm, with bands of scattered scales, dentate at the apex. **Inflorescence** interfoliar, branched to 1 or 2 order(s); peduncle 9–26 cm long, distally 2–5 mm diam., pubescent or with scattered to dense red-brown scales; prophyll 16–27 × 0.6 cm, borne at c. 6 cm above the base of the peduncle, with scattered scales, the apex open for 1–2 cm; peduncular bract inserted at 4–15 cm from the base of the peduncle, 4–11 cm long, with scattered scales, open over its whole length except for a 5 mm beak, deciduous and carried upwards by the lengthening **inflorescence;** non-tubular peduncular bract 1–20 mm, sometimes almost tubular; rachis 1–9 cm long, scaly, with 0–6 branched and 3–16 unbranched branches; rachillae (3–) 6–14 cm long, 1–1.5 mm diam., slightly zigzag, minutely puberulous with a few scales, with distant triads. **Staminate flowers** with sepals 1–1.5 × 1–2.1 mm, the middle one very asymmetric, ciliolate; petals white, 2.2–2.5 × 1.2–1.4 mm, on a 0.4 mm high receptacle; stamens 6, uniseriate, the filaments 0.6–1 mm, thin, anthers 0.7–1.5 × 0.4–0.8 mm, versatile; pistillode 0.6–0.8 × 0.2–0.3 mm. **Pistillate flowers** with sepals 0.9–1.7 × 1–2.4 mm, orbicular, slightly keeled, entire, the innermost largest; petals 2–3.5 × 2.3–3.5 mm, suborbicular, striate, imbricate with brief triangular valvate apices; staminodes 6, 0.2–0.6 mm; pistil 1.5–3.6 × 0.8–2.2 mm. **Fruit** red, ellipsoid, 10–12 × 4–5.5 mm, with an obtuse point; endocarp fibrous, the fibres free or slghtly anastomosing. **Seed** c. 7.5 × 4.5 mm, pointed at the base, rounded at the apex, with a median depression and homogeneous endosperm.

Dypsis corniculata. Close up of crown and inflorescence, Mananara Avaratra (*Beentje & Dransfield* 4820)

NOTE. The type of *N. corniculata* is an inflorescence with a few curved fruits, collected by Boivin on Sainte-Marie, and described and illustrated by Beccari (1912). Jumelle did not mention it in his 1929 treatment of *Neophloga*. It was mentioned as "espèce insuffisamment connu" in Flore de Madagascar (Jumelle & Perrier 1945). In November 1994, just before this book was going to press, JD recollected this at the type locality, and we were able to match it with the distinctive inflorescence of the type. With habitat and leaf details, we have now been able to identify other collections from E Madagascar as *D. corniculata*.

SPECIMENS SEEN. Mananara Avaratra: Antanambe, Oct. 1994 (dead infl.), *Dransfield & Beentje* JD7504 (K, TAN); idem, Oct. 1994 (fl.), *Beentje & Dransfield* 4820 (K, TAN). Mandritsara: Antsiatsiaka, April 1974 (fl., fr.), *Morat* 4489 (P, TAN). Sainte-Marie: Tafondrou forest, March 1847 (fl., y.fr.), *Boivin* s.n. (P, lectotype); Tafondrou forest Dec. 1849/ de ravine tsara April 1851 (y.fr.), *Boivin* s.n. (P, syntype); Kalalao forest, Nov. 1994 (fl.), *Dransfield* JD7522 (K, TAN). Toamasina: Betampona RNI, April 1989 (fl.), *Schatz et al.* 2689 (K); idem, Oct. 1991 (fl.), *Beentje* 4489 (BH, K, MO, P, TAN).

SIMILAR SPECIMEN. *Rakotovao* RN 10610 is a rather scrappy specimen from Manaka (S Zahamena area), collected in January 1960 (fl.) (P). It is probably this species.

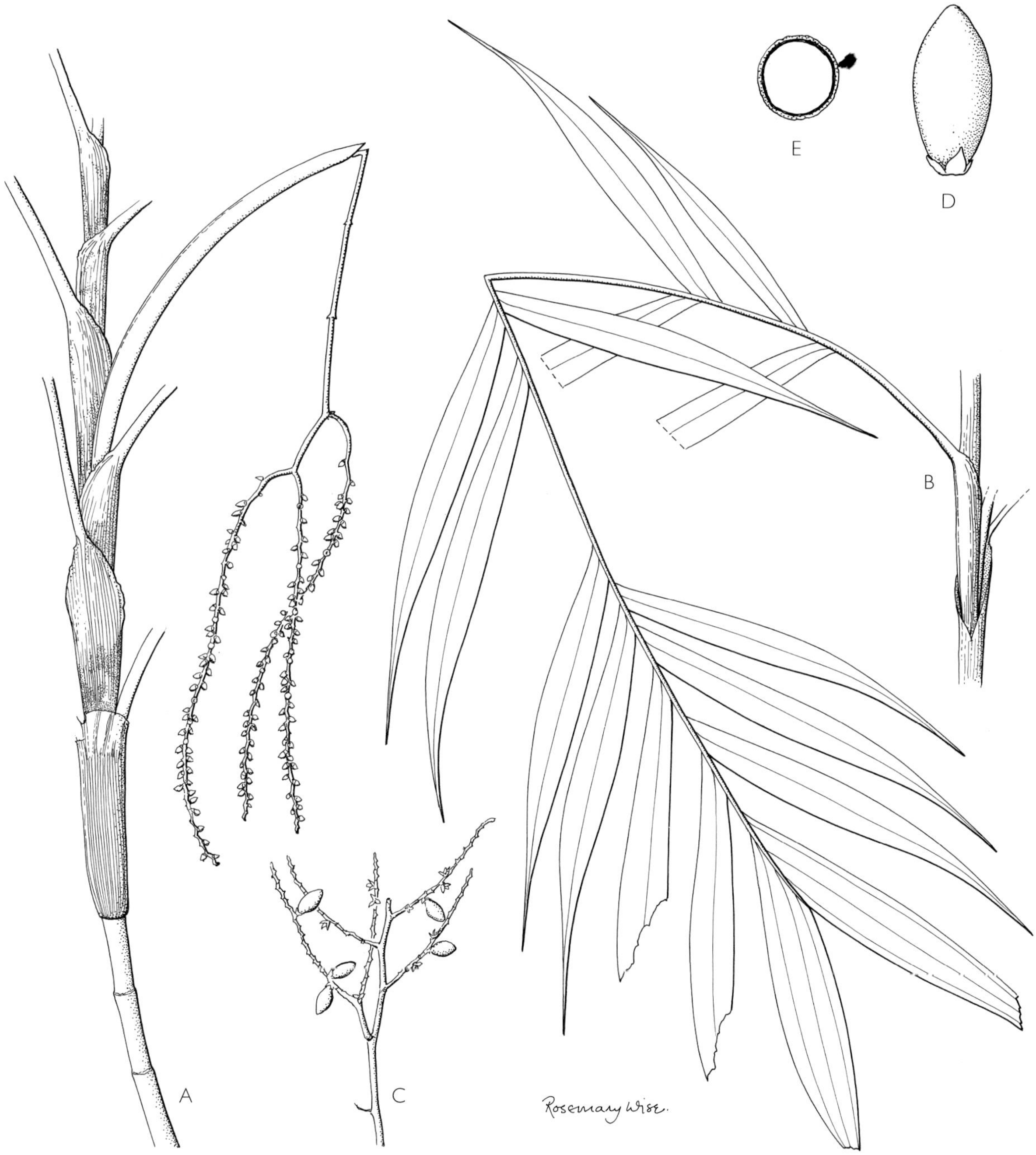

Dypsis corniculata. **A** crown with inflorescence × 3/5; **B** leaf × 3/5; **C** infructescence × 3/5; **D** fruit × 2; **E** fruit in cross section × 2. **A** – **C** from *Schatz et al.* 2689, **D**, **E** from *Morat* 4489. Drawn by Rosemary Wise.

71. DYPSIS THIRYANA

This is not only one of the most attractive small palms from Madagascar but also one of the most easily identified. The small wedge-shaped praemorse leaflets are found only in this species and in *D. trapezoidea*. In the field it can easily be overlooked as the leaflets give the leaves a rather unpalm-like appearance, one more reminiscent of a fern.This would clearly be a wonderful palm to grow as a pot plant; sadly, ripe fruit seem very rarely to be produced in quantity. The name refers to the collector of the type; we have been unable to find any details about him.

DISTRIBUTION. Marojejy and Masoala to Anosibe-an-Ala.

HABITAT. Lowland rain forest, slight to steep slope or ridgetop; 220–900 m.

LOCAL NAMES. *Tsinkiara, Sinkarambolavo maroampototra, Taokonampotatra* (Betsimisaraka).

USES. Meller says the plant is used for thatching; his specimen is all of 60 cm tall, roots included, and we would doubt his statement.

CONSERVATION STATUS. Rare. Spread over a fairly large area, though never common; usually a few individuals per population.

Dypsis thiryana *(Becc.) Beentje & J. Dransf.* **comb. nov.**

SYNONYM:
Neophloga thiryana Becc., Bot. Jahrb. Syst. 38, Beibl. 87: 23 (1906); Palme del Madagascar 24, fig. 17, t. 22 (1912); Jum., Ann. Inst. Bot.-Géol. Colon. Marseille sér. 4, 6 (3): 42 (1929); Cat. Pl. Madagascar, Palmae: 23 (1938); Jum. & H. Perrier, Fl. Madagascar 30: 82, fig. 23 (1945). Type: Madagascar, *Thiry* anno 1904 (P, holotype).

Clustering palm in tufts of 2–4 (solitary according to *Miller & Lowry* 3942, *Benoist* 862, and in the last specimen confirmed by roots). **STEMS** 0.1–1 m high (the type states 5 m, which we find hard to believe), 0.4–0.6 cm diam.; internodes 0.4–4 cm, brown, densely red-scaly; nodal scars 1–2 mm. **LEAVES** 8–10 in the crown, porrect; sheath 6–11 cm long, c. 0.6 cm diam., open for 1–2 cm, distally with dense red-brown and pale white scales, with rounded shoulders or with small triangular bumps to 2 mm high; petiole 1–18 cm long, 1.5–2 mm diam., densely scaly or with scattered scales; rachis 14–30 cm long, in mid-leaf 1–1.5 mm diam., densely scaly or with scattered scales; leaflets 9–14 on each side of the rachis, rich shiny green to dark green, almost regular or (more often) in groups of 2–3, the group interval 1–4 cm, the proximal 2–7 x 0.2–1.2 cm, the median 3–11 x 0.6–1.3 (–2.5) cm, cuneate at the base, with 1 (3) main vein(s), unequally praemorse and dentate at the apex, the distal half prolonged and with a dentate acumen, with some basal scales but glabrescent, rarely with the distal margin scaly,

Dypsis thiryana, in the undergrowth of forest on the Masoala Peninsula.

distal 3–10 x 0.8–3.2 cm, connate for 1.2–3.2 cm, with 4 main veins, truncate and dentate for up to 4 cm. **INFLORESCENCE** interfoliar, erect, branched to 1 order (one unbranched in *Perrier* 17214); peduncle 12–26 cm long, 1.5–2 mm diam., ± glabrous; prophyll 10–17 cm long, to 3 mm wide, borne at up to 5 cm above the base of the peduncle, opening only near the apex, pale brown with scattered scales; peduncular bract inserted at 7–15 cm from the base of the peduncle, 6–11 cm long, opening only in the distal 2–3 cm, pale brown with scattered scales; second tubular peduncular bract often (always?) present at 12–20 cm from the base of the peduncle, 0.9–2.5 cm long, shortly tubular with a long acumen; non-tubular bract sometimes present near the rachis, 1.6 mm long; rachis 0–2.5 (–11.5) cm long, glabrous, with (1–) 2–4 (–8) rachillae; rachillae (2.5–) 6–15 cm long, c. 1 mm diam., glabrous, with distant superficial triads and white flowers. **STAMINATE FLOWERS** with sepals 0.7–1 x 0.8–1 mm, the outer more keeled than the inner, orbicular, ciliolate; petals on a 0.6 mm high receptacle, 1.9–2.2 x 1–1.3 mm, elliptic or ovate, acute, striate; stamens 6, with the antepetalous filaments inserted slightly above the antesepalous ones, filaments connate for 0.15 mm, 0.8–0.9 mm long, thin, anthers dorsifixed and versatile, 1.1–1.2 x 0.4 mm, the locules parallel, slightly apiculate; pistillode c. 0.5 x 0.2 mm. **PISTILLATE FLOWERS** with sepals imbricate, 1–1.3 x 1.3–1.5 mm, orbicular, ciliolate; petals proximally ovate and imbricate, distally triangular, fleshy and valvate, 2.1–3.3 x 1.2–2.6 mm, striate; staminodes 6, 0.2–0.3 mm; ovary 1–2 x 0.8–1.5 mm, ovoid or ellipsoid with pointed apex. **FRUIT** bright red, ellipsoid, 9–11 x 3–5 mm. **SEED** c. 9 x 2–3 mm, with homogeneous endosperm. (Plate on next page).

NOTE. The Tsinkiara on the type has been taken for a locality, but is actually a Betsimisaraka name for any small palm.

Baker briefly mentions the *Meller* specimen in a note under *N. rhodotricha* in J. Linn. Soc. 22: 526 (1887) as distinct, but refrains from naming it.

Perrier 17214 has an unbranched or bifid inflorescence.

SPECIMENS SEEN. Andapa: Marojejy NE, E of Ambalamanasy II, Nov.-Dec. 1948 (fl.), *Humbert & Capuron* 22171 (K, P); Marojejy E, NW of Mandena, Feb. 1989 (fl.), *Miller & Lowry* 3942 (K, P); idem, Nov. 1989 (fl.), *Dransfield et al.* JD6750 (K, TAN). Maroantsetra: Antongil Bay area, Oct. 1912 (fl.), *Perrier* 11978 (P); Sahavary, Feb. 1988 (fl.), *Dransfield et al.* JD6460 (K, TAN); Hiaraka, Oct. 1986 (old infl.), *Dransfield et al.* JD6377 (K, TAN). Mananara Avaratra: 10 km W of Antanambe, April 1992 (y.fr.), *Beentje et al.* 4648 (K, TAN); idem, Oct. 1994 (fr.), *Beentje & Dransfield* 4811 (K, TAN). Soanierana-Ivongo: Soanierana to Antasibe, Dec. 1938 (fl.), *Lam & Meeuse* 5858 (L). Ampasimanolotra: Ambila, March 1951 (bud), *Benoist* 862 (P). Moramanga: Beforon (Beforona), July 1862 (ster.), *Meller* s.n. 31.7.1862 (K). Anosibe-an-Ala: near confluence of Mangoro R and Onive R, Feb. 1923 (y.fr.), *Perrier* 17214 (P).
LOCALITY UNCLEAR: eastern forest, 1904 (fl.), *Thiry* s.n. (Holotype P).

72. DYPSIS TRAPEZOIDEA

A beautiful species known at present only from the isolated hill, Vatovavy, southwest of Mananjary. First collected by Bosser, we had initially included it within the variation of *D. thiryana*, but newly collected specimens show abundant differences between the two. This would be a beautiful ornamental. The species name refers to the leaflets that are almost trapezoid in outline.

DISTRIBUTION. Known only from Vatovavy, Ifanadiana.
HABITAT. Steep rocky slope at foot of cliff, primary lowland forest.
LOCAL NAMES. Not recorded.
USES. Not recorded.
CONSERVATION STATUS. Critical. Single-site locality; site

not protected, except by local fady (taboo); numbers low (estimated at less than a fifty).

Dypsis trapezoidea *J. Dransf.* **sp. nov.**
statura et habitu solitario foliolis praemorsis *D. thiryanae* affinis sed petiolo longiore, foliolis latioribus, inflorescentiae rachidi rachillas breviores aequanti vel excedenti et fructu duplo-majore differt. Typus: Madagascar, Ifanadiana, Vatovavy, *Dransfield et al.* JD7512 (Holotypus K; isotypi AAU, BH, MO, NY, P, TAN).

Solitary slender palm of forest undergrowth. **STEM** to 1.2 m tall, c. 8–12 mm diam. at the base, c. 4–7 mm diam. distally, often conspicuously stilt-rooted at the base, dark grey brown and vertically cracked at the base, dull green distally, internodes 8–18 mm with scattered dark brown scales, nodal scars c. 1–2 mm wide. **LEAVES** c. 7 in crown; sheaths 5–6.5 cm long, c. 1–1.2 cm diam., dull green with scattered evenly spaced thin brown scales, auricles absent; petiole 9–19 cm long, c. 2 mm wide, densely silvery scaly, especially when newly emerged; rachis 21–36 cm, scaly as the petiole; leaflets c. 7–10 on each side of the rachis, mostly arranged in pairs, somewhat trapezoid in outline, with conspicuous praemorse tips, basal leaflets with proximal margin 5.5–6 cm, distal margin 7–9 cm and praemorse margin 0.8–2 cm; mid-leaf leaflets with proximal margin 6.5–8 cm, distal margin 7–11 cm and praemorse margin 2–3 cm; apical leaflet pair much shorter than the rest with proximal margin 3.5–4 cm, distal margin 2.5–3 cm and praemorse margin 2–3.5 cm; leaflets drying pale green, adaxially with scattered brown punctiform scales, abaxially covered with dense white punctiform scales, abaxially also with a band of grey indumentum along the proximal and distal margins. **INFLORESCENCE** interfoliar, much shorter than the leaves, branched to 1 order; peduncle 7–12.5 cm, c. 2 mm wide, densely covered with brown scales; prophyll 9–10 × 0.4–0.6 cm, sparsely scaly; peduncular bract exceeding the prophyll by 4–5 cm; rachis 5–8 cm, densely scaly as the peduncle; rachillae 5–6 (–8), equalling or shorter than the

Dypsis thiryana. Whole plant × 3/5. From *Dransfield et al.* JD6750. Drawn by Rosemary Wise.

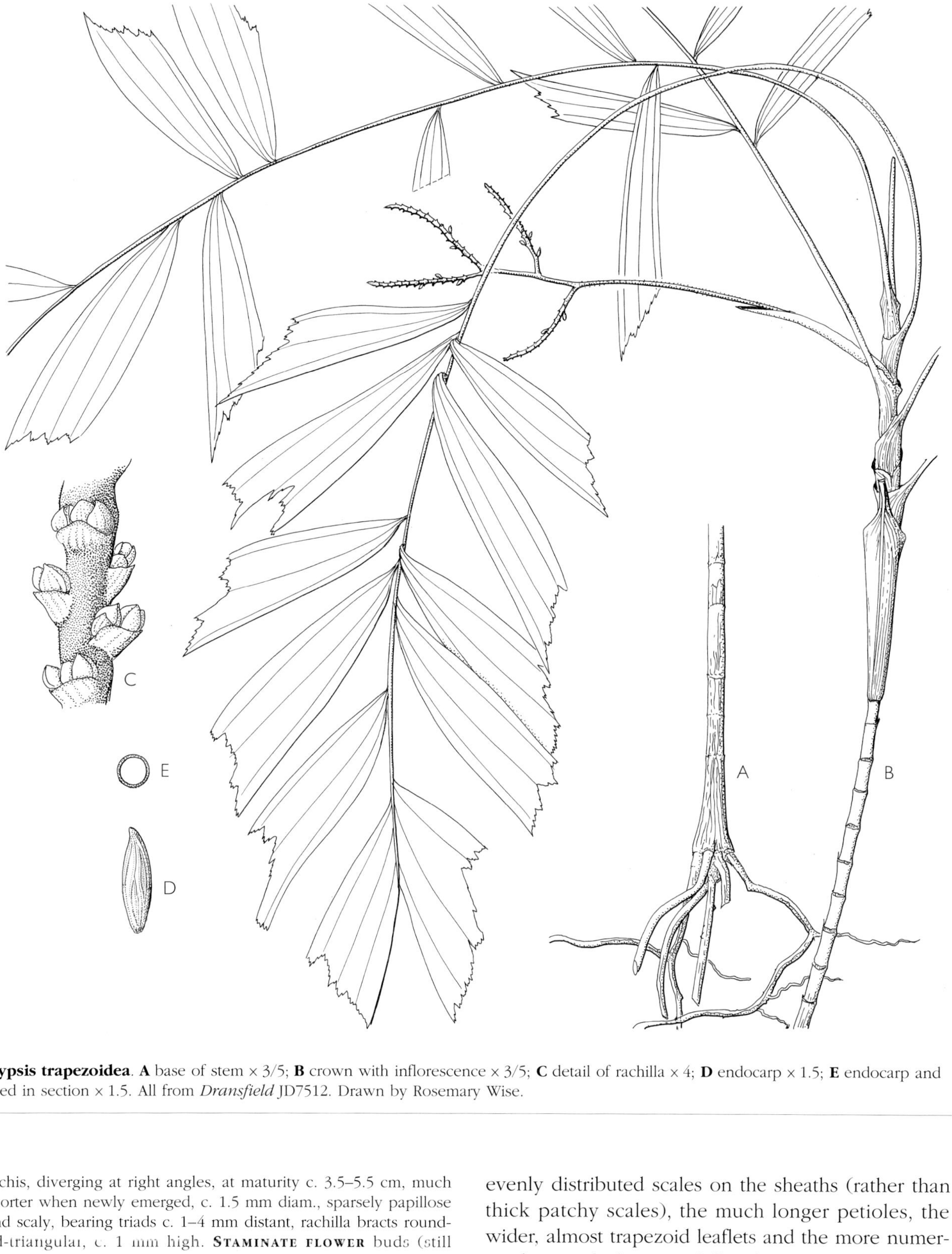

Dypsis trapezoidea. **A** base of stem × 3/5; **B** crown with inflorescence × 3/5; **C** detail of rachilla × 4; **D** endocarp × 1.5; **E** endocarp and seed in section × 1.5. All from *Dransfield* JD7512. Drawn by Rosemary Wise.

rachis, diverging at right angles, at maturity c. 3.5–5.5 cm, much shorter when newly emerged, c. 1.5 mm diam., sparsely papillose and scaly, bearing triads c. 1–4 mm distant, rachilla bracts rounded-triangular, c. 1 mm high. **STAMINATE FLOWER** buds (still immature) rounded, c. 1.2 mm diam.; sepals rounded, imbricate, irregularly keeled, c. 0.8 × 0.8 mm, margins sparsely scaly; petals triangular, 0.9 × 0.7 mm; stamens 6, immature anthers 0.5 × 0.2 mm; pistillode columnar. **PISTILLATE FLOWERS**, only very young buds known. Mature **FRUIT** cherry-red, fusiform, 18 × 6 mm. **SEED** c. 15 × 4 mm, endosperm homogeneous.

NOTE: Because of the praemorse leaflets this species appears very similar to *D. thiryana*. It can be distinguished by the consistently solitary habit, the sparse evenly distributed scales on the sheaths (rather than thick patchy scales), the much longer petioles, the wider, almost trapezoid leaflets and the more numerous but much shorter rachillae that are shorter than the inflorescence rachis (rather than being few, and much longer than the rachis). The mature fruit is also more or less twice the size of that of *D. thiryana*.

SPECIMENS SEEN: Mananjary: Mt. Vatovavy, Nov. 1994 (buds, fr.), *Dransfield et al.* JD7512 (Holotype K; isotypes AAU, BH, MO, NY, P, TAN). idem, Jan. 1964, *Bosser* 18905 (P, TAN).

73. DYPSIS HIARAKAE

A slightly nondescript medium-sized palm from rain forest in the north. The name refers to a village on the west coast of Masoala.

DISTRIBUTION. Manongarivo, Masoala and Mananara Avaratra.
HABITAT. Lowland to submontane rain forest; ridgetop or flat slope; 240–600 m.
LOCAL NAMES. *Sinkiara, Tsirika* (Tsimihety).
USES. Stems used to make blowpipes in Manongarivo.
CONSERVATION STATUS. Rare, possibly vulnerable. Numbers are probably less than a few hundred.

Dypsis hiarakae *Beentje* **sp. nov.**
palma solitaria folio irregulariter pinnato pinnis fasciculatis inflorescentia in 2 vel 3 ordines ramificanti flore staminato 6 staminibus antheris versatalibus endospermio homogeneo distincta. Typus: Madagascar, Ambanizana, Oct. 1986 (y. fr.), *Dransfield et al.* JD6398 (Holotypus K; isotypus TAN).

Solitary palm. **STEM** 3–6 m, 1.2–2.5 cm diam., sometimes stilt-rooted at the base; internodes 1.5–6 cm long, pale green, ± glabrous. **LEAVES** 7–9 in the crown, arching, pinnate; sheath 18–22 cm long, 2/3 closed, pale green with few scattered scales, denser distally, with ragged margins; petiole 15–17 cm long, 5–8 × 2–4 mm diam., adaxially flat, with scattered brown scales; rachis 35–67 cm long, in mid-leaf 2–4 mm wide, with scattered scales; leaflets 12–21 on each side of the rachis, in groups of 2–5, slightly fanned, the group interval 5–12 cm, leaflet interval 0.5–0.9 cm, the proximal 18–24 × 0.3–0.9 cm, median 15–31 × 1.5–5 cm, distal 10–25 × 1.3–4.5 cm, main veins 1–3 (in the distal pair 3–5), with scattered scales on the major and minor veins, and sometimes with long ramenta, apices attenuate, distal pair joined for 1–5 cm, with the apices dentate over a width of 5–10 mm. **INFLORESCENCE** infrafoliar, branched to 2 (rarely 3) orders, erect with numerous arching to hanging rachillae; peduncle 7–37 cm long, proximally c. 12 × 4 mm, distally 3–6 × 3.5–5 mm, densely scaly; prophyll 10–19 cm long, 8–12 mm wide, borne at 1–4 cm above the base of the peduncle, with scattered scales, open in the distal 1–3.5 cm; peduncular bract inserted at 9–14 cm from the base of the peduncle, 14–26 cm long, open for most of its length, beaked for 3 cm, glabrous, rapidly deciduous; rachis 7.5–17 cm long, glabrous or nearly so with 7–10 branched and 5–8 unbranched first order branches, the proximal of these with an axis of up to 7 cm with up to 6 rachillae; rachillae 17–27 cm long, 1–2 mm diam., glabrous to puberulous. **STAMINATE FLOWERS** with sepals 1–1.2 × 1–1.2 mm; petals 2.8–3 × 1.5–2 mm; stamens 6, biseriate, filaments c. 1 mm, anthers 1.3–1.5 × 0.6 mm, parallel and dorsifixed; pistillode c. 1 mm. **PISTILLATE FLOWERS** with sepals 0.9–1.4 × 0.6–1.5 mm; petals 1.5–2.1 × 1.6–2.4 mm; staminodes 6, 0.3–0.5 mm. **FRUIT** red, ellipsoid with a rounded apex, c. 9 × 5 mm. **SEED** 6.5–7 × 3.5 mm, obtuse at both ends, with homogeneous endosperm.

Dypsis hiarakae, collected at Ambalafary (*Beentje et al.* 4578).

NOTE: This species seems closest to *D. confusa*, but is distinct in the long rachillae.

SPECIMENS SEEN: Ambanja: Ambalafary, Feb. 1992 (fr.), *Beentje et al.* 4578 (BH, K, MO, P, TAN); Amtsahakolany Valley, June 1923 (fl.), *Decary* 2139 (K, P). Maroantsetra: Hiaraka, Oct. 1986 (bud), *Dransfield et al.* JD6378 (K, TAN); Ambanizana, Oct. 1986 (y. fr.), *Dransfield et al.* JD6398 (Holotype K; isotype TAN). Mananara

Avaratra: Antanambe, Oct. 1994 (bud), *Dransfield & Beentje* JD7508 (K, TAN).

Similiar specimens. Close, but with significant differences is Midongy Atsimo: Midongy, May 1992 (bud, fr.), *Beentje & Andriampaniry* 4671 (BH, K, MO, P, TAN) from submontane rain forest on slight slopes, c. 500 m.

Clustering palm in tufts of 15–25, the stems 2–6 m tall. **Leaves** with petiole 3.5–10 cm long; leaflets 15–17 on each side of the rachis, median 18–28 × 1.1–2.1 cm, glabrous. **Inflorescence** interfoliar in bud, infrafoliar in fruit, branched to 2 orders; rachis c. 18 cm long, with c. 3 branched and 12 unbranched first order branches; rachillae 9–19 cm long, 1–1.5 mm diam., minutely puberulous. **Staminate flowers** only known from very young buds with 6 stamens, appearing slightly didymous (but possibly because of extreme immaturity). **Fruit** ellipsoid, obtuse, c. 10 × 4.5 mm. **Seed** c. 3.5 mm diam., with homogeneous endosperm.

74. Dypsis confusa

The specific epithet derives from the fact that of the first four specimens we saw two had been named as *Neophloga*, and the other two as *Dypsis* and *Phloga* respectively. It is indeed a palm looking like several other taxa at the same time!

Distribution. Masoala, Mananara and Betampona.
Habitat. Lowland rain forest or peat swamp forest; flat ground or ridge top; 5–300 (–900) m.
Local names. *Tsikara* (Betsimisaraka); *Tsimikara* (Sihanaka).

Dypsis hiarakae. **A** Crown × 2/5; **B** infructescence × 2/5. From *Dransfield et al.* JD6398. Drawn by Rosemary Wise.

USES. Stems used to make blowpipes (fide Cours).
CONSERVATION STATUS. Rare, possibly Vulnerable. Found in several sites, but only one of these is in a protected area. Numbers are thought to be low.

Dypsis confusa *Beentje* **sp. nov.**
D. scottianae superficialiter similis sed foliolis infra squamis dispersis infra tectis petalis floris staminati longioribus differt, a *D. hiarakae* rachillis brevioribus recedit. Typus: Madagascar, Masoala, Antalavia, Feb. 1988, *Dransfield & Staniforth* JD6481 (Holotypus K; isotypi MO, TAN).

Solitary or clustering palm. **STEMS** 3–7 m tall (*Cours* 1904 says 15 m), 1–4 cm diam.; internodes 1–11 cm long, green with some brown scales distally; nodal scars c. 5 mm; crownshaft c. 40 cm long, pale green with brown or reddish scales; wood white, hard. **LEAVES** 5–8 in the crown, pinnate; sheath 12–25 cm long, pale green, densely red-scaly but glabrescent (*Perrier* 15012: with irritant hairs), with rounded, ragged shoulders or with auricles to 1.5 cm long, the outer half open; petiole 3–32 cm long, 3–6 × 2–4 mm diam., with scattered brown scales; rachis 49–86 cm long, in mid-leaf 2.5–5 mm wide, with scattered scales or glabrous; leaflets 11–28 on each side of the rachis, in groups of 2–4 (–8) and in one plane or in slightly ascending fans, group interval 4–14 cm, the proximal 14–37 × 0.2–3.2 cm, median 21–36 × 0.7–2.6 cm, distal 8–20 × 0.2–2.6 cm, connate for 1.5–3 cm, main veins 1 (–3), in the distal pair 2–3, ramenta rarely present, faint reddish glands/scales present on the minor veins, plus occasionally in marginal bands, apices attenuate, unequal, distal pair dentate over a width of 5–6 mm. **INFLORESCENCE** interfoliar, branched to 2 orders, erect; peduncle 20–37 cm long, 2.5–12 × 2–6 mm diam., densely reddish pubescent; prophyll 13–29 cm long, 0.9–1.5 cm wide, borne at 5–14 cm above the base of the peduncle, with scattered scales, open in the distal 1.5–7 cm; peduncular bract inserted at 2–25 cm from the base of the peduncle, 17–24 cm long, green with scattered scales, split over its

Dypsis confusa. **A** crown with inflorescence × 2/5; **B** leaf × 2/5; All from *Dransfield & Staniforth* JD6481. Drawn by Rosemary Wise.

75. Dypsis sahanofensis

This was one of our mysteries, only known from the type (collected in 1911) with very fragmentary material and no description to speak of. Then, just before taking this book to press, we visited Mt Vatovavy, some 70 miles/100 kilometers from the type locality, and we collected a medium-sized palm which turned out to match the type material beautifully. This kind of discovery gives a real thrill!

Distribution. E Madagascar; only known from the type and one recent collection.
Habitat. Rain forest; 315–1400 m.
Local names. Not recorded.
Uses. Not recorded.
Conservation Status. Endangered. The forest on Vatovavy is not protected, except by local fady (taboos), and population numbers were low. If forest clearance in the area continues, the status will become Critical within a few years.

length, with a beak of up to 1.5 cm, circumscissile; rachis 10–36 cm long, scaly, with 8–18 branched and 8–17 unbranched first order branches, the proximal with an axis of up to 8 cm long and with up to 11 second order branches; rachillae yellow to orange, 2.5–15 cm long, c. 1 mm diam., minutely scaly/puberulous to glabrous. **Staminate flowers** with sepals 0.6–1 × 0.7–1.3 mm; petals 1.9–2.5 × 1.2–1.7 mm, orange; stamens 6, 1- to biseriate, filaments 0.9–1 mm, thin, anthers 1.1–1.4 × 0.6–0.7 mm, dorsifixed, not versatile; pistillode 0.6–0.8 × 0.3–0.5 mm. **Pistillate flowers** with sepals 0.8–1 × 1.1–1.3 mm; petals 1.8–2.3 × 1.8–2.2 mm; staminodes 0.2–0.4 mm; pistil only seen as young fruit. **Fruit** orange to red, ellipsoid with a pointed apex, 6–10 × 3–5 mm. **Seed** 3–5 × 2.5–3.5 mm, with homogeneous endosperm.

Note. Most of all this species reminds us of *D. scottiana* from which it differs in the leaflets with scattered scales abaxially, and the longer staminate petals; the rest is gradual: rachis, group interval, leaflets and rachillae are as long as, or slightly longer than specimens of *D. scottiana.* It also resembles *D. hiarakae*, from which it is only distinct by the length of the rachilla.

Moramanga: Mantady, Dec. 1991 (fl.), *Beentje & Andriampaniry* 4545 (K, TAN) keys out as this taxon, but looks quite different. It is clustering with overhanging stems; petiole 0.5–1 cm, leaflets 23–26 on each side of the rachis, almost regular, the median 13–16 × 0.9–1 cm; rachillae 10–13 cm. Similar are: Mandritsara/Andilamena: N of Andilamena, Nov. 1929 (bud), *Perrier* 15012 (P). Ambatondrazaka: Ambatondrazaka, Dec. 1944 (bud), *Homolle* 548 (K, P).

Specimens seen. Maroantsetra: Hiaraka, Oct. 1986 (dead infl.), *Dransfield et al.* JD6383 (K, TAN); Ambanizana, Oct. 1986 (fl.), *Dransfield et al.* JD6386 (K, TAN); idem, Feb. 1989 (fl.), *Schatz & Lowry* 2616 (K, TAN); Andronabe, Feb. 1992 (fl., y.fr.), *Zjhra & Hutcheon* 163 (K); Antalavia, Feb. 1988 (fl.), *Dransfield & Staniforth* JD6481 (Holotype K; isotypes MO, TAN). Mananara Avaratra: Antanambe, April 1992 (y.fr.), *Beentje et al.* 4627 (BH, K, MO, P, TAN). Toamasina: Ambodiriana, Dec. 1944 (bud), *Cours* 1904 (K, P, TAN).

Dypsis sahanofensis, on the isolated hill, Vatovavy (*Beentje & Dransfield* 4835).

Dypsis sahanofensis *(Jum. & H. Perrier) Beentje & J. Dransf.* **comb. nov.**

Synonyms:

Chrysalidocarpus sahanofensis (Jum. & H. Perrier) Jum., Ann. Inst. Bot.-Géol. Colon. Marseille sér. 2, 10 (3): 6 (1922); Ann. Inst. Bot.-Géol. Colon. Marseille sér. 6, 3: 9 (1929); Cat. Pl. Madagascar, Palmae: 11 (1938); Jum. & H. Perrier, Fl. Madagascar 30: 110 (1945). ***Neophloga sahanofensis*** Jum. & H. Perrier, Ann. Inst. Bot.-Géol. Colon. Marseille sér. 3, 1 (1): 32, t. 15–16 (1913). Type: Madagascar, upper Sahanofo R. basin, between Ambohimanga Atsimo and Ambositra, *Perrier* 12060 (Holotype P).

Clustering palm in groups of c. 12. **Stems** 5–6 m long, 3 (–8, fide Perrier) cm diam.; internodes 12–13 cm long. **Leaves** with sheath 20–31 cm long, with sloping shoulders, adaxially dark red-brown, abaxially pale green to mid-brown with some wax and orange-brown caducous scattered scales; petiole 17–18 cm long, proximally c. 1.5 × 0.5 cm, distally c. 1.1 × 0.5 cm, densely scaly on both surfaces, convex on both surfaces with sharp margins; rachis c.1 m long, in mid-leaf 0.5–0.7 cm wide, abaxially with scattered scales, adaxially with a few scales, slightly keeled; leaflets 23–24 on each side of the rachis, grouped to irregular and slightly fanned within the groups, in mid-leaf in groups of 3–5, the groups at intervals of 3–13 cm, the leaflets within the groups at intervals of 0.2–2.5 cm, proximal 26–33 × 0.6–1.4 cm, median 29–38 × 1.7–3 cm, distal 13–23 × 0.5–1.5 cm, glabrous or nearly so, main vein 1, apices attenuate. **Inflorescence** interfoliar, branched to 2 orders; peduncle c. 43 cm, proximally 1.7 × 0.4 cm, distally 1.3 × 0.9 cm, densely reddish-scaly; prophyll 29–44 cm long, borne at 9–25 cm above the base of the peduncle, glabrous or nearly so; peduncular bract inserted at 20–33 cm from the base of the peduncle, c. 34 cm long, deciduous; non-tubular peduncular bract inserted at c. 27 cm from the base of the peduncle, 5 mm

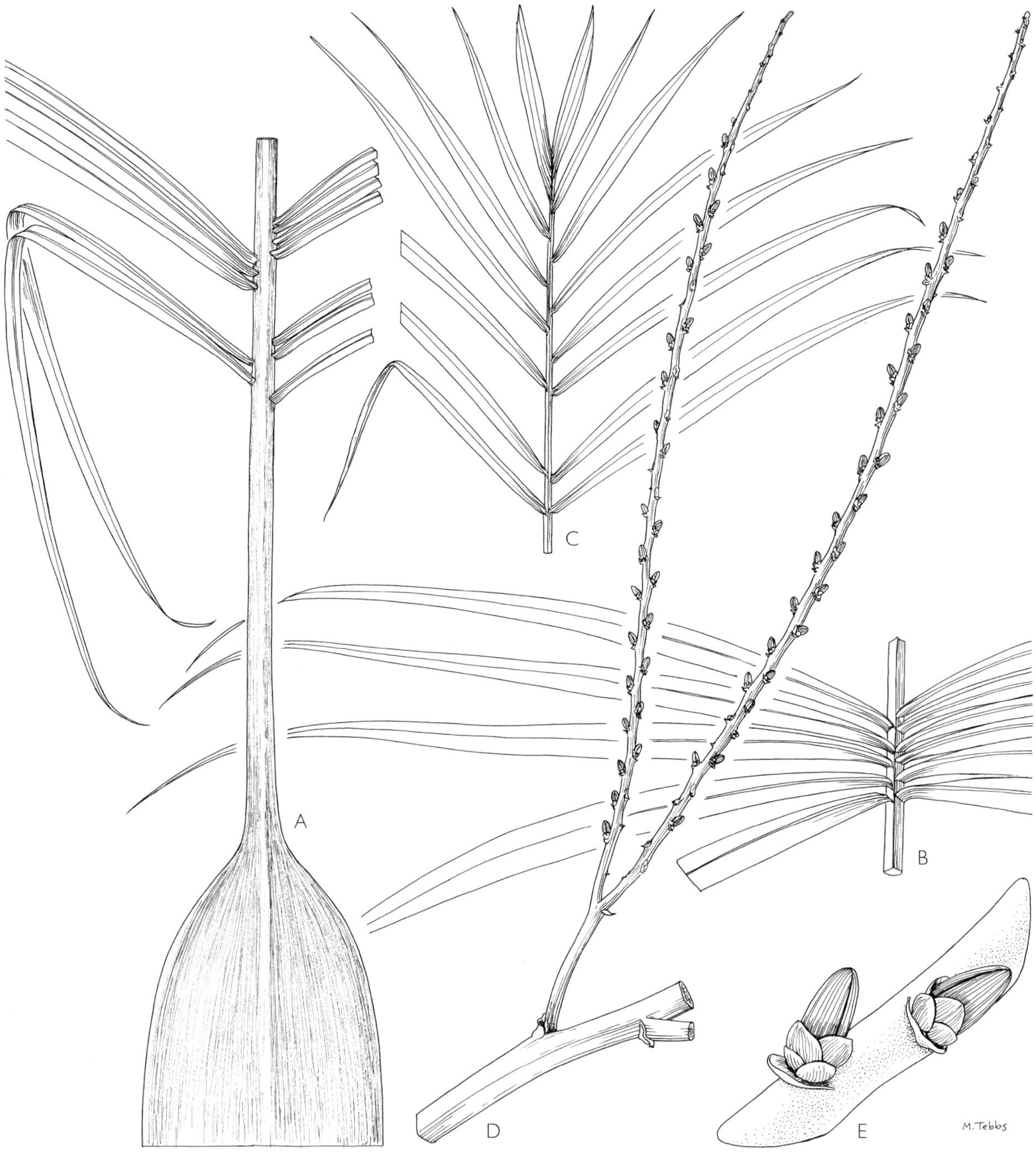

Dypsis sahanofensis. **A** leaf sheath, petiole and basal leaflets × 2/5; **B** mid section of leaf × 2/5; **C** leaf tip × 2/5; **D** part of inflorescence × 2/5; **E** detail of rachilla × 3. All from *Perrier* 12060. Drawn by Rosemary Wise.

long; rachis c. 27 cm, puberulous, with c. 5 branched, and 11–16 unbranched first order branches, the proximal of these with a secondary rachis of up to 5 cm long and 7 x 2.5 mm diam. at base, with up to 3 rachillae; rachillae 19–33 cm long, 1–3.5 mm diam., puberulous, with very distant triads in pits with proud acute rachilla bracts. **STAMINATE FLOWERS** only seen in young bud: sepals to 1.5 x 1.3 mm, keeled and slightly hooded; petals still smaller than the sepals at this stage; stamens still less than 0.3 mm, but clearly 6, anthers probably versatile. **PISTILLATE FLOWERS** with sepals 2.6–3.2 x 2.5–3 mm; petals connate for 0.6–0.8 mm, free for 5.1–5.7 x 3–3.5 mm; staminodes 6, 0.8–1.2 mm long; ovary 5–5.3 x 2–2.2 mm. **FRUIT** not seen.

NOTE. The type at P consists of three sheets. Two sheets of *Perrier* 12049, *Neophloga littoralis*, (sheath and middle part of rachis) were originally excluded by us as not belonging to that taxon. In fact the two pieces not only match those of the type of *D. sahanofensis*, but fit parts of the rachis on the type sheet! There is no doubt that they are the same number; the sheath is figured in Jumelle & Perrier (1913).

SPECIMENS SEEN. Ambositra: upper Sahanofo basin between Ambohimanga du Sud and Ambositra, Nov. 1911 (fl.), *Perrier* 12060 (P; type). Mananjary: Vatovavy, Nov. 1994 (bud), *Beentje & Dransfield* 4835 (K, TAN).

76. DYPSIS ERIOSTACHYS

This handsome little palm, distinctive in its entire-bifid leaf and silky hairy inflorescence, is known only from low crown forest near the summit of the isolated peak of Mount Vatovavy, southwest of Mananjary. The forest on Vatovavy is home for several curious palms, such as *Dypsis basilonga*, *D. trapezoidea* and *D. angusta*. The lower parts of the hill have been largely cleared for shifting cultivation and parts of the exposed northeastern face have at some time been partially destroyed by fire. Nevertheless much of the summit area is covered with fine forest that receives some degree of protection under local beliefs or *fady*. The species name is Greek for woolly spikes, referring to the hairy inflorescences.

DISTRIBUTION. Only known from Vatovavy.
HABITAT. Gentle slope in low canopy forest near summit of hill; 450 m.
LOCAL NAMES. Not recorded.
USES. Not recorded.
CONSERVATION STATUS. Endangered. Single site status; forest only protected by local custom.

Dypsis eriostachys *J. Dransf.* **sp. nov.**
habitu, statura, folio bifido et inflorescentia lanuginosa *D. lanuginosae* superficialiter similis, sed pagina inferiore folii ramentis elongatis, rachillis c. 30 (vice c. 100) pilis inflorescentiis longioribus, non crispatis, floribus staminatis staminibus 6 differt. Typus: Madagascar, Mananjary, Vatovavy, *Dransfield et al.* JD7513 (Holotypus K; isotypi MO, P, TAN).

Slender solitary undergrowth palm to 1.5 m tall. **STEM** 9 mm diam., internodes 13–17 mm long, dark green with pale brown lines, with scattered dark brown caducous scales. **LEAVES** 10 in crown; sheath 8 x 1.7 cm, distally very densely covered with thick red-brown tomentum and scales; auricles ill defined; petiole 1–3 cm, 3 mm wide, densely grey or brown hairy; rachis 23–25 cm; blade entire bifid 32–37 cm long, gradually widening from the base to widest at the tip where 11–13 cm, the two lobes with a broad sinus, and shallowly to deeply lobed apically, adaxially shining, with very inconspicuous punctiform scales, abaxially with minute brown punctiform scales and scattered elongate brown ramenta along a few veins, emerging leaf tinged reddish. **INFLORESCENCE**

erect or spreading, becoming pendulous in fruit, branching to 2 orders; peduncle 25–28 cm long, c. 4 mm wide at the base tapering to 3 mm diam. distally, densely covered with long silky silvery hairs with brown bases; prophyll c. 14 × 0.7 cm, with scattered long silky silvery hairs with dark bases; peduncular bract exceeding the prophyll by 8–16 cm; rachis 16–24 cm, very densely silky hairy; rachillae c. 30, c. 3–5.5 cm long, c. 1.5 mm diam., densely covered in pale brown and silvery silky hairs, triads c. 3 mm apart, rachilla bracts rounded, c. 0.5 mm high, laciniate, edged with hairs. **STAMINATE FLOWER** buds pyramidal, 1.6 × 1.1 mm; sepals rounded, 1.0 × 0.8 mm, margins somewhat erose; petals triangular, 1.3 × 0.8 mm, striate; stamens 6, biseriate, antesepalous with filaments to 0.2 × 0.2 mm, antepetalous with filaments 0.4 × 0.2 mm, anthers elliptic in outline, 0.4 × 0.3 mm; pistillode conical, 0.4 mm high. Immature **PISTILLATE FLOWER** rounded, c. 0.8 mm diam. Submature **FRUIT** fusiform, 16 × 6 mm. **SEED** fusiform, 1.2 × 4.5 mm; endosperm homogeneous; embryo basal.

NOTE. Initially we identified this palm as *Dypsis lanuginosa*, previously known only from a single specimen. On closer examination, however, the newly collected material proved to have six stamens rather than three, and other features such as the long silky more or less straight, rather than twisted, hairs, fewer rachillae, with more distant triads, and the presence of elongate ramenta on the undersurface of the leaf, added evidence that the present species was undescribed.

SPECIMENS SEEN. Mananjary: Vatovavy, Nov. 1994 (fl., fr.), *Dransfield et al.* JD7513 (Holotype K; isotypes MO, P, TAN).

Dypsis eriostachys. **A** crown with leaves and infructescence × 3/5; **B** detail of rachilla with flowers × 3; **C** staminate flower, one petal removed × 25; **D** fruit × 2.5; **E** fruit in cross section × 2.5. All from *Dransfield et al.* JD7513. Drawn by Rosemary Wise.

77. Dypsis lutea

A distinctive undergrowth species with fairly large, entire, pale green leaves and a branched inflorescence. The name *lutea* means golden yellow, and this alludes to the colour of the inflorescence.

Distribution. Masoala and the Moramanga area.
Habitat. Lowland and submontane rain forest; ridge top (?always); 40–1100 m.
Local names. Not recorded.
Uses. Not recorded.
Conservation Status. Critical. There is only one recent collection, from an unprotected area which is being destroyed; the other collections are over eighty years old.

Dypsis lutea *(Jum.) Beentje & J. Dransf.* **comb. nov.**

Synonyms:
Neophloga lutea Jum., Ann. Inst. Bot.-Géol. Colon. Marseille sér. 4, 6 (3): 32 (1929); Ann. Inst. Bot.-Géol. Colon. Marseille sér. 5, 1 (3): 6 (1933); Cat. Pl. Madagascar, Palmae: 21 (1938); Jum. & H. Perrier, Fl. Madagascar 30: 74, fig. 20 (1945). Syntypes: Madagascar, Ambatovola, *Perrier* 12002 (syntype P) & 18391 (syntype P).

Neophloga lutea Jum. var. ***transiens*** Jum. & H. Perrier, Fl. Madagascar 30: 76 (1945). Type: Madagascar, Masoala Peninsula, *Perrier* 11940 (Holotype P), **synon. nov**.

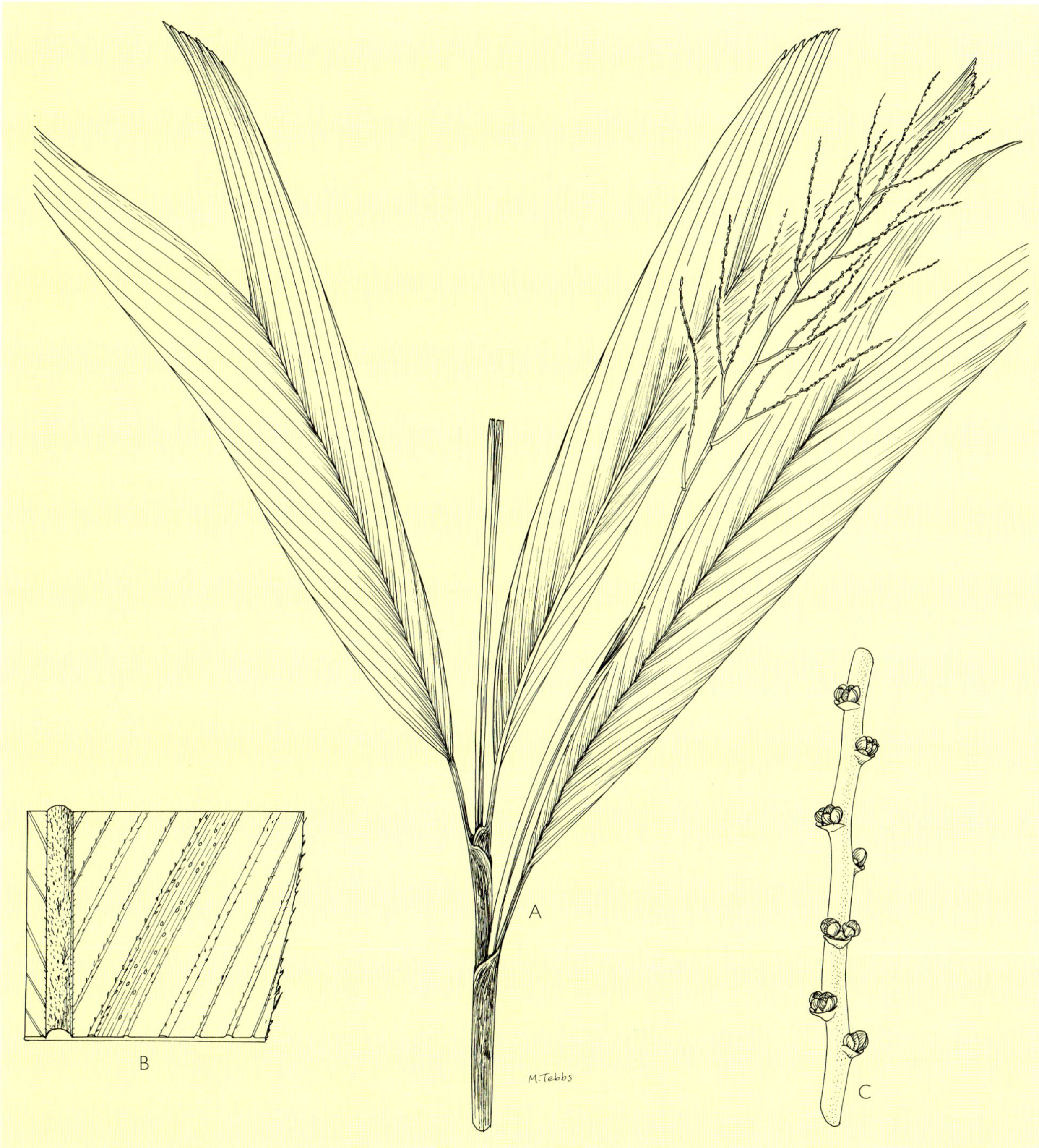

Dypsis lutea. **A** crown × 2.5; **B** detail of undersurface of leaf × 2; **C** detail of rachilla × 2. All from *Dransfield et al.* JD6421.
Drawn by Margaret Tebbs.

Dypsis lutea

Clustering or solitary palm. **STEM** 2.5–3 m high, to 0.8 cm diam., the distal internodes c. 7 cm, blackish brown, covered in laciniate scales. **LEAVES** c. 9 in the crown, entire; sheath 10–17 cm long, c. 1 cm diam., pale green, distally covered in dense red-brown laciniate scales when young, with or without triangular auricles to 3 mm long and laciniate; petiole absent or to 4.5 cm long, c. 1.5 mm diam., with brown laciniate scales mixed with silvery scales; blade obtriangular, 30–55 cm long, pale green, midrib/rachis 21–43 cm long, lobes 9–17 x 2.3–4.2 cm, main veins 7–17, scattered scales on the midrib and to a lesser extent on the veins, apices 5–15 mm wide and dentate, also dentate on the distal outside margins. **INFLORESCENCE** interfoliar, yellow to orange, branched to 1 or 2 orders, 33–77 cm long; peduncle 19–60 cm, 2.5–5 mm diam., densely scaly but glabrescent; prophyll 14–37 cm long, 0.5–1 cm wide, borne at 2.5–9 cm above the base of the peduncle, split for the distal 2.5 cm, with scattered scales; peduncular bract deciduous or persistent, inserted at 12–27 from the base of the peduncle, 13–31.5 cm long, split in the distal 4 cm, with scattered scales; rachis 4–16 cm long, glabrous, without or with up to 5 branched, and 4–11 unbranched first order branches, the proximal of the branched ones with up to 4 rachillae; rachillae 4–17 cm, with dense triads and dark yellow flowers. **STAMINATE FLOWERS** with sepals 1–1.5 x 1–1.4 mm; petals 2.1–2.5 x 1.4–1.6 mm; stamens 6, equal or offset for up to 0.3 mm, filaments 0.8–1 mm long, 0.2–0.3 mm diam., cylindrical with slightly triangular base, anthers 1–1.5 x 0.7–0.8 mm, dorsifixed, versatile, the locules parallel and obtuse; pistillode 0.4–1.1 mm high, 0.3–0.4 mm diam. **PISTILLATE FLOWERS** with sepals 1–1.2 x 1.2–1.6 mm, ciliolate; petals 2.8–3 x 2.2–2.7 mm, ciliolate?; staminodes 6, 0.3–0.6 mm long; gynoecium 2.8–3 mm high, 1.6–2.7 mm diam. **FRUITS** (young) oblong, incurved, attenuate towards an obtuse apex.

NOTE. Var. *transiens* was separated on the base of the peduncular bract being deciduous, not split at the apex, and beaked, whereas var. *lutea* has peduncular bract remaining, opening only at the apex, not beaked. *Perrier* 12002 has one inflorescence which differs strongly from the others on that specimen and looks very much like that of var. *transiens*, including the stouter prophyll, the deciduous peduncular bract and the inflorescence branched to two orders. *Perrier* 18391 shows a peduncular bract with a 1 cm long beak, just as in var. *transiens*. We see no reason to uphold the varieties.

Decary 5064 is from Ivohibe: Ifandana, Sept. 1926 (bud), (P), and is very like *D. lutea*. It is 5 m high; the leaf is 62 cm long, with lobes 42 cm long; the inflorescence is branched to 2 orders, with a rachis of 24 cm with 15 branched and 10 unbranched first order branches; rachillae are 5–9 cm long. The staminate flowers are smaller in all dimensions, but they are in bud.

SPECIMENS SEEN. Maroantsetra/Antalaha: Masoala Peninsula, Nov. 1912 (fl.), *Perrier* 11940 (P, type of var. *transiens*). Moramanga: Ambatovola, Jan. ?1913 (fl.), *Perrier* 12002 (P, syntype); idem, Jan. 1928 (fl.), *Perrier* 18391 (P, syntype); Andasibe, Anevoke, Nov. 1986 (bud), *Dransfield et al.* JD6421 (K, P, TAN).

78. DYPSIS BETAMPONENSIS

A small palm with entire leaves, much-branched inflorescences, and didymous stamens; distinct, and presumably extinct as well. The species name is derived from the type locality.

DISTRIBUTION. Only known from a single collection from Betampona.
HABITAT. Lowland rain forest; c. 100 m (according to Jumelle & Perrier (1945); the type specimen gives no altitude).
LOCAL NAMES. Not recorded.
USES. Not recorded.

Dypsis betamponensis

CONSERVATION STATUS. Presumed extinct: not seen for seventy years. Betampona has been visited by several palm collectors, including HB, in recent years.

Dypsis betamponensis *(Jum.) Beentje & J. Dransf.* **comb. nov.**

SYNONYM:
Neophloga betamponensis Jum., Ann. Inst. Bot.-Géol. Colon. Marseille sér. 4, 6 (3): 33 (1929); Jum. & H. Perrier, Fl. Madagascar 30: 76 (1945). Type: Madagascar, Betampona near Ambodiriana, *Perrier* 17469 (Holotype P).

Small palm to 1 m. **LEAVES** entire, with the sheath c. 11 cm long, 6.5 mm diam., with sparse reddish scales and c. 2 mm high auricles; petiole c. 6 cm long, 2.5–3 mm diam., with reddish scales; lamina 49–52 cm long, midrib 21.5 cm, lobes 30 × 5.7 cm, main veins c. 7, with lines and bands of reddish scales, apices 0.8–1.8 cm wide, long-dentate. **INFLORESCENCE** c. 63 cm, branched to 1 (2) order(s); peduncle c. 41 cm long, c. 2 mm diam., glabrous; prophyll c. 30 cm long, 1 cm wide, borne at 10.5 cm above the base of the peduncle, with scattered pale brown scales; peduncular bract inserted at c. 30 cm from the base of the peduncle, c. 12 cm long; rachis c. 16 cm long, with sparse long (1–2 mm) red curly hairs, especially in the axils of rachillae, with 1 branched and 22 unbranched first order branches, the branched one with 3 rachillae; rachillae reflexed, 3–5.2 cm, c. 1 mm diam., glabrous, with quite dense triads. **STAMINATE FLOWERS** with sepals imbricate, 0.6–0.8 × 0.8–1.1 mm; petals 0.8–0.9 mm high; stamens 6, biseriate, didymous, the filaments 0.2–0.25 mm long and connate at their bases, anthers 0.2–0.35 mm long and wide, in the type often with 4 fertile, 1 sterile and 1 missing or vestigial stamen; pistillode c. 0.2 mm high. **PISTILLATE FLOWERS** only known from very young buds, with sepals imbricate, the petals still enclosed within. **FRUIT** unknown.

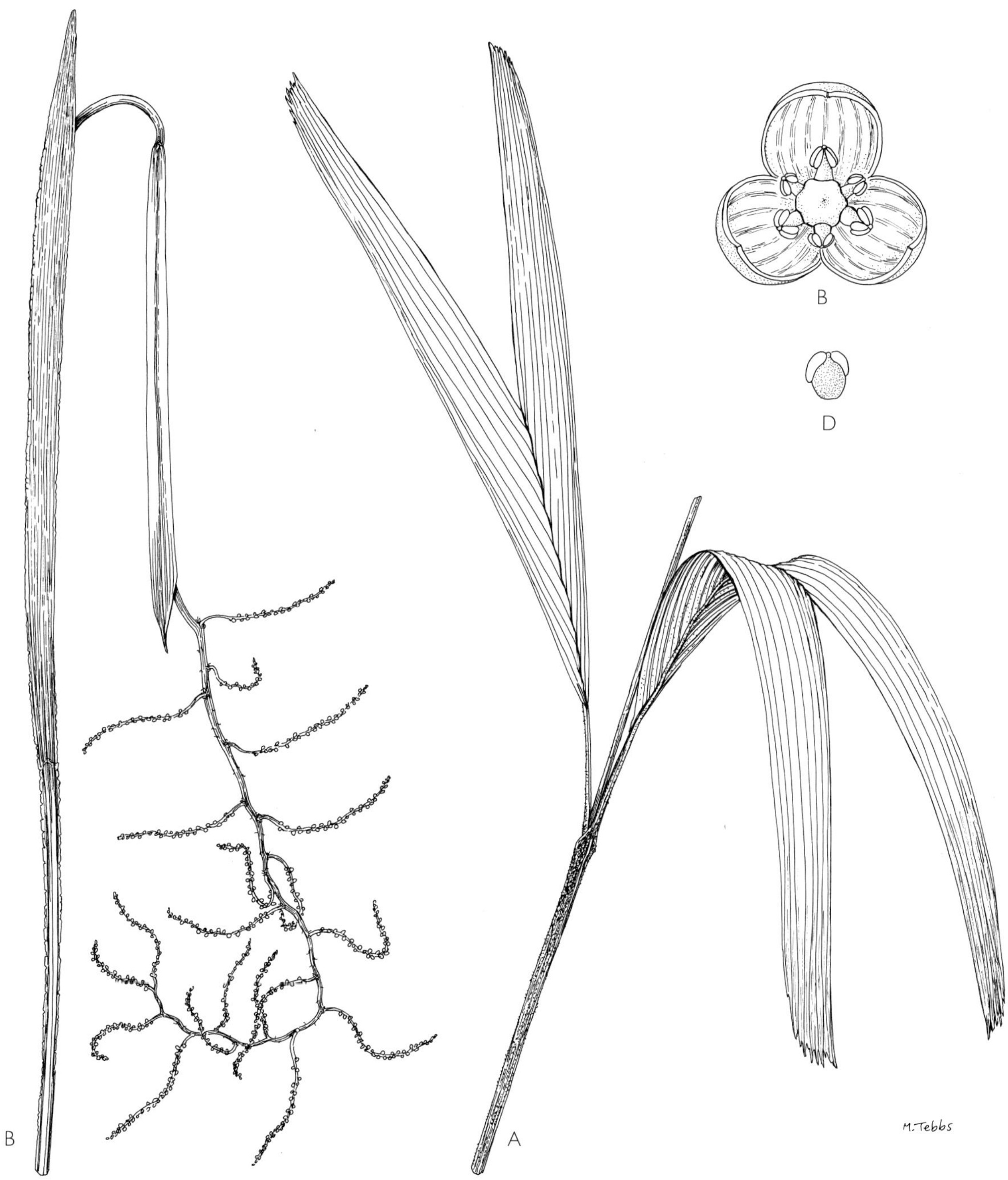

Dypsis betamponensis. **A** crown × 1/2; **B** inflorescence × 1/2; **C** staminate flower × 20; **D** stamen × 30. All from *Perrier* 17469. Drawn by Margaret Tebbs.

NOTE. The didymous stamens are very rare in entire-leaved *Dypsis* species, and only *D. catatiana, D. fanjana* and *D. singularis* are similar in this respect.

SPECIMENS SEEN. Toamasina: Betampona, Dec. 1925 (bud), *Perrier* 17469 (Holotype P).

79. DYPSIS MAHIA

This is a very small species with pencil-thin stems. The small, entire leaves with long lobes are very distinct. *Mahia* is Malagasy for 'slender' and refers to the very slender lobes of the lamina which, curiously and most unusually in the genus, are acuminate rather than dentate.

DISTRIBUTION. Manombo, only known from the type.
HABITAT. Lowland rain forest on thin humus layer over rocky soil; c. 60 m.
LOCAL NAMES. Not recorded.
USES. Not recorded.

CONSERVATION STATUS. Critical. Known from a single site, which is slowly being destroyed by fire, shifting cultivation and logging.

Dypsis mahia *Beentje* **sp. nov.**
palma minima lobis folii integri longis acutis distincta. Typus: Madagascar, Manombo, *Beentje & Andriampaniry* 4792 (Holotypus K; isotypus TAN).

Solitary or with paired stems. **STEMS** to 60 cm, c. 6 mm diam; internodes 3–6 mm, glabrous; nodal scars 1–2 mm. **LEAVES** 8–10 in the crown, entire; sheath 5–5.5 cm long, the outer open, the inner closed, pale brown with scattered reddish scales distally, with laciniate edges, without auricles; petiole 5.5–7 cm long, 2–2.5 mm wide, with scattered scales; lamina entire, 34–38 cm long, connate for 5–7 cm, lobed for 80–85 %, the lobes 1.1–1.8 cm wide; main veins 5, with few small reddish scales; apices long-acuminate, not dentate. **INFLORESCENCE** interfoliar, erect, branched to 1 order; peduncle 15–16 cm long, 1.5–2 mm diam., glabrous; prophyll 11–11.5 cm long, open in the distal 1 cm, with few scattered scales; peduncular bract inserted at c. 8 cm from the base of the peduncle, c. 10.5 cm long, open in the distal 4.5 cm, with few scattered scales; rachis c. 6 cm long, glabrous, with 12 rachillae; rachillae porrect, 3–5 cm long, 0.6 mm diam., glabrous. **STAMINATE FLOWERS** yellow in bud, with sepals 0.5–0.6 × 0.5–0.7 mm; petals 0.7–0.9 × 0.8 mm; stamens 6, uniseriate?, the filaments connate at the base, 0.4 mm long and thin, anthers 0.4 × 0.3 mm,

Dypsis mahia. **A** crown × 1/4; **B** detail of rachilla × 3. All from *Beentje & Andriampaniry* 4792. Drawn by Margaret Tebbs.

dorsifixed, slighttly sagittate. **Pistillate flowers** not seen. **Fruit** not seen.

Note. *D. mahia* has stems and sheaths reminiscent of *D. bernierana*, but is otherwise very different and does not really resemble any other species.

Specimens seen. Farafangana: Manombo, Jan. 1993 (bud), *Beentje & Andriampaniry* 4792 (Holotype K; isotype TAN).

Similar specimen: Antalaha: across Onive R. from Ambatobe, April 1971 (fr.), *Moore* 9916 (P) is a clustering palm to 5 m high, from low rain forest; it is called *Sinkara*, and the stems are used for blowpipes. **Leaves** entire; petiole 11–13 cm; blade 41–43 cm, lobed for 37–39 cm, the lobes 2.3–2.7 cm wide, with scattered scales in bands, apices slightly dentate. **Inflorescence** branched to 2 orders; rachis c. 10 cm, with 7 branched and 11 unbranched first order branches, the proximal of these with a rachis of 2 cm and up to 4 rachillae; rachillae 3.5–5 cm, 1.5 mm diam., with scattered scales or almost glabrous; triads distant. **Fruit** orange-pink, ellipsoid, 10–11 mm, 4–6 mm diam; seed 3.5 mm diam., with homogeneous endosperm.

80. Dypsis pulchella

An attractive undergrowth palmlet with very much the appearance of *D. louvelii*, but with shorter inflorescences and staminate flowers with six rather than three stamens. The species name refers to the pretty appearance of the three known specimens.

Distribution. E Madagascar, Andasibe and lower Mangoro.
Habitat. Rain forest; 300–1000 m.
Local names. Not recorded.
Uses. Not recorded
Conservation Status. Presumed extinct. Not seen for sixty-eight years.

Dypsis pulchella *J. Dransf.* **sp. nov.**
D. louvelii foliis et inflorescentiis superficialiter similis, sed pedunculo breviori floribus staminatis 6 staminibus biseriatis bene distincta. Typus: Madagascar, Mahanoro, lower Mangoro, *Perrier* 18046 (Holotypus P).

Slender ?clustering palm of the forest undergrowth. **Stems** probably not exceeding 1 m tall, 4–7 mm diam., internodes 8–13 mm, with sparse scattered dark brown scales. **Leaves** 5–8 in crown; sheath 3–6 cm long, c. 1 cm diam., covered with scattered dark brown scales, auricles triangular, to 15 × 5 mm, membranous, soon disintegrating; petiole absent or to 2 cm long, c. 2 mm wide, with scattered dark brown scales; blade entire, bifid, plicate, 20–23 cm long, 8–10 cm wide, divided to just under to over half the length, costa 6–11 cm long, the two lobes 12–15 × 2.5–3 cm, adaxially with scattered brown punctiform scales, abaxially densely covered with brown punctiform scales and larger laciniate scales in bands, leaf tips shallowly lobed. **Inflorescence** interfoliar, shorter than the leaves, branched to 1 order; peduncle 13–17 cm long, 1–1.5 mm diam., with scattered dark brown laciniate scales; prophyll inserted c. 2.5 cm above the base of the peduncle, 5–12 cm long, c. 7 mm wide, membranous, with scattered dark brown scales; peduncular bract inserted 3–8 cm above the prophyll insertion, 6–9 cm long, c. 7 mm wide; rachis 2.5–4 cm long, straight or somewhat zigzag, bearing scattered dark brown scales; rachillae 6–9, 2.5–3.5 cm long, c. 1 mm diam., dark brown scaly, triads c. 1–3.5 mm distant, rachilla surface with scattered dark laciniate scales and minutely papillose. **Staminate flower** c. 2.5 × 1.5 mm; sepals rounded, 1.1 × 1.1 mm, irregularly keeled; petals triangular, 1.6 × 1.3 mm, striate; stamens 6, biseriate, antepetalous filaments 0.8 × 0.2 mm, antesepalous filaments 0.5 × 0.2 mm, anthers didymous, the antesepalous c. 0.4 × 0.4 mm, the antepetalous smaller; pistillode scarcely evident. **Pistillate flower** bud immature, c. 1 mm diam. **Fruit**, only immature known, fusiform, 6–7 × 2–3 mm.

Notes. This taxon was included in the folders of *D. louvelii* in the Paris Herbarium. However, the inflorescence has a relatively shorter peduncle and the

sometimes zigzag rachis gives the inflorescence a different appearance, but it is the staminate flowers which immediately separate the taxon from *D. louvelii*. There are six biseriate stamens, in form reminiscent of those of *D. nodifera*. The two collections from the lower Mangoro are clearly conspecific; the collection from Andasibe differs slightly in having a zigzag inflorescence rachis.

SPECIMENS SEEN. Moramanga: Andasibe, Feb. 1925 (fl., fr.), *Perrier* 17216 (P). Mahanoro: Mangoro, Oct. 1927 (fl.), *Perrier* 18045 (P); idem (fl.), *Perrier* 18046 (Holotype P).

81. DYPSIS PLURISECTA

A fairly non-descript species, even though it is only known from the type - but under the microscope a most amazing feature becomes apparent: the stamens have the locules pendulous from a broad connective. These so-called 'geonomoid' stamens are known from very few *Dypsis* species, and resemble those of the genus *Geonoma* from South America (see page 125). The name 'plurisecta' is Latin for 'divided into several segments'.

DISTRIBUTION. Uncertain, possibly Maroantsetra.

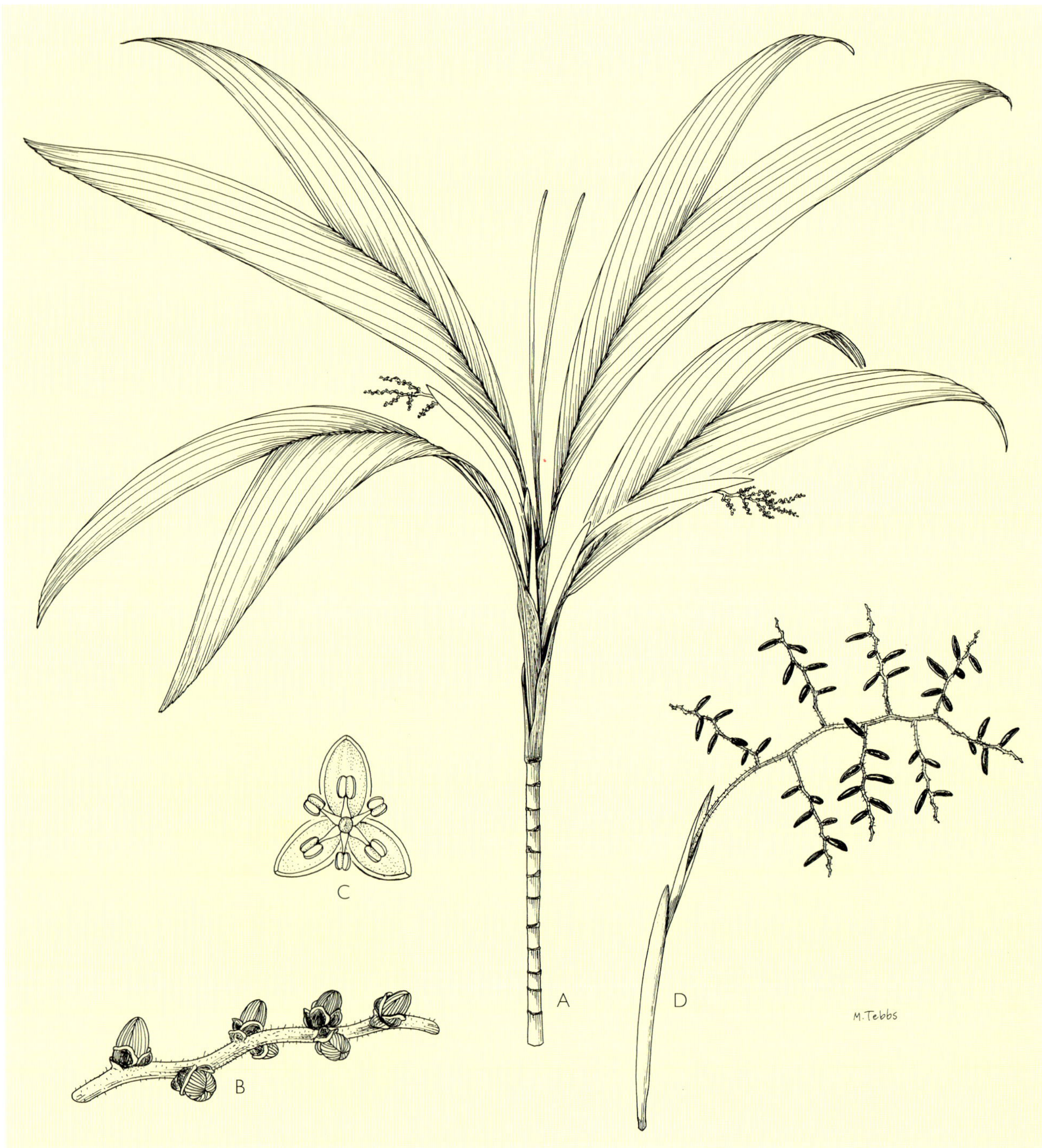

Dypsis pulchella. **A** crown × 1/2; **B** detail of rachilla × 3; **C** staminate flower × 10; **D** infructescence × 1/2. **A** – **C** from *Perrier* 18046, **D** from *Perrier* 17216. Drawn by Margaret Tebbs.

HABITAT. Unknown.
LOCAL NAMES. Not recorded.
USES. Not recorded.
CONSERVATION STATUS. Presumed extinct; not seen for over seventy years.

Dypsis plurisecta *Jum.*
Ann. Inst. Bot.-Géol. Colon. Marseille sér. 3, 6 (1): 35 (1918); Bull. Ac. Malgache 6: 14 (1923); Cat. Pl. Madagascar, Palmae: 14 (1938); Jum. & H. Perrier, Fl. Madagascar 30: 46, fig. 12 (1945). Type: probably from the Maroantsetra area, *Perrier* 11966 (Holotype P).

Habit unknown. Distal internodes 0.5–1 cm long, c. 0.7 cm diam. **LEAVES** with the sheath 5–7 cm long, mostly glabrous but distally with scattered scales; petiole absent or up to 5 mm long; rachis 16–24 cm long, with scattered scales, in mid-leaf 1–1.5 mm diam., with c. 1 mm broad wings between the leaflets; leaflets 5–6 on each side of the rachis, regular or nearly so, opposite or occasionally one pair not opposite, the proximal 5–7 × 0.6–1.1 cm, median 10–15 × 1–2 cm, distal 6–10.5 × 1–2.2 cm, slightly sigmoid with attenuate apices, main veins 2–3, with minute scattered white scales on the minor veins, distal pair joined for 3–4 cm, dentate over c. 1 cm. **INFLORESCENCE** interfoliar, branched to 1 order; peduncle 15–18 cm long, c. 1 mm diam, with scattered scales; prophyll 9–10 cm long, c. 3 mm wide, open in the distal 1 cm; peduncular bract inserted at 7–11 cm from the base of the peduncle, 5–7 cm long, open in the distal 1–2 cm; rachis 5–7 cm long, with c. 9 rachillae; rachillae 2.5–3 cm long, c. 1 mm diam., with scattered scales. **STAMINATE FLOWERS** in bud with sepals 0.5–0.7 × 0.6–0.8 mm; petals c. 1 × 0.8 mm; stamens 6, biseriate, the inner inserted higher up, filaments c. 0.2 mm long, anthers 0.3–0.4 × 0.3–0.4 mm, the locules pendulous from a broad connective (sub-geonomoid); pistillode c. 0.2 mm, conical. **PISTILLATE FLOWERS** unknown. **FRUIT** unknown.

Dypsis plurisecta. **A** crown × 1/2; **B** detail of rachilla × 2. All from *Perrier* 11966. Drawn by Margaret Tebbs.

NOTE. The protologue describes the stamens as three in number, with short filaments, a broad connective, and pendulous locules. All this is quite correct, except for the number of stamens, which is six. In a table on p. 29 of the publication which contains the protologue, there is a description of the leaf sheath anatomy.

This type of stamen also occurs, within *Dypsis*, in *D. fasciculata*, *D. lokohoensis* and *D. thouarsiana*, all species with three stamens.

SPECIMENS SEEN. Probably Maroantsetra ("label lost, without a doubt from the Maroantsetra area"), date unknown (before Aug. 1919), *Perrier* 11966 (Holotype P).

82. DYPSIS ANDAPAE

An entire-leaved species with large leaves and relatively small unbranched inflorescences. The specific epithet derives from the fivondronana, the local administrative unit.

DISTRIBUTION. Marojejy Mountains and Mandritsara area.
HABITAT. Submontane rain forest; steep mid slope; 850–1400 m.
LOCAL NAMES. *Tsingovatrovatra* (Tsimihety, a rather general name for small palms).
USES. Not recorded.
CONSERVATION STATUS. Rare. The distribution of this species is limited.

Dypsis andapae *Beentje* **sp. nov.**
gregem palmunculorum foliis integris inflorescentiis spicatis pertinens *D. coriaceae* et *D. tenuissimae* habitu caespitosa foliis profunde lobatis 6 staminibus affinis sed vagina et lobis folii multo majoribus distincta. Typus: Madagascar, Beamalona, June 1992, *Beentje & Andriampaniry* 4680 (Holotypus K; isotypi MO, P, TAN).

Clustering palm in tufts of 4–6. **STEMS** to 1.2 m tall, 0.8–1.2 cm diam.; distal internodes 4–4.5 cm, glabrous. **LEAVES** 6–10 per crown, arching, entire, bifid; sheath 10–16 cm long, closed for 3/4, with some scattered scales, with 4–10 mm long narrowly triangular auricles; petiole 7–18 cm long, 2–3.5 mm diam., with densely scattered scales, lamina 29–48 cm long, midrib 10–22 cm long, lobes 18–28 x 3.3–7.5 cm, main veins 8–12, with small reddish scattered scales on major and minor veins, apices dentate over a width of 8–15 mm, the outer margins with some teeth as well. **INFLORESCENCE** interfoliar, unbranched; peduncle 22–33 cm long, 1.5–2.5 mm diam., glabrous or minutely puberulous; prophyll 13–22 cm long, 0.7–1.7 cm wide, borne at 1.5–5 cm above the base of the peduncle, open for 1–3 cm, with scattered scales; peduncular bract inserted at 10–15 cm from the base of the peduncle, 9–20 cm long, open for 25–50 %, with scattered scales, sometimes with a beak of up to 5 mm long, tattering; rachilla 13–22 cm long, 1.5–2.5 mm diam., with stellate scales, with dense triads. **STAMINATE FLOWERS** with the free part of the sepals 0.6–1 x 0.5–0.8 mm, the connate part c. 0.5 mm; petals 1.5–2 x 0.7–1.1 mm; stamens 6, biseriate (offset 0.2 mm), filaments 0.5–0.8 mm, anthers 0.6–0.8 x 0.2–0.4 mm, versatile; pistillode < 0.5 mm. **PISTILLATE FLOWERS** with sepals 0.7–1.2 x 0.7–1 mm; petals 1.5–2 x 0.8–1.5 mm; staminodes not seen; ovary c. 0.6–0.8 x 0.5 mm in bud. **FRUIT** orange, ellipsoid with rounded apex, 11–13 x 7–7.5 mm; endocarp with free fibers. **SEED** 9–11.5 x 4.5–5.5 mm, pointed at base and apex, with homogeneous endosperm.

NOTE. In the group of small palms with entire leaves and unbranched inflorescences, closest (through clustering habit, deeply lobed leaf, 6 stamens) to *D. coriacea* and *D. tenuissima*, but distinct by a larger leaf-sheath and much larger leaf lobes.

SPECIMENS SEEN. Andapa: from Ambatoharanana Valley to upper Antsahaberoka R. basin, Nov. 1959 (fl.), *Humbert & Saboureau* 31703 (K, P); Beamalona, June 1992 (fr.), *Beentje & Andriampaniry* 4680 (Holotype K; isotypes MO, P, TAN). Mandritsara: Antsiasiaka, April 1974 (bud), *Morat* 4499 (P).

83. DYPSIS OCCIDENTALIS

A clustering undergrowth palm with pinnate leaves and a long, unbranched inflorescence. *Occidentalis* is Latin for 'western', a name which probably comes from the type locality in the protologue - which we believe is faulty.

DISTRIBUTION. Tsaratanana, Anjanaharibe and Marojejy massifs.
HABITAT. Montane or submontane forest; 450–1400 m.
LOCAL NAMES. None recorded.
USES. None recorded.

CONSERVATION STATUS. Uncertain. These mountains are not well known botanically.

Dypsis occidentalis *(Jum.) Beentje & J. Dransf.* **comb. nov.**

SYNONYM:
Neophloga occidentalis Jum., Ann. Inst. Bot.-Géol. Colon. Marseille sér. 4, 6 (3): 24 (1929); Cat. Pl. Madagascar, Palmae: 21 (1938); Jum. & H. Perrier, Fl. Madagascar 30: 68 (1945). Type: Madagascar, Tsaratanana, *Perrier* 12040 (Holotype P).

Clustering palm in tufts of 6–8. **STEMS** 0.5–2 m high and 7–8 mm diam. **LEAVES** 5–8 in the crown, pinnate; sheath 8–14 cm long, with scattered scales to densely pubescent, and 5 mm long triangular auricles; petiole absent or to 5 cm long, 2–4 mm wide, with scattered scales; rachis 29–40 cm long, in mid-leaf to 3 mm wide, with scattered scales; leaflets 4–13 on each side of the

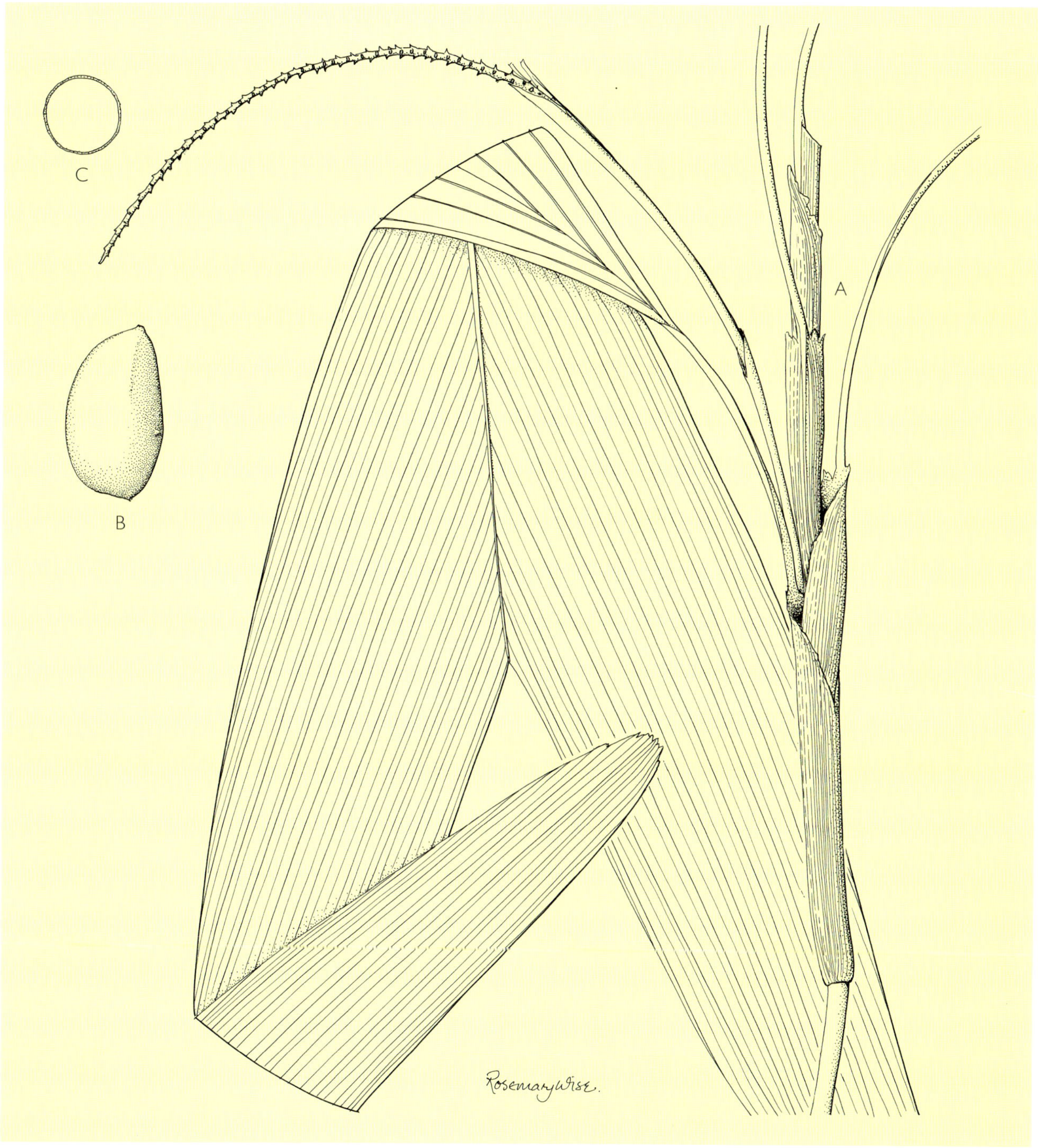

Dypsis andapae. **A** crown showing leaf and inflorescence × 3/5; **B** mature fruit × 2; **C** mature fruit in section × 2. All from *Beentje & Andriampaniry* 4680. Drawn by Rosemary Wise.

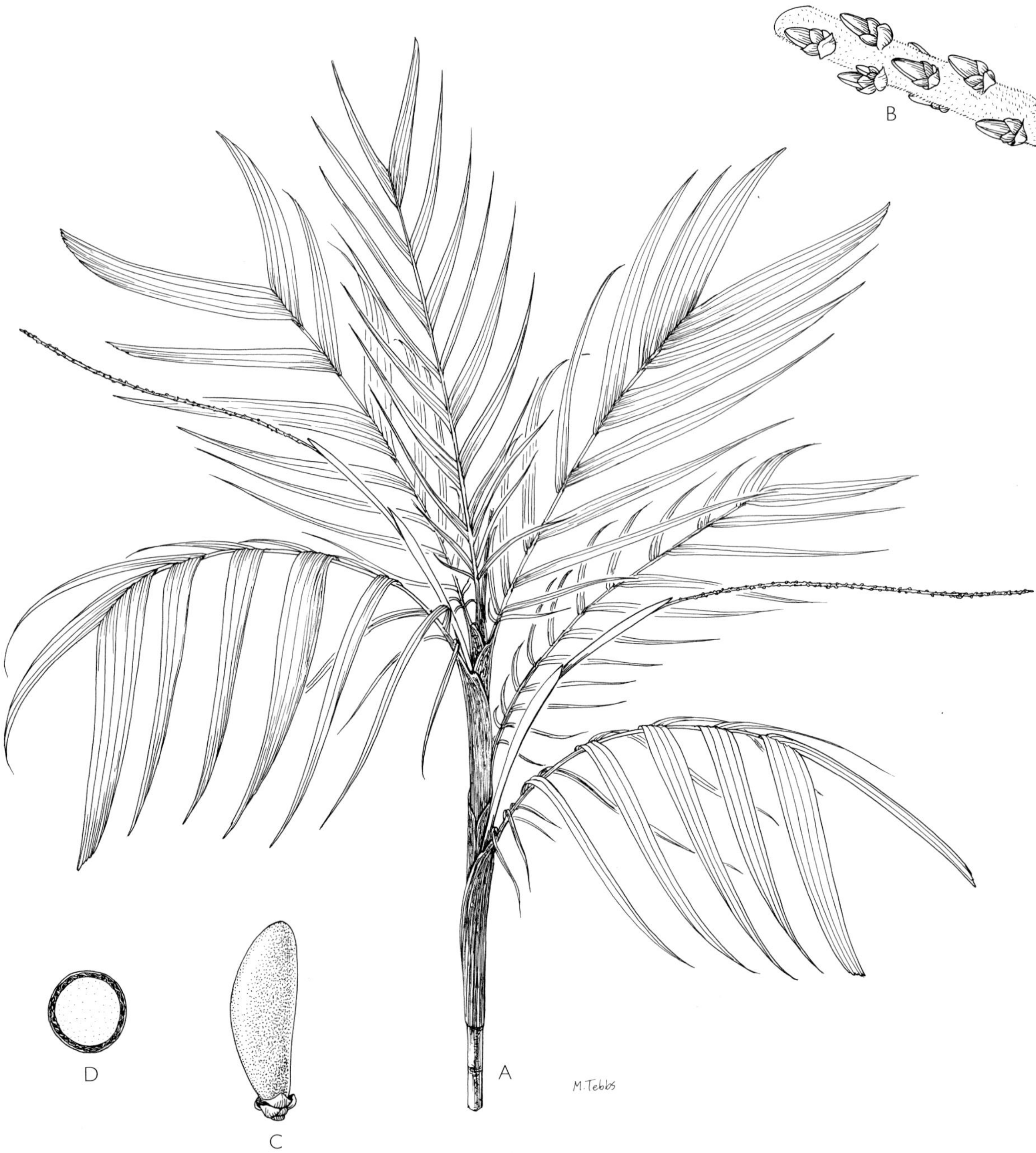

Dypsis occidentalis. **A** crown × 1/8; **B** detail of rachilla × 3; **C** fruit × 3; **D** fruit in cross section × 3. **A** reconstructed from *Humbert* 31703 and *Perrier* 12040, **B – D** from *Decary* 14709. Drawn by Margaret Tebbs.

rachis, ± regular (with few leaflets) to grouped (when more leaflets are present), sigmoid, the proximal 9–27 × 0.2–1 cm, median 15–30 × 1–5 cm, distal 13–21 × 1.9–3.5 cm, the distal pair joined for 2–5 cm, main veins 1, rarely a 2-4-fold leaflet with 3–4 veins, with scattered scales on the minor veins, apices acuminate, the terminal pair dentate over a width of 1–2 cm. **INFLORESCENCE** unbranched, erect; peduncle 22–40 cm long, 1–3.5 mm diam., ?glabrous or minutely puberulous; prophyll 12–23 cm long, 3–9 mm wide, open for 1–7 cm, with scattered scales; peduncular bract inserted at 10–15 cm from the base of the peduncle, 9–31 cm long, open in the distal part and tattering, with a beak of up to 6 mm; rachilla 15–31 cm long, 1–2.5 mm diam., glabrous or minutely puberulous, the triads distant. **STAMINATE FLOWERS** (in bud) with sepals 0.7–0.9 × 0.6–1 mm; petals 1.1–1.4 × 0.6 mm; stamens 6, ?1–seriate, the filaments 0.4–0.7 mm and thin, anthers 0.5–0.7 × 0.2–0.3 mm; pistillode not seen. **PISTILLATE FLOWERS** with sepals 0.6–1.3 × 0.6–1 mm; petals 1.3–1.8 × 0.8–1 mm; staminodes 0.3–0.4 mm high; gynoecium 0.8–1 × 0.3–0.6 mm. **FRUIT** reddish-orange, ellipsoid, 9–10 × 4–5 mm.

NOTE. The protologue of *N. occidentalis* declares that the type specimen comes from the W slopes of Tsaratanana; we read Perriers handwriting on the label of the type as N, as it does not remotely resemble an O (for Ouest).

SPECIMENS SEEN. Ambilobe/Iharana: between Iharana (Vohemar) and Ambilobe, July 1939 (fr.), *Decary* 14709 (P). ?Ambanja, Tsaratanana Mt. area, N side, Nov. 1912 (bud), *Perrier* 12040 (Holotype P). Andapa: NE Marojejy, E of Ambalamanasy II,

Nov./Dec. 1948 (bud), *Humbert & Capuron* 22173 (P); Ambatoharanana Valley to upper Antsahaberoka R., Nov./Dec. 1959 (fl.), *Humbert & Saboureau* 31703b (P); Anjanaharibe W of Beamalona, June 1992 (bud), *Beentje & Andriampaniry* 4684 (K).

84. Dypsis montana

A rare small palm from the high mountains (hence the name). This species slightly resembles *D. occidentalis*, which occurs in the same area.

Distribution. Tsaratanana.
Habitat. Montane forest; c. 1500 m.
Local names. Not recorded.
Uses. Not recorded.
Conservation Status. Unclear. Known from a single site only, but the Tsaratanana is not well known botanically.

Dypsis montana *(Jum.) Beentje & J. Dransf.* **comb. nov.**

Synonym:
Neophloga montana Jum., Ann. Inst. Bot.-Géol. Colon. Marseille sér. 4, 6 (3): 29 (1929); Cat. Pl. Madagascar, Palmae: 21 (1938); Jum. & H. Perrier, Fl. Madagascar 30: 70 (1945). Type: Madagascar, Tsaratanana, *Perrier* 15646 (Holotype P).

Clustering palm to 1 m tall. **Stems** < 4 mm diam. **Leaves** c. 5 in the crown; leaf sheath 7–9 cm long, proximally glabrous, distally with patches of reddish indument, with small triangular auricles 2–5 mm high; petiole 2.5–5 cm long, c. 2 mm diam., with scattered scales; rachis 16–27 cm long, in mid-leaf c. 2 mm wide, with scattered scales; leaflets 3–5 on each side of the rachis, the proximal 10–13 × 0.9–2 cm, median 11–16 × 1.5–2.4 cm, distal pair forming a flabellum 16–22 cm long, connate for 7–8 cm, leaflet interval 1–3 cm, but usually with a large gap just below the terminal flabellum, with scattered scales on the abaxial surface, main veins 1, to 7 on the terminal flabellum. **Inflorescence** interfoliar, unbranched; peduncle 23–27 cm long, densely scaly; prophyll 10–12 cm long, 3–5 mm wide, open only near the apex; peduncular bract inserted at 4.5–13 cm from the base of the peduncle, 7.5–15 cm long, split in the distal 5–9 cm; rachilla 10–14 cm long, 1–2 mm diam., densely stellate-puberulous, with distant triads. **Staminate flowers** with sepals 0.5–0.7 × 0.4–0.6 mm; petals 1–2 × 0.5–1.2 mm; stamens 6, equal, the filaments c. 0.5–0.8 mm long, flattened, anthers 0.6–0.7 × 0.4–0.6 mm, dorsifixed; pistillode 0.7–0.9 × 0.2–0.3 mm, slender and conical. **Pistillate flowers** with sepals 0.5–0.7 × 0.6–0.7 mm; petals 2–2.2 × 1.4–1.5 mm; staminodes c. 0.2 mm; gynoecium c. 1.6–1.7 × 1.4–1.7 mm. **Fruit** unknown. (Plate: see next page).

Specimens seen. ?Ambanja: Tsaratanana, Dec. 1922 (fl.), *Perrier* 15646 (Holotype P); idem, S. Andohanitrema, May 1993 (bud), *Malcomber & Hemingway* 2471 (K).

85. Dypsis bernierana

A charming little palm, quite distinct by its leaf shape and open leaf sheaths. The name refers to the collector of the type, Bernier, about whom very little is known, except that he was French and collected in Madagascar around 1834.

Distribution. Masoala Peninsula and down the coast to Vatomandry.
Habitat. Lowland to submontane rain forest, steep slope; 100–1200 m.
Local names. *Ambosa* (Betsimisaraka, *fide* Bernier).
Uses. Not recorded.
Conservation Status. Vulnerable. This seems to be a rare species, and all over its distribution area the vegetation is under threat from fragmentation and agricultural conversion.

Dypsis montana. **A** crown × 2/5; **B** detail of rachilla × 3. All from *Malcomber & Hemingway* 2471. Drawn by Margaret Tebbs.

Dypsis bernierana *(Baill.) Beentje & J. Dransf.* **comb. nov.**

SYNONYMS:

Haplophloga bernierana Baill., Bull. Soc. Linn. Paris 148: 1171 (1894). Type: Madagascar, Titinga, *Bernier* 41a (Holotype P).

Neophloga bernierana (Baill.) Becc., Bot. Jahrb. Syst. 38, Beibl. 87: 25 (1906); Palme del Madagascar 27, fig. 20, t. 26 (1912); Jum., Ann. Inst. Bot.-Géol. Colon. Marseille sér. 4, 6 (3): 18 (1929); Cat. Pl. Madagascar, Palmae: 19 (1938); Jum. & H. Perrier, Fl. Madagascar 30: 64 (1945).

Solitary dwarf palm. **STEM** to 1 m tall, 6–8 mm diam. distally; internodes 2–10 mm, brown; nodal scars c. 2 mm. **LEAVES** entire, bifid; sheath 3.5–7 cm, open, with scattered scales (denser distally), with fibrous margins; petiole 2–21 cm, 1–1.5 mm diam., with scattered scales; lamina entire, 17–26 cm long, shiny dark green; midrib 5–7.5 cm, lobes 12–19 × 2.3–3.5 cm, main veins 5–6, with scattered scales on the major and minor veins in young leaves, but glabrescent, apex narrowly dentate with 2–3 teeth, sometimes looking acute. **INFLORESCENCE** unbranched; peduncle 10–16 cm long, blackish with dense pubescence; prophyll 4–12.5 cm long; peduncular bract apex to 10–20 cm from the base of the peduncle (insertion point not known); rachilla 5–15 cm long, 1–2 mm diam., with scattered reddish scales; triads distant. **STAMINATE FLOWERS** with sepals 1.2–1.9 × 1.4–1.8 mm, ciliolate, dotted; petals 1.6–2.8 × 1.2–1.6 mm; stamens 6, uniseriate, filaments 0.5–1.3 mm, connate for 0.2 mm, anthers 1.3–1.5 × 0.4–0.5 mm, versatile; pistillode 0.8–1.2 × 0.2–0.3 mm. **PISTILLATE FLOWERS** with sepals 2–2.4 × 1.5–1.8 mm, ciliolate, slightly keeled; petals 2.6–2.8 × 1.3–1.4 mm; staminodes c. 0.4 mm; gynoecium c. 2.3 × 0.8 mm. **FRUIT** red, ellipsoid, 6–13 × 4–5 mm; endocarp fibrous, with almost free fibres. **SEED** with homogeneous endosperm.

Dypsis bernierana, growing in forest undergrowth on the Masoala Peninsula (*Dransfield et al.* JD6364).

Dypsis bernierana. Close-up of fruit (*Dransfield et al.* JD6364).

NOTE. It is probably most closely related to *D. digitata*. The leaf lobes are usually distinctly hooded at their tips.

SPECIMENS SEEN. Maroantsetra: Hiaraka, Oct. 1986 (fl.), *Dransfield et al.* JD6364 (K, P, TAN); without precise locality, May 1975 (y.fr.), *Morat* 4930 (P). Soanierana-Ivongo: Titinga, anno 1834 (fl., y.fr.), *Bernier* 41a (Holotype P). Vatomandry: Mt Takarandonia, Nov. 1927 (fl.), *Perrier* 14124 (P). Locality uncertain ("sentier plus loin que la chûte, 1200 m"), Jan. 1945 (fl.), *Cours* 2479 (K, P).

86. DYPSIS POIVREANA

A small solitary undergrowth species from littoral forest, distinct by its entire, deeply lobed leaves and long unbranched inflorescence. The name refers to the collector of the type, Pierre Poivre (1719–1786).

DISTRIBUTION. Coastal strip around Fenoarivo.
HABITAT. Littoral forest (fide Guillaumet). The type of *D. poivreana* has the description 'palmier aquatique' which seems unlikely.
LOCAL NAMES. *Hovoka* (fide Baron).
USES. Not recorded.

Dypsis bernierana. **A** crown showing inflorescences × 1/2; **B** detail of rachilla in fruit × 1. All from *Cours* 2479.
Drawn by Rosemary Wise.

CONSERVATION STATUS. Critical. Known from only a very small area, in which this kind of vegetation is disappearing rapidly. Not seen in the last twenty-five years.

Dypsis poivreana *(Baill.) Beentje & J. Dransf.* **comb. nov.**

SYNONYMS:
Haplophloga poivreana Baill., Bull. Soc. Linn. Paris 147: 1168 (1894). Type: Madagascar (probably), *Poivre* s.n. (Holotypus P)
Neophloga poivreana (Baill.) Becc., Bot. Jahrb. Syst. 38, Beibl. 87: 24 (1906); Palme del Madagascar 26, fig. 19, t. 25-26 (1912); Jum., Ann. Inst. Bot.-Géol. Colon. Marseille sér. 4, 6 (3): 16 (1929); Cat. Pl. Madagascar, Palmae: 22 (1938); Jum. & H. Perrier, Fl. Madagascar 30: 62, fig. 16 (1945).

Solitary palm to 80 cm. **STEM** c. 8 mm diam.; internodes 2.5–4.5 cm, densely pubescent distally. **LEAVES** c. 8 in the crown, entire; sheath 11–13 cm long, 1/2 to 2/3 closed, with dense scales, auricles 0.7–1 cm; petiole 3–19 cm long, distally 2–3 mm diam., densely scaly or with scattered scales; lamina entire, 24–42 cm long, midrib 5–10 cm long, lobes 19–37 × 2.4–4.5 cm, main veins 5–7, with scattered scales on minor and major veins, apex dentate over a width of 4–12 mm. **INFLORESCENCE** unbranched; peduncle 14–26 cm long, c. 2 mm diam., densely red-pubescent; prophyll 6–13 cm long, 4.5 mm wide, with reddish flaking pubescence, borne at c. 4 cm above the base of the peduncle, open in the distal 1–1.5 cm; peduncular bract inserted at 6–21 cm from the base of the peduncle, 8–14 cm long, with scattered scales, open in the distal 2 cm, sometimes with a 5 mm beak; non-tubular peduncular bract sometimes present, c. 2 mm; rachilla 11–26 cm, 1.5–2 mm diam., densely scaly, with spaced triads. **STAMINATE FLOWERS** with sepals 1.3–1.7 × 0.8–1.8 mm; petals on a 0.2–0.4 mm high receptacle,

Dypsis poivreana. **A** crown × 1/4; **B** detail of rachilla × 4. **A** from *Guillaumet* 2370, **B** from *Geay* 9064. Drawn by Margaret Tebbs.

2–2.6 × 1.3–1.7 mm; stamens 6, biseriate (offset 0.2–0.3 mm), filaments 0.9–1.7 mm and thin (and occasionally connate for c. 0.2 mm), anthers 1.3–1.8 × 0.4–0.6 mm, versatile; pistillode 0.6–1 × 0.3–0.5 mm, pyramidal. **Pistillate flowers** with sepals 1.2–2.2 × 1.6–2.1 mm; petals 3–3.9 × 2.8–3.6 mm; staminodes 0.3–0.9 mm; gynoecium 2.8–4 × 2–2.6 mm, trifid. **Fruit** unknown.

Note. This species slightly resembles *D. digitata* but is distinct in its shorter petioles and closed leaf sheaths; it is close to *D. andapae*, but that species clusters and occurs in high mountains.

Specimens seen. Fenoarivo Atn.: without precise locality, 1909 (fl.), *Geay* 9064 (P); Maningory ferry, Jan. 1964 (fl.), *Peltier & Peltier* 4635 (K, P; some galled flowers present); Tampolo, Feb. 1969 (fl.), *Guillaumet* 2370 (P).
Central Madagascar, without date (fl., y.fr.), *Baron* 2323 (K). Without any locality, without date (around 1750?), *Poivre* s.n. (Holotype P)

87. Dypsis catatiana

This is the commonest small palm of the island, occurring in nearly all East Coast forests over a wide altitude range. It would make a good ornamental, with both the entire-leaved form and that with pinnate leaves being handsome, but will require a humid atmosphere. The name refers to the collector of the type, Louis Domingue Maria Catat (1859–?, who collected in Madagascar in 1889).

Distribution. NE and E Madagascar, between Tsaratanana and Andohahela.
Habitat. Lowland to montane rain forests to bamboo forest, slight to steep mid slope; (150–) 450–1900 m.
Local names. *Sinkaramboalavo* (Betsimisaraka); *Varaotra* (Antanosy).

Dypsis catatiana. Form with entire bifid leaf, Andohahela (*Dransfield et al.* JD6773)

Uses. Not recorded.
Conservation Status. Not threatened. Widespread and common.

Dypsis catatiana *(Baill.) Beentje & J. Dransf.* **comb. nov.**

Synonyms:
Dypsidium catatianum Baill., Bull. Soc. Linn. Paris 148: 1173 (1894). Type: Madagascar, Didy forests, *Catat* 1732 (Holotype P).
Neophloga catatiana (Baill.) Becc., Bot. Jahrb. Syst. 38, Beibl. 87: 25 (1906); Palme del Madagascar 28, fig. 21, t. 27 (1912); Jum., Ann. Inst. Bot.-Géol. Colon. Marseille sér. 4, 6 (3): 20 (1929); Cat. Pl. Madagascar, Palmae: 19 (1938); Jum. & H. Perrier, Fl. Madagascar 30: 66, fig. 17 (1945).

Neophloga indivisa Jum. & H. Perrier, Ann. Inst. Bot.-Géol. Colon. Marseille sér. 3, 1 (1): 29, fig. 9 (1913). Type: E slopes of Mt Andringitra, R. Ihovika, 1000-1200 m, *Perrier* 11974 (Holotype P).

Solitary small or dwarf palm (*Schatz et al.* 1694 is described as colony-forming, to 1.5 m; *Perrier* 12026 and 15985 are said to be clustering). **Stem** 0.2–1 m high, 4–9 mm diam.; internodes 0.5–3 cm, dark green, often with a vertical pale green stripe; nodal scars 0.1–0.2 cm. **Leaves** 4–10 in the crown, porrect to spreading; sheath 3–8 cm long, the outermost often open for c. 50 %, pale green with minute brown or reddish scales, in young leaves with clear triangular ligules 3–6 mm long on each side of the petiole; petiole absent or to 5 cm long, 1.5–2.5 mm diam., slightly channelled adaxially; lamina entire or with (2–) 3–5 (–7) pairs of flat leaflets, shiny medium green, slightly paler on the abaxial surface, young leaves reddish; when *entire* shortly bifid, 14–32 cm long, the midrib 10–22 cm

Dypsis catatiana. **A** crown of form with divided leaves x 2; **B**, **C** stem and leaf sheaths of form with entire bifid leaves x 2; **D** staminate flower x 20; **E** fruit x 2.5. **A**, **E** from *Beentje* 4428, **B** – **D** from *Schatz et al.* 1694. Drawn by Rosemary Wise.

long, the lobes 5–12 x 2–5 cm, with 10–16 main veins, the base cuneate, the apices truncate and dentate, 5–15 mm wide, and dentate on the outside bend, with lines of small reddish and whitish scales on the abaxial midrib and veins; when *pinnately divided* the rachis 9–24 cm long with scattered reddish scales, the individual leaflets flat and at intervals of 1–4.5 cm, the proximal 4–16 (–21) x 0.5–3 cm and connate for 2–6 cm, the median 8–21 x 0.5–3.5 cm and connate for 0.5–8 cm, both the proximal and the median sigmoid, with acuminate apices and 1–6 main veins, the distal leaflets 5–14 x 1.4–3 cm, connate for 2–6 cm, with 5–6 main veins and the apices dentate on the outside bend, all leaflets with small scales on veins. **INFLORESCENCE** interfoliar, erect or spreading and then porrect, unbranched (a single bifurcation seen once in *Guillaumet* 2161, *Perrier* 11998 and *Humbert* 6891), 9–38 cm long; peduncle 6–27 cm long, 1–2 mm diam., with dense minute scales but glabrescent; prophyll 5–15 cm long, borne at 3–6 cm above the base of the peduncle, 2.5–4 mm wide, opening only at the apex, pale brown with scattered scales; peduncular bract inserted at 4–14 from the base of the peduncle, 3–9 cm long, 2.5–4 mm wide, opening only near the apex, pale brown with scattered scales; second peduncular bract often present as a tiny, 1.5–4 mm long, briefly tubular bract situated just above the apex of the first peduncular bract; rachilla 2–14 cm long, 1–2 mm diam., yellow-green to pale yellow, glabrous or with dense minute scales all over, with 20–50 distant superficial triads; flowers yellow-green. **STAMINATE FLOWERS** with imbricate sepals 0.7–1 (–1.8) x 0.6–1 mm, keeled, with membranous margins; receptacle to 0.8 mm high; petals valvate, 1–2 x 0.7–1.3 mm, elliptic, acute, striate; stamens 6, biseriate, didymous, the filaments connate for 0.2–0.4 mm, the antepetalous 0.2–0.5 (–0.8) mm long and narrow, the antesepalous 0.2–0.3 mm and broad, anthers with divergent locules, basifixed, 0.3–0.4 (–0.9) x 0.25–0.4 (–0.5) mm (with 2 sterile anthers in *Jacquemin* H572J; looking atrophied in JD6774);

pistillode 0.2–0.3 mm. **PISTILLATE FLOWERS** with sepals imbricate, 0.8–1.3 × 0.6–1.3 mm, concave, keeled, non-ciliolate; petals proximally imbricate and membranous, distally valvate and fleshy, 1.2–1.8 × 0.9–1.5 mm, concave, elliptic, acute, striate; once (in *Schatz* 1694, a single flower) with a second series of smaller petals inside the outer series, 1.2 mm long; staminodes 6, 0.2–0.4 mm high, from thin to broad and tooth-shaped; ovary 0.8–1.4 × 0.8–1.3 mm, with a low pyramidal stigmatic bump. **FRUIT** deep shiny red, ellipsoid with a slightly pointed apex, 10–15 × 5–9.5 mm; mesocarp c. 2 mm thick, fleshy; endocarp with 22–30 free longitudinal fibres. **SEED** 8.5–10 × 4–5.5 mm, the base pointed, the apex obtuse; endosperm homogeneous. **EOPHYLL** bifid.

NOTE. *N. indivisa* was put into synonymy by Jumelle (1929). Northern populations have generally longer peduncles and more glabrous rachillae, but not consistently so.

Jacquemin H572J has the staminate and pistillate flowers at almost the same stage of development on a single rachilla, which is most unusual in *Dypsis*. *Guillaumet* 2161 has the stamens larger than in all other specimens studied; the measurements of this specimen are included in the description in brackets.

SPECIMENS SEEN. Bealanana: Tsaratanana area, Nov. 1912 (old infl.), *Perrier* 12026 (P); Bealanana to Mangindrano, July 1968 (fl.), *Guillaumet* 2161 (K, P, TAN). Andapa: Andapa to Doana, Andranotsara valley, Oct. 1967 (fl., fr.), *Jacquemin* H572J (P); Marojejy, E of Ambalamanasy II, Dec. 1948 (fl., y.fr.), *Humbert & Capuron* 22119 (K, P); Marojejy, Nov. 1972 (fl.), *Guillaumet* 4037 (TAN), and Dec. 1972 (fr.), *Guillaumet* 4204 (TAN); N slopes of Ambatosoratra, Feb. 1989 (fl.), *Miller* 4257 (TAN). Maroantsetra/Antalaha: Masoala Peninsula, Oct. 1912 (bud), *Perrier* 11948 (P). Ambatondrazaka: Didy forest, Aug. 1889 (fr.), *Catat* 1732 (Holotype P); Zahamena, March 1941 (fr.), *Decary* 16532 (P). Moramanga: Analamazaotra, Sept. 1913 (old infl.), *Perrier* 11996 (P); idem, Dec. 1913 (fr.), *Perrier* 11995 (P); idem, July 1913 (bud), *Perrier* 11998 (P) and 11999 (P); idem, 1912 (fl.), *Viguier & Humbert* 1084 (P); idem, March 1991 (fr.), *Beentje & Raharilala* 4410 (BH, K, MO, TAN); Anranumenabe, Nov. 1986 (bud), *Dransfield et al.* JD6434 (K, TAN); Maromizaha, Feb. 1926 (fr.), *Perrier* 15985 (P); S of Moramanga, Feb. 1930 (fr.), *Decary* 7106 (P), 7161 (P). Anosibe an'Ala: Sandrangato, Dec. 1954 (fl.), *Descoigns* 122a (TAN). Ambositra: Ranomena, July 1992, *Beentje* 4739 (K, MO, TAN). Ifanadiana: Ranomafana, Jan. 1964 (fr.), *Bosser* 18916 (P); idem, Oct. 1987 (fl.), *Schatz et al.* 1693 (P, TAN) and 1694 (K, P, TAN); idem, March 1991 (fr.), *Beentje* 4428 (K, TAN), 4429 (BH, K, MO, P, TAN); idem, Vatoranana to Maharira, March 1992 (fr.), *Malcomber & Rakoto* 1326 (K, P). ?Ivohibe: Ihovika R. (dead infl.), *Perrier* 11974 (P, type of *N. indivisa*). Midongy Atsimo: 24 km S of Midongy, May 1992 (fr.), *Beentje & Andriampaniry* 4669 (K, MO, P, TAN). Befotaka: Mt. Papango, Dec. 1928 (fl.), *Humbert* 6891 (P, TAN). Tolanaro: summit of Marosoui (Marosohihy), Nov. 1928 (bud), *Humbert* 6624 (P, TAN); Bevava col to Bekoho summit, Nov. 1928 (bud), *Humbert* 6425bis (P); Enaniliha, Feb. 1966 (fr.), *Rabevazaha* RN 11521 (P); idem, Dec. 1959 (fl.), RN 10387 *Rakotoson* (K, P); between Saindro Col and Eminiminy, Feb. 1934 (fl., y.fr.), *Humbert* 13999 (K, P); N part of 'chaines anosyennes", Nov. 1971 (bud), *Guillaumet* s.n. (P, TAN); Andohahela, Col Tanatana, Dec. 1989 (fl.), *Dransfield et al.* JD6773 (K, P, TAN) and (fl.) JD6774 (K, TAN); Andohahela, March 1992 (fr.), *Beentje & Andriampaniry* 4594 (BH, K, MO, P, TAN); idem, Dec. 1992 (fr.), *Beentje & Andriampaniry* 4766 (TAN); idem, Itrotroky R, Feb. 1993 (fr.), *Malcomber* 2129 (K, P). Locality not found: Andrianony, Manjarivolo, Nov. 1970 (bud), *Guillaumet* 3511 (P, TAN).

88. DYPSIS CORIACEA

A very attractive small palm with thick shiny leaves. This species was known as 'leather-leaf' among collectors, and the Latin name is a straight translation of this. The entire leaf dries pale green, which is most distinctive.

Dypsis coriacea. An undergrowth palm with very leathery leaves, Sahavary (*Dransfield et al.* JD6459).

Dypsis coriacea. Crown with inflorescences × 2. From *Beentje et al.* 4617. Drawn by Rosemary Wise.

Distribution. Masoala, Maroantsetra and Mananara.
Habitat. Lowland rain forest; on steep slope near stream, or on ridgetops; 200–400 m.
Local names. Not recorded.
Uses. Not recorded.
Conservation Status. Vulnerable. The species has a limited distribution area, most of which is unprotected and under some threat.

Dypsis coriacea *Beentje* **sp. nov.**
gregem palmunculorum 6-staminatorum foliis integris pertinens *D. lucenti* affinis sed rachillis lepidotis et folio glabro multo coriaceo differt. Typus: Madagascar, Sahavary, *Dransfield et al.* JD6459 (Holotypus K; isotypi P, TAN).

Solitary or clustering palm. **Stem(s)** 1–2 m, diam. < 1 cm; distal internodes 2–4.5 cm, pale green with dense to scattered reddish pubescence; nodal scars c. 1 mm. **Leaves** 6–9 in the crown, entire or with 4 leaflets; sheath 4–9 cm long, closed, pale brown with reddish scales and without or with auricles to 8 mm long; petiole 1–12 cm long, channelled, c. 2 mm diam., with scattered scales; lamina *entire*, leathery, rich shiny green and occasionally metallic-tinged, 15–41 × 6–8 cm, lobes 5–12 × 1.5–4 cm; main vein 1, with sparse red and white scales on the midrib, otherwise with few scattered scales to glabrous, the lamina often white-punctate, often with a single deep (to 6 cm) lacuna, in the distal quarter the margins toothed, the teeth to 16 mm long. **Inflorescence** interfoliar, unbranched, spreading; peduncle 8–28 cm long, 2–3 mm diam., glabrous; prophyll 5–16 cm long, 4–6 mm wide, opening in the distal 1cm only, with scattered scales to almost glabrous; peduncular bract inserted at 5–13 from the base of the peduncle, 3–12 cm

long, open in the distal 1–1.5 cm, with scattered scales, quickly deciduous; non-tubular peduncular bract 2–2.5 mm; rachilla 3–26 cm long, 1–2 mm diam., densely pubescent to puberulous, with distant triads. **STAMINATE FLOWERS** with sepals 0.8–1.5 x 0.7–1.6 mm, ciliolate, with scales; petals white, 1.8–2.4 x 1–1.7 mm, striate; stamens 6, uniseriate or very slightly biseriate, filaments 0.8–1.5 mm, thin, anthers 0.8–1.6 x 0.3–0.6 mm, versatile; pistillode 0.7–1 x 0.3–0.5 mm. **PISTILLATE FLOWERS** with sepals 0.8–1.2 x 1.4–1.7 mm; petals 2.1–4 x 1–2.8 mm; staminodes 6, c. 0.5 mm; ovary c. 2 x 1.3 mm. **FRUIT** when young narrowly ovoid, 11–15 x 4.5–6 mm, obtuse, with fleshy mesocarp and fibrous endocarp. **SEED** c. 12.5 x 4 mm, with homogeneous endosperm.

NOTE. The species somewhat resembles *D. lucens* but the leaf veins are glabrous and there are no ramenta; the rachilla is puberulous. It also resembles *D. catatiana*, but has versatile stamens (not didymous ones). There is some variation in the degree of lobing and the length of the rachilla.

SPECIMENS SEEN. Maroantsetra: Hiaraka, Ampanga R., Oct. 1986 (bud), *Dransfield et al.* JD6382 (K, TAN); Sahavary, Feb. 1988 (fl., y.fr.), *Dransfield et al.* JD6459 (Holotype K; isotypes P, TAN); Antalavia, Feb. 1988 (fl.), *Dransfield et al.* JD6482 (K, TAN); Andronabe, feb. 1992 (fl.), *Zjhra & Hutcheon* 120 (K). Mananara Avaratra, Antanambe, April 1992 (fl., y.fr.), *Beentje et al.* 4617 (BH, K, MO, P, TAN).

Dypsis coriacea. Close-up of infructescence (*Dransfield et al.* JD6459).

89. DYPSIS MINUTA

The name of this exquisite little palm, which is restict-ed in its distribution, indicates its size; it is a tiny solitary palm, one of the smallest on the island.

DISTRIBUTION. Masoala and Maroantsetra.

HABITAT. Lowland rain forest; ridgetops; 200–550 m.
LOCAL NAMES. Not recorded.
USES. Not recorded.
CONSERVATION STATUS. Vulnerable. Occurs over a restricted distribution area, without official protection; numbers are thought to be fairly low.

Dypsis minuta *Beentje* **sp. nov.**
D. coriaceae similis sed vagina folii pubescenti inflorescentia arcuata non erecta et folio vivendi non coriaceo differt. Typus: Madagascar, Sahavary, *Dransfield et al.* JD6457 (Holotypus K; isotypi P, TAN).

Solitary palm, stem 15–50 cm. **STEMS** to 5 mm diam.; internodes c. 1–2 cm, densely rusty-red hairy; nodal scars 0.2 mm. **LEAVES** 5–8 in the crown, entire or with several leaflets (fide JD6371, but not apparent from collection); sheath 6–9 cm long, half open, densely red-pubescent, with long-laciniate auricles to 1 cm; petiole 3–18 cm long, 1.5–2 mm diam., densely scaly; lamina entire, 19.5–33 cm long, bright greyish green, midrib 4.5–6 cm, lobes 16–26 × 1.3–2.5 cm, main veins 4–6, with minute reddish scattered scales on major and minor veins, white wax present? or white-punctate, apices narrowly dentate, 2–4 mm wide. **INFLORESCENCE** interfoliar, unbranched (a single bifid rachilla present in JD6457), erect to spreading; peduncle 6–19 cm; prophyll 7–15 cm, pale brown, open in the distal 1 cm; peduncular bract inserted at 6–11.5 cm from the base of the peduncle, 3–12 cm long, glabrous or with scattered scales, open for most of its length; non-tubular peduncular bract often present, 1.5–2 mm long; rachilla 4–12 cm long, 1–2 mm diam., minutely puberulous or glabrous, with distant triads. **STAMINATE FLOWERS** with sepals 0.7–0.9 × 1–1.9 mm, ciliolate; petals 1.7–2 × 1.1–1.2 mm; stamens 6, uniseriate, filaments 1–1.2 mm, anthers 1.3–1.4 × 0.4 mm, versatile;

Dypsis minuta. **A** crown × 2/5; **B** detail of fruiting rachilla × 2.5. **A** from *Dransfield et al.* JD6457, **B** from *Dransfield et al.* JD6483. Drawn by Margaret Tebbs.

pistillode c. 1 × 0.3 mm. **PISTILLATE FLOWERS** white; sepals 0.6–1.2 × 1–1.5 mm, ciliolate; petals 1.8–2.5 × 1.5–1.8 mm; staminodes 0.3–0.4 mm; gynoecium on a 0.2 mm stalk, 1.7–2.3 × 1.2–2 mm. **FRUIT** when young 11–12 × 2–3 mm.

NOTE. Very similar to *D. coriacea* but differs in the pubescent leaf sheath, the curved (not erect) inflorescence and the less leathery leaf, at least when fresh.

SPECIMENS SEEN. Maroantsetra: Hiaraka, Oct. 1986 (bud), *Dransfield et al.* JD6371 (K, P, TAN); Antalavia, Feb. 1988 (y.fr.), *Dransfield et al.* JD6483 (K, TAN); Sahavary, Feb. 1988 (fl., y.fr.), *Dransfield et al.* JD6457 (Holotype K; isotypes P, TAN). No locality, collected before 1867, (fl.), *Lyall* 323 (K, syntype of *N. heterophylla* but not conspecific with the lectotype of that species, *Baron* 486).

90. DYPSIS TENUISSIMA

This is one of the smallest palms in the world, with extremely slender stems. This is reflected in the species name, which is Latin for 'most slender'.

DISTRIBUTION. Only known from Andohahela.
HABITAT. Lowland rain forest, slight mid slope; 500–550 m.
LOCAL NAMES. Not recorded.
USES. Not recorded.
CONSERVATION STATUS. Endangered. Only known

Dypsis tenuissima. **A** whole plant × 1/2; **B** divided leaf × 1/2. **A** from *Beentje & Andriampaniry* 4592, **B** from *Beentje* 4601. Drawn by Margaret Tebbs.

from a single site; numbers seen were less than a hundred. This area is outside the protected area.

Dypsis tenuissima *Beentje* **sp. nov.**
caulibus gracillimis folio minuto tenue multo lobato distincta. Typus: Madagascar, Ezoambo, *Beentje & Andriampaniry* 4592 (Holotypus K; isotypi BH, MO, P, TAN).

Clustering palm in groups of 2–6. **STEMS** 25–50 cm high, 2–4 mm diam.; internodes 1–4 cm. **LEAVES** 4–11 in the crown, entire or pinnate; sheath 2.3–6 cm, closed, with sloping, slightly laciniate shoulders and with scattered scales; petiole 2–4.6 cm, 1–1.5 mm diam., with scattered scales; *entire* leaves 11–19 cm long, midrib 1.7–4.5 cm, the lobes 8–15 × 0.7–1.3 cm (lobed 75–85 %), with 3–5 main veins, narrowly dentate apices, glabrous or with a few proximal scales; *pinnate* leaves with rachis 4.5–6 cm long, with scattered scales; leaflets 2–5 on each side of the rachis, the proximal 9–14 × 0.3–0.6 cm (main veins 1–2), median 9–13 × 0.3–4 cm (interval 0.8–1.2 cm, main vein 1), apices attenuate, distal 11–14 × 0.5–1 cm, connate for 3–4 cm, with 2–3 main veins and narrow dentate apices, glabrous. **INFLORESCENCE** interfoliar, unbranched, 9–16 cm long; peduncle 6–11 cm long, c. 1 mm diam.; prophyll 6–9 cm long, borne at 2.5–4 cm above the base of the peduncle, with few scattered scales especially on the margins (resembling thin ramenta), open at the apex for 0.6–1 cm; peduncular bract inserted at 6–9 cm from the base of the peduncle, deciduous; non-tubular peduncular bract 1–2 mm; rachilla 3.2–4.3 cm long, 1–2 mm diam., with distant triads, glabrous. **STAMINATE FLOWERS** with sepals 0.7–1 × 0.6–1.1 mm, only slightly keeled; petals 2.2–2.5 × 1.2–1.4 mm; stamens 6, uniseriate, filaments c. 0.8 mm and thin, anthers 1.1–1.3 × 0.3–0.5 mm, versatile and parallel; pistillode < 1 mm. **PISTILLATE FLOWERS** with sepals 0.8–1.2 × 1.2–1.4 mm; petals 1.8–2.3 × 1.7–2.2 mm; staminodes 0.3–0.7 mm; ovary 2.1–2.4 × 1.8–2 mm. **FRUIT** ellipsoid, 9.5–10 × 5–5.5 mm, rounded at the apex; endocarp with almost free fibres. **SEED** ellipsoid, 8–9 × 3.5 mm, obtuse at both ends; endosperm homogeneous.

NOTE. Distinct in the very slender stems and the tiny, thin, much-toothed leaf.

SPECIMENS SEEN. Tolanaro: 13 km N of Ezoambo, March 1992 (fl., y.fr.), *Beentje & Andriampaniry* 4592 (Holotype K; isotypes BH, MO, P, TAN); idem, March 1992 (ster.), *Beentje & Andriampaniry* 4601 (K).

91. DYPSIS LUCENS

Only known from the type, which consists of loose pieces of leaf and a few rachillae. These are nevertheless enough to distinguish it from other taxa. The name is Latin for 'shiny', and this presumably refers to the leaf when it is fresh.

DISTRIBUTION. Only known from the type: 'Antongil Bay area'.
HABITAT. Rain forest; c. 400 m.
LOCAL NAMES. Not recorded.
USES. Not recorded.
CONSERVATION STATUS. Probably extinct; not seen since the type was collected over eighty years ago.

Dypsis lucens *(Jum.) Beentje & J. Dransf.* **comb. nov.**

SYNONYM:
Neophloga lucens Jum., Ann. Inst. Bot.-Géol. Colon. Marseille sér. 4, 6 (3): 15 (1929); Cat. Pl. Madagascar, Palmae: 21 (1938); Jum. & H. Perrier, Fl. Madagascar 30: 62 (1945). Type: Madagascar, Antongil Bay, *Perrier* 12032 (Holotype, P).

Palm up to 60 cm. **STEM** 0.7–1 cm diam. **LEAVES** c.10; sheath 6.5–7.5 cm long, 1–1.2 cm diam., partly open, red-brown, covered

Dypsis tenuissima

Dypsis lucens

Dypsis lucens. **A** crown × 1/3; **B** detail of rachilla × 4. All from *Perrier* 12032. Drawn by Margaret Tebbs.

in small dark red-brown scales; petiole absent; lamina entire, shiny, narrowly triangular, 25–30 cm long, midrib 17–19 cm, lobed for 8–12 cm, the lobes 2.4–3.7 cm wide, the apices 5–10 mm wide and toothed, also toothed along the upper outer margins, adaxially with c. 9 prominent veins, abaxially covered in many small scales, intermixed with 2–5 mm long ramenta on the main veins. **Inflorescence** 44–47 cm long, unbranched; peduncle 23–27 cm long, 1.5–2 mm diam., with a few minute scales; prophyll 8–11 cm long, 5–6 mm wide, borne at 0.5–3 cm above the base of the peduncle, opening near the apex only for c. 1 cm, with scattered scales; peduncular bract inserted at 5–8 cm from the base of the peduncle, 13–15 cm long, split for the distal 1 cm, with scattered scales; non-tubular bracts 2, 2–2.5 mm long; rachilla 23–24 cm long, 1–1.5 mm diam., glabrous; triads distant, the rachilla bract small and triangular. **Staminate flowers** only known from an early bud stage; sepals 1.4–1.7 × 1–1.3 mm, ciliolate; petals 2.1–2.4 × 1.2 mm; stamens 6, biseriate (offset 0.2 mm), filaments c. 0.5 mm, anthers 0.5–0.6 mm, still in a very tight cone, the locules parallel. **Pistillate flowers** unknown. **Fruit** and seed unknown.

Note. There is a bit of a mystery regarding the pistillate flowers; Jumelle gives a brief, non-committal description in the protologue which could refer to any pistillate flower in the *Dypsidinae*, but there is none on the type. The specimen has the appearance of *D. concinna*, with an unbranched inflorescence, but there is no petiole.

Specimens seen. Maroantsetra: Antongil Bay area, Oct. 1912 (bud), *Perrier* 12032 (Holotype P).

92. DYPSIS SIMIANENSIS

A very handsome small palm, with entire, very narrow leaves with only a short notch at the tip. We have never seen the fruit, but this species would make an exquisite ornamental. The name comes from the type locality, which was slightly mis-spelt on the label.

DISTRIBUTION. Mananara, Soanierana-Ivongo, Zahamena and Manombo.
HABITAT. Lowland rain forest to rather dry forest; steep to gentle mid slope, soils rocky; 65–250 (–850) m.
NATURAL HISTORY. In Manombo HB photographed an insect visitor to the open male flowers, tentatively identified by Dr. Verdcourt as a Sphecoid wasp; it visited several open staminate flowers and inserted its mouth parts. One of the flowers of *Beentje & Andriampaniry* 4786 is much larger than the others (some 3.5 mm), and it proved to contain a 3 mm long white caterpillar; such galled flowers are of very common occurrence in undergrowth species of *Dypsis*, particularly in those species with three stamens. Male flowers are open when female buds are still minute.
LOCAL NAMES. Not recorded.
USES. Not recorded.
CONSERVATION STATUS. Endangered. Known from a limited number of sites, with the individual populations consisting of few individuals. Recent collections are few; HB has seen a total of less than forty plants.

Dypsis simianensis *(Jum.) Beentje & J. Dransf.* **comb. nov.**

SYNONYM:
Neophloga simianensis Jum., Ann. Inst. Bot.-Géol. Colon. Marseille sér. 4, 6 (3): 13 (1929); Cat. Pl. Madagascar, Palmae: 23 (1938); Jum. & H. Perrier, Fl. Madagascar 30: 61, fig. 15 (1945). Type: Madagascar, Simianona (Simiane) River, *Perrier* 11943 (Holotype P).

Clustering palm in tufts of 2–12 stems. **STEMS** 0.5–2 m long, 0.4–1 cm diam., erect or curving; internodes 0.5–2 cm long, distally with reddish laciniate scales. **LEAVES** 5–15 in the crown, spreading, entire; sheath 4.5–6.6 cm long, pale yellow-brown with scattered reddish scales, densest distally, closed for 60–90%, with oblique opening and minute triangular ligules 2 mm high; petiole 0.5–6.6 cm long, 1.5–2 mm wide, adaxially flat, with a few scattered scales; blade dark green, narrowly triangular, acute at the base, 15–35 x 2.5–4.7 cm, the apex bifid for 2.1–4.2 cm, the apices of the lobes truncate and dentate and 1–1.2 cm wide, only the midrib prominent on both surfaces, with reddish scattered scales, the other veins faint, glabrous, c. 9 on each side of the

Dypsis simianensis, at Manombo (*Beentje* 4518).

rachis. **INFLORESCENCE** interfoliar, erect to spreading, unbranched; peduncle 5.5–24 cm long, c. 1.5 mm diam., proximally with a few scattered scales; prophyll 8–12 cm, borne at 2.5–4 cm above the base of the peduncle, 3 × 2 mm diam., with a few scattered scales mainly on the narrow wings, split only in its distal 5 mm; peduncular bract inserted at 7–8 cm from the base of the peduncle, 4.5–13 cm long, with a few scattered scales, split only in its distal 5 mm; rachilla 5–16 cm long, c. 2 mm diam., glabrous or minutely puberulous, with distant superficial triads; rachilla bract entire, rounded. **STAMINATE FLOWERS** white; sepals 1.3–1.7 × 1.2–1.3 mm, keeled, gibbous, rounded, ciliate; petals 1.7–1.8 × 1.1–1.4 mm, ovate, acute, striate, at anthesis hardly opening and leaving an opening of only 0.6 mm wide for the stamens; stamens 6, uniseriate, the filaments connate for 0.2–0.5 mm, in bud 1 mm long but rapidly lengthening at anthesis to c. 2.5 mm, the antepetalous slightly wider than the antesepalous, anthers 1.2–1.4 × 0.5–0.6 mm, dorsifixed, the locules slightly divergent at the base; pistillode c. 1.5 mm, narrowly ovoid, 0.8 mm diam. **PISTILLATE FLOWERS** unknown except for very young buds. **FRUIT** unknown, except for the fibrous endocarp with almost free fibres. **SEED** c. 8 × 6 mm, rounded at both ends, with homogeneous endosperm.

NOTE. *Cours* 4192 is from higher altitude than the other specimens; it has very short inflorescences with shorter rachillae than usual.

SPECIMENS SEEN. Mananara Avaratra: Antanambe, Oct. 1991 (bud), *Beentje* 4451 (K, MO, P, TAN). Soanierana-Ivongo: Simiane (Simianona) R, Sept. 1912 (fl.), *Perrier* 11943 (P, type). Ambatondrazaka: Bemainty, March 1951 (fl.), *Cours* 4192 (P, TAN). Farafangana: Manombo, Nov. 1991 (bud, seed), *Beentje* 4518 (BH, K, MO, P, TAN); idem, Jan. 1993 (fl.), *Beentje & Andriampaniry* 4786 (BH, K, MO, P, TAN).

Dypsis simianensis. Crown × 3/5. From *Beentje* 4518. Drawn by Rosemary Wise.

93. DYPSIS INTEGRA

The name is Latin for 'entire', since the leaf appears to be completely unlobed on the type, with a rounded apex. On close examination the leaf is seen to be slightly folded near the apex, with a small notch.

DISTRIBUTION. East Coast, between Mananara Avaratra and Vangaindrano.
HABITAT. Lowland rain forest; 50–300 m.
LOCAL NAMES. Not recorded.
USES. Not recorded.
CONSERVATION STATUS. Critical. The only recent collection is from Manombo Forest, which is being destroyed by fire, shifting cultivation and logging. Both of the older sites have been visited, but the species has not been seen there.

Dypsis integra *(Jum.) Beentje & J. Dransf.* **comb. nov.**

SYNONYM:
Neophloga integra Jum., Ann. Inst. Bot.-Géol. Colon. Marseille sér. 4, 6, fasc. 3: 13 (1929); Cat. Pl. Madagascar, Palmae: 21 (1938); Jum. & H. Perrier, Fl. Madagascar 30: 61 (1945). Type: Simiane (Simianona) R., *Perrier* 11944 (Holotype P).

Clustering palm. **STEMS** 20 cm (*Beentje* 4515) to 2 m (*Perrier* 18047), 0.4–0.9 cm diam., internodes 1–1.5 cm long, pale green with dense red-brown scales; nodal scars c.1 mm. **LEAVES** 8–10 in the crown, spreading; sheaths 4.3–6 cm long, open for c. 50%, proximally pale reddish green, distally pale creamy green,

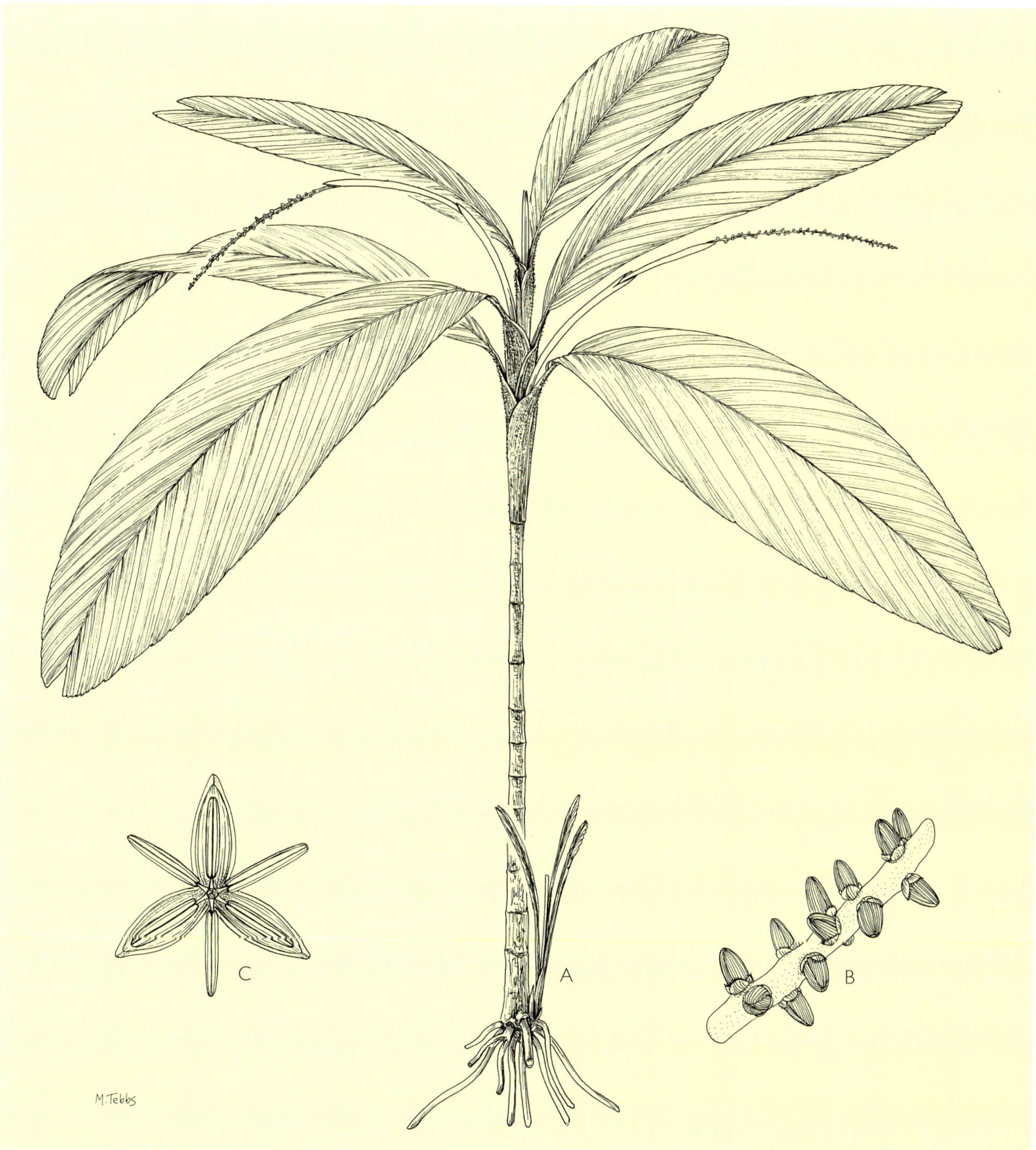

Dypsis integra. **A** whole plant × 2/5; **B** detail of rachilla × 3; **C** open staminate flower × 10. All from *Beentje* 4515. Drawn by Margaret Tebbs.

Dypsis integra, at Manombo (*Beentje* 4515).

with red laciniate scales, very dense distally, with narrowly triangular ligules 3–4 mm long at the base of the petiole; petiole 0.3–2 cm, 2–2.5 mm diam., densely scaly; blade entire, narrowly obovate, 15–33 x 4–7 cm, acute at the base, the margins toothed in the distal third or half, rounded at the apex with a 3–17 mm deep notch, midrib prominent on both surfaces with laciniate scales, other veins faint with minute scattered scales in lines. **INFLORESCENCE** interfoliar, erect to spreading, unbranched; peduncle 12–15 cm long, 1.5–3 x 1.5 mm diam., with scattered laciniate scales; prophyll 7–11 cm, borne at 2–3.5 cm above the base of the peduncle with c. 2 mm wide wings, with a 2-sided lateral split in the distal few cm; peduncular bract inserted at 5–7 cm from the base of the peduncle, 7–12 cm long, with scattered scales, laterally split on one side in the distal 1–2 cm; rachilla 5–11 cm long, 1–2 mm diam., with scattered scales, with slightly distant superficial triads; rachilla bracts entire, rounded, at the margins with reddish laciniate scales. **STAMINATE FLOWERS** with sepals 0.8–1 x 0.8–1 mm, keeled, gibbous, ciliolate to ciliate, with slightly connate base, membranous; petals 2–2.2 x 1–1.3 mm, elliptic, acute, striate, almost free; stamens 6, uniseriate, equal, filaments 0.8 mm long, connate at the base in a 0.2 mm high ring, anthers 1.5–1.6 x 0.5–0.6 mm, dorsifixed, the locules parallel and blunt; pistillode 1.2–1.3 x 0.3 mm. **PISTILLATE FLOWERS** only known from young bud, with sepals imbricate, 1.2–1.3 x 1–1.5 mm, ciliate, rounded; petals valvate, 1.8 x 0.7 mm; ovary 1.6 x 0.7 mm; staminodes c.0.2 mm. **FRUIT** unknown.

NOTE. This species is close to *D. simianensis* and more material might even show the species to be synonymous.

Beentje 4515 differs from the type in the hairiness of the sheath (which is greater in 4515) and the shape of the sheath (more open, and much wider in 4515) as well as in the leaf venation (transverse, sinuous veinlets clearly visible in 4515).

SPECIMENS SEEN. Soanierana-Ivongo: Simiane (Simianona) R., Sept. 1917 (old infl.), *Perrier* 11944 (Holotype P). Mahanoro: lower Mangoro basin, Oct. 1927 (bud), *Perrier* 18047 (P). Farafangana: Manombo, Nov. 1991 (bud), *Beentje* 4515 (K, MO, TAN).

94. DYPSIS DIGITATA

This was thought to be extinct until HB refound it in Manombo Forest, the site of several other palms which were thought to be extinct. The Latin name indicates that the leaflets appear to be in the shape of the fingers of a hand, i.e. almost palmate, which is unique among Madagascar undergrowth palms.

DISTRIBUTION. East Coast between Mananjary and Vangaindrano.
HABITAT. Lowland rain forest; 45–100 m.
LOCAL NAMES. Not recorded.
USES. Not recorded.
CONSERVATION STATUS. Critical. The site of the only recent collection is being destroyed by fire, shifting cultivation and logging. Both the older collection sites are now devoid of forest.

Dypsis digitata *(Becc.) Beentje & J. Dransf.* **comb. nov.**

SYNONYM:
Neophloga digitata Becc., Palme del Madagascar 36 (1914); Jum., Ann. Inst. Bot.-Géol. Colon. Marseille sér. 4, 6 (3): 20 (1929); Cat. Pl. Madagascar, Palmae: 20 (1938); Jum. & H. Perrier, Fl.

Madagascar 30: 66 (1945). Type: Madagascar, Mananjary Province, March-April 1909, *Geay* 8057 (Holotype P).

Small solitary palm, occasionally subcolonial (Beentje). **STEM** very short or 10–40 cm, in *Perrier* 4512 with roots above the surface, 1–1.5 cm diam.; internodes 2–6 mm, brown; nodal scars 1–2 mm. **LEAVES** 7–13 in the crown, erect-arching, often with a few marcescent leaves; sheath 7–13 cm long, 3/4 open, dark reddish with dark scattered scales and ligules 5–10 mm high in young leaves; petiole 5–43 cm long, 1.5–3 mm diam., pale brown with dense to scattered scales; rachis 4–11 cm long, with scattered scales; blade *entire*, 23–41 cm long, midrib 4–11 cm, densely hairy, lobes 19–32 x 1.2–2.9 cm or *pinnate* and then leaflets 2–3 on each side of the rachis, 21–29 x 0.9–2.3 cm (interval 0.6–2 cm), linear, ± equal except that the terminal pair having narrow dentate apices (2–3 teeth only), not acuminate as the more proximal leaflets, main veins 2–4, and often with sinuous transverse veinlets, seemingly glabrous or with scattered scales on the minor veins, or with bands of dense scales on the abaxial surface. **INFLORESCENCE** interfoliar, unbranched, 18–58 cm; peduncle 12–41 cm or more long, densely scaly, 1.5–3.5 mm diam.; prophyll 3–16 x 0.5–0.8 cm, borne at 0.5–2.5 cm above the base of the peduncle, opening near the apex only for 2–6 cm, dark brown, with dense scales; peduncular bract inserted at 2–15 cm from the base of the peduncle, 10–20 cm long, opening in the distal 1–4 cm, sometimes with a closed beak of c. 5 mm long, pale brown with scattered scales; non-tubular peduncular bract sometimes present near the apex of the peduncle, c. 2 cm long; rachilla 6–17 cm long, 1.5–2.5 mm diam., densely puberulous, with very dense, almost continuous triads. **STAMINATE FLOWERS** with sepals 0.8–1.5 x 0.7–1.6 mm; petals (1.2–1.5 mm in bud) 2.8 x 1.5 mm; stamens 6, ± equal, with filaments c. 2.2 mm long and anthers c.

Dypsis digitata, at Manombo (*Beentje* 4512).

1.5 × 0.5 mm, the locules parallel; pistillode c. 1.3 × 0.3 mm. **PISTILLATE FLOWERS** with sepals 1.3–1.8 × 1–2.1 mm; petals 1.8–2.9 × 1.4–2.4 mm; staminodes 0.3–0.7 mm; gynoecium c. 1.9 × 1.3 mm. **FRUIT** red, ovoid or ellipsoid, 11–13 × 5–8 mm, with an obtusely pointed apex; endocarp fibrous, the fibres slightly anastomosing. **SEED** ellipsoid, 10–10.5 × 5–5.5 mm, pointed at both ends; endosperm homogeneous.

NOTE. Very distinct; the open leaf sheath is reminiscent of *D. brevicaulis*.

SPECIMENS SEEN. Mananjary: without further locality except "zone cotière", March/April 1909 (fl.), *Geay* 8056 (P); idem, (y. fr.), *Geay* 8057 (Holotype P). Farafangana: Manombo, Nov. 1991 (bud), *Beentje* 4512 (K). Vangaindrano: Ankararano (loc. not found), May 1919 (fr.), *Perrier* 12625 (P).

95. DYPSIS BREVICAULIS

The leaves of this species appear to come directly from the ground, but there is a small underground stem. The Latin name means 'short-stemmed'.

DISTRIBUTION. East Coast just N of Manantenina and Manafiafy/Sainte-Luce.
HABITAT. Evergreen forest on white sand or laterite; 1–700 m.
LOCAL NAMES. Not recorded.
USES. Not recorded.
CONSERVATION STATUS. Critical. Only known from

Dypsis digitata. **A** habit × 1/6; **B** leaf bases, leaf and inflorescence × 1.5; **C** fruit × 3; **D** endocarp × 3; **E** endocarp and seed in cross section × 3. **A** from *Beentje* 4512, **B** from *Geay* 8056, **C – E** from *Perrier* 12625. Drawn by Margaret Tebbs.

three sites; numbers are extremely low (less than twenty have been seen) and forests in the area are under threat from shifting cultivation and proposed ilmenite mining operations.

Dypsis brevicaulis *(Guillaumet) Beentje & J. Dransf.* **comb. nov.**

SYNONYM:
Neophloga brevicaulis Guillaumet, Adansonia sér.2, 13 (3): 343, figs. 1-2 (1973). Type: Madagascar, Analalava forest N of Manantenina, *Guillaumet* 4003b (Holotype P; isotype K).

Solitary (?) dwarf palm, almost acaulescent (*Humbert* 20671 has 'suffrutex 1 m'). **STEM** largely underground, 5–15 cm, 0.9–2 cm diam. **LEAVES** 5–8, entire; sheath 7–11 cm long, densely covered with reddish scales but glabrescent, with 2 small auricles, almost open in outermost leaves, then with ragged margins; petiole 0–8 cm long, 2–2.5 mm diam., adaxially flat, abaxially concave, with scattered scales; lamina narrowly triangular, 28–150 × 2.7–4.5 cm, with attenuate base, with c. 4 main veins on each side, adaxially with the proximal part of the midrib scaly, abaxially with the midrib scaly, and with minute scattered scales all over, lobed for 3.5–12 cm, the lobes 1.6–2 cm wide and with rather narrow, dentate apex, the outside margin also with one or two indistinct teeth at 6–10 cm from the apex. **INFLORESCENCE** unbranched or rarely branched into 2 rachillae; peduncle c. 13 cm; prophyll c. 13 × 0.3 cm; peduncular bract not seen; rachilla 9–17 cm, c. 1 mm diam., lepidote, with 60–80 triads, these protandrous; rachilla bract wide, concave, rounded-triangular. **STAMINATE FLOWERS** with keeled sepals 1.3–1.5 × 1–1.2 mm, acute, slightly gibbous; petals 2–2.2 × 1.2 mm, ovate, acute; stamens 6, didymous, biseriate, the antepetalous inserted 0.2–0.5 mm higher up, filaments 0.8–0.9 × 0.3 mm, anthers c. 0.4 × 0.6 mm;

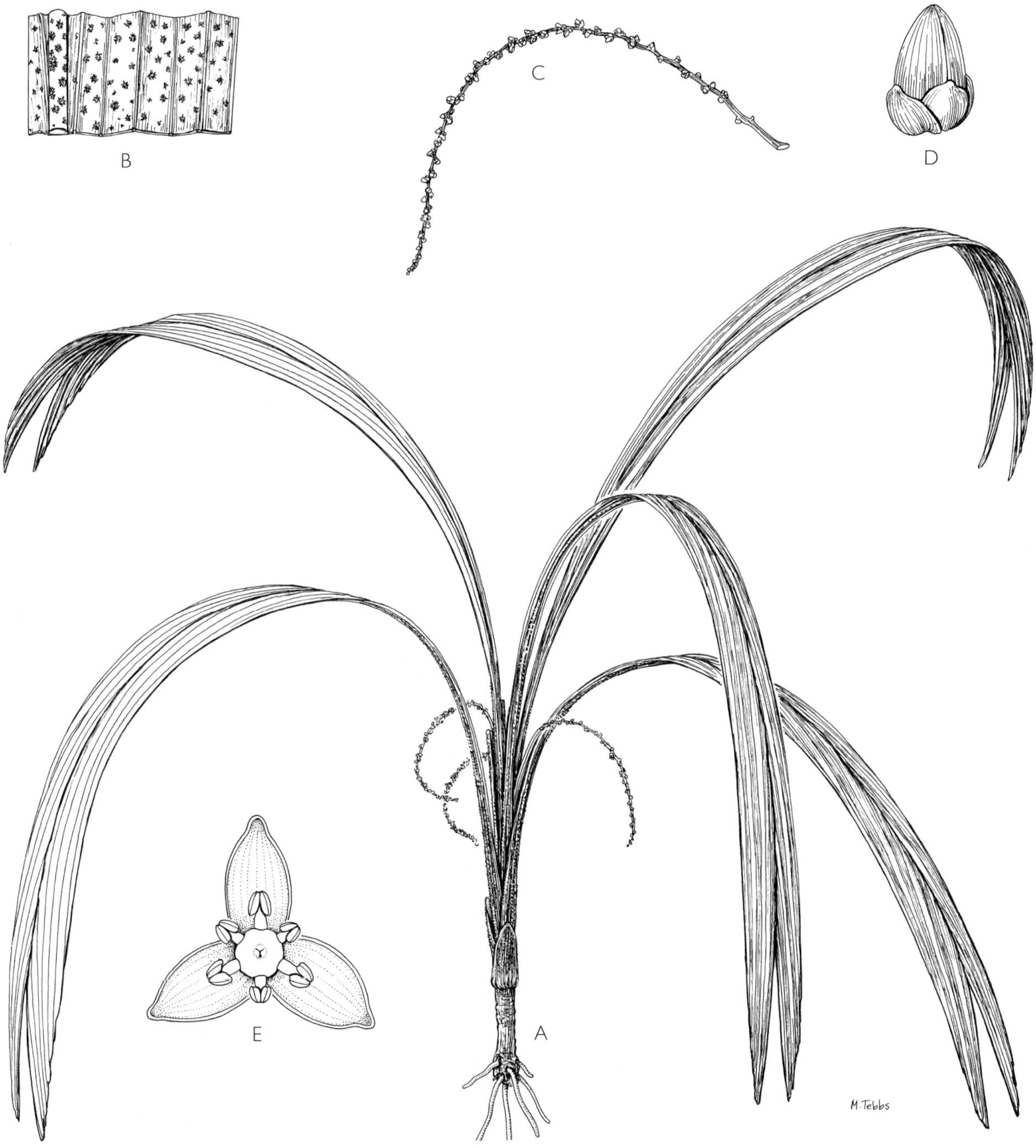

Dypsis brevicaulis. **A** habit × 1/4; **B** detail of undersurface of leaf × 4; **C** rachilla × 1/3; **D** staminate flower bud × 8; **E** open staminate flower × 10. All from *Guillaumet* 4003b. Drawn by Margaret Tebbs.

pistillode 0.8 x 0.5 mm, club-shaped. **PISTILLATE FLOWERS** with cucullate sepals; petals ovate, staminodes 6, very short at the base of an oblong ovary, this trigonous, with connivent stigmas. **FRUIT** unknown.

NOTE. This species has most distinctive long, narrow leaves.

SPECIMENS SEEN. Tolanaro: Analalava forest, Dec. 1971 (fl.), *Guillaumet* 4003b (K, P, type); Mt Vohimavo, N. of Ampasimena, March 1947 (old infl.), *Humbert* 20671 (K, P); St. Luce forest, March 1992 (ster.), *Beentje & Andriampaniry* 4610 (K, TAN).

96. DYPSIS FORFICIFOLIA

A slender solitary or clustered palm of the forest undergrowth. This is the commonest small palm of the forests of the northeast. It is very variable and, as can be seen from the list of synonyms, we are including within its range of variation species that have in the past been recognised as distinct. It appears to flower and fruit abundantly and so there should be little difficulty in obtaining seed. It would certainly make a very handsome ornamental. It could be confused with *D. ambilaensis* but differs in the position of the three stamens (see under *D. ambilaensis*). The species name is derived from the Latin for a pair of shears or scissors, presumably in reference to the forked leaves, reminiscent of an open pair of shears.

DISTRIBUTION. NE Madagascar, apparently quite common.

HABITAT. Coastal lowlands and hill forest; 5–500 m.

LOCAL NAMES. Not recorded.

USES. Not recorded.

CONSERVATION STATUS. So widespread in the northeast of the island as to be at present not threatened.

Dypsis forficifolia *Noronha ex Mart.*
Hist. Nat. Palm. 3 (ed. 1): 180, t. 143. f. 2 (1838); Baill., Bull. Soc. Linn. Paris 147: 1161 (1894); Becc., Bot. Jahrb. Syst. 38, Beibl. 87: 12 (1906); Palme del Madagascar 10 (1912); Jum., Ann. Inst. Bot.-Géol. Colon. Marseille sér. 3, 6 (1): 30, 33 (1918); Bull. Acad. Malgache 6: 9 (1923); Cat. Pl. Madagascar, Palmae: 12 (1938); Jum. & H. Perrier, Fl. Madagascar 30: 39 (1945). Type: Madagascar, *Du Petit Thouars* s.n. (Holotype P).

SYNONYMS:

Dypsis hirtula Mart., Hist Nat. Palm. 3 (ed. 1): 181 (1838); Becc., Palme del Madagascar 13 (1912); Jum., Cat. Pl. Madagascar, Palmae: 13 (1938). Type: Madagascar, "Rocam Bondi", *Poivre s.n.* (Holotype P). **Synon. nov.**

Dypsis littoralis Jum., Ann. Inst. Bot.-Géol. Colon. Marseille sér. 3, 6 (1): 34 (1918); Bull. Acad. Malgache 6: 10 (1923); Cat. Pl. Madagascar, Palmae: 13 (1938); Jum. & H. Perrier, Fl. Madagascar 30: 40 (1945). Type: Madagascar, Mananara, *Perrier* 12056 (Holotype P). **Synon. nov.**

Dypsis masoalensis Jum., Ann. Inst. Bot.-Géol. Colon. Marseille sér. 3, 6 (1): 36 (1918); Bull. Acad. Malgache 6: 15 (1923); Cat. Pl. Madagascar, Palmae: 14 (1938); Jum. & H. Perrier, Fl. Madagascar 30: 48 (1945). Type: Madagascar, Masoala, *Perrier* 12034 (Holotype P). **Synon. nov.**

Dypsis forficifolia. View of crown and inflorescence form with divided leaves, Sahavary (*Dransfield et al.* JD6456).

Dypsis forficifolia, a common forest undergrowth palm in northeast Madagascar, form with entire leaves (*Dransfield et al.* JD6387)

Slender, solitary or clustered forest undergrowth palm. **STEMS** to 4 m tall, c. 7–10 mm diam., internodes 8–30 mm long, often striped when young, dark green near the nodes, pale green between, with scattered to dense caducous dark brown scales, c. 8 × 5 mm. **LEAF** with sheaths 7–9 cm long, 0.7–1.8 cm diam., pale green, rather densely covered in caducous brown scales, auricles usually well developed, triangular; petiole absent or to 12 cm long, 2.5–4 mm wide, abaxially rounded or angled, adaxially grooved or flat; rachis 16–32 cm; blade entire bifid, to 35 × 18 cm, split from one third to one half the length, or with 2–6 leaflets on each side of the rachis; leaflets generally rather broad, occasionally a few narrow leaflets also present, 6–30 × 0.7–8 cm, apical pair with shallow apical lobing, adaxially lamina glabrous, abaxially with scattered bands of caducous, chaffy brown scales, abundant minute punctiform brown scales and scattered large ramenta to 6 mm long; emerging leaf sometimes tinged pink. **INFLORESCENCE** interfoliar, branching to 2, rarely to 3 orders; peduncle 22–50 cm long, c. 2 mm diam. near tip; prophyll 9–25 × 0.4–0.8 cm, with scattered mid brown scales; peduncular bract similar to prophyll but exceeding it, 15–40 × 0.4–0.8 cm; rachis 13–40 cm long, glabrous or sparsely scaly; rachillae from 30 to c. 200, usually diverging at a rather wide but acute angle, occasionally reflexed, 2–4.5 cm long (rarely to 12 cm), 0.5–0.7 mm diam., glabrescent and slightly angled on drying, occasionally with sparse laciniate scales, bearing at anthesis rather distant triads of flowers, c. 1–4 mm distant, each subtended by a smooth entire or shallowly to deeply laciniate rachilla bract c. 1 mm high. **STAMINATE FLOWERS** c. 1 mm diam. at anthesis; sepals c. 0.5 × 0.5 mm, rounded, keeled, somewhat erose margined, shining; petals c. 1 × 0.8 mm, striate; stamens 3, antesepalous, filaments connate in a low ring c. 0.15 mm high, anthers c. 0.25 mm, staminodes absent; pistillode low, conical. **PISTILLATE FLOWERS** c. 1 mm diam., globose; sepals 0.5 × 0.5 mm, broadly imbricate; petals c. 1 × 0.8 mm, striate; staminodes 3 minute; ovary c. 0.8 mm diam. Mature **FRUIT** red, turning black, to c. 15 × 9 mm, ellipsoid. **SEED** 14 × 6 mm, endosperm homogeneous, embryo lateral near the base.

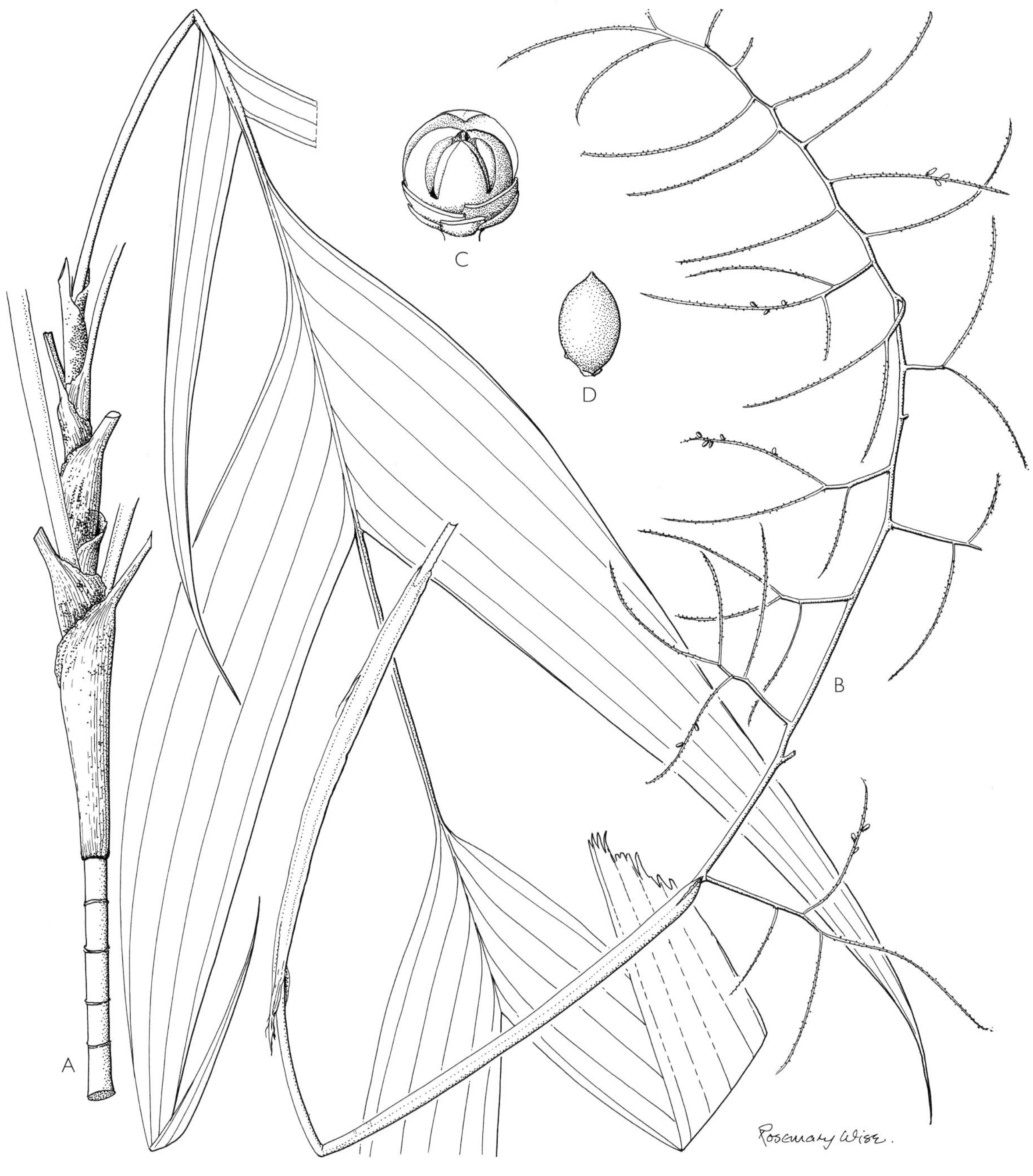

Dypsis forficifolia. **A** crown × 1/2; **B** inflorescence × 1/2; **C** staminate flower, one petal removed × 20; **D** fruit × 1. All from *Beentje et al.* 4633. Drawn by Rosemary Wise.

NOTE. Our interpretation of *D. hirtula* differs from that of Beccari (1910) and Jumelle and Perrier (1945). The holotype of *D. hirtula* is an unnumbered Poivre collection in the Jussieu Herbarium in Paris. Vegetatively this specimen fits within the range of variation of *D. forficifolia*. The inflorescence, however, is branched to three (very rarely four) orders and the rachillae are slender, bearing scattered dark brown laciniate scales and rather distant triads. Viewed in isolation the types of *D. forficifolia* and *D. hirtula* thus appear rather different and can be differentiated using the character of rachilla scaliness utilised by Beccari. However, with the wide range of material now available, illustrating considerable variation in *D. forficifolia,* it is no longer possible to make a clear distinction and we have concluded that *D. hirtula* represents one extreme of the variation of *D. forficifolia*. We have not been able to examine the Majastre collection illustrated by Beccari (1910) as *D. hirtula* but we suspect that this is *D. viridis*. *Perrier* 17466, cited in the Flore de Madagascar as *D. hirtula* and used as the basis of the illustration, is, we believe, *D. viridis* (q.v.). We have also included *D. masoalensis* in synonymy with *D. forficifolia*. Jumelle described the former as differing from the latter in the presence of laciniate rachilla bracts and hairs on the rachillae, and the presence of up to six leaflets on each side of the rachis. The leaf of *D. masoalensis* fits within the range of variation of *D. forficifolia* and the rachillae, similarly, in *D. forficifolia* can vary from being glabrous to with very sparse caducous laciniate scales and the rachilla bracts from being entire to sparsely laciniate. *D. littoralis* fits within the range of variation of *D. forficifolia* and so is included here too.

Plants from near Ambanizana, Masoala Peninsula, tend to have inflorescences branched to three rather than two orders with rather slender and very numerous more or less glabrous rachillae.

Dypsis forficifolia var. *reducta* has staminate flowers with six stamens with didymous anthers, so clearly does not belong here (see under *D. schatzii*).

SPECIMENS SEEN. Maroantsetra: Masoala, Aug. 1912 (buds), *Perrier* 12034 (Holotype of *D. masoalensis* Jum. P); idem, Aug. 1912 (buds), *Perrier* 12037 (P); Ambanizana, Oct. 1986 (fl.), *Dransfield et al.* JD6387 (BH, K, MO, P, TAN); Antalavia, Feb. 1988 (fr.), *Dransfield et al.* JD6470 (K, NY, MO, P, TAN); idem, Nov. 1989 (fr.), *Dransfield et al.* JD6746 (K, TAN); Andronabe, Feb. 1992 (fr.), *Zjhra & Hutcheon* 139 (K, MO, TAN); Andranofotsy River, Sahavary, Feb. 1988 (fr.), *Dransfield et al.* JD6450 (K, MO, NY, P, TAN); idem, Feb. 1988 (fr.), *Dransfield et al.* JD6456 (K, MO, NY, P, TAN); Bay of Antongil, Fananehana, Aug. 1912 (fl.), *Perrier* 12091 (P). Mananara Avaratra: Mananara, littoral forest, Aug. 1912 (fl.), *Perrier* 12056 (Holotype of *Dypsis littoralis* Jum., P); 10 km west of Antanambe, April 1992 (fl.), *Beentje et al.* 4624 (K, TAN); 5 km south of Antanambe, April 1992 (fl., fr.), *Beentje et al.* 4633 (BH, K, MO, P, TAN); idem, April 1992 (fl.), *Beentje et al.* 4637 (K, TAN). Soanierana-Ivongo: Soanierana-Ivongo, Feb. 1969 (fl.), *Guillaumet* 2406 (P). Île Sainte Marie: Tafondrou, March 1847 (fl.), *Boivin* 1704 (FI, P); Forêt de Kalalao, Nov. 1994 (fl., fr.), *Dransfield* JD7525 (K, P, TAN). Without locality: *Du Petit Thouars* s.n. (Holotype P); *Poivre s.n.* (Holotype of *D. hirtula* Mart., P).

97. DYPSIS INTERRUPTA

A strikingly beautiful small palm of the forest undergrowth, known as yet from a single collection from the lowlands in the southeast of the island. The species name refers to the interrupted sequence of otherwise regularly arranged leaflets on each side of the rachis.

DISTRIBUTION. Ifanadiana area.
HABITAT. Hill forest; 510 m.
LOCAL NAMES. Not recorded.
USES. Not recorded.
CONSERVATION STATUS. Critical. The extraordinary forests of Ambinanindrano are gradually being destroyed by shifting cultivation.

Dypsis interrupta *J. Dransf.*, **sp. nov.**
inter species floribus staminatis triandris staminibus antesepalis pistillodio conico foliolis linearibus numerosis in quoque latere rachidis interrupte dispositis distinctissima. Typus: Madagascar, East Coast, Ifanadiana, Ambinanindrano, *Beentje* 4528 (Holotypus K; isotypus TAN).

Solitary palm of forest undergrowth. **STEMS** to 3 m tall, c. 12 mm diam.; internodes 60–70 cm long. **LEAVES** 7 in crown; sheaths forming a well defined crownshaft; sheath 19–20 cm long, 1.5 cm diam., with sparse scattered dark brown scales; low auricles present but apparently soon tattering; petiole 9–12 cm long, c. 4 × 2 mm in cross section, deeply channelled, rather densely covered in discrete dark brown scales; rachis 64–70 cm long, scaly as the petiole; leaflets 24–29 on each side of the rachis, all except the apical pair single-fold, linear, grouped in proximal part of the leaf, regularly arranged distally, regularly arranged within the groups, the leaflet series thus appearing interrupted rather than fascicled, mid-leaf leaflets 15–23 × 0.8–1.5 cm, apical pair slightly shorter, basalmost leaflets 6–8 × 0.4–0.8 cm, leaflet midrib prominent adaxially, texture membranous. **Inflorescence** interfoliar, branched to 2 orders, arching with pendulous rachillae; peduncle 50 cm long, c. 4 mm wide at the base, tapering to 2.5 mm wide, glabrous; prophyll inserted 14.5 cm above the base of the peduncles, 30 × 1.5 cm, membranous, striate, very sparsely scaly; peduncular bract inserted c. 20 cm above the prophyll, otherwise not known; rachis c. 30 cm long, glabrous; rachillae slender, c. 20, 19–30 cm long, glabrous, c. 1.5 mm diam., triads c. 3 mm apart, rachilla bract c. 0.8 mm high, rounded, glabrous. **Flowers** very immature in available specimen. **Staminate flowers** with 3 ?sagittate antesepalous stamens and a conical pistillode. **Fruit** unknown.

Note. This taxon, although known only from a single specimen, is so distinctive that we have considered it important to name it, even though several parts are not represented. Among the species of *Dypsis* with three sagittate antesepalous stamens and a conical pistillode, *D. interrupta* is immediately distinguishable by its numerous linear leaflets arranged in an interrupted series on each side of the rachis. This is clearly a very beautiful species and it would be desirable not only to recollect it to complete our understanding of it as a species, but also to introduce it into cultivation. It was found growing on a steep slope in open forest.

Specimen seen. Ifanadiana: 1 km S of Ambinanindrano, 17–19 Nov. 1991 (fl. buds), *Beentje* 4528 (holotype K; isotype TAN).

Dypsis interrupta. **A** leaf × 2/5; **B** inflorescence × 2/5; **C** open staminate flower × 20. All from *Beentje* 4528. Drawn by Rosemary Wise.

Dypsis interrupta. An elegant palm with grouped leaflets (*Beentje* 4528).

98. DYPSIS LANTZEANA

This is a common palm in the rain forests surrounding the Bay of Antongil. It grows in the forest undergrowth and rarely exceeds about 4 m tall. It could be confused with *D. forficifolia* but can be distinguished by its very hairy rachillae. It is in cultivation in the Palm House of the Royal Botanic Gardens, Kew. This is a most attractive species with its neat habit, young leaves tinged reddish and cherry-red inflorescence branches. The species is named for the collector of the type.

DISTRIBUTION. Northeastern Madagascar, particularly around the Bay of Antongil.
HABITAT. Lowland rain forest; up to 350 m.
LOCAL NAMES. Not recorded.
USES. Not recorded.
CONSERVATION STATUS. Vulnerable. The distribution is limited, and the forests are under some threat from shifting cultivators.

Dypsis lantzeana *Baill.*
Bull. Soc. Linn. Paris 2:1163 (1893); Becc., Bot. Jahrb. Syst. 38, Beibl. 87: 13 (1906); Palme del Madagascar 11 (1912); Jum. in Ann. Inst. Bot.-Géol. Colon. Marseille sér. 3, 6 (1): 30 (1918); Bull. Acad. Malgache 6: 7 (1923); Cat. Pl. Madagascar, Palmae: 13 (1938); Jum. & H. Perrier, Fl. Madagascar 30: 12 (1945) "lanceana". Type: Madagascar, Bay of Antongil, 7 Oct. 1871, *Lantze* s.n. (holotype P).

SYNONYM:
Dypsis lantzeana var. ***simplicifrons*** Becc., Bot Jahrb. Syst. 38, Beibl. 87: 13 (1906); Palme del Madagascar 13 (1912). Syntypes: Madagascar, *Mocquerys* 332, 334 (G-DC).

Dypsis lantzeana. Crown showing pink-tinged emerging leaf, Masoala (*Dransfield et al.* JD6363).

Dypsis lantzeana. **A** crown × 2/5; **B** detail of rachilla × 10. All from *Dransfield et al.* JD6363. Drawn by Rosemary Wise.

Slender solitary or clustered forest undergrowth palm. **Stem** to 4 m tall, 0.7–2 cm diam., internodes 0.8–3 cm long, stem surface with scattered brown scales. **Leaves** 6–15 in crown; sheath 8–9 × 1.5–2 cm, pale green to ivory-coloured, striate, with abundant caducous dark scales, auricles present, to 8 × 4 mm, or not evident; petiole absent or to 10 cm, c. 3.5 mm wide with scattered dark brown scales; rachis 19–47 cm; blade entire bifid to 40 × 15 cm, split to half its length, or leaflets 3–9 on each side of the rachis, generally leaflets somewhat sigmoid, tinged pink when newly emerging, 4.5–26 × 0.5–7 cm, sometimes several very small leaflets crowded at base of rachis, apical leaflets 10–16 × 3.5–6 cm, diverging at a rather wide angle, adaxial surface sparsely dotted with minute brown scales, abaxially with scattered punctiform scales and often with conspicuous ramenta to 20 mm long on ribs. **Inflorescence** interfoliar, branching to 2–3 orders, the branches often somewhat reflexed; peduncle 21–50 cm long, c. 2.5 mm diam. near base; prophyll 10–30 × c. 0.7 cm, membranous, sparsely scaly; peduncular bract exceeding prophyll by 7–25 cm; rachis 21–35 cm; rachillae c. 50 to over 120 in number, 2–6 cm long, c. 0.8–1 mm diam., often reflexed, usually cherry-red when young, densely covered with long branched brown hairs, triads c. 1.5 mm distant, rachilla bract laciniate, it and flowers almost completely obscured by rachilla hairs. **Staminate flowers** spherical, c. 1 × 0.85 mm; sepals imbricate, rounded, irregularly splitting, c. 0.3 × 0.4 mm; petals 1.0 × 0.8 mm; stamens 3 (rarely 2 or reduced to 1), antesepalous, filaments connate in basal 25 mm, free filaments c. 0.3 mm, anthers sagittate, c. 0.1 × 0.1 mm, staminodes absent; pistillode dome-like. **Pistillate flowers** c. 1.2 mm diam.; sepals 0.3 × 0.4 mm, broadly imbricate, irregularly erose; petals striate, triangular c. 1.2 × 1.0 mm; staminodes 6; ovary globular c. 0.7 mm diam. Mature **fruit** ellipsoid, cherry-red ripening dark purple, 10 × 7 mm. **Seed** 7 × 4 mm, endosperm homogeneous, embryo lateral near the base.

NOTE. In the habit, leaf form and inflorescence branching, *D. lantzeana* is very similar to *D. forficifolia*. The rachillae of the present species are densely covered in trichomes, whereas in the latter they are usually glabrescent and only rarely with scattered trichomes. Furthermore, the rachilla bracts in the present species are usually so hairy that the bracts and the subtended flowers are partially obscured. In one remarkable collection, *Dransfield* JD6363, staminate flowers consistently (at least in the Kew duplicate) have only one stamen. This is a particularly tomentose form. It would be interesting to investigate further into this unistaminate condition. In another collection *Henderson et al.* 756, some staminate flowers have only two stamens.

We have pondered at length over the correct orthography of the specific epithet. Baillon originally published "*lantzeana*"; the type annotation and all subsequent authors have used "*lanceana*". We suggest, however, that because Baillon used the spelling Lantze, the collector must have been Lantze rather than Lance (though it is not certain to us whether these are different collectors or the same), and thus favour the former rather than the latter spelling.

SPECIMEN SEEN. Maroantsetra: Bay of Antongil, Oct. 1871 (buds), *Lantze* s.n. (in part - leaf on second sheet = probably *D. fibrosa*) (holotype P); idem, Aug. 1912, *Perrier* 12048 (P); Maroa, 1897 (fl.), *Mocquerys* 332 (type of var. *simplicifrons*, FI); Nosy Mangabe, 1897 (fl.), *Mocquerys* 412 (FI); idem, *Mocquerys* 418 (FI); idem, Feb. 1988 (fl.), *Henderson et al.* 756 (K, MO, NY, P, TAN); idem, April 1988 (fl.), *Schatz & Gentry* 2097 (K, MO, TAN); Masoala, May 1974, *Morat* 4954 (P); Masoala, Hiaraka, Oct. 1986, *Dransfield et al.* JD6363 (BH, K, MO, P, TAN). Mananara Avaratra, 10 km west of Antanambe, Oct. 1991, *Beentje* 4465 (BH, K, MO, P, TAN); idem, April 1992, *Beentje et al.* 4619 (BH, K, MO, P, TAN); 2 km east of Andravolasoa, April 1992 (fr.), *Beentje et al.* 4641 (K, TAN).

99. DYPSIS REMOTIFLORA

Known only from a single collection, this small undergrowth palm has the general appearance of *D. forficifolia*. It can be immediately distinguished by the highly branched inflorescences with extremely slender thread-like branches which carry very few small and distant flowers. The sparse arrangement of the flowers is reflected in the species name.

DISTRIBUTION. East Coast.
HABITAT. Lowland rain forest.
LOCAL NAMES. Not recorded.
USES. Not recorded.
CONSERVATION STATUS. Presumed extinct; not seen for forty years, and the type locality is now devoid of forest.

Dypsis remotiflora *J. Dransf.*, **sp. nov.**
inter species floribus staminatis triandris staminibus antesepalis pistillodio carenti, rachillis gracillimis floribus paucis remotis distinctissima. Typus: Madagascar, East Coast, Ambadikala, *Vigreux* 15417 (Holotypus P).

Solitary (?) or clustered (?) palm. **STEM** to 1 m tall, at least 1 cm diam. (as suggested by diameter of leaf sheaths). Leaf sheaths of unknown length, c. 1 cm diam., striate, bearing abundant dark brown laciniate scales. **LEAF** to 40 cm long, including petiole c. 9 mm long, c. 4 mm wide, broadly triangular in cross section, bearing scattered dark brown scales; rachis 22–29 cm; leaflets 3–4 on each side of the rachis, subopposite or almost alternate (basal leaflets only); basal pair 5–7 x 0.3–0.7 cm, mid-leaf pair to 23 x 5.5 cm, apical pair widely divergent, to 13 x 4.5 cm, somewhat sigmoid; both lamina surfaces with very sparse, minute, brown,

punctiform scales. **INFLORESCENCE** only partially represented in type, branching to 4 orders; prophyll, peduncular bract and peduncle not represented; rachis at least 35 cm long, somewhat sinuous, it and all branches glabrescent; rachillae very numerous, very slender, sinuous, 1.5–4 cm long, c. 0.2 mm diam., each bearing about 3–4 triads only, c. 7 mm distant; rachilla bracts triangular, c. 0.5 mm long. **STAMINATE FLOWER** c. 1.0 × 0.8 mm; sepals ± rounded, imbricate, slightly keeled, 0.5 × 0.6 mm; petals ± elliptic in outline, valvate, 1.0 × 0.7 mm; stamens 3, antesepalous, filaments very short, c. 0.2 × 0.1 mm, connective broad abaxially, anthers 0.3 × 0.1 mm, ± didymous; staminodes absent; pistillode absent. **PISTILLATE FLOWER** slightly smaller than the staminate; sepals imbricate, c. 0.5–0.6 mm; petals basally imbricate, distally valvate, c. 0.7 × 0.4 mm; staminodes 3, minute, dentiform; ovary c. 0.7 × 0.4 mm, ellipsoid, stigma terminal, slightly curved. Mature **FRUIT** ellipsoid, 9.5 × 4 mm. **SEED** ellipsoid, 6 × 3.5 mm; endosperm homogeneous, embryo lateral.

NOTE. This species is remarkable for its extremely slender rachillae with few remote triads of flowers. It is known only from the type. There are two entries for Ambadikala in the US Army Gazetteer for Madagascar, one at 23°19´ S 47°24´ E, the other at 18°15´ S 48°56´ E., of which only the former is recorded on the 1: 500,000 Series maps of Madagascar. This lies in the vicinity of Farafangana.

SPECIMEN SEEN. Farafangana: Ambadikala, Sept. 1954, (fl., fr.) *Vigreux* 15417 (Holotype P).

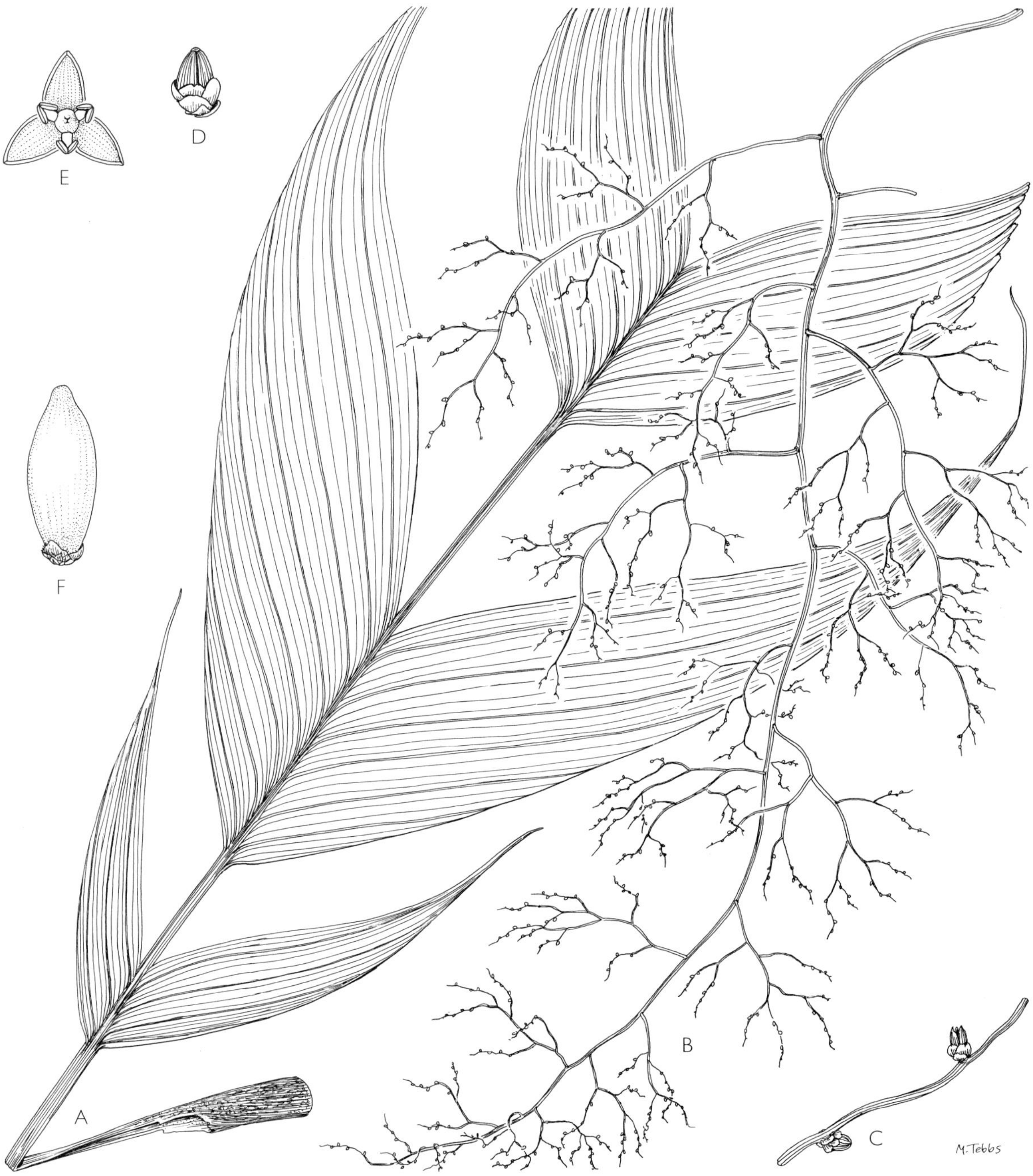

Dypsis remotiflora. **A** leaf × 3/5; **B** inflorescence × 3/5; **C** detail of rachilla × 3; **D** staminate flower bud × 10; **E** opened staminate flower × 15; **F** fruit × 3. All from *Vigreux* 15417. Drawn by Margaret Tebbs.

100. Dypsis louvelii

A neat palmlet of the undergrowth of montane forest, first collected at Analamazaotra, Andasibe (Perinet). This is an easily distinguished species because of its deeply bifid leaf and slender inflorescence with few short rachillae. It has frequently been collected at the type locality where it is no longer very common near to the main trails. Louvel, after whom the palm is named, was a Forest Officer at the old French Forestry Station at Perinet (modern day Andasibe).

Distribution. Central Madagascar, Analamazaotra and environs.
Habitat. Lowland and montane forest, usually in valley bottoms; 300–1100 m.
Local names. Not recorded.
Uses. Not recorded.
Conservation Status. Vulnerable; all recent collections are from one small area.

Dypsis louvelii *Jum. & H. Perrier*
Ann. Inst. Bot.-Géol. Colon. Marseille sér. 3, 1 (1): 21 (1913); Becc., Palme del Madagascar. 59 (1914); Jum., Ann. Inst. Bot.-Géol. Colon. Marseille Fasc. 1: 38 (1918); Bull. Ac. Malgache 6: 19 (1923); Cat. Pl. Madagascar, Palmae: 13 (1938); Jum. & H. Perrier, Fl. Madagascar 30: 24 (1945). Type: Madagascar, Analamazaotra, *Perrier* 11968 (Holotype P).

Solitary slender palm of the forest undergrowth, rarely exceeding 1 m tall. **Stem** 7–10 mm diam., internodes 4–10 mm long, surface glabrous. **Leaves** c. 5–10 in crown, tending to be held ± horizontally; leaf sheath 6–7 long, 1.0–1.4 cm diam., drying dull brown, bearing abundant dark brown scales, the mouth with two short triangular auricle to 4 × 3 mm; petiole absent or very short, not exceeding c. 2 cm; blade entire bifid, rather plicate, or, very rarely, divided into 3 distant narrow leaflets on each side; blade 19–50 cm long with a spread of 4.5–10 cm, the two lobes diverging at a very acute angle, 10–25 × 1.5–4 cm, where divided into leaflets, the leaflets 17–21 × 0.7–0.9 cm, adaxially with sparse punctiform scales, abaxially densely covered with brown punctiform scales. **Inflorescence** interfoliar, arcuate, branching to 1 order, rarely the lowermost 1 or 2 first order branches branched to the second order; peduncle 20–35 × 0.15–0.2 cm, sparsely brown scaly; prophyll sometimes borne up to 2 cm above base of peduncle, 6–10 × 0.3–0.5 cm; peduncular bract exceeding the prophyll, 9–23 × 0.4 cm; rachis usually conspicuously shorter than the peduncle, 5–10 cm; rachillae generally few in number, 9–17, generally somewhat reflexed, 1.3–4 cm long, c. 1 mm diam., sparsely brown scaly, bearing rather densely arranged triads. **Staminate flowers** at anthesis ± ovoid, 1.6 mm long; sepals rounded to broadly triangular, c. 0.7 × 0.7 mm, strongly keeled, inconspicuously dentate at the margins, abaxially slightly papillose, obscurely striate; petals connate in basal 0.25 mm, the lobes triangular, to 1 × 1 mm, abaxially smooth, glabrous; stamens 3 antesepalous, filaments to 0.75 × 0.3 mm, basally joined in the lowermost 0.2 mm, anthers somewhat sagittate, 0.5 × 0.3 mm, introrse; pistillode conical, minute, c. 0.1 mm high. **Pistillate flowers** at anthesis c. 1.6 mm diam.; sepals rounded to broadly triangular, c. 0.7 × 0.7 mm, strongly keeled, inconspicuously dentate at the margins, abaxially slightly papillose, obscurely striate; petals c. 1.5 × 1.5 mm, rounded to broadly triangular; staminodes 3, free, irregularly strap-shaped 0.2–0.4 × 0.2 mm; ovary irregularly reniform, c. 1.2 mm at widest point, stigmas 3, eccentric near the apex, c. 0.1 mm long. Immature fruit narrowly fusiform. Mature **fruit** scarlet, irregularly ellipsoidal, apically somewhat pointed, 14 × 8 mm; epicarp minutely papillose, mesocarp soft, fleshy; endocarp broad fusiform, 12 × 5

Dypsis louvelii, in full fruit, Mantady (*Beentje et al.* 4655).

mm, 0.5 mm thick, with brown, sparsely anastomosing fibres. **Seed** 11 × 4 mm, conforming to the endocarp shape; endosperm homogeneous; embryo lateral.

Note. No specimens were cited in the protologue (Jumelle & Perrier 1913). Of the three specimens cited by Jumelle (1918), we have found only *Perrier* 11968 and thus designate this collection as type.

This is a beautiful, distinctive species, characterised by its dwarf habit, usually with narrow, rather plicate, entire bifid leaves and inflorescences with long arcuate peduncle and short rachis, bearing rather few somewhat reflexed short rachillae. The staminate flowers have three antesepalous stamens with short sagittate anthers, no staminodes and a very small, scarcely discernible pistillode. At present the species seems to have a very restricted distribution.

Specimens seen. Moramanga: Andasibe, Analamazaotra, (fl.), *Perrier* 11968 (holotype P); idem, Feb. 1921 (fl.), *Perrier* 15984 (K, P); idem, Dec. 1932 (fl.), *Leandri* 719 (P); idem, March 1991 (fl., fr.), *Beentje & Raharilala* 4401 (BH, K, MO, P, TAN); Mantady, Dec. 1991 (fl.), *Beentje & Andriampaniry* 4540 (K, MO, TAN); idem, April 1992 (fl., fr.), *Beentje & Dransfield* 4655 (K); idem, Dec. 1992, *Beentje & Andriampaniry* 4772 (K, TAN). Ampasimanolotra: Ct. d'Ambalarondra, Andranampony, April 1951 (fl.), *Cours* 4522 (P). Toamasina: Tampina, au sud de Tamatave, March 1942 (fl.), *Decary* 17691 (K, P). Ambohimanga Atsimo: Route d'Ambohimanga, 1952-1955 (fl.), *Dequaire* 27714 (K, P).

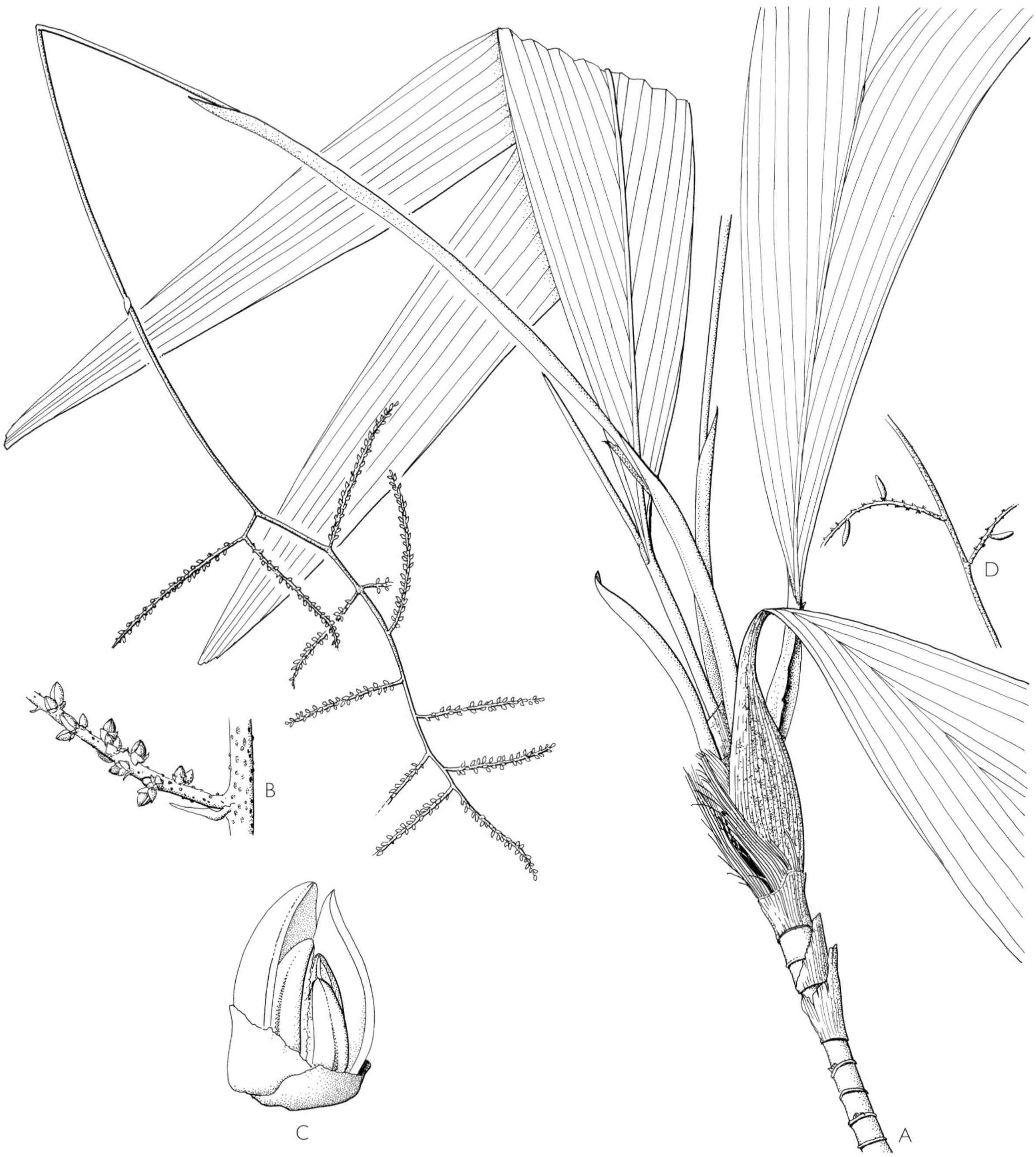

Dypsis louvelii. **A** crown × 3/5; **B** detail of rachilla × 3; **C** staminate flower, one petal removed × 20; **D** fruiting rachillae × 1. All from *Beentje & Raharilala* 4401. Drawn by Rosemary Wise.

101. DYPSIS ZAHAMENAE

At present this species is only known from the upland forests east of Lac Aloatra. It is a small undergrowth palm bearing erect inflorescences with remarkable short, thick, catkin-like branches. First named *Dypsis humbertii* by Jumelle in 1939, after the collector of the type specimen, it has now had to change its name; with the inclusion of all members of *Dypsidinae* in the single genus *Dypsis, D. humbertii* H. Perrier is pre-dated by *Neophloga humbertii* Jum. (1936) (= *Dypsis humbertii* (Jum.) Beentje & J. Dransf. of the present book). The new name refers to the large Nature Reserve of Zahamena where most of the collections have been made.

DISTRIBUTION. Known only from mountains east of Lac Aloatra in Zahamena Reserve.
HABITAT. Humid lower montane forest; 700–1000 m.
LOCAL NAMES. Not recorded.
USES. Not recorded.
CONSERVATION STATUS. Vulnerable; only known from a small, though protected area, with some threat of shifting cultivation to the forests.

Dypsis zahamenae. **A** crown × 2/5; **B** detail of rachilla × 2. All from *Humbert* 17776. Drawn by Margaret Tebbs.

Dypsis zahamenae *J. Dransf.* **nom. nov.**

SYNONYM:
Dypsis humbertii H. Perrier in Humbert, Not. Syst. 8: 46 (1939); Jum. & H. Perrier, Fl. Madagascar 30: 28 (1945). Type: Madagascar, Massif d'Andrangavolo, *Humbert* 17776 (Holotype P). [non *Dypsis humbertii* (Jum.) Beentje & J. Dransf. (*Neophloga humbertii* Jum. 1936)]

Slender undergrowth palm, whether solitary or clustered not known. **STEMS** 0.5–1.2 m tall, 7–8 mm diam., internodes 12–18 mm with scattered chestnut-brown scales. **LEAVES** 6–8 in crown; sheaths 7–9 cm long, 7–13 mm diam., striate, covered in caducous chocolate-brown scales, apically with 2 narrow triangular auricles 7–9 × 3–4 mm, these soon disintegrating; petiole absent or to 7 cm long, 2 mm wide, triangular in cross section, densely covered in soft brown scales; blade entire bifid, 20–35 × 12 cm, with apical cleft not exceeding one third of blade length, or irregularly divided into broad leaflets, up to 3–4 on each side of the rachis; rachis 11–18 cm; leaflets 10–26 × 1.5–4 cm, apical pair (or tips of entire bifid blades) shallowly lobed, c. 1 cm wide; lamina abaxially with abundant caducous brown scales along rachis and main ribs and with abundant punctiform scales, adaxially very sparsely punctiform scaly. **INFLORESCENCE** interfoliar, erect, branching to 1 order; peduncle 11–23 cm long, c. 1.5–2 mm diam.; prophyll c. 10 × 0.5 cm, sparsely brown scaly; peduncular bract similar, exceeding prophyll by up to 6 cm; rachis 3–4 cm, densely brown hairy; rachillae 6–10, short, stout, condensed, usually diverging from rachis at right angles, 10–25 × 3 mm; rachilla bracts very close, conspicuous, rounded, 1.5 × 3 mm, densely covered in laciniate pale grey to brown hairs. **STAMINATE FLOWER** buds pointed, 1.5 × 1 mm; sepals imbricate, rounded, 1 × 1 mm, slightly keeled, margin erose; petals shining, not striate, 1.3 × 0.6 mm; stamens 3 antesepalous, staminodes absent, filaments connate into a tube c. 0.7 mm high, anthers didymous, c. 0.15 × 0.35 mm; pistillode minute. **PISTILLATE FLOWERS** known only in very immature bud. **FRUIT** unknown.

NOTE. This is a very curious species with its highly condensed stubby rachillae. Such rachillae are known elsewhere only in *D. angustifolia* (previously described as variety *angustifolia* of *D. humbertii* Perrier) and in *D. pachyramea*. *D. zahamenae* and *D. angustifolia* have antesepalous stamens and no staminodes whereas *D. pachyramea* has antepetalous stamens and antesepalous staminodes.

SPECIMENS SEEN. Ambatondrazaka: Zahamena, Massif de l'Andrangovalo, Oct. 1937 (fl.), *Humbert & Cours* 17776 (Holotype of *D. humbertii* var. *humbertii*, P); Chutes du Maningory, Dec. 1944 (fl.), *Homolle* s.n. (K, P); between waterfalls and Ambodivoangy, Dec. 1944 (fl.), *Cours* 1848 (K, P); Ambatondrazaka, June 1961 (fl.), *Rakotovao* 11899 (K, P).

102. DYPSIS ANGUSTIFOLIA

A diminutive palm of the forest undergrowth known from Betampona and environs northeast of Toamasina. This is one of three species of *Dypsis* that have highly condensed inflorescences with very short fat flower-bearing branches that appear almost catkin-like. The species name is Latin for having narrow leaves.

DISTRIBUTION. Central part of East Coast lowlands, Betampona and environs.
HABITAT. Lowland rain forest; 400–500 m.
LOCAL NAMES. Not recorded.
USES. Not recorded.
CONSERVATION STATUS. Endangered; occurring in two sites in low numbers.

Dypsis angustifolia *(H. Perrier) Beentje & J. Dransf.* **stat. nov.**

SYNONYM:
Dypsis humbertii H. Perrier var. ***angustifolia*** H. Perrier in Humbert, Not. Syst. 8: 47 (1939); Jum. & H. Perrier, Fl. Madagascar

30: 30: 28 (1945). Type: Madagascar, Toamasina, Ambodiriana, Betampona, *Perrier* 17468 (Holotype P).

Slender, clustering undergrowth palm. **STEMS** to 1 m tall, 6–10 mm diam., internodes 12–25 mm diam., bearing scattered brown scales. **LEAVES** 6–7 in crown; sheaths 9–12 × 0.8 cm, tardily abscising, ± marcescent, striate, with scattered punctiform scales, auricles small, membranous, soon tattering; petiole 4–23 cm, c. 2 mm wide, ± triangular in cross section; blade entire bifid, distinctly plicate on drying, 28–50 cm, deeply cleft to about three quarters of the overall length, the two lobes 20–40 × 1.5–3 cm, occasionally one lobe further divided into 2 narrow leaflets, segment tips shallowly lobed, adaxially with scattered punctiform scales, abaxially paler and with abundant brown punctiform scales. **INFLORESCENCES** interfoliar, erect or curved, branched to 1 order only; peduncle 15–32 cm long, c. 1–2 mm diam., sparsely covered with red scales in exposed portion; prophyll 10–25 × 0.5 cm, membranous, sparsely scaly; peduncular bract inserted far above and exceeding the prophyll by 5.5–9 cm, otherwise similar; rachis 3.5–10 cm, densely brown hairy; rachillae 5–9, inserted at right angles, 1–3 cm long, c. 2 mm diam.; rachilla bracts inconspicuous, c. 0.5 × 1 mm, almost entirely obscured by dense red-brown hairs. **STAMINATE FLOWER** buds c. 1 mm diam., spherical; sepals 3, free, imbricate, rounded, c. 0.6 mm diam., striate, irregularly cleft at margin; petals ± free, longitudinally striate, broad triangular, valvate, 0.8 mm long, 0.8 mm wide at the very base; stamens 3 antesepalous, filaments united in a ring 0.2 mm high, free part 0.2 mm long, anthers didymous, c. 0.2 × 0.1 mm, introrse; pistillode conical, minute. **PISTILLATE FLOWERS** globular; sepals broad, imbricate, 1 × 1 mm, margins erose; petals striate, valvate at tips, irregularly imbricate at base, c. 1.5 × 1.5 mm; staminodes 3, minute; ovary c. 1 mm diam. **FRUIT** unknown.

NOTE. This species is closely related to *D. zahamenae* from which it can be distinguished by the blade being narrow and divided to three quarters rather

Dypsis angustifolia. **A** crown showing inflorescence × 1/2; **B** detail of rachilla × 4. All from *Beentje* 4494. Drawn by Rosemary Wise.

than one third of its length, by the inflorescence rachis being usually more elongate with more distant rachillae, by the rachilla bracts being inconspicuous and almost entirely obscured by hairs, rather than being large and conspicuous, and by the staminate flower buds being rounded rather than pointed with striate rather than smooth shining petals. In all but one collection (one of the two specimens collected under *Beentje* 4494) there is a well developed long petiole, whereas in *D. zahamenae* the petiole is usually very short or absent. Previously considered to be a variety of *D. humbertii* (= *D. zahamenae*) we consider it to be sufficiently distinct to be elevated to specific rank.

Specimens seen. Toamasina: Ambodiriana, Betampona, Dec. 1925 (fl.), *Perrier* 17468 (Holotype P); idem, 25 Oct. 1991 (fl.), *Beentje* 4494 (BH, K, MO, P, TAN); Route from Didy to Ampasimanolotra, received April 1954 (fl.), *Cours* 4926 (P).

103. Dypsis pinnatifrons

This is a handsome palm that has become quite widespread in cultivation outside Madagascar. The great altitudinal and latitudinal range of the species suggests that there may be considerable variability of value for the selection of different strains to suit particular gardening climates.

D. pinnatifrons is superficially very similar in habit, inflorescences and, especially, leaf form to *D. nodifera*. If staminate flowers and/or fruit are available, then there is no difficulty in separating them, for the former has three stamens and a seed with homogenous endosperm while the latter has six stamens and a seed with ruminate endosperm. There do appear to be good differences between the seedlings at about the six-leaf stage, *D. pinnatifrons* with numerous leaflets that are already clearly grouped and divergent within the groups, each leaflet abruptedly narrowed at its tip and *D. nodifera* with entire leaves or with parallel-sided leaflets that are not obviously grouped and divergent. This apparent difference needs confirmation.

The species epithet, *pinnatifrons*, meaning pinnate-leaved, would be appropriate for many a palm in Madagascar but was probably appropriate when it was first published to distinguish the species from the other two species of *Dypsis* known at that time (*D. forficifolia* and *D. hirtula*).

Distribution. Widespread throughout all humid forested areas; one of the commonest palms in Madagascar.
Habitat. Lowland and montane forest, somewhat tolerant of disturbance; to 1000 m.
Local names. *Tsingovatra, Tsingovatrovatra, Ovatsiketry, Ambolo, Hova, Tsobolo.*
Uses. Stems used for making blowpipes.
Conservation Status. Not threatened. Widespread.

Dypsis pinnatifrons. Very robust form on Marojejy. This is the form previously known as *D. sambiranensis* (*Dransfield et al.* JD6765).

Dypsis pinnatifrons *Mart.*
Hist. Nat. Palm. 3 (ed. 1): 180 (1838); Baill., Bull. Soc. Linn. Paris 147: 1162 (1894). Type: Madagascar, *Commerson* s.n. (P).

Synonyms:
Areca gracilis Thouars in Mart., l.c. (non Roxb.). See Note.
Dypsis gracilis Bory ex Mart., Hist. Nat. Palm. 3 (ed. 2): 181 (1845); Becc., Palme del Madagascar 17 (1912); Jum. & H. Perrier, Ann. Inst. Bot.-Géol. Colon. Marseille sér. 3, 1 (1): 23 (1913); Jum., Ann. Inst. Bot.-Géol. Colon. Marseille sér. 3, 6 (1): 36 (1918); Bull. Acad. Malgache 6: 16 (1923); Cat. Pl. Madagascar, Palmae: 12 (1938); Jum. & H. Perrier, Fl. Madagascar 30: 29 (1945). Type: Madagascar, *Perrotet* s.n. (P).
Adelodypsis gracilis (Bory ex Mart.) Becc., Bot. Jahrb. Syst., Beibl. 38: 17 (1906).

Dypsis gracilis var. ***sambiranensis*** Jum. & H. Perrier, Ann. Inst. Bot.-Géol. Colon. Marseille sér. 3, 1 (1): 24 (1913). Type: Madagascar, Lokobe, *Perrier* 18742 (P).
Dypsis sambiranensis (Jum. & H. Perrier) Jum., Ann. Inst. Bot.-Géol. Colon. Marseille sér. 5, 3: 15 (1933); Cat. Pl. Madagascar, Palmae: 14 (1938). **synon. nov.**
Chrysalidocarpus sambiranensis (Jum. & H. Perrier) Jum., Ann. Inst. Bot.-Géol. Colon. Marseille sér. 5, 1: 21 (1933).
Adelodypsis sambiranensis (Jum. & H. Perrier) H.P. Guérin, Ann. Sc. Nat., sér. II, Bot. 10: 28, 37 (1950).

Phloga polystachya Becc. (non Baker), J. Linn. Soc. (Bot.) 29: 62 (1890).

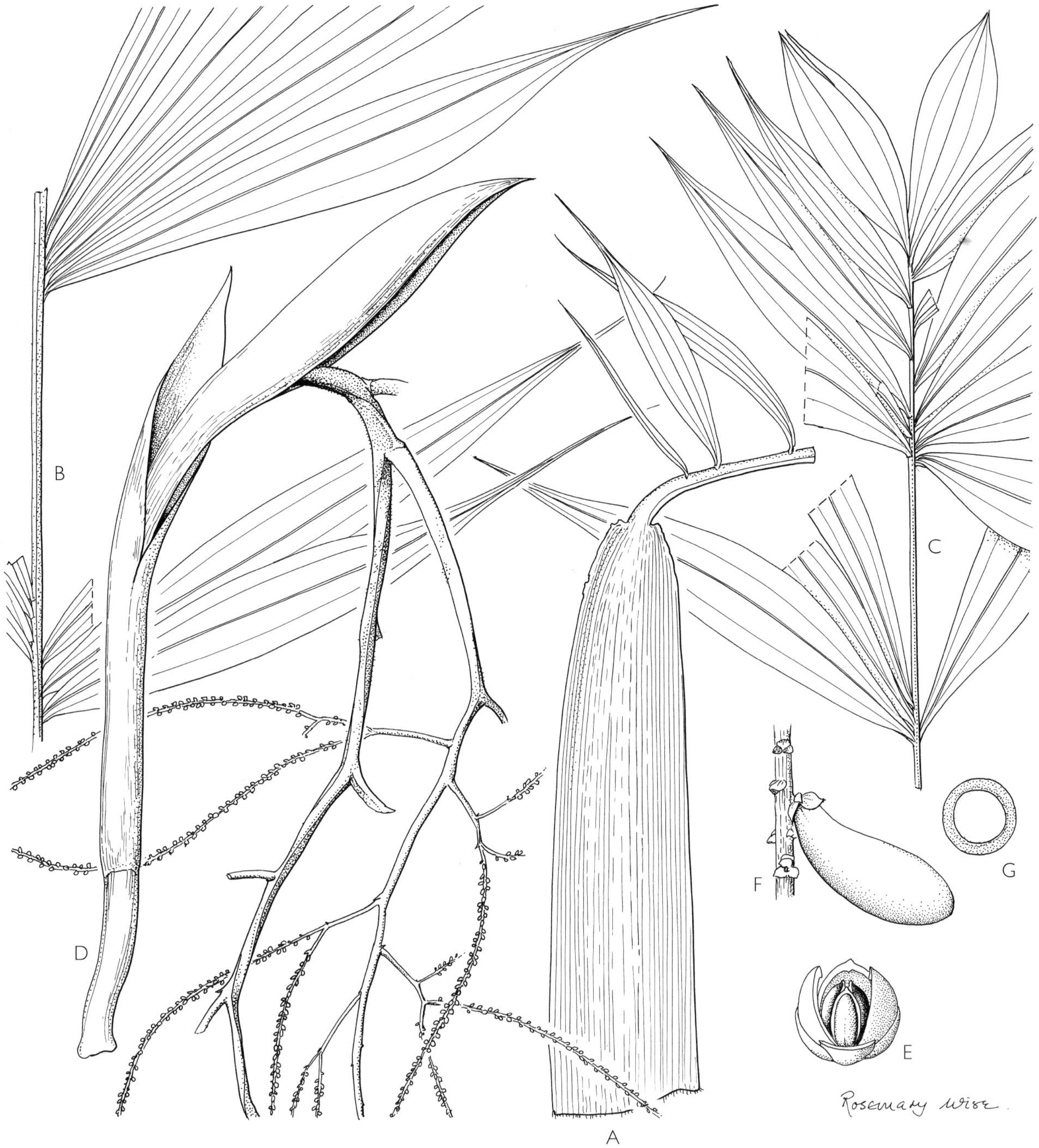

Dypsis pinnatifrons. **A** leaf sheath and lowermost leaflets × 2/5; **B** mid section of leaf × 2/5; **C** leaf tip × 2/5; **D** basal part of inflorescence × 1/3; **E** staminate flower, one petal removed × 9; **F** detail of fruiting rachilla × 3; **G** fruit in cross section × 3. All from *Dransfield et al.* JD6357. Drawn by Rosemary Wise.

Solitary palm of forest undergrowth or middle story. **STEMS** erect, 2.5–12 m tall, 4.5–15 cm diam. at maturity, juvenile stems tending to be more slender and stem increasing in diameter with age, internodes 2–8 cm, very short near crown in old stems, stem surface green and sometimes somewhat mottled when young, becoming grey-brown and often corky-warted and vertically cracked with age, the nodal scars then becoming obscure. **LEAVES** c. 8–16 in crown, spreading, sometimes with a tendency to be marcescent; crownshaft well developed, tending to be rather swollen, green, sometimes tinged bright pink when young; sheaths 25–48 cm long, 6–18 cm diam., lacking distinct auricles, flecked with vertical patches of reddish to chocolate-brown scales and patchy white wax; petiole usually very short or absent, rarely 8–36 cm long, 0.8–3 cm diam., densely covered with caducous pale and dark scales; rachis 75–220 cm, to c. 3 cm wide at base, usually less, densely covered with caducous pale and dark scales; leaflets ± sigmoid, often acuminate in a long drip tip, 22–46 on each side of the rachis, arranged in distant groups of 2–7, the leaflet bases crowded on the rachis and usually strongly swollen to form somewhat corky pulvini, the leaflets fanned within the groups, the proximal leaflet of the group usually shorter than the distal, distal few leaflets usually almost regularly arranged; proximal leaflets (especially when petiole lacking) very short and slender, c. 7 × 0.5 cm, or longer, mid-leaf leaflets 17–49 × 3–7.5 cm, apical leaflets 8–15 × 1–4 cm; emerging leaf often tinged pink, leaflet surfaces glabrous. **INFLORESCENCES** interfoliar, spreading, or somewhat pendulous, shorter than the leaves, c. 80–130 cm long, branching to 3–4 orders; peduncle 35–66 cm long, 1.4–3 cm wide at base, ± crescent-shaped in cross section, densely covered with red-brown indumentum when young; prophyll inserted c. 8–20 cm above the base of the peduncle, 24–47 ×

2–4 cm, with scattered dark red-brown scales; peduncular bract inserted 10–12 cm above the prophyll insertion, 19–35 × 2.5–3.5 cm; rachis 40–55 cm, with red-brown indumentum; rachillae very numerous, usually at least 200, somewhat arched or pendulous, 4–45 cm long, slender, glabrous, c. 1.5 mm diam., increasing in diameter with age, triads c. 1–3 mm apart. **Staminate flower** c. 2 × 1.2 mm; sepals 0.8 × 0.8 mm, ± rounded, irregularly gibbous and keeled; petals 2 × 1 mm, striate; stamens 3, antesepalous, c. 1.3 mm long, filaments 0.9 × 0.4 mm, anthers sagittate, 1.0 × 0.7 mm, thecae 1.0 × 0.3 mm; pistillode pyramidal 0.8 × 0.4 mm. **Pistillate flower** globular; sepals very broad, 1 × 1.5 mm, irregularly explanate, striate, with dark tips; petals 2.5 × 2.5 mm, basally imbricate, with triangular valvate tips; staminodes 3–6, irregularly toothlike; ovary c. 1.5 mm diam. **Fruit** green at first, ?turning brownish at maturity, 14 × 6.5 mm. **Seed** 10 × 4 mm; endosperm homogenous; embryo lateral, near the base.

Note. The typification of the name *Dypsis pinnatifrons* is most complex and open to several interpretations. According to Beccari (1914), Martius' concept of this palm includes several elements, referable to *Phloga polystachya* (our *D. nodifera*), *Vonitra thouarsiana* (our *D. fibrosa*) and *Dypsis gracilis* (see also Dransfield & Moore, 1982) - as Beccari stated - "un grande imbroglio di nomenclatura". Typified on the element Martius noted in passing as *Areca gracilis* Bory non Roxb., an unpublished name on a herbarium sheet, *D. pinnatifrons* is the name applicable to the palm described above.

Inflorescences of *D. pinnatifrons* are usually branched to three orders. In some areas, e.g. Marojejy and Nosy Be, individuals with inflorescences branched to four orders are found. These are often very robust palms, almost twice as large as "normal" *D. pinnatifrons*. Nosy Be is the type locality of *D. sambiranensis* and all modern collections from the island display inflorescences branched to four orders. However, the type specimen, *Perrier* 18742, has inflorescences that appear to be branched to three orders only. Where individuals with inflorescences branched to four orders occur together with individuals with inflorescences branched to three orders, as on Marojejy, the two types of palm look distinctly different, and for some time it appeared that we would maintain two taxa. However, apart from the difference in numbers of orders of branching, there are no obvious differences discernible in the herbarium, and furthermore, the previously conceived restriction of these palms with inflorescences branched to four orders to the northern part of the island does not hold true; one collection from Analamazaotra, *Perrier* 12010, also has inflorescences branched to four orders.We thus feel that it is not possible at present to distinguish two taxa. However, this variation deserves further study in the field.

Dypsis pinnatifrons

Specimens seen. Nosy Be: Ampasindava, Lokobe Forest Reserve, Sept. 1932 (fr.), *Perrier* 18742 (P; Holotype of *D. sambiranensis*); idem, July 1992 (fl.), *Beentje & Andriampaniry* 4696 (BH, K, MO, P, TAN). Ambanja: Ampasindava, path from Ampopo to massif Ambohimarahavavy, *Jacquemin* H524J (P); Manongorivo, 2 km S of Ambalafary, Jan. 1992 (fl.), *Beentje et al.* 4563 (BH, K, MO, P, TAN); Bekolosi, Feb. 1992 (fr.), *Beentje et al.* 4573 (K); Antsatrotro, March 1993 (fl.), *Malcomber et al.* 2280 (K, TAN). Andapa, Marojejy, N of Mandena, Nov. 1989 (fl.), *Dransfield et al.* JD6765 (BH, K, MO, P, TAN); idem, Feb. 1989 (fl.), *Miller & Lowry* 4027 (K, MO). Sambava: 6 km SW of Ambodivoara, Ankatoka R., June 1992 (fl., fr.), *Beentje & Andriampaniry* 4688 (BH, K, MO, P, TAN). Maroantsetra: Masoala, April 1971, *Moore* 9917 (BH, P); south of Maroantsetra, Oct. 1986 (fl, fr.), *Dransfield et al.* JD6357 (BH, K, MO, P, TAN); Nosy Mangabe, Oct. 1987 (fl.), *G.E. Schatz* 1665 (K, MO, P); idem, Feb. 1988 (fl.), *Henderson et al.* 755 (K, MO, NY, P, TAN); idem, Nov. 1989 (fr.), *Carlson* 27 (K, MO, P); Bay of Antongil, 1837, *Richard* 16 (FI, P). Soanierana-Ivongo: Soanierana-Ambohoabe, Dec. 1938 (fl.), *Lam & Meeuse* 5660 (K, L, P); Tanambo, April 1851, *Boivin* s.n. (FI, P). Toamasina, Mahatsara Forest Reserve, Mahavelona, Feb. 1992 (fl.), *Noyes et al.* 939 (K, MO, TAN); road between Toamasina and Fenerive, near PK 83, *Croat* 32533 (MO, P); Ambodirafia, June 1955 (fl.), *Martin* 7257 RN (K, P). Ampasimanolotra: Anivorano Kely and Andrambolahy Kely, April 1951 (fr.), *Cours* 4496 (K, P); Ambila, May 1928 (fl.), *Decary* 6448 (K, P). Moramanga: Andasibe, Analamazaotra, Nov. 1986 (fl.), *Dransfield et al.* JD6415 (BH, K, MO, P, TAN); idem, March 1991 (fr.), *Beentje & Rabarilala* 4402 (K, P, TAN); idem, Nov. 1970 (fl.), *Keraudren* 25334 (K, P); idem, (fl.), *Perrier* 12010 (P); idem, (fl.), *Perrier* 12092 (P); idem, *Service des Eaux et Forêts* 3305 (P); idem, *Moore* 9908 (BH, P); Bemainty, Rahobevava, March 1951 (fl.), *Cours* 4200 (P). Vatomandry: 1903 (fl.), *Bernard* s.n. (P). Vohipeno: Matitana, Baie Mienaru, Aug. 1911 (fl.), *Perrier* 12059 (P). Ifanadiana, Ambinanindrano, Jan. 1993 (fl.), *Beentje & Andriampaniry* 4800 (BH, K, MO, P, TAN). Mananjary: Mananjary, March–April 1909 (fruit), *Geay* 7188 (P); idem, (seedling), *Geay* 7267 (P); idem, (fl.), *Geay* 8053 (P). Farafangana, Vondrozo, Tsararano, Dec. 1963 (fl.), *Bosser* 18553 (K, P). Tolanaro: Manentenina, Forêt de Marovony, 29 Oct. 1990 (fl.), *A. Randrianosolo et al.* 185 (MO, P); 13km N of Ezoambo, March 1992 (fr.), *Beentje & Andriampaniry* 4598 (BH, K, MO, P, TAN); Tolanaro, June-July 1890 (fr.), *Scott Elliot* 2957 (K); idem, Jan. 1962 (fl.), *Rabevazaha* 12183 (P); Réserve Naturelle 11, Nov. 1952 (fl.), *Rakotoniainia* 4884 (K, P); Ampalanana, *Dequaire* 27682 (P). Without locality, April 1890 (fr.), *Scott Eliot* s.n. (FI, K); idem, 1819, *Perrotet* s.n. (FI, P) (syntype of *Dypsis gracilis*); idem, *Chapelier* s.n. (FI, P); idem, *Commerson* s.n. (P) (lectotype of *D. pinnatifrons*).

104. DYPSIS PROCERA

D. procera is an attractive palm of the undergrowth of the very humid forests that surround the Bay of Antongil. The leaf sheaths with their discrete patches of red-brown scales are distinctive. Leaf dissection varies greatly. There are some particularly beautiful forms with entire bifid leaves that would make splendid ornamentals. *D. procera* is cultivated in several Australian collections (and probably elsewhere), where it has been misidentified as *D. hildebrandtii* (see Stewart 1994, p. 97). The species name, *procera*, meaning very tall, or high, seems inappropriate for this undergrowth palm, but it is one of the taller of the palms that belong to *Dypsis* in the restricted sense, rather than the present broad sense.

DISTRIBUTION. Around the Bay of Antongil.
HABITAT. Lowland rain forest, usually on flat land, sometimes on slopes; to 600 m.
LOCAL NAMES. Not recorded.
USES. Not recorded.
CONSERVATION STATUS. Vulnerable. Restricted in distribution in an area where the forest is unprotected and under some threat by shifting cultivators.

Dypsis procera. **A** crown × 1/5; **B** detail of rachilla × 2; **C** staminate flower × 4. **A** from *Dransfield et al.* JD6449, **B**, **C** from *Dransfield et al.* JD6743. Drawn by Margaret Tebbs.

Dypsis procera Jum.
Ann. Inst. Bot.-Géol. Colon. Marseille sér. 3, 6 (1): 33 (1918); Bull. Ac. Malgache 6: 7 (1923); Cat. Pl. Madagascar, Palmae: 14 (1938); Jum. & H. Perrier, Fl. Madagascar 30: 36 (1945). Type: Madagascar, Fananehana, *Perrier* 12089 (Holotype P).

Synonym:
Dypsis longipes Jum., Ann. Inst. Bot.-Géol. Colon. Marseille sér. 3, 6 (1): 37 (1918); Bull. Ac. Malgache 6: 18 (1923); Cat. Pl. Madagascar, Palmae: 13 (1938); Jum. & H. Perrier, Fl. Madagascar 30: 24 (1945). Type: Madagascar, Rantabe, *Perrier* 12030 (Holotype P). **synon. nov.**

Moderate-sized clustering (very rarely solitary) palm of forest undergrowth, tending to form diffuse colonies by few short stolons. **Stems** erect, to 6 m tall, 1–2.5 cm diam., internodes 2–10 cm long, when young green, with scattered caducous brown scales. **Leaves** c. 8 in crown; leaf-sheaths forming a crown-shaft; sheath 17–31 cm long, 1.5–2.5 cm diam., pale green, covered in caducous red-brown scales, these frequently in irregular vertical patches c. 3–22 x 1 mm, leaf sheath mouth with irregularly tattering ligule; petiole rarely very short (3–4 cm), usually 10–25 x 0.5–0.8 cm, bearing scattered, caducous red-brown scales; rachis 38–60 cm long (?rarely more); blade entire bifid or irregularly divided into 2–8 narrow to broad leaflets; basal leaflets 43–80 x 2–15 cm, mid-leaf leaflets 25–50 x 1.2–8.5 cm, apical leaflets 23–55 x 1.5–15 cm; lamina surfaces bearing numerous minute punctiform scales along main and minor veins. **Inflorescence** interfoliar, branched to 2 orders; peduncle 35–46 cm long, 0.5–0.7 cm diam., brown scaly; prophyll 21–36 x 0.5–0.7 cm, borne 5–12 cm above the base of the peduncle, membranous, bearing caducous scales; peduncular bract, rarely preserved, usually abscising before expansion of rachillae from bud, borne at least 12 cm above prophyll, to 16 x 7 cm; rachis 14–35 cm long, scaly as the peduncle, bearing 8–13 first order branches, lower 2–6 branched; rachillae 10–18 in number, 15–50 cm long, 1.2 mm diam., densely covered in red-brown branched trichomes, triads c. 2 mm distant. **Staminate flowers** c. 2 x 1.3 mm; sepals rounded-triangular, 0.8 x 0.7 mm, keeled, glabrous; petals ± elliptic, valvate, striate, 2 x 1.4 mm; stamens 3, antesepalous, filaments 1.1 x 0.4 mm, strap-like, anthers sagittate to subpendulous, 0.9 x 0.5 mm; pistillode conical, to 3 mm. **Pistillate flowers** in very young bud c. 1 mm diam., rounded; sepals broadly imbricate, 1 x 1.1 mm; petals triangular-rounded, striate, 1.1 x 0.5 mm; staminodes 3, minute, irregularly dentiform; immature ovary c. 0.5 x 0.2 mm. Mature **fruit** unknown; immature fruit ellipsoid, 7 x 3 mm, green. **Seed** with homogeneous endosperm.

Note. Jumelle distinguished *Dypsis longipes* from all other species of *Dypsis* sensu stricto by the combination of entire bifid leaf and spicate inflorescence borne on a long peduncle. However, the inflorescence of the holotype clearly displays six scars near the base of the "spike" where branches have been broken off, and there is no doubt that the inflorescence originally bore several long flexuous branches. With this spurious difference removed, *Dypsis longipes* is clearly conspecific with *D. procera*. Furthermore, of the leaf blades in the type one appears to be entire bifid, the other has two leaflets on one side of the rachis; however, in both cases the rachis at the base of the lamina is much narrower than the petioles on the same sheath, suggesting that, far from being whole laminae, the two leaf samples represent the apical portions of two pinnate leaves.

D. procera is a distinctive but rather variable taxon that occurs with some frequency in forest surrounding the Bay of Antongil, in the lowlands and up to altitudes of about 400 m. It is a colonial palm tending to form diffuse clumps by short stolons. Lamina dissection varies from narrowly to broadly and irregularly pinnate, to entire bifid, the entire-bifid forms being particularly striking and beautiful. In most collections the leaf-sheaths bear scales in distinctive irregular vertical patches. *Dransfield* JD6397 differs from other collections in being rather smaller in all its parts, with a short entire bifid leaf and with leaf sheaths that lack the distinctive patches of scales, but are nevertheless scaly; the rachillae are similar to those of *D. procera* and the staminate flowers have three antesepalous stamens and a conical pistillode. This collection is only tentatively included here.

Specimens seen. Maroantsetra: Masoala Peninsula, Hiaraka, Oct. 1986 (fl.), *Dransfield et al.* JD6365 (K, TAN); idem, Oct. 1986 (fl.), *Dransfield et al.* JD6369 (K, TAN); Ambanizana, Oct. 1986 (fl.), *Dransfield et al.* JD6397 (K, TAN) (see note above); Antalavia, Feb. 1988 (fl.), *Dransfield et al.* JD6471 (K, NY, TAN); idem, Nov. 1989 (fl.), *Dransfield et al.* JD6743 (K, TAN); Nosy Mangabe, Feb. 1964 (fl.), *Dransfield* JD6464 (K, TAN); Andranofotsy River, Sahavary, Feb. 1988 (fl.), *Dransfield et al.* JD6455 (K, TAN); idem, Feb. 1988 (fl.), *Dransfield et al.* JD6449 (K, TAN); hill 5 km west of Maroantsetra town, Oct. 1986 (fl.), *Dransfield et al.* JD6361 (K, TAN); Fananehana, Bay of Antongil, Aug. 1912 (fl.), *Perrier* 12089 (Holotype P); Rantabe, Bay of Antongil, Aug. 1912 (fl.), *Perrier* 12030 (Type of *Dypsis longipes* Jum., holotype P). Ampasimanolotra: Ambalarondra, Andranampony, April 1951 (fl.), *Cours* 4537 (K, P).

105. Dypsis paludosa

This species occurs in small pockets of peat swamp developed on white sands behind the beach along the East Coast. Near Mananara Avaratra, it has been collected up to c. 300 m elevation. It is very variable in leaf dissection within populations. Some forms with entire leaves are very beautiful and would make fine ornamentals. The species name is the Latin for marsh-dwelling, referring to the habitat.

Distribution. East Coast: Mananara Avaratra to Ambila-Lemaitso, south of Toamasina, Île Sainte Marie. **Habitat.** Peat swamp forest on white sand behind the coast and in swampy places further inland; to 400 m.

LOCAL NAMES. Not recorded.

USES. Not recorded.

CONSERVATION STATUS. Vulnerable; although widespread, *D. paludosa* occurs in coastal lowland forest that is much threatened.

Dypsis paludosa *J. Dransf.*, **sp. nov.**
palma mediocris, stolonifera vel rare solitaria, foliis plerumque epetiolatis, irregulariter divisis, segmentis plerumque latis, inflorescentia 2-ramosa, rachillis numerosis, crassiusculis, bracteis rotundatis, floribus staminatis triandris, staminibus antesepalis, pistillodio conico. Typus: Madagascar, East Coast, Ampasimanolotra, Ambila-Lemaitso, c. 5 km south of village, *Dransfield et al.* JD6439 (Holotypus K; isotypi BH, MO, TAN);

SYNONYMS:
Adelodypsis boiviniana Becc. (non Baill.), Bot. Jahrb. Syst. 38, Beibl. 87: 17 (1906).
Dypsis boiviniana sensu Becc., (non Baill., 1894), Palme del Madagascar 19 (1913); Jum., Ann. Inst. Bot.-Géol. Colon. Marseille sér. 3. 6(1): 25 (1918); Cat. Pl. Madagascar, Palmae: 12 (1938); Jum. & H. Perrier, Fl. Madagascar 30: 50 (1945).

Clustering (rarely solitary) forest undergrowth palm, tending to form rather open colonies by short stolons. **STEMS** 4–6 (–9) m tall, (1.5) 2.5–5 cm diam.; internodes 1–6 cm long, grey-brown near base, green near the crown, when young covered rather densely with caducous red-brown scales. **LEAVES** 9–12 in crown, tending to be porrect, giving the crown a characteristic shuttlecock appearance; crownshaft well developed, c. 30 cm long; sheaths 21–27 cm long, 6–10 cm wide when split and flattened out, abaxially rather densely covered with red-brown stellate scales, scattered or arranged in longitudinal lines, auricles irregularly triangular, to 2 × 0.7–1.5 cm, soon tattering; petiole absent or very short, rarely to 7.5 cm, 10–16 mm wide, adaxially flat or shallowly channelled, abaxially slightly angled, scaly as the leaf sheath; rachis 41–114 cm long, tapering from up to 16 mm wide at base, adaxially channelled near the base, abaxially somewhat angled, abaxially bearing sparse to dense caducous dark brown or pale scales with dark brown punctiform bases; blade coriaceous, very variable, tending to be ± cuneate in outline, entire bifid, or basally entire and with 1–3 distal leaflets or irregularly or regularly pinnate with 2–13 leaflets, when entire, blade to 90 cm long with lobes to 70 × 15 cm, when split into leaflets, leaflets very varied in length and width, mid-leaf leaflets 30–70 × 1.4–5 cm, basal leaflets sometimes very short and slender, 10 × 0.5 cm, all but the apical pair acuminate, apical pair shallowly lobed; leaflets on drying tending to have reflexed margins; adaxial blade surface with very sparse punctiform scales, abaxial surface with scattered brown punctiform scales. **INFLORESCENCE** interfoliar, rarely becoming infrafoliar with age, shorter than the leaves, branching to 2 orders; peduncle 15–50 cm long, 7–15 mm diam., densely rusty hairy; prophyll (8) 15–43 × 1–3 cm, inserted 3–14 cm above the base of the peduncle, sparsely dark brown scaly; peducular bract 20–30 × 1.5–3 cm, inserted 19–28 cm above the base of the inflorescence, similar to prophyll; rachis 12–25 cm, sparsely to densely dark brown hairy; rachillae numerous (c. 50), 15–35 cm, spreading to pendulous, c. 2.5–3 mm diam. when fresh, shrinking on drying, glabrous to rather densely covered in dark brown hairs; rachilla bracts, entire, low, c. 0.5 × 1.5 mm tending to form shallow pits, c. 1.5–2 mm apart. **STAMINATE FLOWERS** in immature bud c. 1.1 × 0.8 mm; sepals 1 × 0.8 mm, irregularly imbricate; petals 0.8 × 0.5 mm; stamens 3, antesepalous, anthers sagittate, 0.5 × 0.3 mm; pistillode conical, c. 0.3 mm high. **PISTILLATE FLOWERS** at anthesis c. 3 × 2.5 mm; sepals 1.8 × 1.8 mm, irregularly explanate, imbricate, splitting; petals striate, 3 × 2.5 mm, basally imbricate, apically triangular, valvate, somewhat reflexed at anthesis; staminodes 6, minute, dentiform, c. 0.3 × 0.1 mm; ovary irregularly globose, 2.5 × 2.5 mm, stigmas 3, slightly reflexed. **FRUIT** ellipsoid, sometimes somewhat fusiform, c. 18 × 9 mm in immature but probably full grown state, endocarp with coarse fibres. **SEED** 11 × 5 mm; endosperm homogeneous.

Dypsis paludosa, in the Mananara Avaratra Biosphere Reserve (Photo: B. Rogers).

NOTE. For an account of the interpretation of *Dypsis boiviniana*, see under that species. *D. paludosa* is the palm that Beccari (1912, 1914) interpreted as *D. boiviniana* based on the mixed gathering by Boivin from Île Sainte Marie. When we made collections of a dypsid from Ambila-Lemaitso, we thought that we had at last refound the palm interpreted as *D. boiviniana* by Beccari and the Flore de Madagascar (1945). However, the new material has glabrescent or completely glabrous rather than hairy rachillae. A collection made by Perrier (12054 in P) from near Mananara is a close match for the original Boivin three-stamened plant, and has hairy rachillae. In leaf it is indistinguishable from the Ambila-lemaitso taxon and also from a sterile collection in P made by Boivin (No 1706 from Tafondrou, Île Sainte Marie). This Boivin collection almost certainly represents the leaves of the specimen with the second order hairy rachillae, that Beccari interpreted as being the type of *D. boiviniana*; the labels on the sheets and state "Forêt de Tafondrou, 1849, Sainte Marie de Madagascar" and are undoubtedly original, rather than new labels written by Beccari. Furthermore, this taxon survives to this day on Île Sainte Marie. The great similarity in

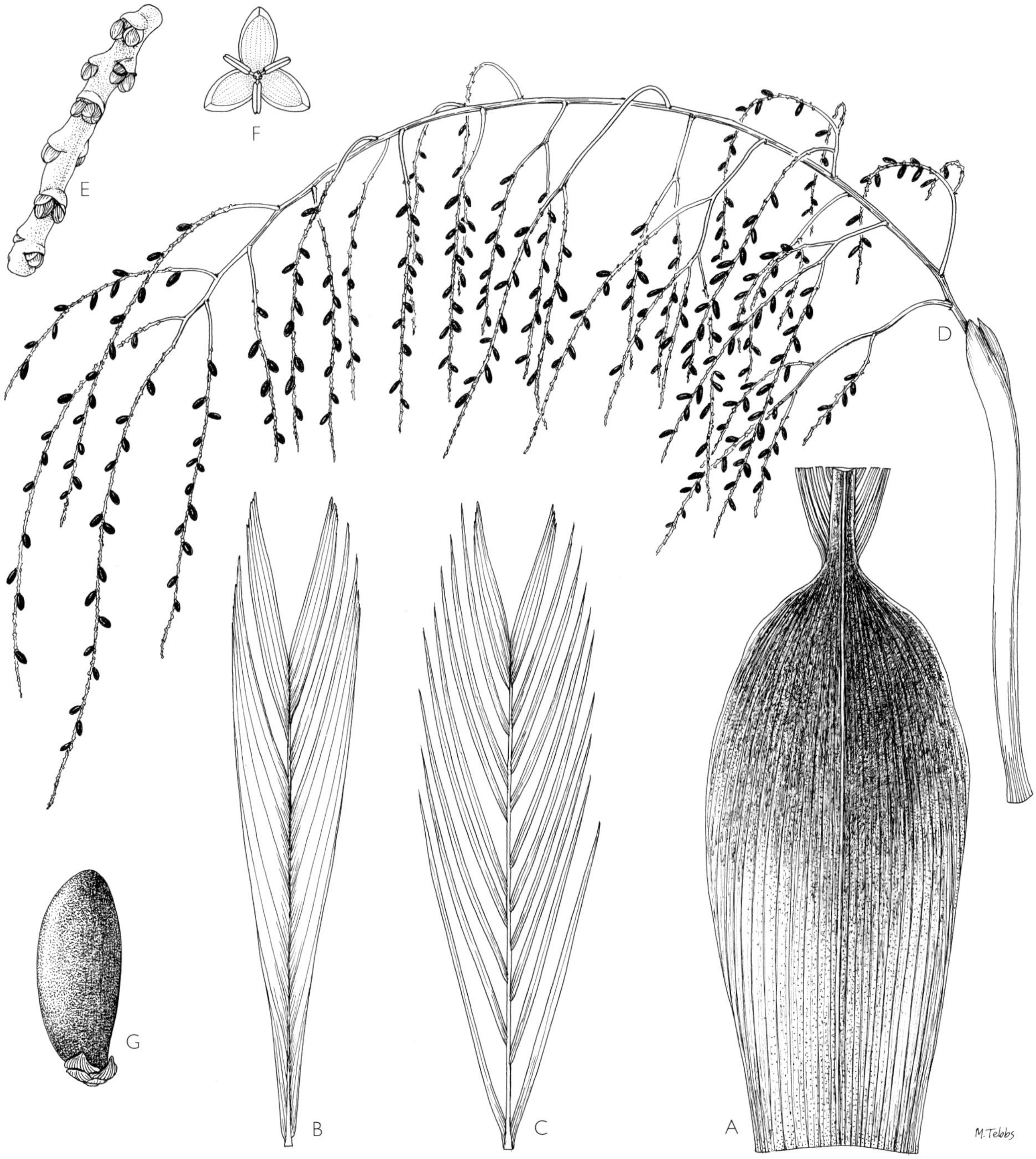

Dypsis paludosa. **A** leaf sheath × 2/5; **B**, **C** two leaves to show variation × 1/8; **D** infructescence × 1/5; **E** detail of rachilla × 1.5; **F** staminate flower × 10; **G** fruit × 2. **A**, **C**, **G** from *Dransfield et al.* JD6494, **B**, **D**, **F** from *Dransfield et al.* JD6439, **E** from *Dransfield et al.* JD6438. Drawn by Margaret Tebbs.

form, texture and size of the leaves of the Ambila-Lemaitso taxon, *Perrier* 12054 and *Boivin* 1706 and the similarity in the form of the inflorescences, suggest to us that the hairiness of the rachillae is probably a variable feature, and we are thus including all these collections in the same species that we describe as new and name D. *paludosa*. A collection (*Dransfield* JD7524), made in November 1994 in the Forêt de Kalalao, about 20 km north of Tafondrou, matches the early specimens of Boivin, but is in very young inflorescence bud. The inflorescence branches to two orders and the rachillae are hairy although very young. It is interesting to record that true *D. boiviniana* was found growing nearby.

Specimens seen. Mananara Avaratra: Mananara, Sept. 1912 (buds), *Perrier* 12054 (P); Antanambe, Oct. 1991 (fl.), *Beentje* 4462 (BH, K, MO, P, TAN); idem, April 1992 (fl.), *Beentje & Dransfield* 4620 (BH, K, MO, P, TAN); idem, April 1992 (fl.), *Beentje & Dransfield* 4647 (BH, K, MO, P, TAN); idem, Oct. 1994, (bud), *Dransfield & Beentje* JD7503 (K, TAN). Île Sainte Marie: Tafondrou, 1849 (sterile), *Boivin* 1706 (P); said to be Ravin-tsara (but probably same as *Boivin* 1706, mixed with type of *Dypsis boiviniana* Baillon), *Boivin* s.n. (P); Forêt de Kalalao, Nov. 1994 (buds), *Dransfield* JD7524 (K, TAN). Ampasimanolotra: Ambila-Lemaitso, c. 5 km south of village, Nov. 1986 (buds), *Dransfield et al.* JD6438 (K, TAN), JD6439 (Holotype K; isotypes BH, MO, TAN); c. 6 km south of village, March 1988 (fr.), *Dransfield et al.* JD6492 (K, TAN), JD6493 (K, TAN), JD6494 (K, TAN); 8.6 km south of village, Sept. 1991 (fl.), *Beentje* 4443 (BH, K, MO, P, TAN); 7.2 km south of village, Sept. 1991 (fl.), *Beentje* 4448 (K, TAN).

106. Dypsis mirabilis

There is nothing particularly unusual about the vegetative morphology of this species. It is a single-stemmed palm of the undergrowth of moderate size, with leaves that are divided into few broad leaflets, a habit common among undergrowth dypsids. The species name means wonderful, which may seem inappropriate for a rather ordinary looking palm. However, this species has most unusual staminodes in the staminate flower, and it is this rather obscure feature that caused us wonder.

Distribution. Marojejy and environs.
Habitat. Lowland forest in valley bottoms; 90–200 m.
Local names. Not recorded.
Uses. Not recorded.
Conservation Status. Probably endangered. The distribution area is very small, and numbers are thought to be very low.

Dypsis mirabilis *J. Dransf.*, **sp. nov.**
palma solitaria, folio foliolis paucis latis, inflorescentia 2-ramosa, rachillis pendulis, inter species floribus staminatis triandris staminibus antesepalis, staminodiis antepetalis ad pistillodium adnatis distincta. Typus: Madagascar, Marojejy, *Dransfield* JD6771 (Holotypus K; isotypus TAN).

Solitary undergrowth palm. **Stem** to 2.5 m tall, to 15 mm diam., internodes c. 40 mm long. **Leaves** c. 7 in crown; sheath 24 cm long, c. 25 mm diam., longitudinally striate, sparsely scaly, the sheath

Dypsis mirabilis. View of crown (*Beentje & Andriampaniry* 4687).

mouth with an irregularly tattering brown membranous margin; petiole c. 11 cm, adaxially channelled, abaxially ± rounded, covered with abundant caducous dark brown scales; rachis c. 85 cm; blade irregularly divided into 3–4 broad long acuminate leaflets on each side of the rachis, basal leaflets c. 80 × 11 cm, mid-leaf leaflets 70 × 4–7 cm, apical leaflets 52 × 7–8.5 cm, shallowly lobed distally, leaflet texture rather thin in dried specimens, glabrous, short transverse veinlets evident. **INFLORESCENCE** interfoliar, branching to 2 orders, to c. 115 cm long; peduncle 45 cm long, densely covered with dark red indumentum; prophyll and peduncular bract lacking in available material; rachis c. 70 cm long, covered with indumentum as the peduncle; rachillae ± pendulous, very numerous (c. 100), 12–16 cm long, c. 2 mm diam., glabrous, tending to be narrow just above the triad insertion, triads 1–3 mm apart, rachilla bract c. 0.5 × 2 mm. **STAMINATE FLOWERS** at anthesis rounded, orange, c. 1.8 mm diam.; sepals 1.0 × 0.9 mm, apically erose, keeled; petals rounded triangular, 1.5 × 1.3 mm, striate; stamens 3, antesepalous, emerging slightly between the petals at anthesis, filaments terete, 1.1 mm long, c. 0.5 mm wide at base, tapering to 0.3 mm wide at tip, anthers didymous, 0.4 × 0.1 mm; staminodes antepetalous, filaments adnate to the central pyramidal pistillode forming a composite structure c. 1.1 × 0.7 mm, the filaments c. 1.1 × 0.2 mm, free at the very tip, empty anthers c. 0.1 × 0.1 mm, free pistillode c. 0.1 mm high. **PISTILLATE FLOWERS** known only in very immature bud; sepals in fruit triangular, c. 1 × 1 mm; petals in fruit triangular, striate, c. 2 × 2 mm; staminodes 6. Immature **FRUIT** ovoid, green, 12 × 7 mm; endocarp composed of brown fibres. **SEED** 9 × 5 mm; endosperm homogeneous.

NOTE. The antesepalous stamens have didymous anthers (unusual among species with three stamens

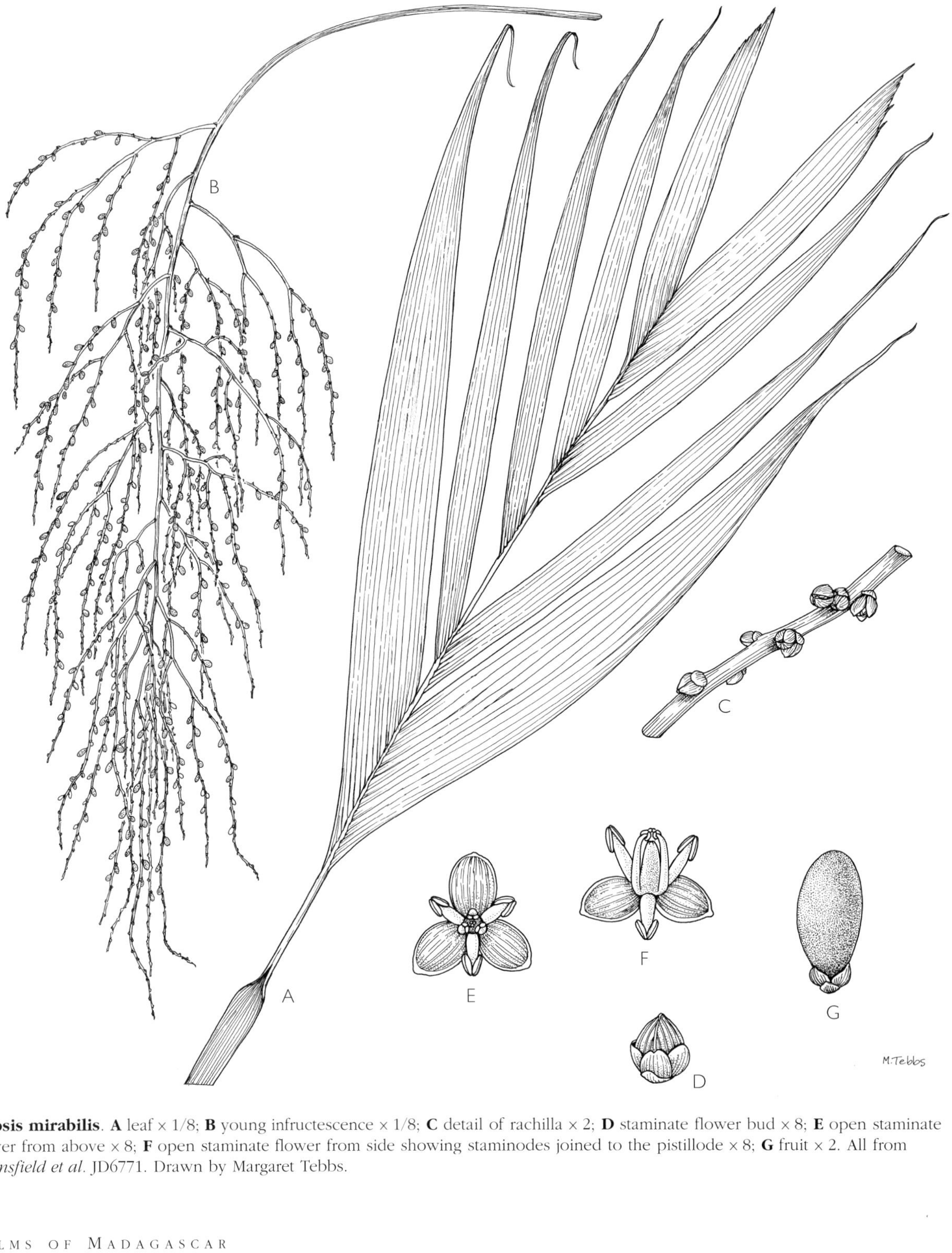

Dypsis mirabilis. **A** leaf × 1/8; **B** young infructescence × 1/8; **C** detail of rachilla × 2; **D** staminate flower bud × 8; **E** open staminate flower from above × 8; **F** open staminate flower from side showing staminodes joined to the pistillode × 8; **G** fruit × 2. All from *Dransfield et al.* JD6771. Drawn by Margaret Tebbs.

where, if the stamens are antesepalous, they are usually sagittate), but the most unusual feature is that there are three well developed staminodes that are adnate by their filaments to the pistillode, thus forming a large pyramidal body in the centre of the flower. The relationships of this species are obscure. *Beentje & Andriampaniry* 4687 has very young flower buds; nevertheless, three large antesepalous stamens and three minute central fused staminodes are visible. It is possible that *Casimir* 11731 (K, P) from Maroambihy, Sambava, also belongs here; inflorescence structure and vegetative features are similar, but no staminate flowers are available to confirm the identity.

Specimen seen. Andapa: Marojejy, Mandena, headwaters of Bangouabe, east of Jean Poste Juvance, Nov. 1989 (fl., fr.), *Dransfield* JD6771 (holotype K; isotype TAN); 6 km SW of Ambodivoara, Ankatoka River valley, June 1992 (fl. buds), *Beentje & Andriampaniry* 4687 (K, MO, P, TAN).

107. Dypsis fasciculata

This species occurs in lowlands near the coast; recently made collections are for the most part from forest developed on white sands, poor in nutrients, but the palm will also grow on laterite soils. When growing vigorously this can be quite a handsome species but it is usually rather untidy, with narrow grouped leaflets. Superficially it resembles a depauperate form of *Dypsis nodifera*. The species name refers to the grouped leaflets.

Distribution. NE Madagascar, from Antalaha and Marojejy to Betampona.
Habitat. Lowland forest near the coast, often on white sands, rare inland; 5–225 m.
Local names. Not recorded.
Uses. Not recorded.
Conservation Status. Vulnerable. This is quite widespread but occurs in coastal forests that are often affected by fire.

Dypsis fasciculata *Jum.*
Ann. Inst. Bot.-Géol. Colon. Marseille sér. 3, 6 (1): 36 (1918); Bull. Acad. Malgache 6: 18 (1923); Cat. Pl. Madagascar, Palmae: 12 (1938); Jum. & H. Perrier, Fl. Madagascar 30: 39 (1945). Type: Madagascar, Antalaha, 50 m alt., Nov. 1912, *Perrier* 12042 (?P - not found).

Solitary or clustering palm of the forest undergrowth. **Stems** 3–6 m tall, c. 15–40 mm diam., internodes to 40 mm long near base, c. 15 mm long in mature stems near the crown, green with scattered dark brown scales. **Leaves** c. 8 in crown; crownshaft well developed; sheaths 13–24 × 1.5–2.5 cm, densely covered with dark red-brown scales, these sometimes in vertical patches; auricles sometimes present, soon tattering; petiole 8–35 cm long, 5–10 x 3–5 mm in cross section, bearing abundant caducous chocolate-brown scales; rachis 70–90 cm or more long; leaflets 11–23 on each side of the rachis, conspicuously grouped in 2s–6s (usually 3s–4s), c. 10–20 cm between the groups, the proximal few leaflets usually very slender and short (to 16 x 0.3), mid-leaf leaflets 20–47 x 1–4 cm, including the long attenuate drip tip, apical leaflets 10–20 x 2.5–6 cm; leaflet surfaces glabrous. **Inflorescence** interfoliar, shorter than the leaves, branching to 2 orders; peduncle to c. 50 cm long, 0.8–1.5 cm wide at base, tapering to c. 4 mm wide distally, densely covered in red-brown scales; prophyll inserted c. 6–11 cm above base, 10–40 × 0.8–1.2 cm, sparsely dark scaly; peduncular bract inserted to 14 cm above prophyll, exceeding the prophyll tip by c. 8–11 cm; rachis 20–46 cm, basally scaly as the peduncle, distally glabrescent; rachillae spreading or pendulous, 16–30 in number, 20–c. 50 cm long, c. 1.5–2 mm diam. at anthesis, increasing to c. 3 mm diam. in fruit, glabrous, rachilla bracts rounded, very inconspicuous, triads c. 2 mm distant, in shallow pits. **Staminate flowers** c. 2 × 1.5 mm; sepals c. 1.4 × 0.6 mm, irregularly split and keeled; petals c. 1.6 × 1.2 mm, striate; stamens 3 antesepalous, filaments broad 1.0 × 0.6 mm, anther thecae free, ± pendulous, c. 0.7 × 0.2 mm; pistillode pyramidal, 0.7 × 0.4 mm. **Pistillate flowers** c. 2.5 × 1.9 mm; sepals rounded, c. 1.1 × 1.1, striate, irregularly erose; petals rounded triangular, irregularly imbricate, c. 2 × 1.8 mm; staminodes 3 (4), minute, toothlike, 0.2 × 0.1 mm; ovary rounded to top-shaped, 2 × 1.8 mm. **Fruit** green (?still immature), 14 × 7.5 mm. **Seed** 10 × 4 mm; endosperm homogeneous; embryo lateral near the base.

Note. The interpretation of this name has caused particular problems because we have been unable to locate the type, *Perrier* 12402, from Antalaha, which, we assume, should be in Paris. Using Jumelle's imperfect description, we have here applied the name to a taxon from forest near the Bay of Antongil, further to the south from Antalaha, that keys out to *D. fasciculata* and which more or less fits the protologue. Perhaps most significant is Jumelle's description of the stamen form. He describes the thecae as being pendulous, a most unusual stamen type in *Dypsis*; those collections cited here that have staminate flowers also have stamens of this form. Should the type ever be found, then the conspecificity of the newly collected material will have

Dypsis fasciculata

to be checked. We have also recently made two collections of this taxon on Île Sainte Marie, one (sterile) from the Forêt de Kalalao, near to the historic locality Tafondro where Boivin collected, and one (fertile) from the Forêt d'Ambohidena (see notes under *Dypsis thouarsiana*).

As interpreted here, this is a distinctive, easily distinguished species, with fascicled leaflets and inflorescences branched to two orders, with lax spreading or pendulous rachillae and staminate flowers with only three stamens with pendulous and divergent thecae. *Miller* 3327 from Marojejy differs from the other collections cited in being rather small and slender in all its parts. Neverthess it seems to belong here.

SPECIMENS SEEN. Andapa: Marojejy, along trail to Marojejy East from Mandena, Oct. 1988 (fl.), *Miller* 3327 (MO). Maroantsetra, Masoala Peninsula, Antalavia, Feb. 1988 (fl.), *Dransfield et al.* JD6466 (K, TAN); idem, Feb. 1988 (fl, young fruit), *Dransfield et al.* JD6472 (K, TAN). Mananara Avaratra: Mananara, Aug. 1917 (fl.), *Perrier* 12045 (P); 5 km south of Antanambe, April 1992 (fl., fr.), *Beentje & Dransfield* 4636 (BH, K, MO, P, TAN). Île Sainte Marie: Forêt de Kalalao, Nov. 1994 (sterile), *Dransfield et al.* JD7521 (K, TAN); Forêt d'Ambohidena, Nov. 1994 (buds), *Dransfield et al.* JD7528 (K, TAN). Toamasina: Betampona, near Ambodiriana, Dec. 1925 (fl.), *Perrier* 17477 (P).

Dypsis fasciculata. **A** leaf sheath and petiole with lowermost leaflet × 2/5; **B** mid section of leaf × 2/5; **C** leaf tip × 2/5; **D** inflorescence × 2/5; **E** staminate flower, one petal removed × 20; **F** part of rachilla in young fruit × 2/5. All from *Beentje & Dransfield* 4636. Drawn by Rosemary Wise.

108. DYPSIS LOKOHOENSIS

A slender palmlet of the undergrowth of montane forest. Because of its inflorescence structure and staminate flower size and shape, this species has the general appearance of members of the genus that were previously included in *Neophloga,* and had previously been misidentified as *D. (Neophloga) lutea* and *Neophloga lanceolata*. However, there are only three stamens in the staminate flower, and they are of an extraordinary form. The slender filaments end in a bifid connective, each arm reflexed and bearing a single short pendulous anther. The overall form thus approaches that of the stamens of some members of the New World palm genus *Geonoma* (Tribe *Geonomeae*). The Lokoho River, after which the palm is named, drains the whole southern half of the massif of Marojejy.

DISTRIBUTION. Madagascar, only known from Marojejy.
HABITAT. Humid lowland and lower montane rain forest on steep slopes; 400–1200 m.
LOCAL NAMES. Not recorded.
USES. Not recorded.

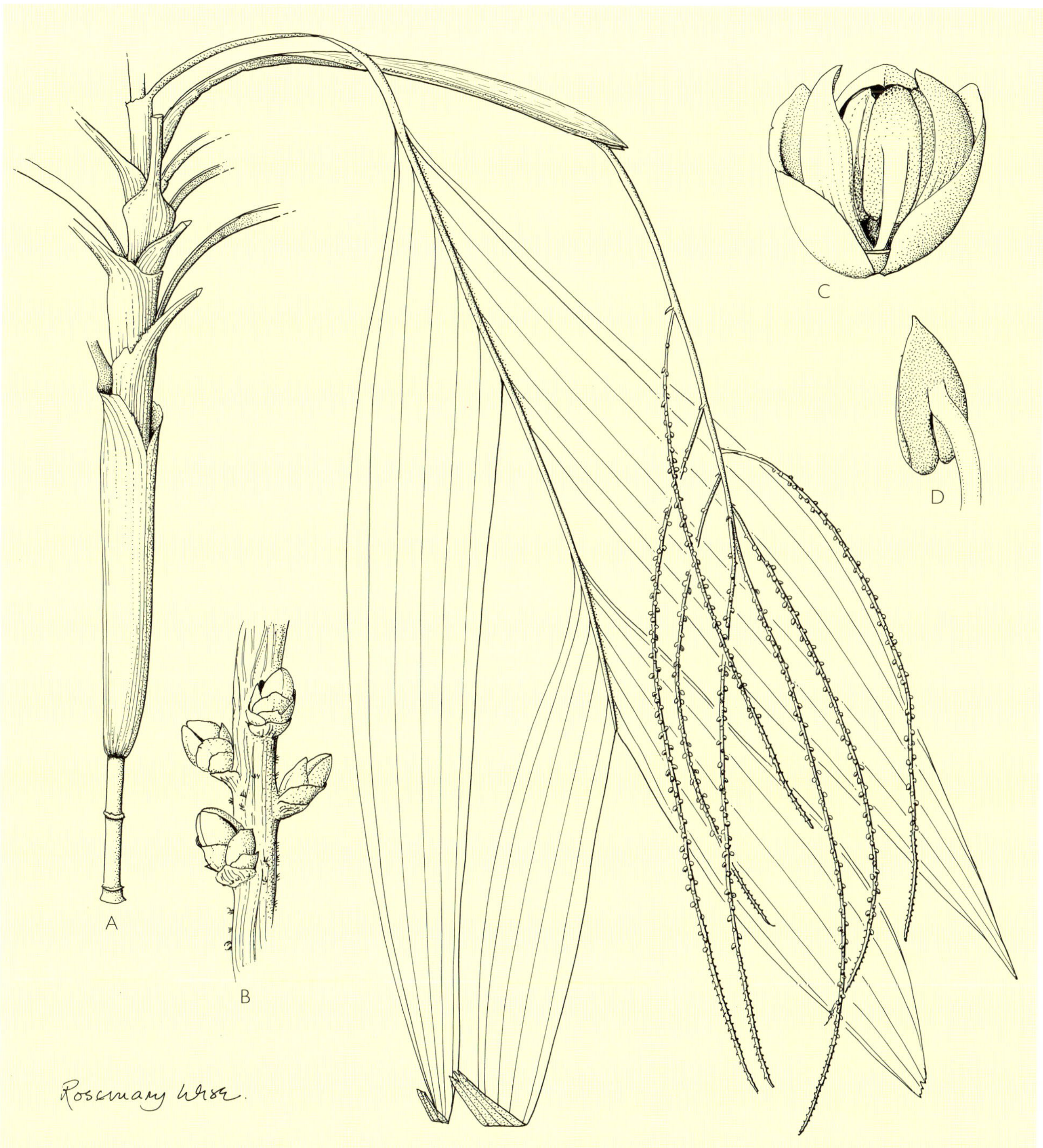

Dypsis lokohoensis. **A** crown × 3/5; **B** detail of rachilla × 6; **C** open staminate flower × 15; **D** stamen × 15. All from *Dransfield et al.* JD6759. Drawn by Rosemary Wise.

CONSERVATION STATUS. Vulnerable. Only occurs in a restricted area, though there is some protection.

Dypsis lokohoensis *J. Dransf.* **sp. nov.**
Palma gracilis, foliis integro-bifidis vel pinnatis, inflorescentia 1- (rare 2) ramosa, rachillis paucis (4–9), pendulis, floribus staminatis triandris, staminibus antesepalis, connectivo valde bifido antheris pendulis distinctissima. Typus: Madagascar, Andapa, Marojejy, *Dransfield et al.* JD6759 (Holotypus K; isotypi P, TAN).

Slender clustering undergrowth palm. **STEM** to 3 m tall, 6–15 mm diam., internodes 5–14 mm long with scattered caducous brown scales, nodal scars prominent. **LEAVES** 4–9 in crown; sheaths 6–12 cm long, c. 7–15 mm diam., longitudinally striate and bearing scattered caducous dark brown scales, sheath mouth lacking auricles; petiole 2–9 cm long, c. 1–3 mm diam., ± triangular in cross section, bearing sparse scattered dark brown scales; lamina entire bifid or divided into 3–9 subregular to grouped leaflets on each side of the rachis, rachis (or costa) 8–33 cm long, sparsely dark brown scaly; entire bifid lamina to 20–24 cm, divided to 65%, with lobes 14–15 cm long, 3.9–4.4 cm wide at widest point, tapering gradually to shallowly lobed tips, the apical cleft to 13 cm deep; pinnate leaf with leaflets subregular or grouped irregularly, somewhat sigmoid, the basal 4–30 × 0.3–4.5 cm, the subapical to 14–25 × 2–4 cm, the apical pair 10–17 × 1.8–6 cm, apically shallowly lobed; both lamina surfaces bearing minute brown punctiform scales along main ribs and veins. **INFLORESCENCES** interfoliar at first, apparently becoming infrafoliar, ± porrect, branching to 1 (or very rarely 2) orders; peduncle 12–32 cm long, 1–2 mm diam., sparsely scaly; prophyll 7–25 × 0.4–0.7 cm, apically irregularly laciniate, abaxially sparsely scaly; peduncular bract 18–20 × 0.5–0.7 cm, deciduous; rachis 3–6 cm, densely scaly when young; rachillae 4–9, somewhat curved or flexuous, 7–23 cm, densely scaly when young; triads borne ± superficially, c. 1 mm distant, rachilla bracts minute. **STAMINATE FLOWERS** c. 2 × 2 mm; sepals rounded triangular, abaxially keeled, c. 1 × 1 mm; petals 2 × 1.3 mm, ovate-oblong, with triangular tips, striate; stamens 3, antesepalous, filaments elongating to 2 × 0.3 mm, pale, connectives tanniniferous, dark, bifid with pendulous arms to 0.5 mm, anthers pendulous and divergent on the connective arms, 0.4 × 0.2 mm; staminodes absent; pistillode conical, c. 0.5 mm high. **PISTILLATE FLOWERS** ovoid, c. 1.8 × 1.6 mm; sepals 1.1 × 0.9 mm, imbricate, rounded; petals 1.2 × 0.8 mm, imbricate basally, triangular valvate distally; staminodes 3, minute, dentiform; ovary 1.1 × 0.4 mm, ellipsoid. **FRUIT** only known in very immature state.

NOTE. The two specimens that make up the Humbert collection differ greatly in leaf dissection, yet they both bear the same collector's number. In one the leaf blade is entire bifid, in the other it is divided into irregularly arranged leaflets. However, details of indument and texture are identical. *Cours* 3243 from Ambatosoratra, is more robust; it has a pinnate leaf with subregular leaflets. Ambatosoratra lies at 14° 34′ S 49° 44′ E on the eastern slopes of the Marojejy Massif. The most recent collections have all been made on the eastern slopes of Marojejy, and vary in size and leaf dissection. *Dransfield et al.* JD6396 (K, TAN) from near Ambanizana on the Masoala Peninsula keys to this species but does not have the same texture; unfortunately, the available material does not have flowers and it is not possible to match the specimen with more certainty.

SPECIMENS SEEN. Andapa: Marojejy, trail to Marojejy Est from Mandena, above Camp 2, Nov. 1989 (fl.), *Dransfield et al.* JD6751 (K, TAN); (fl.), JD6752 (K, TAN); (fl.), JD6759 (Holotype K; isotypes P, TAN); Valley of the Lokoho, Mt. Ambodilaitra, north of Andranomiforitra and Belambo, March 1949 (fl.), *Humbert* 23275 (P); Ambatosoratra, Jan. 1949 (fl.), *Cours* 3243 (K, P); Col d'Andapa, Route de Sambava, April 1970 (fl.), *Bosser* 20114 (K, P).

Dypsis lokohoensis

109. DYPSIS THOUARSIANA

INSUFFICIENTLY KNOWN SPECIES

Dypsis thouarsiana *Baill.*
Bull. Soc. Linn. Paris 2:1163 (1894). Type: Madagascar, *Du Petit Thouars* s.n. (P).

The interpretation of the name *Dypsis thouarsiana* has been one of the most intriguing problems in Madagascar palm nomenclature. The name was published by Baillon (1894a) in his paper entitled "Les palmiers malgaches à petites fleurs". Loosely translated, his protologue reads "The species that we name *D. thouarsiana* takes after the preceding one (*D. lantzeana*) in the deep division of the leaves and by the confluence of the terminal segments, despite the lower ones being separate; but it is also equally reminiscent of *D. pinnatifrons* in the narrow rigid elongate segments, nerved like the leaves of *Gladiolus*, linear lanceolate and nearly half a metre long. In the plant of Du Petit Thouars and in a variety which we attribute to the same species, called Vounouthre or Talanouc by the natives of northeast Madagascar (*Boivin* 1709, 1709[2], Sainte Marie, Tafondrou), the inflorescence has a short stocky compressed peduncle, and long divisions which are two to four decimetres, rather thick, rigid and carrying three-flowered glomerules. In the male flowers, the only ones that are well developed, there are only three stamens and obtuse concave strongly imbricate sepals, the base of which is prolonged on the outside in a sort of solid obtuse spur, equalling a fifth of the total height of the sepal and running into a short obtuse keel along the dorsal median line".

Thus three collections are cited in this protologue. The collection made by Du Petit Thouars, without locality, consists of leaf fragments and what appears to be a first order branching system of a very young inflorescence; floral bracteoles have developed, and three imbricate sepals are partially developed, within which are three very small valvate structures which we interpret as petals.

Stamens, however, have not yet developed. *Boivin* 1709 consists of two leaves while *Boivin* 1709² consists of a first order branching system of an inflorescence, still young, but with more floral details available. Within the petals are three antesepalous stamens; these have widely divergent anther thecae that seem to be pendulous from the tip of the filament, a most unusual stamen form, though not unique to this taxon (see also *D. lokohoensis* and *D. fasciculata*). The material is so incomplete that we cannot be certain that the leaves in *Boivin* 1709 are from the same species as the inflorescence in *Boivin* 1709². The Du Petit Thouars collection seems to be conspecific with the Boivin leaf collections. A fourth collection, probably made by Boivin, is annotated "talanouac" but consists of leaves only, matching those of the other collections.

The leaves of the three collections described above are relatively small; they have distinctive broad multifold basal and apical segments and single fold mid-leaf segments. There is a long petiole and at its base a short entire ligule. It must be said that these leaves are reminiscent of the plant previously named *Vonitra fibrosa* that has long been called *Vonitra thouarsiana*, so this may go some way to explain why Beccari based his *Vonitra thouarsiana* on Baillon's name (while preparing his description from a collection made by the Rev. Baron). *Boivin* 1709² also carries the name "vounouthre" - i.e., *vonitra*, the consistently applied vernacular name for this important and common palm. Yet there are only three stamens in *Boivin* 1709², as described by Baillon and clearly evident in the specimen, while in Beccari's new genus *Vonitra* there are six biseriate stamens. It is most surprising that Beccari did not mention this anomaly. Could it be that the four early collections represent more than one taxon? The leaves could be interpreted as juvenile *D. fibrosa* leaves. Only in *D. fasciculata* do rachilla and stamen form approach those of *Boivin* 1709², but this species has inflorescences branched to two orders; the inflorescence fragment in *Boivin* 1709² could represent the whole branched portion of an inflorescence branched to two orders, or, more likely (because of "rachis" is flattened on one side) a whole first order branching system of an inflorescence branched to three orders. It is possible that the inflorescence in the Du Petit Thouars collection represents a fragment of a very young *D. fibrosa* inflorescence, but the stamens that would allow identification have not yet developed.

Further collecting on Île Sainte Marie may sort out the problems, but at present we have a species, *Dypsis thouarsiana,* that appears to have been typified by Baillon on the Du Petit Thouars collection (the Boivin collections are mentioned as a variety) that are too young to show the diagnostic three stamens described by Baillon. The two numbered Boivin collections may not even belong to the same taxon, as one consists of leaves while the other of inflorescence fragments, and these are clearly not conspecific with *Dypsis fibrosa.*

110. DYPSIS PERRIERI

D. perrieri is a rather massive stocky, litter-trapping palm of the lower strata of rain forest. In habit it is reminiscent of other litter-trapping species such as *Masoala madagascariensis, Ravenea albicans*, and *Dypsis marojejyi*, having a rather short stem and large leaves that do not fall neatly from the stem. It is easily distinguished when in flower because of the large torpedo-like peduncular bract, densely covered in thick red tomentum. In shape this bract is very reminiscent of that of *Beccariophoenix madagascariensis*, but the flowers, fruit and thick tomentum are very different. Found on Marojejy and around the bay of Antongil, this species has recently been found near Antanambe, south of Mananara Avaratra.The name refers to the collector of the type, and of many other exciting palms, Joseph Marie Henry Alfred Perrier de la Bâthie (1873–1958), whose many excellent collections have given the world a wealth of information on many common, rare, and several now possibly extinct species.

DISTRIBUTION. Marojejy, Masoala and Mananara Avaratra.

HABITAT. Moist forest, on steep slopes, near waterfalls on rocks or in valley bottoms; 150–800 m.

LOCAL NAMES. *Besofina* (Betsimisaraka, meaning big ears, because of the large sheath auricles); *Menamosona* (Betsimisaraka, meaning red back, because of the red tomentum on the sheath and bracts), *Kase.*

USES. Good palm-heart.

CONSERVATION STATUS. Vulnerable. Despite its fairly large distribution area, this species is not common in any of its sites; the fact that the palm-heart is eaten is a contributing factor to its status.

Dypsis perrieri *(Jum.) Beentje & J. Dransf.* **comb. nov.**

SYNONYMS:
Antongilia perrieri Jum., Ann. Inst. Bot.-Géol. Colon. Marseille sér. 4, 6 (2): 19 (1928); Cat. Pl. Madagascar, Palmae: 7 (1938); Jum. & H. Perrier, Fl. Madagascar 30: 127, fig. 34 (1945). Type: Madagascar, Maroantsetra-Mananara, *Perrier* 11946 (Holotype P).

Dypsis perrieri, growing on the slopes of Marojejy.

Chrysalidocarpus auriculatus Jum., Ann. Inst. Bot.-Géol. Colon. Marseille sér. 5, 1 (1): 25 (1933); Cat. Pl. Madagascar, Palmae: 8 (1938); Jum. & H. Perrier, Fl. Madagascar 30: 117 (1945), **synon. nov**. Type: Madagascar, Masoala, Marambo Hills, *Perrier* 11942 (Holotype P).

Chrysalidocarpus ruber Jum., Ann. Inst. Bot.-Géol. Colon. Marseille sér. 5, 1 (1): 23 (1933); Cat. Pl. Madagascar, Palmae: 11 (1938); Jum. & H. Perrier, Fl. Madagascar 30: 100 (1945), **synon. nov**. Type: Madagascar, Masoala, *Perrier* 11941 (Holotype P).

Massive, squat solitary palm. **Trunk** 2–8 m high, 20–30 cm diam., near the crown c. 12 cm diam. Trunk with marcescent leaves and persistent bases of leaf sheaths; internodes c. 4 cm; nodal scars dotted with fibre remains, c. 2 cm wide. Base of crown litter-accumulating. **Leaves** 12–20 in the crown, porrect; sheath c. 1 m long, abaxially densely reddish-tomentose to floccose, turning fibrous and desintegrating with age, with auricles 3–12 cm long; apparent petiole 40–160 cm long, proximally c. 5 × 2.5 cm diam., distally 3–3.5 × 2–2.7 cm, the margins often with a slight wing formed by the rein of the proximal leaflets, deeply channelled with sharp margins, abaxially densely reddish-tomentose to reddish-brown scaly, glabrescent, dark green under the tomentum; rachis c. 3–3.5 m long, proximally channelled, in mid-leaf 1.3–1.8 × 2.3 cm diam., distally keeled, with patches of pale white, pale brown or dark red tomentum of peltate laciniate scales; leaflets 45–50 on each side of the rachis, regular, rigid and plicate, adaxially dark green, abaxially bright green, the proximal 80–148 × 1.2–3.4 cm, the most proximal ones inserted at different levels, median 69–107 × 3–5.5 cm (interval [0.5 -] 4–6.5 cm), distal 21–60 × 1–3 cm, the distal pair joined for 3–6 cm, main veins 3–7, apices acuminate and unequally bifid, with scattered minute scales on the minor veins, occasionally with large

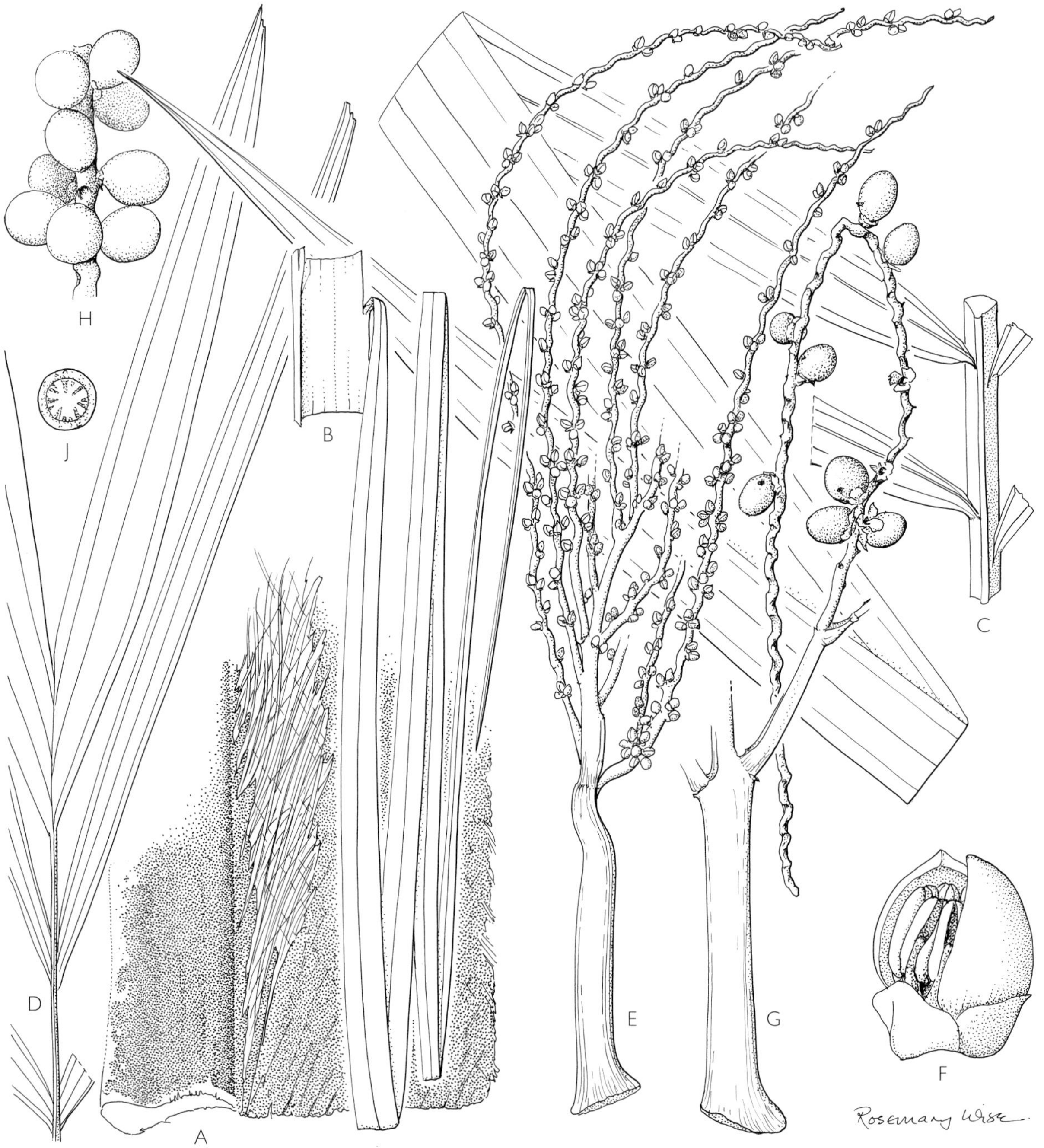

Dypsis perrieri. **A** leaf sheath and fibrous margins × 2/5; **B** proximal part of leaf with lowermost leaflet × 2/5; **C** mid section of leaf × 2/5; **D** leaf tip × 2/5; **E** first order inflorescence branch × 2/5; **F** staminate flower, one petal removed × 8; **G** first order branch of infructescence × 2/5; **H** fruiting rachilla × 2/5; **J** fruit in cross section × 2. All from *Dransfield et al.* JD6749. Drawn by Rosemary Wise.

brown or red ramenta 5–15 mm long on the abaxial midrib, once (in the type) with dense silvery hairs on the abaxial midrib. **INFLORESCENCE** interfoliar to infrafoliar, branched to 2 or 3 orders, spreading to pendulous, to 2 × 1.5 m; peduncle 57–>100 cm long, proximally 8–9 × 2.5 cm, distally c. 5.5 × 3.5 cm, glabrous in young fruit; prophyll erect, 40–60 cm long, c. 5.5 cm wide, borne at c. 3 cm above the base of the peduncle, rufous brown, sometimes rotting away early; peduncular bract inserted at c. 8 cm from the base of the peduncle, circumscissile and carried upwards by the lengthening inflorescence, woody, 80–150 cm long, 3.5–4 cm diam., with a 9–10 cm long beak, densely reddish tomentose, splitting over its whole length except for the distal 16–20 cm; non-tubular peduncular bracts in the distal part of the peduncle, 4–5 × 2 cm; rachis 30–40 cm long, glabrous, with 12–25 branched and 10 unbranched first order branches, the proximal of these with a rachis to 36 cm long, up to 2.2 × 1.1 cm diam. proximally, and with up to 16 second order branches; rachillae spreading or pendulous, 15–50 cm long and 2–6 mm diam.; triads distant, in slight pits, with a narrow, obtuse rachilla bract. **STAMINATE FLOWERS** cream, with sepals 1.5–3 × 1.6–3.2 mm, the outermost smallest, proximally gibbous, keeled, ovate, rounded, ciliolate; petals 4–5.2 × 2.8–3.8 mm, elliptic, acute; stamens 6, 2-seriate, the antepetalous more adaxial and inserted slightly higher up, filaments 1.3–2 mm long with slightly triangular bases, anthers 2.3–3.5 × 1–1.6 mm, dorsifixed, versatile, the locules slightly divergent proximally and obtuse or apiculate, slightly unequal, with a wide (–0.8 mm) dark-coloured connective; pistillode conical, 1–1.4 × 0.5–1.1 mm. **PISTILLATE FLOWERS** with sepals 2.2–4 × 4.5–6.5 mm, concave and enveloping through up to 230°, broadly elliptic, rounded with a fleshy, solid abaxial bulge, ciliolate; petals 5–7.5 × 5–8 mm, concave and enveloping to c. 270°, in fruit 9.7–11 mm wide, with membranous margins, broadly elliptic; staminodes 6, flat, triangular, obtuse, 0.8–1.4 mm high; gynoecium asymmetrical, 3.5–4.5 mm × 2.8–4 mm. **FRUIT** ellipsoid, dull greenish brown, 15–19 × 12–16 mm, rounded or shortly stalked at the base, rounded at the apex; mesocarp up to 2 mm thick; endocarp very fibrous, with few anastomizations. **SEED** slightly obovoid or ellipsoid, 14–16 × 11–12 mm, pointed at the base, rounded at the apex, with a subbasal depression corresponding to the embryo; endosperm ruminate, with the intrusions dense, irregular, up to 1.3 mm thick but usually c. 0.3 mm thick and up to 5.5 mm deep.

Dypsis perrieri. Detail of crown with inflorescence enclosed by the red-hairy peduncular bract and infructescence in almost mature fruit, Marojejy (*Dransfield et al.* JD6749).

NOTE. Jumelle (1928b) described this as a new genus, based on the sagittate anthers and the very fibrous endocarp of the fruit. On examining the type the first character is not correct – the locules are parallel. The second character is more difficult to evaluate, since there are no fruits present in the type.

The types of *C. auriculatus* and *C. ruber* are clearly conspecific; the fruit mentioned in the protologue of *Chrysalidocarpus auriculatus* are of doubtful provenance – they have homogeneous endosperm, so they are unlikely to belong to *D. perrieri*. These fruits were the reason why the taxon was included in *Chrysalidocarpus*; but they are not present on the P sheets of this number.

SPECIMENS SEEN. Andapa: Marojejy, Dec. 1972 (bud), *Guillaumet* 4202 (TAN); idem, NW of Mandena, Oct. 1988 (fl.), *Miller et al.* 3435 (K, MO, P, TAN); idem, N of Mandena, Nov. 1989 (fl., fr.), *Dransfield et al.* JD6749 (K, P, TAN). Antalaha: Marambo, Oct. 1912 (bud), *Perrier* 11942 (P, type of *C. auriculatus*); Masoala Peninsula, without date (fl.), *Perrier* 11941 (P, type of *Chrysalidocarpus ruber*). Maroantsetra: Antalavia, Feb. 1988 (bud), *Dransfield et al.* JD6475 (K, P, TAN); between Maroantsetra and Mananara, without date (fl.), *Perrier* 11946 (Holotype P).

SIGHT RECORDS. Mananara Avaratra: Antanambe (*Dransfield & Beentje*).

111. DYPSIS MOOREI

A massive undergrowth litter-trapping palm, known only from the type; named for Hal Moore (1917–1980), who discovered it.

DISTRIBUTION. Masoala Peninsula.
HABITAT. Lowland rain forest with low canopy; c. 50 m?
LOCAL NAMES. *Maroala* (Betsimisaraka).
USES. Not recorded.
CONSERVATION STATUS. Endangered; only known from a single collection from an area which is known to be under severe threat of degradation.

Dypsis moorei *Beentje* **sp. nov.**
D. perrieri affinis sed petiolo multo longiore rachide multo breviore inflorescentia glabra semel ramificanti pedunculo graciliore sepalis petalisque floris staminati majoribus differt. Typus: Madagascar, Antalaha: across Onive R from Ambatobe, *Moore* 9918 (Holotypus P; isotypi BH, K).

Solitary palm. **TRUNK** c. 1 m high, covered in old leaf bases. **LEAVES** ascending to erect; sheath fibrous; petioles green, c. 3.5 m long, distally c. 2 × 1.6 cm, channelled, densely scaly; rachis c. 1.4 m long, in mid-leaf c. 1.5 × 1.2 cm, glabrous, possibly slightly waxy; leaflets c. 54 on each side of the rachis, regular, directed forward (Moore), green on both surfaces, the proximal 61–66 × 1.4–1.7 cm, median 80–81 × 3.7–3.8 cm (interval 4–5.5 cm), distal 10–46 × 1.1–2.3 cm, main veins 3, with minute reddish scattered scales on the minor veins, apices attenuate or acute. **INFLORESCENCE** interfoliar, branched to 1 order, erect, c. 1.5 m

Dypsis moorei. **A** basal leaflets × 1/3; **B** mid section of leaf × 1/3; **C** leaf tip × 1/3; **D** distal part of peduncle × 1/3; **E** peduncular bract × 1/3; **F** tip of inflorescence × 1/3; **G** detail of rachilla × 2. All from *Moore* 9918. Drawn by Margaret Tebbs.

long; peduncle c. 90 cm long, c. 2.6 × 1.6 cm diam. medially, distally 1.7 × 1.1 cm, densely scaly; prophyll not seen; peduncular bract caducous, inserted high on the peduncle (*fide* Moore), c. 1.3 m long, green, drying brown; rachis 18–30 cm long, densely scaly; rachillae stiff, 42–51 cm long, c. 7 mm diam., densely scaly, slightly zigzag with distant triads. **Staminate flowers** greenish white; sepals 3.3–4.5 × 4–4.9 cm, keeled, gibbous, imbricate; petals 5.7–7 × 3.5–4 mm; stamens 6, biseriate (offset c. 1 mm), filaments connate at the base for c. 0.2 mm, free for c. 2 mm, anthers 2.7–3 × 1.6–2.2 mm, latrorse and versatile, with parallel locules and a wide black connective; pistillode c. 1.3 × 1.2 mm. **Pistillate flowers** with sepals 3.5–6 × 6–8 mm, the innermost widest; petals c. 8 × 7.5 mm in bud; staminodes 6, 0.8–1 mm high, flat; ovary slightly stalked, c. 6.5 mm × 2.8 mm. Mature **fruit** ± spherical, to 25 × 22 × 22 mm, stigmatic remains lateral or slightly below the equator; endocarp strongly fibrous. **Seed** 22 × 15 × 18 mm, endosperm deeply ruminate, embryo subbasal. **Eophyll** deeply bifid.

Note. This species is somewhat near *D. perrieri* but differs in the petiole being much longer, the rachis much shorter, the inflorescence branching to 1 order only, with glabrous axes; the peduncle is much more slender; the staminate sepals and petals are larger.

Specimens seen. Antalaha: across Onive R from Ambatobe, April 1971 (fl.), *Moore* 9918 (Holotype P; isotypes BH, K).

112. Dypsis dransfieldii

This species is distinctive in its rather open clustering habit, fibrous leaf sheaths that do not form a well defined crownshaft (but lacking pendulous piassava) and the surprisingly long peduncle that projects beyond the leaves. Such inflorescences are reminiscent of those of the group of *Dypsis* species that were formerly included in *Vonitra*. In fact *D. dransfieldii*, as an intermediate between *Vonitra* and *Neodypsis*, was one of the first pieces of evidence we discovered for the general blurring of generic boundaries within *Dypsidinae*, which has lead to the recognition of the single large genus *Dypsis*. Curious rather than beautiful, this palm is a distinctive feature of coastal white sand forests along the western side of the Masoala Peninsula and has been found nowhere else.

Distribution. NE Madagascar: Masoala Peninsula.
Habitat. Littoral forest, on steep or level slope; 2–20 m.
Local names. None recorded.
Uses. None recorded.
Conservation Status. Endangered. Single site status, in a fragile habitat.

Dypsis dransfieldii *Beentje* **sp. nov.**
a *D. nossibense* foliorum foliolorumque numero, pedunculi longitudine differt. Typus: Madagascar, Antalavia, *Dransfield et al.* JD6735 (K holotypus; isotypi P, TAN).

Clustering palm in clumps of 3–5, basally with short stilt roots. Mature **stems** unbranched, 6–8 m tall, 7–8 cm diam.; wood very hard; internodes c. 10 cm (2 cm near the crown); bark dull reddish brown, distally with red-brown tomentum, ringed with

Dypsis dransfieldii. Inflorescences in bud and at staminate anthesis (*Dransfield et al.* JD6469).

Dypsis dransfieldii. **A** leaf sheath opened out × 1/2; **B** distal part of petiole with lowermost leaflets × 1/2; **C** mid section of leaf × 1/2; **D** leaf tip × 1/2; **E** mid portion of inflorescence rachis × 1/2; **F** staminate flower, one petal removed × 10; **G** fruit × 2; **H** endocarp and seed in cross section × 1.5. **A – G** from *Dransfield et al.* JD6735, **H** from *Dransfield et al.* JD6468. Drawn by Rosemary Wise.

close leaf scars. **LEAVES** 6–12 per crown, porrect, marcescent; sheaths with thick reddish brown tomentum, 36–48 cm, proximally c. 16 cm wide when flattened, deeply channelled, distally 0.5–1.8 × 0.6–1.2 cm, with pale brown fibrous tattering, margins 2.5–3 cm wide, red-brown tomentose; on the opposite side from the petiole with a long, mid-brown, 12–16 cm fibrous tongue; petiole absent, but pseudopetiole appearing after disintegration of sheath material, 15–30 cm, channelled, straw-brown with scattered minute scales; rachis 1.3–1.7 m, puberulous or with flaking grey-brown tomentum, in mid-leaf 0.7–1.2 × 0.5–0.7 cm and keeled; leaflets 33–34 on each side of the rachis, regular, slightly sigmoid, proximal leaflets 28–57 × 0.4–1.8 cm, median 52–62 × 1.9–3.2 cm (leaflet interval 4–5 cm), distal 15–39 × 0.5–2.9 cm, with bifid apices, main veins 3–5, midrib prominent adaxially, apices attenuate, glabrous. Sucker shoots with bifid leaves; young leaves tinged pink. **INFLORESCENCE** interfoliar, branched to two orders, erect, 2–2.6 m, projecting beyond the leaves, dull reddish brown; peduncle 141–200 cm, round in cross-section, proximally 1.2–2 × 0.5–1.5 cm, green overlaid with red-brown tomentum, glabrescent, distally 0.7–1.8 × 0.5–0.8 cm, pinkish; prophyll 72–91 cm long, borne at 7–16 cm above the base of the peduncle, cylindrical, split only near the apex, with flaking red-brown tomentum; peduncular bract pale cream, inserted at 30–44 cm above the base of the peduncle, 124–126 cm long, with scattered scales, beaked for c. 8.5 cm, abscising and carried upwards by the lengthening inflorescence; non-tubular peduncular bracts inserted at c. 89 cm (1.5–6.5 × 0.7–1.6 cm), 102 cm (1 cm); rachis 30–45 cm, angular in parts, with c. 15 branched and 14 unbranched first order branches, these flattened, swollen and 6 × 2–3 mm at base; rachillae 18–32 cm, slightly flattened, some-

what pitted, 1.2–2 mm diam. at anthesis, 2.5–3 mm diam. in fruit, covered with dense simple scales; flowers cream-coloured. **STAMINATE FLOWERS** with sepals 0.8–1.4 × 1.4–2.2 mm, unequal, keeled, hooded; petals 2.4–2.5 × 1.6–1.8 mm, ovate, acute; stamens 6, in 1 series (only on large magnification a slight distinction in 2 series apparent), the filaments 0.5–0.8 mm, anthers 1.6–1.8 × 0.4–0.6 mm, dorsifixed, the locules parallel; pistillode 0.8–1 mm high, c. 0.4 mm diam. **PISTILLATE FLOWERS** only known in bud stage, with sepals c. 1.5 × 1.8 mm, petals c. 1.3 × 1.2 mm, gynoecium c. 1 × 0.7 mm. **FRUIT** ovoid, 15–20 × 12–14 mm, with persistant sepals 2–3 mm and persistant petals 3.5–6 × 5 mm, fruit verrucose when mature. **SEED** c. 13 × 9 mm; endocarp with loose fibrous strands 10–22 mm long, the basal ones longest; endosperm ruminate for 2–2.5 mm. **EOPHYLL** bifid, with scaly petiole; scale leaves 2 (respectively 2 and 5–7 cm long).

NOTE. This species differs from *D. nossibensis* in the number of leaves, the number of leaflets, and the size of the peduncle.

SPECIMENS SEEN. Maroantsetra: Antalavia, Feb. 1988 (fr.), *Dransfield et al.* JD6468 (K, P, TAN); idem, Feb. 1988 (fr.), *Dransfield et al.* JD6469 (K, TAN); idem, Nov. 1989 (fl.), *Dransfield et al.* JD6735 (K holotype; isotypes P, TAN); idem, Nov. 1989 (seedling), *Dransfield et al.* JD6736 (K, TAN).

Dypsis dransfieldii. View of the crown showing long peduncle, Antalavia (*Dransfield et al.* JD6469).

113. Dypsis nossibensis

This clearly belongs to the 'Vonitra' group, though it sometimes lacks the distinctive piassava (leaf sheath fibre) on the trunk. The inflorescence has the bulbous bases to the first order branches that are so distinctive for this group, but there are only one or two inflorescences per tree, and the trunk does not branch, which is unlike several other species of this group. The name refers to the island of Nosy Be.

DISTRIBUTION. NW Madagascar, only known from Lokobe forest.
HABITAT. Moist forest; steep mid slope; alt. 100–185 m.
LOCAL NAMES. None recorded.
USES. Wood used in construction.
CONSERVATION STATUS. Critical. Only known from a single forest, in which several palm species, including this one, are felled for planks. Total numbers seen were less than twenty-five.

Dypsis nossibensis *(Becc.) Beentje & J. Dransf.* **comb. nov.**

SYNONYMS:
Chrysalidocarpus nossibensis Becc., Bot. Jahrb. Syst. 38, Beibl. 87: 34 (1906); Palme del Madagascar 41, t. 40 (1914); Jum., Cat. Pl. Madagascar, Palmae: 10 (1938), as doubtful species. Type: Madagascar, Nosy Be, Lokobe, March 1851, *Boivin* s.n. (Holotype P).
Vonitra nossibensis (Becc.) Perrier, Fl. Madagascar 30: 130 (1945).

Vonitra loucoubensis Jum., Rev. Bot. Appl. 1922: 161 (1922); Ann. Inst. Bot.-Géol. Colon. Marseille sér. 5, 1: 15 (1927); Jum., Cat. Pl. Madagascar, Palmae: 26 (1938). Type: Madagascar, Nosy Be, Lokobe, *Perrier* 12028 (Holotype P). NOTE: this name is predated by *Chrysalidocarpus lucubensis* Becc. from 1906, which is a synonym of *C. madagascariensis* Becc.

Dypsis nossibensis, growing at Lokobe showing fibres and 'tongue' (*Beentje & Andriampaniry* 4698).

Solitary palm. **TRUNK** 5–10 m, not branching, 7–15 cm diam., bare and then grey and conspicuously ringed, or with sheath remnants, with (over distal 2 m of trunk) or without a disintegrated fibrous mass; base of trunk with aerial roots; internodes c. 7 cm.; base of crown bulbous, 8 cm diam.; wood hard. **LEAVES** c. 18 in the crown, porrect to arching and held on edge in distal half of the leaf, with the leaflets in one plane; sheath 28–50 cm, 6–9 cm wide, more distally with a central woody part of 3–4 cm wide with two fibrous "wings" of c. 4.5 cm wide, with short fine brown tomentum with fibrous tattered margins, and opposite the petiole with a long brown tongue, 100–120 × 5–5.5 cm; petiole 39–40 cm long, proximally 1.5–2.7 × 1.5–2 cm, distally 1.7–2.1 × 1–1.2 cm, convex or channelled with rather sharp margins and a thin grey to red-brown tomentum; rachis c. 2.8 m, in mid-leaf 0.8–1.5 cm wide and keeled, with red-brown to grey tomentum but glabrescent; leaflets 62–64 on each side of the rachis, regular, mid-green, glabrous, the proximal 62–86 × 1.1–2.4 cm, median 52–78 × 3.3–5 cm (interval 3–4 cm), distal 16–44 × 0.9–3.5 cm, apices attenuate, main veins 5–7, distinct, midrib prominent adaxially, ramenta few, spaced, rather small, on the abaxial midrib of median leaflets. **INFLORESCENCE** erect, only a single stage present at one time, branched to 2 orders, 106–150 cm, with spreading or semi-pendulous rachillae; peduncle 75–84 cm long, proximally 4.2–6 × 2.4–3 cm, distally 2.5–2.7 × 1.6–2 cm, green with short reddish tomentum; prophyll 45–64 × 6.5–8 × 2–3 cm, borne at c. 6 cm above the base of the peduncle, 2-winged, split abaxially; peduncular bract inserted at 15–20 cm, abscising and carried upwards by the lengthening bud, 102–106 × 9.5–18 cm, beaked and closed for 5.5–9 cm, glabrous and pale red-brown abaxially, yellow-brown adaxially; incomplete peduncular bracts 3–6 × 36 mm; rachis 31–37 cm, proximally 3.5–5 × 2.5–3 cm diam., with 13–17 branched and 13–16 unbranched first order branches, these puberulous, with bulbous bases and proximally 1.5–2 × 0.4–1 cm; rachillae 25–56 cm, 3–5 mm diam., angular in bud and slightly zigzag, proximally with dense stellate scales, more distal with fewer scales, pitted; rachilla bracts c. 2 mm. **STAMINATE FLOWERS** slightly trigonous; sepals unequal, 3–4 × 4–7 mm, the innermost the largest, fleshy, keeled, hooded, broadly ovate, rounded, with a minute fringe of tiny hairs; petals purplish white, 4.4–4.5 × 5 mm, fleshy, ovate, slightly acute, inserted on the edge of a fleshy receptacle c. 3 mm high and 4 mm diam.; stamens 6, slightly bi-seriate with the antepetalous stamens inserted slightly higher and the anthers shorter; filaments 2.5–3 mm, connate in a basal ring around the pistillode, anthers dorsifixed, the locules parallel, 2.8–3.2 × 1.8–2 mm; pistillode 1.7–1.8 × 1.7 mm, trigonous. **PISTILLATE FLOWERS** trigonous at base, distinctly asymmetrical, 6 mm high in bud; sepals 4.5–6 × 6–7 mm, unequal, fleshy, keeled, hooded, fringed, imbricate, the innermost largest; petals purplish white, 7–8.3 × 7 mm, broadly ovate, acute, hooded; gynoecium 6–8 × 4.5–6.5 mm, asymmetrical, the distal part partly split in 3, stigmas short, pyramidal; staminodes 6, 0.4–0.7 mm, flat. **FRUIT** not seen in mature state, young fruit 11 × 13 mm, asymmetric. **EOPHYLL** bifid; seedling with 2 scale leaves.

NOTE. HB believes he has seen two trees of this species on the Antsahampano peninsula on the mainland near Nosy Komba, but as he was on public transport at the time was unable to make sure. This would not change the conservation status, as the vegetation of this peninsula is also under threat.

The type of *V. nossibensis* consists of an inflorescence and a very young leaf. We have omitted the measurements of the leaf from the above description.

SPECIMENS SEEN. Nosy Be: Lokobe, March 1851 (fl.), *Boivin* s.n. (P, type of *Chrysalidocarpus nossibensis*); idem, anno 1913 (fl.), *Perrier* 12028 (P, type of *V. loucoubensis*); Lokobe forest near Ampasindava, July 1992 (ster.), *Beentje & Andriampaniry* 4695 (K, TAN); idem, July 1992 (fl.), *Beentje & Andriampaniry* 4698 (K); near Pointe Lokobe, Feb. 1992 (fl.), *Birkinshaw* 124 (K); idem, June 1992 (fl., y.fr., seedling), *Beentje & Birkinshaw* 4704 (BH, K, MO, P, TAN).

Dypsis nossibensis. **A** leaf sheath × 2/5; **B** petiole and basal leaflets × 2/5; **C** mid section of leaf × 2/5; **D** leaf tip × 2/5; **E** first order inflorescence branch × 2/5; **F** detail of rachilla × 3/5; **G** staminate flower, one petal removed × 10; **H** young fruit × 2; **J** fruit in cross section × 2; **K** seedling × 1. All from *Beentje & Birkinshaw* 4704. Drawn by Rosemary Wise.

114. Dypsis crinita

This is one of the commonest palms of valley bottoms of rain forests in the lowlands of north and north-east Madagascar. It is particularly abundant as a riverside tree along the fast-flowing rocky rivers that flow out into the Bay of Antongil. It is a robust palm, immediately distinguishable by its tall, dichotomously branched trunks, abundant piassava (leaf sheath fibre), liver-coloured young leaves and long inflorescences, branched to two orders, usually with many inflorescences in each crown. Seedlings of this species must frequently be submerged by flood water, and in this respect behave as rheophytes, although when the trees are mature, the crowns tower well above the water level. In deep humid valleys, this palm can often be found in large numbers, adding a very special beauty to the riverside scenery. An epiphytic orchid (*Eulophiella*) can often be found growing in the piassava of big old plants of *D. crinita*. It seems likely that this palm is in cultivation, as it is easily accessible to seed collectors and produces large quantities of fruit, but we know of no critically identified plants in collections. The name means 'with tufts of long hairs', presumably a reference to the leaf sheath fibre.

Distribution. NW and NE Madagascar: Manongarivo, Masoala Peninsula and Mananara Biosphere Reserve.
Habitat. Streamsides; alt. 200–250 m; may grow as a rheophyte.
Local names. *Vonitra* (general).
Uses. Heartwood used against children's cough (Manongarivo), piassava used as an oil filter (Manongarivo).
Conservation Status. Rare. Though fairly restricted in distribution, numbers seem to be in the thousands rather than hundreds.

Dypsis crinita *(Jum. & H. Perrier) Beentje & J. Dransf.* **comb. nov.**

Synonym:
Vonitra crinita Jum. & H. Perrier, Agr. Prat. Pays Chauds 10, 85: 293 (1910); Jum., Ann. Inst. Bot.-Géol. Colon. Marseille sér. 3, 1 (1): 9, fig. 4 (1913); Jum., Ann. Inst. Bot.-Géol. Colon. Marseille sér. 4, 5: 13 (1927); Cat. Pl. Madagascar, Palmae: 25 (1938); Jum. & H. Perrier, Fl. Madagascar 30: 134 (1945). Type: none indicated, but locality Manongarivo Massif, alt. 200 m, probably *Perrier* 12052 (Holotype P; ?isotype K).

Clustering palm in tufts of 2–4. **Trunk** 4–15 m, usually branched 2–3 times, rarely unbranched, grey, conspicuously ringed, 12–20 cm (Perrier: to 35 cm) diam., slightly inflated at the base, c. 10 cm diam. near the crown, distally covered in sheath fibers; internodes 2–3 cm, nodal scars 0.5 cm; wood pink, rather soft. **Leaves** 12–15 in each crown, arching, held on edge in the distal half; sheath 54–56 cm long, proximally c. 5 cm wide, with very fibrous margins, red-brown tomentose, the tongue opposite the

Dypsis crinita. Juveniles growing as rheophytes, Mananara Avaratra

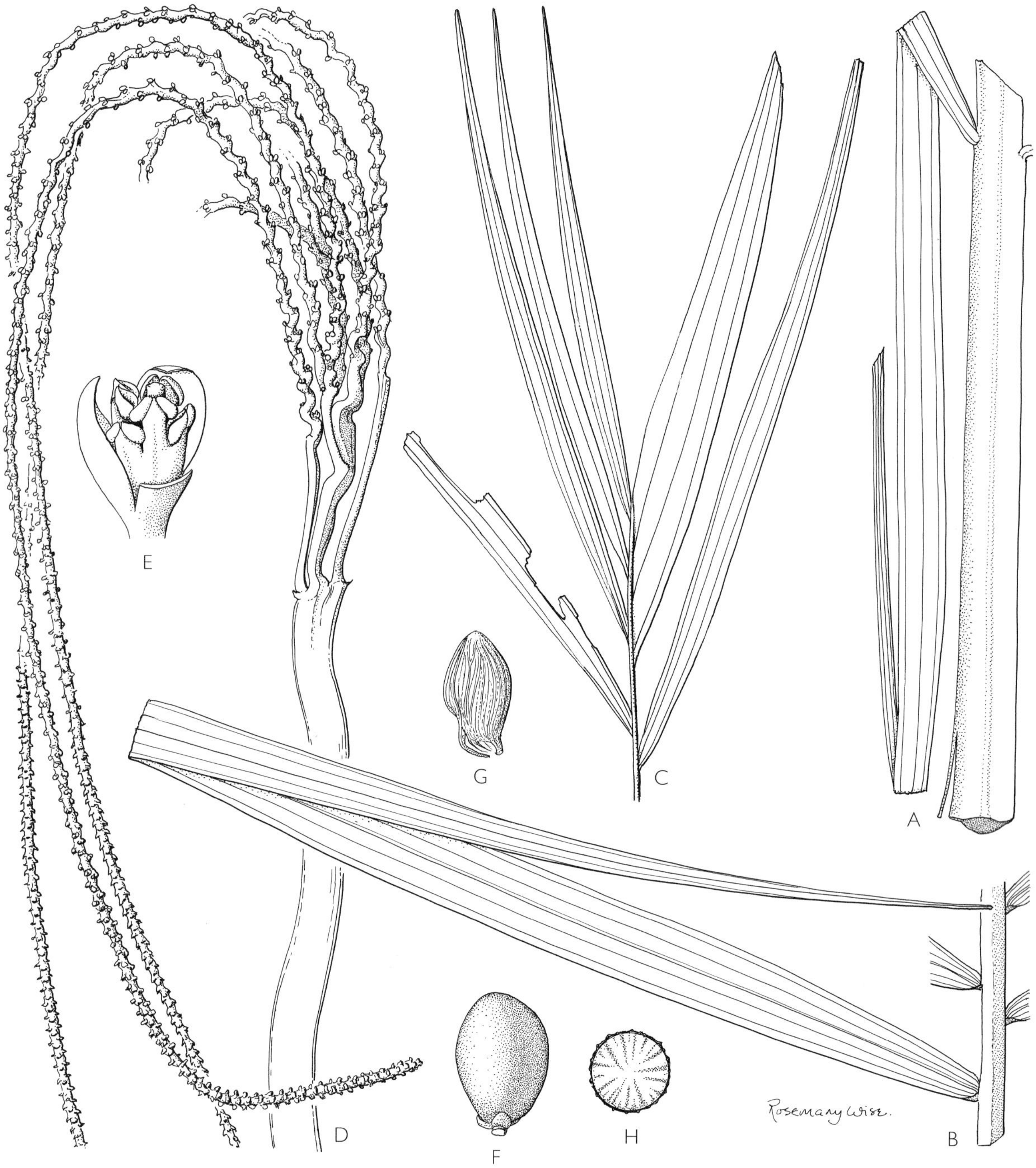

Dypsis crinita. **A** proximal part of leaf with lowermost leaflet × 3/5; **B** mid section of leaf × 3/5; **C** leaf tip × 3/5; **D** first order inflorescence branch × 3/5; **E** staminate flower, one petal removed × 10; **F** fruit × 1; **G** seed enveloped by fibrous endocarp × 1; **H** seed in cross section × 1. **A – D**, **F – H** from *Dransfield et al.* JD6745, **E** from *Beentje et al.* 4577. Drawn by Rosemary Wise.

petiole said by Perrier to be very long; petiole 90–100 cm long, proximally c. 2 × 1.3 cm diam., distally 1.2–2 × 0.6–1 cm, channelled with sharp edges, abaxially with patches of red hairs; rachis 2–3 m long, in mid-leaf 0.6–0.9 × 0.5 cm diam., strongly keeled; leaflets regular, 51–60 on each side of the rachis, bright green, stiff, slightly sigmoid, the proximal 53–78 × 0.7–1.8 cm, the median 54–75 × 2.3–3 cm, distal 11–32 × 1.2–2.7 cm, the terminal pair joined for c. 4 cm and toothed at the apex, main veins 5–7, prominent adaxially, few large red-brown ramenta on abaxial midrib, apices attenuate. **INFLORESCENCE** interfoliar, among the fibrous mass, many per tree, erect, branched to 2 orders; peduncle 126–180 cm long, proximally c. 2.5 cm diam., distally 2.2 cm diam., with reddish laciniate scales, glabrescent; prophyll 52–59 cm long, 3.7 cm wide; peduncular bract inserted at c. 15 cm from the base of the peduncle, c. 200 cm long, beaked for 7–11 cm, abscising and carried upwards by the lengthening inflorescence; rachis c. 40 cm long, with c. 10 branched and 10 unbranched first order branches, these with bulbous bases and more distally 6 × 2.5 mm diam.; rachillae pendulous, very many, 45–64 cm long, 2–3 mm diam., with many small stellate scales; triads spaced, more distally replaced by pairs. **STAMINATE FLOWERS** with sepals very unequal, the outermost smallest, keeled, hooded, proximally gibbous, 2–2.3 × 2.5–3.5 mm, with minute fringe of hairs on margins; petals purplish at anthesis, equal, ovate, acute, c. 2.5 × 1.8 mm; stamens biseriate, filaments 1.3–1.5 mm, anthers 0.3–0.5 mm, basifixed; pistillode c. 1.5 mm, bottle-shaped. **PISTILLATE FLOWERS** with sepals unequal, hooded, keeled, 1.8–2.8 × 4–6 mm, imbricate; petals imbricate, orbicular, rounded, 3.5–3.8 × 4–6 mm, the innermost enveloping the gynoecium for 270°; ovary c. 3.5 × 3.3 mm, asymmetrical; staminodes 6, flat, tooth-shaped, 0.5–1 mm high. **Fruit** slightly ovoid, green turning purple-black, 20–24 × 17–18 mm; endocarp fibrous. **Seed** 15–17 × 11–13 mm; endosperm with dense deep ruminations reaching the centre of the seed.

Note. No specimens were mentioned in the protologue; the species was said to grow at 200 m in the Manongarivo. Beccari studied a sheet of the species given to him by Jumelle, and says (Beccari 1910) that it came from Ananalave in Manongarivo. Jumelle and Perrier (1913) suddenly state the altitude as 1,200 m, and this altitude is repeated in Jumelle (1927a), which also cites *Perrier* 10052 as the sole specimen; Jumelle and Perrier (1945) repeats the altitude, and the specimen is again cited, but is now said to have come from the Ampasimena Peninsula at the W base of Manongarivo. This conforms to a sheet in Paris, *Perrier* 12052, which agrees with the description. To the west of Manongarivo there is the Ampasindava Peninsula, with a small headland called Ampasimena; in the peninsula the altitude does not rise above 800 meters.

Specimens seen. Ambanja: Ampasimena Peninsula, Dec. 1909 (fl., fr.), *Perrier* 12.052 (Holotype P, see note); Manongarivo, Ambalafary, Feb. 1992 (fl.), *Beentje et al.* 4577 (BH, K, MO, P, TAN). Maroantsetra: NE of Antalavia, Nov. 1989 (bud, fr.), *Dransfield et al.* JD6745 (K, TAN).
A sheet at Kew, *Perrier* s.n., NW Madagascar, without further locality, Dec. 1909 (fl., fr.), is presumably a duplicate of the type.
Sight record: Mananara Biosphere Reserve (*Dransfield & Beentje*, 1994).

Dypsis crinita
(x: sight records)

115. Dypsis utilis

This is an upland palm, closely related to *D. crinita*, but even larger in all its parts, and with inflorescences branched to three rather than two orders. It has been seen near Analamazaotra (in fact there is a much photographed specimen opposite the station hotel at Andasibe; see photo below) and at Ranomafana. Sometimes cited as being critically endangered, this may not be so rare as was originally thought. The Latin name means 'useful', a reference to the piassava produced by the leaves and formerly exported for rope production.

Distribution. E Madagascar.
Habitat. Streamsides in swamp forest or moist forest, gregarious; alt. 950–1000 m.
Local names. *Vonitra* or *Vonitrandrano* (water vonitra).
Uses. Jumelle (1927a) states that the palm-heart is eaten, and that the fruit pulp is sugary and also consumed.
Conservation Status. Vulnerable. Restricted to only a few sites; numbers thought to be low.

Dypsis utilis. A mature palm at Andasibe.

Dypsis utilis *(Jum.) Beentje & J. Dransf.* **comb. nov.**

SYNONYM:
Vonitra utilis Jum., Compt. Rend. Acad. Paris 164: 921 (1917); Rev. Bot. Appl. 1922: 161 (1922); Ann. Inst. Bot.-Géol. Colon. Marseille sér. 5, 1: 16 (1927); Cat. Pl. Madagascar, Palmae: 26 (1938); Jum. & H. Perrier, Fl. Madagascar 30: 131, figs. 35, 36 (1945). Type: Madagascar, no locality mentioned, probably (from Jumelle 1927) Analamazaotra, *Perrier* 12005 (Holotype P).

Robust palm, solitary or more often several in a clump. **TRUNK** 6–17 m high, branching dichotomously 2–3 times at 9–11 m above the ground, the branches closely parallel, 25–37 cm diam., near crown c. 18 cm diam.; internodes 7–9 cm; bark light brown; wood soft, pinkish-white, moist; crown untidy with pale brown pendulous fibre. **LEAVES** 13–14 per crown, with the dead leaves marcescent, 4–5 m, erect to porrect, often held on edge; sheath fibrous, c. 80 cm long, brown-grey tomentose, with fibrous wings c. 4.3 cm wide each, with opposite the petiole a light brown fibrous tongue c. 50 cm long; petiole 45–100 cm (to 185 cm in young trees), proximally c. 5 × 2.5 cm, distally 2–4.2 × 1.1–1.5 cm, flat adaxially or slightly convex, convex abaxially, with sharp edges, with scattered scales, glabrescent; rachis 2.8–3.8 m, 1.5–1.9 × 1–1.3 cm in mid-leaf, keeled, straw-brown, with scattered pale or dark red scales; leaflets 65–73 on each side of the rachis, in one plane, regular, bright green, stiff, attenuate, the proximal ones 64–82 × 1.2–2 cm, median 77–92 × 3.1–4.8 cm (interval 4–4.5 cm), distal 18–47 × 1–1.8 cm, ramenta few, red-brown, quite large, abaxial on the midrib, main veins 2 on each side of the midrib and main veins, prominent abaxially, young leaves reddish. **INFLORESCENCE** interfoliar to infrafoliar, up to 20 per tree (in two crowns) in all stages from closed and erect to

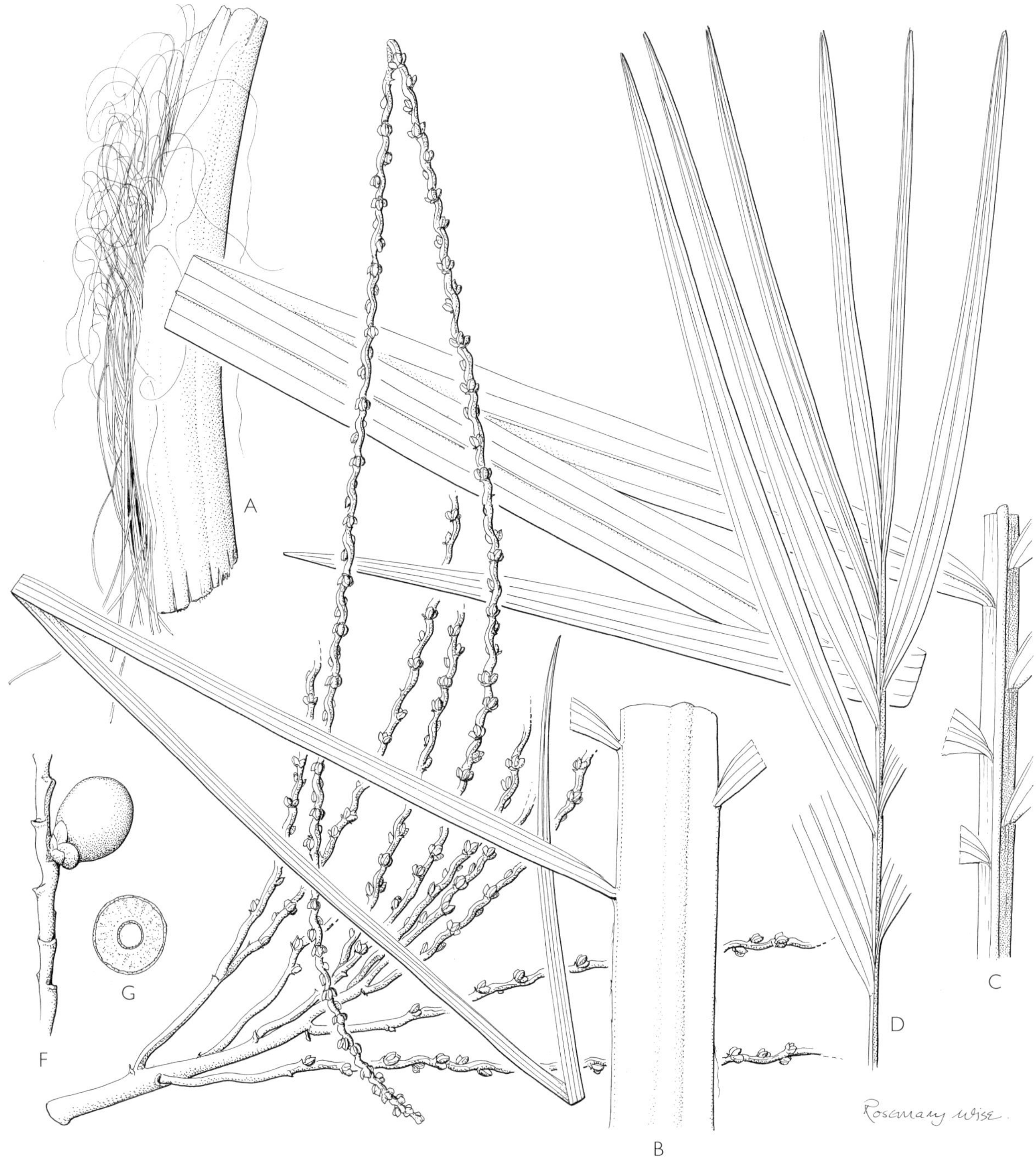

Dypsis utilis. **A** distal part of leaf sheath × 1/2; **B** basal part of leaf × 1/2; **C** mid section of leaf × 1/2; **D** leaf tip × 1/2; **E** part of inflorescence × 1/2; **F** detail of rachilla and fruit × 3/5; **G** fruit in cross section × 3/5. All from *Beentje & Raharilala* 4417. Drawn by Rosemary Wise.

Dypsis utilis
(x: sight records)

old fruiting stage (pendent); branching to 3 orders (rarely to 2 orders?); peduncle 125–135 cm, proximally 3.3 x 2.5 cm, distally 2.8 x 2 cm, green, glabrous, curved; prophyll dark brown, 25–70 cm long, 6.5 x 2.5 cm diam., split at the apex; peduncular bract 154–191 x 8.4 cm, green turning mid-brown, adaxially red-brown and smooth, densely scaly but soon glabrescent abaxially, glabrous adaxially, beaked for c. 7 cm, abscising and carried upwards by the lengthening inflorescence; rachis c. 62 cm long, somewhat angled, the branches bulbous at their bases; first order branches (number unclear) proximally 1.4–1.5 x 0.5–0.7 cm; rachillae pendulous, coral-pink in bud but turning pale green, 69–91 cm long, 2.5–5 mm diam., glabrous or nearly so; triads distant, spirally arranged, slightly sunken in pits; buds yellow-green to reddish brown, slightly trigonous. **STAMINATE FLOWERS** purplish or reddish at anthesis, with sepals 2–2.2 x 3.5–5 mm, unequal, the outermost smallest, fleshy, hooded; petals 2.7–2.8 x 1.8 mm, fleshy, ovate, acute; stamens 6, biseriate, didymous, the outermost inserted slightly lower than the inner, filaments 0.8–1 mm, anthers dorsifix, 0.6–0.8 mm; pistillode c. 1.5 x 0.8 mm, bottle-shaped, showing stamen indentations. **PISTILLATE FLOWERS** with sepals 3–3.5 x 3.5–8.5 mm, unequal, hooded, the innermost largest and enveloping the bud for some 270°; petals 3.5–4.5 x 5–7 mm, orbicular, imbricate, unequal, the innermost largest; ovary 3.5–3.8 x 2.5–3 mm, slightly asymmetric with an indistinct trigonous apex; staminodes 0.6–0.8 mm, flat, tooth-shaped. **FRUIT** dull green turning purplish brown or black, globose-ellipsoid or obovoid, 17–28 x 14–20 mm, with persistent petals c. 6 mm long; endocarp fibrous. **SEED** c. 23 mm x 16 mm, pointed at the base, rounded at the apex, with ruminate endosperm, the ruminations many, slightly irregular, and almost reaching the middle of the seed.

NOTE. This is the most robust member of the "Vonitra" group.

SPECIMENS SEEN. Moramanga: Analamazaotra (fl.), *Perrier* 12005 (Holotype P); idem, Feb. 1924 (fl.), *Perrier* 16067 (P); idem, Oct. 1963 (fl.), *Moore* 9005 (BH, TAN); idem, March 1991 (ster.), *Beentje & Rabarilala* 4409 (K, TAN); Maromiza, March 1991 (fl., fr.), *Beentje & Rabarilala* 4417 (BH, K, MO, P, TAN).
SIGHT RECORD. A specimen from Ranomafana was brought to us by Dan Turk and was definitely this species (because of the length of the leaf and the branching of the inflorescence).

116. DYPSIS FIBROSA

This is one of the most widespread of species in the "Vonitra" group, occurring throughout the north-west and eastern rain forest belt. In habitat it differs from the much larger *D. crinita* that often grows in nearby valley bottoms and riversides, while the present species will grow on ridges and slopes. The sheaths produce abundant piassava, leaf sheath fibre, and hence the species name. This species has been cultivated in many botanical gardens and private collections where the combination of neat dark green leaves that are flushed red when newly emerged and attractive brown fibre make it particularly ornamental. Its ability to branch dichotomously adds to its appeal.

DISTRIBUTION. NW and E Madagascar.
HABITAT. Moist upland forest or coastal hill forest on steep slopes or less often on ridge tops, also in littoral or peatswamp forest overlying white sand at low altitudes; alt. 5–800 m.
LOCAL NAMES. *Vonitra* (widespread), *Vonitrambohitra* (mountain vonitra, fide Jumelle), *Ravimbontro* (Nosy Mangabe).

Dypsis fibrosa, surviving in a cleared area, Sahavary.

USES. Leaf extensively used for thatching (Masoala), inflorescences sold as brushes (Masoala). Formerly one of the main piassava producers (30–50 francs a kg in 1951).

CONSERVATION STATUS. Not threatened. Widespread.

Dypsis fibrosa *(Wright) Beentje & J. Dransf.* **comb. nov.**

SYNONYMS:

Dictyosperma fibrosum Wright, Kew Bull. (1894): 359. Type: Madagascar, *Proctor Brothers* s.n. (Holotypus K).

Vonitra fibrosa (Wright) Becc., Agric. Colon. 5: 322 (1911); Becc., Palme del Madag.: 8, figs. 3, 4 (1912).

V. thouarsiana Becc. (non *Dypsis thouarsiana* Baill.), Bot. Jahrb. Syst. 38, Beibl. 87: 18 (1906); Jum. & H. Perrier, Ann. Inst. Bot.-Géol. Colon. Marseille sér. 3, 1 (1): 5, figs. 2, 3 (1913); Jum., Rev. Bot. Appl. 1922: 160 (1922); Ann. Inst. Bot. Géol. Colon. Marseille sér. 5, 1, fasc. 1: 8 (1927); Cat. Pl. Madagascar, Palmae: 26 (1938); Jum. & H. Perrier, Fl. Madagascar 30: 132, fig. 36: 3, 37 (1945). Type: Central Madagascar, anno 1883?, *Baron* 2319 (Holotype K). **synon. nov.**

Solitary or clustering palm, when clustering in groups of 2–6. **TRUNK** 3–9 m, branched once or twice (rarely three times) a few meters above the ground, rarely unbranched, the branches closely parallel, 5–18 cm diam.; distal part of the trunk covered in fibrous piassava; base swollen, sometimes with surface roots resembling stilt roots; bark pale brown to grey, ringed, internodes 0.8–2 cm; wood hard, white. **LEAVES** 8–25 in each crown, occasionally with up to 8 marcescent leaves; leaves arching, held on edge in the distal half; sheath 40–60 cm long, red-brown floccose, proximally 10–12 cm wide, more distally with a central woody part and a fibrous part together with a 30–34 cm long pale brown tongue opposite the

Dypsis fibrosa. **A** proximal part of leaf with lowermost leaflets × 1/2; **B** mid section of leaf × 1/2; **C** leaf tip × 1/2; **D** tip of peduncular bract × 1/2; **E** first order inflorescence branch × 1/2; **F** fruiting rachillae × 1/2. All from *Beentje* 4449. Drawn by Rosemary Wise.

petiole becoming tattered and so producing the piassava clothing the upper part of the trunk; petiole 40–170 cm long, proximally 1.2–2.6 × 0.8–1.5 cm, distally 0.9–1.6 × 0.8–1 cm, with red-brown patches of tomentum but glabrescent, adaxially slightly convex or channelled, with sharp edges; rachis 1.4–2 m long, in mid-leaf 0.9–1 cm wide and keeled, with red-brown patches of tomentum but glabrescent; leaflets regular, 34–51 on each side of the rachis, in one plane, dull dark green (red in young leaves), the proximal 45–82 × 0.8–2.5 cm, median 46–71 × 2.6–4.3 cm (108 × 7.3 cm in *Perrier* 14097), distal 10–42 × 0.5–2.3 cm, acute, the terminal pair in young plants wide (up to 5 cm) and connate for up to 25 % of their length, main veins 5–7, the midrib prominent adaxially, glabrous, apices unequally attenuate. **INFLORESCENCE** interfoliar, erect in bud, porrect-arching in flower and fruit, branched to 3 orders (once to 4 orders); peduncle 70–94 (–150) cm long, proximally 1.5–2.5 × 0.9–1.8 cm, distally 1–1.7 × 0.7–1 cm, green, glabrous; prophyll 39–56 cm, reddish-tomentose, glabrescent; peduncular bract 103–188 cm, 6–8 cm wide, splitting only near its apex, beaked for 2–12 cm, coriaceous, inserted at 17–20 cm from the base of the peduncle, abscising and carried upwards with the lengthening inflorescence, brown with patches of red-brown pubescence; rachis 37–60 cm long, green, glabrous, with 15–17 branched and 6–14 unbranched branches; first order branches flattened, 0.8–1.3 × 0.3–0.5 cm, with basal swelling; rachillae arching to almost pendulous (4.5–) 17–53 cm long (up to 78 cm in fruit) and 0.1–0.2 cm diam. (up to 0.5 cm in fruit), glabrous, green to red-brown, with spaced triads in slight pits. **FLOWERS** orange in bud, yellow at anthesis. **STAMINATE FLOWERS** slightly trigonous, narrowed near the base, with sepals 1–1.4 × 1.4–2.3 mm, broadly ovate, hooded; petals 1.5–1.8 × 1.3 mm, hooded; stamens 6, in 2 series, didymous, densely appressed against the pistillode, filaments 1–1.4 × 0.5 mm, anthers medifixed or dorsifixed with almost globose thecae, c. 0.3 × 0.3 mm; pistillode bottle-shaped, with indentations conforming to the stamens, c. 1.3 × 0.4–0.5 mm in diam. **PISTILLATE FLOWERS** globose, with sepals 1.2–1.8 × 2.3–2.8 mm, broadly ovate, hooded; petals suborbicular, 2.1–2.8 × 1.8–2.4 mm; ovary asymmetrical, c. 1.8 × 1.6 mm, topped by an indistinct trigonous apex; staminodes c. 0.2 mm high, dentiform. **FRUIT** black, obovoid to almost globose, 20–30 × 18–25 mm; persistent sepals c. 3 × 3–4 mm, persistent petals c. 5 × 7 mm; mesocarp fleshy, c. 5 mm thick; endocarp fibrous, with up to 25 mm long fibres. **SEED** 20–23 × 15–18 mm, ellipsoid, pointed at the apex; endosperm with ruminations 3–4 mm deep. **EOPHYLL** bifid, germination adjacent-ligular.

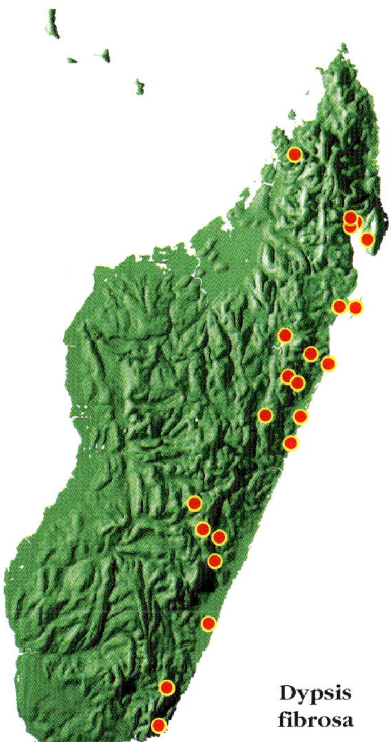

Dypsis fibrosa

SPECIMENS SEEN. Ambanja: Manongarivo, Bekolosi, Dec. 1992 (fr.), *Malcomber et al.* 1972 (K). Maroantsetra: Antalavia, Apr. 1988 (fl.), *Gentry & Schatz* 62180 (K, MO, P); W of Maroantsetra, Oct. 1963 (fl.), *Moore* 9009 (P, TAN); 5 km W of Maroantsetra, Oct. 1986 (bud), *Dransfield et al.* JD6360 (K, P, TAN); Nosy Mangabe, April 1989 (fl.), *B. DuPuy* MB146 (K, TAN); Hiaraka, Oct. 1986 (fl., fr.), *Dransfield et al.* JD6373 (K, TAN). Sainte Marie: Kalalao forest, Nov. 1994 (bud), *Dransfield et al.* JD7523 (K, TAN). Soanierana-Ivongo: Andasibe, Dec. 1938 (bud), *Lam & Meeuse* 5862 (K, L). Ambatondrazaka: Maningory Falls, Dec. 1944, *Homolle* 547 (P, probably a very young plant). Toamasina: Betampona, Oct. 1991 (bud, fr.), *Beentje* 4498 (BH, K, MO, P, TAN); Toamasina, (leaf, seed), *Proctor Bros.* s.n. (K). Moramanga: Analamazaotra, (fl., fr.), *Perrier* 12001 (P). Ampasimanolotra: Andrambolahy kely to Andranampony, April 1951 (fl.), *Cours* 4511 (K, P, TAN); Anivoranokely, Sept. 1954 (fl.), *Vigrence* 15462 (P); 5 km S of Ambila-Lemaitso, Nov. 1986 (fr.), *Dransfield et al.* JD6440 (K, P, TAN); idem, Sept. 1991 (fl., fr.), *Beentje* 4449 (BH, K, MO, P, TAN). Vatomandry: without precise location, Nov. 1927 (fl.), *Perrier* 14097 (P). Ifanadiana: Ambohimanga rd. (fl.), *Dequaire* 27702 (P); 34km E of Ranomafana, March 1991 (bud), *Beentje* 4439 (BH, K, MO, P, TAN). Mananjary: Mt Vatovavy, Aug. 1911 (fl.), *Perrier* 12053 (P). Farafangana: Manombo, Nov. 1991 (bud), *Beentje* 4519 (BH, K, MO, P, TAN). Vangaindrano: Analalava, Dec. 1971 (fl.), *Guillaumet* 4017 (TAN). Tolanaro: Manantenina, Marovony, Oct. 1990 (infl.), *Randrianasolo et al.* 185 (K); NW of Ste Luce, Oct. 1989 (fr.), *McPherson et al.* 14218 (P). Without locality, Central Madagascar, s.d. (fl.), *Baron* 2319 (K, type of *V. thouarsiana* sensu Beccari).

CULTIVATED: Sri Lanka, Peradeniya Royal Bot. Gard., July 1986 (fr.), *Rutherford & Bandara* R 136 (K).

NOTE. The fruit is said to be eaten by wild pig, and we have seen signs which seem to confirm this. Though Beccari (1906) thought he was making a new combination based on *Dypsis thouarsiana* when coining the name *Vonitra thouarsiana*, the descriptions, both of the new genus and of the species, were based on *Baron* 3190, quite distinct from the types of *Dypsis thouarsiana*. Beccari saw the types of *Dypsis thouarsiana* and thought the leaf was a young one, possibly a seedling, of his new taxon; he also considered the inflorescence of *Dypsis thouarsiana* as too young too analyse. The taxa, however, are clearly distinct, the types of *Dypsis thouarsiana* having three most peculiar stamens and leaves with 3–4 leaflets on each side of the rachis (see under 109. *Dypsis thouarsiana*).

117. DYPSIS ANTANAMBENSIS

This is a new species in the "Vonitra" group; it thrives on steep slopes and on ridge tops in low-canopy forest on ultramafic rock. It tends to branch near ground level and have several more or less equal rather slender trunks that are clothed in long persistent short piassava fibre. The leaves are distinctive having rather short leaflets that diverge from the rachis at an acute angle and tend to be held rather stiffly. The name is derived from the type locality. As far as we know, this species is not in cultivation.

DISTRIBUTION. Only known from one small area in the Mananara Avaratra Biosphere Reserve.
HABITAT. Rather open rain forest, on ultramafic soils on steep mid slopes and ridge tops; 250–290 m.
LOCAL NAMES. Not recorded.
USES. Not recorded.
CONSERVATION STATUS. Endangered. Despite its occurrence in a protected area, the number of individuals seems to be less than fifty.

Dypsis antanambensis *Beentje* **sp. nov.**
inter species inflorescentia semel ramosa *D. pusillae* Beentje & J. Dransf. accedens sed statura majore, foliolis pluribus, inflorescentia longiore ab ea recedens. Type: Madagascar, Antanambe, *Beentje, Dransfield & Andriampaniry* 4629 (Holotypus K, isotypi BH, MO, TAN).

Palm apparently basally branched to give clumps of up to 3 otherwise unbranched aerial stems. **STEMS** up to 4 m high, covered for their whole length with a dense fibrous layer c. 12 cm diam., without layer c. 5.5 cm diam., internodes 1.5–2.5 cm, bark conspicuously stepped and ringed; wood brown, quite hard. **LEAVES** c. 12 in the crown, porrect, with stiff leaflets; sheath c. 48 × 3 cm, pale brown with white bloom and orange-red scales, with many marginal fibres; petiole c. 66 cm long, proximally 1.1–2 × 1.4–1.6 cm diam., distally 1–1.6 × 0.8–1.1 cm, red-brown, margins sharp, abaxially convex with patches of brown scales, adaxially channelled; rachis c. 2.5 m, in mid-leaf keeled, 0.8–1.1 × 0.6–0.9 cm; leaflets c. 49 on each side of the rachis, regular, stiff, erect, those on opposite sides of the rachis at an angle of c. 90° with each other, the proximal 36–37 × 0.4–0.7 cm, some with long pendulous reins, median 52–53 × 2.8 cm (leaflet interval 4 cm), distal 15–38 × 0.4–1.9 cm; leaflets glabrous, the apices unequally acuminate, bifid for 3–4 cm with one tip 5–7 cm longer, main veins 5–7, midrib prominent adaxially. **INFLORESCENCE** interfoliar, spreading, branched to 1 order, in bud c. 225 cm long; peduncle c. 160 cm, distally 1.2 × 0.9 cm diam., with dense small red-brown scales; prophyll c. 131 × 2.2–3 cm, cylindrical, 2-keeled, split only at the apex; peduncular bract 102–122 cm (inserted at c. 100 cm), deciduous and carried upwards by the lengthening inflorescence, glabrous, ?beaked for c. 6 cm; rachis c. 41 cm, with c. 20 branches and scattered red-brown scales; rachillae 34–60 cm long, 3.5–5 × 2

Dypsis antanambensis. **A** base of petiole and sheath × 1/2; **B** distal part of petiole and basal leaflets × 1/2; **C** mid section of leaf × 1/2; **D** leaf tip × 1/2; **E** basal part of inflorescence rachis × 1/2; **F** detail of rachilla × 3; **G** seedling × 1. All from *Beentje et al.* 4629. Drawn by Rosemary Wise.

mm diam., somewhat pitted, bare at the base, densely covered in tiny stellate scales, with distant triads, more distally with paired flowers only; rachilla bracteoles c. 3 × 0.6 mm. **STAMINATE FLOWERS** with sepals 1.4–1.6 × 1.4 mm; petals (in bud) 1.7–1.8 mm long; stamens 6, biseriate, the outer (antesepalous) series sessile, the inner series with filaments c. 0.6 mm long, anthers c. 0.6 × 0.4 mm, dorsifixed, locules parallel. **PISTILLATE FLOWERS** with sepals 1.4–1.6 × 1.4 mm; petals (in bud) 1.7–1.8 mm long; gynoecium c. 0.8 × 0.5 mm; staminodes c. 0.2 mm. **FRUIT** unknown, except for: endocarp very fibrous, 15–25 × 12–18 mm. **SEED** with deeply ruminate endosperm, the ruminations many and up to 5 mm deep. **EOPHYLL** pinnate.

NOTE. Among the species of the Vonitra-group this species can be distinguished by its short, stiff leaflets, the inflorescence branched to 1 order resembling *D. pusilla* but is much larger, has a greater number of leaflets and a larger inflorescence.

SPECIMENS SEEN. Mananara Avaratra: Antanambe, April 1992 (bud), *Beentje et al.* 4629 (Holotype K; isotypes BH, MO, P, TAN, type); idem, Oct. 1994 (sd.), *Beentje & Dransfield* 4809 (K, TAN).

118. DYPSIS PUSILLA

This is the smallest species of the "Vonitra" group. It is an attractive palm of the undergrowth in valley bottoms. The stems are almost always unbranched, and often solitary, and densely clothed with very short piassava. It can be distinguished from other species in the group by the short usually unbranched stems, the short piassava, the delicate curving leaflets and the inflorescence that is branched to one order only. Its rather dainty appearance is reflected in the species name. As far as we know, this species is not in cultivation, but it would clearly be an attractive subject for a small garden.

DISTRIBUTION. Masoala Peninsula and Mananara Biosphere Reserve.

Dypsis antanambensis, growing in forest on ultramafic rock, Mananara Avaratra *(Beentje et al.* 4629).

HABITAT. Littoral forest on deep humus near the beach, riverine in moist forest, or on humus banks in moist forest on steep hillside; 1–220 m.

LOCAL NAMES. *Vonitra* (Betsimisaraka).

USES. Not recorded.

CONSERVATION STATUS. Vulnerable. With a fairly small distribution area and low numbers of individuals per population, this species might move into the Endangered category in the near future, if shifting cultivation continues apace in its native habitat.

Dypsis pusilla *Beentje* **sp. nov.**

D. fibrosae affinis insignis statura pusilla, trunco raro ramificanti, inflorescentia brevi semel ramificanti. Typus: Madagascar, Antalavia, *Dransfield et al.* JD6474 (Holotypus K; isotypus TAN).

Solitary or clustered palm. **STEM** 0.8–2 m, usually unbranched, sometimes dichotomous; bare stem 4–5 cm diam., but usually covered with very thick layer of persistent fibrous leafbases and then c. 12 cm diam.; fibres coarse, short, not pendulous. Trunk with basal surface roots and occasionally with aerial roots at 40–50 cm from ground; bark brown, internodes 3–10 mm long. **LEAVES** 15–16 in the crown, up to 2 m long, porrect with arching rachis; sheath 22–26 × 6–7.5 cm, with rather dense red or blackish laciniate scales, auricles (up to 5 cm long) and margins disintegrating quickly; petiole 75–100 cm long, c. 1 × 0.6 cm (proximal), × 0.6 × 0.4–0.5 cm (distal), flattened or slightly convex adaxially, convex to keeled abaxially, green to purplish with dense red laciniate scales; rachis 78–94 cm, in mid-leaf 3–4 mm wide, angular, with

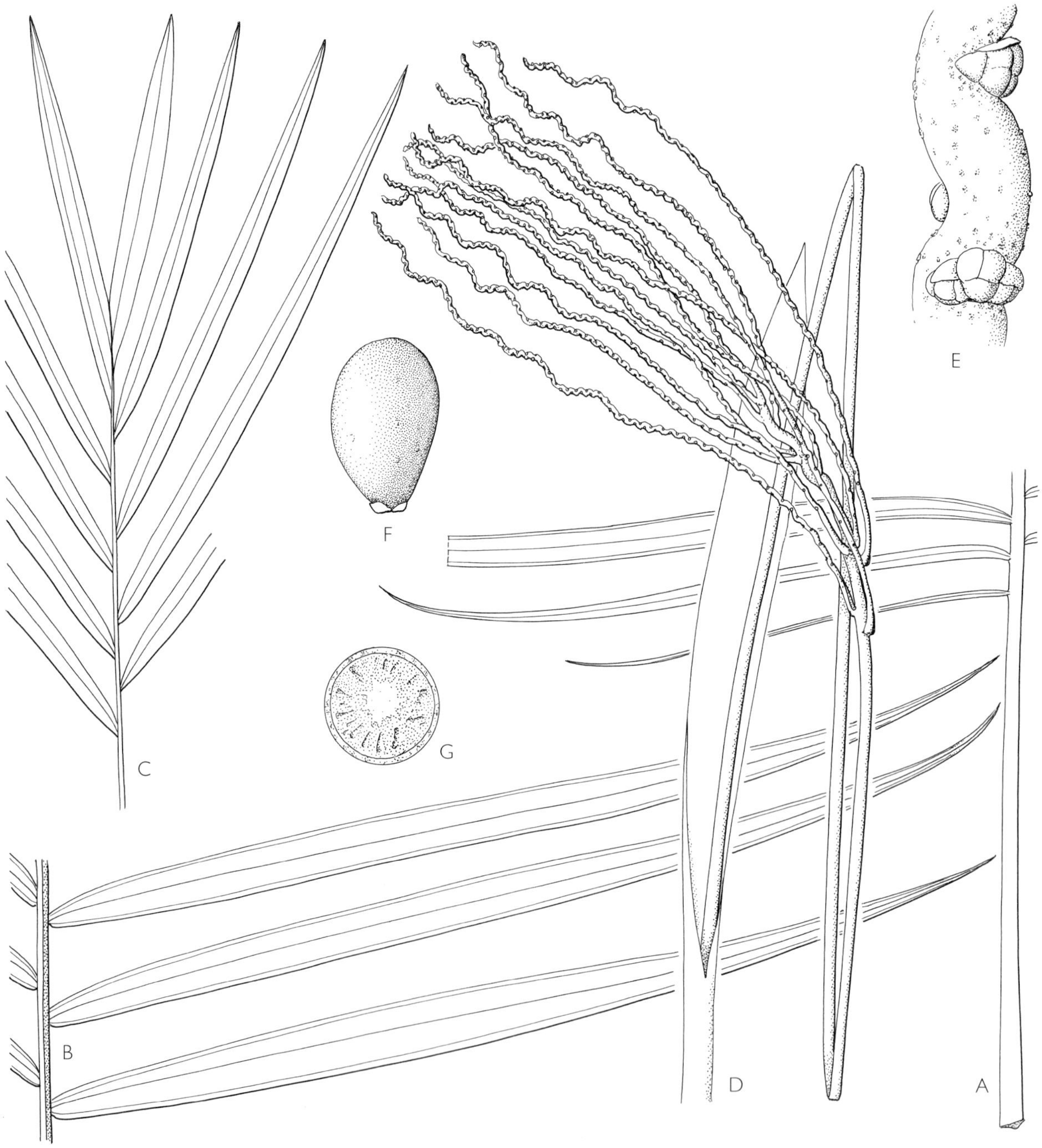

Dypsis pusilla. **A** proximal part of leaf with lowermost leaflet × 1/2; **B** mid section of leaf × 1/2; **C** leaf tip × 1/2; **D** inflorescence × 1/2; **E** detail of rachilla × 8; **F** fruit × 2; **G** seed in cross section × 2. **A** – **C** from *Dransfield et al.* JD6474, **D** – **G** from *Beentje* 4468. Drawn by Rosemary Wise.

scattered scales; leaflets 25–28 on each side of the rachis, stiff, proximally at an angle of 90° with the leaflets on the opposite side of the rachis, more distal in one plane, regular, glossy dark green, proximal ones 19–41 × 0.3–2.5 cm (most proximal ones short and narrow), median 38–51 × 1.8–3.2 (interval 2.5–3 cm), distal 13–31 × 0.8–2.5 cm, the apical pair hardly connate, apices attenuate and bifid, main veins 3, with the midrib prominent adaxially, leaflets glabrous or with a few scattered scales near the base. **INFLORESCENCE** interfoliar, erect, branching to one order (in *Beentje* 4651 with the most proximal rachilla bifurcate); peduncle 47–132 cm, proximally 5–6 × 4 mm, distally 3–5 × 2–4 mm, with dense reddish laciniate scales but soon glabrescent; prophyll 10–34 cm long, 2-keeled, with a few scattered reddish scales, cylindrical, split at apex only, soon disintegrating at the apex; peduncular bract inserted at 7–14 cm from base, 30–48 cm long, pale brown with patches of dense reddish scales, circumscissile but carried up and covering the rachillae until expansion, beaked for 9–10 cm; rachis 7–15 cm, with 17–22 branches; rachillae green, occasionally with a bulbous base, 12–32 cm long, 2–2.5 mm diam., covered in minute stellate scales; triads spirally arranged, quite close, in pits. **STAMINATE FLOWERS** globose in bud; sepals 0.6–1 × 1–1.3 mm; petals 1.7–1.8 × 1.2–1.5 mm, ovate, acute; stamens 6, in 2 series, didymous, filaments 0.4–0.8 mm long, anthers 0.4–0.5 × 0.5–0.6 mm, rounded; pistillode 0.2–0.9 × 0.4–1 mm, hexagonal with central trifid apex. **PISTILLATE FLOWERS** with sepals c. 0.5 mm high, 1 mm wide, petals 1.4–1.8 × 1.2–1.3 mm; ovary 1–1.4 mm high, 1.1–1.2 mm diam., slightly asymmetrical; staminodes 0.3 mm high. **FRUIT** purple to black, obovoid, rounded, 13–19 × 9–15 mm; mesocarp 1.2 mm thick, rather fleshy; endocarp fibrous. **SEED** 15–16 × 9–12 mm, with endosperm ruminations 3–4 mm deep.

NOTE. Distinct by its small stature, its trunk which rarely branches and its short inflorescence branched to only one order; only very few inflorescences are present at the same time, which is unusual in the Vonitra group.

Dypsis pusilla. HB provides scale, Mananara Avaratra.

SPECIMENS SEEN. Maroantsetra: Antalavia, Feb. 1988 (y.fr.), *Dransfield et al.* JD6467 (K, TAN); idem, Feb. 1988 (y.fr.), *Dransfield et al.* JD6474 (K, P, TAN); Ambanizana, Nov. 1992 (bud, fr.), *Schatz et al.* 3394 (K). Mananara Avaratra: Antanambe, Oct. 1991 (bud), *Beentje* 4461 (K, MO, P, TAN); idem, Oct. 1991 (fr.), *Beentje* 4468 (BH, K, TAN); idem, April 1992 (fr.), *Beentje et al.* 4651 (K, TAN).

119. DYPSIS AQUATILIS

INSUFFICIENTLY KNOWN SPECIES

This species is known from a single incomplete specimen. The collector described it as growing beside a river and in the water. The inflorescence seems to have a long peduncle and is somewhat reminiscent of those of members of the "Vonitra" group. We are tempted to suggest that this is a stemless rheophyte and those with a mind to try to refind it should perhaps look for a palm with a habit reminiscent of *Chamaedorea cataractarum*.

DISTRIBUTION. Manantenina.
HABITAT. "Riverside and in the water" (Guillaumet).
USES. Not recorded.
CONSERVATION STATUS. Uncertain. Probably Endangered, since it seems to be restricted to a single area.

Dypsis aquatilis *Beentje* **sp. nov.**
Palma acaulescens verosimiliter rheophytica *D. fibrosae* affinis folio pinnata pinnis squamas dispersas ramentas ferentibus inflorescentia pedunculo longo terete in 2 ordines ramificanti distincta. Typus: Madagascar, Tolanaro: Manantenina, *Guillaumet* 4003a (Holotype P).

"Trunk very short, if not non-existant" (Guillaumet). **LEAVES:** sheath unknown; petiole part on type 25 cm long, proximally 2 cm diam., deeply channelled, rapidly tapering to the distal diam. 0.8 cm and slightly channelled; rachis c. 8 mm diam. in upper/mid-leaf; leaflets regular, close, the median 21–23.5 × 1.2–1.3 cm, the distal 10–18 × 0.4–1.1 cm, main vein 1, apices unequally acute, indument: adaxial midrib as well as margins with small dark ramenta, abaxial minor veins with dense but scattered minute scales. **INFLORESCENCE** branched to 2 orders; peduncle at least 21.5 cm and probably considerably longer, 5 × 3 mm diam., glabrous; bracts at least 3 major ones, opening near the apex only, the most distal tubular bract 5.5 cm long; non-tubular peduncular bract c. 1 cm long; rachis (one seen) 27.5 cm long, with 12 branches, the proximal 4 or 5 bifurcate; rachis bracts proximally c. 5 mm long; rachillae 17–24 cm long, 1.5 mm diam., glabrous, with spaced triads. **STAMINATE FLOWERS** with sepals 1.5–1.8 × 1–2 mm; petals 1.8–2.4 × 1.6–1.7 mm, acute to obtuse; stamens 6, didymous, uniseriate, the filaments 1–1.4 × 0.5–0.6 mm, flat and slightly triangular, anthers 0.4–0.5 mm high, 0.7–0.8 mm wide, dorsifixed; pistillode 1.7–2 × 0.8 mm. **PISTILLATE FLOWERS** unknown. **FRUIT** unknown.

NOTE. Distinct from all other species in the "*Vonitra*" group by the leaves with scattered scales and much ramenta, as well as by their small size.

SPECIMENS SEEN. Tolanaro: Manantenina, Dec. 1971 (fl.), *Guillaumet* 4003a (P).

Dypsis aquatilis. **A** petiole and basal leaflets × 2/5; **B** distal part of leaf × 2/5; **C** inflorescence × 2/5; **D** detail of rachilla × 2. All from *Guillaumet* 4003a. Drawn by Margaret Tebbs.

120. DYPSIS NODIFERA

This is one of the commonest and most widespread of all Madagascar palms, occurring from sea level up to nearly 1500 m. It is a very attractive single-stemmed species with a slender trunk and leaves with strongly grouped leaflets. In fact it appears uncannily like *D. pinnatifrons*, and, without staminate flowers and/or fruit it is impossible to tell them apart. The two species can also grow in the same habitat and area, to add to the confusion, but generally, the present species is the smaller and more slender of the two. Young seedlings, however, seem to be easily identified (see notes under *D. pinnatifrons*). The exposed forms of *D. nodifera* with very narrow leaflets can be strikingly beautiful. *D. nodifera* is undoubtedly in cultivation, where it will be impossible to distinguish from *D. pinnatifrons* until it flowers.

DISTRIBUTION. NW, E and SE Madagascar.
HABITAT. Moist forest, on steep or slight mid slopes, or littoral forest on white sand and then on the flat; alt. 5–1440 m.

Dypsis nodifera, growing in the Parc Botanique et Zoologique de Tsimbazaza, Antananarivo.

LOCAL NAMES. *Ovana* (Betsimisaraka); *Bedoda* (Betsimisaraka); *Sincaré* = probably *Tsinkiara*, a general name for small palms. *Tsirika, Tsingovatra* (Betsimisaraka; these are fairly general palm names).
USES. Hollowed out stems are used as blowpipes.
CONSERVATION STATUS. Not threatened. Widespread and common.

Dypsis nodifera *Mart.*
Hist. Nat. Palm. 3: 312 (1849). Type: Madagascar, E coast, *Bojer* s.n. (Holotype P).

SYNONYMS:
Phloga nodifera (Mart.) Salomon, Die Palmen: 109 (1887).
NOTE. The author of the combination is often cited as Noronha ex Salomon, even by Salomon himself. This is not correct: Salomon is the author of the combination, and Noronha had nothing to do with it.

Phloga polystachya Noronha ex Baker in J. Linn. Soc. 22: 527 (1887) **nomen invalid.**; Baill., Bull. Soc. Linn. Paris 148: 1175 & 150: 1185 (1895); Becc., Bot. Jahrb. Syst. 38, Beiblatt 87: 10 (1906); Palme del Madagascar 3, fig.1, t. 1 (1912); Jum. & H. Perrier, Ann. Inst. Bot.-Géol. Colon. Marseille sér. 3, 1 (1): 2, pl. 1, fig. 1 (1913); Jum., Cat. Pl. Madagascar, Palmae: 23 (1938); Jum. & H. Perrier, Fl. Madagascar 30: 124, fig. 33 (1945).
NOTE. The name *Phloga polystachya* was published as a *nomen nudum* by Noronha in Prodromus Phytologicus, in Du Petit-Thouars, Mélanges de Botanique et des Voyages (1811); there was no description of any kind, but Noronha gives its local name as "Ou Van"; the same local name for this species was given to HB in 1991. Martius, in 1849, cites *Phloga polystachya* Noronha MSS as a synonym of his own *Dypsis nodifera* Mart., and Moore (1963) therefore cited the name as *Phloga polystachya* Noronha ex Mart. Hooker (1887) cited the name *Dypsis nodifera* Mart. under the description of *Phloga*, but did not make the new combination.

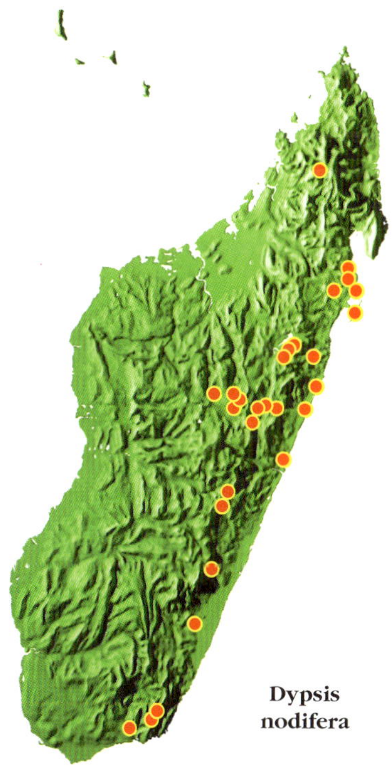

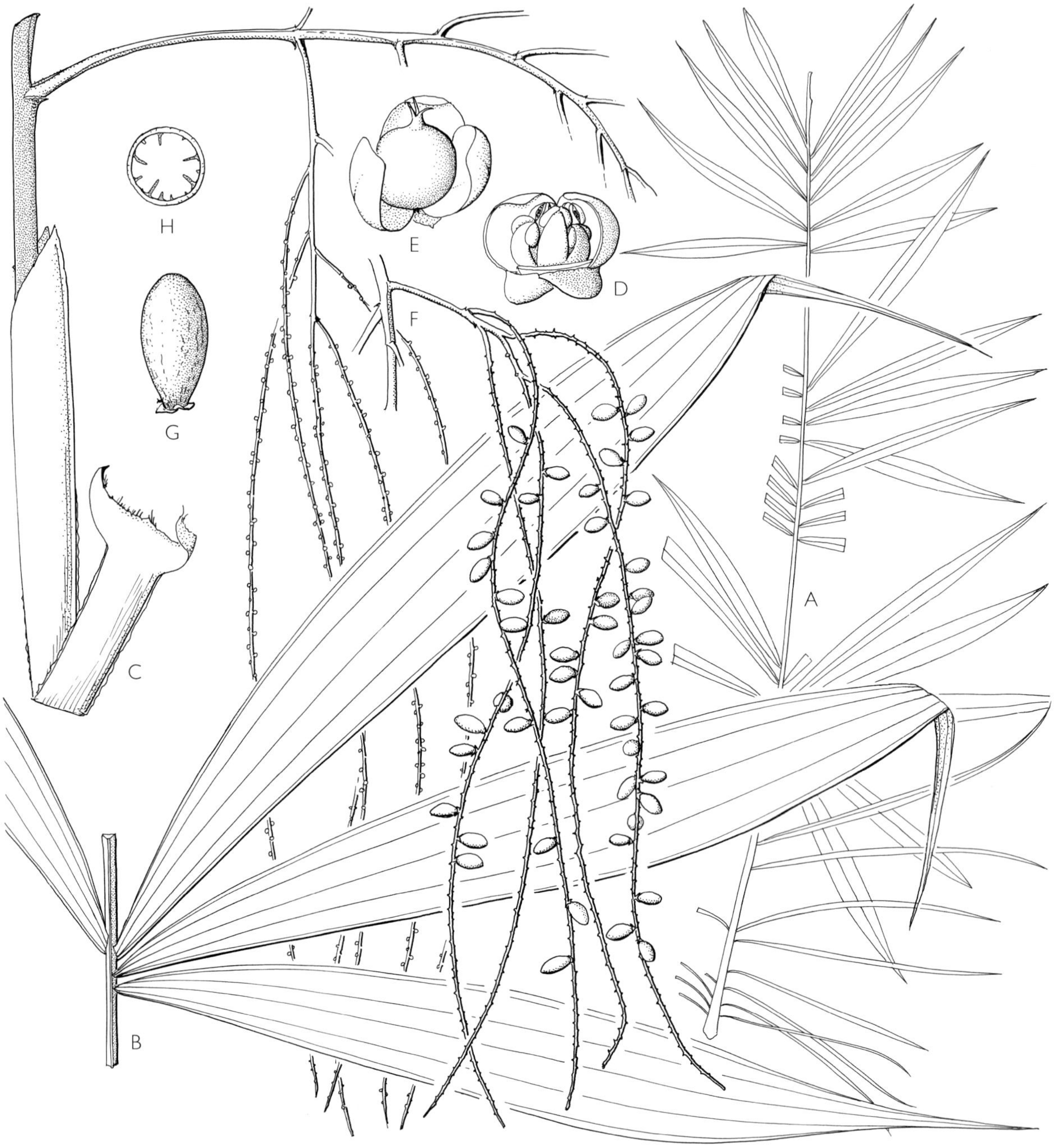

Dypsis nodifera. **A** whole leaf × 1/8; **B** mid section of leaf × 1/2; **C** basal part of inflorescence × 1/2; **D** staminate flower × 10; **E** pistillate flower × 10; **F** part of infructescence × 1/2; **G** fruit × 2; **H** fruit in cross section × 2. **A**, **F** – **H** from *Dransfield et al.* JD6776, **B** from *Beentje* 4497, **D** from *Beentje* 4446, **E** from *Beentje* 4441. Drawn by Rosemary Wise.

The first to make the combination under *Phloga* was Baker (1887); it is curious that Baker also described *Dypsis polystachya*, with a different type, two pages earlier, but did not compare the two. *Phloga polystachya* is an invalid name, as it is a synonym of *Dypsis nodifera*.

Dypsis polystachya Baker in J. Linn. Soc. 22: 525 (1887). Type: Madagascar, without locality, *Baron* 1287 (Holotype K).

Phloga polystachya Noronha ex Baker var. ***stenophylla*** Becc., Bot. Jahrb. Syst. 38, Beibl. 87: 11 (1906); Palme del Madagascar 5 (1912); Jum. & H. Perrier, Fl. Madagascar 30: 124 (1945). Syntypes: Central Madagascar, *Baron* 2880 (?P, lectotype), *Baron* 4560 (P–not seen there), **synon. nov.**

Solitary (but sometimes subcolonial) palm. **STEM** 2–10 m tall, 1.2–6 cm diam., 1–3 cm diam. near the apex, sometimes procumbent with only the distal part erect; stilt roots sometimes present; internodes 3.5–10 cm long, dull grey to distally red-brown; wood hard; crownshaft pale green with reddish flecks, 25–35 cm long. **LEAVES** 6–12 in the crown (to 17 and 3-ranked according to Dorr), erect to porrect, slightly arching; sheath 12–30 cm long, 1.5–1.6 cm across, closed, pale green or grey-green, sometimes tinged with pink, with scattered (distally denser) red-purple scales, without obvious ligules or (more often) with dark triangular auricles 0.5–5 cm high; petiole absent or up to 28 cm long, proximally 5–12 × 3–6 mm across, distally 3.5–10 × 2.5–4 mm across, yellow-green, adaxially channelled or flat, less often slightly convex, densely scaly; rachis 24–75 cm long, in mid-leaf 3–5 × 2–3.5 mm, densely scaly on both surfaces but adaxially glabrescent; leaflets (12–) 23–59 on each side of the rachis, in groups of 2–6, the groups at intervals of 5–15 cm, the leaflets within the groups 0.2–0.8 cm apart, twisted and fanned within the groups with their apices pendulous, within the groups the leaflets increasing in size from proximal to distal, often with swellings 2–4 mm across at the insertion point on the rachis, the distal often reflexed towards the leaf base, the proximal 14–35 x 0.1–1.8 cm, the median 12–37 x 1.2–4.5 cm, the distal 6.5–20.5 x 0.4–3.5 cm, proximal and median narrowly ovate, long-acuminate, dark green, with 1–7 main veins and minute scattered reddish scales, larger scales on the margins and occasionally some medium-sized (3–4 mm) ramenta in the proximal part, distal pair joined at the base for 0.3–3.3 cm, with 1–5 main veins and dentate narrow apices. **INFLORESCENCE** interfoliar to infrafoliar, 20–90 cm long, branched to 3 orders, less often to 2 or 4 orders, porrect to arching with pendulous rachillae, protandrous; peduncle (5 -) 13–32 cm long, proximally 10–15 x 5–8 mm across, distally 5–9 x 4–6 mm, densely scaly; prophyll 7–33 cm long, 0.9–3 cm wide, borne at 2–6 cm above the base of the peduncle, glabrous, opening near the apex only; peduncular bract inserted at 4–11 cm from the base of the peduncle, soon deciduous, 12–26 cm long, opening over its entire length except for the 0.5–2.5 cm long beak, adaxially rich red-brown, abaxially pale brown with some scattered scales; non-tubular peduncular bract inserted at 12–18 cm from the base of the peduncle, triangular, 0.2–1.8 cm long; rachis (7–) 13–30 cm long, densely scaly but glabrescent, with rachis bracts to 12 x 8 mm, with 8–14 branched and 6–10 unbranched first order branches, the proximal ones basally 5–12 x 3–5 mm across; rachillae (7–) 12–34 cm long, c. 1 mm across, puberulous but glabrescent, with spaced superficial triads; buds red, flowers white, pink or cherry-red. **STAMINATE FLOWERS** with sepals imbricate, 0.8–1.2 x 1–1.6 mm, proximally keeled and gibbous, concave, with ragged membranous margins; petals red, with fleshy apex, 0.8–1.6 x 1.7–2.3 mm, broadly ovate, rounded with small apiculus; stamens 6, didymous, biseriate, with the antepetalous stamens inserted higher up and more central than the antesepalous ones, filaments 0.5–1 x 0.5–0.7 mm and triangular, anthers 0.3–0.4 x 0.5–0.6 mm, the locules dorsifixed, proximally divergent, obtuse, not versatile; pistillode invisible or nearly so. **PISTILLATE FLOWERS** with sepals 1.2–1.5 x 1.5–1.9 mm, broadly elliptic; petals white, elliptical with broad membranous wings, distally fleshy, 1.4–2.2 x 2–2.5 mm; staminodes six, minute; gynoecium 1–1.5 x 1.3–1.5 mm, asymmetric, the style arms up to 0.6 mm long. **FRUIT** ellipsoid, green (always?), 8–10 x 5–8 mm, rounded at both ends; endocarp fibrous, with somewhat anastomosing fibres. **SEED** ellipsoid, c. 7.5 x 5.5 mm, with rounded ends; endocarp deeply (> 50 %) ruminate.

Dypsis nodifera. Detail of immature fruit (*Dransfield et al.* JD6436) (Photo: D.N. Cooke).

NOTE. Although the generic name was validly published by J.D. Hooker in Bentham & J.D. Hooker, Genera Plantarum 3: 877, 909 (1883), the only species mentioned under *Phloga* is *Dypsis nodifera* Mart., and under the Code, Article 33.1, Ex. 2, this does not qualify as a new combination.

The species bears an extraordinary resemblance to *Dypsis pinnatifrons*, but can be distinguished by the six stamens (not three) and the ruminate endosperm (not homogeneous, as in the *Dypsis*).

Plants out in the open often have much narrower leaves, and forms with extremely narrow leaves have been split off as var. *stenophylla*; we believe this is merely an adaptation to bright light, and there are intermediate forms; therefore, we have relegated this variety to synonymy.

The flowers were visited by flies in *Dransfield et al.* JD6436. *Overdorff* reports that leaves and fruit are eaten by *Lemur fulvus* and *L. rubriventer*.

Variability can be quite large locally: in *Beentje* collections at Betampona several trees standing close together were collected, and inflorescences within this population were branched to two or three orders; the prophyll varied in length from 12 to 33 cm. But *Beentje & Dransfield* 4808 (Mananara Avaratra: Antanambe, Oct. 1994, y.fr.) is so different that it is only tentatively included here. The rachis is slightly longer, but the leaflets are not as tightly grouped and certainly not ovate; they are also less fanned.

SPECIMENS SEEN. Ambanja: source of Sambirano R., (fl.), *Last* anno 1889–1890 (K). Mananara-Avaratra: Antanambe, 1989 (fr.), *Morat et al.* 8518 (P). Soanierana-Ivongo: Titinga (Tintingue), without date (fl.), *Bernier* 40 (P); Soanierana to Andasibe, Dec. 1938 (bud), *Lam & Meeuse* 5859 (L), 5860 (L). Sainte-Marie: Manambo heights, April 1851 (fl.), *Boivin* s.n. (P); between Tsarahac (?) and Sirambe, Nov. 1850 (bud), *Boivin* s.n. (P). Ambatondrazaka: Nonokambo, Aug. 1937 (fl.), *Herb. Jard. Bot. Tana* 2839 (P); Manaka E, April 1961 (fl.), *Rakotovao* RN 11850 (K, P). Toamasina: Betampona, Oct. 1991 (y.fr.), *Beentje* 4496 (K), 4497 (K); Tampina, Dec. 1938 (fl.), *Lam & Meeuse* 6045 (L). Ampasimanolotra: Ambila-Lemaitso, Feb. 1924 (fl.), *Perrier* 16040 (P); idem, May 1928 (fl.), *Decary* 6378 (P); 5–9 km S of Ambila-Lemaitso, Nov. 1986 (fl., y.fr.), *Dransfield et al.* JD6436 (K, P, TAN); idem, March 1988 (fl., fr.), *Henderson et al.* 759 (K); idem, June 1989 (bud, fr.), *Phillipson* 1860 (K, P, TAN); idem, Sept. 1991 (fl., fr.), *Beentje* 4441 (BH, K, MO, P, TAN), 4446 (BH, K, MO, P, TAN). Moramanga: S of Moramanga, Feb. 1930 (fr.), *Decary* 7143 (P); Mantady, March 1991 (fl.), *Beentje & Rabarilala* 4403 (BH, K, MO, P, TAN), 4405 (BH, K, TAN); idem, Oct. 1993 (fl.), *van Nek* 2000 (WAG), & (y.fr.) 1997 (WAG); Analamazaotra, Aug. 1862 (fl., y.fr.), *Meller* s.n. 21.8.1862 (K); idem, July 1938 (fl.), *Herb. Jard. Bot. Tana* 3778 (K, P); idem, Nov. 1986 (bud), *Dransfield et al.* JD6414 (K, TAN); Fanovana, May 1968 (bud), *Rakotozafy* 620 (P); Rahobevava, March 1951 (fl.), *Cours* 4296 (P). Ambohidratrimo: Ambohidratrimo forest, sine die (bud, y. fr.), *Le Myre de Vilers* s.n. (P). Manjakandriana: Ambatolaona, Ankeramadinika, June 1962 (bud), *Bosser* 16139 (K, P) and Jan. 1964 (fr.), *Bosser* 18983 (P, TAN); E of Ambatolaona, Nov. 1912 (fl., y.fr.), *Viguier & Humbert* 1250 (P); Andrangaloaka (Andrangalaoka), Nov. 1880 (y.fr.), *Hildebrandt* 3717 (K, P); idem, (y.fr.), *Parker* comm. Aug. 1881 (K); Mandraka, Feb. 1985 (fr.), *Dorr et al.* 3716 (K). Anosibe an Ala: Sandrangato–Anosibe, Nov. 1952 (y.fr.), *Capuron & Lerandri* 1681 (P). Marolambo: Andasibe, Onive R., Nov. 1914 (bud), *Perrier* 12076 (P). Mahanoro: 6 km S of Mahanoro, Oct. 1991 (bud, fr.), *Beentje* 4505 (BH, K, MO, P, TAN). Ambositra: Ambohimitombo, Dec. 1894 (bud), *Forsyth Major* 603 (K). Ifanadiana: Ranomafana N.P. near Ambatolahy, Dec. 1986 (fr.), *Nicoll* 207 (K, TAN); idem, Duke Primate Centre, Oct. 1987 (fr.), *Overdorff* 7 (K, P). Vondrozo: Vondrozo, (fr.), *Decary* 4854 (K, P, TAN). Amboasary Atsimo: Behara, Feb. 1967 (y.fr.), *Serv. For. Mad.* 26477 SF (K, P). Tolanaro: Andohahela, Col Tanatana, Dec. 1989 (fl., fr.), *Dransfield et al.* JD6776 (K, P, TAN); Andohahela, R. Itrotroky, Feb. 1993 (bud), *Malcomber et al.* 2120 (K, P).

Locality not found: Ambatovy, Feb. 1965 (y.fr.), *Peltier & Peltier* 5165 (P).
Central Madagascar, without precise locality: 1883 (bud), *Baron* 508 (K, P); Oct. 1882 (y.fr.), *Baron* 1287 (K, type of *Dypsis polystachya*); (ster.), *Baron* 2880 (P, type of var. *stenophylla*, not seen); (fl.), *Baron* 4340 (K, P); (bud, fr.), *Baron* 4465 (K, P); (bud), *Baron* s.n. (K). Madagascar, without any locality: ? 1887 (y.fr.), *Baron* 5072 (K); 1926/1927 (fr.), *Decary* 4854 (K, P, TAN).

121. Dypsis thermarum

This is clustering undergrowth palm that is abundant in the undergrowth in deep valleys of the Ranomafana National Park. It can easily be distinguished by its few narrow and long leaflets and the densely hairy short inflorescences. It forms rather untidy clumps at the foot of steep slopes. The species name is the Latin for "of the hot springs", a translation of the Malagasy *ranomafana*.

Distribution. Only known from Ranomafana National Park.
Habitat. Humid montane rain forest, on steep slopes; 800–1400 m.

Dypsis thermarum. Whole plant in undergrowth at Ranomafana.

Dypsis thermarum. Detail of inflorescences *(Dransfield & Beentje* JD7511).

LOCAL NAMES. *Fanikara* (Tanala).
USES. Stems used for making traps to catch crayfish.
CONSERVATION STATUS. Rare, probably not at present under threat, despite being geographically restricted.

Dypsis thermarum *J. Dransf.*, **sp. nov.**
inflorescentia gregem specierum floribus staminatis triandris staminibus antepetalis, staminodiis antesepalis alternantibus similis, sed floribus staminatis 6 staminibus biseriatis, antheris didymis differt, *D. angustae* similis sed inflorescentia rachillis numerosioribus longioribus differt, *D. anovensi* similis sed textura folii forma inflorescentiae diversa differt. Typus: Madagascar, Fianarantsoa, Ranomafana National Park, Nov. 1994 (fl.), *Dransfield & Beentje* JD7511 (Holotypus K; isotypi BH, MO, P, TAN).

Slender solitary or clustering undergrowth palm, forming small clumps of up to 8 stems. **STEMS** to 2 m tall, 4–9 mm diam., internodes 10–42 mm long, dark green with pale yellow-green vertical striping, glabrescent, or with very sparse pale brown scales. **LEAVES** c. 6 in crown; sheaths 6–8 cm long, 0.6–1 cm diam., sparsely brown scaly, whitish or pale yellow-green, forming a well-defined crownshaft, auricles present, rather narrow, 6–11 × 4–5 mm; petiole 4–14 cm long, 2 mm diam., sparsely scaly; rachis 8–14 cm; leaflets 2–5 on each side of the rachis, ± equal, mostly composed of a single fold, occasionally with up to 4 folds, sometimes slightly irregular, 15–35 × 0.4–2.5 cm, adaxially lamina with scattered pale brown punctiform scales, abaxially with abundant pale brown punctiform scales and scattered bands of caducous dark brown scales. **INFLORESCENCE** branching to 1, very rarely to 2 orders, much shorter than the leaves; peduncle 10–16 cm long, c. 2 mm wide; prophyll 9–12 × 0.5–0.7 cm, rather densely covered in reddish brown indumentum; peduncular bract similar to prophyll,

Dypsis thermarum. **A** stem base and suckers × 3/5; **B** crown × 3/5; **C** detail of rachilla × 4; **D** staminate flower, one petal removed × 20. **A** from *Beentje* 4420, **B – D** from *Dransfield et al.* JD7511. Drawn by Rosemary Wise.

exceeding it by 0.5–4 cm; rachis 5–10 cm, c. 2 mm diam., very densely covered in long laciniate dark red-brown hairs; rachillae 11–28, tending to diverge ± at right angles, short and rather stocky, 1.5–4 cm long, c. 1.5 mm diam., increasing slightly in diameter in fruit, somewhat angular, densely covered in red-brown laciniate hairs; rachilla bracts c. 2 mm apart, short with numerous marginal laciniate hairs. **Staminate flowers** rounded, c. 1.5 mm diam.; sepals rounded, irregularly keeled, 0.9 × 0.9 mm; petals triangular, striate, 1.2 × 1.4 mm; stamens 6, biseriate, antesepalous filaments 0.4 × 0.3 mm, antepetalous 0.5 × 0.3 mm, anthers didymous, 0.3 × 0.3 mm; pistillode absent. **Pistillate flowers** globular, c. 1.3 mm diam.; sepals 1 × 0.4 mm, imbricate, keeled; petals 1.4 × 1.3 mm, broadly imbricate at the base, striate; staminodes 3 or more, irregular, minute; ovary c. 1 mm diam., with slightly eccentric stigmas. **Fruit** orange at maturity, fusiform, 11 × 4 mm. **Seed** 7 × 3.5 mm; endosperm homogeneous, embryo lateral near the base.

Note. This species appears to be a common palm of the forest undergrowth on steep slopes in the National Park at Ranomafana. It resembles *D. angusta* but can be distinguished by the staminate flowers having six rather than three stamens. The leaflets are also of a different texture. *D. anovensis*, known only from its type, is superficially similar, but has an inflorescence of very different form. Collections made in November 1994 (*Dransfield & Beentje* JD7511) have flowers at staminate anthesis, and these flowers appear to have six fertile stamens. Despite this, the affinities of the palm remain with the species formerly included in *Trichodypsis*.

Specimens seen. Fianarantsoa: Ranomafana National Park, Nov. 1994 (fl.), *Dransfield & Beentje* JD7511 (Holotype K; isotypes BH, MO, P, TAN); idem, July 1992 (fl.), *Beentje & Beentje* 4734 (BH, K, MO, P, TAN); idem, Dec. 1986, *Marion Nicoll* 203 (K, MO, P, TAN); idem, March 1992, *Malcomber & Schatz* 1309 (K, MO, P, TAN); idem, April 1993, *Malcomber et al.* 2446 (K, MO, TAN); idem, March 1991 (fl., fr.), *Beentje* 4420 (K, BH, MO, P, TAN).

122. Dypsis angusta

This is a small palm of the undergrowth of low crown forest occurring on the isolated peak of Vatovavy, at the edge of the coastal plain south of Mananjary. We refound it in November 1994. The Latin name means narrow, presumably in reference to the narrow leaflets.

Distribution. East Coast, only known from the Ifanadiana area and Farafangana.
Habitat. Rain forest; 45–500 m.
Local names. Not recorded.
Uses. Not recorded.
Conservation Status. Endangered. Only known from three sites, in which population numbers are low. None of these sites is protected.

Dypsis angusta *Jum.*
Ann. Inst. Bot.-Géol. Colon. Marseille sér. 3, 6 (1): 34 (1918); Bull. Acad. Malgache 6: 12 (1923); Cat. Pl. Madagascar, Palmae: 12 (1938); Jum. & H. Perrier, Fl. Madagascar 30: 43 (1945). Type: Mananjary, Mt Vatovavy, Aug. 1915, *Perrier* 11976 (holotype P).

Slender clustering undergrowth palm. **Stems** to 2 m or more tall, 5–10 mm diam., internodes to 9 cm long near the stem base, decreasing to 0.8–1.2 cm long near the crown, basally bearing abundant chocolate-brown caducous scales, glabrescent distally; crown with c. 4–6 leaves; leaf-sheaths 5–7 cm long, forming a crownshaft, striate, covered with caducous chocolate-brown scales, auricles scarcely developed or to 10 × 2 mm, membranous. **Leaf** to 43 cm long including petiole; petiole 3–12 cm long, 1.5–3 mm wide, ± triangular in cross-section, bearing sparse caducous pale and chocolate-brown scales; rachis 10–29 cm long, bearing 2–3 or 7–9 (see notes below) narrow, ± approximate, regularly arranged leaflets on each side; leaflets 9–30 × 0.7–2.1 cm, drying dark or pale

Dypsis angusta. Inflorescence, Vatovavy (*Dransfield & Beentje* JD7514)

greenish brown, composed of 1–4 folds, acuminate, adaxially with sparse minute punctiform scales, abaxially minutely roughened and with scattered punctiform scales. **INFLORESCENCE** interfoliar, shorter than the leaves, sparsely branched to 1 (–2) orders, basal 1 or 2 first order branches sometimes branched; peduncle 7–18 cm long, c. 2.5 mm diam., densely covered in caducous, dark brown laciniate scales; prophyll 12–16 × 0.7–1cm, bearing scattered pale or dark brown scales; peduncular bract 8–10 × 0.7–1 cm, similar to prophyll; rachis 4–9 cm long, c. 2.5 mm diam., densely scaly as the peduncle; rachillae 2–20, rather short and stocky, 2–7 cm long, c. 1.5–2.5 mm diam. bearing rather dense chocolate-brown scales; triads c. 1 mm distant in newly emerged inflorescence, eventually c. 3 mm distant. **STAMINATE FLOWERS** c. 1 mm diam.; sepals broad, rounded, imbricate, somewhat striate, c. 0.7 × 0.7 mm; petals rounded triangular, 1 × 0.8 mm; stamens 3, antepetalous, filaments 0.8 × 0.3 mm, anthers 0.4 × 0.2 mm; staminodes tooth-like, 0.3 × 0.2 mm; pistillode very small. **PISTILLATE FLOWER** bud immature, c. 1 × 1 mm; sepals rounded, imbricate, c. 1 × 1 mm; petals triangular, c. 1 × 0.6 mm; staminodes 3, minute; immature ovary c. 0.4 mm diam. Mature **FRUIT** unknown.

NOTE. The type, *Perrier* 11976, from Vatovavy, has leaves with two to three leaflets on each side of the rachis. A recent collection from Vatovavy (*Dransfield et al.* JD7514) has an inflorescence larger than that of the type and with up to 20 rachillae. Collections made at Manombo (*Beentje* 4511) have eight to nine pairs of leaflets. We are including the latter collection in *D. angusta* because of the manifest similarities in inflorescence structure and leaf texture and colour. All four collections cited are from the central part of the east coast, yet they are rather heterogeneous, and we are not completely certain that they all represent one species. Until more material is collected this will remain a poorly known species.

SPECIMENS SEEN. Mananjary: Mt. Vatovavy, Aug. 1911 (fl.buds), *Perrier* 11976 (holotype P); idem, Nov. 1994 (fl.), *Dransfield et al.* JD7514 (K, TAN). Farafangana: Manombo Reserve Special, Nov. 1991 (dead infl.), *Beentje* 4511 (K, TAN). Fianarantsoa: Amfamahirana, Nov. 1963 (fl.), *Bosser* 18300 (K, P). (Plate: see next page).

123. DYPSIS ANOVENSIS

This very poorly known species has been collected only once. Until it is recollected its relationships will remain obscure. The species name reflects its only known locality.

DISTRIBUTION. Only known from the Anove River near Soanierana-Ivongo.
HABITAT. Rain forest at low elevation.
LOCAL NAMES. Not recorded.
USES. Not recorded.
CONSERVATION STATUS. Presumed extinct. Not seen for more than eighty years. Most of the forest has disappeared from the Anove River mouth.

Dypsis anovensis *J. Dransf.* **nom. nov.**

SYNONYM:

Dypsis linearis Jum., Ann. Inst. Bot.-Géol. Colon. Marseille sér. 3. 6: 35 (1918); Bull. Acad. Malgache 6: 14 (1923); Cat. Pl. Madagascar, Palmae: 13 (1938); Jum. & H. Perrier, Fl. Madagascar 30: 44 (1945). Type: Madagascar, R. Anove, *Perrier* 12066 (Holotype P) (non *Neophloga linearis* Becc., 1906).

Slender forest undergrowth palm to 2 m tall. **STEM** c. 7 mm diam. (not exceeding 3 cm, fide Jumelle). **LEAVES** c. 6 in crown (7–8, fide Jumelle); sheaths 6.5 cm long, c. 7 mm diam., covered in caducous chocolate-coloured scales; petiole 8–15 cm long, c. 2.5 mm wide near base, covered in a thick layer of caducous pale brown scales; rachis 9–10 cm; leaflets 3–4 on each side of the rachis, ± regularly arranged, drying pale green, up to 2.5 cm distant, the longest to 25 cm long, c. 4–12 mm wide, mostly composed of a single fold, both surfaces with scattered punctiform scales, abaxially with broad bands of caducous brown scales. **INFLORESCENCE** interfoliar, in bud much shorter than the leaves, but one dead inflorescence axis suggesting eventually as long as leaves (to 45 cm fide Jumelle), branching to 1 order only; peduncle 21 cm; prophyll to 8 × 0.4 cm, membranous; peduncular bract exceeding prophyll by 9.5 cm, otherwise similar (exceeding prophyll by 23 cm fide Jumelle); rachis c. 5 cm (8–12 cm, fide Jumelle) with rather dense brown scales; rachillae probably c. 14, c. 2 cm long, c. 0.5 mm diam., rather sparsely scaly, triads c. 0.5 mm apart in immature inflorescence, rachilla bract entire, inconspicuous. **STAMINATE FLOWERS** in very young bud, all parts at an early state of development; stamens 3 antepetalous.

NOTE. The need for the new name, *Dypsis anovensis*, is a consequence of the inclusion of *Neophloga* in *Dypsis*.

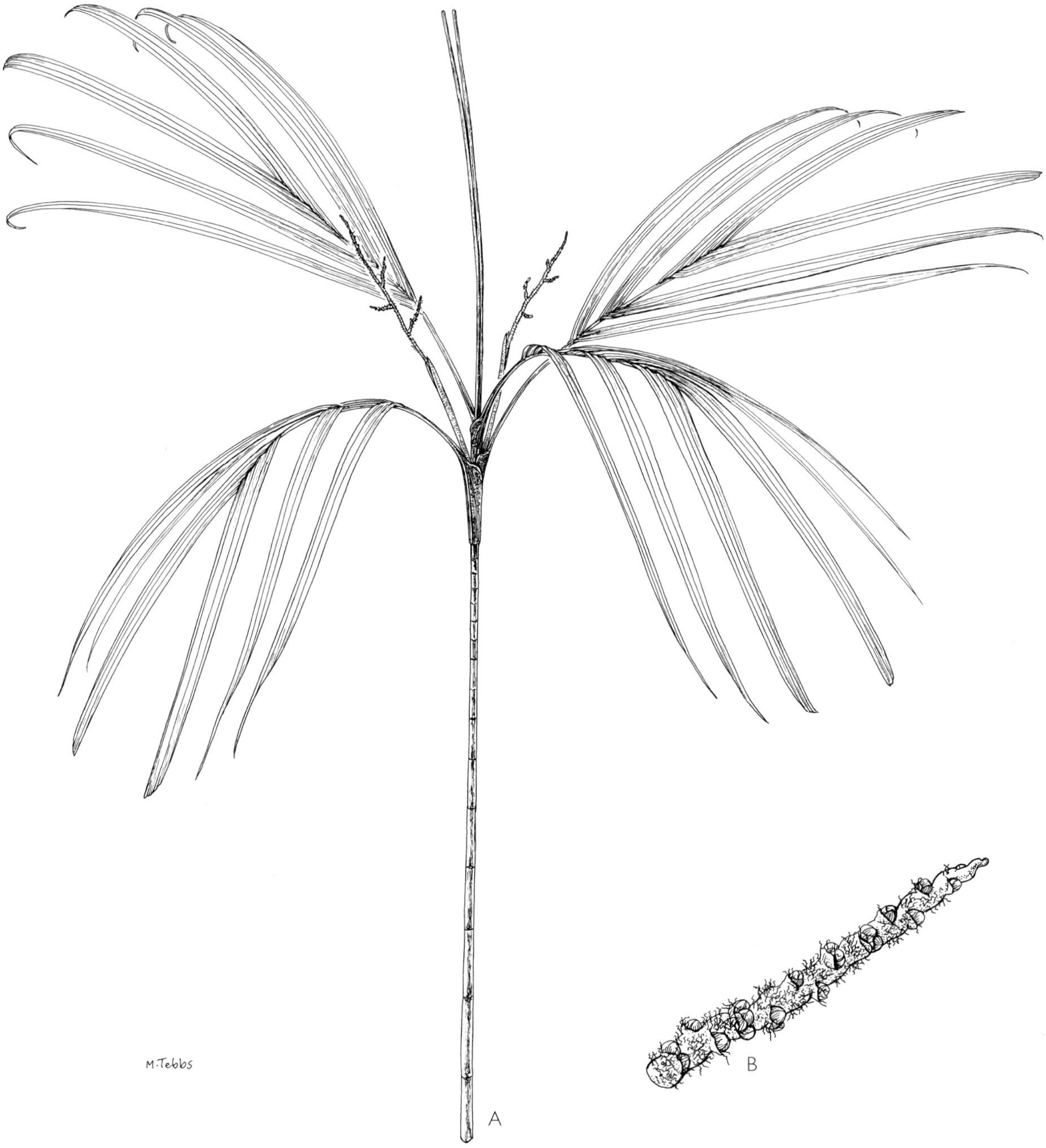

Dypsis angusta. **A** whole plant × 1/3; **B** detail of rachilla × 3. All from *Perrier* 11976. Drawn by Margaret Tebbs.

Dypsis anovensis is known only from its type. The combination of slender, linear leaflets and small sparsely branched inflorescence is paralleled by *D. cookei* from high elevation forest on Marojejy; however, the leaflet texture of the latter is different and there are consistently more leaflets on each side of the rachis, that dry dark rather than pale green. Some of the dimensions mentioned by Jumelle suggest that he had more mature inflorescences at hand than the one preserved on the type sheet in P.

SPECIMEN SEEN. Soanierana-Ivongo: near to River Anove, Sept. 1912 (buds), *Perrier* 12066 (holotype P).

124. DYPSIS AMBILAENSIS

This is a small palm of the undergrowth of white sand forest developed on raised beaches behind the East Coast. It appears superfically to be almost identical to *D. forficifolia*. Unfortunately, there are no easy ways to identify the two species, without dissecting the flowers and observing the stamens. However, we have noted that the two species appear to be geographically separated. Ambila-Lemaitso, the type locality, is the root of the species name.

DISTRIBUTION. East Coast, Tampina to Ambila-

Dypsis anovensis. **A** crown × 2/5; **B** detail of rachilla × 3. All from *Perrier* 12066. Drawn by Margaret Tebbs.

Lemaitso, south of Toamasina.

HABITAT. Coastal forest on white sands at low elevation.

LOCAL NAMES. Not recorded.

USES. Not recorded.

CONSERVATION STATUS. Endangered. Known only from the coastal white sand forests south of Toamasina, forests that are of very limited extent and much prone to damage by fire, as well as to development.

Dypsis ambilaensis *J. Dransf.* **sp. nov.**
habitu foliisque *D. forficifoliae* superficialiter similis sed floribus staminatis triandris staminibus antepetalis staminodiis antesepalis alternantibus differt. Typus: Madagascar, Ampasimanolotra, Ambila-Lemaitso, *Dransfield* JD6496 (Holotypus K; isotypus TAN)

Slender, clustering undergrowth palmlet, 1–3 m tall. **STEMS** c. 5–8 mm diam., internodes 7–18 mm long, green when young, densely covered with caducous dark brown to black scales, old internodes ± blackish. **LEAVES** c. 7–8 in crown; leaf-sheaths 6–8 × 0.9–1.5 cm, striate, sparsely covered with caducous black scales near base, densely covered distally, leaf sheath mouth with 2 triangular, membranous auricles 3–5 × 3–4 mm; leaf c. 25–40 cm long; petiole absent or to 7 × 0.2–0.4 cm; rachis 15–22 cm long, abaxially covered with blackish scales; blade entire bifid, to 22 × 11 cm, the apical lobes to 13 × 5.5 cm, or divided into 2–4 pairs of narrow to broad, ± sigmoid leaflets, the narrowest at the base, the mid-leaf leaflets usually the broadest, leaflets ranging from 7–20 × 0.5–4 cm, the apical pair (or the two lobes in the

Dypsis ambilaensis, a palm of coastal forest on white sand, Ambila-Lemaitso (*Dransfield et al.* JD6444).

Dypsis ambilaensis. **A** crown with inflorescence × 1/2; **B** staminate flower showing antepetalous stamens × 10. All from *Dransfield et al.* JD6496. Drawn by Rosemary Wise.

entire, bifid leaf) shallowly lobed distally, adaxial surface glabrous, abaxial surface with bands of caducous brown scales, numerous punctiform scales and sometimes with long narrow ramenta along main ribs. **Inflorescence** interfoliar, only slightly exceeding leaves, branching to 2 orders; prophyll 12–25 × 0.4–0.5 cm, bearing sparse to dense laciniate red-brown scales or glabrescent; peduncle 17–25 × 0.2 cm bearing scattered laciniate dark red-brown scales; peduncular bract exceeding the prophyll by c. 2–6 cm, otherwise similar; rachis 9–26 cm, sparsely to rather densely covered in red-brown laciniate scales; first order branches 10–27, at least about a half of them branching to the second order; rachillae 16–c. 55, slender, 3–5.5 cm, c. 0.7 mm diam., glabrous, bearing triads c. 2–3 mm distant, each subtended by a low rounded rachilla bract, the bract usually with laciniate red-brown hairs to 1 mm long along margin. **Staminate flowers** at anthesis c. 1 mm diam., red, with slight fecal smell; sepals unequal, broad, imbricate, 0.5 × 0.4–0.7 mm; petals rounded triangular, striate, 0.8 × 0.6 mm; stamens 3, antepetalous, filaments connate at the very base, free portion ± triangular, 0.3 × 0.3 mm, anthers didymous, 0.3 × 0.1 mm, staminodes minute, c. 0.1 × 0.1 mm; pistillode a low dome. **Pistillate flowers** globular, c. 1.1 mm diam.; sepals irregularly rounded, imbricate c. 0.5 × 0.5 mm; petals broadly triangular c. 1 × 1 mm, valvate, imbricate only at the base, striate; staminodes 5–6, irregular, c. 0.15 × 0.1 mm; ovary rounded, c. 1 mm diam., stigmas 3, thread-like, c. 0.15–0.2 mm. Immature **fruit** curved; mature fruit somewhat curved, red, 10–15 × 4.5–5 mm, striate when dry. **Seed** 8 × 3.5 mm; endosperm homogeneous, embryo lateral, c. 2 mm above base.

Note. This species is superficially very similar to *D. forficifolia* and can only be separated with certainty if staminate flowers are available. In *D. ambilaensis* there are three antepetalous stamens and three staminodes (i.e., this species belongs to the previously recognised *Dypsis* § *Trichodypsis*) while in *D. forficifolia* the stamens are antesepalous and there are no staminodes (i.e., this species belongs to *Dypsis* § *Dypsis*). So far, *D. ambilaensis* has been found only in coastal forest on white sands near Ambila-Lemaitso and Tampina, south of Toamasina. It is particularly confusing that *D. forficifolia* can grow in an apparently identical forest type along the coast further to the north, near Mananara.

A few of the staminate flowers on the specimen *Dransfield et al.* JD6444 appear to have six fertile stamens (see also *Dypsis thermarum*).

SPECIMENS SEEN. Toamasina: Tampina, Forest Reserve near railway, Dec. 1938 (fl.), *Lam & Meeuse* 6040 (L). Ampasimanolotra: Ambila-Lemaitso, forest west of lagoon on white sand, March 1988 (fl.), *Dransfield* JD6496 (Holotype K; isotype TAN); Nov. 1986 (fl.), *Dransfield* JD6444 (K, TAN); idem, Feb. 1924 (fl., fr.), *Perrier* 16039 (P); idem, May 1928 (fl., fr.), *Decary* 6480 (P).

125. DYPSIS GLABRESCENS

A small solitary or clustering palm of the undergrowth of lowland forest in the north-east of the island. The species epithet, Latin for becoming glabrous (lacking hairs) refers to the branches of the inflorescence. In some populations, the rachillae are quite hairy when they are newly expanded.

DISTRIBUTION. Madagascar, only known from Île Sainte Marie, Mananara Avaratra and Betampona.
HABITAT. Rain forest; tending to occur in valley bottoms. 50–600 m.
LOCAL NAMES. Not recorded.
USES. Not recorded.
CONSERVATION STATUS. Endangered; despite being known from three localities, all populations are small. Only in Betampona does the palm occur within a reserve.

Dypsis glabrescens *(Becc.) Becc.*
Palme del Madagascar 16 (1913); Jum., Ann. Inst. Bot.-Géol. Colon. Marseille sér. 3, 6 (1): 30 (1918); Bull. Acad. Malgache 6: 11 (1923); Cat. Pl. Madagascar, Palmae: 12 (1938); Jum. & H. Perrier, Fl. Madagascar 30: 43 (1945).

Dypsis glabrescens, growing in the Mananara Avaratra Biosphere Reserve.

Synonym:
Trichodypsis glabrescens Becc., Bot. Jahrb. Syst., Beibl. 38: 15 (1906). Type: Madagascar, Île Sainte Marie, Tafondrou, *Boivin* 1709 (holotype P).

Slender solitary or clustering undergrowth palmlet to 3 m tall. **Stem** 7–10 mm diam., internode to 9–30 mm, glabrous. **Leaves** 5–6 in crown; sheath 6–8 cm long, 8–11 mm diam., striate, drying reddish brown, with scattered to dense caducous scales, auricles membranous, 12–16 × 2–3 mm; petiole 1–12 cm, 2–2.5 mm wide near base; rachis 9–19 cm; blade entire bifid, to 33 cm long, split to almost two thirds the length or with 2–4 leaflets on each side of the rachis, diverging at an acute angle, rather close, the longest at the base to 26 × 1.5–2.8 cm, apical leaflets to 20 × 1.5 cm, apical leaflet tips somewhat contracted, blunt, only very shallowly lobed, adaxial surface with sparse minute brown punctiform scales, abaxially drying pale green, with abundant minute brown punctiform scales and bands of caducous brown hairs. **Inflorescence** interfoliar, shorter than the leaves, branching to 1 order; peduncle 18–23 cm long, c. 2 mm diam., with scattered dark brown trichomes; prophyll 6–11 × 0.4 cm, dull brown, with scattered caducous scales; peduncular bract similar, but exceeding the prophyll by 7–9 cm; rachis 6–11 cm, sparsely to densely covered with caducous brown trichomes, eventually glabrescent; rachillae c. 14–25, diverging at a ± acute or right angle, 4–7 cm long, with scattered dark brown trichomes or almost glabrous, but covered in numerous minute papillae or smooth (*Beentje* 4488), triads c. 1–3 mm distant, rachilla bracts inconspicuous, entire. **Staminate flowers** c. 1.1 mm diam.; sepals c. 0.7 × 0.8 mm, rounded, broad imbricate, erose at margin, keeled; petals at anthesis very fleshy, striate, c. 1.5 × 1.6 mm, basally very briefly joined, the lobes gaping slightly; stamens 3, antepetalous, alternating with 3 antesepalous staminodes, anthers didymous, c. 0.2 × 0.1 mm, pistillode conical, minute. **Pistillate flower** bud rounded, c. 1 mm diam.; sepals erose margined, broadly imbricate, c. 0.8 ×

Dypsis glabrescens. **A** crown, showing inflorescence with galled staminate flowers × 2/5; **B** staminate flower × 20. All from *Beentje & Dransfield* 4631. Drawn by Rosemary Wise.

0.9 mm; petals valvate apically, strongly imbricate at base, fleshy, 1.2 × 1 mm; staminodes 6 (fide Beccari); ovary c. 0.8 mm diam. **FRUIT** cherry-red at maturity, ovoid, 13–15 × 10–11 mm. **SEED** 10 × 6 mm, endosperm homogenous.

NOTE. This species resembles *D. mocquerysiana* in its small stature and entire-bifid or sparsely divided leaf and the position of the three stamens. It differs, however, in the inflorescence that has a much shorter rachis bearing many fewer rather uniform and relatively long rachillae that generally diverge at an acute angle; the rachillae are usually minutely papillose. *Perrier* 12037, cited in Flora of Madagascar under *D. glabrescens*, has antesepalous stamens, so cannot belong to *D. glabrescens*. It is included under *D. forficifolia*. A palm from Betampona, represented by *Beentje* 4488 matches *D. glabrescens* in all its parts except for lacking the minute papillosity of the rachillae of *D. glabrescens*. At first we thought that the specimens from Mananara Avaratra represented an undescribed taxon, but then a new collection from the type locality of *D. glabrescens* has illustrated a wider range of variation into which the Mananara collections fit. *Razanaparany* RN10018 (K, P) from nearby Ambodiriana, has similar rachillae to *Beentje* 4488, but has a robust leaf with many scales and of quite different texture and form and with a very much larger inflorescence. We have not been able to match this incomplete collection.

SPECIMENS SEEN. Mananara Avaratra: 10 km west of Antanambe, April 1992 (fl.), *Beentje & Dransfield* 4631 (BH, K, MO, P, TAN); idem, Oct. 1994 (fl., fr.), *Dransfield & Beentje* JD7501 (K, TAN). Île Sainte Marie: Tafondrou, July 1849 (fl.), *Boivin* 1709 (Holotype P; isotype fragment FI); Forêt de Kalalao, Nov. 1994 (dead infl.), *Dransfield et al.* JD7530 (K, TAN). Toamasina: Betampona, Oct. 1991 (fl.), *Beentje* 4488 (BH, K, MO, P, TAN).

126. DYPSIS LAEVIS

This species, known only from its type, is distinctive in its epetiolate leaves with minute basal leaflets and long lax inflorescence with glabrous axes. The species name is Latin for smooth, referring to the lack of hairs on the inflorescence branches.

DISTRIBUTION. Central east coastal Madagascar, known only from Manombo near Farafangana.
HABITAT. Lowland forest.
LOCAL NAMES. Not recorded.
USES. Not recorded.
CONSERVATION STATUS. Critical. The small protected area where this species occurs is severely threatened with illegal logging and shifting cultivation.

Dypsis laevis *J. Dransf.* **sp. nov.**
inter species floribus staminatis triandris staminibus antepetalis, staminodiis antesepalis alternantibus *D. ambilaensi* affinis, sed forma folii et inflorescentia longa laxa glabra differt. Typus: Madagascar, East Coast, Farafangana, Manombo, *Beentje & Andriampaniry* 4785 (Holotypus K; isotypus TAN).

Slender, solitary forest undergrowth palm. **STEM** to 3 m tall, c. 15 mm diam., internodes 15 - 30 mm, green with scattered dark red-brown to blackish scales. **LEAVES** 9 in crown, porrect; sheath 8–9 cm long, c. 13 cm diam., pale green with scattered red-brown scales, auricles membranous, rounded-triangular, c. 8 × 8 mm, disintegrating; petiole absent; rachis 26–38 cm, c. 4 mm wide at the very base, with scattered caducous dark scales; leaflets 5–7 on each side of the rachis, the proximal 1–4 very small and slender, 1–9 × 0.1–1.0 cm, mid-leaf leaflets broad, somewhat irregular in arrangement, 9–20 × 1.8–7 cm, somewhat sigmoid, acuminate in a long drip-tip, apical pair to 12 × 4 cm, joined for 5–7 cm along

rachis, conspicuously lobed along the margins, adaxial surface glabrous, abaxial surface with minute scattered red-brown punctiform scales. **Inflorescence** longer than the leaves, branched to 2 orders, all axes reddish brown, glabrous; peduncle 37–46 cm long, c. 2–3 mm diam.; prophyll borne c. 2 cm above base of peduncle, 22 cm long, c. 1 cm wide, very sparsely and minutely punctate scaly; peduncular bract borne c. 9 cm above base of peduncle, to 26 cm long, similar to prophyll, exceeding it by 14 cm; rachis 30 cm; branches rather lax and distant; rachillae c. 45 in number, 6–7.5 cm long, 0.7–0.8 mm diam., triads c. 1 mm apart, rachilla bracts inconspicuous. **Staminate flower** still immature, c. 0.7 mm diam.; sepals 0.3 × 0.3 mm, rounded, smooth, keeled, margins erose; petals striate, c. 0.6 × 0.5 mm; stamens still very young, 3, antepetalous, alternating with 3 antesepalous staminodes; pistillode minute. **Pistillate flower** c. 1.2 mm diam.; sepals broadly imbricate, 5 × 7 mm, margins erose; petals striate, 1 × 0.8 mm, basally imbricate; staminodes 6, strap-shaped, minute; ovary gibbous c. 0.8 mm diam., stigmas ± lateral. Immature **fruit** curved, 6 × 3 mm. **Seed** 5 × 1.5 mm; endosperm homogeneous, embryo basal.

Note. The position of the three stamens allies this species with those species previously included in *Dypsis* § *Trichodypsis*. It perhaps most resembles *D. ambilaensis* but differs in the leaf form and in the long, lax, glabrous inflorescences.

Specimen seen. Farafangana: Manombo Forest, Jan. 1993 (fl., fr.), *Beentje & Andriampaniry* 4785 (Holotype K; isotype TAN).

Dypsis laevis. **A** crown × 1/4; **B** detail of rachilla × 4; **C** open staminate flower × 30; **D** fruit × 3. All from *Beentje & Andriampaniry* 4785. Drawn by Margaret Tebbs.

127. DYPSIS VIRIDIS

This is a most attractive slender clustering palm. Superficially it resembles a very slender form of *Dypsis forficifolia*, but the arrangement of the stamens in the present species is quite different. However, it can easily be distinguished, even when not in flower, by the numerous very slender stems in the clump, that are usually palest green, each internode marked with a vertical darker green stripe and by the irregularly divided leaflets of a thin dull green. We assume that the species name *viridis* (Latin - green) refers to the pale green colour that the plants assume on drying in the herbarium.

DISTRIBUTION. NE Madagascar from the Bay of Antongil to Soanierana-Ivongo.
HABITAT. Lowland and hill rain forest on ridges; up to 400m.
LOCAL NAMES. Not recorded.
USES. Not recorded.
CONSERVATION STATUS. Vulnerable. At present the palm is known from very few localities and only at Antanambe is it within a protected area.

Dypsis viridis, an elegant clump-forming palm of the forest undergrowth, Mananara Avaratra.

Dypsis viridis *Jum.*
Ann. Inst. Bot.-Géol. Colon. Marseille sér. 3, 6 (1): 35 (1918); Bull. Ac. Malgache 6: 13 (1923); Cat. Pl. Madagascar, Palmae: 14 (1938); Jum. & H. Perrier, Fl. Madagascar 30: 42 (1945). Type: Madagascar, Bay of Antongil, *Perrier* 12057 (holotype P).

Slender clustering forest undergrowth palmlet. **STEMS** to 1.5 m tall, 3–8 mm diam., internodes 12–25 mm., very pale ivory-coloured but often with a vertical green stripe. **LEAVES** c. 6–7 in crown; sheath 5–6 cm long, 4–8 mm diam., pale ivory-coloured, very sparsely scaly, auricles not developed; petiole 3–7 cm long, c. 2 mm wide near the base, very sparsely scaly; rachis c. 11 cm long; leaflets drying pale green, 2–7 on each side of the rachis, rather irregular, small, 7–20 × 1–3.5 cm, abaxially with scattered bands of caducous scales, otherwise glabrescent. **INFLORESCENCE** branching to 1 (2) orders; peduncle 12–41 cm long, glabrescent, usually shorter than the leaves but occasionally longer; prophyll 8–24 × 0.2–0.3 cm, very sparsely scaly; peduncular bract exceeding the prophyll by 4–12 cm, otherwise similar; rachis 5–8 cm, sparsely caducous scaly; rachillae 7–11, spreading, 1.2–6 cm long, c. 0.5 mm diam., somewhat angular, minutely brown papillose, ± glabrous; triads c. 1–1.5 mm distant, rachilla bracts inconspicuous. **STAMINATE FLOWERS** c. 1.1 × 1 mm; sepals 0.5 × 0.8 mm, broad, rounded, imbricate, somewhat erose at margin, keeled; petals striate, 0.9 × 0.8 mm; stamens 3, antepetalous, alternating with 3 low triangular antesepalous staminodes, the filaments connate in basal 0.3 mm, anthers didymous 0.2 × 0.2; pistillode conical, minute. **PISTILLATE FLOWERS** known only in immature bud, c. 0.8 mm diam.; sepals broadly imbricate c. 0.5 × 0.5 mm; petals striate, c. 0.6 × 0.6 mm; other parts very immature. Ripe **FRUIT** cherry-red, ellipsoid, 10 × 5 mm. **SEED** 7 × 5 mm, endosperm homogeneous.

NOTE. This is one of the smallest and most slender of the three-staminate species of *Dypsis*. It is a very dainty palm. Although *Perrier* 12031 is labelled as the type in Paris, Jumelle & Perrier (1945) explicitly cited *Perrier* 12057 as type. Both were cited as syn-

Dypsis viridis. **A** crown × 3/5; **B** detail of rachilla × 6; **C** staminate flower, one petal removed × 30. All from *Beentje et al.* 4621. Drawn by Rosemary Wise.

types in the protologue. Two further collections may be referable to this species. *Perrier* 17466, from Ambodiriana, collected in November 1925 (P) is vegetatively very similar to the type of *D. viridis* but has an entire bifid leaf. It was identified by Jumelle as *D. hirtula* and used as the basis of the illustration of this species in Jumelle & Perrier (1945). It does not match the type of *D. hirtula*. Furthermore there are no staminate flowers to allow closer identification. *D. hirtula* has three antesepalous stamens while in *D. viridis* the three stamens are antepetalous. *Cours* 2513 from forest south of Mangabe, collected in January 1950 (K, P) is also similar to *D. viridis* in the position of the three stamens and in vegetative characters. The inflorescence has, however, more numerous and finer rachillae. It is tentatively determined as *D. viridis*.

Specimens seen. Maroantsetra: Bay of Antongil, Aug. 1912, *Perrier* 12057 (holotype P); Fananehana, Aug. 1912, *Perrier* 12031 (P). Mananara Avaratra: 10km west of Antanambe, April 1992 (fl.), *Beentje et al.* 4621 (BH, K, MO, P, TAN), Oct. 1994 (fl., fr.), *Dransfield & Beentje* JD7505 (BH, K, TAN). Soanierana-Ivongo: Soanierana to Antasibe, Dec. 1938 (fl.), *Lam & Meeuse* 5857 (L).

128. DYPSIS HILDEBRANDTII

Despite having been collected by almost every botanist to visit the forest at Analamazaotra, *D. hildebrandtii* remains the commonest palm of the forest undergrowth in the montane forest there. The plant cultivated as *D. hildebrandtii* in Australia illustrated by Stewart (1994) is *D. procera*. True *D. hildebrandtii* is cultivated in the Palm House at Kew; it is one of the smallest palms of Madagascar and is altogether a very pretty plant. Occurring as it does in montane forest, one might expect it to be relatively tolerant of cool growing conditions. The species is named for the veteran plant collector, J.M. Hildebrandt (1847–1881).

DISTRIBUTION. Central Madagascar, in montane rain forest.
HABITAT. Montane forest, hill slopes and ridges; usually 700–1000 m, rarely as low as 300 m.
LOCAL NAMES. *Tsirika* (Merina).
USES. Not recorded.
CONSERVATION STATUS. Vulnerable. Restricted to the Moramanga area, but probably occurs in the forests more to the south of Anosibe-an-Ala, which are botanically not well known.

Dypsis hildebrandtii, the commonest small palm at Analamazoatra (*Dransfield et al.* JD6413).

Dypsis hildebrandtii *(Baill.) Becc.*
Palme del Madagascar 14 (1913); Jum. & H. Perrier, Ann. Inst. Bot.-Géol. Colon. Marseille sér. 3. 1: 22 (1913); Jum., Ann. Inst. Bot.-Géol. Colon. Marseille sér. 3, 6 (1): 32 (1918); Bull. Acad. Malgache 6: 3 (1923); Cat. Pl. Madagascar, Palmae: 13 (1938); Jum. & H. Perrier, Fl. Madagascar 30: 32 (1945).

SYNONYM:
Trichodypsis hildebrandtii Baill., Bull. Soc. Linn. Paris 2: 1165 (1894); Hist. des Plantes 13: 370 (1895). Becc., Bot. Jahrb. Syst. 38, Beibl. 87: 14 (1906). Type: Madagascar, Imerina Orientalis, *Hildebrandt* 3829 (G-BOIS, G-DC, K, M, P).

Slender, solitary or clustering palmlet of the forest undergrowth. **STEMS** to c. 2 m tall, rarely to 4 m, 3–13 mm diam., internodes 4–32 mm long, pale to dark green, with a sparse to dense covering of laciniate dark brown scales. **LEAVES** 4–10 in crown; sheaths 5–10 x 0.4–1.5 cm, longitudinally striate, sparsely covered in caducous red-brown scales, forming a well- or ill-defined crownshaft, the sheath mouth with two triangular, acute or acuminate auricles to 7 x 6 mm; petiole absent or very short, with the blade somewhat decurrent along its edges, to 30 x 1.5–3 mm, very rarely longer, sparsely brown scaly; rachis 6–30 cm; blade entire bifid or with 2 (or rarely–4) leaflets on each side of the rachis, occasionally leaves of two types borne in the same crown; entire bifid blade 12–21 x 4–9 cm with an apical notch 4–10 cm; leaflets usually broad, 5.5–22 x 1.2–3 cm, occasionally 1–2 slender leaflets at the very base of the rachis, 5–15 x 0.3–0.7 cm; adaxial surface of leaf with minute punctiform scales, abaxial surface with minute punctiform scales, bands of scattered caducous laciniate scales and occasionally with a few ramenta on main veins; emerging leaves tinged reddish, leaves drying dark brown. **INFLORESCENCES** interfoliar, occasionally also infrafoliar on the same plant, branching to 2 orders; peduncle 10–29 cm long, c. 2 mm in cross section, densely brown laciniate hairy in exposed parts; prophyll 10–35 x 0.4–0.8 cm, with scattered grey and brown laciniate scales; peduncular bract similar to the prophyll, usually exceeding the prophyll be about 2–5 cm; rachis 11–33 cm, rarely 6.5–8 cm, densely covered in grey and brown, curled laciniate scales, the rachis and rachillae together ± elongate triangular in outline; rachillae usually numerous, c. 20–50 or more, rarely as few as 12, 1.5–4 cm, in populations with few rachillae, the rachillae sometimes longer (to 7 cm), c. 0.6 mm diam., usually diverging at an acute angle and curved towards the tip, covered with pale or dark brown laciniate scales; rachilla bracts c. 3.5 mm apart, rounded, c. 0.6 mm, with laciniate margins. **STAMINATE FLOWERS** rounded, to 1.2 × 1 mm at anthesis; sepals broad triangular, keeled, 0.5 × 0.7 mm, ciliate-margined; petals c. 0.8 × 0.8 mm, striate; stamens 3, antepetalous, with 3 minute staminodes, filaments united in a ring c. 0.3 mm high, anthers ± didymous, 0.2 × 0.3 mm; pistillode minute. **PISTILLATE FLOWER** in young bud globular; sepals broad, ± rounded-triangular, imbricate, striate, 0.8 × 0.7 mm, the margins ciliate; petals c. 0.8 × 0.6 mm, imbricate with valvate tips; staminodes 3, minute, dentiform; ovary c. 0.4 × 0.3 mm. Mature **FRUIT** red, 10 × 5.5 mm, ellipsoid to fusiform; mesocarp c. 0.4 mm thick; endocarp thin, fibrous. **SEED** 8 × 3.5 mm; endosperm homogeneous, embryo lateral near the base.

NOTE. It is easily distinguishable by its small stature, slender stems, leaves bifid or with two to three (rarely more) broad leaflets and its inflorescence that is branched to two orders and bears numerous very hairy rachillae. The collections from Vatomandry, the lower part of the Mangoro Basin and Ambatovola made by Perrier (the last four collections cited above) differ from typical *D. hildebrandtii* in having few rather robust rachillae in the inflorescence. They are all from relatively low elevations; however, although they present a rather distinctive facies, we have found no clear distinguishing features.

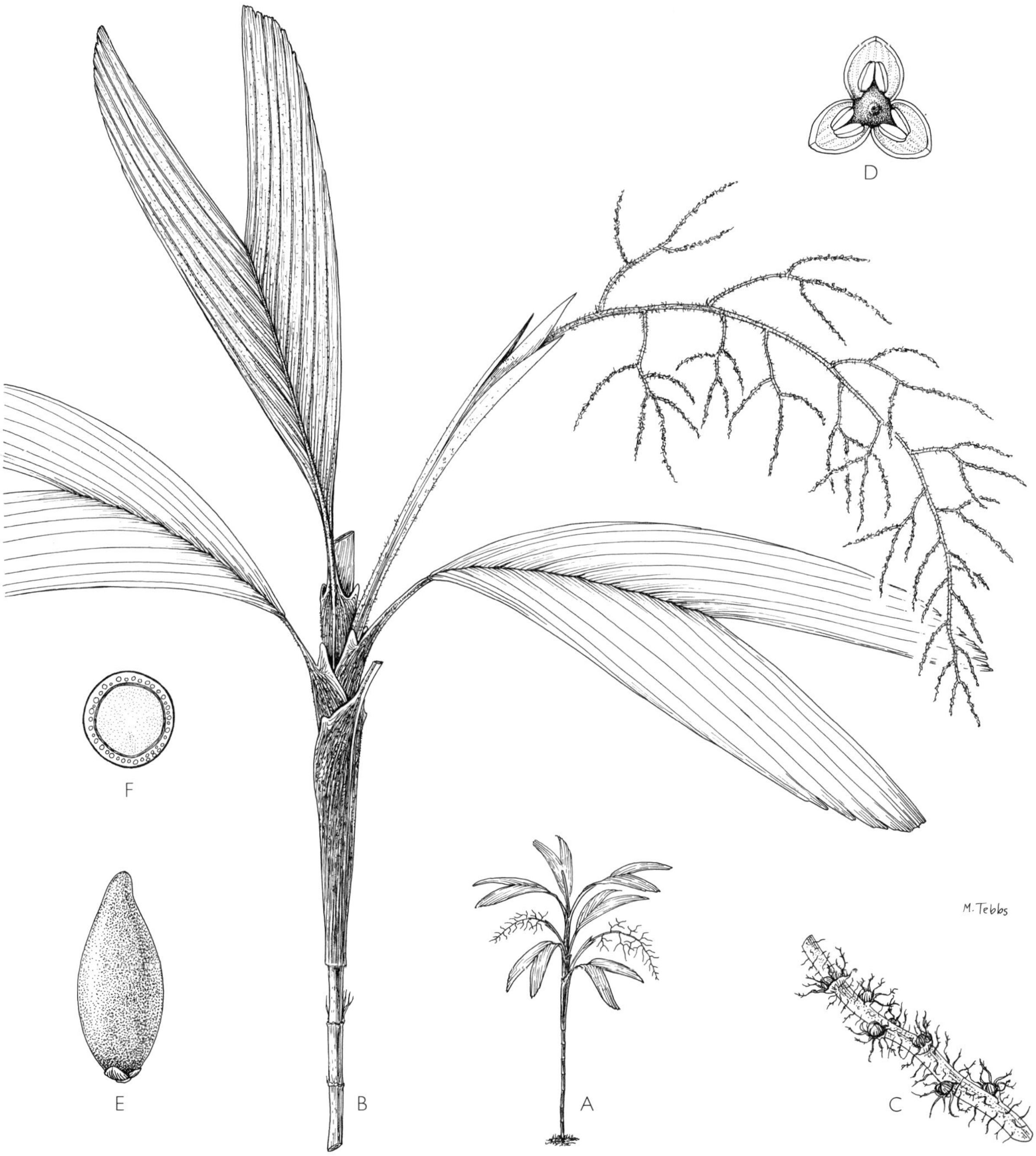

Dypsis hildebrandtii. **A** habit × 1/10; **B** crown with inflorescence × 1/2; **C** detail of rachilla × 5; **D** open staminate flower × 15; **E** fruit × 3; **F** fruit in cross section × 3. **A**, **B**, **E**, **F** from *Dransfield et al.* JD6487, **C**, **D** from *Dransfield et al.* JD6413. Drawn by Margaret Tebbs.

SPECIMENS SEEN. Moramanga: Analamazaotra, Feb. 1912 (fl.), *Perrier* 11956 (P); idem, Feb. 1912 (fl.), *Perrier* 11957 (P); idem, Feb. 1926 (fl.), *Perrier* 15987 (K, P); idem, Feb. 1925 (fl.), *Perrier* 17215 (P); idem, Dec. 1932 (fl.), *Leandri* 724 (P); idem, Nov. 1938 (fl.), *Lam & Meeuse* 5312 (K, L, P); idem, Oct. 1912 (fl.), *Viguier & Humbert* 1058 (P); idem, Sept. 1951 (fl.), *Benoist* 1117 (P); idem, Oct. 1963 (fl.), *Moore* 9909 (BH, P); idem, May 1980 (fl.), *Raketamalata* s.n. (P); idem, 1000 m, Nov. 1986 (fl.), *Dransfield et al.* JD6413 (BH, K, MO, P, TAN); idem, Nov. 1986 (fl.), *Dransfield et al.* JD6427 (BH, K, MO, P, TAN); idem, Nov. 1986 (fl.), *Dransfield et al.* JD6428 (BH, K, MO, P, TAN); idem, March 1988 (fl.), *Dransfield et al.* 6487 (K, NY, P, TAN); idem, Aug. 1992 (fl.), *Beentje & Beentje* 4743 (K); idem, June 1938 (fl.), without collector 3745 (P); Mantady, March 1991 (fr.), *Beentje & Rabarilala* 4404 (BH, K, MO, P, TAN); idem, March 1991 (fl.), *Beentje & Rabarilala* 4407 (BH, K, TAN); idem, Dec. 1991 (fl.), *Beentje & Andriampaniry* 4549 (K, TAN); idem, July 1987 (fl.), *Phillipson* 2120 (K, MO, P, TAN); Moramanga, Feb. 1959 (fl.), *H.-J. Schlieben* 8129 (K); 17 km south on road to Anosibe an'Ala, Oct. 1991 (fl.), *Beentje* 4507 (BH, K, MO, P, TAN); idem, Oct. 1991 (fl.), *Beentje* 4509 (K, TAN); Mantasoa, Jan. 1889 (fl.), *Catat* (P); Forêt de Sandrangato, March 1942 (fr.), *Decary* 17851 (P). Anosibe An'Ala, 16 July 1968 (fl.), *LTFT* 26762-SF (P); Ankarahara, July 61 (fl.), *Peltier* 3262 (P). Ambatovola, Jan. 1912 (fl.), *Perrier* 11989 (P). Mahanoro, Bassin du Mangoro, Oct. 1927 (fl.), *Perrier* 18044 (P); idem, *Perrier* 18048 (P); idem, *Perrier* 18392 (P). Vatomandry, Sept. 1921 (fl.), *Perrier* 14265 (P); idem, Sept. 1921 (fl.), *Perrier* 14266 (P).

LOCALITY UNCERTAIN: Imerina Orientalis, Jan. 1891 (fl.), *Hildebrandt* 3829 (Holotype P; isotypes G-BOIS, FI, G-DC, K, M); Central Madagascar, Oct. 1882 (fl.), *Baron* 1490 (K); received Nov. 1885 (fl.), *Baron* 3911 (K, P); *Myre de Vilers* s.n. (P); "chiefly from northwest Madagascar", received Sept. 1887 (fl.), *Baron* 5082 (K).

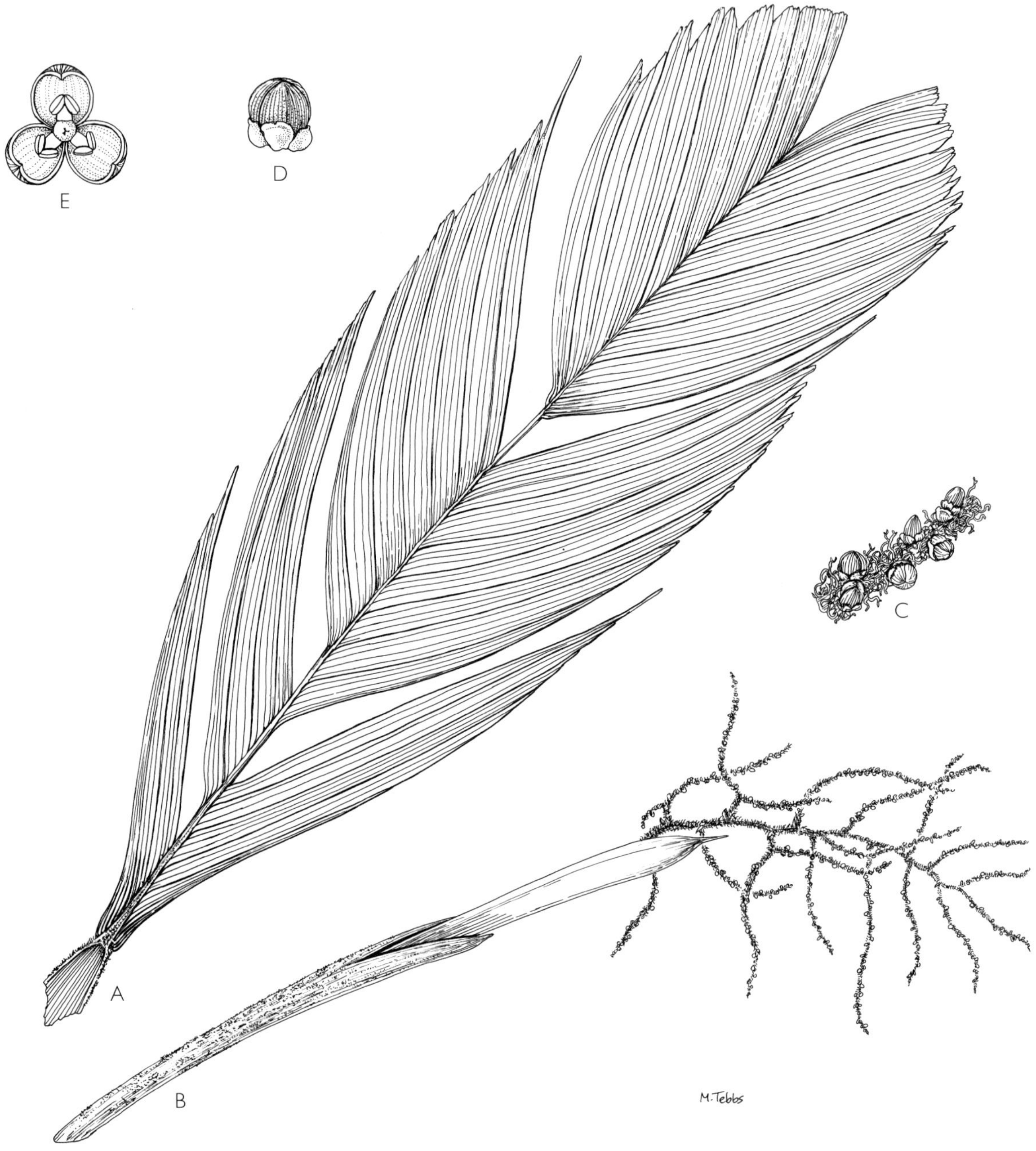

Dypsis bosseri. **A** leaf × 2/5; **B** inflorescence × 2/5; **C** detail of rachilla × 4; **D** staminate flower bud × 6; **E** open staminate flower × 6. All from *Bosser* 16972. Drawn by Margaret Tebbs.

129. Dypsis bosseri

Known only from a single collection, this is a small palm of the forest undergrowth. It most resembles *D. hildebrandtii* but is larger in all its parts and has a rather congested inflorescence. It is named for the collector of the type, Jean Bosser.

Distribution. Known only from forest west of Mahavelona.
Habitat. Lowland forest.
Local names. Not recorded.
Uses. Not recorded.
Conservation Status. Possibly extinct; not collected since the original collection, and most of the forest at Mahavelona has now disappeared.

Dypsis bosseri *J. Dransf.* **sp. nov.**
inter species floribus staminatis triandris staminibus antepetalis, staminodiis antesepalis alternantibus, *D. hildebrandtii* similis sed habitu multo majore, folio robustiore epetiolato inflorescentia congesta rachillis pilis ferruginosis dense tectis differt. Typus: Madagascar, East Coast, Mahavelona, *Bosser* 16972 (Holotypus P).

Slender forest undergrowth palm to 2.5 m tall. **Stem** not preserved in available specimen. **Leaf** sheath dimensions not known, surface very densely covered in thick ferruginous scales, auricles apparently absent; petiole absent; rachis 37–40 cm, c. 5 mm wide at the base, adaxially with scattered ferruginous scales, abaxially rather densely covered with caducous ferruginous scales; leaflets 4–5 on each side of the rachis, the lowermost pair very short, 5 × 0.5 cm, sometimes only partly separated from the second pair, second pair of leaflets 14–19 × 2–2.5 cm, third pair to 25 × 6 cm, apical pair to 15 × c. 8 cm, joined for 12.5–14 cm along the rachis, with an apical notch to 5 cm deep, the apical margins deeply lobed to 5 mm, occasionally deeper, adaxial surface glabrous, abaxially with scattered brown punctiform scales and bands of scattered brown scales. **Inflorescence** branched to 2 orders, relatively stout; peduncle 21 cm long; prophyll 16 × 1 cm, membranous, with scattered red-brown scales; peduncular bract similar, exceeding the prophyll by c. 6.5 cm; rachis 11 cm long, c. 2.5 mm diam., densely covered with red-brown trichomes to c. 1 mm long; rachillae c. 28, 2.5–6 cm long, c. 0.8 mm diam., very densely covered in red-brown trichomes, triads c. 2 mm apart, rachilla bracts to 0.5 mm, laciniate. **Staminate flowers** c. 1.5 mm diam.; sepals rounded, c. 0.5 × 0.5 mm, keeled, margins erose; petals triangular, striate, c. 1.5 × 1 mm; stamens 3, antepetalous, c. 0.5 mm high, anthers c. 0.2 × 0.1 mm, didymous, staminodes triangular, antesepalous, c. 0.2 × 0.2 mm; pistillode conical, minute. Other parts not known.

Note. We have described this new species, even though it is known only from a single specimen, because it seems so distinctive. The combination of robust epetiolate leaf with few broad and very close segments and the stocky inflorescence, with axes all densely covered in thick ferruginous hairs is distinctive. Perhaps it most resembles robust forms of *D. hildebrandtii*, but is more massive in all its parts.

Specimen seen. Toamasina: Mahavelona, Dec. 1962 (fl.), *Bosser* 16972 (Holotype P).

130. Dypsis furcata

There is only one herbarium specimen of this slender palmlet. In spite of that, the form of leaf appears to be so distinctive and unusual that there should be no difficulty in identifying it, should it be refound. This must have been a very beautiful palm. The species name refers to the deeply forked leaf.

DISTRIBUTION. Central East Coast of Madagascar.
HABITAT. Rain forest at low elevation.
LOCAL NAMES. Not recorded.
USES. Not recorded.
CONSERVATION STATUS. Presumed extinct. Not collected since the type collection was made more than sixty-five years ago.

Dypsis furcata *J. Dransf.* **sp. nov.**
inter species floribus staminatis triandris staminibus antepetalis, staminodiis antesepalis alternantibus, foliis valde furcatis, inflorescentia 2-ramosa distincta. Typus: Madagascar, East Coast, Mangoro, *Perrier* 18051 (Holotypus P).

Clustering slender forest undergrowth palm to 1–2 m tall. **STEM** 7 mm diam., internodes 13–16 mm long, bearing scattered dark brown scales. **LEAVES** probably c. 5 in crown, entire deeply bifid, to 35 cm long; sheaths 6–8.5 cm, c. 1.2 cm diam., striate, rather densely covered with red-brown scales and branched trichomes, auricles triangular, 7 × 6 mm; petiole 7–13 cm long, c. 3.5 mm wide at the base, with abundant caducous dark brown scales; costa 2 cm only; blade deeply bifid, the two lobes 27–35 × 2–3 cm, tapering to the narrow tips, apical margin shallowly and irregularly lobed, adaxial surface with sparse scattered brown punctiform scales, abaxial surface with abundant brown punctiform scales and bands of elongate pale trichomes with brown bases. **INFLORESCENCE** interfoliar at first, later infrafoliar, branched to 2 orders; peduncle 23–30 cm, c. 3 mm wide at the base; prophyll inserted c. 4–5.5 mm above base, 17–21 × 0.8 cm, membranous, sparsely scaly; peduncular bract similar to prophyll, exceeding it by 5–7 cm; rachis when fully expanded 16–26 cm, densely covered in red-brown branched trichomes; most (not just basalmost) first order

Dypsis furcata. **A** crown × 2/5; **B** inflorescence and leaf sheath × 3/5; **C** detail of rachilla × 3; **D** staminate flower bud × 4; **E** open staminate flower × 4. All from *Perrier* 18051. Drawn by Margaret Tebbs.

branches branched to a further order; rachillae c. 70, 1.5–3 cm long, c. 0.8 mm diam., with scattered red-brown branched trichomes, triads c. 0.8–1 mm apart, rachilla bracts 0.25 × 0.5 mm, rounded, with entire margins. **Staminate flowers** bud 0.6 mm diam.; sepals 0.4 × 0.4 mm, margins erose; petals 0.5 × 0.4 mm, triangular, striate; stamens 3, antepetalous, alternating with 3 minute antesepalous staminodes; pistillode minute. Persistent striate petals at base of very immature fruit to 1 × 1.1 mm. Other parts unknown.

Note. Known from a single collection, this is nevertheless a very distinctive species, with deeply divided leaf blades and densely hairy inflorescences with numerous rachillae.

Specimens seen. Mahanoro: Lower Mangoro, forest remnants, Oct. 1927 (fl.), *Perrier* 18051 (Holotype P).

131. Dypsis lanuginosa

Represented by a single herbarium specimen collected in 1922. We know little about this undergrowth palmlet. Its most distinctive feature is the densely woolly inflorescence with a large number of rachillae. Superficially it resembles *D. eriostachys*, but the latter has staminate flowers with six rather than three stamens. The species name refers to the woolly hairs on the inflorescence.

Distribution. Lower Mangoro.
Habitat. Lowland forest; 300 m.

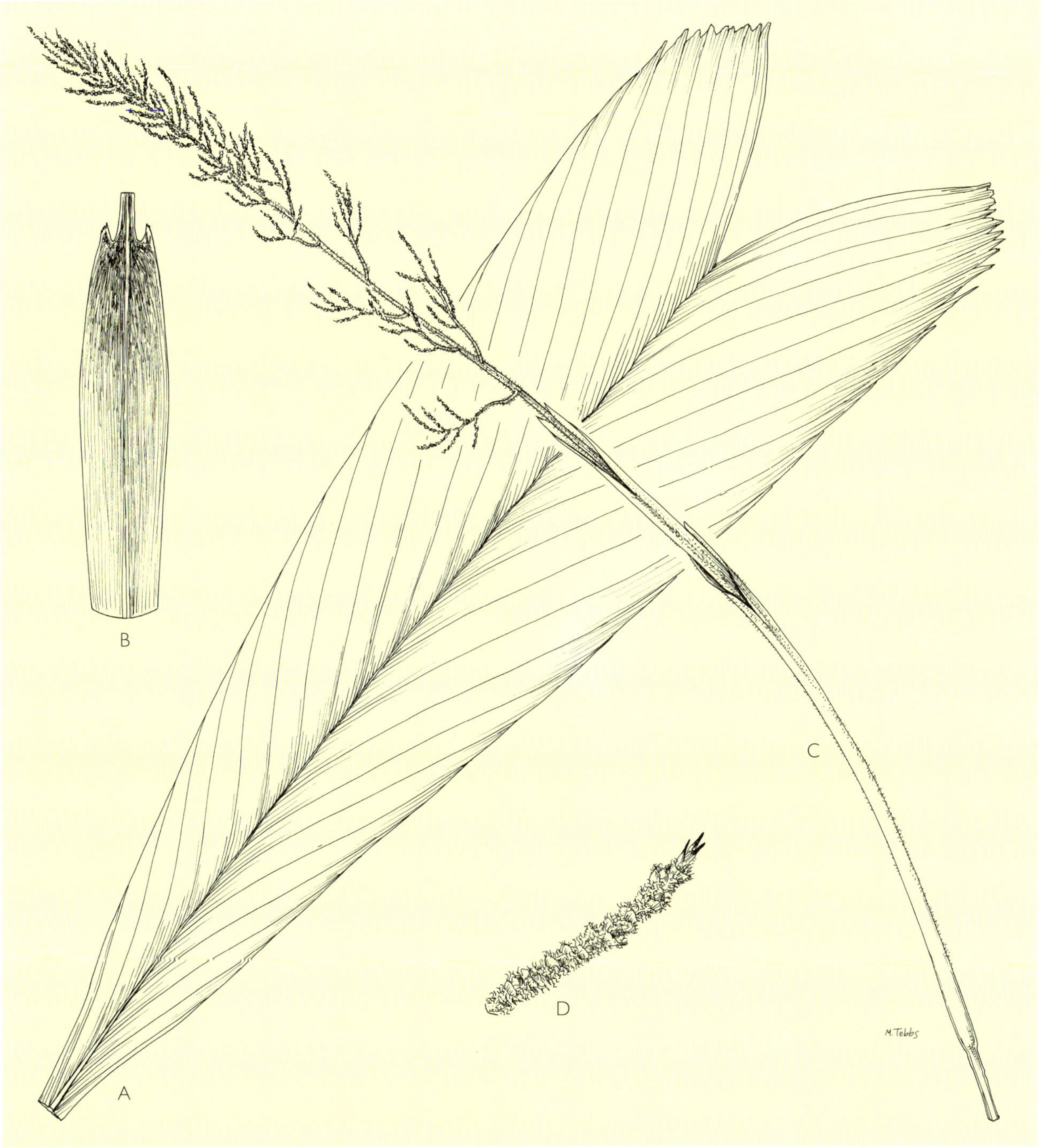

Dypsis lanuginosa. **A** leaf × 2/5; **B** leaf sheath × 2/5; **C** inflorescence × 2/5; **D** detail of rachilla × 3. All from *Perrier* 18050. Drawn by Margaret Tebbs.

Local names. Not recorded.
Uses. Not recorded.
Conservation Status. Presumed extinct. Not collected for more than seventy years. The forests of the lower Mangoro River have now disappeared.

Dypsis lanuginosa *J. Dransf.* **sp. nov.**
gregem specierum floribus staminatis triandris staminibus antepetalis, staminodiis antesepalis alternantibus pertinens, folio magno integro-bifido inflorescentia lanuginosa distincta; *D. lantzeanae* superficialiter similis sed positione staminum omnino diversa. Typus: Madagascar, lower basin of Mangoro, *Perrier* 18050 (Holotypus P).

Slender clustering undergrowth palm to 4 m tall. **Stem** 13 mm diam., but said to reach 40 mm diam., internodes 24–27 mm long, covered in dark brown caducous scales. **Leaf** with sheath c. 15 cm long, c. 2 cm diam., distally very densely covered with thick red-brown tomentum and scales; auricles ill-defined; petiole absent; rachis 38–44 cm; blade entire bifid or with very small basal leaflets, otherwise entire-bifid, probably undulate and bullate; basal leaflets where present 7–11 × 0.4–1 cm, blade to 48 cm long, gradually widening from c. 4 cm near the base to widest at the tip where 18 cm, the two lobes with a broad sinus, and shallowly to deeply lobed apically, adaxially with minute brown punctiform scales, abaxially minutely roughened and with irregular bands of caducous dark brown scales. **Inflorescence** probably erect, branching to 2 orders; peduncle 40 cm long, c. 4 mm wide at the base tapering to 3 mm diam. distally, loosely shaggy hairy; prophyll inserted at least 3.5 cm above the base, 28 × 0.7 cm, with scattered dark and pale stellate and shaggy hairs; peduncular bract exceeding the prophyll by 9 cm; rachis 25 cm; rachillae numerous, probably at least 100, c. 2–2.5 cm long, c. 1 mm diam., densely covered in pale brown shaggy hairs, triads c. 1–1.5 mm apart, rachilla bracts rounded, c. 0.5 mm high, obscured by hairs. Immature **staminate flower** buds c. 0.5 mm diam.; stamens 3, antepetalous. Other parts unknown.

Dypsis lanuginosa

Note. Although known from a single collection, this appears to be a distinctive species. The leaves are entire bifid or with two very small basal leaflets, and otherwise entire bifid, and there is no petiole. Drying of the specimen suggests that in the living state the leaf was probably beautifully undulate and bullate. The shaggy-hairy inflorescence is similar to that of *D. lantzeana* but instead of having three antesepalous stamens, the three stamens are antepetalous.

Specimen seen. Mahanoro: lower basin of Mangoro, forest vestiges, Oct. 1922 (fl.), *Perrier* 18050 (Holotype P).

132. Dypsis mocquerysiana

An attractive small solitary palm of the undergrowth of very humid rain forests in the lowlands surrounding the Bay of Antongil. *D. mocquerysiana* has been grown at Kew since 1986, where it flowers regularly and once produced two fruits. It is a very neat palm with short stems, with tidy leaf scars, and, usually, entire bifid leaves, somewhat reminiscent of *D. louvelii*, but rather broader. The finely and highly branched inflorescence, however, is quite different

Dypsis mocquerysiana, in forest undergrowth, Sahavary (*Dransfield et al.* JD6454).

from that of the latter species. Mocquerys, after whom the palm is named, was an early plant collector in the north-east of Madagascar.

DISTRIBUTION. NE Madagascar: Masoala and Mananara Avaratra.
HABITAT. Lowland rain forest, usually in humid valley bottoms; 50–400 m.
LOCAL NAMES. Not recorded.
USES. Not recorded.
CONSERVATION STATUS. Vulnerable. Occurs in a restricted area, where shifting cultivation is increasing.

Dypsis mocquerysiana *(Becc.) Becc.*
Palme del Madagascar 15 (1913); Jum., Cat. Pl. Madagascar, Palmae: 14 (1938); Jum. & H. Perrier, Fl. Madagascar 30: 33 (1945).

SYNONYMS:
Trichodypsis mocquerysiana Becc., Bot. Jahrb. Syst. 38, Beibl. 87: 15 (1906). Type: Madagascar, Bay of Antongil, *Mocquerys* 333 (Holotype G-DC; isotype FI).

Dypsis mananarensis Jum., Ann. Inst. Bot.-Géol. Colon. Marseille sér. 3, 6 (1): 32 (1918) as "manaranensis"; Bull. Ac. Malgache 6: 6 (1923); Cat. Pl. Madagascar, Palmae: 14 (1938); Jum. & H. Perrier, Fl. Madagascar 30: 34 (1945); Type: Madagascar, Mananara, *Perrier* 12067 (holotype P) (lectotype chosen here) **synon. nov.**

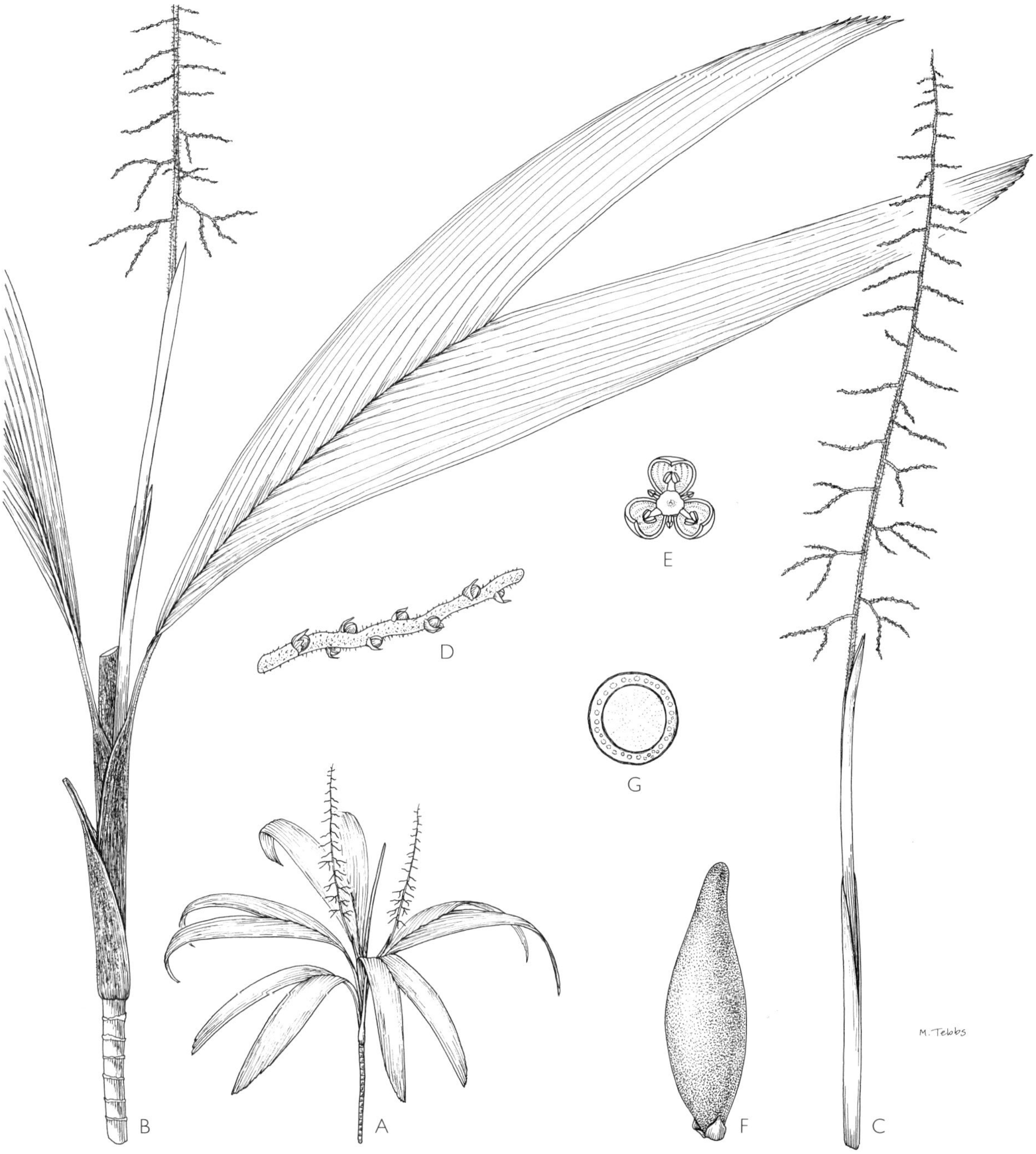

Dypsis mocquerysiana. **A** habit × 1/10; **B** crown × 2/5; **C** inflorescence × 2/5; **D** detail of rachilla × 3; **E** staminate flower × 20; **F** fruit × 3; **G** fruit in cross section × 3. **A – C** from *Dransfield et al.* JD6454, **D**, **F**, **G** from *Dransfield et al.* JD6368, **E** from *Cooke* 95. Drawn by Margaret Tebbs.

Slender solitary (or clustering, fide Jumelle) palm of the forest undergrowth, not exceeding 2 m tall. **STEM** 7–9 mm diam.; internodes c. 7 mm diam., stem surface with scattered caducous scales. **LEAVES** 4–8 in crown, tending to be held ± horizontally; sheath to 10 cm long, 1 cm diam., densely covered with thick caducous dark brown indumentum, the sheath mouth sometimes with two short irregular triangular auricles (1–3 cm, fide Jumelle 1923); petiole absent or very short, not exceeding 6 cm long, c. 4 × 3 mm in section, densely covered in caducous dark brown indumentum as the sheath; rachis (or costa) 8–19 cm long, tapering along its length, bearing caducous dark brown indumentum as the petiole; blade entire bifid or with 2 broad leaflets on one or both sides of the rachis; blade to 50 cm long, bifid for 50% to almost 80% of the length, the two lobes (or apical lobes) diverging from the rachis at an angle of about 10–15°, held flat or ± cucullate, somewhat plicate, to 30 × 3.5–5.5 cm, adaxially glabrous, abaxially with minute, scattered punctiform dark brown scales and a few bands of caducous laciniate scales, and usually with conspicuous dark ramenta on major veins, 2–5 mm long. **INFLORESCENCE** protandrous, erect at anthesis, held above the leaves, bent down in fruit, branching to 1–2 orders; peduncle 17–37 cm long, c. 1.5–3 mm diam., bearing abundant caducous chocolate-brown laciniate scales when young in exposed parts; prophyll sometimes borne 2–3 cm above base of peduncle, 9–18 × 0.4–0.8 cm, very sparsely scaly; peduncular bract borne 2–5 cm above the prophyll, 17–27 × 0.7–0.8 cm, similar to prophyll; rachis usually elongate, 9–22 cm long, bearing numerous short first-order branches, the basal few branched further to the second order, the whole branching portion of the inflorescence ± narrow rectangular in outline, axes sparsely to very densely covered in shaggy dark brown scales; rachillae usually numerous (c. 90), short, rarely exceeding 4 cm long, c. 1 mm diam., surface covered with minute punctiform scales and larger shaggy dark brown scales; triads 1–1.5 mm distant; rachilla bracts c. 0.5–1 mm, triangular, entire or long laciniate. **STAMINATE FLOWERS** c. 0.7–1 mm diam.; sepals 0.5–0.7 mm long, irregularly triangular or rounded, keeled, margins entire, or laciniate; petals 0.8 × 0.5 mm, striate; stamens 3, antepetalous, filaments 0.2 × 0.1 mm, basally connate in a ring with staminodes, anthers c. 0.2 mm, didymous, staminodes clublike, c. 0.1 mm; pistillode a low dome. **PISTILLATE FLOWERS** c. 1.5 mm diam.; sepals imbricate, c. 0.5 mm, margins laciniate; petals triangular, 1.3 × 1 mm, striate; staminodes 6, irregularly dentiform, minute; ovary (post anthesis) c. 1.5 mm diam.. Mature **FRUIT** bright red, 13 × 5.5 mm, epicarp shiny, mesocarp thin fleshy, endocarp thin with few scattered longitudinal fibres. **SEED** 11 × 4 mm; endosperm homogeneous; embryo lateral near the base.

NOTES. Beccari cited two syntypes, *Mocquerys* 333 and *Mocquerys* 419. As he used the latter for his "*Dypsis mocquerysiana dissecta*", we interpret this as selecting *Mocquerys* 333 as representing the typical facies (i.e., the type) of the species.

In the protologue of *D. mananarensis*, Jumelle cited two collections: *Perrier* 12067 and *Perrier* 12064; of these the former is selected here as lectotype, as it is the more complete collection. Of the six leaves represented in the two syntypes, five are entire bifid, while the sixth, in *Perrier* 12064, has a blade split on one side of the rachis to give two broad leaflets. The plate in the Flora of Madagascar is thus most misleading, as it represents a leaf with four segments, and the description also mentions that the blade has most often four segments. On the label attached to the lectotype, *Perrier* 12067, Perrier has written "Parait differer de D11 (No.11963)"; we have found no trace of this collection. Jumelle cited the type locality as "Manarana" and the specific epithet as "*manaranensis*". The misspelling of the specific epithet was later corrected to "*mananarensis*" on the type sheet *Perrier* 12067.

Jumelle distinguished *D. mananarensis* from *D. mocquerysiana* on the dissection of the blade, decurrence of segments and the absence of fimbriate bracts subtending the flower clusters (as illustrated by Beccari). However, even within populations of *D. mocquerysiana* on the Masoala Peninsula, the degree of fimbriation of the rachilla bracts varies greatly from strongly fimbriate to almost lacking fimbriation, and there seem to be no significant differences between dissected-bladed *D. mocquerysiana* and *D. mananarensis*.

Lam & Meeuse 5861, from Soanierana-Antasibe, differs from other collections of *D. mocquerysiana* in being rather large, in having a petiole about 6 cm long, in the leaf lacking ramenta on the undersurface and in the rachillae bearing rather longer shaggier hairs. We have only tentatively included it here. *Heim* s.n. has ramenta on the undersurface of the leaves, but has petioles about 6 cm long, a relatively short inflorescence and the the rachilla hairs are shaggier than those found in typical *D. mocquerysiana*. Furthermore, the collection was made in the Matitanana, far to the south of the main area of distribution around the Bay of Antongil. This collection is not complete and is only tentatively included here.

SPECIMENS SEEN. Maroantsetra: Masoala Peninsula, Maroa, Bay of Antongil, (fl.), *Mocquerys* 333 (Holotype G-DC, isotype FI); Nosy Mangabe, (fl.), *Mocquerys* 419 (G-DC, FI); Hiaraka, Oct. 1986 (fl., fr.), *Dransfield et al.* JD6368 (BH, K, MO, TAN); Antalavia, April 1988 (fl.), *Gentry & Schatz* 62172 (K, MO); Andranofotsy River, Sahavary, Feb. 1988 (fl., fr.), *Dransfield et al.* JD6454 (K, TAN). Mananara Avaratra: Mananara, Aug. 1912 (fl.), *Perrier* 12067 (lectotype of *D. mananarensis*, P); idem, *Perrier* 12064 (P). Tentatively included specimens: Soanierana-Ivongo: Antasibe, Dec. 1938 (fl.), *Lam & Meeuse* 5861 (L) (see note above); Matitanana, forêt de Tsianovoka, Ampenetra, Sept. 1934 (fl.), *Heim* s.n. (P) (see notes).

CULTIVATED: ex Masoala, Ambanizana, Royal Botanic Gardens, Kew, Sept. 1991 (fl.), *Cooke* 95 (K).

133. DYPSIS COOKEI

This is a beautiful little palm, distinctive in its rather thick metallic blue green leaves with narrow leaflets. Leaf texture and colour are somewhat reminiscent of that of the Mexican palm, *Chamaedorea metallica*. Clearly this would make a very handsome ornamental but, sadly, it is known from only one locality (Marojejy) and has been collected once only. It is named for David Cooke, who is in charge of the Palm House at Kew, and has joined two Kew expeditions to Madagascar, helping greatly in the collection of palms and contributing his very special brand of humour and companionship.

DISTRIBUTION. Marojejy.
HABITAT. Humid lower montane forest on steep slope; 1100 m.
LOCAL NAMES. Not recorded.
USES. Not recorded.
CONSERVATION STATUS. Endangered. Although this species occurs within the Marojejy Nature Reserve, it is known from but a single population consisting of less than twenty plants.

Dypsis cookei *J. Dransf.* **sp. nov.**
inter species floribus staminatis triandris staminibus antepetalis, staminodiis antesepalis alternantibus, foliolis unicostatis coriaceis metallico-atroviridibus inflorescentia 2–3-ramosa rachillis dense pruinosis distincta. Typus: Madagascar, Reserve Naturelle Marojejy, along trail to summit of Marojejy Est, north of Mandena, *Dransfield et al.* JD6761 (Holotypus K; isotypi MO, P, TAN).

Slender clustering palmlet to 2 m tall, often less. **STEMS** green, c. 6 mm diam.; internodes 15–17 mm long, glabrous, shining dark green when fresh, becoming striate on drying. **LEAVES** 6–7 in the crown; leaf-sheaths pale green, tinged purple when young, smooth when fresh, becoming striate on drying, bearing caducous, minute scattered punctiform scales, otherwise glabrous, c. 7 cm long, two minute auricles present at the base of the petiole (?always); leaf without sheath 25–35 cm long; petiole 12–20 mm long, c. 1.5 mm wide, subtriangular in section, bearing scattered brown scales; rachis 15–20 cm, bearing scattered scales as on the petiole; leaflets 6–8 on each side of the rachis, very narrow, dark blue-green with an almost metallic sheen, mostly composed of one fold but occasion-

Dypsis cookei. A beautiful palm of the undergrowth of mountain forest, Marojejy.

ally apical pair and the basal leaflets composed of two folds, proximal leaflets 12–16 × 0.6–1.1 cm, median leaflets 17–20 × 0.7–0.8 cm, distal leaflets 12–14 × 1.2–1.4 cm; lamina surfaces glabrous apart from very sparse scattered minute punctiform brown scales. **INFLORESCENCE** interfoliar, c. 38 cm long, only slightly protruding beyond the leaves, crimson-purple throughout, branching to 2–3 orders with c. 28 rachillae, the whole branched part of the inflorescence more or less triangular in outline when pressed; peduncle glabrous, c. 15–20 cm long; prophyll c. 12 cm long, peduncular bract exceeding the prophyll by about 8 cm, the tip reaching to the level of the basalmost rachilla, the two bracts drying brown and bearing scattered dark brown scales, especially along the margins; rachis c. 10.5 cm; rachillae c. 25, 2–4 cm long, the longest near the base, c. 1 mm diam. when fresh, bearing triads 1–2 mm apart, the rachilla surface densely pruinose. **STAMINATE FLOWER** bud c. 0.7 mm diam.; sepals 3 rounded, gibbous, c. 0.5 mm diam., abaxially minutely and densely papillose; petals broadly triangular, c. 0.5 mm long, c. 0.5 mm wide, obscurely striate; stamens 3, antepetalous, anthers didymous, c. 0.25 mm long, the filaments triangular c. 0.2mm long; pistillode pyramidal, c. 3 mm high. **PISTILLATE BUDS** very immature, c. 0.5 mm diam., rounded. Other parts not seen.

NOTE. This species is easily distinguished by its dark metallic green unicostate leaflets of a rather thick, coriaceous texture. The whole plant has a rheomorphic appearance but in fact grows far from water, on a steep slope in montane rain forest.

SPECIMEN SEEN. Andapa: Reserve Naturelle Marojejy, along trail to summit of Marojejy Est, north of Mandena, above Camp 2, Nov. 1989 (buds), *Dransfield et al.* JD6761 (Holotype K, isotypes MO, P, TAN).

Dypsis cookei. **A** crown × 1/2; **B** detail of rachilla × 2; **C** opened staminate flower × 20. All from *Dransfield et al.* JD6761. Drawn by Rosemary Wise.

Dypsis cookei. Detail of inflorescence (*Dransfield et al.* JD6761).

134. DYPSIS BEENTJEI

This beautiful species is immediately recognizable by its stemless habit and dark green bifid leaves, each with a central pale yellow line. So far it is known only from an area of forest developed on supposed ultramafic rock in the Biosphere Reserve south of Mananara Avaratra. Here it grows in a valley bottom, in damp hollows and along small streams. It would make a handsome ornamental.

DISTRIBUTION. East Coast.
HABITAT. Rain forest, waterlogged alluvial flat beside river, ultramafic bedrock; c. 250 m.
LOCAL NAME. Not recorded.
USES. Not recorded.
CONSERVATION STATUS. Endangered. Although known from a single population, this does occur within the Biosphere Reserve. However, we have seen no more than about thirty plants.

Dypsis beentjei *J. Dransf.* **sp. nov.**
acaulescens folio bifido atroviride, linea media albo-lutea, inflorescentia sparse divaricate ramosa floribus staminatis triandris, staminibus antepetalis staminodiis antesepalis alternantibus. Typus: Madagascar, Mananara Avaratra, 10 km west of Antanambe, *Beentje et al.* 4626 (Holotypus K; isotypi BH, MO, P, TAN).

Clustering acaulescent palm with stems subterranean and procumbent, forming tufts in the forest undergrowth. **STEM** c. 4 × 1.4 cm, dull brown, bearing robust roots c. 2 mm diam. Leaves c. 9 in crown, ± erect; sheath 7–9 cm long, 2.5–3 cm wide at the base when opened out, pale creamy brown, tinged reddish when fresh, drying reddish brown, with a conspicuous central pale yellow line, the sheath striate on drying, densely covered with caducous dark brown scales;

Dypsis beentjei. Detail of infructescence (*Beentje et al.* 4626).

petiole to 55 cm long or more, c. 4 mm wide, ± triangular in cross section, pale creamy brown, covered in caducous dark brown scales; lamina narrow-triangular, entire-bifid, to 60 cm long, c. 10 cm wide across the two tips, apically cleft to 19 cm, the lobes to 3.5 cm wide at the base, tapering to the shallowly lobed tip, 0.5 cm wide, the lamina base long decurrent on the petiole, lamina leathery in texture, adaxially dark shining green when fresh, with scattered minute punctiform dark brown scales, the mid-line pale cream-coloured, abaxially matt, bearing rather dense apparently caducous dark brown scales of varying size with laciniate margins, particularly along the ribs. **INFLORESCENCE** interfoliar, erect at first, becoming curved, branching to two orders, ± rectangular in outline; peduncle to 25 cm long, 3 mm diam., rounded in cross-section, densely covered in dark brown laciniate-margined scales; prophyll to 8 × 0.5 cm, membranous, soon tattering, bearing scattered dark brown scales; peduncular bract long exceeding the prophyll, to 18 × 0.5 cm, membranous, soon tattering, scaly as the prophyll; rachis to 9 cm long, scaly as the peduncle; lowermost 3–4 branches branched with one branch of the second order; rachillae 15–20, 3–5 cm long, c. 6 mm diam., somewhat zigzag, the triads 2–6 mm distant, each subtended by a lacinate rachilla bract to 1.5 × 0.5 mm, the rachilla surface partially obscured by dark brown laciniate scales. **STAMINATE FLOWERS** at anthesis pale green, depressed spherical, 2 × 2.5 mm; sepals rounded and gibbous at first, to c. 1 × 1 mm, then becoming explanate and splitting irregularly into 3 sections; corolla spherical, tubular at the base, lobes gaping at anthesis; stamens 3 free, antepetalous, alternating with 3 antesepalous staminodes; staminodes wedge-shaped, succulent, 0.9 × 0.8 mm; fertile stamens with succulent filaments 1.0 × 0.5 mm, connective dark, 0.3 × 0.3 mm, anthers didymous, 0.5 × 0.3 mm; pistillode pyramidal c. 0.4 mm high. **PISTILLATE FLOWER** buds very immature, c. 1 mm diam. Mature **FRUIT** scarlet, irregularly ovoid to ellipsoid, to 17 × 10 mm; epicarp glabrous; mesocarp soft, fleshy; endocarp ellipsoid, 12 × 6 mm, c. 0.2 mm thick, with longitudinal brown fibres, irregularly anastomosing. **SEED** irregularly ellipsoid, 9–10 × 5 mm, shallowly longitudinally grooved; endosperm homogeneous, embryo lateral.

Dypsis beentjei. **A** habit × 2/5; **B** detail of rachilla × 2; **C** infructescence × 2/5. All from *Beentje et al.* 4626. Drawn by Rosemary Wise.

Note. This unusual species is immediately recognisable by its acaulescent habit and relatively large entire bifid dark green leaves, each with a central pale yellow band. The inflorescence is sparsely and divaricately branched and is partially hidden among leaf litter on the ground. Apart from being obviously allied to other species with three antepetalous stamens, its relationships are not clear.

Specimens seen. Mananara Avaratra: 10 km west of Antanambe, Oct. 1991 (fl.), *Beentje* 4475 (K, TAN); idem, April 1992 (fl.,fr.), *Beentje & Dransfield* 4626 (Holotype K; isotypes BH, MO, P, TAN).

135. Dypsis pachyramea

This is one of the smallest members of the genus. It is an attractive species with mid green, deeply plicate leaves and short, squat inflorescences hidden among the leaf bases. It is found in very humid lowland forest on the western side of the Masoala Peninsula. The species name is derived from the Greek for thick branches.

Distribution. Masoala Peninsula.
Habitat. Lowland forest, on steep slopes and in humid valley bottoms; to 400 m.
Local names. Not recorded.
Uses. Not recorded.
Conservation Status. Vulnerable. Only known from two sites.

Dypsis beentjei. Whole plant, showing the pale stripe along the midrib, Mananara Avaratra (*Beentje et al.* 4626).

Dypsis pachyramea *J. Dransf.* **sp. nov.**
habitu et forma inflorescentia congesta rachillis brevibus crassisque *D. angustifoliae* superficialiter similis sed gregem specierum floribus staminatis triandris staminibus antepetalis, staminodiis antesepalis alternantibus pertinens. Typus: Madagascar, Masoala, Hiaraka, *Dransfield et al.* JD6362 (Holotypus K; isotypi BH, MO, P, TAN).

Slender clustering undergrowth palmlet occurring in rather dense stands. **STEMS** to 50 cm tall, usually less, decumbent, rooting at the ground and sprouting at the rooting point, 7–8 mm diam., in the upper part obscured by marcescent leaf sheaths, internodes 7–23 mm long; nodal scars c. 2 mm wide, stem surface covered in scattered chestnut-brown scales. **LEAVES** c. 8 in the crown; sheaths tubular, 60 x 9–14 mm, striate when dry, apically with 2 short triangular auricles 4.5 x 3 mm, these soon disintegrating, the sheath surface covered in dense caducous fimbriate-margined chestnut-brown scales; petiole absent or to c. 9 cm long, triangular in cross section, 2 mm wide, densely covered in fimbriate-margined caducous chestnut-brown scales; blade narrow triangular, entire bifid, plicate, matt mid green, 39–43 cm, with costa 20–23 cm long, the lobes 17–20 cm long, 3.5–4.0 cm wide at the base, tapering to 1 cm wide at the shallowly dentate tips; costa bearing abundant caducous chestnut-brown scales, adaxial blade surface with scattered brown punctiform scales both on and between the ribs, abaxial surface with caducous fimbriate scales along ribs and scattered punctiform brown scales between ribs. **INFLORESCENCE** interfoliar, erect or sharply reflexed, much shorter than the leaves, 7–15 cm, branched to one order only; peduncle to 10 cm long, 3 mm diam. distally, densely covered with fimbriate dark brown caducous scales; prophyll inserted to 10 mm above the base of the peduncle, 46 × 6 mm, tattering and bearing caducous dark brown scales; peduncular bract inserted c. 2 cm above the prophyll insertion, 40 x 6 mm, tattering and scaly as the prophyll; rachis to 3 cm; rachillae 5–10, thick, short, 10–25 × 3 mm in liquid-preserved specimen, the longest being the continuation of the rachis, rachilla surface obscured by dense fimbriate scales. Triads borne congested ± throughout the rachilla length, partially obscured by the scales and partially sunken in pits, the rachilla bracts rounded, c. 0.5 × 1 mm, forming the lower lips of the pits. **STAMINATE FLOWERS** at anthesis c. 1.1 mm, spherical; sepals rounded, c. 0.8 mm diam., margins irregularly dentate or entire; petals ± free except at the very base, broadly triangular, 1 × 1 mm, striate, glabrous; androecium with 3 antepetalous stamens, filaments forming a fleshy ring c. 0.4 mm high, anthers didymous, 0.2 mm, ± sessile, introrse, connective c. 0.1 mm, staminodes scarcely visible as 3 minute triangular lobes positioned between the anthers; pistillode scarcely visible. **PISTILLATE FLOWERS** at anthesis 2 × 1.5 mm, briefly stalked; sepals ovate, margins irregularly dentate, 1.2–1.5 × 0.9 mm; petals 1.5 × 1 mm, free, with thick triangular striate valvate tips, basally weakly imbricate; staminodes 6, irregularly triangular, 0.1–0.75 mm, irregularly grouped at one side of the ovary or separated, if grouped then connate; ovary spherical, c. 1.1 mm diam., stigmas eccentrically apical, to 0.2 mm, basally connate. Immature **FRUIT** green, fusiform, mature fruit cherry-red, ellipsoid, 17 × 7 mm; epicarp smooth, mesocarp 1–2 mm thick, endocarp 0.4 mm thick with sparse anastomosing fibres. **SEED** ellipsoid, 14 × 5 mm, embryo lateral.

NOTE. This species was at first confused with *Dypsis humbertii* var. *angustifolia* (= *D. angustifolia*). The inflorescences of both taxa are remarkably similar and unlike other species of *Dypsis* in the few short fat rachillae densely covered with hairs that obscure the flowers. There are also similarities in habit and form of the leaf. However, stamens in the new taxon are antepetalous, while in *D. angustifolia* they are antesepalous.

SPECIMENS SEEN. Maroantsetra: Masoala Peninsula, Hiaraka, Oct. 1986 (fl.), *Dransfield et al.* JD6362 (Holotype K, isotypes BH, MO, P, TAN); Andranofotsy River, Sahavary, swamp at headwaters of Sahafotra River, Feb. 1988 (fl.), *Dransfield et al.* JD6451 (K, MO, NY, P, TAN).

Dypsis pachyramea, in the undergrowth of forest on the Masoala Peninsula (*Dransfield et al.* JD6362).

Dypsis pachyramea. Detail of crown with infructescence, Masoala (Photo: B. Du Puy & D. Du Puy).

136. DYPSIS MONOSTACHYA

This is a curious species that superficially looks much more like the species of *Dypsis* formerly included in the genus *Neophloga*. However, there are clearly only three stamens, and these being opposite the petals, this taxon belongs to the grouping formerly included in *Dypsis* § *Trichodypsis*. Initially we had confused *Dypsis spicata* with this species; for differences, see under the latter. The species name is derived from the Greek for a single spike.

DISTRIBUTION. Bay of Antongil and Mandritsara.

HABITAT. Lowland and montane forest; to 900 m.

LOCAL NAMES. Not recorded.

USES. Not recorded.

CONSERVATION STATUS. Not known. Presumed extinct, although the known sites are rarely visited by collectors.

Dypsis monostachya *Jum.*
Ann. Inst. Bot.-Géol. Colon. Marseille sér. 3, 6 (1): 29, 36 (1918); Bull. Ac. Malgache 6: 15 (1922-23); Cat. Pl. Madagascar, Palmae: 14 (1938); Jum. & H. Perrier, Fl. Madagascar 30: 46 (1945). Type: Madagascar, Rantabe, Bay of Antongil, *Perrier* 12047 (Holotype P).

Clustering slender palmlet to 1m tall. **STEM** 5–7 mm diam., internodes 7–25 mm long with scattered scales, nodal scars c. 1 mm

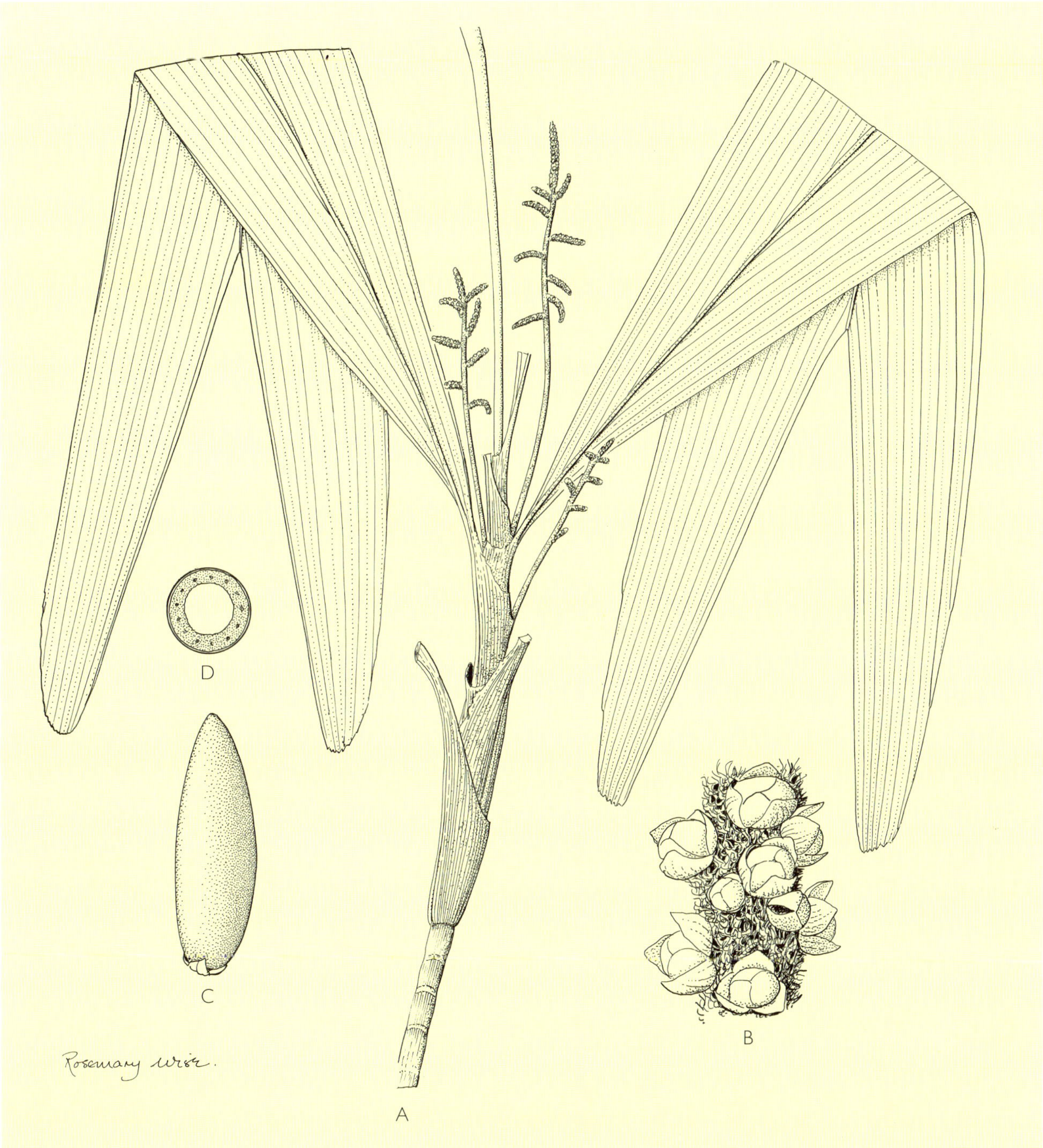

Dypsis pachyramea. **A** crown × 1/2; **B** detail of rachilla × 6; **C** fruit × 2; **D** fruit in cross section × 2. All from *Dransfield et al.* JD6451. Drawn by Rosemary Wise.

wide. **LEAVES** c. 5 in crown; sheaths 6–8 × 0.5–0.8 cm, longitudinally striate, sparsely to densely covered with caducous dark red-brown scales, membranous by the mouth, sometimes tattering but lacking distinct auricles; petiole lacking or to 7.5 cm long, c. 1–2 mm wide, triangular in cross section, bearing sparse to dense caducous dark brown scales; lamina divided into c. 6 narrow leaflets on each side of the rachis, rachis 8–28 cm, sparsely scaly; leaflets 15–22 × 0.5–1.2 cm, the apical pair only slightly broader than the proximal leaflets, lamina surfaces with minute punctiform brown scales borne on pale thickened bases, denser on the abaxial surface than on the adaxial, leaflets drying very pale green. **INFLORESCENCE** spicate, interfoliar, only slightly shorter than the leaves or exceeding them, 24–45 cm long, erect, but with the flower-bearing portion sometimes pendulous; prophyll 6–9.5 × 0.4–0.6 cm; peduncular bract 6.5–18 × 0.4 cm, both prophyll and peduncular bract sparsely scaly; peduncle 13–24 cm long; spike 9–21 × 0.15–0.2 cm; triads rather sparse in proximal portion, more densely arranged distally, rachilla bract c. 1 mm, rounded to apiculate with laciniate scales, a distinctive patch of chestnut-brown scales present on axis just above the triads. **STAMINATE FLOWERS** c. 1.4 mm high; sepals c. 0.6 mm long, keeled; petals valvate, ± elliptical, 1 × 0.6 mm; stamens 3, antepetalous, alternating with 3 antesepalous triangular staminodes, together borne on a short androecial tube ring to 0.4 mm high, connective ± triangular, anthers subdidymous, pistillode not seen. **PISTILLATE FLOWER** rounded; sepals imbricate, rounded, c. 0.8 × 0.8 mm; petals rounded-triangular, 1.2 × 1 mm; staminodes 3, minute, dentiform; ovary c. 1 mm diam., post anthesis. **FRUIT** unknown.

NOTE. As circumscribed here, *D. monostachya* is known from two separate populations. The type was collected at Rantabe at the southern end of the Bay of Antongil. The second collection was made further south but at relatively high altitude at Andilamena, Mandritsara. This latter collection differs from the

Dypsis monostachya. **A** crown × 2/5; **B** detail of rachilla × 3. All from *Perrier* 12047. Drawn by Margaret Tebbs.

type in having rather thick scattered scales on the leaf sheath and in its inflorescences which are longer and have the flower bearing portion apparently pendulous. Otherwise the two collections are very similar.

SPECIMENS SEEN. Maroantsetra: Rantabe, Bay of Antongil, Aug. 1912 (fl.), *Perrier* 12047 (Holotype P). Mandritsara: north of Andilamena, April 1923 (fl.), *Perrier* 15015 (P).

137. DYPSIS SPICATA

This species is one of the smallest in the genus. It appears very similar to *Dypsis catatiana* and can only be separated with certainty if staminate flowers are available: *D. catatiana* has six stamens while the present species has only three. It is a very attractive palmlet. The species name comes from the Latin for spicate, bearing a spike, in reference to the unbranched inflorescence.

DISTRIBUTION. Marojejy and environs.
HABITAT. Humid lower montane forest on steep slopes; 850–1000 m.
LOCAL NAMES. Not recorded.
USES. Not recorded.
CONSERVATION STATUS. Rare. Occurs in a small area, though partly protected in the Marojejy Special Reserve.

Dypsis spicata *J. Dransf.*, **sp. nov.**
gregem specierum floribus staminatis triandris staminibus antepetalis, staminodiis antesepalis alternantibus pertinens, inflorescentia spicata *D. monostachyae* affinis, sed foliis multo minoribus, integro-bifidis vel 2 foliolis, inflorescentia pilis castaneis juxta flores carentibus differt. Typus: Antsiranana, Réserve Naturelle de Marojejy, along trail to Marojejy Est, *Dransfield et al.* JD6764 (Holotypus K; isotypus TAN).

Solitary slender palmlet to 1 m tall. **STEM** 3–7 mm diam., internodes 5–15 mm long with scattered scales, nodal scars c. 1 mm wide. **LEAVES** 5–8 in crown; sheaths 4–7.5 cm long, 0.5–0.8 cm diam., longitudinally striate, with very sparse caducous dark red-brown scales, membranous by the mouth, sometimes tattering but lacking distinct auricles; petiole 1–3.5 cm long, c. 1 mm wide, triangular in cross section, bearing sparse caducous dark brown scales;

Dypsis spicata. Form with few broad leaflets (*Dransfield et al.* JD6748).

lamina entire bifid or divided into 2 broad leaflets on each side of the rachis, rachis (or costa) 11–15 cm, sparsely scaly; entire bifid lamina 17–20 cm long, the two apical lobes 4.5–7 × 2.5–3.7 cm; where blade divided into leaflets, leaflets 6–9 × 1.5–2 cm, the distal pair not conspicuously broader than the proximal leaflets, lamina surfaces with minute punctiform brown scales borne on pale thickened bases, denser on the abaxial surface than on the adaxial; leaflets drying dark grey on upper surface, chestnut-brown on under surface. **INFLORESCENCE** spicate (?always–see below), interfoliar, erect at first, becoming pendulous in fruit, shorter than the leaves, 14–21 cm long; peduncle to 13 cm long; prophyll 8–11 × 0.2–0.4 cm, peduncular bract 6.5–9 × 0.2–0.4 cm, both prophyll and peduncular bract sparsely scaly; spike 3–9 × 0.15–0.25 cm, elongating somewhat after anthesis; triads rather sparse in proximal portion, more densely arranged distally, rachilla bract c. 1 mm, rounded to apiculate with laciniate scales, rachilla ± glabrous. **STAMINATE FLOWERS** c. 0.7 mm high; sepals c. 0.4 mm long, keeled; petals ± elliptical, 0.5 × 0.4 mm; stamens 3, antepetalous, alternating with 3 antesepalous triangular staminodes, together borne on a short androecial tube ring to 0.2 mm high, connective ± triangular, anthers subdidymous, pistillode not seen. **PISTILLATE FLOWERS** rounded; sepals imbricate, rounded, c. 1 × 1 mm; petals imbricate, rounded-triangular, 1.2 × 1 mm; staminodes 3, minute, dentiform; ovary c. 1 mm diam., post anthesis. Mature **FRUIT** cherry-red, glistening, ellipsoid, c. 12 × 7 mm. **SEED** fusiform, 12 × 3.5 mm; endosperm homogeneous.

Dypsis spicata. Close-up of fruits (*Dransfield et al.* JD6764).

NOTE. We had misidentified the plants from Marojejy and environs as *Neophloga catatiana* (= *Dypsis catatiana*); there is a quite astonishing similarity between the two taxa. However, the three antepetalous stamens alternating with staminodes, borne in an androecial ring are quite different from the six stamens of *Dypsis catatiana* and its relatives. At first we included these collections with *D. monostachya*, but later decided that the collections from Marojejy represent an undescribed taxon. *D. monostachya* is distinguished by it larger leaves that are divided into about six narrow leaflets on each side of the rachis, that dry pale green; in *D. spicata* the leaf is undivided or divided into two broad leaflets on each side of the rachis and the lamina dries dark grey on the upper surface, and chestnut-brown on the lower surface. In *D. monostachya* the peduncle is much longer than in *D. spicata* and in the former there is a distinctive cluster of shining chestnut-brown scales on the inflorescence axis, just above and adaxial to the triads. Such hairs are absent in the latter species. *D. monostachya* remains a rather poorly known taxon.

One further collection should be mentioned here. This is *Homolle* 518 (K, P), collected in the environs of Lac Alaotra, on the banks of the Maningory between Menasaka and Ambodiriana in December 1944 (fl.). It has an entire bifid leaf and seems very similar to *D. spicata*. However, the inflorescence bears a single branch at the base of the spike. We tentatively name it as *D. spicata*.

SPECIMENS SEEN. Andapa: Réserve Naturelle de Marojejy, along trail to Marojejy Est, Feb. 1989 (fl., fr.), *Miller & Lowry* 3951 (K, MO, TAN); idem, Nov. 1989 (fl.), *Dransfield et al.* JD6747 (K, TAN); Nov. 1989 (fl.), *Dransfield et al.* JD6748 (K, TAN); idem, Nov. 1989 (fl.), JD6753 (K, TAN); idem, Nov. 1989 (fl., fr.), JD6764 (Holotype K; isotype TAN); 4km west of Beamalona, Anjanaharibe, June 1992 (fl.), *Beentje & Andriampaniry* 4681 (K, TAN); idem, June 1992 (fl.), *Beentje & Andriampaniry* 4685 (K, TAN).

138. DYPSIS ACAULIS

A very distinctive stemless palm, known only from a single herbarium specimen collected by Perrier de la Bâthie on the Masoala Peninsula. This is one of the very few palms in Madagascar that has whitish undersides to the leaves. This combined with the habit, the undivided blade and the spicate inflorescence should make it easy to identify. Despite this, it has not been refound. The species name (Latin for stemless) refers to the habit of this palm.

DISTRIBUTION. NE Madagascar, Masoala Peninsula.

HABITAT. Lowland rain forest; 40 m.
LOCAL NAMES. Not recorded.
USES. Not recorded.
CONSERVATION STATUS. Presumed extinct. Not collected for more than eighty years.

Dypsis acaulis *J. Dransf.* **sp. nov.**
acaulescens, procumbens, foliis bifidis pagina inferiore laminae indumento griseo tecta, inflorescentia spicata, longe pedunculata, floribus staminatis triandris staminibus antepetalis bene distincta. Typus: Madagascar, Antalaha, Masoala, Marambo, *Perrier* 12044 (Holotypus P; isotypus K).

Acaulescent forest undergrowth palm. **STEM** probably solitary, procumbent, 5 cm long, c. 18 mm diam., circular in cross section, drying pale buff; roots cylindrical, c. 3 mm diam. at insertion. **LEAVES** entire,

Dypsis spicata. **A** crown with entire bifid leaf × 3/5; **B** crown with divided leaves × 3/5; **C** staminate flower, one petal removed × 30. **A** from *Dransfield et al.* JD6753, **B**, **C** from *Dransfield et al.* JD6748. Drawn by Rosemary Wise.

bifid; sheath open to the base, 6 × 3 cm (opened out at the base), pale brown covered with reddish brown tomentum and dark brown punctiform scales; petiole c. 26 × 0.4 cm, triangular in cross section, adaxially shallowly channelled, the margins quite sharp, ± glabrescent, pale buff or with sparse reddish brown tomentum; costa of leaf 18 cm long, outer leaf margins 45 cm long, inner margins c. 28 cm, the two lobes 4.5–5.5 cm wide, the tips irregularly dentate or subpraemorse; lamina with c. 8–9 major adaxial folds, adaxial surface drying grey, transverse veinlets conspicuous but short, abaxially covered in dense grey indumentum and red-brown punctiform scales. **Inflorescence** spicate to c. 22 cm; peduncle c. 13 cm, c. 2 mm diam.; prophyll tightly tubular, 9 × 0.3 cm; peduncular bract c. 15 × 0.3 cm, both covered in scattered brown lacerate scales; spike to 9 × 0.3–0.4 cm, cylindrical, densely covered in thick red-brown tomentum entirely obscuring bracts and flowers, only very faintly showing impressions of bracts; bracts scarcely 1 mm high forming pits. **Staminate flowers** 0.7 mm high; sepals 3, c. 0.25 mm, margins ciliate, chaffy, buff-coloured; petals 3, dark, c. 0.7 × 0.3 mm, longitudinally striate; stamens 3, antepetalous, anthers didymous, staminodes 3, dentiform, between the fertile stamens; pistillode minute, pyramidal. **Fruit** (lost), said to be bright red, fusiform, 20 × 6 mm. (Plate: see next page).

Note. Known only from material with very young staminate flower buds.

Specimens seen. Antalaha: Masoala, Marambo, towards Antalaha, Aug. 1912 (buds), *Perrier* 12044 (Holotype P; isotype K).

139. Dypsis canescens

Insufficiently known species

Distribution. NW Madagascar, only known from the type.
Habitat. Forest on sandstone; c. 40 m.
Local names. Not recorded.
Uses. Not recorded.
Conservation status. Presumed extinct; not seen for more than fifty years.

Dypsis canescens *(Jum. & H. Perrier) Beentje & J. Dransf.* **comb. nov.**

Synonym:
Chrysalidocarpus canescens Jum. & H. Perrier, Ann. Inst. Bot.-Géol. Colon. Marseille sér. 3, 1 (1): 38, t. 19 (1913); Jum., Ann. Inst. Bot.-Géol. Colon. Marseille sér. 10, 3: 12 (1922); Jum., Cat. Pl. Madagascar, Palmae: 8 (1938); Jum. & H. Perrier, Fl. Madagascar 30: 106 (1945). Type: Madagascar, Sambirano forest at Ambaliha, *Perrier* s.n. (?P, not seen)

Description from literature: Trunk 4–8m, straight, ringed, < 10 cm diam., base somewhat thickened in a cone. **Leaves** 8–12, ascending; sheath glabrous, membranous, without ligule; petiole absent; rachis c. 2 m long; leaflets regular, recurved, 3.5–4 cm distant, proximal 60–65 × 0.6–0.7 cm, median 75–80 × 1.5 cm, some brownish ramenta on the midrib, adaxially white-mealy when young. **Inflorescence** at most branched to 2 orders; peduncle 60 cm; prophyll borne at 13 cm above the base of the peduncle, 40 cm long, split only at apex; peduncular bract deciduous, inserted at 30 cm above the base of the peduncle, 45 cm long, conical at the apex, split along its length; rachillae often with paired or solitary staminate flowers due to abortion of the pistillate ones. **Staminate flowers** with keeled and slightly spurred sepals, glabrous, ovate, obtuse, 2 × 2 mm; petals ovate, acute, 3 × 2.2 mm; filaments c. 1.3 mm; ovary rudiment cylindrical, 1.8 mm, about as long as the stamens. **Pistillate flowers** unknown. **Fruit** unknown.

Note. In the protologue the species is compared (as is *Chrysalidocarpus brevinodis*) with *D. lutescens* with the following difference: absence of petiole.

The type locality is either Ambaliha on the Ampasindava Peninsula, as seems to be indicated in the protologue, or another Ambaliha (not found on map or in Gazetteer) on the left bank of the Sambirano River.

The fact that it is compared with *D. lutescens* makes us think the leaflets might be regular, but unfortunately we cannot be certain.

Specimens seen. None. Despite a thorough search, we have not found the type specimen.

140. Dypsis ramentacea

This pretty little palm is known from a single collection from forest undergrowth in the lowlands near to Mananara Avaratra. The species name refers to the scales (ramenta) on the undersurface of the leaves.

Distribution. Mananara, Antanambe.
Habitat. Lowland rain forest; 115 m.
Local names. Not recorded.
Uses. Not recorded.
Conservation Status. Critical; found in a tiny patch of forest near agricultural land; individuals seen less than twenty-five.

Dypsis ramentacea *J. Dransf.* **sp. nov.**
habitu inflorescentiaque *D. louvelii* superficialiter similis sed petiolo longiori, pagina inferiori laminae ramentas ferenti rachillis inflorescentiae papillosis differt (floribus ignotis). Typus: Madagascar, Mananara, Antanambe, *Beentje* 4458 (Holotypus K; isotypi BH, MO, P, TAN).

Slender solitary forest undergrowth palmlet to 70 cm tall. **Stems** c. 4–5 mm diam., when dry, when fresh to 8 mm diam., internodes 4–12 mm long, sparsely and minutely scaly. **Leaves** 9–10 in crown, apparently marcescent, crownshaft poorly defined; sheath 5–6 cm long, 1 cm diam., rather densely covered in caducous dark red-brown indumentum; auricles triangular, to 5 × 3 mm, membranous, soon tattering; petiole 9–15 cm long, c. 2 mm wide, ± triangular in section, bearing scattered dark brown scales; lamina entire 25–31 cm long, deeply bifid, the rachis 8–9.5 cm, the two lobes to 18–22 × 3.5 cm, diverging at an acute angle, the lobes shallowly toothed, adaxial lamina surface with

Dypsis acaulis. Whole plant × 1/3. From *Perrier* 12044. Drawn by Margaret Tebbs.

sparse, minute brown punctiform scales, abaxial surface with conspicuous, abundant dark brown ramenta on all major veins and scattered red-brown punctiform scales. **INFLORESCENCES** 25–30 cm long, arching or semi-horizontal, branching to 1 order; peduncle to 21 cm, 1 mm diam., covered in discrete elevated dark brown scales; prophyll to 15 × 0.4 cm, membranous, straw-coloured, with sparse dark scales; peduncular bract exceeding the prophyll by 7 cm, similar to prophyll; rachis 5–6 cm; rachillae 8–10 in number, 2–3.5 cm long, c. 0.6 mm diam., with dark brown papillae, triads 1–3 mm distant. **FLOWERS** said to be red in bud, opening white, not available. Mature **FRUIT** bright red, ovoid, 13 × 8 mm; seed 1–2 × 6 mm; endosperm homogeneous, embryo lateral near the base.

NOTE. Although no flowers have been available for study, the habit and inflorescence structure suggests that this is probably a species of *Dypsis* with three stamens and is certainly undescribed. Because it is so distinctive, we have named it *D. ramentacea*, even though staminate flowers that would suggest relationships are missing. The inflorescence is reminiscent of that of *D. louvelii* but the rachillae are distinctive in the presence of dark papillae. The undersurface of the leaf bears abundant ramenta, whence the specific epithet. Such ramenta are also present in *D. mocquerysiana* but this species has a very different erect inflorescence with many rachillae, and the leaves are usually epetiolate. There is also a superficial similarity with *D. glabrescens* but in the latter ramenta are absent, there is no petiole and the rachillae are shorter.

SPECIMENS SEEN. Mananara Avaratra: Antanambe, 1.5 hours walk upstream from first ferry north of Antanambe, Oct. 1991 (fr.), *Beentje* 4458 (Holotype K; isotypes BH, MO, P, TAN).

Dypsis ramentacea, growing south of Mananara Avaratra (*Beentje* 4458).

DOUBTFUL NAMES

Areca madagascariensis Mart. **nomen dubium.**
Hist. Nat. Palm. 3: 179, t. 158: V, 1–5 (1837). Syntypes: Madagascar, Arbre a sel, Areque singe, *Poivre* s.n., *Michaux* s.n., both types not found at P or elsewhere.

NOTE: The epithet clashes with, and predates, *Chrysalidocarpus madagascariensis*, but *Areca madagascariensis* must be a **nomen dubium**.

Phloga microphoenix Baill. **nomen dubium.**
Bull. Soc. Linn. Paris 150: 1185 (1895). Syntypes: Madagascar, Île Sainte Marie, Tafondrou forest, *Boivin* 1703 (not seen), 1704 (P).

NOTE. We are very doubtful about this name. *Boivin* 1704 is *Dypsis forficifolia* Mart.; however, Baillon describes the leaflets as many, regular, linear-lanceolate, 40 × 2 cm; this sounds most unlike *Phloga nodifera*, or *Dypsis forficifolia* for that matter. Beccari (1906, 1914) says the type consists of a young *Phloga nodifera* inflorescence and the leaves of another palm, probably *Dypsis boiviniana*. He cites *Boivin* 1704 under *Dypsis forficifolia*, so presumably saw nr. 1703.

Dypsis vilersiana Baill. **nomen dubium.**
Bull. Soc. Linn. Paris 147: 1165 (1894). Type: Central Madagascar, *Le Myre de Vilers* s.n. (P, type; not seen, said to be an inflorescence only, described as having 3 stamens).

NOTE. Beccari 1906 says this is a young inflorescence of *Phloga polystachya*.

Dypsis ramentacea. Crown, showing inflorescence × 2/5. From *Beentje* 4458. Drawn by Rosemary Wise.

Neophloga andasiensis Jum. & H. Perrier **nomen nudum** in Martelli, Nuov. Giorn. Bot. Ital. n.s. 41: 717 (1935).

Note. We have no idea to which palm this name refers, but, as a *nomen nudum* it can safely be ignored.

Haplophloga comorensis Baill.
Bull. Soc. Linn. Paris 148: 1171 (1894). Type: ? , said to be from Mayotte.

Note. We have seen no material of this species and Baillon's description provides very little information.

Haplophloga loucoubensis Baill.
Bull. Soc. Linn. Paris 148: 1171 (1894). Type: ? , said to be from Nosy Be.

Note. We have seen no material of this species and Baillon's description provides very little information.

SUBTRIBE LEMUROPHOENICINAE

Lemurophoenicinae *J. Dransf.* **subtribus nova**
ad subfamiliam *Arecoideas*, tribum *Areceas* pertinens, a ceteris subtribubus pseudomonomeris structura fructus differt; facie externa florum subtribui *Dypsidinis* accedens sed floribus masculis 52–59 vice 6 stamina ferentibus et fructu suberoso-verrucato endocarpio basin tumorem globuliformem cordiformem ferenti endospermio superficialiter ruminato et embryone apicali recedit. Genus solum madagascariense *Lemurophoenix* J. Dransf.

Belonging to Subfamily *Arecoideae*, tribe *Areceae*, but differing from all other pseudomonomerous subtribes in fruit structure; external form of the flowers similar to that of members of *Dypsidinae*, but the staminate with 52–59 rather than 6 stamens; fruit with a corky-warted pericarp, the endocarp bearing a basal heart-shaped button, the endosperm superficially ruminate and the embryo apical. Only genus, and type: *Lemurophoenix* J. Dransf. (Madagascar).

X. LEMUROPHOENIX

J. Dransfield, Kew Bull. 46: 61 (1991). Type: *L. halleuxii* J. Dransf.

Massive solitary unarmed monoecious pleonanthic tree palm. **TRUNK** bare, ringed with leaf scars. Crownshaft well developed. **LEAF** reduplicately pinnate; sheath tubular, bearing wax and scales; petiole rather short, channelled adaxially, rounded or ridged abaxially, densely covered with caducous chocolate-brown scales; rachis adaxially somewhat channelled near the base, ridged distally, abaxially rounded or flattened, scaly as the petiole; leaflets regular, numerous, linear-lanceolate, long acuminate except near the tip where bifid; adaxial leaflet surface glabrous, abaxial bearing a few large dark brown ramenta near the base on the main vein and more numerous small ramenta on secondary veins, pale brown peltate scales abundant on all veins; transverse veinlets not visible. **INFLORESCENCE** infrafoliar, branched to 3 orders, the whole inflorescence exposed long before anthesis, protandrous; peduncle moderate in length; prophyll splitting along one side; peduncular bract longer than the prophyll; first-order branches widely spreading, the basal few branched to the third order, the distal branched to the second order or unbranched; rachillae numerous, elongate, pendulous or spreading, somewhat swollen, with flowers partially embedded in shallow pits; rachilla bracts rather obscure, forming the lower lip to the pits; floral bracteoles minute, included in the pits. **FLOWERS** borne in triads of a central pistillate and two lateral staminate for about two-thirds the rachilla length, and in pairs of staminate flowers in the distal third. **STAMINATE FLOWER** in bud ± bullet-shaped; sepals 3, ± distinct, minutely connate at the base, imbricate, strongly keeled and gibbous; petals 3, ± distinct in bud, valvate, boat-shaped, adaxially grooved, glabrous, later the floral receptacle greatly enlarging carrying the petal bases above the calyx, the petals becoming reflexed by a swollen pulvinus at the petal bases; stamens 52–59, borne on the dome-shaped receptacle, filaments terete, straight or contorted in bud, rarely filaments partially connate, anthers frequently rather irregular in outline due to close-packing in the bud, basally sagittate, medifixed, latrorse; pollen elliptic, monosulcate with perforate tectate exine; pistillode columnar, hidden among the filament bases. **PISTILLATE FLOWER** known only in immature bud; sepals 3, distinct, imbricate, unequal, rounded; petals 3, distinct, basally strongly imbricate, with triangular valvate tips; staminodes 10–12, minute, tooth-like or strap-shaped, distributed evenly around the gynoecium; gynoecium pseudomonomerous, ovoid, stigmas apical, as yet scarcely developed, ovule heminanatropous, basally attached. **FRUIT** large, usually borne in abundance, globose, the epicarp cracked polygonally into low corky warts, stigmatic remains basal; mesocarp rather spongy, easily separable from the endocarp; endocarp spherical, with a basal heart-shaped pale brown button. **SEED:** endosperm very shallowly and sparsely ruminate; embryo apical. **GERMINATION** adjacent-ligular; eophyll bifid.

Lemurophoenix halleuxii. View into the crown of a fruiting individual, David Cooke providing scale (*Dransfield et al.* JD6402).

DISTRIBUTION. Monotypic; endemic to Madagascar.

LEMUROPHOENIX HALLEUXII

This majestic palm, probably the grandest palm of the whole island, occurs in a steep-sided valley below the long ridge leading eastwards from the village of Sahavary, northeast of Maroantsetra. The valley is at about 350–450 m above sea level. There is a great abundance of palms in the forest here, yet there is little peculiar about the habitat. Furthermore *Lemurophoenix* does not occur in similar valleys nearby. As one drops down from the crest of the ridge, suddenly one becomes aware of immense palms in the forest canopy in the valley bottom, the huge greyish-pink crownshafts and vast bunches of fruit standing out amidst the surrounding dull green. There are about 30 mature individuals and about 20 juveniles in various stages of development. Beneath the mature trees lie thick carpets of bare endocarps, either newly fallen or rotting. Seedlings are distinctive but very scarce. It appears that there is little effective dispersal and very limited regeneration. Recently a single mature tree has been discovered in a deep valley further south on the Masoala Peninsula.

As one of the most beautiful and spectacular palms of Madagascar, *Lemurophoenix halleuxii*, the red-lemur palm, has been much sought after. Seeds are harvested regularly from the Sahavary locality, most of them exported to palm enthusiasts and nurserymen.

However, it has proved to be difficult to grow so far. Growing conditions that mimic the natural habitat would seem to be the most likely to lead to success. The species was named for Dominique Halleux who was among the team of people who first located the palm in the mid 1980s.

DISTRIBUTION. Masoala Peninsula and environs.
HABITAT. Growing on steep slopes and in a deep narrow valley in primary forest at 200–450 m elevation.
LOCAL NAME. *Hovitra vari mena* (Betsimisaraka).
USES. Not recorded.
CONSERVATION STATUS. Endangered. Only known from two populations; numbers very low (c. 50); not protected.

Lemurophoenix halleuxii *J. Dransf.*
Kew Bull. 46: 62 (1991). Type: Madagascar, Maroantsetra, *Dransfield et al.* JD6453 (Holotype K; isotypes BH, MO, NY, P, TAN).

Massive solitary unarmed monoecious pleonanthic tree palm. **TRUNK** bare to 20 m tall, c. 1 m diam. at very base, c. 50 cm d.b.h., tapering somewhat distally, ringed with leaf scars. Crownshaft well developed, to 1.5 m long, c. 50 cm diam. **LEAF** with sheath greyish-pink when newly exposed, bearing white wax and dark brown scales; petiole c. 25 cm long, c. 10 cm wide,

Lemurophoenix halleuxii, towering above the valley bottom, Sahavary.

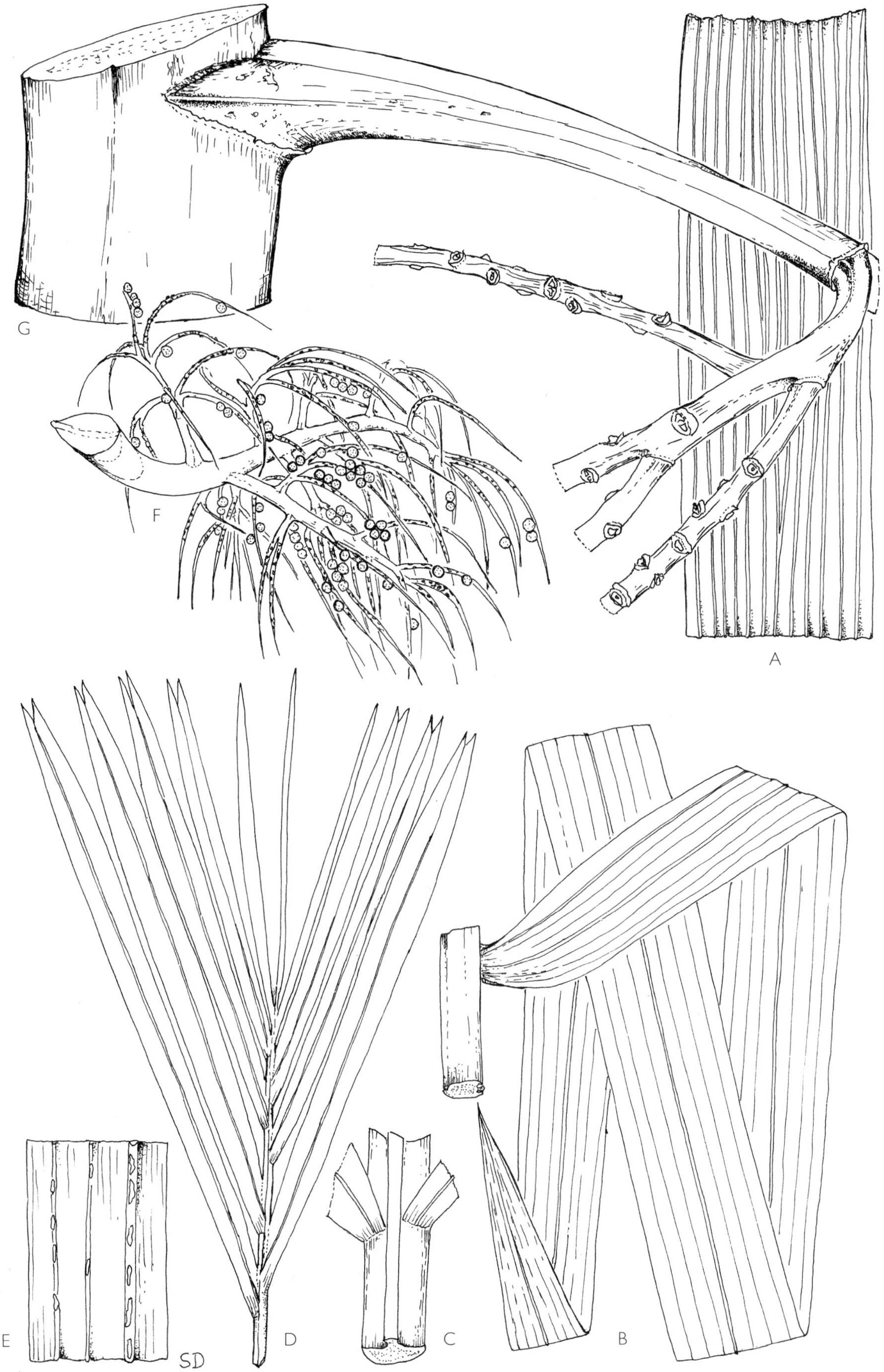

Lemurophoenix halleuxii. **A** detail of leaf sheath × 3/4; **B** leaflet from mid leaf, abaxial view × 1/3; **C** detail of rachis at mid leaf, adaxial view × 1/2; **D** leaf tip × 1/3; **E** detail of lower surface of leaflet × 4; **F** infructescence, greatly reduced; **G** tip of peduncle and proximal first-order branch × 3/4. **A** from *Dransfield et al.* JD6453, **B – G** from *Dransfield et al.* JD6402. Drawn by Soejatmi Dransfield.

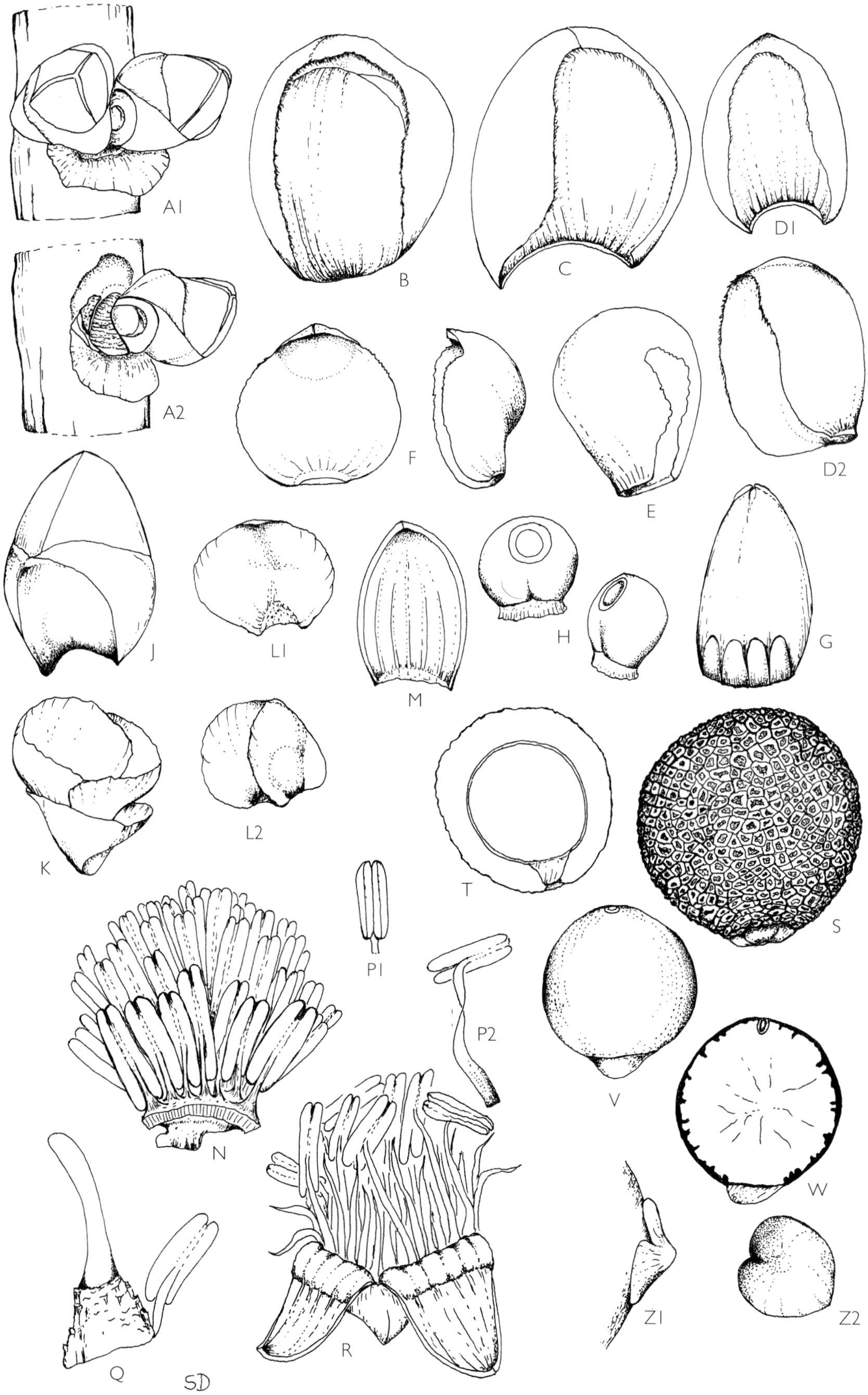

Lemurophoenix halleuxii. **A1** triad × 4; **A2** triad, one staminate flower removed × 4; **B** pistillate flower bud × 12; **C, D1**, **D2**, **E** sepals of pistillate flower bud × 12; **F** two views of petal of pistillate flower × 12; **G** ovary with four staminodes × 20; **H** two views of ovule × 60; **J** staminate flower bud × 5; **K**, **L1**, **L2** sepals of staminate flower × 5; **M** petal of staminate flower bud × 5; **N** androecium of staminate flower bud × 10; **P1** stamen from bud × 10; **P2** stamen from open flower × 4; **Q** pistillode and one stamen × 10; **R** open staminate flower × 5; **S** fruit × 3/4; **T** vertical section of immature fruit × 3/4; **V** endocarp × 3/4; **W** vertical section of endocarp and seed × 3/4; **Z1**, **Z2** two views of basal excrescence on endocarp × 1.5. **A – V**, **Z** from *Dransfield et al.* JD6453, **W** from *Dransfield et al.* JD6402. Drawn by Soejatmi Dransfield.

Lemurophoenix halleuxii. Close-up of corky-warted fruit (*Dransfield et al.* JD6742).

Lemurophoenix halleuxii. Crown in full fruit, Sahavary (Photo: D.N. Cooke).

c. 5 cm thick, channelled adaxially, rounded or ridged abaxially, densely covered with caducous chocolate-brown scales; rachis to 4.25 m long, adaxially somewhat channelled near the base, ridged distally, abaxially rounded or flattened, scaly as the petiole; leaflets c. 60 on each side of the rachis, regularly arranged, linear-lanceolate, long acuminate except near the tip where bifid, the proximal to 65 × 2 cm, mid-leaf leaflets to 95 × 6 cm, decreasing abruptly to 17 × 0.7 cm near the leaf tip, adaxial leaflet surface glabrous, abaxial bearing a few large dark brown ramenta near the base on the main vein and more numerous small ramenta on secondary veins, pale brown peltate scales abundant on all veins; transverse veinlets not visible; emerging rachis crimson, expanding blade flushed red. **INFLORESCENCE** massive, to 2 m long, infrafoliar, branched to 3 orders, the whole inflorescence exposed long before anthesis, protandrous; peduncle c. 50 cm long, grossly swollen and winged to 35 cm wide at the base, decreasing to 10 × 4.5 cm; prophyll to 90 × 30 cm, inserted c. 15 cm above base of peduncle, abaxially bright crimson, cream-coloured within, splitting along one side; peduncular bract to 120 × 30 cm, inserted c. 10 cm above prophyll, otherwise similar; first-order branches c. 12–15, widely spreading, the basal few devoid of branches in the basal 20–25 cm, to 4–5 cm thick, then branched to the third order, the distal first-order branches branched to the second order or unbranched; rachillae c. 110, elongate, pendulous or spreading, to 40 cm long, c. 8–9 mm diam., cream-coloured when newly emerged, becoming green, somewhat swollen, with flowers partially embedded in shallow pits c. 6–10 mm apart, rachilla surface minutely papillose and scaly; rachilla bracts rather obscure, 2 × 8 mm, forming the lower lip to the pits; floral bracteoles minute, included in the pits. **FLOWERS** borne in triads of a central pistillate and two lateral staminate for about two-thirds the rachilla length, and in pairs of staminate flowers in the distal third. **STAMINATE FLOWERS** in bud ± bullet-shaped, greenish brown, c. 8 × 5 mm; sepals c. 3.5 × 6 mm, strongly keeled and gibbous; petals c. 7 × 3 mm, pale brown; stamens 52–59, borne on the dome-shaped receptacle, filaments terete, 2–4 × 0.1 mm, straight or contorted in bud, anthers c. 2 × 0.5 mm; pistillode columnar, c. 1.5 × 0.2 mm. **PISTILLATE FLOWERS** known only in immature bud, c. 4.5 mm diam.; sepals c. 3.5 × 4.5 mm; petals c. 3.5 mm long; staminodes 0.4 × 0.2 mm; gynoecium c. 2.5 × 1.5 mm. **FRUIT** chestnut brown, c. 50 mm diam. at maturity, usually borne in abundance, globose, the epicarp cracked polygonally into low corky warts less than 1 mm high, 2–5 mm across, mesocarp c. 8 mm thick, whitish, rather spongy, easily separable from the endocarp; endocarp spherical, c. 33 mm diam., with a basal heart-shaped pale brown button 11 × 11 × 7 mm. **SEED** endosperm very shallowly and sparsely ruminate. **EOPHYLL** bifid.

SPECIMENS SEEN. Maroantsetra, Andranofotsy River, Sahavary, hills east of village, Andilampananina, Oct. 1986 (fr.), *Dransfield et al.* JD6402 (BH, K, MO, NY, P, TAN); idem, Feb. 1988 (fl., fr.), *Dransfield et al.* JD6453 (Holotype K; isotypes BH, MO, NY, P, TAN); Masoala Peninsula, Nov. 89 (fr.), *Dransfield et al.* JD6742 (K, TAN).

SUBTRIBE MASOALINAE

Masoalinae *Beentje & J. Dransf.*, **subtribus nova**
ad subfamiliam *Arecoideas* tribum *Areceas* pertinens, palmae monoecae robustae, solitariae, brevicaules, foliis grandis, reduplicatis, corona foliorum detritifera, inflorescentia interfoliacea staminata vel pistillata vel bisexuali, prophyllo vagina folii includenti, bractea pedunculari prophyllo simile, bracteis peduncularibus incompletis numerosis, fructu vestigio stigmatis terminalis distincta. Subtribus madagascariense, duobus generibus *Masoala* et *Marojejya*. Typus: *Masoala*.

Belonging to subfamily *Arecoideae* tribe *Areceae*. Robust, solitary, short-stemmed litter-trapping palms with large reduplicate leaves; inflorescences interfoliar, staminate, pistillate or bisexual, the prophyll hidden by the leaf sheaths, peduncular bract similar to the prophyll, numerous incomplete peduncular bracts also present; fruit with terminal stigmatic residue. Two genera in Madagascar, *Masoala* and *Marojejya*. Type: *Masoala*.

Dransfield & Uhl (1986a) placed *Marojejya* in the same subtribe as *Sclerosperma* from West Africa (*Sclerospermatinae*), while *Masoala* was unplaced, apart from being included in tribe *Areceae*. When new material of *Masoala* became available, it was immediately obvious that *Marojejya* had much in common with *Masoala* and that the two genera, quite distinct from each other in the sexuality of the inflorescences, nevertheless were probably related. We then found material of *Marojejya insignis* with inflorescences bearing flowers of both sexes and material of an undescribed taxon that appeared to be intermediate between the two genera, and thus emphasising their similarity. This latter taxon, described here as *Masoala kona*, we now believe to be more similar to *Masoala madagascariensis* than to either species of *Marojejya* and so have decided to maintain the two genera, while emphasising their presumed relationship by accommodating them in a new subtribe. We believe that the inclusion of *Marojejya* in *Sclerospermatinae* was unjustified, based on similarities in habit and the condensed inflorescence that are superfical rather than natural.

These are wonderful palms, rather local in their distribution; where they occur, they are often accompanied by other unusual palms. The great litter-trapping crowns, often with snaking zigzag tree roots from neighbouring dicotyledonous trees penetrating the litter in the crowns, are most conspicuous.

KEY TO MASOALINAE

1. Inflorescences bisexual, branching to 2 orders, elongate, lax, long-exserted from the leafsheath (?*kona* sometimes 1 order); peduncle elongate; rachis longer than the peduncle; petals free (briefly connate in *M. kona*); pistillode as large as the stamens; sepals broadly imbricate; fruit symmetrical, with apical stigmatic remains .... **Masoala**

   Inflorescences unisexual (but pistillate with abortive staminate flowers), congested, ± hidden by leaf bases, branching to 1 order; peduncle short; rachis shorter than the peduncle; staminate flowers with petals connate in lower 1/3 and adnate to the receptacle; pistillode minute; sepals narrow in staminate flower; fruit asymmetrical, with lateral or sub-basal stigmatic remains .... **Marojejya**

XI. MASOALA

Jum., Ann. Inst. Bot.-Géol. Colon. Marseille sér. 5, 1 (1): 8 (1933); Jum. & H. Perrier, Fl. Madagascar 30: 50 (1945); N.W. Uhl & J. Dransf., Genera Palmarum 478 (1987). Type: *Masoala madagascariensis* Jum.

Robust, solitary, unarmed, pleonanthic, monoecious palms. **TRUNK** erect, short, covered with the remains of the leaf sheaths. **LEAVES** large, reduplicately pinnate, marcescent; sheaths tubular, distally attenuate, the margins smooth, without ligules or auricles; petiole relatively short, adaxially deeply channelled, abaxially rounded; rachis adaxially keeled, abaxially rounded or flattened; leaflets numerous, linear, single-fold or multi-fold, the midrib strong, the tips briefly bifid, adaxially glabrous, abaxially minutely dotted, transverse veinlets obscure. **INFLORESCENCE** solitary (occasionally multiple), interfoliar, branching to 2 orders; peduncle elongate, semicircular; prophyll inserted some distance above the base of the peduncle, large, flattened, narrow, elliptical, beaked, strongly 2-keeled, coriaceous, splitting longitudinally along the abaxial face, persistent; peduncular bracts 1–3, similar to the prophyll but not 2-keeled, relatively large, open; rachis longer than the peduncle, bearing spirally arranged, short triangular bracts each subtending a first order branch; proximal first order branches with a short bare portion bearing 1–2 branches; rachillae ± straight, rather thick, elongate, bearing either triads or solitary pistillate flowers proximally and paired or solitary staminate flowers distally, each subtended by a low triangular bract; floral bracteoles conspicuous, triangular. **STAMINATE FLOWERS** symmetrical; sepals 3, distinct, imbricate, coriaceous, triangular, keeled; petals 3, valvate, distinct or proximally briefly connate, triangular, longer than the sepals, coriaceous; stamens 6, filaments

Masoala madagascariensis, growing in rather open forest on the eastern foothills of Marojejy (*Dransfield* JD6770).

KEY TO MASOALA

1. Leaves pinnate, with ± equal leaflets .... 1. **M. madagascariensis**

   Leaves irregularly divided, the basal segments large and multifold .... 2. **M. kona**

slender, distinct, anthers elongate, dorsifixed, latrorse, slightly versatile, the locules slightly divergent; pistillode cylindrical or slightly conical. **PISTILLATE FLOWERS** ovoid, larger than the staminate; sepals 3, distinct, broadly imbricate, triangular, keeled, coriaceous, striate; petals 3, distinct, imbricate with valvate tips, triangular, coriaceous, striate; staminodes 6, minute; gynoecium ovoid, stigmas 3, large, triangular; ovule 1. **FRUIT** subglobose with pronounced terminal stigmatic remains; mesocarp fleshy; endocarp fibrous, adherent to the seed. **SEED** with homogeneous endosperm and basal embryo. **EOPHYLL** bifid.

DISTRIBUTION. Two species, endemic to Madagascar.

1. MASOALA MADAGASCARIENSIS

Until 1986, M. *madagascariensis* was known only from very incomplete material preserved in the Paris Herbarium, material that gave little indication of its appearance and relationships. In Genera Palmarum, *Masoala* was included within the Arecoid palms in an uncertain position, but as the book went to press, new material was collected. We have now seen this remarkable palm in several localities, from the Marojejy Massif southwards to the Bay of Antongil. *M. madagascariensis* is a squat but rather massive litter-trapping palm of the most humid forest.

DISTRIBUTION. Marojejy, Masoala and Mananara.
HABITAT. Lowland rain forest; dry hillside to swampy valley bottom, occasionally on ultramafic soils; 200–420 m.
LOCAL NAMES. *Kase, Hovotralanana, Mandanozezika* (Betsimisaraka).
USES. Used for thatch; palm-heart edible.
CONSERVATION STATUS. Vulnerable. The species is only known from three recent sites, and numbers in all sites were low.

Masoala madagascariensis *Jum.*
Ann. Inst. Bot.-Géol. Colon. Marseille sér. 5, 1 (1): 8 (1933); Cat. Pl. Madagascar, Palmae: 16 (1938); Jum. & H. Perrier, Fl. Madagascar 30: 52, fig. 14 (1945). Type: Madagascar, Masoala Peninsula, Marambo, Oct. 1912, *Perrier* 11938 (Holotype P).

Solitary palm. **TRUNK** 3.5–10 m, 30–35 cm diam. when covered in sheath remnants, 14–20 cm diam. for clean trunk; internodes 2–5 cm, pale brown; nodal scars faint; wood hard. **LEAVES** 20–31 per crown, porrect to spreading, held in shuttlecock mode, stiff, 3–4 m long, tending to be held on edge, marcescent, litter-accumulating at base, with aerial roots penetrating the litter; sheath 45–50 cm long,

Masoala madagascariensis. Inflorescence (*Dransfield et al.* JD6390).

Masoala madagascariensis. Detail of crown showing broad reins (*Dransfield et al.* JD6390).

open, rather undulate, bright green, glabrous or with scattered brown indument, with large lateral auricles to 2 m long (JD6738) running into the reins or scarcely auricled (JD6770), long-attenuate into glabrous green apparent petiole, to 80 cm long or absent, proximally c. 25 cm wide, distally 5.5 × 2.5 cm diam., flat or slightly convex adaxially, with somewhat embedded scattered scales; rachis c. 3.9 m long, in mid-leaf c. 1.4 × 2.2 cm diam., keeled, abaxially with scattered small red glands/scales; leaflets mostly regular, plane, stiff, ± porrect, 55–70 on each side of the rachis, bright green, the most proximal pendulous and in groups of 4, proximal 71–117 × 1–4.5 cm, median 65–88 × 2.7–5.6 cm (interval 3–6 cm), distal 28–66 × 2.2–2.8 cm, main veins 4 rather faint, next to a clear midrib, abaxially on the midrib with a few large (6–12 mm long) red-brown laciniate ramenta, and on the minor veins with scattered small red glands, apices attenuate and slightly bifid, the distal pair joined for 3–5 cm and with acute to dentate apices less than 1 cm wide. **Inflorescence** interfoliar, arching and then erect, 0.7–1.3 m long, branching to 2 orders at base, with arching rachillae; multiple inflorescence primordia once observed (JD6738); peduncle 50–120 cm long, 2.5–8 × 1.3–4 cm diam., densely red-pubescent; prophyll 50–98 × 9–16 cm, borne at 8–10 cm above the base of the peduncle, bright pale green, tomentose when young, with scattered brown scales when older; peduncular bract 80–104 cm long, c. 8 cm wide, apparently circumscissile, not or hardly beaked, with few scattered scales; non-tubular peduncular bracts at regular intervals, spirally inserted, the proximal 17–22 × 6.5–8 cm, the 4 to 5 subsequent ones 0.5–7 × 3–7 cm and varying from narrowly triangular to a small ridge; rachis c. 90 cm long with 15–35 first order branches, these proximally 1.2–3 × 0.6–1.4 cm, with up to 7 second order branches; rachillae (9–) 20–43 cm long, 9–12 mm diam. when fresh, as little as 6 mm diam. when dried, glabrous or nearly so, with triads in pits, the young flowers green. **Staminate flowers** with the sepals 2.8–3.8 × 2–3.3 mm, keeled, acute; petals in bud 2.8–5 × 2.6–3.8 mm, fleshy; stamens slightly biseriate (offset 0.3 mm) with the antesepalous filaments flatter and to 0.8 mm wide, and the antesepalous ones triangular in cross-section (to 1.2 mm wide), filaments 1–2 mm long (connate at the base for up to 0.5 mm), anthers 2.3–3.5 × 1–1.2 mm, dorsifixed, with parallel locules, latrorse; pistillode cylindrical, 2.3–3.5 × 0.4–1.2 mm. **Pistillate flowers** with the sepals 5–6.5 × 4.5–7 mm; petals 6–8.5 × 7–9.5 mm; staminodes minute, 6 (no sign of ring), c. 0.3 mm high and flat; pistil at anthesis 11–18 × 6–10.5 mm, the trifid apex 2–3.5 mm high. **Fruit** green when young, when mature yellowish brown, subglobose with pronounced terminal stigmatic boss, 24–25 × 18–19 mm; mesocarp fleshy, 1.5–2 mm thick; endocarp

Masoala madagascariensis. Detail of pistillate flowers just after fertilisation (*Dransfield et al.* JD6770).

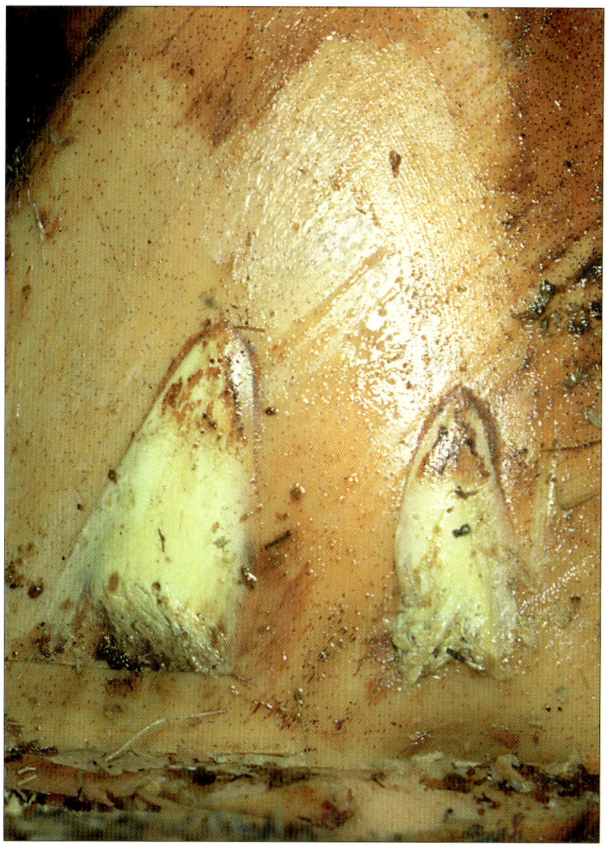

Masoala madagascariensis. Abortive multiple inflorescence buds (*Dransfield et al.* JD6738).

fibrous, thick, adherent to the seed, forming a hard layer around it, with parallel, free fibers reaching to the apex of the style boss. **SEED** depressed globose, 10–11 × 12–15 mm with homogeneous endosperm and basal embryo.

NOTE. In the type there are rachillae which are proximally almost exclusively pistillate (a few triads intermixed), and distally exclusively staminate; other rachillae in the type are exclusively staminate from the very base.

In JD6738 there are sometimes 2–3 inflorescence buds in a leaf axil, but only one seems to develop. We are uncertain about the protogyny or protandry of this species. *Beentje* 4472 has pistillate buds slightly larger than the staminate buds in the same triad, JD6390 with staminate buds of the same size as *Beentje* 4472 but almost non-existent pistillate buds; triads have been compared at the same distance above the base of the rachilla.

SPECIMENS SEEN. Andapa: Bangouabe R. headwaters, Nov. 1989 (fl., y.fr.), *Dransfield et al.* JD6770 (K, P, TAN). Antalaha: Marambo, Oct. 1912 (fl.), *Perrier* 11938 (P, type). Maroantsetra: Antalavia, Oct. 1986 (bud), *Dransfield et al.* JD6390 (K, P, TAN); idem, Nov. 1989 (bud), *Dransfield et al.* JD6738 (K, P, TAN) and JD6740 (K, P, TAN). Mananara Avaratra: Antanambe, Oct. 1991(fl.), *Beentje & Andriampantry* 4472 (BH, K, MO, P, TAN); idem, April 1992 (fr.), *Beentje et al.* 4632 (K).

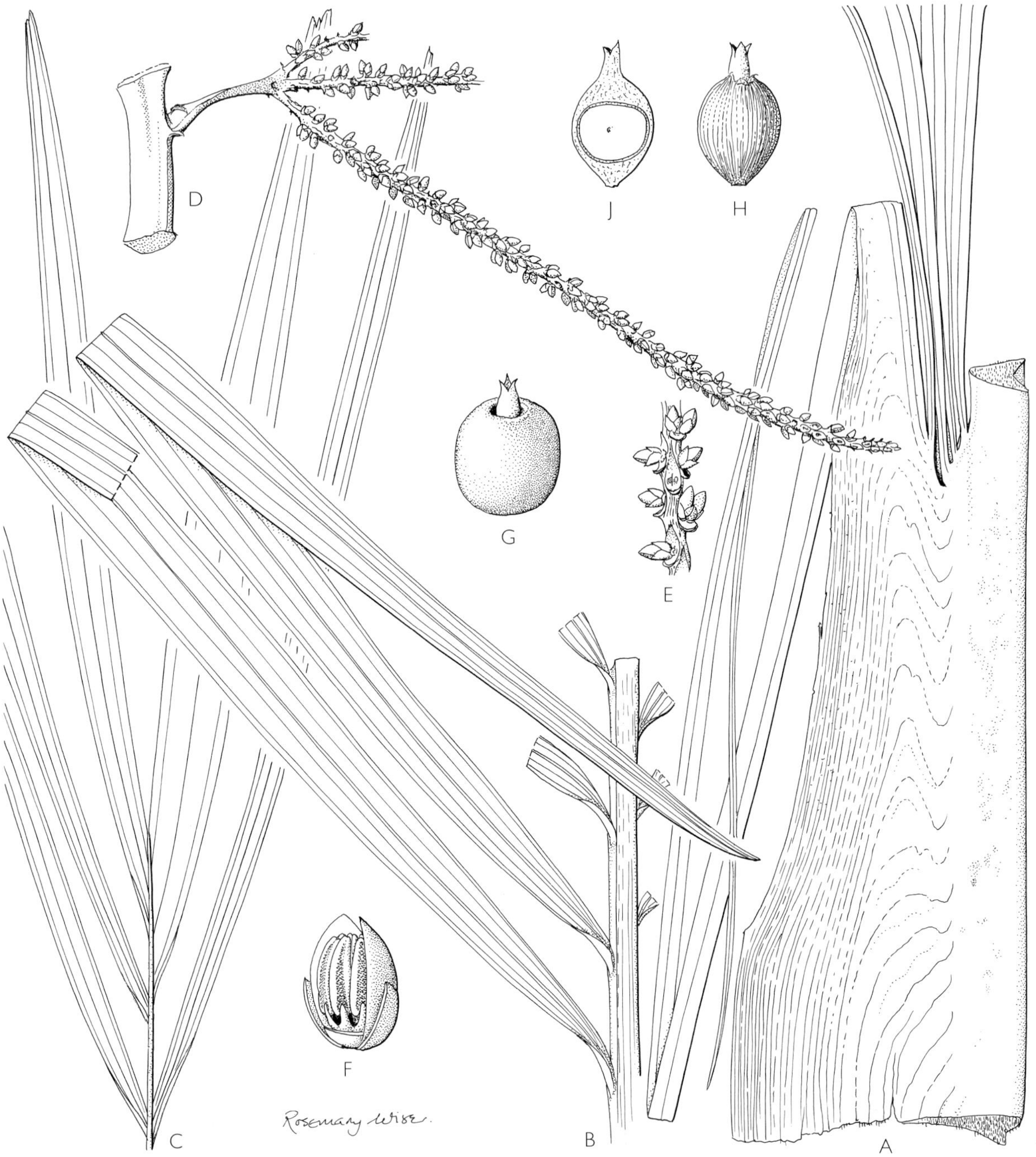

Masoala madagascariensis. **A** leaf base, rein and basal leaflets × 1/3; **B** mid section of leaf × 1/3; **C** leaf tip × 1/3; **D** first order branch of inflorescence × 1/3; **E** detail of rachilla × 1; **F** staminate flower, one petal removed × 4; **G** fruit × 1; **H** endocarp × 1; **J** endocarp in vertical section × 1. **A** from *Dransfield et al.* JD6390, **B** – **F** from *Dransfield et al.* JD6770, **G** – **J** from *Beentje et al.* 4632. Drawn by Rosemary Wise.

2. Masoala kona

A new species with a leaf resembling that of *Marojejya insignis*, but in the inflorescence clearly congeneric with *Masoala madagascariensis*, though it is much smaller than that species. The name derives from the local name for this plant.

Distribution. Ifanadiana area.
Habitat. Mid-altitude low-canopy rain forest; steep to slight mid slopes or near ridge crests, on sandy/quartz soils; 450–550m.
Local names. *Kona, kogne* (Tanala)
Uses. A leaf of this taxon, when stuck in a bamboo pole is thought by some to ward off thunder-clouds.
Conservation status. Endangered. The species is confined to a small area; populations are small, and the forest in this area is disappearing steadily, due to shifting cultivation.

Masoala kona *Beentje* **sp. nov.**
A *M. madagascariensi* statura minore foliis irregulariter pinnatis segmentis proximalibus latis, inflorescentia paucis rachillis floribus pistillatis semper proximalibus et floribus staminatis semper distalibus differt. Typus: Madagascar, Ifanadiana: 3 hrs walk E of Tsaratanana, March 1991, *Beentje & Andriampaniry* 4437 (Holotypus K; isotypi BH, MO, P, TAN).

Masoala kona, growing near Ifanadiana (*Beentje & Andriampaniry* 4437).

Masoala kona. Inflorescence (*Beentje & Andriampaniry* 4524).

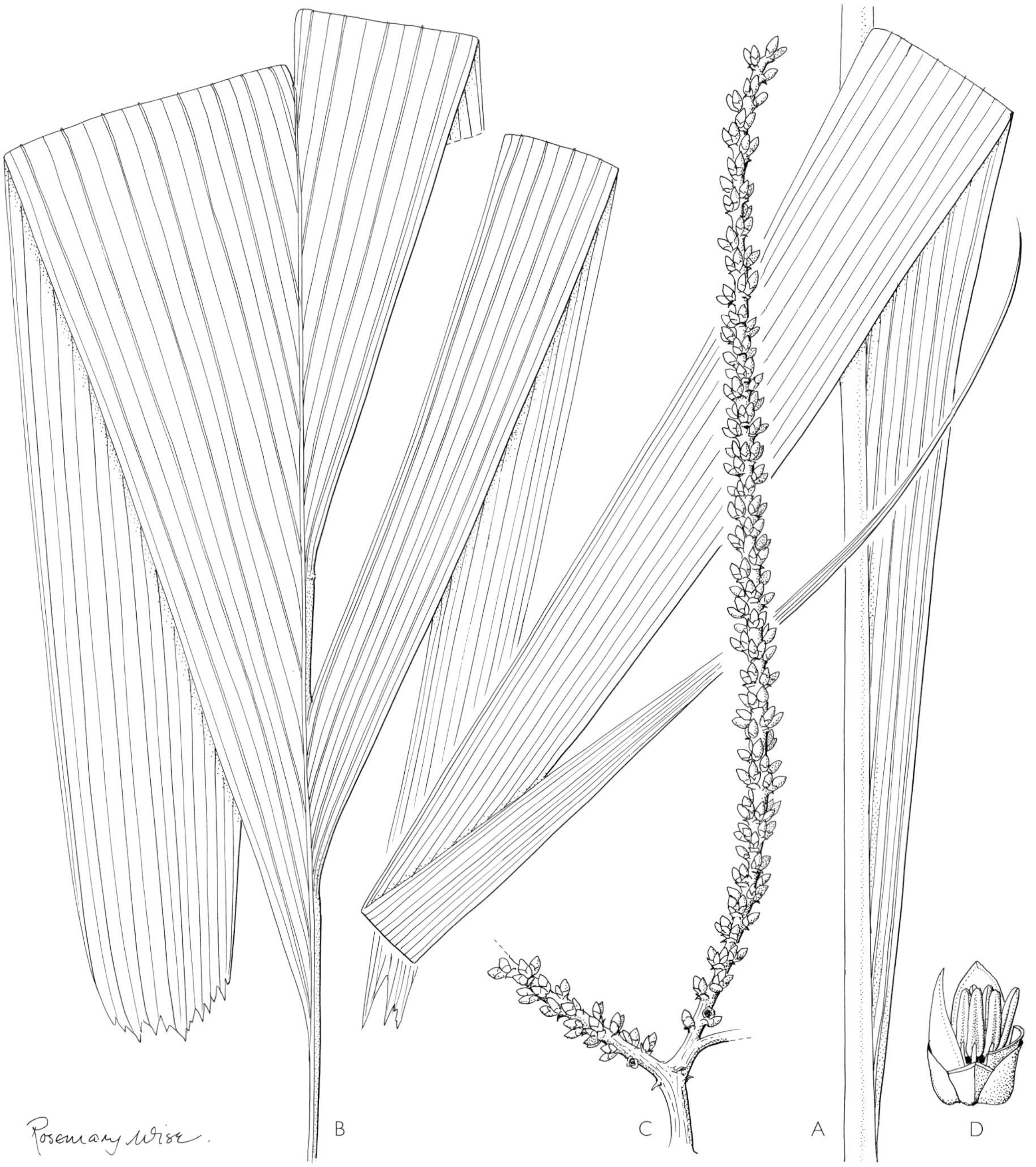

Masoala kona. **A** basal leaflets × 3/5; **B** leaf tip × 3/5; **C** part of inflorescence with staminate flower buds × 3/5; **D** staminate flower, one petal removed × 4. **A**, **B** from *Beentje & Andriampaniry* 4437, **C, D** from *Beentje & Andriampaniry* 4525. Drawn by Rosemary Wise.

Solitary palm. **TRUNK** 1.2–9 m, the whole or in older trees the upper part with persistent leaf bases, c. 35 cm diam. when covered in old leaf bases, 13–20 cm diam. when bare; internodes 1–2 cm long, dark brown, nodal scars 1–2 cm high; wood hard. **LEAVES** spirally arranged, marcescent, 13–17 living ones in the crown, erect, plus 5–15 dead ones; sheath litter-accumulating, c. 28 × 24 cm, abaxially yellow-brown to red-brown with dense red-brown tomentum, adaxially yellow-brown; auricles and petiole absent, though an apparent petiole of up to 30 cm may be present; proximal part of leaf litter-accumulating; rachis 2.8–4.5 m long, proximally 3–4.6 × 1–3 cm diam. and channelled with a central ridge, in mid-leaf c. 1.5 cm wide and keeled, with scattered scales abaxially; leaflets 6–15 on each side of the rachis, the proximal large and multi-fold, 165–250 × 13–24 cm and attached to the rachis for up to 1.5 m, the more distal ones irregularly alternating between thin single-fold and wide multi-fold leaflets, 73–130 × 2.5–16 cm (interval 6–12 cm), distal pair 55–88 × 4–16 cm, connate for 8–18 cm, main veins 3–5 in single-fold leaflets, with scattered scales on minor veins, apices in distal pair long-dentate over 1.5–4 cm, in others acute. **INFLORESCENCE** interfoliar, erect, 0.7–1.2 m long, branching to 1 or 2 order(s), possibly occasionally multiple (personal observation under difficult circumstances, HB 4437), protogynous (HB 4525); peduncle 22–56 cm long, 2.2–3 × 1.3–2.2 cm diam., with dense red-brown scales; prophyll c. 32 cm long, c. 9 cm wide, with scattered scales; peduncular bracts 3 (always?) 20–62 cm long, 7–8 cm wide, with dense red-brown scales and a beak of up to 2 cm, the most proximal inserted only a few mm from the base of the peduncle and sub-woody; rachis 27–30 cm long, with 0–3 branched and c. 10 unbranched first order branches; rachillae dark brown, 22–60 cm long, with pistillate flowers occupying the proximal half

Masoala kona. **A** mid section of leaf × 3/5; **B** part of peduncle, rachis and first order inflorescence branch × 3/5; **C** rachilla with pistillate flowers × 3/5; **D** pistillate flower × 2. **A** from *Beentje & Andriampaniry* 4437, **B** from *Beentje & Andriampaniry* 4525, **C**, **D** from *Beentje & Andriampaniry* 4437. Drawn by Rosemary Wise.

to two-thirds of its length and single or paired staminate flowers the distal part; the interface between the two sexes mixed, not sharp, and occasionally with some triads here; also most pistillate buds at an early stage appearing to be flanked by staminate buds which later abort. **STAMINATE FLOWERS** yellow, the sepals 4.5–5.5 × 2.8–3.8 mm, keeled, acute, with ragged edges; petals connate for 2–3.3 mm, free parts 5–6 × 3.2–4.6 mm, with fleshy and thickened proximal margins; stamens and pistil connate for the proximal 2–3 mm with the petals; stamens equal, the filaments 2.2–3 × 1.2–1.3 mm, fleshy with wider base, angular, anthers 3–3.7 × 1.2–1.4 mm, dorsifixed, latrorse, slightly versatile, the locules slightly divergent; pistillode 4.5–5 × 1.5–1.6 mm. **PISTILLATE FLOWERS** with 4 bracts, 2.8–6 × 3–6 mm; between innermost bracts and sepals there is a ring of up to 2 mm high, thin, ?glandular hairs; sepals 7–10 × 5.5–7.5 mm, not keeled; petals 11–18 × 6–14 mm, with laciniate margins; staminodes 6, connate for the proximal 0.1–0.3 mm, triangular and flat, 0.6–1 mm high; ovary 12–15 × 5–6 mm, trifid for c. 3 mm, the inner surfaces papillose. **FRUIT** only seen empty, ellipsoid, 25–40 × 12–14 mm, topped by the persistent style/stigma mass. **EOPHYLL** deeply bifid; scale leaves two.

SPECIMENS SEEN. Ifanadiana: 3 hrs walk E of Tsaratanana, March 1991 (bud), *Beentje & Andriampaniry* 4437 (Holotype K; isotypes BH, MO, P, TAN); idem, Nov. 1991 (fl.), *Beentje & Andriampaniry* 4524 (K, MO, TAN); idem, Nov. 1991 (fl.), *Beentje & Andriampaniry* 4525 (K, MO, P, TAN); Ambinanindrano, Jan. 1993 (old infl.), *Beentje & Andriampaniry* 4799 (K, MO, P, TAN); idem, Nov. 1994 (old infl.), *Beentje & Dransfield* 4839 (K, TAN).

SUBTRIBE MASOALINAE

XII. MAROJEJYA

Humbert, Mém. Inst. Sc. Madag. sér. B. 6: 92 (1955); N.W. Uhl & J. Dransf., Genera Palmarum 477 (1987). Type: *M. insignis* Humbert.

Stout, solitary, unarmed, pleonanthic, monoecious palms. **STEM** erect, short, often covered with the remains of the leaf sheath. **LEAVES** massive, entire, variously dissected, or reduplicately pinnate; sheath tubular, distally attenuate, with or without auricles; petiole absent or present, adaxially channelled, abaxially rounded; rachis adaxially channelled but more distally keeled, abaxially rounded; blade undivided or divided into single- to multi-fold leaflets, abaxially minutely dotted, transverse veinlets obscure. **INFLORESCENCE** unisexual (but see below), hidden among the leaf bases and the trapped litter therein, branching to 1 order; staminate and pistillate inflorescences basically similar but the staminate with more numerous slender and longer branches; peduncle large, elliptical in cross-section; prophyll tubular, strongly flattened, 2-keeled, splitting longitudinally, thinly coriaceous; basal peduncular bract similar to the prophyll but smaller, subsequent peduncular bracts numerous, incomplete, gradually decreasing in size, spirally arranged, acute to acuminate; rachis shorter than the peduncle, bearing spirally arranged rachillae, each subtended by a conspicuous, triangular, acute bract; staminate rachillae stout, catkin-like, somewhat flexuous; rachilla bracts conspicuous, paired, narrow triangular, subtending densely crowded, paired staminate flowers except near the base and the apex where the staminate flowers solitary, rarely with a few pistillate flowers near the base; floral bracteoles 2, well-developed, acute, with ciliate margins. **STAMINATE FLOWERS** somewhat asymmetrical due to close packing; sepals 3, free, unequal, narrow, ovate, keeled, chaffy, with ciliate margins, tending to be widely separated; petals 3, ± boat-shaped, valvate, coriaceous, connate basally and there adnate to the receptacle; stamens 6, filaments basally connate, the free parts flattened, tapering, elongate, inflexed at the tip, anthers medifixed, versatile, latrorse; pollen elliptic, monosulcate, with foveolate tectate exine; pistillode small, 3-lobed. **PISTILLATE RACHILLAE** shorter, thicker, and fewer than the staminate, bearing crowded, spirally arranged, triangular bracts forming the lower lips of shallow pits, each pit including 3 membranous bracteoles, 2 very small abortive staminate flowers and a large solitary pistillate flower. **PISTILLATE FLOWERS** much larger than the staminate, obpyriform, somewhat asymmetrical; sepals 3, distinct, somewhat chaffy, ovate with triangular tips, ± striate; petals 3, distinct, larger than and similar to the sepals with short, triangular, valvate tips; staminodes 6, narrow, triangular; gynoecium gibbous, unilocular, uniovulate, gradually tapering to 3 large, triangular, recurved stigmas, ovule large, pendulous, campylotropous. **FRUIT** asymmetrically globular, perianth persistent, stigmatic remains forming a large subapical, lateral or subbasal beak; epicarp smooth, mesocarp thin and granular, endocarp composed of several layers of broad, soft anastomosing fibres closely adhering to the seed. **SEED** irregularly rounded, flattened or ± kidney-shaped, smooth or grooved, subapically attached, endosperm homogeneous; embryo basal, opposite the fruit attachment. **GERMINATION** adjacent-ligular; eophyll bifid, with or without a petiole.

DISTRIBUTION. Two species, endemic to Madagascar.

A mysterious palm, called *konabe* in Betsimisaraka, has been seen by us in Mananara Avaratra Biosphere Reserve, but only as young plants. These are said to be very rare. The taxon was certainly different from the two mentioned here (it might even be an aberrant *Masoala*; it has some similarities with *Masoala kona*) but the material is too incomplete to describe formally. (See photo below).

Konabe, an as yet unidentified palm, possibly a species of *Marojejya*, Mananara Avaratra.

Marojejya insignis, growing on the slopes of Marojejy (*Dransfield et al.* JD6754).

KEY TO MAROJEJYA

1. Leaves entire (sometimes tattered with age); seed grooved and ridged .... 1. **M. darianii**
   Leaves dissected; seed smooth .... 2. **M. insignis**

1. MAROJEJYA DARIANII

Ever since its discovery in 1983 this has been one of the most sought-after of all Madagascar palms. Tales of the presence of a huge entire-leafed palm growing in the hills near Maroantsetra began circulating in the early 1980s, and through dogged persistence Californian palm enthusiast Mardy Darian, with the help of Jean Gerard and Dominique Halleux, tracked it down. JD can remember the excited phone calls from Mardy Darian describing it. The first description that appeared in Principes (Dransfield & Uhl 1984a) was based on fragments and photographs, sent by Mardy Darian. These could not prepare one for the sheer size and beauty of the palm as it grows in a small peat swamp in the hills above Sahavary. This locality is now well known and large quantities of seed have been exported. We do not know of the fate of most of these, but some juvenile plants are growing well in private collections in Queensland, Australia. Although it is so spectacular, *Marojejya darianii* is undoubtedly a difficult palm to grow for most of us. It is also one of the most threatened palms of Madagascar. We have heard recent rumours of the presence of this palm in forest near Toamasina, but we have seen neither photographs nor material to verify this.

DISTRIBUTION. Only known from a single site near Maroantsetra.
HABITAT. Upland swamp in valley bottom; 400–450 m.
LOCAL NAMES. *Ravimbe* ("big leaf", Betsimisaraka).
USES. Not recorded.
CONSERVATION STATUS. Critical. Known only from a single site, which is not protected.

Marojejya darianii *J. Dransf. & N.W. Uhl*
Principes 28, 4: 151, figs. 1-3 and cover (1984). Type: Madagascar, Maroantsetra, *Darian* s.n. (Holotype BH; isotype K).

Solitary, monoecious medium-sized tree palm of great beauty. **TRUNK** 8–15 m tall, 15–35 cm diam., when young covered in leaf bases and then appearing wider, when older becoming bare with

Marojejya darianii, view into the crown of an exceptionally tall individual (*Dransfield et al.* JD6399).

Marojejya darianii, growing near Sahavary.

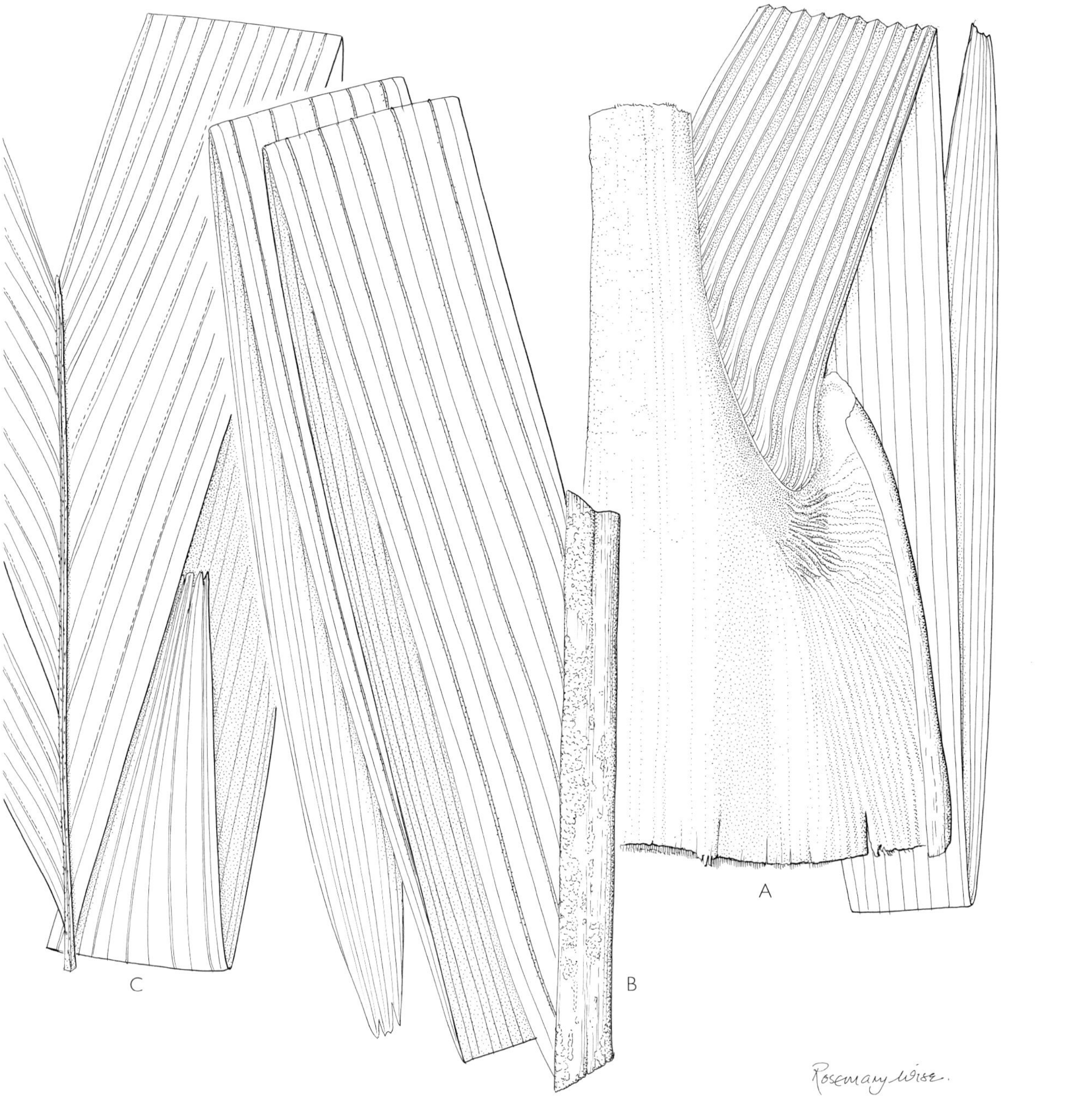

Marojejya darianii. **A** base of leaf blade and auricle × 2/5; **B** mid section of leaf × 2/5; **C** leaf tip × 2/5. All from *Dransfield et al.* JD6399. Drawn by Rosemary Wise.

warty lumps, pale grey-brown, very obscurely ringed; internodes short, with scattered short spine-like adventitious roots. **Leaves** 20–30 in the crown, stiffly erect, entire, bifid, pinnately ribbed, becoming tattered with age; sheath bright green with 2 conspicuous large rounded auricles, 10–12 cm wide; petiole absent; rachis c. 15 cm wide near base, spongy in texture, abaxially covered in white indument, adaxially glabrous; blade 3.5–5 m, bifid in apical 20–50 cm or sub-praemorse, proximally with margins decurrent into the sheath, gradually widening, 1–1.2 m wide at widest point at 66%, multi-fold, each fold with a distinct midrib and faint veins, margins finely serrate, adaxially glabrous, abaxially with abundant pale floccose scales on ribs, with scattered minute dot-like scales on blade. **Inflorescence** with peduncle c. 50 cm long; peduncular bract and prophyll membranous. **Staminate inflorescence** with peduncle distally c. 3 × 2.8 diam.; rachis c. 8 cm long, with c. 35? closely packed erect rachillae; rachilla catkin-like, c. 18–25 cm long, 7–8 mm across, apparently bearing flowers to the tip; bracts c. 2 × 1.75 mm, spirally arranged, congested, horizontally inserted, joined laterally and forming pits c. 2 mm across and deep, the free tips apiculate, distally pointing, the exposed part densely tomentose except at the tip, the tip extending between the staminate flowers; floral bracts minute. **Staminate flowers** c. 3.5 × 2 mm; bract long-acuminate, the acumen c. 1.5 mm long; sepals 1.3–2.4 × 1–1.3 mm, acute, slightly keeled, the margins very finely serrate, 1 sepal usually slightly wider than the others; petals connate at the base for c. 0.8 mm, ± boat-shaped, 1.8–2 × 1.5–1.7 mm; stamens with filaments awl-shaped, long and slender, c. 2–2.5 mm, united at the base with the pistillode, the antesepalous inserted lower than the antepetalous, anthers medifixed, c. 2.5 × 1 mm, latrorse, ± exserted at anthesis; pistil-

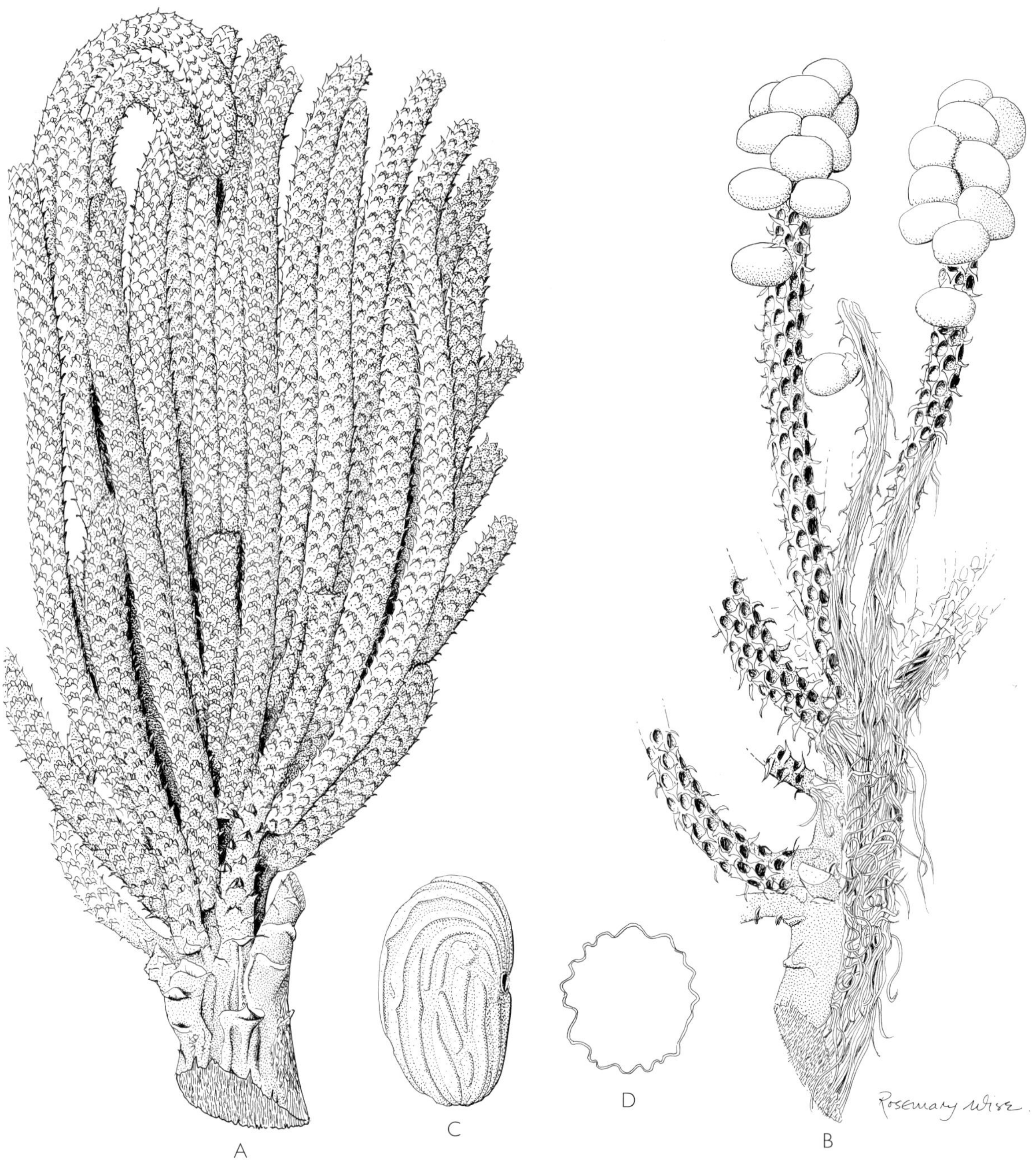

Marojejya darianii. **A** distal part of staminate inflorescence × 1/2; **B** part of infructescence × 1/2; **C** seed × 2; **D** seed in cross section × 2. **A**, **B** from *Dransfield et al.* JD6399, **C**, **D** from *Darian s.n.* Drawn by Rosemary Wise.

lodes 3, irregularly joined, c. 0.75 mm long. **PISTILLATE INFLORESCENCE** with peduncle c. 50 cm long, ± circular in cross-section, distally 5 × 3.5 cm diam.; bracts not available, but some small (c. 2 × 1.5 cm) non-tubular bracts near the apex of the peduncle; rachis 6–10 cm long, with c. 48 closely set, catkin-like rachillae, most erect but the proximal first spreading and then more distally curving upwards; rachillae ± equal, 10–16 cm long, 0.9–1.3 cm diam., bearing spirally arranged bracts united laterally and basally to form pits; pits c. 4.5 mm across; rachilla bracts triangular, apiculate, c. 3 × 4 mm, at first apparently closing the pit, becoming reflexed; abortive staminate flowers 2, very small, concealed within the pit, lateral to the pistillate, floral bracteoles 2–6, triangular, 2–4 × 2–3.5 mm; sepals (in young fruit) 4–5 × 3.5–4.4 mm; petals (in young fruit) 7–7.5 × 5–6.5 mm; staminodes flat, triangular, ?possibly basally connate for < 0.2 mm, 1.2–1.7 mm high. **FRUIT** pink when young, turning green and then red, obovoid, to 20–25 × 15–22 mm; endocarp fibrous, the fibres anastomosing and densely set. **SEED** obovoid, 20–23 × 12–15 × 10–12 mm, the surface covered with deep, broad, mostly longitudinal, anastomosing grooves, with rounded base and obtuse apex; endosperm homogeneous with slight intrusions corresponding to the grooves in the seed-coat; embryo lateral to the hilum. **EOPHYLL** bifid, epetiolate, with long-decurrent blade margins.

SPECIMENS SEEN. Maroantsetra: Sahavary, anno 1983 (seed), *Darian* s.n. (Holotype BH; isotype K); idem, Oct. 1986 (fl., yfr.), *Dransfield et al.* JD6399 (K, P, TAN); idem, Feb. 1988 (fr.), *Dransfield et al.* JD6452 (K, P, TAN); 10 km N of Sahavary, April 1989 (fr.), *Du Puy & Du Puy* s.n. (K, TAN).

2. MAROJEJYA INSIGNIS

This is one of the grandest palms of Madagascar. First discovered in 1949 on the spectacular mountain massif of Marojejy, it has now been found to occur throughout the length of the eastern rain forests. There is even a handsome individual growing in the Parc de Tsimbazaza in Antananarivo opposite the Herbarium building. This is a rather massive, litter-trapping palm. The species name is Latin for 'outstanding' or 'remarkable'.

DISTRIBUTION. Widespread, from Marojejy to Andohahela.

HABITAT. Lowland to mid-altitude rain forest; ridgetop or steep slopes, (70–) 350–1150 m.

LOCAL NAMES. *Menamoso, Beondroka* (Tsimihety); *Maroalavehivavy* (the female *Beccariophoenix*), *Betefoka, Besofina, Hovotralanana, Mandanzezika* (Betsimisaraka); *Fohitanana, kona* (Tanala).

USES. Palm-heart eaten.

CONSERVATION STATUS. Vulnerable. Despite its wide distribution, the size of individual populations is small, and we estimate the total number of individuals at less than two thousand. Cutting for palm-heart continues.

Marojejya darianii. Detail of auricle and base of leaf blade (*Dransfield et al.* JD6452).

Marojejya insignis. View of the litter-trapping crown, Andohahela (*Dransfield et al.* JD6778).

Marojejya insignis *Humbert*

Mém. Inst. Sc. Madag. sér. B, 6: 94, fig.1, pl. 28 (1955). Type: Madagascar, E slopes of Marojejy Mts, *Humbert* 23632 (P, holotype; K isotype).

Solitary tree palm. **TRUNK** 2–8 m, 14–40 cm diam. (with sheath remnants up to 50 cm), obscurely ringed; distal part (c. 1 m) of trunk, covered in sheath/petiole remnants, often with zigzag roots of other plants visible; internodes brown, 1.5–2.5 cm, nodal scars c. 1 cm. Superficial roots sometimes present. **LEAVES** 15–20, spirally arranged, 4–5 m long, held porrect in shuttlecock mode, marcescent; sheath glabrous, 40–94 cm long, attenuate towards the blade, without auricles or with clear auricles to 6 × 5 cm, abaxially densely red-tomentose to almost glabrous, bright green to cherry-red, distally (together with the proximal part of the lamina) litter- and water-accumulating with aerial roots penetrating the litter, soon splitting opposite the petiole, the margins often fibrous, the margins with conspicuous robust parallel veins, those of the auricles sinuous or arching; (apparent?) petiole 0–143 cm, adaxially channelled with sharp edges to flat, 4–9 × 3–6.5 cm diam., in older leaves with scattered scales; rachis 3–5.8 m long, in midleaf 1.5–3 cm wide and keeled, abaxially with scattered chocolate-coloured to reddish scales (up to 5 × 1 mm) in shallow depressions or ± glabrous; *either* with the lamina entire in proximal quarter, multi-fold, outer margin 1.5–2 m long, 20–30 cm wide and more distally regularly pinnate, with (22, Amby) 30–60 leaflets on each side of the rachis, *or* with the leaf regularly pinnate with 59–84 leaflets on each side of the rachis, and nearly all leaflets single-fold, stiff or slightly arching, in one plane, green, the proximal 36–80 × 0.4–2.6 (–7.5) cm (the basal two occasionally 2-fold), the median 75–120 × 2–5 (–15) cm (interval 2–7.5 cm), distal 28–52 × 1–7 cm, connate for 2.5–12 cm, main veins 1–10, fold interfaces with medium-sized reddish scales (floccose whan young, with minute reddish dots or not visible

Marojejya insignis. **A** leaf base, rein and basal leaflets × 2/5; **B** mid section of leaf × 2/5; **C** leaf tip × 2/5; **D** fruit × 2; **E** seed in cross section × 2. **A – C** from *Dransfield et al.* JD6778, **D**, **E** from *Humbert* 23632. Drawn by Rosemary Wise.

in older leaves), with scaly margins, otherwise glabrous or with several 4–5 mm long ramenta on the proximal midrib, apices slightly falcate, acute to attenuate, in the distal pair dentate. **INFLORESCENCES** interfoliar, hidden among the leaf sheaths, occasionally a pistillate inflorescence with the odd rachilla with the distal few centimeters with staminate flowers; protandrous (*Beentje* 4533). **STAMINATE INFLORESCENCE** with creamy yellow bracts; peduncle 5–17 cm long, 2.5–4 × 0.7–2.5 cm diam., glabrous to densely reddish-tomentellous; prophyll 30–35 × 8–10 cm, membranous, oblong-obovate, with small scattered scales, split along its 2 keels, subacute or apically beaked for c. 2 cm; first peduncular bract 41–46 × 9–10 cm, with a beak 2–4 cm, subacute or beaked for 2–4 cm, with scattered scales or with flaking pubescence; non-tubular peduncular bracts 9–16, decreasing in size from base to apex of the peduncle, 1.5–35 × 1–6 cm; rachis 5–22 cm long, densely reddish-tomentellous with 20–66 densely packed erect or porrect rachillae; rachillae 5–21 cm long, 8–15 mm diam., each subtended by acute or acuminate bracts to 20 × 7.5 mm, at the very base with a bare part 1–2 cm long and c. 15 × 9 mm diam. with several empty triangular bracts to 4 mm long, the main part densely set with pairs, distally with densely packed solitary flowers; flowers sunk in pits, flanked by long-acuminate bracteoles 1.5–3 × 1 mm. **STAMINATE FLOWERS** crimson; sepals linear, 2.3–3.8 × 0.5–1.7 mm, only very slightly imbricate, very slightly connate at the very base; petals proximally connate into a tube with stamens and pistillode for 1.5–2 mm, the free lobes valvate, ovate, acute, 2–3.6 × 1.4–2.6 mm; filaments connate into a ring for 0.2–0.6 mm, free for 1–2.7 mm, the antepetalous inserted slightly higher than the antesepalous and also slightly wider, apically inflexed, anthers 1.6–3 × 0.4–1.2 mm, the locules parallel, latrorse, mucronate, versatile; pistillode 0.3–0.8 mm long, irregularly 3-lobed. **PISTILLATE INFLORESCENCE** interfoliar, to 30 cm, branched to 1 order, pinkish at anthesis, with creamy yellow bracts; peduncle 8–27 × 5 × 3.5 cm (> 50 cm in JD6775), dense-

Marojejya insignis. **A** staminate inflorescence, prophyll and first peduncular bract removed × 2/5; **B** pistillate inflorescence, prophyll and first peduncular bract removed × 2/5. All from *Dransfield et al.* JD6404. Drawn by Rosemary Wise.

ly reddish-tomentellous; prophyll 26–35 x 6.5–15 cm, with small scattered scales, split over its length, adnate for 3 cm; first peduncular bract inserted at c. 8 cm from the base of the peduncle, 34–45 x 5–15 cm, with a beak 2–4 cm or not beaked at all, with scattered scales; non-tubular peduncular bracts 9–15, decreasing in size from base to apex of the peduncle, 2–31 x 1.2–7 cm; rachis 8–18 cm, densely reddish-tomentellous, with 16–50 rachillae; rachillae 5–12 cm long, with a short bare base to 1.5 cm long and 7–13 x 4 mm diam. with a few narrowly triangular sterile bracts, the rest c. 12 mm diam., very densely set with solitary pistillate flowers; bracteoles two, 3–6 x 3–4 mm, acute, very similar to the sepals. **PISTILLATE FLOWERS** green to cream-coloured; bracteoles 2–6 x 2.3–6 mm, the largest very similar to the sepals; sepals imbricate, slightly asymmetrical (more so in fruit), broadly ovate with slightly thickened, darker apex, the margins ciliolate, 5–8 x 4.5–7.5 mm, with slightly ragged edges; petals broadly ovate with small fleshy triangular apices, 7–16 x 5–12 mm; staminodes slender, 1–1.7 mm, sometimes connate for up to 0.2 mm, in fruit connate with the fruit base and carried upwards for up to 1 mm; pistil 9–12 x 4–6 mm, styles to 2 mm long, papillose adaxially. **FRUIT** bright to dark plum red to purplish turning black, irregularly obovoid, 18–21 x 11–20 x 10–14 mm, with a pronounced style boss to 5 mm high, this either subapical to almost sub-basal; mesocarp fleshy, 1–1.5 mm thick; endocarp fibrous, 14–16 x 12–16 mm, the fibres parallel near the base, anastomosing distally, without a beak. **SEED** subglobose, 11–14 x 10–15 x 9–10 mm, with homogeneous endosperm; embryo basal. **EOPHYLL** bifid, with a long petiole (> 13 cm).

Marojejya insignis
(x: sight records)

NOTE. This palm, described as late as 1955, remained very poorly known until it was recollected by Harold Moore in 1971. Since then, numerous collections of *Marojejya*-like palms have been made from the Marojejy mountains in the north to Andohahela in the south. At first, every collection from a new area appeared to differ in some way from the type, as well as from each other. Features such as the presence or absence of a petiole, combined with the presence or absence of auricles, and the dissection of the leaf have proved to be extraordinarily variable. We even found such variation within populations. For a time we felt there were at least two separate taxa, but upon careful examination of the now ample material we have been forced to admit that all this variation occurs within a single taxon, *M. insignis*.

SPECIMENS SEEN. Andapa: Mt Beondroka, March 1949 (ster.), *Humbert* 23424 (K, P); Marojejy East peak slopes, March 1949 (fr.), *Humbert* 23632 (K, L, P, type); idem, Oct. 1988 (old infl.), *Miller et al.* 3508 (K); idem, Nov. 1989 (y.fr.), *Dransfield et al.* JD6754 (K, TAN). Maroantsetra: Sahavary, Oct. 1986 (bud), *Dransfield et al.* JD6404 (BH, K, MO, P, TAN); idem, Feb. 1988 (y.fr.), *Dransfield et al.* JD6462 (K, P); idem, Feb. 1988 (old infl.), *Dransfield et al.* JD6484 (K, TAN). Toamasina: Ankiririryra Forest, March 1971 (fl.), *Moore* 9901 (P); Betampona, Oct. 1991 (fl.), *Beentje* 4499 (BH, K, MO, P, TAN); idem, *Beentje* 4500 (BH, K, MO, P, TAN). Moramanga: Anranomenabe, Nov. 1986 (ster.), *Dransfield et al.* JD6432 (K, P, TAN); Mantady, Dec. 1991 (fr.), *Beentje & Andriampaniry* 4547 (BH, K, MO, P, TAN); idem, *Beentje & Andriampaniry* 4551 (BH, K, MO, P, TAN). Ifanadiana: Ambinanindrano, Nov. 1991 (fl., fr., seedling), *Beentje* 4533 (BH, K, MO, P, TAN); idem, July 1992 (fr.), *Beentje & Andriampaniry* 4733 (K); idem, Jan. 1993 (y.fr.), *Beentje & Andriampaniry* 4804 (K). Manakara: Amby, May 1992 (old infl.), *Beentje & Andriampaniry* 4663 (K, MO, TAN); idem, July 1992 (fl.), *Beentje & Andriampaniry* 4725 (K, MO, TAN). Tolanaro: Andohahela, Dec. 1989 (y.fr.), *Dransfield et al.* JD6775 (K, P, TAN); idem, (old infl.), *Dransfield et al.* JD6778 (K, TAN).

SIGHT RECORDS. Andranomifotitra valley (reported in protologue), Ambatasoratra slopes (reported). Mananara Avaratra (*Beentje & Dransfield*).

Marojejya insignis. Infructescence (*Dransfield et al.* JD6775).

Marojejya insignis. Detail of crown showing a pistillate and two dead staminate inflorescences (*Beentje* 4533).

Marojejya insignis. Detail of staminate rachillae (*Dransfield et al.* JD6404).

Marojejya insignis. Detail of pistillate rachillae (*Dransfield et al.* JD6404).

TRIBE COCOEAE
SUBTRIBE BECCARIOPHOENICINAE
XIII. BECCARIOPHOENIX

Jum. & H. Perrier, Ann. Fac. Sc. Marseille 23 (2): 34, fig. 1 & 2 (1915); N.W. Uhl & J. Dransf., Genera Palmarum 483 (1987). Type: *B. madagascariensis* Jum. & H. Perrier.

Robust, solitary, unarmed, pleonanthic, monoecious tree palm. **LEAVES** massive, pinnate, marcescent; sheath tubular with a large auricle, later disintegrating into fibers; petiole absent; rachis adaxially channelled near the base, abaxially rounded, distally with 2 lateral grooves; leaflets single-fold, regularly arranged, acute, adaxially glabrous, abaxially often slightly waxy, with conspicuous transverse veinlets. **INFLORESCENCE** solitary, interfoliar, branching to 1 (–2) orders; peduncle massive, elliptic in cross-section; prophyll inserted at the base of the peduncle, thick and coriaceous, persistent, disintegrating into coarse interwoven fibers, strongly 2-keeled, tubular, with a solid beak, splitting briefly at the tip; peduncular bract inserted at the apex of the peduncle, longer than the prophyll, very thick and woody, tubular with a solid beak, splitting longitudinally and circumscissile at the insertion point, leaving a collar-like scar, adaxially smooth, abaxially longitudinally grooved; rachis very short, bearing crowded, spirally arranged, short, triangular and acuminate bracts each subtending a first order branch; rachillae straight, rigid, rather thick, each with a large swelling at the base, proximally with a short bare portion, distally bearing (sub-) distichous triads (near the very base occasionally in groups of 2 pistillate and 2 staminate flowers) or near the very apex paired or solitary staminate flowers, each subtended by a low triangular bract; floral bracteoles well developed, broad, rounded, striate, rather coriaceous. **STAMINATE FLOWERS** relatively very large, covered in white wax, subsymmetrical; sepals 3, distinct, keeled, imbricate, coriaceous, rather short; petals 3, distinct except at the very base, valvate, strongly coriaceous to woody, ± boat-shaped, longer than the sepals, somewhat striate, adnate at the very base to the floral axis; stamens 18–21, filaments short, slender at the base, adnate to the floral axis, anthers elongate, erect, ± basifixed, sometimes irregularly sagittate; pollen elliptic, monosulcate, with rugulate, perforate, tectate exine; pistillode absent or minute. **PISTILLATE FLOWERS** slightly larger than the staminate, covered in thin wax; sepals 3, distinct, broadly imbricate, hooded, strongly coriaceous, ± striate; petals 3, distinct, imbricate, thinly coriaceous, with very brief valvate triangular tips; staminodes connate in a brief irregularly toothed ring; gynoecium tri-locular, tri-ovulate, obpyriform, stigmas 3, tightly appressed in bud, ovule form unknown. **FRUIT** relatively large, 1-seeded, ± ovoid with a triangular beak, the perianth segments persisting as a cupule; epicarp purplish with scattered brown tomentum, smooth, mesocarp rather dry with abundant longitudinal fibers, the outer fine, the inner broad and flattened, easily separated from the endocarp, endocarp woody, relatively thin, marked with 3 pores below the middle, 1 opposite the embryo. **SEED** broadly ovoid, attached near the base with a broad hilum, with numerous anastomosing raphe branches, endosperm deeply ruminate; embryo lateral below the equator. Germination adjacent-ligular; eophyll entire, lanceolate.

DISTRIBUTION. Monotypic, endemic to Madagascar.

Beccariophoenix madagascariensis. Individual growing in coastal white-sand forest (*Beentje* 4758).

BECCARIOPHOENIX MADAGASCARIENSIS

A truly wonderful palm, beautiful, somewhat mysterious, and totally special. When this tree is in bud (and it often is) there seem to be torpedoes poking out of the crown; these are the extraordinary peduncular bracts at the tips of the long peduncles. A very rare species. The genus is most appropriately named for O. Beccari. Until the mid 1980's the palm was virtually unknown and had been assumed to be extinct. The story of its rediscovery is told in Principes

Beccariophoenix madagascariensis. A mature palm, felled by a hunter for its apex, Mantady (1986).

Beccariophoenix madagascariensis. Young plant in cultivation in Queensland; note the windows in the distal part of the leaf blade.

(Dransfield 1988). After the publication of this article, enthusiasts turned their attention to obtaining seeds, and soon large quantities were being exported from Madagascar. Seedlings from these introductions are already growing well around the world. In fact, young plants with their 'windowed' leaves can be remarkably handsome!

DISTRIBUTION: Mantady and SW Madagascar.
HABITAT. Submontane rain forest, ridgetop and sides of crest, (100–) 900–1200 m; white sand forest, c. 20 m.
LOCAL NAMES. *Manarano, Manara, Maroala* (Andasibe); *Sikomba* (Antanosy).
USES. Young leaflets much sought after for the manufacture of "Manarano" hats, formerly exported in quantity; therefore species rare. Used in house construction. Palm-heart eaten.
CONSERVATION STATUS. Critical. At Mantady the cutting down of this beautiful tree continues; in July 1992 nine mature trees were seen that had been felled for their palm-heart. At the last count there were less than 20 mature trees left in Mantady, as well as some 20 in the southern population on white sand, in an area threatened by strip-mining.

Beccariophoenix madagascariensis. Fruiting rachilla (*Dransfield et al.* JD6433).

Beccariophoenix madagascariensis. Inflorescence after the peduncular bract has fallen (*Dransfield et al.* JD6433).

There are unconfirmed reports of another population near Toamasina where seed is said to be harvested for export.

Beccariophoenix madagascariensis *Jum. & H. Perrier*
Ann. Fac. Sc. Marseille 23 (2): 35, fig. 1 & 2 (1915); Jum., Cat. Pl. Madagascar, Palmae: 7 (1938); Jum. & H. Perrier, Fl. Madagascar 30: 121, fig. 32 (1945); J. Dransf., Principes 32: 59–68 (1988). Type: Madagascar, Analamazaotra, alt. 900 m, "Manarano", *Perrier* 12014 (Holotype P; isotype FI, P).

Solitary palm. **TRUNK** 2–12 m, 24–35 cm diam.; internodes 2–8 cm, very pale grey or pale brown, nodal scars 1.5–2.5 cm, stepped distally (to 5 cm proud); wood hard. **LEAVES** 11–30 in the crown, straight, 3.5–5 m; sheath 80–160 cm long, 20–26 cm in diameter, green to brown with a thin grey to rich brown tomentum, with large fibres at right angles, with 2 large triangular, slightly obtuse auricles, decomposing into large greyish fibres; petiole absent, but leaf base fibres disintegrating to give a 'false petiole' to 80 cm long; rachis 3.5–3.7 m long (to 6 m in young plants), yellow-green, proximally 8.5 × 5 cm diam., in mid-leaf c. 2.5 cm wide; leaflets 100–130 on each side of the rachis, regular, stiff and spreading, not very rigid, rich shiny green, abaxially with white waxy bloom, proximal 44–100 × 1–1.5 cm, median 110–176 × 2.7–4.5 cm, distal 20–40 × 0.5–1.7 cm, transverse veinlets very conspicuous, brown ramenta conspicuous on the abaxial midrib, scattered red or pale brown scales present (but sometimes difficult to see) on the minor veins. **INFLORESCENCE** massive, interfoliar, several per tree, c. 120 cm; peduncle thick, 83–100 cm long, distally 10 × 6 cm in diameter; prophyll 45–90 cm long, c. 25 cm wide, borne at the base of the peduncle, two-keeled, brown, persistent; peduncular bract inserted at the apex of the peduncle on a thickened collar c. 3 cm wide, 68–80 cm, c. 14 cm in diameter, woody, thick (2–3 cm), beaked for c. 8 cm, splitting except at apex, circumscissile and deciduous, adaxially

Beccariophoenix madagascariensis. Rachillae in bud (*Dransfield et al.* JD6433).

Beccariophoenix madagascariensis. Inflorescence in bud (*Dransfield et al.* JD6433).

Beccariophoenix madagascariensis. Open inflorescence (*Beentje* 4758).

Beccariophoenix madagascariensis. **A** Proximal part of leaf showing lowermost leaflets × 1/2; **B** mid section of leaf × 1/2; **C** leaf tip × 1/2; **D** tip of peduncle and base of rachis, showing peduncular bract scar × 1/2; **E** tip of inflorescence rachis in section × 1/2; **F** tip of rachilla with staminate flowers × 1/2; **G** staminate flower, one petal removed × 3; **H** fruiting rachilla × 1/2; **J** endocarp × 1/2; **K** endocarp and seed in section × 1/2. **A – E**, **H** from *Beentje & Andriampaniry* 4537, **F**, **G**, **J**, **K** from *Dransfield et al.* JD6433. Drawn by Rosemary Wise.

peach-rose to pale cream and slightly spongy, abaxially brown, occasionally with deep grooves while still in bud; second peduncular bract 21 cm long, thick and woody; rachis 5.5–10 cm long, with 31–46 first order branches; rachillae 43–62 cm, 5–8 mm across, each with a grossly swollen basal pulvinus; proximally with a few quadrads, medially with triads, distally with pairs or solitary staminate flowers; rachilla bracts narrow, rounded; flowers covered in white wax, otherwise yellowish, tinged with red. **Staminate flowers** oblong, 7 mm wide; sepals 3 mm, keeled; petals 5 × as long as the sepals, 15 × 7 mm, ovate, woody, with small incurved mucro; filaments 1.5-2 mm, anthers 9 mm. **Pistillate flowers** with 2 bracteoles, one small and ovate, the other much larger, forming a cupule; bud conical, 11 mm wide; sepals concave, 13 × 9 mm; petals less coriaceous, c. 11 × 11 mm; ovary irregularly globular. **Fruit** purple-brown, brown-tomentose, ovoid, 35 × 25 mm, somewhat beaked with an acute tip, at the base with persistent perianth which is slightly accrescent; pulpa 3-4 mm thick; endocarp hard, woody, 1.5 mm thick.

Specimens seen. Moramanga: Analamazaotra area, without date (fl.), *Perrier* 12014 (Holotype P; isotype FI); Mantady, November 1986 (fl., fr.), *Dransfield et al.* JD6433 (AAU, BH, K, MO, P, TAN); Mantady, Dec. 1991 (fl., fr.), *Beentje & Andriampaniry* 4537 (BH, K, MO, P, TAN). Tolanaro: Manampanihy valley near Ampasimena, March 1947 (fl., y.fr.), *Humbert* 20572 (K, P); Ste. Luce, Dec. 1992 (y.fr.), *Beentje* 4758 (K, TAN).

SUBTRIBE BUTIINAE

XIV VOANIOALA

J. Dransf. Kew Bull. 44: 192 (1989). Type: *Voanioala gerardii* J. Dransf.

Solitary, monoecious palms. **TRUNK** erect, very conspicuously stepped and ringed with oblique leaf scars. **LEAVES** pinnate, cleanly abscising; leaf-sheath tubular at first, fibrous, apparently soon disintegrating to leave a massive elongate rectangular leaf base, forming an apparent petiole with sparsely fibrous margins; true petiole absent; rachis ± rectangular in cross-section in the mid-leaf region; leaflets numerous. **INFLORESCENCE** solitary, apparently protandrous, branching to 1 order; prophyll tubular, 2-keeled, fibrous, remaining hidden among the leaf bases; peduncular bract bright green and strictly tubular in bud, later splitting longitudinally, flattening and becoming somewhat cowl-like, abaxially deeply and longitudinally grooved, bearing scattered brown scales on the ridges between the grooves, adaxially smooth, glabrous, pale cream-coloured; peduncle ± circular in cross-section; rachis bearing spirally arranged rachillae, each subtended by a small triangular rachis bract; rachillae numerous, most with a basal bare portion. **STAMINATE FLOWERS** asymmetrical, broadly or narrowly triangular in outline; sepals 3, distinct, slightly to strongly imbricate at the base, triangular, acute to acuminate, membranous, glabrous; petals 3, distinct, unequal, valvate, glabrous, thinly coriaceous except at the thick angular tips, broadly and irregularly triangular-ovate, with acute or acuminate tips, abaxially smooth, adaxially marked with the impressions of the stamens and papillose near the thick tips; stamens 12 (–13), filaments subulate, very short to moderate in length, anthers basifixed, basally sagittate, apiculate at the tips, latrorse; pollen circular or elliptic in polar view, monosulcate. **PISTILLATE FLOWERS** only known as buds, much larger than the staminate, irregularly triangular; sepals 3, distinct, unequal, strongly imbricate, broadly ovate with triangular keeled tips, coriaceous, glabrous, the margins minutely toothed; petals 3, distinct, longer than the sepals, basally irregularly imbricate, conspicuously valvate at the triangular tips, abaxially with scaly indumentum towards the apex, adaxially strongly papillose towards the tip; staminodal ring high with 9 irregular triangular teeth, 0.1–0.5mm; gynoecium syncarpous, tricarpellate, triovulate, stigmas 3, angled, papillose and scaly, ovules with axile placentation. Immature **FRUIT** green, covered with dense chestnut-brown scaly indumentum. Mature fruit 1-seeded, somewhat irregularly ellipsoid, tipped with a short beak and stigmatic remains; epicarp purplish-brown, densely covered with brown scaly indumentum; mesocarp with an outer fibrous zone just below the epicarp, and an inner fleshy zone; endocarp ± ellipsoid, apically pointed, basally truncate, very heavily thickened, pale brown when fresh, becoming grey with age. **SEED** irregularly ellipsoid, filling the endocarp cavity, laterally attached with a narrow irregular hilum, endosperm homogeneous but irregularly intruded by the endocarp protuberances, very hard, white, with a narrow irregular central lacuna; embryo basal, top-shaped, positioned opposite an endocarp pore. **GERMINATION** remote-tubular; eophyll entire, lanceolate.

DISTRIBUTION. Monotypic, endemic to Madagascar.

Voanioala gerardii. View into the crown of a mature fruiting palm (*Dransfield et al.* JD6737).

VOANIOALA GERARDII

An extraordinary palm, occurring as scattered trees in the depths of the forested Masoala Peninsula. The discovery of the palm and the first scientific collection are described in Dransfield (1989b, 1992b). The forest coconut is a robust palm with a conspicuously 'stepped' trunk. Endocarps and partially rotted fruit carpet the ground beneath the only mature trees known and, as seedlings have not been observed away from the base of the trees, it seems that there is little effective dispersal at the present day. As in *Satranala decussilvae*, we suggest that the extraordinarily hard sculptured endocarp is an adaptation by a now extinct animal, such as the elephant bird, *Aepyornis*.

Seed collected in October 1986 germinated at Kew in February 1987. From this batch of seedlings, Margaret Johnson (1989) counted the chromosomes of *Voanioala* and discovered that there are at least 596 chromosomes. This quite extraordinary number, at that time the highest recorded not only in the palms, but in all the monocotyledons, has been confirmed by Röser (1994) who counted over 600 in another sample from the same batch of seed, grown at Fairchild Tropical Garden in Florida. The genus is named after the local name, while the specific epithet honours Jean Gerard, one of the discoverers.

DISTRIBUTION. Masoala Peninsula.
HABITAT. Primary forest rich in palms and pandans in swampy valley bottom and on gentle slopes at c. 400 m.
LOCAL NAMES. *Voa-nio-ala* (forest coconut, Betsimisaraka).
USES. Cut for palm-heart.
CONSERVATION STATUS. Critical. *Voanioala gerardii* seems to be a very rare palm; less than ten trees are known to exist in the wild. Unless the superb primary rain forests of the Masoala Peninsula can be effectively protected against further destruction, and palms in particular safeguarded from destructive exploitation for palm cabbage, then the chances of survival for this remarkable and beautiful palm are slim indeed.

Voanioala gerardii *J. Dransf.*
Kew Bull. 44: 195 (1989); Principes 36: 124–127 (1992). Type: Madagascar, Masoala Peninsula, Antalavia, Oct. 1986, *Dransfield et al.* JD6389 (Holotype K; isotypes AAU, BH, MO, NY, P, TAN, US).

Robust solitary unarmed pleonanthic monoecious tree palm. **TRUNK** 15–20 m tall, basally with a large root boss to c. 1 m diam., c. 35 cm diam., distally the stem bare, very conspicuously "stepped" and ringed with oblique leaf scars c. 10 cm distant, the distal part of the internode projecting c. 5 cm outwards from the proximal part of the following internode. **LEAVES** c. 15–20 in crown, c. 5 m long, cleanly abscising; leafsheath tubular at first, fibrous, apparently soon disintegrating to leave a massive elongate rectangular leaf base, forming an apparent petiole c. 150 × 30 cm, c. 8–10 cm thick, with sparsely fibrous margins, abaxially densely covered with caducous brown indumentum; leaf base suddenly contracting into rachis, true petiole absent, the rachis

Voanioala gerardii. Infructescence held by David Cooke (*Dransfield et al.* JD6737).

Voanioala gerardii. Seedlings beneath a mature tree, Masoala Peninsula.

Voanioala gerardii. Detail of peduncular bract and flower buds (*Dransfield et al.* JD6389).

Voanioala gerardii. Mature fruit (*Dransfield et al.* JD6737).

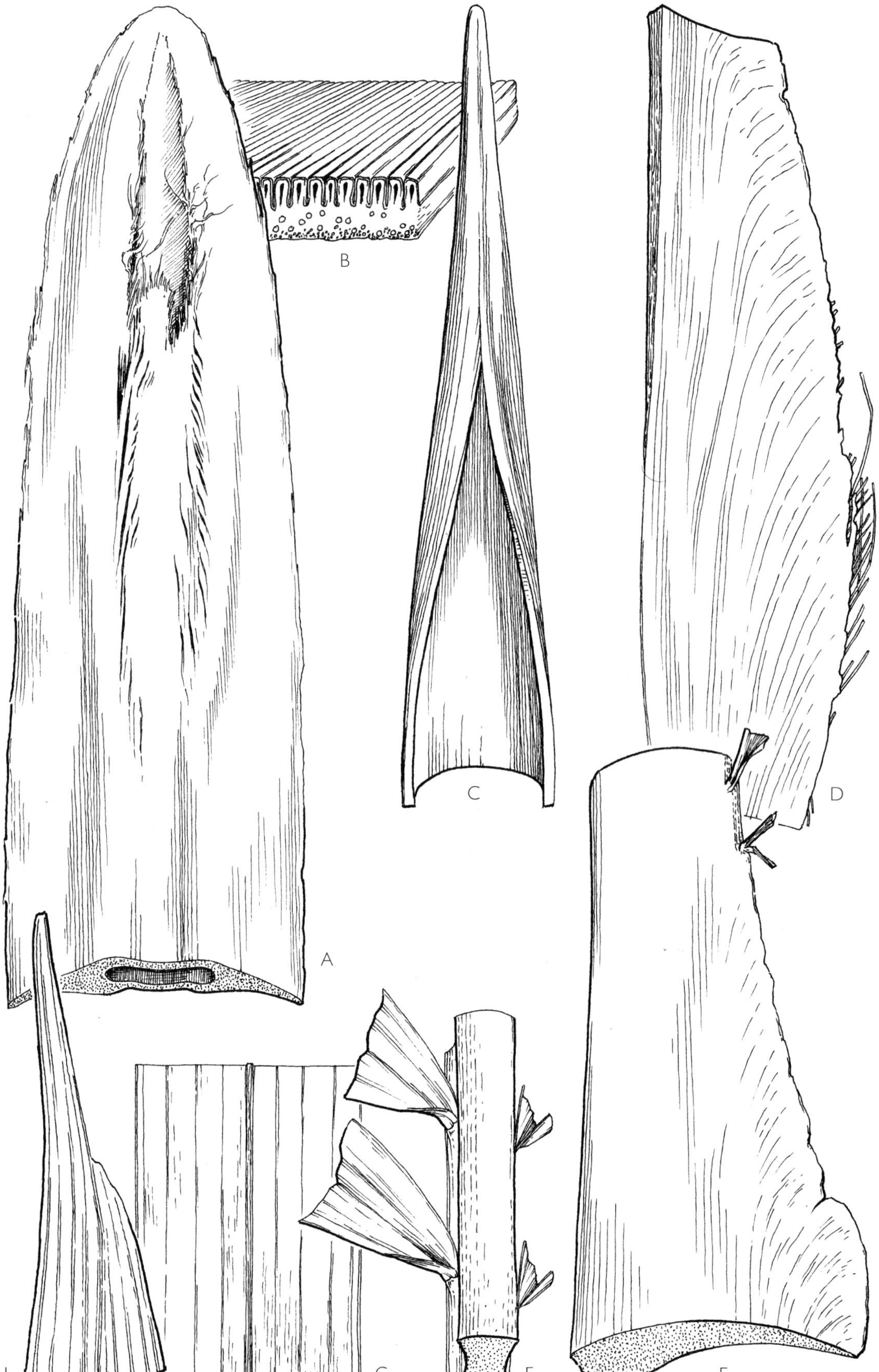

Voanioala gerardii. **A** prophyll, abaxial view × 2/5; **B** detail of peduncular bract to show deep grooves × 2; **C** tip of peduncular bract × 1/2; **D** portion of leaf base × 1/2; **E** leaf base and insertion of lowermost leaflets × 1/2; **F** mid section of rachis with leaflets bases, abaxial view × 1/2; **G** detail of leaflet × 1; **H** leaflet tip × 1. All from *Dransfield et al.* JD6389. Drawn by Eleanor Catherine

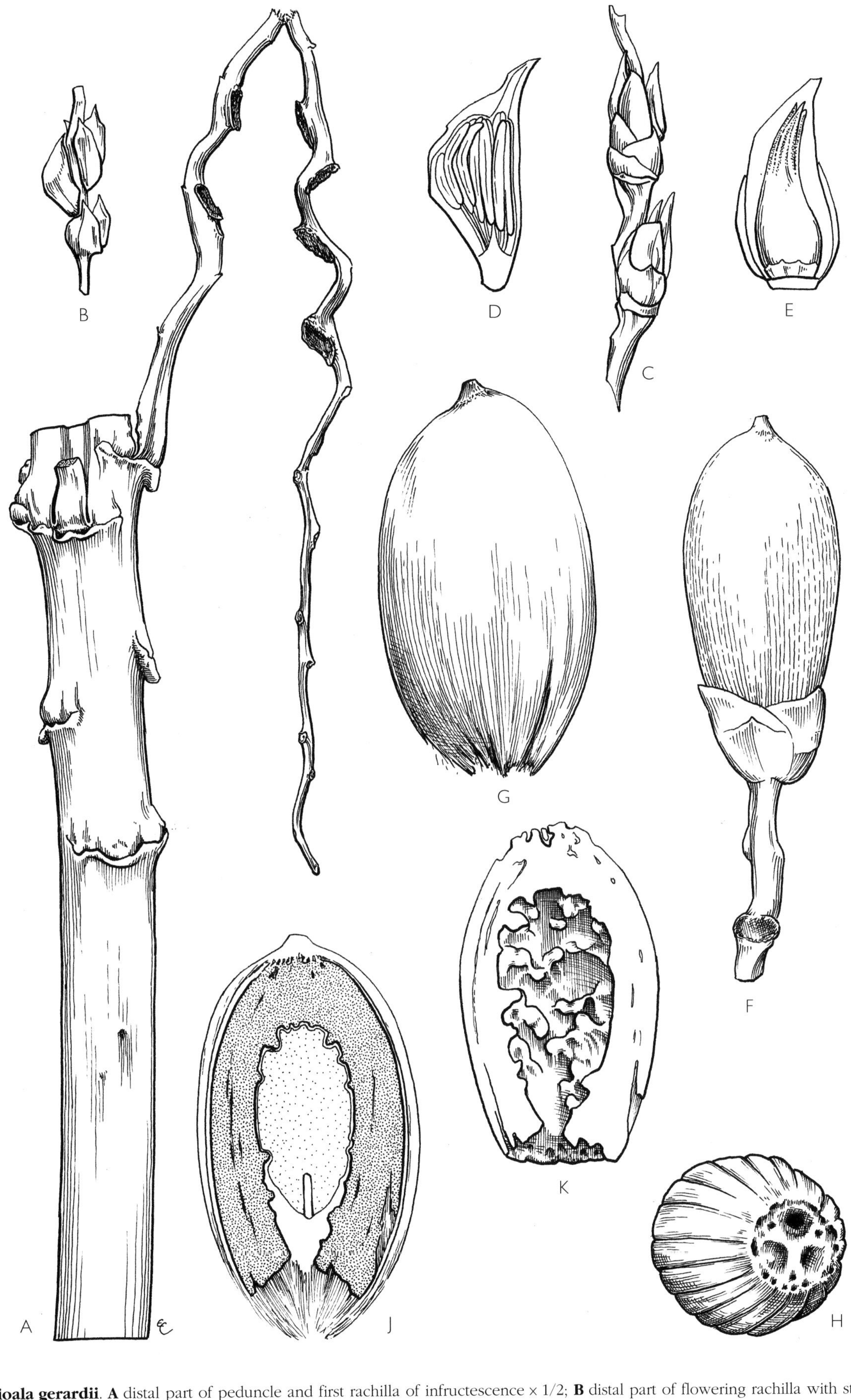

Voanioala gerardii. **A** distal part of peduncle and first rachilla of infructescence × 1/2; **B** distal part of flowering rachilla with staminate flower buds × 1; **C** proximal part of flowering rachilla with triads × 1; **D** staminate flower bud × 3; **E** pistillate flower bud, part of calyx and corolla removed × 3; **F** young fruit × 1; **G** mature fruit × 4/5; **H** bare endocarp viewed from proximal end to show three pores × 4/5; **J** mature fruit in vertical section × 4/5; **K** endocarp sectioned to show irregular intrusions × 4/5. All from *Dransfield et al.* JD6389. Drawn by Eleanor Catherine.

Voanioala gerardii. Whole fruit, endocarp, and cross and vertical sections (*Dransfield et al.* JD6389).

± rectangular in cross section in the mid-leaf region, 4 × 3 cm, abaxially densely covered with caducous brown indumentum as the leaf base; leaflets c. 70 on each side of the rachis, regularly arranged, rather stiff, scarcely pendulous, very coriaceous, concolorous, shining mid green when fresh, drying pale, mid-leaf leaflets c. 150 × 7 cm, unevenly bilobed at the tips, mid vein prominent adaxially, abaxially bearing a few brown ramenta near the base, c. 8 longitudinal veins besides the mid vein, transverse veinlets obscure but lamina minutely transversely striate, portion of leaflet exposed in the sword leaf bearing caducous chocolate scales, thin wax also present on both surfaces. **INFLORESCENCES** to c. 1.5 cm long, interfoliar, erect in bud, later horizontal; peduncle c. 90 cm long, circular in cross section, 4–5 cm diam., pale cream-coloured at anthesis, becoming green in fruit, brown scaly when newly emerged; prophyll tubular, 2-keeled, c. 70 × 13 cm, fibrous, remaining hidden among the leaf bases, bearing caducous brown scales; peduncular bract c. 120 × 18 cm, bright green and strictly tubular in bud, later splitting longitudinally, flattening and becoming somewhat cowl-like, abaxially deeply and closely longitudinally grooved, bearing scattered brown scales on the ridges between the grooves, adaxially smooth, glabrous, pale cream-coloured; rachis c. 60 cm long; rachillae c. 60, those near the base longest, to c. 50 cm, decreasing in length towards tip of inflorescence, most with a basal bare portion 2–5 cm long, c. 7 mm diam. near the base, decreasing to 1.5 mm diam. near the tip, the rachillae bearing 0–7 triads near the base and paired or solitary staminate flowers distally, the flower groups spirally arranged, or becoming somewhat distichous by close-packing, c. 5–10 mm apart. **STAMINATE FLOWERS** asymmetrical, broadly or narrowly triangular in outline, c. 10–12 × 7–9 mm, creamy-yellow just before anthesis, the whole inflorescence smelling sweetly; sepals c. 3–4 × 4 mm, distinct, slightly to strongly imbricate at the base, triangular, acute to acuminate, membranous, glabrous; petals 9–19 × 3–6 mm., unequal, glabrous, thinly coriaceous except at the thick angular tips, broadly and irregularly triangular-ovate, with acute or acuminate tips, abaxially smooth, adaxially marked with the impressions of the stamens and papillose near the thick tips; stamens with filaments subulate, 0.5–2.5 × 0.1 mm, anthers 9 × 1 mm, basifixed, basally sagittate, apiculate at the tips, latrorse. **PISTILLATE FLOWERS** only known as buds, irregularly triangular, c.18 × 10 mm; sepals 8–12 × 10 mm, unequal, strongly imbricate, broadly ovate, with triangular, keeled tips, coriaceous, glabrous, the margins minutely toothed; petals 15 × 8 mm, basally irregularly imbricate, conspicuously valvate at the triangular tips, abaxially with scaly indumentum towards the apex, adaxially strongly papillose towards the tip; staminodial ring c. 1.2 mm high with 9 irregular, triangular teeth, 0.1–0.5 mm; gynoecium c. 4 mm diam. **FRUIT** green when immature, turning rich red-brown when ripe, 7–8 × 4–5 cm, covered with dense chestnut-brown scaly indumentum, one-seeded, somewhat irregularly ellipsoid, tipped with a short beak and stigmatic remains; epicarp purplish-brown, densely covered with brown scaly indumentum; mesocarp with an outer fibrous zone just below the epicarp, and an inner fleshy zone; endocarp ellipsoid, apically pointed, basally truncate, very heavily thickened, pale brown when fresh, becoming grey with age, very deeply and irregularly longitudinally grooved, with 3 very deep basal impressions each with a central germination pore, in section the body of the endocarp traversed by longitudinal irregular vertical canals and fibres, inner surface of the endocarp with numerous irregular rounded excrescences intruding into the cavity. **SEED** irregularly ellipsoid, 4 × 2 cm, filling the endocarp cavity, laterally attached with a narrow irregular hilum, endosperm homogeneous but irregularly intruded by the endocarp protruberances, very hard, white, with a narrow, irregular central lacuna. **EOPHYLL** and leaf 2 entire, lanceolate, c. 30 × 7 cm, leaves 3 and 4 bifid.

SPECIMENS SEEN. Maroansetra, Masoala Peninsula, Antalavia, 17 Oct. 1986 (fl, fr.), *Dransfield et al.* JD6389 (Holotype K; isotypes AAU, BH, MO, NY, P, TAN, US) & (seedling) JD6391 (BH, K, MO, NY, P, TAN); Nov. 1989 (fr.), *Dransfield et al.* JD6737 (K, TAN).

XV. COCOS

Cocos nucifera *L.*, the coconut (*voanibo*), is widely planted throughout the more humid lowlands of Madagascar, where it is one of the most important village crops. It is very variable in Madagascar, and may represent one of the first crop plants to have been introduced to the island by the early immigrants from Indomalaysia.

SUBTRIBE ELAEIDINAE

XVI. ELAEIS

Jacq., Select. Stirp. Amer. Hist. 280 (1763); J. Dransf. & N.W. Uhl, Genera Palmarum 514 (1987). Type: *Elaeis guineensis* Jacq.

Robust solitary pleonanthic monoecious palm. **TRUNK** erect with tardily abscising leaf-bases. **LEAVES** large, reduplicately pinnate; sheath fibrous; true petiole absent; lower part of leaf rachis with marginal fibre spines and short spines developed from the midribs of the basal leaflets; leaflets numerous, slightly irregular, held in more than one plane, composed of a single fold except near the leaf tip where sometimes compound, especially in juveniles, leaflet tips entire. **INFLORESCENCE** interfoliar, solitary, short and condensed, usually unisexual, usually several adjacent axils producing inflorescences of the same sex, followed by a series of the other sex, branching to 1 order; peduncle short, usually hidden among the leaf bases; prophyll short, tubular at first, soon disintegrating into a mass of fibres, the larger fibres spine-like; peduncular bract inserted some distance from the prophyll, distally disintegrating into a mass of fibres; subsequent peduncular bracts small, not tubular; rachis very short; rachillae crowded together, the staminate more slender than the pistillate, rachilla tips of both sexes spine-like. **STAMINATE FLOWERS** borne in pairs sunken in crowded pits in the rachilla and only slightly protruding from the pits at anthesis; sepals 3, free, imbricate, membranous; petals 3, free, ovate, ± equalling the sepals; stamens 6, exserted at anthesis, filaments broad fleshy, united laterally to form a tube with 6 short reflexed tips bearing rectangular versatile anthers; pollen elliptic or triangular, monosulcate, trichotomosulcate, or tetrachotomosulcate, with reticulate tectate exine; pistillode columnar, trifid, small. **PISTILLATE FLOWERS** much larger than the staminate, solitary in the axils of conspicuous bracts with spine-like tips; sepals 3, free, imbricate; petals 3, free, imbricate; staminodes forming a low tanniniferous ring; pistil ovoid, trilocular, triovulate, with 3 apical slightly reflexed fleshy stigmas. **FRUIT** 1- (rarely more) seeded, ovoid but usually irregular from close-packing; epicarp smooth shiny, mesocarp oily, fibrous, brightly coloured, endocarp stony with 3 apical pores. **SEED** rounded; endosperm homogenous, with or without a central cavity; embryo subapical, opposite one of the pores. **GERMINATION** adjacent-ligular; eophyll entire, lanceolate.

DISTRIBUTION. Two species, one endemic to S and C America, the other widespread in the more humid parts of Africa and Madagascar.

ELAEIS GUINEENSIS

The oil palm can be found throughout the humid lowland areas in the east of the island, and even in towns on the plateau. Often planted as an ornamental, there are also small commercial plantations, for example near Toamasina and Sambava. It also occurs in the northwest, where Jumelle and Perrier suggested it was native; however, whether the oil palm is native or not in Madagascar is open to debate. We have never seen it in the wild in Madagascar.

DISTRIBUTION. Widespread throughout the more humid parts of Africa, the distribution much influenced by man. In Madagascar doubtfully native, ?naturalised in the northwest and east in humid areas, particularly near villages.
HABITAT. Village margins, valley bottoms., up to c. 500 m as a naturalised plant.
LOCAL NAME. *Tsingilo* (cited in Fl. Madagascar).
USES. The most important perennial oil crop of the tropics, also used for a multiplicity of other purposes; in Madagascar, of limited importance.
CONSERVATION STATUS. Not threatened.

Elaeis guineensis *Jacq.*
Jacq., Select. Stirp. Amer. Hist. 280 (1763); Jum. & H. Perrier, Ann. Inst. Bot.-Géol. Colon. Marseille Sér. 3, 1 (1): 78 (1913); J. Dransf., Flora Tropical East Africa, Palmae: 50 (1986). Type: t.172 in Select. Stirp. Amer. Hist. (1763);

SYNONYMS:
Elaeis guineensis var. ***madagascariensis*** Jum. & H. Perrier, Les Matières Grasses 6 (1911). Type: not indicated.

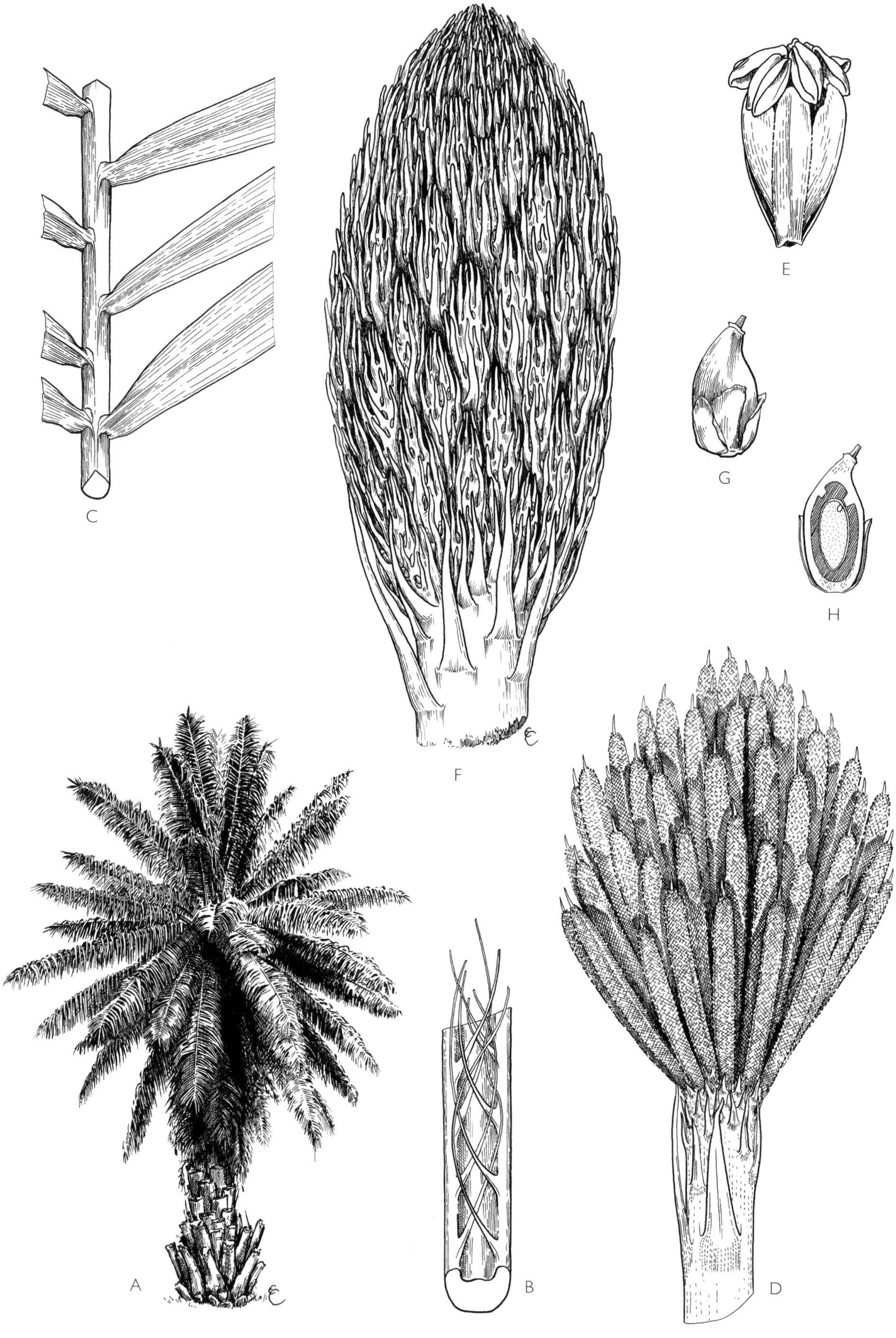

Elaeis guineensis (from African material). **A** habit × 1/100; **B** leaf base × 2/3; **C** mid section of leaf × 2/3; **D** staminate inflorescence × 1/3; **E** staminate flower × 12; **F** pistillate inflorescence × 1/3; **G** fruit × 2/3; **H** fruit in vertical section × 2/3. **A** from several photographs, **B, C** from *Drummond & Hemsley* 3341, **D – F** from material cult. in West Africa. **G, H** from *Maggs* 1. Drawn by Eleanor Catherine.

Elaeis madagascariensis (Jum. & H. Perrier) Becc., Palme del Madagascar 55 (1914). Lectotype: without locality, date or collector, sent by Jumelle to Beccari (FI).

Solitary robust palm. **TRUNK** grey to 20 m tall, swollen at the base, 30–50 cm diam., covered in remains of leaf bases when young, but in high rainfall areas often obscured by epiphytes. **LEAVES** c. 40–50 in crown, to 7.5 m long in well grown subadults; leaf base long persistent with coarse brown fibres and upward pointing fibre spines to 35 × 5 mm, ± 10 mm distant, confined to the sheath margins; apparent petiole to 1.25 m long, c. 20 cm wide at base, distally armed with bulbous-based spines to 4 cm long, 1–5 cm distant, formed from the basal leaflets; leaflets 100–150 on each side of the rachis, inserted irregularly in two planes, to 120 × 8 cm. **STAMINATE INFLORESCENCE** with peduncle 15–20 cm long, 5 cm or more diam.; rachillae c. 50, crowded, 10–20 cm long, 1–2 cm diam., with spine-like tip to 1 cm. **STAMINATE FLOWERS** with sepals 2 × 1.5 mm; petals 2 × 1.5 mm; staminal tube to 2.5 mm long, anthers reflexed 1.5 mm long. **PISTILLATE INFLORESCENCE** more massive than the staminate, with much shorter rachillae, spine-like tips to 2 cm long. **PISTILLATE FLOWERS** with sepals 10 × 4 mm; petals similar to sepals; staminodal ring to 1 mm; ovary c. 5 mm diam., styles and stigmas to 10 mm long. **FRUIT** usually 1-seeded, partially enclosed by enlarged calyx and corolla tipped by the stigmatic remains, variable, in wild type c. 3 × 2 cm, bright orange, sometimes dusky violet in exposed parts, mesocarp bright orange, endocarp 2–5 mm thick. **SEED** c. 2 × 1.5 cm.

NOTES. Small populations in the west of the island were separated by Jumelle and Perrier as a distinct variety, *E. guineensis* var. *madagascariensis*, and Beccari went further, elevating this variety to specific rank as *E. madagascariensis,* based on what we believe to be spurious differences in the position of the anthers and fruit size (anthers reflexed in African populations, erect in Madagascar, fruit smaller in Madagascar populations). However, *E. guineensis* in Africa is itself very variable, and current thinking (Rajanaidu, pers. com.) is that the taxon in Madagascar fits within the range of variation found in Africa. In the absence of good fertile specimens from the wild in Madagascar, we have based the above description on the wild type African oil palm.

SPECIMENS EXAMINED. Mahabo: Valley of Sakeny, near Malaimbandy, Oct. 1933 (sterile), *Humbert* 11385 (P).
Without locality, date or collector, sent by Jumelle to Beccari (FI) (Lectotype of *E. madagascariensis*).

GLOSSARY

(ADAPTED FROM UHL & DRANSFIELD 1987)

Abaxial – the side of an organ that faces away from the axis that bears it, e.g., the undersurface of a leaf, or the outer surface of a tubular bract.

Abcission – separation.

Acanthophyll – a spine derived from a leaflet.

Acaulescent – lacking a visible stem.

Acuminate – tapering to a point.

Acute – distinctly and sharply pointed but not drawn out.

Adaxial – the side of an organ that faces towards the axis that bears it, e.g., the upper surface of a leaf, or the inner surface of a tubular bract.

Adjacent-ligular – in germination, the new shoot developing next to the seed, enclosed by a ligule.

Adnate – attached to, as when two different types of organ are joined together.

Anastomose – to unite, usually forming a network.

Androecium – collective term for all the stamens.

Antepetalous – opposite the petals.

Antesepalous – opposite the sepals.

Anther – the part of the stamen that contains pollen (Fig. 1).

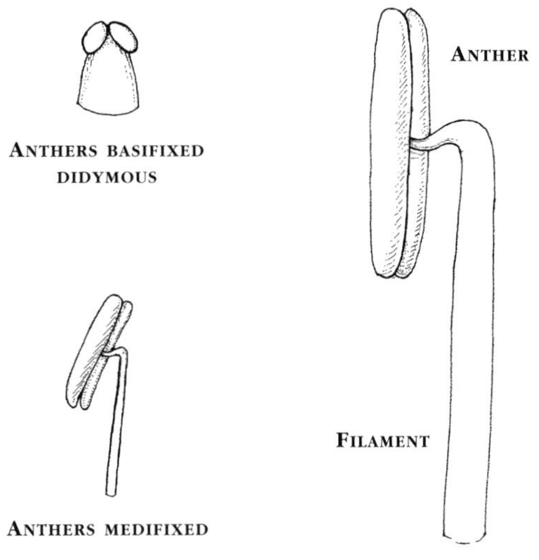

Figure 1

Anthesis – time of flower opening when pollen is shed, or stigmas are receptive.

Apiculate – bearing a short, sharp but not stiff point.

Apocarpous – with free carpels.

Armed – bearing some form of spines.

Auricle – an ear-like lobe.

Axillary – borne in an axil, the angle between the stem and a leaf or bract.

Basifixed – attached at the base.

Bifid – divided into two usually equal parts.

Bract – modified leaf, associated with an inflorescence.

Bracteole – a small bract borne on a flower stalk.

Caducous – dropping off early.

Caespitose – clustered.

Calyx – the outermost or lowermost whorl of floral organs, the sepals.

Carpel – the single unit of the gynoecium.

Catkin-like – used to describe a cylindrical rachilla when flowers are densely crowded.

Ciliate – bearing a fringe of hairs.

Cincinnus – a flower cluster wherein each successive flower arises in the axil of the bracteole on the preceding flower stalk.

Circumscissile – falling off neatly and leaving a circular scar.

Clustered – with several stems in the same individual.

Connate – united, when organs of the same kind are joined together.

Connective – the part of the stamen that connects the anthers to the filament.

Coriaceous – leathery.

Corolla – the second whorl of flower organs, the petals.

Costa – a rib, usually used for the midrib of the whole leaf rather than of leaflets.

Costapalmate – shaped like the palm of a hand and having a short midrib or costa.

Crown – the cluster of leaves borne at the tip of the stem.

Crownshaft – a conspicuous cylinder formed by tubular leaf sheaths at the tip of the stem (Fig. 2).

Cupule – a small cup-like structure.

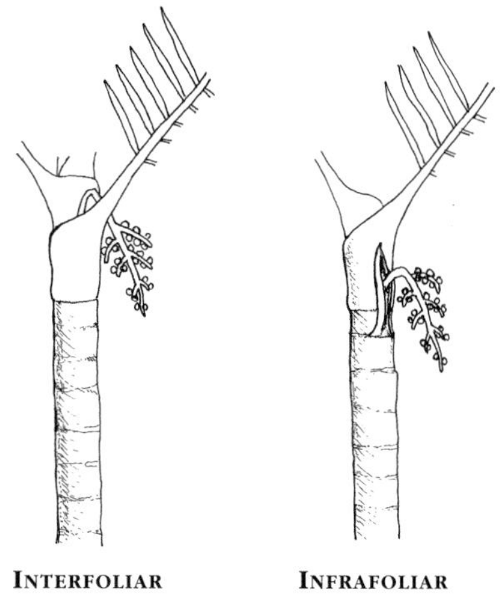

Figure 2

Deciduous – shed periodically, falling.

Decumbent – reclining, but with the apex ascending.

Dentate – toothed.

Dichotomous – equally forking (Fig. 3).

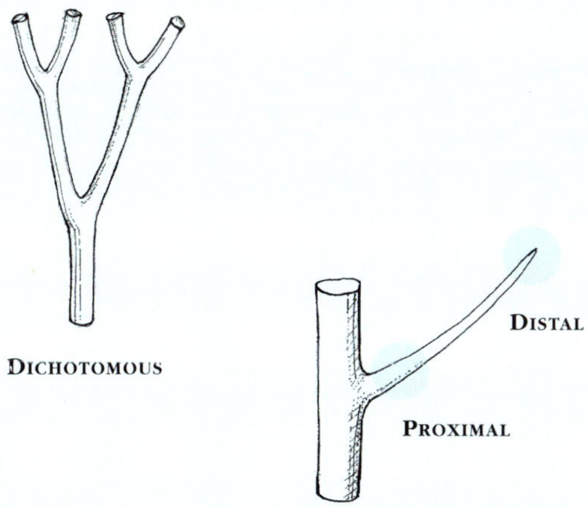

Figure 3

Didymous – of anthers where connective is almost absent.

Digitate – like fingers.

Dimorphic – with two different forms.

Dioecious – when staminate and pistillate flowers are borne on different plants.

Distal – farthest from point of attachment (Fig. 3).

Distant – widely separated.

Distichous – arranged in two ranks.

Divaricate – spread widely.

Dorsifixed – attached by the abaxial side.

Dyad – a pair.

Elliptic – oblong, with regularly rounded ends.

Embryo – the rudimentary plant in a seed.

Endocarp – the innermost layer of the fruit wall.

Endosperm – in palms the nutritive body of the seed.

Entire – (in palm leaves) not divided into leaflets.

Eophyll – in palm seedlings, the first leaf with a blade.

Epicarp – the outermost layer of the fruit wall.

Epipetalous – borne on the petals.

Extrorse – of anthers, opening away from the centre of the flower.

Fibrous – composed of, or including, fibres.

Filament – in stamens, the stalk that carries the anthers.

Filamentous – thread-like.

Flabellate – fan-shaped.

Gibbous – irregularly swollen, with one side more swollen than the other.

Glabrous – smooth, without hairs of scales.

Glaucous – covered with a blue-grey bloom.

Gynoecium – the ovule-bearing innermost part of the flower.

Hapaxanthic – describing shoots that flower and then die (see pleonanthic).

Hastula – a flap of tissue borne at the insertion of the blade on the petiole in a palmate leaf.

Hermaphrodite – of flowers, bearing both stamens and gynoecium.

Homogeneous – uniform, the same throughout (usually of endosperm).

Imbricate – overlapping.

Indumentum – any covering of hairs or scales.

Induplicate – V-shaped in cross section.

Inflexed – bent inwards.

Inflorescence – branching system that bears the flowers, including all its bracts and branches.

Infrafoliar – borne below the leaves.

Infructescence – as inflorescence, but in fruit.

Interfoliar – borne among the leaves.

Internode – the space or part of a stem between the attachment of two leaves.

Introrse – of anthers, opening towards the centre of the flower.

Keeled – bearing a ridge or keel.

Knee – a swelling on the leaf sheath, where there is a distinct bend.

Lamina – the blade of a leaf.

Lanceolate – narrow, tapering at both ends.

Latrorse – of anthers, opening sideways.

Leaflets – as used in this book, divisions of a pinnate leaf.

Ligule – distal projection of a leaf sheath.

Linear – several times longer than broad, narrow.

Locule – the cavity in which the ovule is borne.

Marcescent – withering before being shed.

Medifixed – attached at the middle (Fig. 1).

Mesocarp – the middle layer of the fruit wall.

Moderate – medium-sized.

Monoecious – describing a plant bearing both staminate and pistillate flowers.

Node – the point of attachment of a leaf on the stem.

Obovoid – egg-shaped, but broader distally.

Obpyriform – pear-shaped, but attached at the broader end.

Ovary – the ovule-bearing part of the gynoecium.

Ovoid – egg-shaped, broader at the base.

Ovule – the young or developing seed.

Palmate – shaped like the palm of a hand.

Papillose – bearing minute rounded projections.

Pedicel – a flower stalk.

Peduncle – the lower unbranched part of an inflorescence, the inflorescence stalk (Fig. 4).

Peduncular bracts – empty bracts borne on the peduncle between the prophyll and the first rachis bracts.

Perianth – collective term for sepals and petals.

Persistent – remaining attached for a long time.

Petal – one unit of the inner floral envelope or corolla.

Petiole – the stalk of the leaf (Fig. 5).

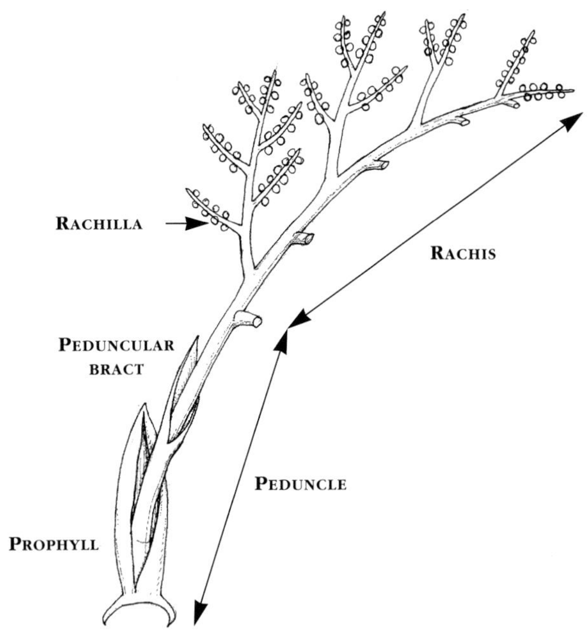

Figure 4

Pinna(ae) – a leaflet (or leaflets) of a pinnate leaf.

Pinnate – feather-like.

Pistil – see gynoecium.

Pistillate – bearing a pistil (gynoecium), female.

Pistillode – a sterile gynoecium.

Pit – a cavity formed by the rachilla and united bracts.

Pleonanthic – describing shoots that flower continually, not dying after flowering (see hapaxanthic).

Plicate – pleated, folded.

Plumose – softly feathered, with leaflets arranged in different planes.

Praemorse – jaggedly toothed as if bitten.

Procumbent – of stems, tending to bend over and grow parallel to the ground.

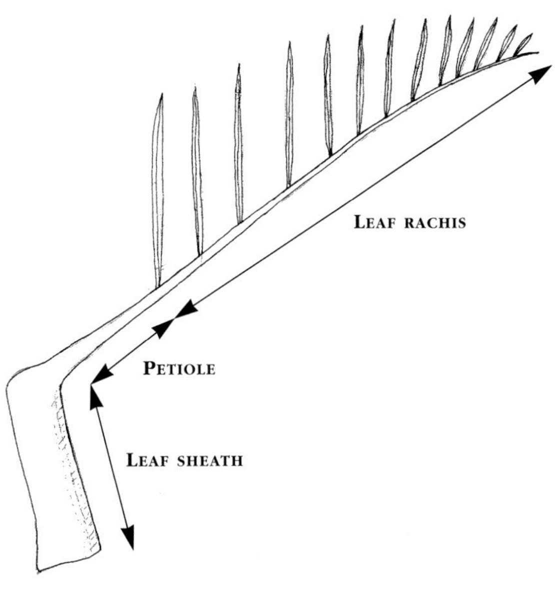

Figure 5

Prophyll – the first bract borne on an inflorescence (Fig. 4).

Protandrous – of flowers or inflorescences, pollen shed before stigmas receptive.

Protogynous – of flowers of inflorescences, stigmas receptive before pollen is shed.

Proximal – nearest to the attachment.

Pseudomonomerous – appearing to be of one member, but actually of several, used to describe a gynoecium with one fertile carpel, but with parts of two other carpels present.

Punctiform – dot-like.

Pyrene – a seed-like body formed by a hard layer of endocarp surrounding each seed (used only where more than one seed regularly develops).

Pyriform – pear-shaped.

Rachilla – the branch that bears the flowers (Fig. 4).

Rachilla bracts – bracts borne on a rachilla.

Rachis – the axis of a leaf beyond the petiole, or the axis of an inflorescence beyond the peduncle (Fig. 4).

Rachis bracts – bracts subtending first-order branches in the inflorescence.

Ramenta – rather thin elongate scales with ragged edges.

Receptacle – the central axis of a flower to which the floral organs are attached.

Reduplicate – of leaflets, L-shaped in cross-section.

Rein – a narrow marginal strip which is usually shed when a pinnate leaf unfolds.

Remote-ligular – in germination, the young plant connected to the seed by the elongated tubular cotyledonary stalk, and bearing a ligule.

Remote-tubular – in germination, the young plant connected to the seed by the elongated tubular cotyledonary stalk, but lacking a ligule.

Rheophyte – a plant adapted to growing within the flood-zone of a river.

Rhizome – underground stem.

Root boss – a swelling at the base of the stem, from which the roots arise.

Rugose – wrinkled.

Ruminate – of endosperm, darkly streaked due to the infolding of seed coats (Fig. 6).

Sagittate – shaped like an arrow head.

Scale leaf – a reduced leaf.

Segment – a division of a palmate or costapalmate leaf.

Sepal – a single part of the outermost whorl of floral envelopes, the calyx.

Sessile – lacking a stalk.

Sheath – basal part of the leaf that is usually tubular, but often splits.

Sigmoid – S-shaped.

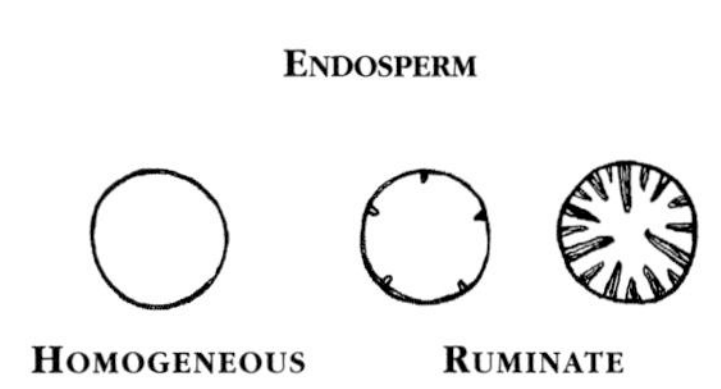

Figure 6

Sinuous – wavy.

Solitary – single-stemmed, not clustered.

Spicate – spike-like, the inflorescence unbranched (Fig. xxxx).

Spinule – very small spine.

Stamen – the male organ of the flower, consisting of a filament and anthers that contain pollen.

Staminate – bearing stamens, male.

Staminode – a sterile stamen, often greatly modified in form

Stem – the main axis.

Stigma – the pollen receptor on the gynoecium.

Stilt-roots – oblique roots at the stem base but above ground, that prop up the stem.

Striate – lined.

Style – the part of the carpel between the ovary and the stigmas.

Sub – as a prefix, meaning nearly, almost, or somewhat.

Subulate – awl-shaped.

Sympodial – composed of superposed branches, lacking a single main axis.

Syncarpous – with united carpels.

Terete – circular in cross-section.

Testa – the seed coat.

Thecae – the locules of an anther.

Tomentose – densely covered with short hairs.

Triad – a group of three flowers composed of a central pistillate flower, flanked by two lateral staminate flowers.

Trilocular – of a gynoecium, with three chambers or locules.

Triovulate – bearing three ovules.

Truncate – as though cut off nearly straight across.

Ultramafic – rock rich in iron and manganese.

Unarmed – without any spines.

Undulate – wavy.

Unilocular – of a gynoecium, with one chamber.

Uniovulate – with one ovule.

Valvate – meeting exactly without overlapping.

Ventricose – swollen.

Versatile – of anthers, attached near the centre and movable on the filament.

BIBLIOGRAPHY

Adany, A. J., Birkinshaw, C. R. & Andrews, J. R. (1994). Illegal palm felling in Lokobe Reserve, Madagascar. Principes 38(4): 204–210.

Audru, J. & Andriatsarafara, F.R. 1986. Le satramira chez les Antandroy de la region de Miadana (*Hyphaene shatan* Boj. - Palmae), une providence ou une peste végétale. IEMVT. Maisons-Alfort. pp 53.

Baillon, H.E. (1894a). Les palmiers malgaches à petites fleurs. Bull. Mens. Soc. Linn. Paris 2 (147): 1161–1168.

— (1894b). Les palmiers malgaches à petites fleurs (cont'd). Bull. Mens. Soc. Linn. Paris 2 (148): 1171–1176.

— (1895). Les palmiers malgaches à petites fleurs (cont'd). Bull. Mens. Soc. Linn. Paris 2 (150): 1185 - 1186.

Baker, J.G. (1887). Further contributions to the flora of Madagascar. J. Linn. Soc. (Bot.) 22: 441–537.

Battistini, R. & Richard-Vindard, G. (eds.) (1972). Biogeography and ecology in Madagascar. Monographiae Biologicae 21. Junk, Den Haag.

Beccari, O. (1890). Rivista monografica delle specie del genere *Phoenix* Linn. Malesia 3: 345–416.

— (1906). Palmarum Madagascariensium Synopsis. Engl. Bot. Jahrb. 38, Beibl. 87: 1–41.

— (1913). Studio sui *Borassus* e descrizione di un genere nuovo asiatico di Borasseae. Webbia 4: 293–385.

— (1912–1914) Palme del Madagascar. Istituto Micrografica Italiano, Firenze. pp 1–32 appeared in 1912, pp. 33–59 in 1914 (Stafleu & Cowan).

Beccari, O. & Pichi-Sermolli, R.E.G. (1956). Subfamiliae Arecoidearum Palmae Gerontogeae. Webbia 11: 1– 187.

Beentje, H.J. (1993a). The days I didn't find lakatra. Principes 37(1): 4–11.

— (1993b). A new aquatic palm from Madagascar. Principes 37(4): 197–202.

— (1994a). The days I found lakatra, and trouble. Principes 38(2): 73–79.

— (1994b). A monograph of *Ravenea* (*Palmae*: *Ceroxyloideae*). Kew Bull. 49(4): 623–671.

Beentje, H.J. & Dransfield, J. (in press). Generic delimitation in the palms of Madagascar. In Proc. AETFAT Conference, Wageningen 1994.

Bernardi, L. (1974). Notulae ad *Sindroa*... Candollea 29: 163–171.

Bouché C.D. (1878). (Untitled passage). Monatsschrift des Vereines zur Beforderung des Gartenbaues in den Konigleges Preussischen Staaten fur Gaertnerei und Pflanzenkunde April 1878: 197.

Burkill, I.H. (1935). A dictionary of the economic products of the Malay Peninsula. Government Printer, Singapore.

Davison, G.W.H. 1988. Endau Rompin: A Malaysian Heritage. Malayan Nature Society, Kuala Lumpur, 221 pp.

Dransfield, J. (1986). Flora of tropical East Africa: Palmae. A.A. Balkema, Rotterdam.

— (1988). *Beccariophoenix madagascariensis*. Principes 32: 59–68.

— (1989a). Speciation patterns in the palms of Madagascar. pp. 153–171 in Holm-Nielsen L.B., Nielsen I.C. & Balslev H. (eds.) - Tropical Forests - botanical dynamics, speciation and diversity. Academic Press, London.

— (1989b). Searching for a forest coconut in Madagascar. In: Hepper, F.N. (ed.). Plant hunting for Kew. Royal Botanic Gardens, Kew, Her Majesty's Stationery Office, London. pp 51–60.

— (1992a). Observations on rheophytic palms in Borneo. Bull. Inst. Fr. Études Andines 21(2): 415–432.

— (1992b). *Voanioala*, the forest coconut. Principes 36(3): 124–127.

Dransfield, J. & Beentje, H.J. (1995). *Satranala* (*Coryphoideae*: *Borasseae*: *Hyphaeninae*), a new palm genus from Madagascar. Kew Bull. 50(1): 85–92.

Dransfield J. & Uhl N.W. (1984a). A magnificent new palm from Madagascar. Principes 28(4): 151–154.

— (1984b). *Halmoorea*, a new genus from Madagascar, with notes on *Sindroa* and *Orania*. Principes 28(4): 163–167.

— (1986a). An outline of a classification of palms. Principes 30(1): 3–11.

— (1986b). *Ravenea* in the Comores. Principes 30(4): 156–160.

Ferguson, I.K., Dransfield, J. & Flawn, I. (1987). A review of the pollen morphology and systematics of the genera *Ravenea* and *Louvelia* (*Ceroxyleae*: *Ceroxyloideae*: *Palmae*). J. Palyn. 23: 65–72.

Ferguson I.K. & Harley M.M. (1993). The significance of new and recent work on pollen morphology in the Palmae. Kew Bull. 48: 205–243.

Fisher, J.B. (1973). Unusual branch development in the palm, *Chrysalidocarpus*. Bot. J. Linn. Soc. 66: 83–95.

Fisher J.B. & Moore Jr. H.E. (1977). Multiple inflorescences in the palms (Arecaceae): their development and significance. Bot. Jahrb. Syst. 98: 573–611.

Goodman, S.M. (1993). A reconnaissance of Île Sainte Marie, Madagascar: The status of the forest, avifauna, lemurs and fruit bats. Biol. Conserv. 65: 205–212.

Guillaumet J.-L. (1973). Une nouvelle espèce de palmier nain de Madagascar. Adansonia sér. 2, 13: 341–349.

Hodel, D.R. (1993). A tale of two (three?) *Neodypsis*. The Palm Journal: 8–18.

Hooker, J.D. (1887). Palmae. In Bentham, G. & Hooker, J.D., Genera Plantarum 3 (2): 870–948.

— (1884). *Ravenea hildebrandtii*. Botanical Magazine 110: tab. 6776.

Humbert, H. (1955). Une merveille de la nature à Madagascar. Première exploration du massif du Marojejy et de ses satellites. Mém. Inst. Scient. Madagascar B, 6: 1–210.

Humbert, H. & Cours Darne, G. (1965). Notice de la carte Madagascar. Carte Internationale du Tapis Vegetal. Trav. Sect. Sci. Tech. Inst. Fr. Pondichéry. Hors Sér. 6.

Jenkins, M. (ed.). 1987. Madagascar. An environmental profile. IUCN, Gland, Switzerland and Cambridge, U.K. pp 374.

Johnson, M. (1989). An unusually high chromosome number in *Voanioala gerardii* (*Palmae*: *Arecoideae*: *Cocoeae*: *Butiinae*) from Madagascar. Kew Bull. 44(2): 207–210.

Jumelle, H. (1922). Les *Chrysalidocarpus*, palmiers de Madagascar. Ann. Inst. Bot.-Geol. Colon. Marseille sér. 2, 10 (3): 5–32.

— (1923). Les *Dypsis*, palmiers de Madagascar. Bull. Acad. Malgache. Nouvelle Sér. 6: 1–20.

— (1924a). Les *Neodypsis*, palmiers de Madagascar. Compt. Rend. Acad. Sc. Paris 179: 448–450.

— (1924b). Les *Neodypsis*, palmiers de Madagascar. Ann. Inst. Bot.-Geol. Colon. Marseille sér. 4, 2 (2): 5–32.

— (1924c). *Neodypsis* et *Chrysalidocarpus*, palmiers de Madagascar. Compt. Rend. Acad. Sc. Paris 179: 249–251.

— (1927a). Les *Vonitra*, palmiers de Madagascar. Ann. Inst. Bot.-Geol. Colon. Marseille sér. 4, 5 (1): 5–19.

— (1927b). *Ravenea* et *Louvelia*, palmiers de Madagascar. Ann. Inst. Bot.-Geol. Colon. Marseille sér. 4, 5 (1): 21–51.

— (1928a). Les *Neophloga*, palmiers de Madagascar. Ann. Inst. Bot.-Geol. Colon. Marseille sér. 4, 6 (3): 5–47.

— (1928b). Un nouveau genre Malgache de palmiers. Ann. Inst. Bot.-Geol. Colon. Marseille sér. 4, 6 (2): 17–19.

— (1933a). Nouveaux palmiers de Madagascar. Ann. Inst. Bot.-Geol. Colon. Marseille sér. 5, 1 (1): 5–29.

— (1933b). Nouvelles observations sur quelques palmiers Malgaches. Ann. Inst. Bot.-Geol. Colon. Marseille sér. 5, 1 (3): 5–19.

— (1938). Catalogue des plantes de Madagascar. Palmae. G. Pitot, Tananarive.

Jumelle, H. & Perrier de la Bâthie, H. (1912). Un nouveau genre de palmiers de Madagascar. Compt. Rend. Acad. Sc. Paris 155: 410–411.

— (1913). Palmiers de Madagascar. Ann. Inst. Bot.-Geol. Colon. Marseille sér. 3, 1: 1–91.

— (1945). Palmiers. In Humbert, H. (ed.) - Flore de Madagascar et des Comores 30e famille. Imprimerie officielle, Antananarivo.

Koechlin, J., Guillaumet, J.-L. & Morat, P. (1974). Flore et Végétation de Madagascar. J. Cramer, Vaduz. pp 687.

Martius, C.F.P. von (1823–50). Historia Naturalis Palmarum. Leipzig, Germany.

Moore, H.E. (1963). An annotated checklist of cultivated palms. Principes 7 (4): 119–182.

— (1965). Palm hunting around the world. Principes 9(1): 13 - 29.

— (1973). The major groups of palms and their distribution. Gentes Herb. 11(2): 27–140.

Neumann O. (1878). (Untitled passage). Monatsschrift des Vereines zur Beforderung des Gartenbaues in den Konigleges Preussischen Staaten fur Gaertnerei und Pflanzenkunde April 1878: 323–324.

Nicoll, M. & Langrand, O. (1989). Madagascar: Revue de la conservation et des aires protégées. WWF, Gland, Switzerland. pp 374.

Paulian, R. (1984). Introduction to the mammals. pp 151–154. In Jolly, A., Oberlé, P. & Albignac, R. Key environments: Madagascar. Pergamon Press, Oxford, U.K.

Röser, M. (1994). Pathways of caryological differentiation in palms (*Arecaceae*). Pl. Syst. Evol. 189: 83–122.

Steenis, C.G.G.J. van (1981). Rheophytes of the world. Sijthoff & Noordhoff, Alphen aan der Rijn, Netherlands. pp 405.

Stewart, L. (1994). A guide to palms and cycads of the world. Cassell, London. pp.246.

Tomlinson, P.B. (1969). The anatomy of the vegetative organs of *Juania australis* (*Palmae*). Gentes Herb. 10: 412–424.

Uhl, N.W. (1969). Floral anatomy of *Juania*, *Ravenea*, and *Ceroxylon* (*Palmae-Arecoideae*). Gentes Herb. 10: 394–411.

Uhl, N.W. & Dransfield, J. (1987). Genera Palmarum. A classification of palms based on the work of H.E. Moore Jr. L.H. Bailey Hortorium and the International Palm Society, Allen Press, Kansas.

Uhl, N.W., Dransfield, J., Davis, J.I., Luckow, M.A., Hansen, K.S. & Doyle, J.I. (in press). Phylogenetic relationships among palms: cladistic analysis of morphological and chloroplast DNA restriction site variation. In Rudall, P. J., Cribb, P.J. , Cutler, D.F. & Humphries, C.J. (eds.). Monocotyledons: systematics and evolution. Royal Botanic Gardens, Kew.

Wendland, H. (1878). Beiträge zur Kenntniss der Palmen. Bot. Zeit. 36(8): 114–118.

Zanoni, T. & Read, R.W. (1986). Hybrid in *Chrysalidocarpus*. Principes 30(1): 35–37.

INDEX TO SPECIMENS SEEN

COLLECTOR	NR	GENUS	SPECIES
Aubert du Petit		*Dyps*	*forfi*
Aubert du Petit		*Dyps*	*pinna*
Barnett	455	*Dyps*	*baron*
Baron	486	*Dyps*	*heterop*
Baron	502	*Dyps*	*decip*
Baron	508	*Dyps*	*nodif*
Baron	1286	*Dyps*	*heterop*
Baron	1287	*Dyps*	*nodif*
Baron	1490	*Dyps*	*hilde*
Baron	2319	*Phoe*	*recli*
Baron	2319a	*Dyps*	*fibro*
Baron	2322	*Dyps*	*laste*
Baron	2323	*Dyps*	*poivr*
Baron	2880	*Dyps*	*nodif*
Baron	3002	*Dyps*	*heterop*
Baron	3248	*Dyps*	*heterop*
Baron	3270	*Dyps*	*baron*
Baron	3271	*Dyps*	*decip*
Baron	3413	*Dyps*	*conci*
Baron	3911	*Dyps*	*hilde*
Baron	4340	*Dyps*	*nodif*
Baron	4465	*Dyps*	*nodif*
Baron	4509	*Dyps*	*baron*
Baron	4776	*Phoe*	*recli*
Baron	4839	*Phoe*	*recli*
Baron	5071	*Dyps*	*heterop*
Baron	5072	*Dyps*	*nodif*
Baron	5082	*Dyps*	*hilde*
Baron	5923	*Hyph*	*coria*
Baron	6068	*Dyps*	*baron*
Baron	6724	*Phoe*	*recli*
Baron	<1.1882	*Rave*	*madag*
Baron	<10/820	*Dyps*	*conci*
Baron	<12.1881	*Dyps*	*nodif*
Baron	<1889	*Dyps*	*decip*
Baron	8/81	*Dyps*	*decip*
Beentje	4401	*Dyps*	*louve*
Beentje	4402	*Dyps*	*pinna*
Beentje	4403	*Dyps*	*nodif*
Beentje	4404	*Dyps*	*hilde*
Beentje	4405	*Dyps*	*nodif*
Beentje	4407	*Dyps*	*hilde*
Beentje	4408	*Dyps*	*heterop*
Beentje	4409	*Dyps*	*utili*
Beentje	4410	*Dyps*	*catat*
Beentje	4411	*Dyps*	*jumel*
Beentje	4412	*Dyps*	*baron*
Beentje	4414	*Dyps*	*baron*
Beentje	4417	*Dyps*	*utili*
Beentje	4418	*Dyps*	*jumel*
Beentje	4419	*Rave*	*robus*
Beentje	4420	*Dyps*	*therm*
Beentje	4424	*Dyps*	*linea*
Beentje	4425	*Dyps*	*ambol#*
Beentje	4426	*Dyps*	*confu#*
Beentje	4428	*Dyps*	*catat*
Beentje	4429	*Dyps*	*catat*
Beentje	4434	*Rave*	*madag*
Beentje	4435	*Rave*	*madag*
Beentje	4436	*Dyps*	*ambol*
Beentje	4437	*Maso*	*kona*
Beentje	4439	*Dyps*	*fibro*
Beentje	4440	*Rave*	*lakat*
Beentje	4441	*Dyps*	*nodif*
Beentje	4443	*Dyps*	*palud*
Beentje	4445	*Dyps*	*arena*
Beentje	4446	*Dyps*	*nodif*
Beentje	4448	*Dyps*	*palud*
Beentje	4449	*Dyps*	*fibro*
Beentje	4450	*Rave*	*madag*
Beentje	4451	*Dyps*	*simia*
Beentje	4452	*Dyps*	*laste*
Beentje	4453	*Dyps*	*fanja*
Beentje	4454	*Dyps*	*fanev*
Beentje	4455	*Oran*	*longi*
Beentje	4456	*Rave*	*sambi*
Beentje	4457	*Dyps*	*sp.*
Beentje	4458	*Dyps*	*ramen*
Beentje	4459	*Dyps*	*fanja*
Beentje	4460	*Dyps*	*lutes*
Beentje	4461	*Dyps*	*pusil*
Beentje	4462	*Dyps*	*palud*
Beentje	4463	*Dyps*	*forfi*
Beentje	4464	*Dyps*	*oligo*
Beentje	4465	*Dyps*	*lantz*
Beentje	4466	*Oran*	*ravak*
Beentje	4467	*Dyps*	*tsarav*
Beentje	4468	*Dyps*	*pusil*
Beentje	4472	*Maso*	*madag*
Beentje	4474	*Satr*	*decus*
Beentje	4475	*Dyps*	*beent*
Beentje	4478	*Dyps*	*sp.*
Beentje	4485	*Dyps*	*bejof*
Beentje	4486	*Rave*	*drans*
Beentje	4487	*Dyps*	*schat*
Beentje	4488	*Dyps*	*glabr*
Beentje	4489	*Dyps*	*corni*
Beentje	4491	*Rave*	*sambi*
Beentje	4493	*Oran*	*longi*
Beentje	4494	*Dyps*	*angustifolia*
Beentje	4496	*Dyps*	*nodif*
Beentje	4497	*Dyps*	*nodif*
Beentje	4498	*Dyps*	*fibro*
Beentje	4499	*Maro*	*insig*
Beentje	4500	*Maro*	*insig*
Beentje	4501	*Rave*	*sambi*
Beentje	4502	*Rave*	*sambi*
Beentje	4503	*Rave*	*sambi*
Beentje	4504	*Rave*	*sambi*
Beentje	4505	*Dyps*	*nodif*
Beentje	4506	*Dyps*	*mananj*
Beentje	4507	*Dyps*	*hilde*
Beentje	4509	*Dyps*	*hilde*
Beentje	4510	*Dyps*	*nause*
Beentje	4511	*Dyps*	*angusta*
Beentje	4512	*Dyps*	*digit*
Beentje	4513	*Dyps*	*singu*
Beentje	4515	*Dyps*	*integ*
Beentje	4516	*Dyps*	*nause*
Beentje	4517	*Dyps*	*integ*
Beentje	4518	*Dyps*	*simia*
Beentje	4519	*Dyps*	*fibro*
Beentje	4521	*Dyps*	*mananj*
Beentje	4522	*Oran*	*trisp?*
Beentje	4523	*Rave*	*lakat*
Beentje	4524	*Maso*	*kona*
Beentje	4525	*Maso*	*kona*
Beentje	4526	*Rave*	*madag*
Beentje	4527	*Rave*	*madag*
Beentje	4528	*Dyps*	*inter*
Beentje	4530	*Rave*	*madag*
Beentje	4532	*Rave*	*drans*
Beentje	4533	*Maro*	*insig*
Beentje	4534	*Dyps*	*baron*
Beentje	4535	*Dyps*	*conci*
Beentje	4536	*Rave*	*louve*
Beentje	4537	*Becc*	*madag*
Beentje	4538	*Dyps*	*linea*
Beentje	4539	*Dyps*	*conci*
Beentje	4540	*Dyps*	*louve*
Beentje	4544	*Dyps*	*baron*
Beentje	4545	*Dyps*	*confu#*
Beentje	4546	*Rave*	*madag*
Beentje	4547	*Maro*	*insig*

COLLECTOR	NR	GENUS	SPECIES
Beentje	4549	*Dyps*	*bilde*
Beentje	4550	*Dyps*	*conci*
Beentje	4551	*Maro*	*insig*
Beentje	4552	*Hypb*	*coria*
Beentje	4553	*Dyps*	*madag*
Beentje	4554	*Dyps*	*madag*
Beentje	4555	*Bism*	*nobil*
Beentje	4558	*Rave*	*sambi*
Beentje	4559	*Dyps*	*andri*
Beentje	4560	*Rave*	*sambi*
Beentje	4561	*Dyps*	*madag*
Beentje	4562	*Dyps*	*rivul*
Beentje	4563	*Dyps*	*pinna*
Beentje	4569	*Rave*	*robus*
Beentje	4570	*Rave*	*sambi*
Beentje	4571	*Dyps*	*andri*
Beentje	4572	*Rave*	*sambi*
Beentje	4573	*Dyps*	*pinna*
Beentje	4574	*Dyps*	*pilul*
Beentje	4575	*Dyps*	*laste*
Beentje	4576	*Rave*	*robus*
Beentje	4577	*Dyps*	*crini*
Beentje	4578	*Dyps*	*biara*
Beentje	4579	*Dyps*	*madag*
Beentje	4580	*Dyps*	*laste*
Beentje	4581	*Dyps*	*rivul*
Beentje	4582	*Rave*	*sambi*
Beentje	4583	*Dyps*	*ampas*
Beentje	4585	*Dyps*	*pinna*
Beentje	4586	*Dyps*	*onila*
Beentje	4587	*Rave*	*rivul*
Beentje	4588	*Rave*	*glauc*
Beentje	4589	*Rave*	*glauc*
Beentje	4590	*Rave*	*xerop*
Beentje	4591	*Dyps*	*mcdon*
Beentje	4592	*Dyps*	*tenui*
Beentje	4594	*Dyps*	*catat*
Beentje	4596	*Dyps*	*malco*
Beentje	4598	*Dyps*	*pinna*
Beentje	4599	*Dyps*	*scott*
Beentje	4600	*Rave*	*robus*
Beentje	4601	*Dyps*	*tenui*
Beentje	4603	*Dyps*	*malco*
Beentje	4604	*Rave*	*robus*
Beentje	4605	*Rave*	*kroci*
Beentje	4607	*Dyps*	*scott*
Beentje	4608	*Dyps*	*scott*
Beentje	4609	*Dyps*	*saint*
Beentje	4610	*Dyps*	*brevi*
Beentje	4611	*Rave*	*music*
Beentje	4612	*Rave*	*music*
Beentje	4613	*Dyps*	*mananj*
Beentje	4614	*Dyps*	*decar*
Beentje	4615	*Dyps*	*ambos*
Beentje	4616	*Oran*	*longi*
Beentje	4617	*Dyps*	*coria*
Beentje	4618	*Dyps*	*fanev*
Beentje	4619	*Dyps*	*lantz*
Beentje	4620	*Dyps*	*palud*
Beentje	4621	*Dyps*	*virid*
Beentje	4622	*Dyps*	*oligo*
Beentje	4623	*Rave*	*julie*
Beentje	4624	*Dyps*	*forfi*
Beentje	4625	*Rave*	*sambi*
Beentje	4626	*Dyps*	*beent*
Beentje	4627	*Dyps*	*confu*
Beentje	4628	*Satr*	*decus*
Beentje	4629	*Dyps*	*spnov*
Beentje	4630	*Rave*	*albic*
Beentje	4631	*Dyps*	*conva*
Beentje	4632	*Maso*	*madag*
Beentje	4633	*Dyps*	*forfi*
Beentje	4636	*Dyps*	*fasci*
Beentje	4637	*Dyps*	*forfi*
Beentje	4638	*Dyps*	*oligo*
Beentje	4639	*Dyps*	*mango*
Beentje	4640	*Dyps*	*jumel*
Beentje	4641	*Dyps*	*lantz*
Beentje	4643	*Rave*	*albic*
Beentje	4644	*Rave*	*albic*
Beentje	4645	*Dyps*	*ovobo*
Beentje	4646	*Dyps*	*subra*
Beentje	4647	*Dyps*	*palud*
Beentje	4648	*Dyps*	*thiry*
Beentje	4651	*Dyps*	*pusil*
Beentje	4652	*Rave*	*latis*
Beentje	4653	*Dyps*	*conci*
Beentje	4654	*Dyps*	*pilul*
Beentje	4655	*Dyps*	sp.
Beentje	4657	*Rave*	*latis*
Beentje	4658	*Dyps*	*decip*
Beentje	4659	*Dyps*	*orope*
Beentje	4660	*Rave*	*robus*
Beentje	4661	*Rave*	*julie*
Beentje	4662	*Rave*	*robus*
Beentje	4663	*Maro*	*insig*
Beentje	4665	*Rave*	*robus*
Beentje	4666	*Rave*	*madag*
Beentje	4667	*Rave*	*madag*
Beentje	4668	*Rave*	*robus*
Beentje	4669	*Dyps*	*catat*
Beentje	4670	*Dyps*	*malco#*
Beentje	4671	*Dyps*	*biara#*
Beentje	4672	*Dyps*	*prest*
Beentje	4673	*Dyps*	*prest*
Beentje	4674	*Rave*	*sambi*
Beentje	4675	*Rave*	*sambi*
Beentje	4676	*Dyps*	*lutes*
Beentje	4677	*Rave*	*robus*
Beentje	4678	*Oran*	*longi*
Beentje	4679	*Dyps*	*nause*
Beentje	4680	*Dyps*	*andap*
Beentje	4681	*Dyps*	*spica*
Beentje	4684	*Dyps*	*occid*
Beentje	4685	*Dyps*	*spica*
Beentje	4686	*Rave*	*robus*
Beentje	4687	*Dyps*	*mirab*
Beentje	4688	*Dyps*	*pinna*
Beentje	4690	*Dyps*	*lutes*
Beentje	4691	*Phoe*	*recli*
Beentje	4692	*Phoe*	*recli*
Beentje	4693	*Dyps*	*madag*
Beentje	4694	*Hypb*	*coria*
Beentje	4695	*Dyps*	*nossi*
Beentje	4696	*Dyps*	*pinna*
Beentje	4697	*Dyps*	*ampas*
Beentje	4698	*Dyps*	*nossi*
Beentje	4699	*Rave*	*sambi*
Beentje	4700	*Dyps*	*ampas*
Beentje	4701	*Dyps*	*ampas*
Beentje	4704	*Dyps*	*nossi*
Beentje	4705	*Rave*	*sambi*
Beentje	4706	*Oran*	*longi*
Beentje	4707	*Bora*	*sambi*
Beentje	4708	*Bora*	*sambi*
Beentje	4709	*Dyps*	*onila*
Beentje	4711	*Rave*	*glauc*
Beentje	4712	*Rave*	*glauc*
Beentje	4713	*Dyps*	*onila*
Beentje	4714	*Rave*	*rivul*
Beentje	4715	*Rave*	*rivul*
Beentje	4716	*Dyps*	*baron*
Beentje	4717	*Rave*	*robus*
Beentje	4719	*Rave*	*julie*
Beentje	4722	*Rave*	*julie*
Beentje	4723	*Rave*	*madag*
Beentje	4724	*Rave*	*madag*
Beentje	4725	*Maro*	*insig*
Beentje	4726	*Rave*	*lakat*
Beentje	4727	*Dyps*	*mananj*
Beentje	4728	*Dyps*	*mananj*
Beentje	4729	*Dyps*	*ifana*
Beentje	4730	*Oran*	*longi*
Beentje	4731	*Dyps*	*ambol*
Beentje	4732	*Rave*	*drans*
Beentje	4733	*Maro*	*insig*
Beentje	4734	*Dyps*	*therm*
Beentje	4735	*Dyps*	*confu#*
Beentje	4737	*Dyps*	*nodif*
Beentje	4739	*Dyps*	*catat*
Beentje	4740	*Dyps*	*heterop*
Beentje	4741	*Rave*	*madag*
Beentje	4742	*Dyps*	*ambos*
Beentje	4743	*Dyps*	*bilde*
Beentje	4744	*Dyps*	*jumel*
Beentje	4745	*Rave*	*louve*

COLLECTOR	NR	GENUS	SPECIES
Beentje	4751	*Rave*	*xerop*
Beentje	4753	*Rave*	*xerop*
Beentje	4754	*Rave*	*xerop*
Beentje	4756	*Rave*	*music*
Beentje	4758	*Becc*	*madag*
Beentje	4759	*Dyps*	*scott*
Beentje	4760	*Dyps*	*saint*
Beentje	4761	*Rave*	*kroci*
Beentje	4766	*Dyps*	*catat*
Beentje	4768	*Dyps*	*onila*
Beentje	4769	*Rave*	*louve*
Beentje	4770	*Dyps*	*conci*
Beentje	4771	*Dyps*	*baron*
Beentje	4772	*Dyps*	*louve*
Beentje	4776	*Rave*	*latis*
Beentje	4777	*Dyps*	*linea*
Beentje	4782	*Dyps*	*scott#*
Beentje	4783	*Rave*	*julie*
Beentje	4784	*Rave*	*julie*
Beentje	4785	*Dyps*	*laevi*
Beentje	4786	*Dyps*	*simia*
Beentje	4787	*Rave*	*julie*
Beentje	4788	*Dyps*	*elega*
Beentje	4789	*Rave*	*robus*
Beentje	4792	*Dyps*	*mahia*
Beentje	4793	*Rave*	*lakat*
Beentje	4794	*Dyps*	*inter*
Beentje	4796	*Dyps*	*fibro*
Beentje	4797	*Dyps*	*ambol*
Beentje	4798	*Dyps*	*fibro*
Beentje	4799	*Maso*	*kona*
Beentje	4800	*Dyps*	*pinna*
Beentje	4804	*Maro*	*insig*
Beentje	4805	*Rave*	*sambi*
Beentje	4808	*Dyps*	*nodif#*
Beentje	4810	*Satr*	*decus*
Beentje	4811	*Dyps*	*thiry*
Beentje	4812	*Rave*	*albic*
Beentje	4814	*Rave*	*albic*
Beentje	4815	*Satr*	*decus*
Beentje	4818	*Rave?*	*drans?*
Beentje	4819	*Dyps*	*bovom*
Beentje	4820	*Dyps*	*corni*
Beentje	4827	*Dyps*	*bovom*
Beentje	4835	*Dyps*	*saban*
Beentje	4839	*Maso*	*kona*
Benoist	862	*Dyps*	*thiry*
Benoist	1117	*Dyps*	*bilde*
Bernard		*Dyps*	*pinna*
Bernardi	11646	*Rave*	*bilde*
Bernardi	11750	*Dyps*	*lance*
Bernier	40	*Dyps*	*nodif*
Bernier	41	*Dyps*	*berni*
Bernier	41.a	*Dyps*	*berni*
Bewsher	34	*Rave*	*bilde*
Birkinshaw	105	*Dyps*	*madag*
Birkinshaw	124	*Dyps*	*nossi*
Birkinshaw	136	*Rave*	*sambi*
Bogner	186	*Dyps*	*linea*
Boiteau 1936?		*Dyps*	*decip*
Boivin	1703	*Dyps*	*nodif*
Boivin	1704	*Dyps*	*forfi*
Boivin	1704a	*Dyps*	*fibro*
Boivin	1706	*Dyps*	*palud*
Boivin	1707	*Dyps*	sp.
Boivin	1709	*Dyps*	*glabr*
Boivin	2282	*Hyph*	*coria*
Boivin	2283	*Phoe*	*recli*
Boivin	3100	*Phoe*	*recli*
Boivin		*Dyps*	*nossi*
Boivin		*Dyps*	*pinna*
Boivin		*Dyps*	*thou*
Boivin		*Dyps*	sp.
Boivin		*Dyps*	sp.
Boivin 1847		*Dyps*	*corni*
Boivin anno1850		*Dyps*	*oligo*
Boivin april 1851		*Dyps*	*nodif*
Boivin Mar 1851		*Dyps*	*nossi*
Boivin mar1851		*Dyps*	*nossi*
Boivin s.n.		*Dyps*	*palud*
Bojer sn		*Dyps*	*nodif*
Bory de St Vincent		*Dyps*	sp.
Bosser	2165	*Dyps*	sp.
Bosser	2166	*Dyps*	*bilde*
Bosser	3506	*Hyph*	*coria*
Bosser	7278	*Dyps*	sp.
Bosser	15946	*Dyps*	*heterop*
Bosser	16139	*Dyps*	*nodif*
Bosser	16826	*Dyps*	*lutes*
Bosser	16972	*Dyps*	*bosse*
Bosser	18297	*Rave*	*nana*
Bosser	18300	*Dyps*	*angusta*
Bosser	18553	*Dyps*	*pinna*
Bosser	18599	*Dyps*	*mcdon#*
Bosser	18867	*Rave*	*glauc*
Bosser	18905	*Dyps*	*trape*
Bosser	18916	*Dyps*	*catat*
Bosser	18983	*Dyps*	*nodif*
Bosser	19739	*Dyps*	*conci*
Bosser	19853	*Dyps*	*heterop*
Bosser	20114	*Dyps*	*lokoh?*
Bosser	20367	*Dyps*	*baron*
Campenon, 1887		*Dyps*	sp.
Capuron	1681	*Dyps*	*nodif*
Capuron Dec 1953		*Maro*	*insig*
Carlson	27	*Dyps*	*pinna*
Casimir RN	11731	*Dyps*	*mirab#*
Catat	1732	*Dyps*	*catat*
Catat		*Dyps*	*bilde*
Catat Jan1889		*Dyps*	sp.
Chapelier		*Dyps*	*fibro*
Chapelier		*Dyps*	*pinna*
Chapelier		*Rave*	*madag*
Cheek	2380	*Dyps*	*madag*
Cloisel	108	*Dyps*	*scott*
Commerson		*Dyps*	*comme*
Commerson		*Dyps*	*pinna*
Commerson s.n.		*Dyps*	*forfi*
Cons.Res.Nat.Pa	4185	*Dyps*	*rivul*
Cours	1848	*Dyps*	*zaham*
Cours	1904	*Dyps*	*confu*
Cours	1945	*Rave*	*sambi*
Cours	2312	*Dyps*	*conci#*
Cours	2437	*Dyps*	*linea*
Cours	2479	*Dyps*	*berni*
Cours	2513	*Dyps*	*virid#*
Cours	3234	*Dyps*	*cerac*
Cours	3243	*Dyps*	*lokoh*
Cours	3299	*Dyps*	*cours*
Cours	3342	*Dyps*	*baron*
Cours	3576	*Dyps*	*pumil*
Cours	3831	*Dyps*	sp.
Cours	4058	*Dyps*	*jumel*
Cours	4061	*Dyps*	*baron*
Cours	4192	*Dyps*	*simia*
Cours	4200	*Dyps*	*pinna*
Cours	4234	*Dyps*	*linea*
Cours	4296	*Dyps*	*nodif*
Cours	4297	*Dyps*	*baron*
Cours	4384	*Rave*	*robus*
Cours	4410	*Dyps*	*conci#*
Cours	4494	*Dyps*	*canal*
Cours	4496	*Dyps*	*pinna*
Cours	4510	*Rave*	*julie*
Cours	4511	*Dyps*	*fibro*
Cours	4512	*Rave*	*sambi*
Cours	4522	*Dyps*	*louve*
Cours	4537	*Dyps*	*proce*
Cours	4540	*Dyps*	sp.
Cours	4542	*Dyps*	*cours*
Cours	4713	*Dyps*	*conci*
Cours	4926	*Dyps*	*angustifolia*
Cremers	1419	*Dyps*	*conci*
Cremers	2082	*Dyps*	sp.
Cremers	2292	*Rave*	*glauc*
Cremers	3788	*Dyps*	*madag*
Croat	31012	*Phoe*	*dacty*
Croat	31918	*Dyps*	*decar*
Croat	32268	*Dyps*	sp.
Croat	32483	*Dyps*	*lutes*
Croat	32533	*Dyps*	*pinna*
Curtis	116	*Dyps*	*curti*
Curtis	117	*Dyps*	*?*

COLLECTOR	NR	GENUS	SPECIES
Darian s.n.,1983	sn	Maro	daria
De Lastelle 1841		Dyps	laste
Decary	2111	Hyph	coria
Decary	2139	Dyps	hiara#
Decary	4854	Dyps	nodif
Decary	5064	Dyps	luteu#
Decary	5081	Dyps	sp.
Decary	5466	Dyps	sp.
Decary	6303	Dyps	lutes
Decary	6378	Dyps	nodif
Decary	6448	Dyps	pinna
Decary	6480	Dyps	ambil
Decary	7082	Dyps	conci
Decary	7106	Dyps	catat
Decary	7143	Dyps	nodif
Decary	7161	Dyps	catat
Decary	7218	Dyps	conci
Decary	7556	Dyps	onila
Decary	9303	Dyps	decar
Decary	10294	Dyps	decar
Decary	10296	Dyps	decar
Decary	10299	Dyps	decar
Decary	10729	Dyps	scott
Decary	14220	Dyps	bonsa#
Decary	14709	Dyps	occid
Decary	15535	Phoe	recli
Decary	15889	Dyps	decip
Decary	16503	Dyps	heterop
Decary	16532	Dyps	catat
Decary	16672	Dyps	heterop
Decary	16680	Dyps	bonsa#
Decary	16683	Dyps	heterop
Decary	17691	Dyps	louve
Decary	17851	Dyps	hilde
Decary	18222	Dyps	conci
Decary 13jun1939		Dyps	bonsa#
Dequaire	27682	Dyps	pinna
Dequaire	27702	Dyps	fibro
Dequaire	27714	Dyps	louve
Dequaire	27728	Dyps	conci#
Descoigns	174	Dyps	hilde
Descoings	122	Dyps	catat
Descoings	122	Dyps	hilde
Descoings	1759	Dyps	sp.
Dorr	2917	Dyps	sp.
Dorr	3532	Dyps	baron
Dorr	3619	Rave	sambi
Dorr	3714	Dyps	sp.
Dorr	3716	Dyps	nodif
Dorr	3729	Dyps	baron
Dorr	3891	Dyps	linea
Dorr	3892	Dyps	linea
Drake	679	Hyph	coria
Dransfield	6357	Dyps	pinna
Dransfield	6358	Dyps	laste
Dransfield	6359	Dyps	laste
Dransfield	6360	Dyps	fibro
Dransfield	6361	Dyps	proce
Dransfield	6362	Dyps	pachy
Dransfield	6363	Dyps	lantz
Dransfield	6364	Dyps	berni
Dransfield	6365	Dyps	proce
Dransfield	6366	Dyps	proce
Dransfield	6367	Dyps	oligo
Dransfield	6368	Dyps	mocqu
Dransfield	6369	Dyps	proce
Dransfield	6370	Dyps	heterop
Dransfield	6371	Dyps	minut
Dransfield	6372	Rave	drans
Dransfield	6373	Dyps	fibro
Dransfield	6374	Rave	drans
Dransfield	6375	Oran	longi
Dransfield	6376	Oran	ravak
Dransfield	6377	Dyps	thiry
Dransfield	6378	Dyps	hiara
Dransfield	6379	Rave	robus
Dransfield	6380	Rave	robus
Dransfield	6381	Rave	robus
Dransfield	6382	Dyps	coria
Dransfield	6383	Dyps	confu
Dransfield	6384	Rave	drans
Dransfield	6385	Rave	drans
Dransfield	6386	Dyps	confu
Dransfield	6387	Dyps	forfi
Dransfield	6388	Dyps	coria
Dransfield	6389	Voan	gerra
Dransfield	6390	Maso	madag
Dransfield	6391	Voan	gerra
Dransfield	6396	Dyps	lokoh#
Dransfield	6397	Dyps	proce
Dransfield	6398	Dyps	hiara
Dransfield	6399	Maro	daria
Dransfield	6400	Oran	trisp
Dransfield	6401	Oran	ravak
Dransfield	6402	Lemu	halle
Dransfield	6403	Maro	insig
Dransfield	6404	Maro	insig
Dransfield	6405	Dyps	hejof
Dransfield	6407	Dyps	lutes
Dransfield	6408	Rave	sambi
Dransfield	6409	Rave	sambi
Dransfield	6411	Dyps	conci
Dransfield	6412	Dyps	jumel
Dransfield	6413	Dyps	hilde
Dransfield	6414	Dyps	nodif
Dransfield	6415	Dyps	pinna
Dransfield	6416	Dyps	conci
Dransfield	6417	Dyps	nodif
Dransfield	6418	Rave	louve
Dransfield	6419	Rave	louve
Dransfield	6420	Rave	louve
Dransfield	6421	Dyps	lutea
Dransfield	6422	Rave	robus
Dransfield	6423	Dyps	nodif
Dransfield	6424	Rave	lakat
Dransfield	6425	Rave	madag
Dransfield	6426	Dyps	baron
Dransfield	6427	Dyps	hilde
Dransfield	6428	Dyps	hilde
Dransfield	6429	Dyps	jumel
Dransfield	6430	Rave	madag
Dransfield	6431	Rave	madag
Dransfield	6432	Maro	insig
Dransfield	6433	Becc	madag
Dransfield	6434	Dyps	catat
Dransfield	6435	Dyps	conci
Dransfield	6436	Dyps	nodif
Dransfield	6437	Dyps	psamm
Dransfield	6438	Dyps	palud
Dransfield	6439	Dyps	palud
Dransfield	6440	Dyps	fibro
Dransfield	6441	Dyps	lutes
Dransfield	6442	Bism	nobil
Dransfield	6443	Bism	nobil
Dransfield	6444	Dyps	ambil
Dransfield	6445	Rave	sambi
Dransfield	6446	Dyps	orope
Dransfield	6447	Dyps	baron#
Dransfield	6448	Rave	sambi
Dransfield	6449	Dyps	proce
Dransfield	6450	Dyps	forfi
Dransfield	6451	Dyps	pachy
Dransfield	6452	Maro	daria
Dransfield	6453	Lemu	halle
Dransfield	6454	Dyps	mocqu
Dransfield	6455	Dyps	proce
Dransfield	6456	Dyps	forfi
Dransfield	6457	Dyps	minut
Dransfield	6458	Oran	ravak
Dransfield	6459	Dyps	coria
Dransfield	6460	Dyps	thiry
Dransfield	6461	Dyps	hejof
Dransfield	6462	Maro	insig
Dransfield	6463	Dyps	tsarav
Dransfield	6464	Dyps	proce
Dransfield	6465	Dyps	fanev
Dransfield	6466	Dyps	fasci
Dransfield	6467	Dyps	pusil
Dransfield	6468	Dyps	drans
Dransfield	6469	Dyps	drans
Dransfield	6470	Dyps	forfi
Dransfield	6471	Dyps	proce
Dransfield	6472	Dyps	fasci
Dransfield	6473	Dyps	hovom#
Dransfield	6474	Dyps	pusil
Dransfield	6475	Dyps	perri

COLLECTOR	NR	GENUS	SPECIES
Dransfield	6478	*Dyps*	*cauda*
Dransfield	6479	*Oran*	*longi*
Dransfield	6481	*Dyps*	*confu*
Dransfield	6482	*Dyps*	*coria*
Dransfield	6483	*Dyps*	*minut*
Dransfield	6484	*Maro*	*insig*
Dransfield	6485	*Rave*	*louve*
Dransfield	6486	*Rave*	*louve*
Dransfield	6487	*Dyps*	*bilde*
Dransfield	6488	*Dyps*	*conci*
Dransfield	6489	*Dyps*	*conci*
Dransfield	6490	*Dyps*	*jumel*
Dransfield	6491	*Rave*	*robus*
Dransfield	6492	*Dyps*	*palud*
Dransfield	6493	*Dyps*	*palud*
Dransfield	6494	*Dyps*	*palud*
Dransfield	6495	*Dyps*	*psamm*
Dransfield	6496	*Dyps*	*ambil*
Dransfield	6735	*Dyps*	*drans*
Dransfield	6736	*Dyps*	*drans*
Dransfield	6737	*Voan*	*gerra*
Dransfield	6738	*Maso*	*madag*
Dransfield	6739	*Dyps*	*tokor*
Dransfield	6740	*Maso*	*madag*
Dransfield	6741	*Dyps*	*cauda*
Dransfield	6742	*Lemu*	*balle*
Dransfield	6743	*Dyps*	*proce*
Dransfield	6744	*Dyps*	*bovom*
Dransfield	6745	*Dyps*	*crini*
Dransfield	6746	*Dyps*	*forfi*
Dransfield	6747	*Dyps*	*spica*
Dransfield	6748	*Dyps*	*spica*
Dransfield	6749	*Dyps*	*perri*
Dransfield	6750	*Dyps*	*thiry*
Dransfield	6751	*Dyps*	*lokoh*
Dransfield	6752	*Dyps*	*lokoh*
Dransfield	6753	*Dyps*	*spica*
Dransfield	6754	*Maro*	*insig*
Dransfield	6755	*Dyps*	*decus*
Dransfield	6756	*Dyps*	*oreop*
Dransfield	6757	*Rave*	*sambi*
Dransfield	6758	*Rave*	*sambi*
Dransfield	6759	*Dyps*	*lokoh*
Dransfield	6760	*Dyps*	*tsarav*
Dransfield	6761	*Dyps*	*cooke*
Dransfield	6762	*Dyps*	*tsarav#*
Dransfield	6763	*Dyps*	*heterop*
Dransfield	6764	*Dyps*	*spica*
Dransfield	6765	*Dyps*	*pinna*
Dransfield	6766	*Dyps*	*pilul*
Dransfield	6767	*Rave*	*robus*
Dransfield	6768	*Rave*	*robus*
Dransfield	6769	*Dyps*	*baron*
Dransfield	6770	*Maso*	*madag*
Dransfield	6771	*Dyps*	*mirab*
Dransfield	6772	*Dyps*	*decar*
Dransfield	6773	*Dyps*	*catat*
Dransfield	6774	*Dyps*	*catat*
Dransfield	6775	*Maro*	*insig*
Dransfield	6776	*Dyps*	*nodif*
Dransfield	6777	*Dyps*	*linea*
Dransfield	6778	*Maro*	*insig*
Dransfield	6779	*Dyps*	*malco*
Dransfield	6780	*Rave*	*robus*
Dransfield	6782	*Dyps*	*mcdon*
Dransfield	7501	*Dyps*	*conva*
Dransfield	7502	*Dyps*	*subra*
Dransfield	7503	*Dyps*	*palud*
Dransfield	7504	*Dyps*	*corni*
Dransfield	7505	*Dyps*	*virid*
Dransfield	7506	*Dyps*	*fanja*
Dransfield	7508	*Dyps*	*hiara*
Dransfield	7509	*Dyps*	*boivi*
Dransfield	7510	*Maro*	sp.
Dransfield	7511	*Dyps*	*therm*
Dransfield	7512	*Dyps*	*trape*
Dransfield	7513	*Dyps*	*erios*
Dransfield	7514	*Dyps*	*angusta*
Dransfield	7515	*Dyps*	*scand*
Dransfield	7516	*Raph*	*farin*
Dransfield	7518	*Rave*	*robus*
Dransfield	7519	*Dyps*	*boivi*
Dransfield	7520	*Rave*	*robus*

COLLECTOR	NR	GENUS	SPECIES
Dransfield	7521	*Dyps*	*corni*
Dransfield	7522	*Dyps*	*corni*
Dransfield	7523	*Dyps*	*fibro*
Dransfield	7524	*Dyps*	*forfi*
Dransfield	7525	*Dyps*	*palud*
Dransfield	7526	*Dyps*	*sanct*
Dransfield	7527	*Dyps*	*sanct*
Dransfield	7528	*Dyps*	*fasci*
Dransfield	7530	*Dyps*	*glabr*
Du Petit Thouar		*Dyps*	*thou*
Du Petit Thouar		*Dyps*	*thoua*
Du Petit Thouars		*Dyps*	*nodif?*
Du Puy	146	*Dyps*	*fibro*
Du Puy	146	*Dyps*	*fibro?*
Du Puy	149	*Dyps*	*lant?*
Du Puy	150	*Dyps*	sp.
Du Puy	152	*Oran*	*longi*
Du Puy	152	*Oran*	*ravak*
Du Puy	160	*Dyps*	*last?*
Du Puy	161	*Oran*	*ravak*
Du Puy	176	*Dyps*	*rivul*
Du Puy	180	*Dyps*	*last?*
Du Puy	217	*Dyps*	*baron*
Du Puy	222	*Dyps*	*baron*
Du Puy	256	*Bora*	*madag*
Du Puy	410	*Dyps*	*madag*
Du Puy	450	*Rave*	*glauc*
Du Puy	543	*Dyps*	*scott*
Du Puy	575	*Rave*	*xero*
Du Puy	673	*Dyps*	*onila*
Du Puy	674	*Hyph*	*coria*
Du Puy	714	*Dyps*	*madag*
Du Puy	791	*Bora*	*madag*
Du Puy	807	*Rave*	*sambi*
Du Puy	s.n.	*Maro*	*daria*
Dumetz	503	*Dyps*	*scott*
Dumetz	527	*Dyps*	*scott*
Dumetz	616	*Dyps*	*scott*
Dumetz	618	*Dyps*	*scott*
Dumetz	632	*Rave*	*samb*
Forsyth Major	603	*Dyps*	*nodif*
Forsyth Major	604	*Dyps*	*heterop*
Forsyth Major	605	*Dyps*	*heterop*
Forsyth Major	606	*Dyps*	*linea*
Geay	7188	*Dyps*	*pinna*
Geay	7267	*Dyps*	*pinna*
Geay	8051	*Dyps*	sp.
Geay	8052	*Dyps*	sp.
Geay	8053	*Dyps*	*pinna*
Geay	8055	*Phoe*	*recli*
Geay	8056	*Dyps*	*digit*
Geay	8057	*Dyps*	*digit*
Geay	9064	*Dyps*	*poivr*
Gentry	11685	*Dyps*	sp.
Gentry	52551	*Dyps*	*pinna*
Gentry	62172	*Dyps*	*mocqu*
Gentry	62179	*Dyps*	*cauda*
Gentry	62180	*Dyps*	*fibro*
Goudon		*Dyps*	*pervi*
Goudot		*Dyps*	*comme*
Goudot		*Dyps*	sp.
Grandidier july 1898		*Phoe*	*recli*
Grandidier Sept 1866		*Hyph*	*coria*
Greenway	1488	*Dyps*	*pemba*
Greve	154	*Phoe*	*recli*
Guillaumet	2161	*Dyps*	*catat*
Guillaumet	2215	*Dyps*	*bilde*
Guillaumet	2370	*Dyps*	*poivr*
Guillaumet	2404	*Dyps*	*oligo*
Guillaumet	2405	*Dyps*	sp.
Guillaumet	2406	*Dyps*	*forfi*
Guillaumet	2512	*Dyps*	*heterop*
Guillaumet	2526	*Dyps*	sp.
Guillaumet	2527	*Dyps*	*arena*
Guillaumet	2528	*Dyps*	*boivi#*
Guillaumet	2531	*Dyps*	*fanev*
Guillaumet	3511	*Dyps*	*catat*
Guillaumet	3901	*Dyps*	*comme*
Guillaumet	3980	*Dyps*	*mcdon*

COLLECTOR	NR	*GENUS*	*SPECIES*
Guillaumet	4003a	*Dyps*	*aquat*
Guillaumet	4003b	*Dyps*	*brevi*
Guillaumet	4017	*Dyps*	*fibro*
Guillaumet	4030	*Dyps*	*baron*
Guillaumet	4037	*Dyps*	*catat*
Guillaumet	4059	*Dyps*	*pumil*
Guillaumet	4105	*Dyps*	*cours*
Guillaumet	4106	*Dyps*	*cours*
Guillaumet	4108	*Dyps*	*andri*
Guillaumet	4111	*Dyps*	sp.
Guillaumet	4145	*Dyps*	sp.
Guillaumet	4172	*Rave*	*madag*
Guillaumet	4196	*Dyps*	sp.
Guillaumet	4201	*Dyps*	sp.
Guillaumet	4202	*Dyps*	*perri*
Guillaumet	4204	*Dyps*	*catat*
Guillaumet	4210	*Dyps*	sp.
Guillaumet 12/1971	sn	*Dyps*	*madag*
Guillaumet nov.71	sn	*Dyps*	*catat*
Heim		*Dyps*	*mocqu#*
Henderson	755	*Dyps*	*pinna*
Henderson	756	*Dyps*	*lantz*
Henderson	757	*Raph*	*farin*
Henderson	759	*Dyps*	*nodif*
Herb J.Bot.Tana	2839	*Dyps*	*nodif*
Herb J.Bot.Tana	3091	*Dyps*	*decip*
Herb J.Bot.Tana	3778	*Dyps*	*nodif*
Herb J.Bot.Tana	4267	*Phoe*	*recli*
Herb J.Bot.Tana	4847	*Rave*	*madag*
Herb J.Bot.Tana 12/46		*Dyps*	sp.
Herb J.Bot.Tana 8/37	sn	*Dyps*	*bonsa#*
Herb Lac Alaotr	4234	*Dyps*	*?fasc*
Herb Lac Alaotr	4410	*Dyps*	sp.
Herb Lac Alaotr=Cours	3342	*Dyps*	*baron*
Hildebrandt	2918	*Dyps*	*madag*
Hildebrandt	3052	*Hyph*	*coria*
Hildebrandt	3304	*Phoe*	*recli*
Hildebrandt	3717	*Dyps*	*nodif*
Hildebrandt	3829	*Dyps*	*bilde*
Hildebrandt	3974	*Dyps*	*decip*
Hildebrandt	4132	*Rave*	*madag*
Hildebrandt s.n	sn	*Bism*	*nobil*
Hodel	1144	*Dyps*	*lepto*
Homolle	518	*Dyps*	*spica#*
Homolle	547	*Dyps*	*fibro*
Homolle	548	*Dyps*	*confu#*
Homolle	3437	*Dyps*	*linea*
Homolle 1944	sn	*Dyps*	*zaham*
Hull s.n. Oct 1993		*Dyps*	*caba?*
Hull s.n. Oct 1993		*Dyps*	*lance*
Humbert	2822	*Dyps*	*onila*
Humbert	2856	*Hyph*	*coria*
Humbert	4868	*Rave*	*madag*
Humbert	5715	*Dyps*	*decar*
Humbert	5817	*Dyps*	*humbe*
Humbert	5958	*Dyps*	*scott*
Humbert	6232	*Rave*	*robus*
Humbert	6283	*Dyps*	*onila*
Humbert	6354	*Rave*	*nana*
Humbert	6539	*Dyps*	*onila*
Humbert	6624	*Dyps*	*catat*
Humbert	6891	*Dyps*	*catat*
Humbert	7025	*Dyps*	*onila*
Humbert	11298	*Phoe*	*recli*
Humbert	11653	*Dyps*	*onila*
Humbert	12109	*Dyps*	*linea*
Humbert	12248	*Dyps*	*onila*
Humbert	12458	*Bism*	*nobil*
Humbert	12491	*Rave*	*xerop*
Humbert	13017	*Rave*	*xerop*
Humbert	13698	*Rave*	*xerop*
Humbert	13904	*Dyps*	*linea*
Humbert	13905	*Dyps*	*onila*
Humbert	13977	*Dyps*	*malco#*
Humbert	13999	*Dyps*	*catat*
Humbert	14031	*Rave*	*robus*
Humbert	14032	*Dyps*	*malco*
Humbert	14136	*Dyps*	*malco#*
Humbert	14155	*Dyps*	*decar*
Humbert	17776	*Dyps*	*zaham*
Humbert	17908	*Dyps*	*jumel*

COLLECTOR	NR	*GENUS*	*SPECIES*
Humbert	18432	*Dyps*	*onila*
Humbert	19032	*Dyps*	*onila*
Humbert	19245	*Dyps*	*onila*
Humbert	20572	*Becc*	*madag*
Humbert	20671	*Dyps*	*brevi*
Humbert	21947	*Rave*	*nana*
Humbert	22119	*Dyps*	*catat*
Humbert	22171	*Dyps*	*thiry*
Humbert	22173	*Dyps*	*occid*
Humbert	22287	*Dyps*	*andri*
Humbert	22558	*Dyps*	*cours*
Humbert	22756	*Dyps*	*pumil*
Humbert	23159	*Dyps*	*cours*
Humbert	23275	*Dyps*	*lokoh*
Humbert	23287	*Dyps*	*baron*
Humbert	23377	*Dyps*	*onila*
Humbert	23391	*Dyps*	*laste*
Humbert	23424	*Maro*	*insig*
Humbert	23463	*Dyps*	*cours*
Humbert	23490	*Rave*	*nana*
Humbert	23493	*Dyps*	*bonsa*
Humbert	23559	*Dyps*	*andri*
Humbert	23632	*Maro*	*insig*
Humbert	23679	*Dyps*	*pumil*
Humbert	23682	*Dyps*	*bonsa*
Humbert	23682	*Rave*	*drans*
Humbert	23695	*Rave*	*nana*
Humbert	23729	*Dyps*	*bonsa*
Humbert	23799	*Dyps*	*pumil*
Humbert	24305	*Dyps*	*andri*
Humbert	24766	*Dyps*	*heterom*
Humbert	25281	*Dyps*	*heterom*
Humbert	28705	*Rave*	*glauc*
Humbert	31702	*Dyps*	*decus*
Humbert	31703	*Dyps*	*andap*
Humbert	31725	*Dyps*	*heterom*
Humbert	31737	*Dyps*	sp.
Humbert	31883	*Dyps*	*andri*
Humbert	32891	*Dyps*	*madag*
Humbert	32920	*Dyps*	*madag*
Humbert	11385bis	*Elae?*	*guine*
Humbert	17908.b	*Dyps*	*heterop*
Humbert	31703.b	*Dyps*	*occid*
Humbert	6425.b	*Dyps*	*catat*
Humbert No/De 1950		*Raph*	*farin*
Humblot	338	*Dyps*	*humbl*
Humblot	395	*Rave*	*bilde*
Humblot	1338	*Dyps*	*humbl*
Humblot	1347	*Dyps*	*lance*
Humblot	1395	*Rave*	*bilde*
Humblot	1608	*Rave*	*bilde*
Jacquemin	524	*Dyps*	*pinna*
Jaquemin	572J	*Dyps*	*catat*
Keraudren	25334	*Dyps*	*pinna*
Kirk		*Rave*	*bilde*
Lam	5312	*Dyps*	*bilde*
Lam	5659	*Dyps*	*soani*
Lam	5660	*Dyps*	*pinna*
Lam	5857	*Dyps*	*virid*
Lam	5858	*Dyps*	*thiry*
Lam	5859	*Dyps*	*nodif*
Lam	5860	*Dyps*	*nodif*
Lam	5861	*Dyps*	*mocqu*
Lam	5862	*Dyps*	*fibro*
Lam	5863	*Dyps*	*laste*
Lam	6040	*Dyps*	*ambil*
Lam	6045	*Dyps*	*nodif*
Lam	6135	*Hyph*	*coria*
Lam	6136	*Bism*	*nobil*
Lantz 1871		*Dyps*	*lantz*
Last 1889-1890		*Dyps*	*nodif*
Lastelle anno1841		*Dyps*	*pilul*
Le Myre de Vilers		*Dyps*	*heterop*
Le Myre de Vilers		*Dyps*	*heterop*
Le Myre de Vilers		*Dyps*	*bilde*
Le Myre de Vilers		*Dyps*	*nodif*
Leandri	386	*Dyps*	*madag*

COLLECTOR	NR	GENUS	SPECIES
Leandri	717	*Dyps*	*conci*
Leandri	719	*Dyps*	*louve*
Leandri	720	*Dyps*	*conci*
Leandri	721	*Dyps*	*jumel*
Leandri	724	*Dyps*	*bilde*
Leandri	1104	*Dyps*	*madag*
Leandri	1659	*Dyps*	*conci*
Leandri	2717	*Dyps*	*madag*
Leandri	2988	*Dyps*	*madag*
Leandri	2997	*Dyps*	*madag*
Leandri	3923	*Rave*	*glauc*
Lyall	323	*Dyps*	*minut*
Lyne	100	*Dyps*	*pemba*
Malcomber	883	*Rave*	*sambi*
Malcomber	885	*Dyps*	*andri*
Malcomber	976	*Dyps*	*baron*
Malcomber	1309	*Dyps*	*therm*
Malcomber	1326	*Dyps*	*catat*
Malcomber	1333	*Dyps*	*linea*
Malcomber	1401	*Rave*	*sambi*
Malcomber	1542	*Dyps*	*malco*
Malcomber	1972	*Dyps*	*fibro*
Malcomber	2120	*Dyps*	*nodif*
Malcomber	2129	*Dyps*	*catat*
Malcomber	2280	*Dyps*	*pinna*
Malcomber	2309	*Dyps*	*pilul*
Malcomber	2446	*Dyps*	*therm*
Malcomber	2448	*Rave*	*madag*
Malcomber	2471	*Dyps*	*monta*
Martin RN	7257	*Dyps*	*pinna*
McPherson	14218	*Dyps*	*fibro*
Meller 2.8.1862		*Dyps*	*rhodo*
Meller 31.7.186	sn	*Dyps*	*thiry*
Meller Aug. 1862	sn	*Dyps*	*nodif*
Miller	3327	*Dyps*	*fasci*
Miller	3435	*Dyps*	*perri*
Miller	3508	*Maro*	*insig*
Miller	3509	*Dyps*	*decus*
Miller	3510	*Dyps*	*oreop*
Miller	3512	*Dyps*	*andri*
Miller	3566	*Dyps*	*cours*
Miller	3941	*Dyps*	*acumi*
Miller	3942	*Dyps*	*thiry*
Miller	3951	*Dyps*	*spica*
Miller	3975	*Dyps*	*oreop*
Miller	4027	*Dyps*	*pinna*
Miller	4185	*Dyps*	*decus*
Miller	4257	*Dyps*	*catat*
Miller	4378	*Dyps*	*oreop*
Miller	4391	*Rave*	*sambi*
Miller	4491	*Dyps*	*laste*
Miller	4493	*Dyps*	*tsarav#*
Moore	6101	*Dyps*	*cabad*
Moore	6517	*Dyps*	*cabad*
Moore	9005	*Dyps*	*utili*
Moore	9009	*Dyps*	*fibro*
Moore	9010	*Rave*	*sambi*
Moore	9014	*Dyps*	sp.
Moore	9018	*Dyps*	*lutes*
Moore	9019	*Rave*	*sambi*
Moore	9024	*Rave*	*sambi*
Moore	9025	*Dyps*	sp.
Moore	9026	*Dyps*	*madag*
Moore	9027	*Dyps*	sp.
Moore	9028	*Rave*	*moore*
Moore	9029	*Rave*	*robus*
Moore	9901	*Maro*	*insig*
Moore	9908	*Dyps*	*pinna*
Moore	9909	*Dyps*	*bilde*
Moore	9913	*Dyps*	*conci*
Moore	9916	*Dyps*	*mahia#*
Moore	9917	*Dyps*	*pinna*
Moore	9918	*Dyps*	*moore*
Moore	9919	*Dyps*	*tsarav#*
Moore	9920	*Dyps*	*psamm#*
Moore	9921	*Oran*	*trisp*
Moore	9923	*Dyps*	*decip*
Moore	9968	*Dyps*	sp.
Moore	10118	*Oran*	*longi*
Moore	10119	*Dyps*	*mana?*
Moquerys	332	*Dyps*	*lantz*

COLLECTOR	NR	GENUS	SPECIES
Moquerys	333	*Dyps*	*mocqu*
Moquerys	334	*Dyps*	*lantz*
Moquerys	412	*Dyps*	*lantz*
Moquerys	418	*Dyps*	*lantz*
Moquerys	419	*Dyps*	*mocqu*
Morat	445	*Dyps*	*louve*
Morat	4489	*Dyps*	*corni*
Morat	4499	*Dyps*	*andap*
Morat	4590	*Dyps*	*baron#*
Morat	4591	*Rave*	*sambi*
Morat	4930	*Dyps*	*berni*
Morat	4954	*Dyps*	*lantz*
Morat	8518	*Dyps*	*nodif*
Nicoll	203	*Dyps*	*therm*
Nicoll	204	*Dyps*	*?fasc*
Nicoll	207	*Dyps*	*nodif*
Nicoll	342	*Rave*	*madag*
Nicoll	425	*Dyps*	sp.
Nicoll	439	*Dyps*	*bejof*
Nicoll	493	*Dyps*	*graci*
Nicoll	494	*Dyps*	sp.
Noronha sn		*Dyps*	*pinna*
Noyes	922	*Dyps*	*lutes*
Noyes	939	*Dyps*	*pinna*
Overdorff	7	*Dyps*	*nodif*
Parker <8.1881	sn	*Dyps*	sp.
Parker Aug.1881		*Dyps*	*nodif*
Peltier	3262	*Dyps*	*bilde*
Peltier	4635	*Dyps*	*poivr*
Peltier	5165	*Dyps*	*nodif*
Perrier	145	*Dyps*	*madag*
Perrier	679	*Hyph*	*coria*
Perrier	1541	*Dyps*	*canal*
Perrier	10052	*Dyps*	*crini*
Perrier	11926	*Dyps*	*angusta*
Perrier	11935	*Dyps*	*heterom*
Perrier	11936	*Dyps*	*ankai*
Perrier	11937	*Oran*	*longi*
Perrier	11938	*Maso*	*madag*
Perrier	11939	*Rave*	*albic*
Perrier	11940	*Dyps*	*lutea*
Perrier	11941	*Dyps*	*perri*
Perrier	11942	*Dyps*	*perri*
Perrier	11943	*Dyps*	*simia*
Perrier	11944	*Dyps*	*integ*
Perrier	11946	*Dyps*	*perri*
Perrier	11948	*Dyps*	*catat*
Perrier	11952	*Hyph*	*coria*
Perrier	11956	*Dyps*	*bilde*
Perrier	11957	*Dyps*	*bilde*
Perrier	11958	*Rave*	*rivul*
Perrier	11960	*Rave*	*madag*
Perrier	11961	*Dyps*	*pilul*
Perrier	11962	*Dyps*	*schat#*
Perrier	11966	*Dyps*	*pluri*
Perrier	11968	*Dyps*	*louve*
Perrier	11970	*Dyps*	*rivul*
Perrier	11972	*Dyps*	*conci*
Perrier	11974	*Dyps*	*catat*
Perrier	11975	*Rave*	*robus*
Perrier	11976	*Dyps*	*angusta*
Perrier	11977	*Dyps*	*procu*
Perrier	11978	*Dyps*	*thiry*
Perrier	11985	*Dyps*	*lance*
Perrier	11986	*Dyps*	*lance*
Perrier	11987	*Dyps*	*lance*
Perrier	11988	*Dyps*	*jumel*
Perrier	11989	*Dyps*	*bilde*
Perrier	11994	*Dyps*	*linea#*
Perrier	11995	*Dyps*	*catat*
Perrier	11996	*Dyps*	*catat*
Perrier	11998	*Dyps*	*catat*
Perrier	11999	*Dyps*	*catat*
Perrier	12001	*Dyps*	*fibro*
Perrier	12002	*Dyps*	*lutea*
Perrier	12003	*Dyps*	*heterop*
Perrier	12004	*Dyps*	*pilul*

COLLECTOR	NR	GENUS	SPECIES
Perrier	12005	*Dyps*	*utili*
Perrier	12008	*Rave*	*latis*
Perrier	12010	*Dyps*	*pinna*
Perrier	12018	*Dyps*	*baron*
Perrier	12021	*Rave*	*louve*
Perrier	12022	*Rave*	*lakat*
Perrier	12023	*Dyps*	*laste*
Perrier	12026	*Dyps*	*catat*
Perrier	12028	*Dyps*	*nossi*
Perrier	12030	*Dyps*	*proce*
Perrier	12031	*Dyps*	*virid*
Perrier	12032	*Dyps*	*lucen*
Perrier	12033	*Dyps*	*oligo*
Perrier	12034	*Dyps*	*forfi*
Perrier	12037	*Dyps*	*forfi*
Perrier	12038	*Dyps*	*fanja?*
Perrier	12039	*Dyps*	*curti*
Perrier	12040	*Dyps*	*occid*
Perrier	12041	*Rave*	*sambi*
Perrier	12042	*Dyps*	*fasci*
Perrier	12043	*Dyps*	*curti*
Perrier	12044	*Dyps*	*acaul*
Perrier	12047	*Dyps*	*monos*
Perrier	12048	*Dyps*	*lantz*
Perrier	12049	*Dyps*	*mango*
Perrier	12050	*Dyps*	*mango*
Perrier	12051	*Dyps*	*boivi#*
Perrier	12052	*Dyps*	*crini*
Perrier	12053	*Dyps*	*fibro*
Perrier	12054	*Dyps*	*palud*
Perrier	12055	*Dyps*	*oreop*
Perrier	12056	*Dyps*	*forfi*
Perrier	12057	*Dyps*	*virid*
Perrier	12058	*Dyps*	*lutes*
Perrier	12059	*Dyps*	*pinna*
Perrier	12060	*Dyps*	*saban*
Perrier	12061	*Rave*	*glauc*
Perrier	12064	*Dyps*	*mocqu*
Perrier	12065	*Dyps*	*soani*
Perrier	12066	*Dyps*	*anove*
Perrier	12067	*Dyps*	*mocqu*
Perrier	12068	*Maro*	*insig*
Perrier	12069	*Bora*	*sambi*
Perrier	12070	*Dyps*	*amban*
Perrier	12071	*Dyps*	*ambol*
Perrier	12072	*Dyps*	*tanal*
Perrier	12073	*Dyps*	*mananj*
Perrier	12074	*Dyps*	*onila*
Perrier	12076	*Dyps*	*nodif*
Perrier	12077	*Bism*	*nobil*
Perrier	12079	*Dyps*	*onila*
Perrier	12080	*Dyps*	*laste*
Perrier	12082	*Dyps*	*baron*
Perrier	12083	*Dyps*	*onila*
Perrier	12087	*Dyps*	*nause*
Perrier	12088	*Dyps*	*basil*
Perrier	12089	*Dyps*	*proce*
Perrier	12090	*Dyps*	*boivi#*
Perrier	12091	*Dyps*	*forfi*
Perrier	12092	*Dyps*	*pinna*
Perrier	12099	*Dyps*	*decip*
Perrier	12499	*Dyps*	*onila*
Perrier	12625	*Dyps*	*digit*
Perrier	13292	*Dyps*	*arena*
Perrier	13649	*Rave*	*glauc*
Perrier	14097	*Dyps*	*fibro*
Perrier	14124	*Dyps*	*berni*
Perrier	14158	*Dyps*	*mananj*
Perrier	14202	*Dyps*	*presto*
Perrier	14265	*Dyps*	*bilde*
Perrier	14266	*Dyps*	*bilde*
Perrier	14839	*Dyps*	*onila*
Perrier	14969	*Dyps*	*jumel*
Perrier	14980	*Dyps*	*oreop*
Perrier	15012	*Dyps*	*confu#*
Perrier	15015	*Dyps*	*monos*
Perrier	15104	*Dyps*	*heterop*
Perrier	15265	*Dyps*	*tsaratan*
Perrier	15266	*Dyps*	*heterom*
Perrier	15413	*Dyps*	*canal*
Perrier	15414	*Dyps*	*ligul*
Perrier	15646	*Dyps*	*monta*
Perrier	15647	*Dyps*	*oreop*

COLLECTOR	NR	GENUS	SPECIES
Perrier	15800	*Dyps*	*onila*
Perrier	15801	*Dyps*	*acumi*
Perrier	15802	*Dyps*	*lepto*
Perrier	15803	*Dyps*	*rivul*
Perrier	15883	*Dyps*	*baron*
Perrier	15982	*Dyps*	sp.
Perrier	15983	*Dyps*	*conci*
Perrier	15984	*Dyps*	*louve*
Perrier	15985	*Dyps*	*catat*
Perrier	15986	*Dyps*	*jumel*
Perrier	15987	*Dyps*	*bilde*
Perrier	15988	*Dyps*	*arena*
Perrier	15989	*Dyps*	*baron*
Perrier	15990	*Dyps*	*baron*
Perrier	15991	*Rave*	*sambi*
Perrier	16039	*Dyps*	*ambil*
Perrier	16040	*Dyps*	*nodif*
Perrier	16061	*Dyps*	*baron*
Perrier	16067	*Dyps*	*utili*
Perrier	16069	*Rave*	*robus*
Perrier	16070	*Rave*	*sambi*
Perrier	16076	*Dyps*	*oreop*
Perrier	16100	*Dyps*	*madag*
Perrier	16226	*Cocos*	*nucif*
Perrier	16227	*Dyps*	*ankai*
Perrier	16548	*Dyps*	*onila*
Perrier	16550	*Rave*	*glauc*
Perrier	16580	*Rave*	*rivul*
Perrier	16834	*Rave*	*madag*
Perrier	17214	*Dyps*	*thiry*
Perrier	17215	*Dyps*	*bilde*
Perrier	17216	*Dyps*	*pulche*
Perrier	17217	*Dyps*	*conci*
Perrier	17467	*Dyps*	*schat*
Perrier	17468	*Dyps*	*angustifolia*
Perrier	17469	*Dyps*	*betam*
Perrier	17470	*Dyps*	*pervi*
Perrier	17473	*Rave*	*sambi*
Perrier	17474	*Dyps*	*cerac*
Perrier	17477	*Dyps*	*fasci*
Perrier	17554	*Dyps*	*madag*
Perrier	17570	*Dyps*	*madag*
Perrier	17872	*Hyph*	*coria*
Perrier	18042	*Dyps*	*mango*
Perrier	18043	*Dyps*	*procum*
Perrier	18044	*Dyps*	*bilde*
Perrier	18045	*Dyps*	*pulche*
Perrier	18046	*Dyps*	*pulche*
Perrier	18047	*Dyps*	*integ*
Perrier	18048	*Dyps*	*bilde*
Perrier	18049	*Dyps*	*elega?*
Perrier	18050	*Dyps*	*lanug*
Perrier	18051	*Dyps*	*furca*
Perrier	18390	*Dyps*	*heterop*
Perrier	18391	*Dyps*	*lutea*
Perrier	18392	*Dyps*	*bilde*
Perrier	18399	*Dyps*	*heter*
Perrier	18446	*Dyps*	*heter*
Perrier	18653	*Rave*	*xerop*
Perrier	18654	*Dyps*	*decar*
Perrier	18730	*Dyps*	*ampas*
Perrier	18742	*Dyps*	*pinna*
Perrier	18870	*Dyps*	*baron*
Perrier	15266.b	*Dyps*	*heterom*
Perrier 8.12.1909		*Dyps*	*crini*
Perrier s.n.		*Dyps*	*canes*
Perrier s.n.		*Dyps*	*madag*
Perrier s.n.		*Phoe*	*recli*
Perrotet an1819		*Dyps*	*pinna*
Perville A1841		*Dyps*	*pervi*
Peyrieras		*Rave*	*madag*
Phillipson	1767	*Hyph*	*coria*
Phillipson	1860	*Dyps*	*nodif*
Phillipson	2120	*Dyps*	*bilde*
Phillipson	2235	*Dyps*	*madag*
Phillipson	3406	*Dyps*	*onila*
Phillipson	3961	*Dyps*	*scott*
Phillipson	3991	*Rave*	*xerop*
Poisson	70	*Bism*	*satra*
Poivre		*Dyps*	*poivr*
Poivre s.n.		*Dyps*	*hirtu*
Pool Comm.4/1876	sn	*Dyps*	*conci*
Proctor Bros,1890		*Dyps*	*fibro*

COLLECTOR	NR	*GENUS*	*SPECIES*
Raberazaha	12183	*Dyps*	*pinna*
Rabevazaha RN	11521	*Dyps*	*catat*
Rabevohitra	1888	*Dyps*	*scott*
Rabevohitra	1918	*Dyps*	sp.
Rabevohitra	1932	*Dyps*	*scott*
Rabevohitra	2208	*Dyps*	*comme*
Raketamalata		*Dyps*	*bilde*
Rakotoson RN	10387	*Dyps*	*catat*
Rakotovao RN	9600	*Dyps*	*procum*
Rakotovao RN	10610	*Dyps*	*corni*
Rakotovao RN	11859	*Dyps*	*baron*
Rakotovao RN	11899	*Dyps*	*zaham*
Rakotovao RN	12115	*Dyps*	*cf. conci*
Rakotovao SF/RN	11850	*Dyps*	*bonsa#*
Rakotovao SF/RN	11850	*Dyps*	*nodif*
Rakotozafy	216	*Dyps*	sp.
Rakotozafy	620	*Dyps*	*nodif*
Randrianasolo	185	*Dyps*	*pinna*
Randrianasolo	185 K	*Dyps*	*fibro*
Randrianasolo	185 P	*Dyps*	sp.
Razafindratsira	sn	*Dyps*	sp.
Razafindratsira		*Dyps*	*laste*
Razafindratsira 1986		*Dyps*	*lepto*
Razafindratsira anno 89		*Becc*	*madag*
Razafindratzina	sn	*Rave*	*xerop*
Richard	16	*Dyps*	*pinna*
Richard	64	*Dyps*	sp.
Richard	144	*Hyph*	*coria*
RN	6402	*Dyps*	*madag*
RN	8050	*Dyps*	sp.
RN	10018	*Dyps*	*manan*
RN	11019	*Dyps*	*baron*
RN Rakotoniainia	4884	*Dyps*	*pinna*
RN Razafindrako	3601	*Rave*	*glauc*
Schatz	1665	*Dyps*	*pinna*
Schatz	1680	*Dyps*	sp.
Schatz	1693	*Dyps*	*catat*
Schatz	1694	*Dyps*	*catat*
Schatz	1858	*Dyps*	sp.
Schatz	2097	*Dyps*	*lantz*

COLLECTOR	NR	*GENUS*	*SPECIES*
Schatz	2415	*Bora*	*mada?*
Schatz	2612	*Dyps*	sp.
Schatz	2616	*Dyps*	*confu*
Schatz	2687	*Dyps*	*schat*
Schatz	2689	*Dyps*	*corni*
Schatz	2897	*Dyps*	*bonsa*
Schatz	3394	*Dyps*	*pusil*
Schlieben	8129	*Dyps*	*bilde*
Schlieben	11214	*Rave*	*bilde*
Scott Elliot	2615	*Dyps*	*scott*
Scott Elliot	2957	*Dyps*	*pinna*
Scott Elliot4/1	sn	*Dyps*	*pinna*
SF	3305	*Dyps*	*pinna*
SF	26477	*Dyps*	*nodif*
SF	26668	*Dyps*	*linea*
StJohn	26542	*Rave*	*moore*
StJohn	26543	*Rave*	*bilde*
StJohn	26544	*Rave*	*bilde*
Thiry A1904		*Dyps*	*thiry*
van Nek	1997	*Dyps*	*nodif*
van Nek	2000	*Dyps*	*nodif*
Veitch anno 189	sn	*Dyps?*	sp.
Vigrence	15462	*Dyps*	*fibro*
Vigreux	15417	*Dyps*	*remot*
Viguier	1058	*Dyps*	*bilde*
Viguier	1084	*Dyps*	*catat*
Viguier	1250	*Dyps*	*nodif*
Williams	177	*Dyps*	*pemba*
Willing	64	*Rave*	*xerop*
Zjhra	120	*Dyps*	*coria*
Zjhra	139	*Dyps*	*forfi*
Zjhra	163	*Dyps*	*confu*
Zjhra	203	*Dyps*	*bonsai*

INDEX TO LOCAL NAMES

INDEX TO SCIENTIFIC NAMES

(ACCEPTED NAMES OF SPECIES TREATED IN DETAIL IN BOLD, SYNONYMS IN ITALICS; SPECIES MENTIONED IN PASSING IN ROMAN)
(MAIN REFERENCE IN **BOLD** NUMERALS, ILLUSTRATIONS IN *ITALICS*)